李崇银院士论文选集

Selected Works of Academician Li Chongyin

Li Chongyin（李崇银）

科学出版社

北　京

图书在版编目（CIP）数据

李崇银院士论文选集/李崇银著. —北京：科学出版社，2019.6
ISBN 978-7-03-061277-9

Ⅰ. ①李… Ⅱ. ①李… Ⅲ. ①大气科学–文集 Ⅳ. ①P4-53

中国版本图书馆 CIP 数据核字（2019）第 098706 号

责任编辑：朱　瑾 / 责任校对：郑金红
责任印制：吴兆东 / 封面设计：无极书装

科学出版社 出版
北京东黄城根北街 16 号
邮政编码: 100717
http://www.sciencep.com

北京虎彩文化传播有限公司 印刷
科学出版社发行　各地新华书店经销
*
2019 年 6 月第　一　版　开本：787×1092　1/16
2019 年 6 月第一次印刷　印张：40
字数：980 000

定价：588.00 元

Prof. Li Chongyin (李崇银)

Photos

A photo taken in 1980-1981 as a visiting scholar during the collaborative research of the Department of Atmospheric Sciences at the University of Illinois, USA.

Visiting the hydraulic experimental base of Professor Yashino (first from right) of the University of Tsukuba in Japan in 1985.

Visiting Badaling of the Great Wall with R. S. Lindzen (middle), the academician of the American Academy of Sciences, and Professor J. Egger from German in June 1986.

A photo taken with Prof. A. B. Pittok when Li was invited to visit Australia DAS/CSIRO (1991.11-1992.4).

A photo taken with Prof. H. Lim (first from left), Chng Lak Seng (second from left) and Dr. Chiyu Tik (first from right) when Li was invited to visit Singapore in 1992.

A photo taken with Prof. T. Asai (right) and Prof. T. Nitta (left) as a memento during the Asian Monsoon Workshop in Japan in 1995.

In the CLIVAR-SSG meeting in May 1999 at Southampton Oceanography Centre, UK. (first from right is Prof. Li Chongyin)

A photo taken with Dr. A. D. Moura, the director of International Research Institute for Climate and Society (IRI), US, when visiting IRI with the Chinese CLIVAR delegation in 1998.

Exchanging greetings with R. Dickinson (right), an academician of the American Academy of Sciences and the American Academy of Engineering at an international conference in 2000.

Making a report at Nansen Environmental and Remote Sensing Center (NERSC) when invited to visit Norway in July 2013.

Making a report when attending an academic seminar abroad in 2014.

Forword

Prof. Li Chongyin, an academician of Chinese Academy of Sciences (CAS) and well-known scientist in atmospheric science and climate dynamics, was born in Daxian (new calld Dazhou City), Sichuan Province on 15 April 1940, the Han nationality. After graduation from the Department of Applied Geophysics, University of Science and Technology of China in 1963, he has been engaged in atmospheric science research all along in the Institute of Atmospheric Physics (IAP), Chinese Academy of Sciences. He was the director of Atmospheric Circulation and Geophysical Fluid Mechanism Laboratory (1984-1994) and vice director of the LASG (State Key Laboratory of Numerical Modeling for Atmospheric Sciences and Geophysical Fluid Dynamics) (1985-1993), vice chairman of Academic Committee in the IAP/CAS (1988-2013), chairman of Dynamical Meteorology Committee in the Chinese Meteorology Society (1987-2006), vice president of the Chinese Meteorology Society (2006-2010) and chairman of the Chinese Climate Research Committee (2004-2012). He was also the member of ICDM/IAMAS (1991-2002), the member of CLIVAR Scientific Steering Group/WCRP (1997-1999), the member of Asia-Australian Monsoon Panel/CLVAR (1995-2001) and the member of ICCL/IAMAS (2003-2016).

Prof. Li Chongyin has achieved systematic and innovative achievements in the frontier fields of atmospheric science and climate, such as in tropical meteorology, atmospheric low frequency oscillation and its dynamics, the sea-air interactions (ENSO cycle, IOD and PIOAM) and climate variability. Particularly, he has made outstanding contributions to push the development of tropical meteorology, dynamic meteorology and climatic dynamics. In the studies on atmospheric low frequency oscillation, he systematically demonstrated the characteristics of atmospheric low frequency oscillation, advanced the theory and dynamical mechanism on the atmospheric low frequency teleconnection / low frequency remote response. In the studies on dynamics of the atmospheric intraseasonal oscillation, he advanced and consummated the conditional instability of the second kind (CISK) wave theory of tropical intraseasonal oscillation (a. k. a. MJO). He is the first one in the world to indicate that the convection heating feedback (a. k. a. CISK) is an important mechanism to excite tropical intraseasonal oscillation. Later on, he showed the importance of the CISK-Rossby wave in driving tropical intraseasonal oscillation as well as the CISK-Kelvin wave. In the studies on ENSO cycle dynamics, he not only indicated the important effects of the ENSO on the temperature and precipitation in Eastern China, and the activity of typhoon over the Western Pacific, but also got more important innovative achievements in the occurrence mechanism of

El Niño. He revealed the important impact of anomalous East Asian winter monsoon on the occurrence of ENSO. Furthermore, he also indicated that the premonition of an El Niño event is in the equatorial western Pacific, and the subsurface ocean temperature anomaly in the warm pool and its eastward propagation play a key role in occurrence of the ENSO. Thus, a new concept on essence of the ENSO is advanced: The ENSO is exactly a cycle of subsurface ocean temperature anomalies in the tropical Pacific driven by zonal wind anomalies over the equatorial center-western Pacific, which mainly results from anomalous East Asian winter monsoon. He indicated the interactions between the MJO and ENSO at first. He also achieved a series of achievements in tropical meteorology, the typhoon dynamics and interdecadal climate variability, such as the influences of ENSO and MJO on the typhoon activity, the process of Asian summer monsoon onset and the impact of MJO, the mechanisms of interdecadal climate variations, and so on.

Prof. Li Chongyin has published 21 scientific treaties and over 450 scientific papers (about 220 in English, which are arranged as appendixs) by himself and cooperators (the most are his students) at home and abroad. In this symposium, some selected partial papers in English are collected, which involves research fields on atmospheric intraseasonal oscillation and dynamics, the ENSO, IOD and PIOAM, monsoon and tropical cyclone, climate variability, the oceanic wave activity and renewable energy resources over the sea, and the near space environment.

LASG, Institute of Atmospheric Physics,
Chinese Academy of Sciences

Contents

1 Atmospheric Intraseasonal Oscillation and Dynamic Mechanism

2 The ENSO, IOD and PIOAM (Pacific–Indian Ocean Associated Mode)

3 Monsoon and Tropical Cyclone

4 Climate Variability

5 The Oceanic Wave Activity and Renewable Energy Resource over the Sea

6 Stratosphere and Near Space Environment

1

Atmospheric Intraseasonal Oscillation and Dynamic Mechanism

The CISK-Overstability Convection

Li Chongyin (李崇银)

Institute of Atmospheric Physics, Academia Sinica, Beijing

Abstract A kind of the conditional instability of the second kind (CISK)—overstability convection presented in the hurricane model is investigated in this paper. The structure of this unstable mode and the basic conditions to excite it are also shown for discussion.

In the tropical atmosphere, the cumulus convection could excite the CISK oscillatory unstable mode through the CISK mechanism, and the cumulus friction (cumulus momentum mixing) gives a fundamental effect to produce this overstability convection.

Some meteorologists assumed that the waves in the lower stratosphere in the tropical atmosphere may be caused by the forcing of the condensation heating. The present investigation shows from the dynamic angle that the CISK-overstability convection excited by the cumulus convection heating, especially by the cumulus momentum mixing, through the CISK mechanism is a fundamental factor to drive the waves in the equatorial stratosphere and the oscillatory phenomena in tropical atmosphere.

Ⅰ. Introduction

Both the theoretic studies and the analyses on the observational data indicate that there are mixed Rossby-gravity waves and Kelvin waves in the lower stratosphere and upper troposphere in the tropics[1-6]. For the mixed Rossby-gravity wave, the zonal wave-length is about 10,000 km, the vertical wavelength is 5 km, and the period is about 4—5 days (the phase speed is 23 m/sec). While for the Kelvin wave, the zonal wavelength is about 30,000 km, the vertical wavelength is 10 km, and the period is about 10—20 days (the phase speed is 25 m/sec).

The discovery of these waves in the lower stratosphere in the tropics is looked upon as one of the fundamental advances in the dynamic meteorology. Some investigations have been made, but a lot of problems should be studied further. For example the formation mechanisms of the mixed Rossby-gravity waves and Kelvin waves have not yet been found satisfactorily. At first, meteorologists suggested that those waves may be caused by the condensational heating and are related to the cumulus convection. Then, J. R. Holton[7] completed a numerical experiment of the genesis of waves in the equatorial stratosphere with a diagnostic model; he assumed that there is a periodic condensation heating in the limited region in the tropics. The disturbances were caused in the equatorial stratosphere and similar to the mixed Rossby-

Received July 24, 1982; revised January 4, 1983.

gravity waves and the Kelvin waves. But why the condensation heating is periodic in the tropics? Holton did not discuss this problem. An assumption may be proposed naturally, i.e. whether the condensation heating can lead to produce an oscillatory unstable mode through some physical processes.

A theoretic investigation of the physical property and the formative condition of the overstability convection has been made[8], and it indicates that the overstability convection is also present in a hurricane model. In this paper, the overstability convection produced in the hurricane model, named the conditional instability of the second kind (CISK)-overstability convection, is studied. We will understand the structures and the fundamental factors which cause them. Further more, we will give a theoretical explanation of some periodic oscillations in the tropical atmosphere at the base of this CISK-overstability convection.

Ⅱ. Numerical Model

In consideration of the promotion between the cumulus convection and the synoptic scale depression in the tropics, J. G. Charney and A. Eliassen developed the conditional instability of the second kind (CISK), and gave a theoretical explanation to the genesis and development of hurricane [9]. In the CISK theory, Ekman pumping is the key to the convergence of moisture, it may be named Ekman-CISK. The deep cumulus convection over the tropical ocean not only provides the condensational heating but also leads to the vertical transportation of the horizontal momentum. Recently, meteorologists have paid much attention to the cumulus momentum transportation. E. K. Schneider and R. S. Lindzen describe this effect as the cumulus friction[10], i.e.

$$F = \frac{1}{\rho}\frac{\partial}{\partial z}[M(V - V_c)] \tag{1}$$

where M is the cloud mass flux, V_c and V are respectively the horizontal velocity of the cloud air and the environmental air. Since the development of the deep cumulus is usually rapid, the vertical velocity is sufficiently large. Then the drag forces will have no enough time to change V_c, and V_c will therefore be approximately conserved Recently, Mak[11] has indicated that the cumulus friction, or called the cumulus momentum mixing, can create a secondary circulation and excite the conditional instability of the second kind as Ekman pumping; it is named the cumulus momentum mixing-conditional instability of the second kind (CMM-CISK).

We introduce both the Ekman pumping and the cumulus momentum mixing into the CISK theory. For simplification, we assume that the top of the Ekman layer is at the same height as the cumulus base is. We may then write the perturbation governing equations in a cylindriral-pressure coordinate system as follows.

$$\frac{\partial v}{\partial t} = -fu - \frac{\partial}{\partial p}[M(v - v_B)] \tag{2}$$

$$\frac{\partial \phi}{\partial r} = fv \tag{3}$$

$$\frac{\partial \phi}{\partial p} = -\frac{RT}{p} \tag{4}$$

$$\frac{\partial (ru)}{\partial r} + r\frac{\partial \omega}{\partial p} = 0 \tag{5}$$

$$\frac{\partial T}{\partial t} - \frac{pS}{R}\omega = \frac{Q}{c_p} \tag{6}$$

where u, v and ω respectively refer to the radial, azimuthal and p-vertical velocity components; ϕ and T are the geopotential and temperature respectively, v_B is the azimuthal velocity at the cumulus base; Q is the condensational heating rate;

$$S \equiv -\frac{R}{p}\left(\frac{\partial \bar{T}}{\partial p} - \frac{R\bar{T}}{pc_p}\right)$$

is the static stability.

For analysis sake, we introduce two new variables ξ and ψ defined by

$$\xi = rv \tag{7}$$

$$u = \frac{1}{r}\frac{\partial \psi}{\partial p}, \omega = -\frac{1}{r}\frac{\partial \psi}{\partial r} \tag{8}$$

where ξ is the relative angular momentum and ψ is the meridional stream function. Eqs. (1)—(5) can be reduced to the following two equations:

$$\frac{\partial \xi}{\partial t} = -f\frac{\partial \psi}{\partial p} - \frac{\partial}{\partial p}[M(\xi - \xi_B)] \tag{9}$$

$$\frac{\partial^2 \xi}{\partial t \partial p} = \frac{S}{f}r\frac{\partial}{\partial r}\left(\frac{1}{r}\frac{\partial \psi}{\partial r}\right) - \frac{R}{fp}r\frac{\partial}{\partial r}\left(\frac{Q}{c_p}\right). \tag{10}$$

The radial boundary conditions are:

$$\lim_{r\to\infty}\frac{\psi}{r} = 0, \tag{11}$$

$$\lim_{r\to\infty}\frac{\xi}{r} = 0,\ \lim_{r\to\infty}\frac{\psi}{r} = 0. \tag{12}$$

The vertical boundary conditions are

at p_{top},

$$\omega = 0,\ \psi = 0, \tag{13}$$

at p_{bottom},

$$\omega_B = -\frac{Kp_B}{2f}\frac{1}{r}\frac{\partial (rv_B)}{\partial r}, \text{ or}$$

$$\psi_B = \frac{Kp_B}{2f}\xi_B, \tag{14}$$

where $K = \left(\frac{D_E}{H}\right)f\sin 2\alpha$, H is the scale height, $D_E = \sqrt{2v/f}$, v is the viscous coefficient, and α is the surface wind cross-isobar angle.

As usual, the convective condensational heating is depicted in parameterization form. Since the two physical processes are included, both the frictional convergence and the secondary circulation caused by the cumulus momentum mixing can make a contribution to the release of the latent heat. We therefore write the condensational heating into two parts. One of them is caused by the Ekman pumping and parameterized by the vertical velocity ω_B at the top of Ekman layer according to Charney's work; another is caused by the cumulus momentum mixing and parameterized by the vorticity at a reference layer as that used in [11] and [12]. Thus, the heating by cumulus convection may be written as follows

$$Q = c_p \varepsilon \eta(p) \left[M_* \frac{1}{r} \frac{\partial r v_*}{\partial r} + N \frac{1}{r} \frac{\partial r v_B}{\partial r} \right] \tag{15}$$

where $\eta(p)$ is the heating profile, ε and N are coefficients.

Substitution of (15) into (10) leads to the following equation

$$\frac{\partial^2 \xi}{\partial t \partial p} = \frac{s}{p} r \frac{\partial}{\partial r}\left(\frac{1}{r}\frac{\partial \psi}{\partial r}\right) - \frac{H}{f^2}\left[M_* r \frac{\partial}{\partial r}\left(\frac{1}{r}\frac{\partial \xi_*}{\partial r}\right) + N r \frac{\partial}{\partial r}\left(\frac{1}{r}\frac{\partial \xi_B}{\partial r}\right)\right], \tag{16}$$

Where $H \equiv \dfrac{\varepsilon R f \eta}{p}$ can be considered as a heating parameter.

The dependence of ψ and ξ on r and p is separable and their radial structures must be the same. We may seek a model solution for ψ and ξ in the forms of

$$\psi(r,p,t) = \varphi(r) E(p) e^{\sigma_t}$$

and

$$\xi(r,p,t) = \varphi(r) A(p) e^{\sigma_t}. \tag{17}$$

The radial structure function $\varphi(r)$ must satisfy the following equation

$$r \frac{d}{dr}\left(\frac{1}{r}\frac{d}{dr}\right)\varphi + \lambda^2 \varphi = 0 \tag{18}$$

with the boundary conditions corresponding to (11) and (12), namely

$$\lim_{r \to 0} \frac{\varphi}{r} = 0, \ \lim_{r \to \infty} \frac{\varphi}{r} = 0. \tag{19}$$

The solution of $\varphi(r)$ is

$$\varphi(r) = r J_1(\lambda r) \tag{20}$$

where $J_1(x)$ is the first-order Bessel function of the first kind, and λ^2 is a constant.

Substituting (17) into (9) and (10) and making use of (18), we obtain the following governing equations for $E(p)$ and $A(p)$:

$$\sigma A = -f \frac{dE}{dp} - \frac{d}{dp}[M(A - A_B)], \tag{21}$$

$$\sigma f \frac{dA}{dp} = -\lambda^2 S E + \frac{\lambda^2 H}{f}[M_* A_* + N A_B]. \tag{22}$$

The boundary conditions corresponding to (13) and (14) are

$$E(p_t) = 0, \quad E(p_B) = \frac{K p_B}{2f} A(p_B). \tag{23}$$

Thus, (21), (22) and (23) constitute an eigenvalue-eigenfunction problem. The eigenvalue σ and the vertical structure can be obtained according to the solution of the above problem.

Ⅲ. The Structure of the CISK-Overstability Convection

The numerical solutions to 15-level model described in the above section show that the heating by cumulus convection produces not only a steady unstable mode but also an oscillatory mode of instability. To compare this with the results obtained, the variations of the growth rate of the steady unstable mode with the length scale of the disturbance are shown in Fig. 1. The Ekman-CISK curve expresses the Ekman-CISK process and is similar to the results of the CISK. The CMM-CISK expresses the cumulus momentum mixing-CISK process. It is obvious that the cumulus momentum mixing can excite the CISK as the Ekman pumping can. The combined CISK expresses the combined processes of the Ekman pumping and the cumulus momentum mixing. The growth rates of the unstable mode and their relations with the heating profiles have been investigated in [13]. Therefore, we mainly show the structures of the CISK-overstability convection here.

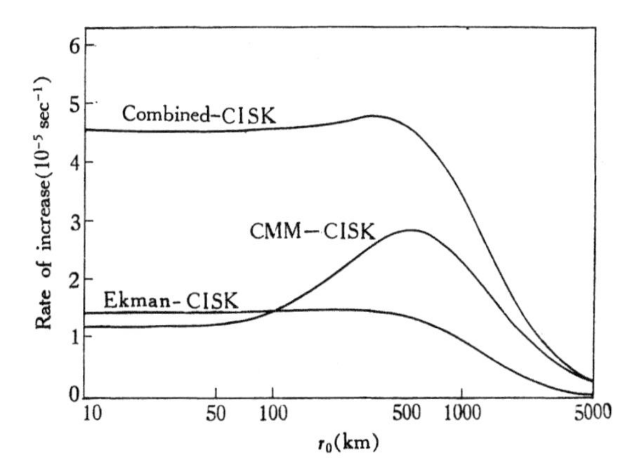

Fig. 1. Variations of growth rate with length scale (for steady unstable mode).

The distribution of the vertical velocity of the CISK-oscillatory unstable mode is shown in Fig. 2. It is very clear that the vertical velocity has a periodic variation with time. The maximum of the vertical velocity is located in the middlelower troposphere (750 Mb). The vertical velocity has the same sign in the vertical direction, but its distribution is not in axial symmetry.

The time-height sections of the temperature and the tangential velocity are respectively shown in Figs. 3 and 4. It can be shown that except the periodic variation, both the temperature and the tangential velocity vertically have opposite signs in the upper layer from the lower layer. There is a cyclonic disturbance (or the positive temperature variation) in the upper layer if there is an anticyclonic disturbance (or the negative temperature variation) in the lower layer, and *vice versa*. According to the comparison between Figs. 3 and 4, we can also see that the positive temperature perturbation is basically associated with the cyclonic circulation, but the negative temperature perturbation with the anticyclonic circulation. This matching structure shows that the cyclonic. disturbance has a property with warming core as the tropical cyclone.

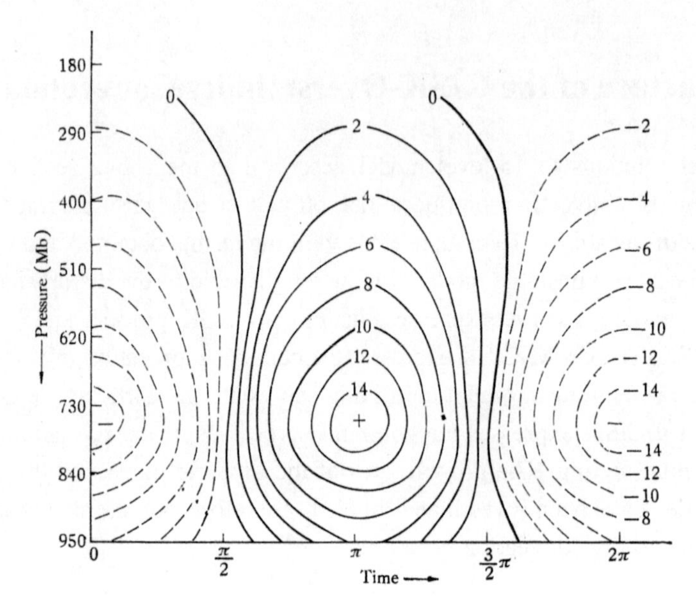

Fig. 2. Vertical velocity section of CISK-oscillatory unstable mode.

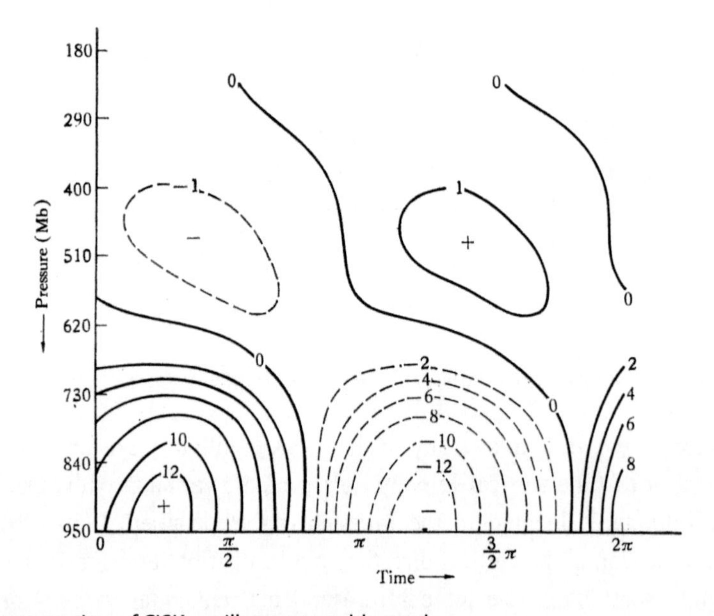

Fig. 3. Temperature section of CISK-oscillatory unstable mode.

Comparing Fig. 2 with Fig. 3, it is very clear that the rise is present during the temperature increasing in the middlelower troposphere, and the sinking is present during the temperature decreasing there. This implies that the disturbance satisfies the condition, for which the warmer air rises and colder air sinks in the middle-lower troposphere. So the available potential energy will be transformed into the kinetic energy of the disturbance. Thence the disturbance can get an unstable development.

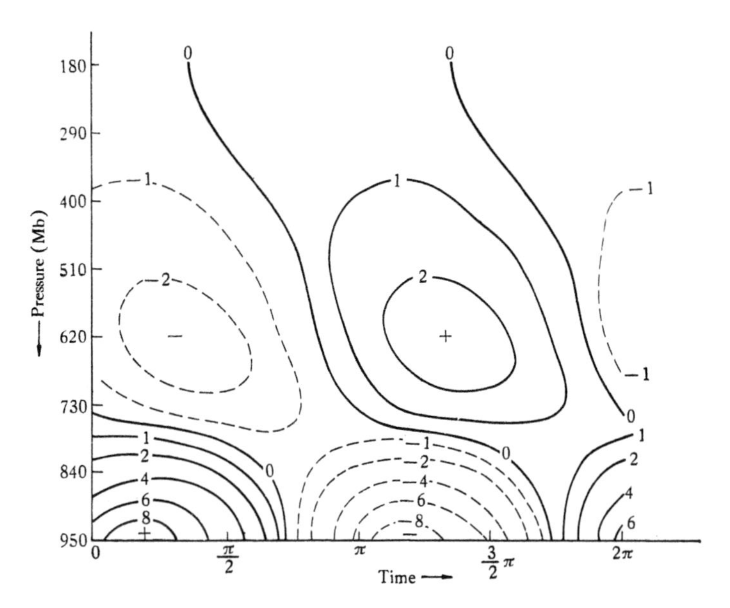

Fig. 4. Tangential velocity section of CISK-oscillatory unstable mode.

The variation of the kinetic energy of the disturbance shows that the kinetic energy has a periodic change and increases from a period to another. This means that the kinetic energy is produced. According to the energetics, it is clear that the disturbance can get an unstable development.

Ⅳ. On the Formative Conditions of the CISK-Overstability Convection

In order to further understand the formative conditions of the CISK-overstability convection, the theoretic analysis is necessary. We use a simple 2-level model to find the analytic solution. In the 2-level model, (21) and (22) can be written as follows.

$$\sigma A_1 + \frac{f}{\Delta} E_2 + \frac{M_2}{\Delta}(A_2 - A_4) = 0 \tag{24}$$

$$\sigma A_3 + \frac{f}{\Delta}(E_2 - E_4) - \frac{M_2}{\Delta}(A_2 - A_4) = 0 \tag{25}$$

$$\sigma f \frac{A_3 - A_1}{\Delta} + \lambda^2 S_2 E_2 - \frac{\lambda^2 H_2}{f} A_4 (M_4 + N) = 0 \tag{26}$$

where Δ=400 mb is the pressure interval. To obtain (24) to (26), we have already assumed that $E(p_t) = 0$, $M(p_t) = 0$ at the top of the model atmosphere, and $A_* = A_B = A_4$, where $p_4 = 950$ mb is the bottom of the model atmosphere.

For simplification, we assume that the tangential velocity just varies linearly from p_4 to p_3 i.e., $A_4 \approx aA_3$, where a is a coefficient. Thus, we can obtain from (23):

$$E_4 = KbA_3, \tag{27}$$

where

$$b=\frac{ap_4}{2f}.$$

It is not difficult to obtain the following equation from (24) to (26)

$$(2+\delta S_2)\sigma^2+\left[\frac{fb\delta S_2}{\Delta}K+\frac{fb}{\Delta}K+\frac{\delta S_2 a}{\Delta}M_2-\frac{\delta a}{\Delta}H_2(M_4+N)\right]\sigma \\ +\frac{S_2\delta fbM_2K}{2\Delta^2}=0, \tag{28}$$

where $\delta=\lambda^2\Delta^2/f^2$. Solving the algebraic equation (28). we get $\sigma(\sigma=\sigma_r\pm i\sigma_I)$, where

$$\sigma_r=\frac{H_2\delta a(M_4+N)-bfK(S_2\delta+1)-S_2\delta aM_2}{2(2+\delta S_2)\Delta}, \tag{29}$$

$$\sigma_I=\sqrt{\frac{2(2+S_2\delta)M_2bS_2\delta fK-[fbK(1+S_2\delta)+S_2\delta aM_2-H_2\delta a(M_4+N)]^2}{2(2+\delta S_2)\Delta}}. \tag{30}$$

For the unstable mode, the growth rate $a_r>0$ must be satisfied. The physical meanings of the terms in (29) are obvious. The first term represents the influence of the convective condensation heating due to the Ekman pumping and the cumulus momentum mixing. They make a positive contribution to the growth rate. The second and third terms represent respectively the weakening effect due to the surface boundary friction and the cumulus friction. Charney et al. have indicated in their CISK theory that the heating contribution due to the Enman pumping is greater than the frictional dissipation due to boundary layer and the conditional instability of the second kind just can be excited. Therefore, the physical nature expressed in (29) is the same as the one in the classical CISK theory. But the Ekman pumping and the cumulus momentum mixing are included together here.

Only when σ_I is real number and non-nought, can the oscillatory unstable mode be produced. This means a requirement for

$$2(2+S_2\delta)bS_2\delta fKM_2-[fb(1+S_2\delta)K+S_2a\delta M_2-H_2\delta a(M_4+N)]^2>0. \tag{31}$$

In order to satisfy the inequality (31). $M_2>0$ is the necessary condition. In other words, the cumulus momentum mixing (or the cumulus friction) is a necessary condition to excite the CISK - oscillatory instability. There is no oscillatory unstable mode if the steady unstable mode as the cumulus momentum mixing is neglected (using $M_2=M_4=0$). This analytic solution is consistent with the calculated result in the multi-level model. In this case, we did not get the overstability convection either when the effect of the cumulus friction is not considered. The Ekman pumping alone cannot excite the overstability convection in the multi-level model. This seems to be inconsistent with the inequality (31). The reason is that we used the approximation which is associated with the vertical velocity at the top of the boundary layer to get the analytic solution.

According to the inequality (31), we can also find that, the production of the oscillatory unstable mode is related to the heating, Coriolis parameter and the static stability. These are confirmed by computing in the multi-level model. For example, the profile of the static stability could influence the presence of the oscillatory unstable mode. When the static stability does not change with height, it is unfavourable to excite the oscillatory instability. The influence of the profile of the heating is as that noticed in [13]. If there is much heating in the lower troposphere, the presence of the oscillatory instability is more favourable.

We know that the excitation of overstability convection is related to the Rayleigh number R, Taylor number T and the Prandtl number P. R mainly reflects the heating of fluid, and T and P mainly reflect the rotating and viscosity of fluid respectively. The oscillatory instability is possible only when R is sufficiently large, as long as T is not too large and P is rather small (but $P \neq 0$). For the CISK model used in this work, the convective condensation heating and Coriolis parameter f are equivalent to the Rayleigh number and the Taylor number respectively. Introducing the cumulus friction into model is well matched that the certain Prandtl number is considered. Therefore, it is well consistent with the result obtained in the previous studies that the oscillatory unstable mode is possible only when the cumulus momentum mixing is introduced. Thus, we could con clude that the convective condensation heating, the rotation of the earth and the cumulus friction are the basic factors to produce the CISK-overstability convection. Especially, the cumulus friction not only has a dissipating effect but also transports the momentum and the heating. Through this transportation, the periodic phenomenon of the convection activity is excited.

Ⅴ. Discussion

Both the analysis and the numerical calculation mentioned above show that the convection heating and the cumulus friction excite not only the steady convection but also the overstability convection based on the CISK mechanism. In order to discuss the relation between the overstability convection and the waves in the tropical atmosphere, we will analyze their periods. As indicated above the CISK-overstability convection depends highly on the profile of the heating, which is not fixed in the real atmosphere. We can just discuss some typical cases. Assuming that the vertical distribution of the convection heating has a maximum at the upper troposphere, the middle troposphere and the lower troposphere respectively, the calculated results show that (i) when the maximum heating is at the lower troposphere, the period of the oscillatory instable mode with wavelength 9300 km is 4.4 days, and the period is 6.1 days for the oscillatory unstable mode with wavelength 24,500 km; (ii) when the maximum heating is at the middle/upper troposphere, the disturbance with wavelength 9300 km is not the oscillatory unstable mode but the disturbance with wavelength 24500 km is the oscillatory unstable mode, and the periods are 16.6 and 28 days respectively.

In the tropical atmosphere, the oscillatory period of the Kelvin wave is about 10—20 days, and it is about 4—5 days for the mixed Rossby-gravity wave. Even though the calculated results are difficult to compare with the observed ones (since the calculation is just for some special cases), we can still deem that the Kelvin waves and the mixed Rossby-gravity waves can be driven by the CISK-overstability convection. Through the CISK mechanism, what kind of the periodic wave can be excited by the convection heating and the cumulus friction? This depends on the vertical distribution of the convection heating to a great extent. As the heating is maximum at lower troposphere, it is favorable to produce the overstability convection corresponding to the Kelvin waves. It is favorable to produce the overstability convection corresponding to the mixed Rossby-gravity waves if the heating is maximum at the middle-upper troposphere.

The tropical atmosphere is often in the moist condition and instability situation. The cumulus convection is very frequent. But the condensation heatings are quite different since the intensities of the convection are not the same. In general, the deep cumulus convection can make a maximum heating in the middle-upper troposphere. But there is a maximum heating in the middle-lower troposphere for the normal convection. The mixed Rossby-gravity waves and the Kelvin waves may be the reactions of the above two convection processes respectively. The analysis from the observational data shows that the wave-length and the period have a wider range for the Kelvin waves and the mixed Rossby-gravity waves. This supports the above conclusion from another angle. Because the different cumulus convections can make the condensation heating with the different intensities and vertical distributions and then excite the oscillatory unstable mode with the different wave-lengths and periods.

Ⅵ. Concluding Remarks

The wave in the tropical atmosphere is a major issue. In this paper, a preliminary discussion is made and a theoretic mechanism to produce the mixed Rossby-gravity waves and the Kelvin waves is proposed. Although some problems remain to be investigated further, the following conclusions are significant.

1. The convection heating is able to excite the CISK oscillatory unstable mode (or called the CISK-overstability convection) through the CISK mechanism. This CISK-overstability convection could drive the mixed Rossby-gravity waves and Kelvin waves, bringing about other oscillatory phenomena in the tropical atmosphere.

2. Except the condensation heating, the cumulus friction (the cumulus momentum mixing) produced by convection is also important. It gives a fundamental effect to excite the CISK-overstability convection. If there is not the cumulus friction in the model, the steady unstable mode can just be obtained but the oscillatory unstable mode cannot.

3. The CISK-overstability convection is excited based on the CISK mechanism. They have the property like the CISK unstable mode has.

References

[1] Matsuno, T., J. Meteor. Soc. Japan, 44(1966), 25-43.

[2] Yanai, M. & Maruyama, T., *ibid.* 44(1966), 291-294.

[3] Lindzen, R. S. *Mon. Wea. Eev.*, 95(1967), 441-451.

[4] Wallace, J. M. & Kousky, V. E., *J. Atmos. Sci.*, 25(1968), 900-907.

[5] Holton, J, R. & Lindzen, R. S., *Mon. Wea. Rev.*, 96(1968), 385-386.

[6] Chang, C. P., *J. Atmos. Sci.*, 27(1970), 133-138.

[7] Holton, J. R., *ibid.*, 29(1972), 368-375.
[8] 麦文建, 李崇银, 中国科学, (B), 1982, 758-768.
[9] Charney, J. G. & Eliassen, A., *J. Atmos. Sci.*, 21(1964), 68-75.
[10] Schneider, E. K. & Lindzen, R. S., *J. Geophys. Res.*, 81(1976), 3158-3160.
[11] Mak. M. K., *13th Technical Conference on Hurricanes and Tropical Meteorology*, December 1.5, 1980, Miami Beach.
[12] Mak. M. K., *Tellus*, 33(1981), 531-537.
[13] 李崇银, 大气科学, 7(1983), 第 3 期, 260-268.

Actions of Summer Monsoon Troughs (Ridges) and Tropical Cyclones over South Asia and the Moving CISK Mode

Li Chongyin (李崇银)

Institute of Atmospheric Physics, Academia Sinica, Beijing

Abstract The analyzed wind field during MONEX shows that the monsoon troughs (or ridges) over South Asia have 30—50 day oscillation and slowly propagate northward. Simultaneously, the tropical cyclones in South Asia could disperse energy to the east.

Considering the feedback effect of the cumulus convection and the influence of the shearing basic flow, we introduce the vertical shearing basic flow into the conditional instability of the second kind (CISK) model and obtain a moving CISK mode. Under the Summer Monsoon in South Asia, the meridional propagation speed (northward) is about 0.6 latitude per day and the oscillation period is about 34 days for this CISK mode. And it has an easterly group velocity. The above-mentioned characteristics of this mode are similar to the analyzed results of the data during MONEX. Therefore, it can be initially concluded that the CISK mode caused by the shearing basic flow and the feedback effect of the cumulus convection seems to be an important mechanism driving the monsoon troughs (ridges) and the tropical cyclones over South Asia in summer.

Ⅰ. Introduction

The summer monsoon over South Asia is the strongest and the most typical in the world. The actions of the summer monsoon can seriously influence the weather and climate in South Asia region (including the southern part of China). Therefore, meteorologists pay great attention to the study on the summer monsoon over South Asia all along.

On the actions of summer monsoon, the "active" monsoon and "break" monsoon are known very well. In fact, the "active" and "break" monsoons are associated with the actions of monsoon troughs (ridges). Thus, understanding the action law and dynamic mechanism of the monsoon trough (ridge) is of great significance.

In monsoon trough, there are often the monsoon depressions over Indian subcontinent and the tropical cyclone over the Bay of Bengal. They usually cause severe weather (heavy rain)

Received January 11, 1984; revised March 6, 1985.

in these regions. Therefore, the investigation on the action laws of the tropical cyclone and monsoon depression over South Asia is important.

MONEX provided a lot of observation data for the study of summer monsoon over South Asia. Based on the analysis of the observation data during summer monsoon season in 1979, Krishnamurti and Subrahmanyam[1] noticed that the monsoon troughs (ridges) have a 30—50 day periodic oscillation, and held that the 30—50 day mode exists during summer monsoon season over South Asia. Murakami et al.[2] also confirmed the existence of 40—50 day oscillation during summer monsoon season over South Asia through the analysis of MONEX data. Simultaneously, they pointed out the phenomena of eastward dispersion of energy of tropical cyclones and monsoon depressions over South Asia. This indicates that the wave with eastward propagation of group velocity exists in this region. Then, are the above-mentioned two phenomena caused by the same wave? How this wave is excited? These are questions which have not been investigated and need to be answered.

The CISK is regarded as a fundamental mechanism for the genesis and maintenance of the tropical cyclone. Some investigations on the monsoon depression indicate that the formation mechanism of the monsoon depression is mainly the barotropic instability, while the development and maintenance of the monsoon depression are dependent on CISK. Therefore, it is necessary to consider CISK mechanism when the action of the tropical cyclone (or monsoon depression) over South Asia during summer monsoon season is discussed. In addition, there exists an outstanding vertical shearing. in the wind field over South Asia during the summer. So, the dynamic effects of the vertical shearing wind should be considered when the action of the summer monsoon system is studied.

Ⅱ. Results of Analysis

According to the analysis of wind field data, Krishnamurti and Subrahmanyam confirmed the existence of a low frequency mode oscillation with a period range of 30—50 days during the MONEX summer. They observed the steady meridional propagation of a train of troughs and ridges from the equator to about 30°N. The meridional scale of this mode is around 3000 km, and its meridional speed of propagation is ~0.75° latitude per day. The amplitude of the wind for this mode is about 3—6 ms^{-1}. Based on Ref. [1], the positional variations of troughs (dashed lines) and ridges (solid lines) over South Asia during the period from May 2 to June 9, 1979 are shown in Fig. 1. It is clear that the monsoon trough moves from about 5°N to near 27°N during the period of May 2—22, and the monsoon ridge propagates from about 3°N to near 23°N during the period from May 20 to June 9. Therefore, the slower northward moving of monsoon trough (ridge) is undoubted during MONEX summer, and the transformation time between trough and ridge (half-period of oscillation) is about 20 days (Fig. 1).

Fig. 2 is the time-latitude sections of the zonal mean winds at 850 mb, averaged between 60° and 150°E. These U' and V' have been filtered by using 45 day filter. The periodic variations of the wind field and its slower northward moving from the equator are all clear.

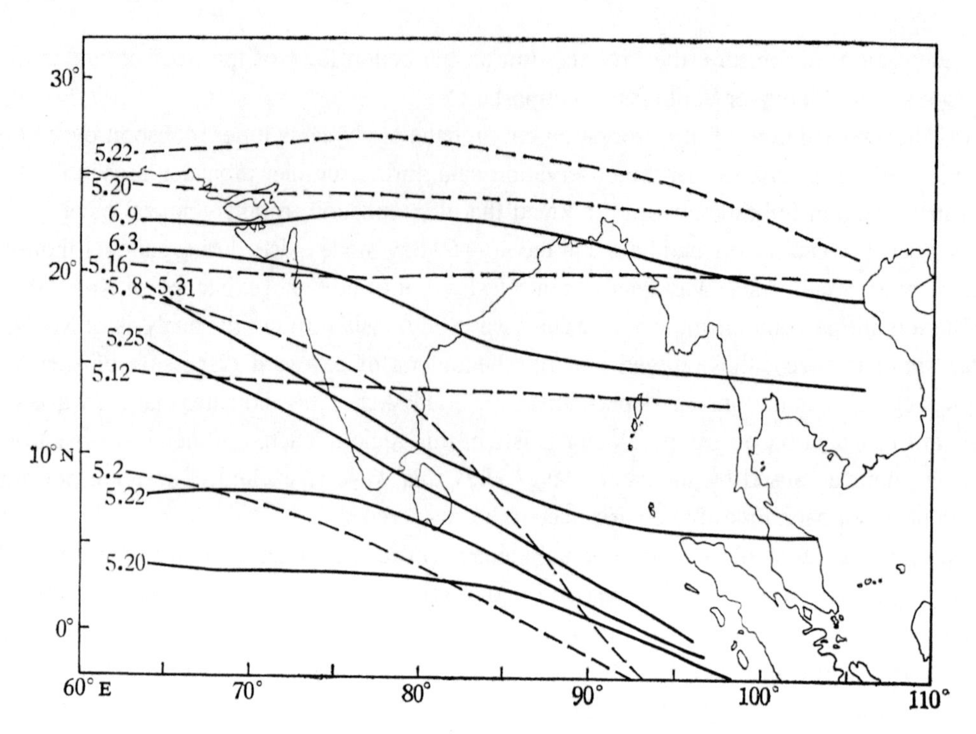

Fig. 1. The positions of monsoon troughs (dashed lines) and ridges (solid lines) over South Asia during the period from May 2 to June 9, 1979.

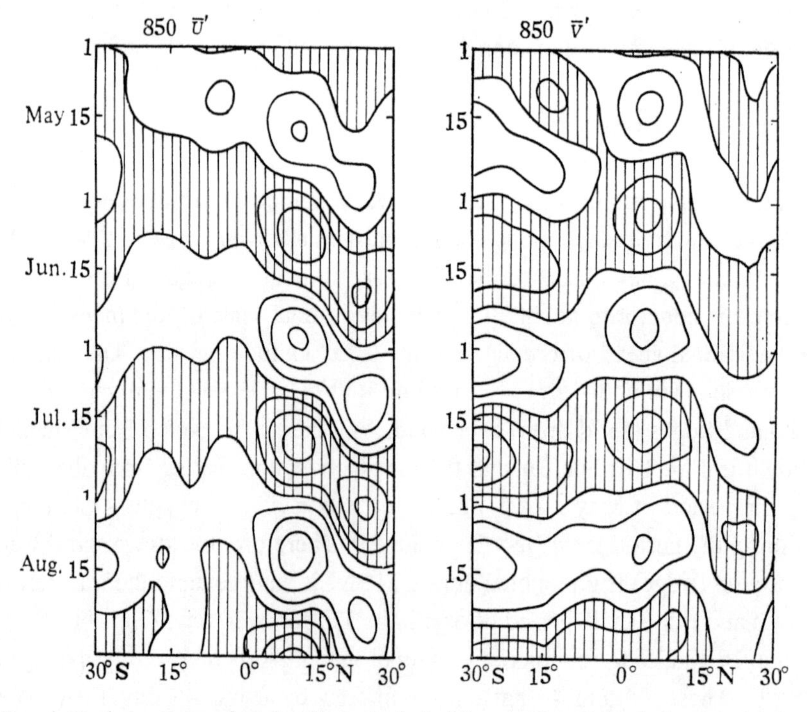

Fig. 2. The time-latitude sections of 45 day filtered winds (U' and V') at 850 mb (from Ref. [2]).

The tropical cyclones and monsoon depressions are important component parts of the monsoon trough. They move not only westward in the monsoon trough, but also northward with the propagation of monsoon trough. Besides, Murakami et al. indicated that the development of the tropical cyclone over South Asia should generate another tropical cyclone in east by south direction In other wards, tropical cyclones can disperse energy east-southeastward. The propagation direction of the group velocity is opposite to the phase velocity of tropical cyclone and monsoon depression. The time-longitude sections of U' and Φ' at 850 mb along 15°N over South Asia during summer in 1979 are shown in Fig. 3. It can be seen that there exist 40—50 day oscillations in the wind field and geopotential field, and the same period exists in the activities of the tropical cyclones (numbered in Fig. 3). Simultaneously, the successive appearance of the tropical cyclone reflects the existence of eastward group velocity. The speed of energy dispersion is about 4—8 longitudes per day. Fig. 3 shows that three perturbations exist in May, The first perturbation ("1") is a tropical cyclone which developed in the equatorial Bay of Bengal on May 7 and reached the east coast of India; the second one ("2")was generated over Indo-China on May 16; and the third one ("3") formed near the Philippines on May 20. Thus, it is obvious that the tropical cyclones propagate eastward in form, i.e. the energy of the cyclones is dispersed eastward (the eastward group velocity). Three similar events occurred during the periods: June 16—July 6, August 1—21 and September 17—28.

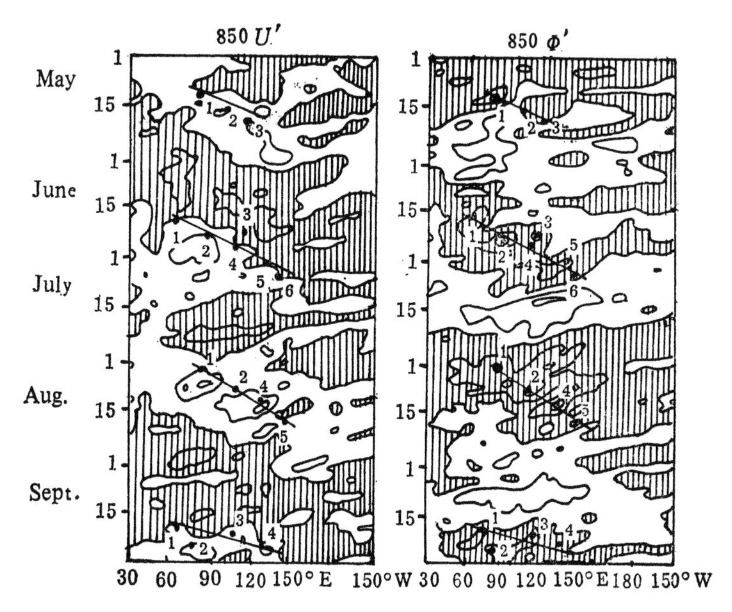

Fig. 3. Time-longitude sections of 850 mb U' and Φ' along 15°N over South Asia during summer in 1979. Shaded parts denote anomaly easterlies (left) and negative (right), respectively.

III. Moving CISK Mode Caused by Vertical Wind Shearing

As we know, the genesis and development of the tropical cyclones and the development and maintenance of the monsoon depressions all are dependent on CISK. However, there are stronger easterlies in the upper troposphere and stronger westerlies in the lower troposphere over South Asia during summer monsoon season. Therefore, the vertical shearing basic flow must influence CISK. And the actions of the tropical cyclones and monsoon depressions are also influenced.

The tropical cyclone and monsoon depression can be regarded as an axial symmetry system. The axi-symmetrical coordinates are used in the classic CISK theory. To introduce the basic flow, we apply a plane-symmetrical coordinates instead of the cylindrical one. Since wind includes x-component (u) and y-component (v), for the case of $d\bar{u}/dp$, according to Ref. [3-5], the disturbance equations can be written

$$\frac{\partial u}{\partial t}+\omega\frac{d\bar{u}}{dp}=fv, \tag{1}$$

$$\frac{\partial \phi}{\partial y}=-fu, \tag{2}$$

$$\frac{\partial \phi}{\partial p}=-\frac{RT}{p}, \tag{3}$$

$$\frac{\partial v}{\partial y}+\frac{\partial \omega}{\partial p}=0, \tag{4}$$

$$\frac{\partial T}{\partial t}+v\frac{\partial \bar{T}}{\partial y}-\frac{pS}{R}\omega=\frac{Q}{C_p}, \tag{5}$$

where u, v and ω are respectively x-, y- and p-direction velocities; ϕ and T are geopotential and temperature respectively; f is the Coriolis parameter (constant); R is the gas constant; C_p is the specific heat at constant pressure; $S=-\frac{R}{p}\left(\frac{\partial \bar{T}}{\partial p}-\frac{R\bar{T}}{pC_p}\right)$, static instability parameter; Q is the convective condensation heating rate for the unit mass air ; $\bar{u}$ and $\bar{T}$ are the basic quantities and they should satisfy the relationship

$$\left.\begin{aligned}\frac{\partial \bar{\phi}}{\partial y}&=-f\bar{u},\\ \frac{\partial \bar{\phi}}{\partial p}&=-\frac{R\bar{T}}{p}.\end{aligned}\right\} \tag{6}$$

For the two-level model, the convective condensation heating is introduced following the example of the Charney's parameterization[6],

$$Q_2=-\frac{\mu L}{2\Delta}(\bar{q}_{s3}-\bar{q}_{s1})\left(\omega_2+\frac{1}{2}\omega_4\right), \tag{7}$$

where L is the mean latent heat of condensation; q_s is the saturation value of the specific humidity; Δ=450 mb is the pressure interval; μ is a heating parameter.

Assume that the basic flow is linear variation with altitude, i.e. $d\bar{u}/dp=$ constant. From Eqs. (1)—(7), we can obtain the following equation

$$\frac{\partial^2 \omega_2}{\partial y^2}+2\frac{d\bar{u}}{dp}\frac{f}{S_2\Delta}\frac{\partial}{\partial y}(\omega_3-\omega_1)+\frac{f^2}{\Delta^2 S_2}(\omega_4-2\omega_2)=\mu H\frac{\partial^2}{\partial y^2}\left(\omega_2+\frac{1}{2}\omega_4\right) \tag{8}$$

where

$$H=\frac{RL}{2C_p S_2 p_2 \Delta}(\bar{q}_{s3}-\bar{q}_{s1})$$

is an atmospheric state parameter. For the tropical atmosphere, one can take H=1.1 approximately.

The investigation completed by Krishnamurti shows that the vertical distribution of the vertical velocity in monsoon depression is similar to that in tropical cyclone[7]. In general, there is the maximum ω-velocity between 300—400 mb. Thus, the approximate relationship $\omega_3-\omega_1=\frac{1}{2}(\omega_4-\omega_2)$ could be used. Introducing the stream function ψ and assuming $\psi=\psi e^{\sigma t}$, the equation relative to ψ_2 can be obtained from (8):

$$\frac{\partial^2 \psi_2}{\partial y^2}+\frac{d\bar{u}}{dp}\frac{f}{S_2\Delta}\frac{\partial}{\partial y}(\psi_4-\psi_2)+\frac{f^2}{\Delta^2 S_2}(\psi_4-2\psi_2)=\mu H\frac{\partial^2}{\partial y^2}\left(\psi_2+\frac{1}{2}\psi_4\right) \tag{9}$$

Because of the friction effect of boundary layer, the vertical motion caused by Ekman pumping can be introduced on the basis of the Charney's study. And the following relationship can be obtained

$$\psi_4=\frac{K}{K+\sigma}\psi_2, \tag{10}$$

where K is the friction coefficient of boundary layer. For the tropical atmospheric boundary layer, we can take $K=1.72\times10^{-6}\ \text{s}^{-1}$.

(10) is introduced into Eq. (9). Taking $\psi_2=Ae^{ily}$ and since $\sigma=\sigma_r+i\sigma_I$, we can obtain the following two equations:

$$[2(\sigma_r+K)-\mu H(2\sigma_r+3K)]l^2-2l\varepsilon\sigma_1+\eta(2\sigma_r+K)=0, \tag{11}$$

$$2(1-\mu H)\sigma_I l^2+2l\varepsilon\sigma_r+2\eta\sigma_I=0. \tag{12}$$

Thus, the growth rate (σ_r) and the frequency (σ_I) of the CISK mode can be written as follows:

$$\sigma_r=\frac{[(3\mu H-2)l^2-\eta]K}{2[(1-\mu H)l^2+\eta]+\dfrac{l^2\varepsilon^2}{(1-\mu H)l^2+\eta}}, \tag{13}$$

$$\sigma_{Iy}=-\frac{l\varepsilon\sigma_r}{(1-\mu H)l^2+\eta}, \tag{14}$$

where $\eta=2f^2/\Delta^2 S^2$; $\varepsilon=f\dfrac{d\bar{u}}{dp}/S_2\Delta$, and ε describes the effect of vertical shearing basic flow; l is the wavenumber in y-direction.

According to Formulae (13) and (14), if ε=0, then σ_I=0, and the standing CISK mode just exists. In other words, the convective condensational heating feedback should produce an immovable CISK in the case without vertical shearing basic flow. But if $\varepsilon\neq0$, then $\sigma_I\neq0$, in general, the moving CISK mode should appear. In other words, when the vertical shearing basic flow exists, the feedback effect of the cumulus convection can cause a moving CISK mode.

Analogously, considering $d\bar{v}/dp$ effect, then taking k instead of l and exchanging u and v each other, we can still obtain

$$\sigma_r = \frac{[(3\mu H - 2)k^2 - \eta]K}{2[(1-\mu H)k^2 + \eta] + \dfrac{k^2\varepsilon^2}{(1-\mu H)k^2 + \eta}}, \tag{15}$$

$$\sigma_{Ix} = -\frac{\varepsilon k \sigma_r}{(1-\mu H)k^2 + \eta}, \tag{16}$$

where k is the wavenumber in x-direction; $\varepsilon = f\dfrac{d\bar{v}}{dp}/S_2\Delta$. If the wavenumbers are same in x- and y-directions ($k = l$) and $\dfrac{d\bar{u}}{dp} = \dfrac{d\bar{v}}{dp}$, then Formula (13) is just the same as (15), This is an inevitable outcome and indicates that the treatment in this section is reasonable.

Ⅳ. Results of Model Atmosphere and Discussion

The development and maintenance of the tropical cyclone and monsoon depression are dependent on CISK. The moving CISK mode must influence the action of tropical cyclones and monsoon depressions. According to the wave theory in the atmosphere, from (14) we can respectively obtain the phase speed C_y and the group velocity C_{gy} as follows:

$$C_y = -\frac{\sigma_{Iy}}{l} = \frac{\varepsilon\sigma_r}{(1-\mu H)l^2 + \eta}, \tag{17}$$

$$C_{gy} = -\frac{d}{dl}\sigma_{Iy} = C_y\left[1 - \frac{2l^2(1-\mu H)}{(1-\mu H)l^2 + \eta}\right] + \frac{l\varepsilon}{(1-\mu H)l^2 + \eta}\frac{d\sigma_r}{dl}. \tag{18}$$

Based on the CISK theory, when the disturbance has developed into the tropical cyclone, the growth rate of the tropical system (with scale k_c, l_c) should be the maximum. Then $(d\sigma_r/dl)_{l_c} \approx 0$ can be taken approximately. We, therefore, obtain

$$C_y \approx \frac{f\sigma_r}{S_2\Delta[(1-\mu H)l^2 + \eta]}\frac{d\bar{u}}{dp}, \tag{19}$$

$$C_{gy} \approx C_y\left[1 - \frac{2l^2(1-\mu H)}{(1-\mu H)l^2 + \eta}\right]. \tag{20}$$

Analogously, it is obtained from (16) that

$$C_x \approx -\frac{f\sigma_r}{S_2\Delta[(1-\mu H)k^2 + \eta]}\frac{d\bar{v}}{dp}, \tag{21}$$

$$C_{gx} \approx C_x\left[1 - \frac{2k^2(1-\mu H)}{(1-\mu H)k^2 + \eta}\right]. \tag{22}$$

From the analyzed expressions of the phase speed and the group velocity for the CISK mode —— Formulae (12) — (22), it is obviously shown that:

1) In general, the following inequality is tenable:

$$1<\frac{2l^2(1-\mu H)}{(1-\mu H)l^2+\eta}\left(\text{or}\frac{2k^2(1-\mu H)}{(1-\mu H)k^2+\eta}\right)<2,$$

therefore, the group velocity (C_g) is smaller than the phase speed for the CISK mode and they have the opposite propagation directions. In other words, if the CISK modes propagate westward, their energy should be slowly dispersed eastward.

2) During the summer monsoon season over South Asia, there is $d\bar{u}/dp \geqslant d\bar{v}/dp>0$ generally. Therefore, the CISK modes here propagate northwestward and the group velocity propagates southeastward since $1-2k^2(1-\mu H)/[(1-\mu H)k^2+\eta]<0$ in general.

The above-mentioned results are in accord with the analyzed data by Murakami et al.

For the atmospheric basic state dicing the summer monsoon season over South Asia, we can take $f = 0.377 \times 10^{-4}\,\text{s}^{-1}$ (the value at 15°N), $s_2 = 0.032\text{m}^2\text{s}^{-2}\text{mb}^{-2}$, $\mu = 0.865$. Since the diameter of the convective cloud region in the tropical cyclone is about 800—1000km in general, then $k = l = 3.49 \times 10^{-6}\text{m}^{-1}$. Based on the distribution of mean wind during the summer monsoon season over South Asia, $\frac{d\bar{u}}{dp}\approx 3.54\times10^{-2}\text{ms}^{-1}\text{mb}^{-1}$ and $\frac{d\bar{v}}{dp}\approx 1.23\times10^{-2}\text{ms}^{-1}\text{mb}^{-1}$ could be accepted. By using the above-mentioned parameters, the following results can be obtained:

$$C_y = 2.7(\text{km/h}),\ C_x = -0.94(\text{km/h}),$$

$$C_{gy} = -0.40(\text{km/h}),\ C_{yx} = 0.14(\text{km/h}),$$

and the period of the meridional propagation of the CISK mode is about 34 days.

According to the analyzed wind field a 850mb during MONBX, Krishnamurti et al. showed that the monsoon troughs (or ridges) have 30—50 day oscillation phenomena during the summer monsoon season over South Asia, And the monsoon trough (or ridge) line propagates northward with about 0.75 latitude per day. On the basis of the present study, the convective condensational heating through CISK mechanism can produce a moving CISK mode under the effect of the vertical shearing basic flow. For the summer atmospheric condition over South Asia, this mode propagates northward with a speed of 0.6 latitude per day and about 34 days oscillation period. These are similar to the results of the analyzed data. Therefore, It can be preliminarily concluded that 30—50 day oscillation during the summer monsoon season over South Asia is driven by the slow, northward moving CISK mode.

It is shown in [2] that the tropical cyclones have energy dispersion phenomena and the group velocity is about 3—8 longitude per day southeastward. These results seem to be similar to the above-mentioned simple computations of the general tendency. The disturbances propagate west by north and their energies disperse east by south. But there exist some differences. Because the model and parameters used are all simple, this cannot reflect the real atmospheric motion state during the summer monsoon season over South Asia. In addition, it may be the other reason that the vertical shearing of flow is only introduced into our theory, but the results do not include the influences of the basic flow on the movement of disturbances. According to the "guide" principle of the movement of tropical cyclones, the speed and direction of the tropical cyclone movement are controlled by the "guide" flow and the mean winds at 500 mb can represent the guide flow. There is weaker easterlies at 500 mb in general during the summer monsoon season over South Asia. Then, the speed and movement direction of the tropical cyclones (or monsoon depressions) in the model will be close to that in real atmosphere.

The group velocity computed in this paper is smaller than and different from the result obtained by Murakami et al. But the following analysis should be reasonable: the development and maintenance of the tropical cyclone depend on CISK. The energy comes from the release of latent heat and the supply of the water vapour is mainly through the frictional convergence of the boundary layer. Therefore, the transport of disturbance energy must be influenced by the lower troposphere flow. Simultaneously, it has been proved in other study that the cyclonic shearing of basic flow could strengthen CISK and is favourable for the development of tropical cyclone[8]. During the summer monsoon season over South Asia, the cyclonic shearing of basic flow in the lower troposphere is mainly caused by each enhancement of monsoon. Thus, since the group velocity of the CISK mode exerts influence of mean westerly in lower troposphere on energy transport, the velocity of energy dispersion will be basically close to that obtained by Hurakami et al.

During the summer season over South Asia, since the southwest monsoon in the lower troposphere is stronger and there is stonier easterlies caused by South Asia High in the upper troposphere, the vertical shearing of basic flow is stronger and steady. Then, the CISK mode is the strongest and the periodic oscillation of monsoon troughs (or ridges) and the energy dispersion phenomena of tropical cyclones are the most obvious in South Asia as compared with other places in the world. This has been proved by the MONEX data analysis. Based on the results of this study, it can be inferred that the ITCZ over West Pacific and the South China Sea also has long-period oscillation phenomena. But, since the vertical shearing of basic flow is weaker there, the CISK mode may be weaker relatively and has different oscillation period. Of course, these inferences remain to be confirmed by using observation data.

MONEX data analysis and the results in this paper show that the tropical cyclones and monsoon depressions in monsoon trough could disperse energy eastward by south. And this is favourable for the development of new disturbance. But it should be pointed out that the energy dispersion is just one possibility to cause the new disturbance and the genesis of tropical disturbance has other mechanism and is controlled by different conditions.

V. Concluding Remarks

In this paper, we introduce the vertical shearing of basic flow which is the most obvious in South Asia during the summer monsoon season into CISK model, and obtain a moving CISK mode. The meridional speed and oscillation period of this mode is about 0.6° latitude per day and 34 days, respectively. These are similar to the characters of the 30—50 day mode noticed by Krishnamurti et al. through the analyzed MONEX data. In another study, it has been shown that the 30—50 day mode is the most obvious over South Asia[9]. This is also conformable to the theoretical analysis in this paper. Since South Asia is located in the tropics, CISK often exists and the vertical shearing of the mean winds is the most obvious there. Thus, moving CISK mode is the most active in South Asia during the summer monsoon season. It can be therefore priliminarily concluded that 30—50 day oscillation over South Asia during the summer monsoon season is driven by moving CISK mode which is caused by the convective condensational feedback and the vertical shearing of basic flow.

The group velocity is opposite to the phase for the CISK mode. Considering the influence of mean westerlies in the lower troposphere on the energy transport, the energy dispersion speed close to Munakami's can be obtained. This shows that the energy dispersion of tropical cyclone in South Asia is also associated with the CISK mode.

A dynamic explanation for 30—50 day oscillation phenomena and the eastward dispersion of energy of tropical cyclone over South Asia during the summer monsoon season is given for the first time in this study. But, since the model is too simple, the theoretical results close to the real atmosphere completely is difficult to obtain. The detailed theoretical analysis and numerical simulation need to be further studied.

References

[1] Krishnamurti, T. N. & Subrahmanyam, D., The 30—50 day mode at 850 mb during MONEX, *J. Atmos. Sci.*, 39(1982), 1088-1095.

[2] Murakami, T., Nakazawa, T. & J. He, On the 40—50 day oscillations during the 1979 Northern Hemisphere Summer, Part I: phase propagation, *J. Meteor. Soc. Japan*, 62(1984), 440-468.

[3] Shukla, J., CISK-barotropic-baroclinic instability and the growth of monsoon depression, *J. Atmos. Sci.*, 35(1978), 495-508.

[4] Mak, M. & Jim Kao, C. Y., An instability study of the onset-vortex of the southwest monsoon, 1979, *Tellus*, 34(1982), 358-368.

[5] 李崇银, 垂直切变基本气流中的 CISK, 大气科学, 7(1983), 427-431.

[6] Charney, J. G. & Eliassen, A., On the growth of the hurricane depression, *J, Atmos. Sci.*, 21 (1964), 68-75.

[7] Krishnamurti, T. N. et al., Study of a monsoon depression (Ⅰ), synoptic structure, *J. Meteor, Soc. Japan*, 53(1975), 227-240.

[8] 李崇银, 环境流场对台风发生发展的影响, 气象学报, 41(1983), 275-284.

[9] Krishnamurti, T. N. & Sulochana Gadgil, On the structure of the 30 to 50 day mode over the globe during FGGE Tellus, Series A, 37A(4)(1985), 336-360.

An Observational Study of the 30-50 Day Atmospheric Oscillations Part Ⅱ: Temporal Evolution and Hemispheric Interaction Across the Equator

Li Chongyin (李崇银), Zhou Yaping (周亚萍)

LASG, Institute of Atmospheric Physics, Academia Sinica, Beijing

Abstract In this part, the temporal evolution and interaction across the equator of 30-50 day oscillation in the atmosphere are investigated further. The annual variation of 30-50 day oscillation is quite obvious in the mid-high latitudes. In the tropical atmosphere, the obvious interannual variation is an important property for temporal evolution of 30-50 day oscillation. The low-frequency wavetrain across the equator over the central Pacific and central Atlantic area, the movement of the long-lived low-frequency system across the equator and the meridional wind component across the equator will obviously show the interaction of 30-50 day oscillation in the atmosphere across the equator.

Ⅰ. Introduction

In Part Ⅰ (Li et al., 1990), some characteristics in relation to the structure and propagation of 30-50 day oscillation have been shown. We focused on the differences of the structure and propagation of 30-50 day oscillation in the mid-high latitudes from that in the tropical atmosphere. It is clear that the 30-50 day oscillation in the mid-high latitudes has typical barotropics structure and mainly propagates westward.

The 30-60 day oscillations have been regarded as a global low-frequency system. But the study in relation to the temporal variation and interaction across the equator is not quite ample yet. Based on the analysis of 30-50 day motion fields at 850 hPa and 200 hPa for 1980-1984 BCMWF data and FGGE (1979) data, Mehta and Krishnamurti (1988) initially investigated the interannual variability of 30-50 day oscillation. They found that the northward propagation speed of the trough and ridge systems over Indian summer monsoon region is different in 1979, 1982 and 1983 from that in 1980, 1981 and 1984. The amplitude of the divergent wave was found to be variable from one year to the next. But it is very clear that the characteristics of interannual variation of 30-60 day oscillation, especially its relationship with ENSO event are not clearly understood; and the characteristics of annual variation are not sufficiently

Received December 21, 1990; revised March 18, 1991.

studied too.

The possible relation of the quasi-40 day oscillation of the summer monsoon in East Asia to the atmospheric circulation in Australia has been indicated by analyzing the data in May-August, in 1980 (Chen et al., 1982). The meridional propagation of the quasi-40 day oscillation was studied by He (1990) and Li (1990). But the propagational feature of 30-50 day oscillation across the equator from one hemisphere to another is not studied very well.

According to the analyses using the ECMWF data, temporal evolution of 30-50 day oscillations and their hemispheric interaction across the equator will be investigated further in this part.

II. Annual Variation of 30-50 Day Oscillation in the Atmosphere

The 30-50 day oscillation is a universal feature of the atmospheric motion. As well as the regional differences, the temporal variation of 30-50 day atmospheric oscillation is also very important. In this section, we focus on the annual variation of 30-50 day oscillation in the atmosphere.

Fig.1 shows the kinetic energies of 30-50 day oscillation at 850 hPa along 55°S latitude averaged in January and July, 1981, respectively. It is very clear that the kinetic energy in July is much larger than that in January. This means the 30-50 day atmospheric oscillation in winter is much stronger than that in summer along 55°S.

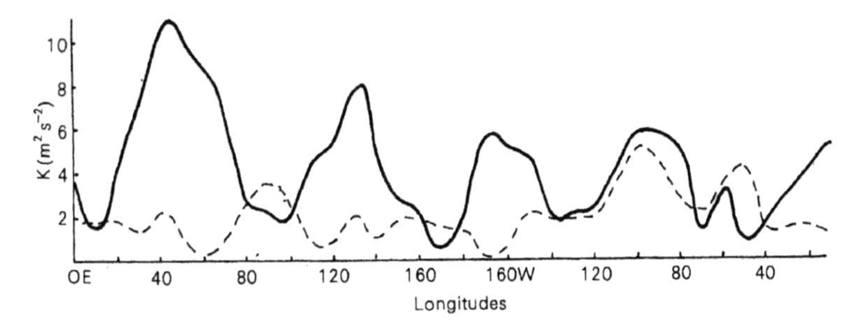

Fig. 1. Longitudinal distribution of kinetic energy at 850 hPa for 30-50 day oscillation in January (dashed line) and July (solid line), 1981 along 55°S latitude.

Is the annual variation a general feature of 30-50 day oscillation, as shown in Fig.1 that it is stronger in winter and weaker in summer? In order to investigate this question, the temporal variation of 30-50 day band-pass filtered u^2 at 200 hPa in the region (30°-50°N, 80°-180°E) is given in Fig.2. It is able to represent the temporal variation of 30-50 day oscillation in the mid-high latitudes. We can find in Fig.2 that the maximum u^2 is in the wintertime and the minimum u^2 in the summertime. In other words, the 30-50 day oscillations in the mid-high latitudes of Northern Hemisphere are stronger in the wintertime but weaker in the summertime.

In other regions of the mid-high latitudes, the 30-50 day band-pass filtered u^2 has a similar temporal variation as shown in Fig.2 (figures are not given). Therefore, according to the results shown in Fig.1 and Fig.2, it can be suggested that the 30-50 day oscillations in the mid-high latitudes have obvious annual variation and they are stronger in the wintertime but weaker in the summertime.

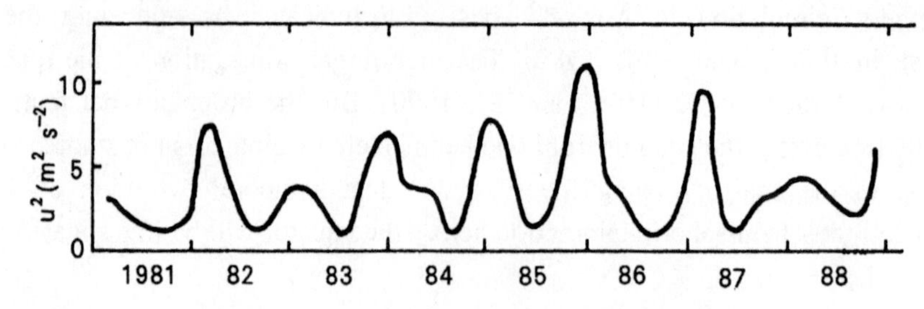

Fig. 2. Temporal variation of 30-50 day band-pass filtered u^2 at 200 hPa in (30°—50°N, 80°—180°E) region.

It is different from obvious annual variation in the mid-high latitudes that the 30-50 day oscillations in the tropical atmosphere do not have significant annual variation. Fig.3 shows the longitudinal distribution of mean kinetic energy at 500 hPa for 30-50 day oscillation along 10°S—10°N latitudes in January and July, 1981. We can clearly see that kinetic energies of 30-50 day oscillation in the tropical atmosphere have nearly the same magnitude except in South Asia monsoon region. In other words, the 30-50 day oscillations in the most tropical area have the same strength and no obvious annual variation. In South Asia region, there is clear seasonal variation of 30-50 day oscillation and this seasonal variation is different from that in the mid-high latitudes, i.e., the 30-50 day oscillation is stronger in summer but weaker in winter. This seasonal variation feature of 30-50 day oscillation in South Asia is a significant reflection of the monsoon activity. In summer, there are stronger summer monsoon and the cumulus convection in South Asia. So that the stronger 30-50 day atmosphere oscillation can be excited.

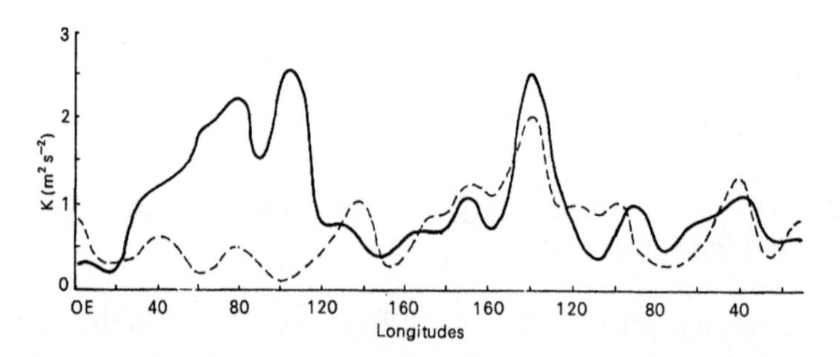

Fig. 3. Longitudinal distribution of kinetic energy at 500 hPa for 30-50 day oscillation in January (dashed line) and July (solid line), 1981 along 10°S—10°N latitudes.

Ⅲ. Interannual Variation of 30-50 Day Oscillation in the Atmosphere

Mehta and Krishnamurti (1988) analyzed the zonal winds at 850 hPa over the Arabian Sea in the period 1979—1984 and showed that the 30-50 day oscillations were stronger during

1979 and 1983. The temporal-longitude sections of the 200 hPa velocity potential, averaged between 30°S and 30°N for 1979-1984, showed more obvious eastward propagating waves during 1979 and during October 1982 to October 1983. Because 1982-1983 was the El Niño year and 1979 was characterized by weak warm anomalies of SST in the eastern equatorial Pacific Ocean, they suggested a possible connection between the 30-50 day oscillation and El Niño activity.

According to the analyses of interannual variations of 30-50 day oscillation before and after 1982-1983 and 1986-1987 El Niño events, the connection between the 30-50 day oscillation and El Niño event will be revealed more clearly. The temporal variation of u^2 of 30-60 day oscillation at 200 hPa along the equatorial middle-western region is given in Fig.4. It is very clearly shown that the seasonal variation of 30-50 day oscillations in the tropics is not obvious, with clear difference from that in the mid-high latitudes. And the interannual variation of 30-50 day atmospheric oscillations in the tropics is closely related to El Niño event. We can see in Fig.4 that there are the maximum u^2 over the middle-western equatorial Pacific Ocean in the Spring of 1982 and Spring-Summer of 1986, before the occurrence of El Niño event. Lau and Chan (1987) have suggested that the increasing amplitude and the decreasing frequency of 30-50 day atmospheric oscillation over the equatorial Pacific Ocean can excite El Niño event. The observational analysis in this paper gives an obvious affirmation to Lau's perspective.

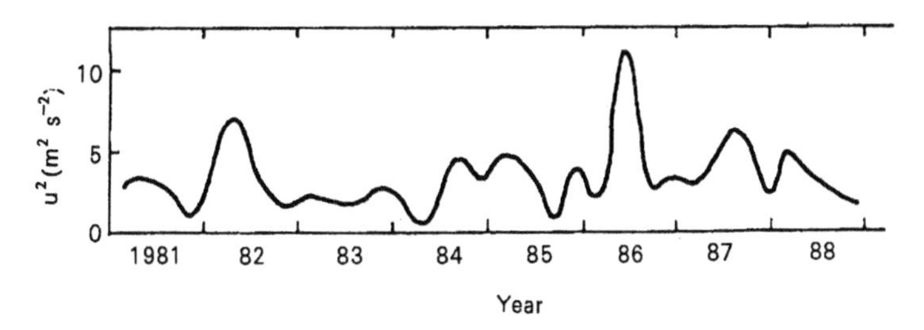

Fig. 4. Temporal variation of 30-50 day band-pass filtered u^2 at 200 hPa in (10°S—10°N, 110°—180°E) region.

There is stronger 30-50 day atmospheric oscillation over the mid-western equatorial Pacific prior to the El Niño event. What process could enhance the 30-50 day atmospheric oscillation over the mid-western Pacific? Our studies have indicated that the anomalously stronger winter monsoon (cold waves) in East Asia in the wintertime prior to the El Niño event is capable of enhancing the cumulus convection over the middle-western equatorial Pacific and then enhancing the 30-50 day oscillation there (Li, 1989; Li, 1990). Comparing Fig.2 with Fig.4, it can be seen that there are stronger 30-50 day oscillation u^2 in 30°—50°N latitudes of East Asia-western Pacific area in the spring of 1982 and the winter of 1985—86, prior to the enhancement over the middle-western equatorial Pacific, although not as obviously as those in the tropics. In other words, the stronger 30-50 day atmospheric oscillations over East Asia mid-high latitudes in the wintertime propagate with the cold wave activity into the middle-western Pacific area and enhance the 30-50 day atmospheric oscillations over there, then the El Niño event could be excited.

Ⅳ. Interaction of 30-50 Day Oscillation Across the Equator

Based on the 1979 FGGE IIIb data in (30°S—30°N, 30°E—150°W) area, the transfer of sensible and latent heat and momentum for the quasi-40 day oscillation in a cross-equatorial meridional section is investigated (He, 1988). It suggested an interaction of 30-50 day oscillation across the equator. But it is not understood what are the avenues for the interaction of 30-50 day oscillation across the equator? According to the analyses in the ECMWF data, some avenues for the interaction of 30-50 day atmospheric oscillation across the equator are explored.

There are 30-50 day low-frequency wavetrains in both the Northern Hemisphere and the Southern Hemisphere. We will indicate further that the 30-50 day low-frequency wavetrains are able to cross the equator in some places and produce the cross-equatorial wavetrains. These cross-equatorial low-frequency wavetrains obviously show an interaction of 30-50 day oscillations across the equator because the low-frequency energy will propagate along the wavetrains from one hemisphere to another.

In Part I (Li and Wu, 1990), the low-frequency wavetrains in the Northern Hemisphere are noted. There we focus on the cross-equatorial parts of low-frequency wavetrains, especially over the central equatorial Pacific. Fig. 5 shows the correlation coefficients of the filtered geopotential height at 500 hPa based on the reference points (115°E, 45°N) and (150°W, 40°N), respectively. According to the distributions of the correlation coefficients, it is very clear that there are two major cross-equatorial wavetrains over the central equatorial Pacific. Because the correlation coefficients over the central equatorial Pacific area are quite large, the cross-equatorial low-frequency wavetrains are obviously existential. Although it is no more obvious than that over the central equatorial Pacific area, the cross-equatorial wavetrain can also be seen over the central equatorial Atlantic area.

The long lasting low-frequency systems ("storms") have been found to propagate meridionally northwards from the equator during the Northern winter season (Krishnamurti et al., 1985). Are there the long lasting low-frequency systems to propagate across the equator? We have indicated the existence of the long-lived vortices in Part I. In fact, the low-frequency vortex, shown in Fig.6 of Part I, propagates from the Northern Hemisphere to the Southern Hemisphere over the central Pacific area. In that figure, the cyclonic vortex "C" located near 25°N on January 4, 1982 has already moved near the equator on January 24. Then that cyclonic vortex propagated into the Southern Hemisphere. Fig.6 presents the isolated wind fields of the oscillations at 200 hPa from February 28 to March 15, 1982. It is also shown that a cyclonic vortex marked with "C" propagates from Northern to Southern Hemisphere over the central equatorial Pacific area. In the same time, a cyclonic vortex marked with "D" in the Southern Hemisphere obviously propagates into the Northern Hemisphere over the central equatorial Atlantic area.

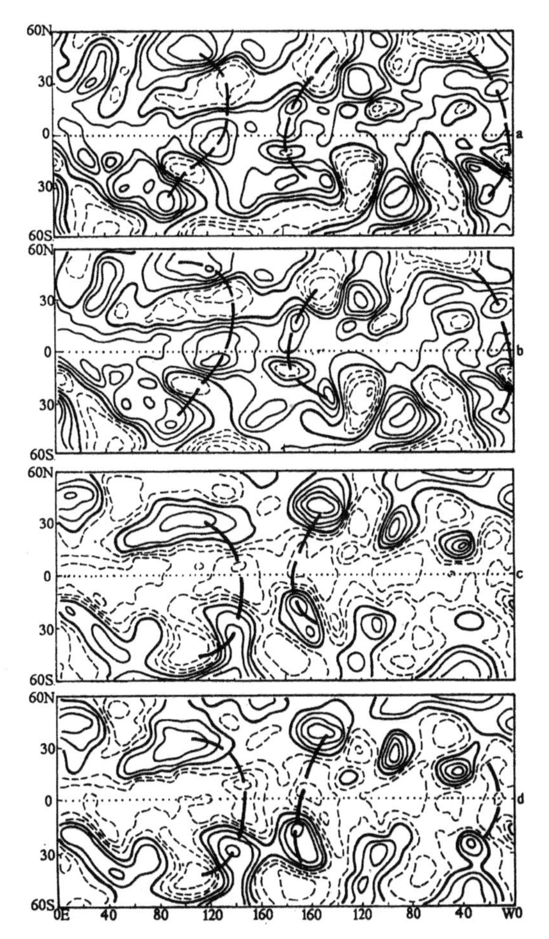

Fig. 5. Distributions of the correlation coefficients of the filtered geopotential height at 500 hPa. The contour intervals are 0.2. a. The reference point at (115°E, 45°N); 3 days lag b. The reference point at (115°E, 45°N); 6 days lag c. The reference point at (150°W, 40°N); 3 days lag d. The reference point at (150°W, 40°N); 6days lag.

It is clearly presented that the cross-equatorial propagation of the low-frequency vortex is an important avenue for the interaction of 30-50 day atmospheric oscillations between the two hemispheres. The central Pacific area and central Atlantic area are major regions for the cross-equatorial propagation of the low-frequency vortex, especially in the wintertime.

Fig.7 presents longitudinal distributions of the meridional wind components of 30-50 day oscillation at 200 hPa along the equator for the 1st, 3rd, 5th and 7th phases, respectively. The slow eastward propagation along the equator shows the basical property of 30-50 day oscillation in the tropical atmosphere. It is shown in Fig.7 that the cross-equatorial meridional winds along the equator are oscillational and the cross-equatorial effects are able to exist at any place. The longitudinal distributions of v^2 at 200 hPa along the equator averaged for 8 phases in the summertime and wintertime are given in Fig.8. We can find that there are three major cross-equatorial effect regions along the equator in the wintertime, i.e., the central Pacific area, the eastern Atlantic and western African area and the western Indian Ocean; there are two major cross-equatorial effect regions in the summertime, i.e., the eastern Pacific area and the central Atlantic area.

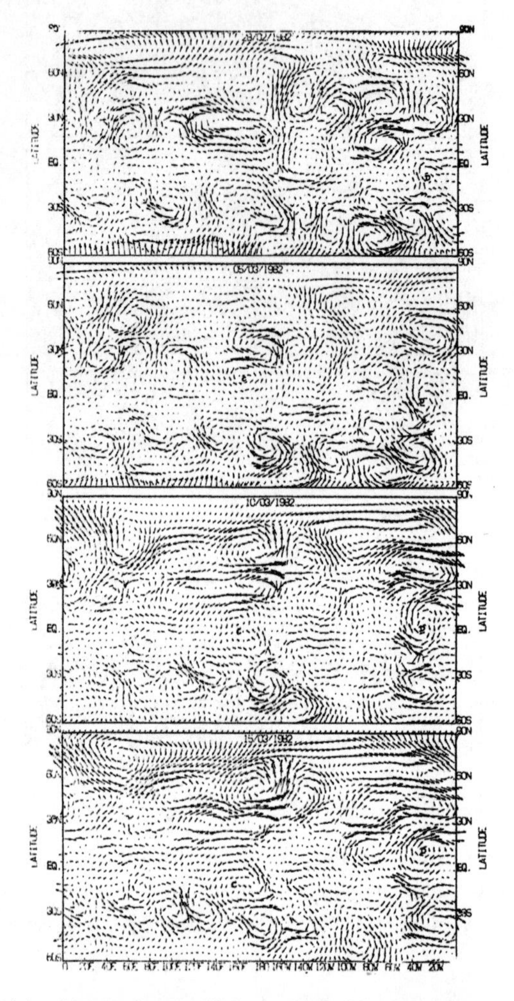

Fig. 6. The isolated wind fields of 30-50 day oscillations at 200 hPa from February 28 to March 15, 1982.

The discussions in this section show that the interaction of 30-50 day oscillations across the equator has three avenues. They are the cross-equatorial low-frequency wavetrains, the cross-equatorial low-frequency vortex and the cross-equatorial meridional winds. And, the major regions for the cross-equatorial interaction of 30-50 day oscillations are over the central-eastern Pacific and the central Atlantic.

It is well-known that there are three main avenues for cross-equatorial air. They are located in 50°—60°E, 105°E and 150°E regions along the equator. Specially, the famous Somali jet which crosses the equator through 50°—60°E avenue, is a significant circulation system for the Southern Hemisphere affecting the Northern Hemisphere. But for the 30-60 day oscillations, the above usual avenues are not apparent. On the contrary, the more important avenues are located over equatorial central Pacific and central Atlantic. This indicates that the cross-equatorial effect of 30-50 day oscillation has remarkable differences from that of general circulation.

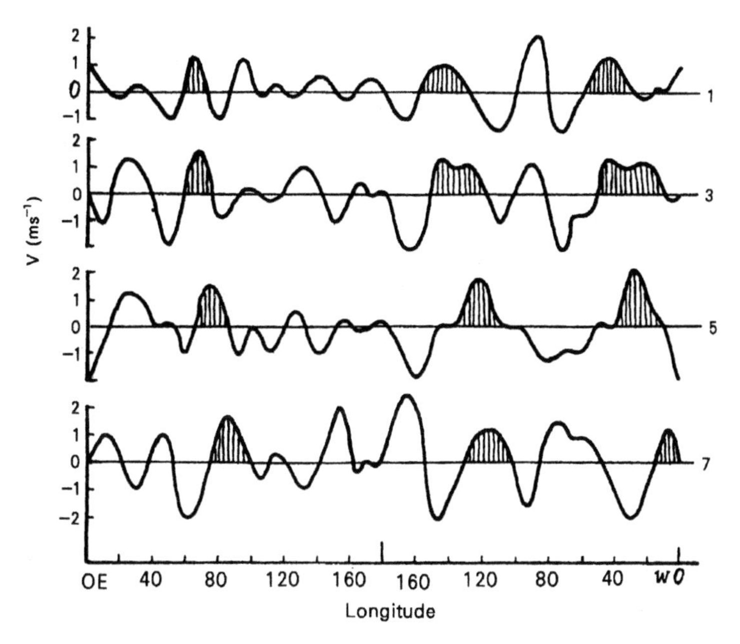

Fig. 7. Longitudinal distributions of meridional wind components of 30-50 day oscillation at 200 hPa along the equator for the wintertime in 1980. Numerals 1, 3, 5 and 7 represent 4 oscillation phases in total 8 phases, respectively.

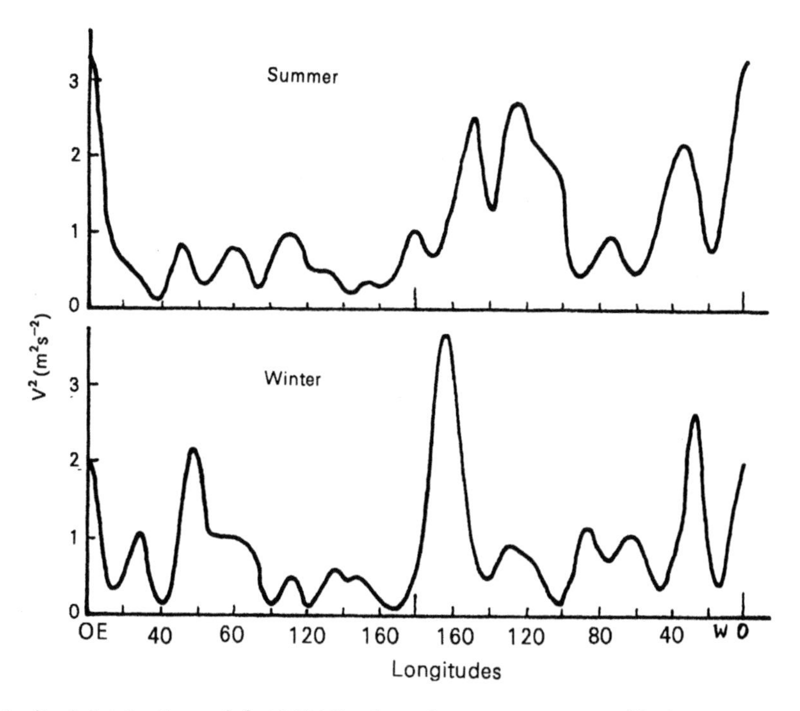

Fig. 8. Longitudinal distributions of v^2 at 200 hPa along the equator averaged 8 phases.

Ⅴ. Concluding Remarks

Based on the analyses in ECMWF data, some properties in relation to the temporal evolution and hemispheric interaction of 30-50 day oscillation across the equator are discussed in this part. Especially, the following conclusions should be indicated.

1. The 30-50 oscillations in the atmosphere have annual variation property. It is more obvious in the mid-high latitudes. In general, the 30-50 oscillations are stronger in the wintertime and weaker in the summertime for the mid-high latitude area. And, they are stronger in the summertime and weaker in the wintertime for the tropical atmosphere.

2. The interannual variation of 30-50 day atmospheric oscillation in the tropics is very clear and it is related to the El Niño event. There are stronger 30-50 day atmospheric oscillations over the mid-western equatorial Pacific prior to the El Niño event.

3. The cross-equatorial effect of 30-50 day atmospheric oscillation has three major avenues: the cross-equatorial low-frequency wavetrains, the cross-equatorial propagation of the low-frequency vortex and the cross-equatorial meridional wind. The cross-equatorial effects of 30-50 day atmospheric oscillation mainly occur over the central Pacific and the eastern Atlantic-western Africa. Sometimes, there are obvious cross-equatorial meridional wind of 30-50 day oscillation over the western equatorial Indian Ocean.

This study was supported in part by National Natural Science Foundation of China.

References

Chen Longxun and Jin Zuhui (1982), On the interaction of circulation between the two hemispheres in the East Asia monsoon circulation system during summer, *Proceedings of the Symposium on the Summer Monsoon in South Asia*, People's Press of Yunnan Province, 218-231.

He Jinhai (1988), The transfer of physical quantities in QDPO and its relation to the interaction between the NH and SH circulations, *Adv. in Atmos. Sci.*, 5: 97-106.

He Jinhai (1990), Discussion of meridional propagation mechanism of quasi-40 day oscillation, *Adv. in Atmos. Sci.*, 7: 78-86.

Krishnamurti, T. N. and S. Gadgil (1985), On the structure of 30-50 day mode over the globe during FGGE, *Tellus*, 37A, 336-360.

Lau, K.M. and P.H. Chan (1987), The 40-50 day oscillation and the El Niño-southern oscillation: A new perspective, *Bull, Amer, Meteor. Soc.*, 67: 533-534.

Li Chongyin (1989), The frequent activities of stronger aerotroughs in East Asia in wintertime and the occurrence of the El Niño event, *Scientia Sinica*, Series B, 32: 976-985.

Li Chongyin (1990), Intraseasonal oscillation in the atmosphere, *Chinese Journal of Atmospheric Sciences*, 14: 35-52.

Li Chongyin (1990), Interaction between anomalous winter monsoon in East Asia and El Niño event, *Adv. in Atmos. Sci.*, 7: 36-46.

Li Chongyin and Wu Peili (1990), An observational study of the 30-50 day atmospheric oscillations Part I: Structure and propagation, *Adv. in Atmos. Sci.*, 7: 294-304.

Mehta, A.V., and T. N. Krishnamurti (1988), Interannual variability of the 30-50 day wave motions, *J. Meteor. Soc. Japan*, 66: 535-548.

Global Atmospheric Low-Frequency Teleconnection*

Li Chongyin[1] (李崇银), Zhang Qin[2] (张勤)

[1] LASG, Institute of Atmospheric Physics, Academia Sinica, Beijing
[2] Department of Meteorology, Nanjing Institute of Meteorology**, Nanjing

Abstract An analytic study in relation to the action of 30-60 day low-frequency oscillation in the global atmosphere is completed by using ECMWF grid-point data (1980—1988). An important feature is shown, i. e. there are obvious 30-60 day low-frequency teleconnection patterns in the global atmosphere. In the Northern Hemisphere, there are mainly Eurasia-Pacific (EAP) and Pacific-North America (PNA) low-frequency teleconnection patterns, and there are mainly Australia-South Africa (ASA) and Round-South America (RSA) patterns in the Southern Hemisphere. Meanwhile, the EAP and PNA low-frequency wave trains in the Northern Hemisphere join and interact with the ASA and RSA low-frequency wave trains in the Southern Hemisphere.
Keywords atmospheric intraseasonal oscillation, low-frequency teleconnection, wave train.

Ⅰ. Introduction

The term "atmospheric teleconnection" first appeared in 1969, Bjerknes referred to the correlation between the far-apart atmospheric circulation variations as the atmospheric teleconnection[1]. Actually its discovery dates back to as early as 1897 when Hildebrandson pointed out the teleconnectivity in atmospheric motion which was named "southern oscillation (SO)" later by Walker[2], which represents teleconnection of the surface pressure variations between the southeastern Pacific and the southern Indian Ocean. In the early 1980s, Wallace and Gutzler produced various teleconnection patterns of the atmospheric circulation changes in the Northern Hemisphere[3]. The "great circle" theory presented by Hoskins and Karoly provides the atmospheric teleconnection with dynamical explanations[4]. Consequently, the atmospheric teleconnectivity has become the important theory in understanding the atmospheric circulation and weather/climate variations in the 1980s.

Another important aspect concerning the changes in atmospheric circulation and climate is atmospheric intraseasonal (30-60 day) oscillation. A series of studies have not only revealed

Received November 30, 1990.
*Project supported in part by the National Natural Science Foundation of China.

** In 2004, it was renamed Nanjing University of Information Science & Technology.

the structural features and behavior of intraseasonal oscillations in the tropical atmosphere[5, 6] but also pointed out the existence, the characteristic structure and behavior of the extratropical counterpart[7, 8]. Thus, the intraseasonal oscillations have been affirmed to be a universal feature of the global atmospheric motion.

Intraseasonal oscillations are different in character and closely related for the tropical and the extratropical atmosphere. But what are their prominent characteristics on a global basis? In this paper, the analysis shows that there exists a pronounced low-frequency teleconnection with fairly stable patterns and wave trains in the atmospheric intraseasonal oscillation on a global basis.

Ⅱ. Data and Their Processing

The data used in this work are ECMWF seven-layer grid-point analysis 1980-1988, we focus on the geopotential height data at 500 hPa for the investigation.

Like other studies in this aspect, the present work employs the band-pass filtering method to isolate 30-60 day oscillation. The central frequency of the band-pass filter corresponds to a 45 day oscillation period. The response function of the filter is shown in Fig. 1.

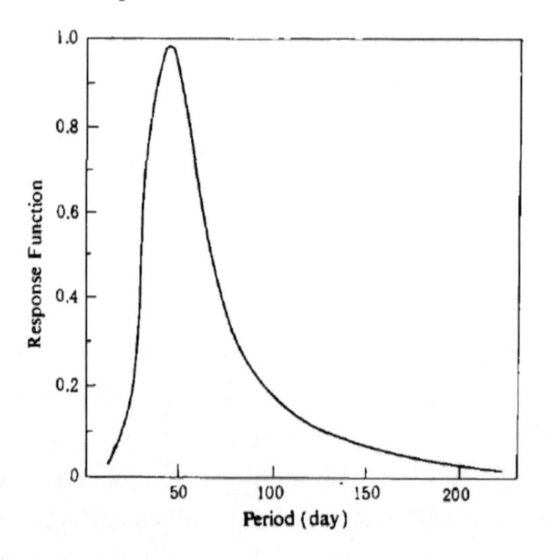

Fig. 1. The response function of the 30-60 day band-pass filter.

The 30-60 day band-pass filtered geopotential height fields at 500 hPa are able to represent the horizontal structures and the behavior of 30-60 day oscillations in quite wide senses. In order to reveal the features of atmospheric low-frequency teleconnection, the authors used a similar method adopted by Wallace in studying the atmospheric teleconnection. For various base points, the correlation coefficients to all grid points in the globe are computed. The analysis of the correlation coefficient field can clearly show the characteristics of the global atmospheric low-frequency teleconnection.

III. The Patterns of Atmospheric Low-Frequency Teleconnection

To reveal the existence and features of global atmospheric low-frequency teleconnection the authors calculate separately the correlation coefficients for zero, ±4 and ±8 day lag correlation based on 10 referential points (10°E, 45°N; 80°E, 45°S; 110°E, 50°N; 140°E, 20°N; 150°E, 15°S; 160°W, 5°S; 140°W, 40°N; 110°W, 30°S; 70°W, 45°N; 30°W, 20°S). The global distributions of the correlation coefficients clearly show that there exists a very significant teleconnection in the 30-60 day atmospheric oscillation for any type of lag and any base points. Additionally, these teleconnection patterns are fairly steady in the global context and show no significant change because of the different base points. Fig. 2 shows the global distributions of contemporary correlation coefficients for the base points 140°E, 20°N; 160°W, 5°S; 70°W, 45°N, respectively, with other points. The basic features of the 30-60 day atmospheric low-frequency teleconnection are very obvious. In the Northern Hemisphere, there are evident Eurasia-Pacific (EAP) and Pacific-North America (PNA) teleconnection patterns. The two patterns and their corresponding low-frequency wave trains are fully consistent with the results of Ref. [9] and extended the Eurasia (EU) wave train which was indicated by K. M. Lau[10] to the tropical western Pacific area.

From Fig. 2, we can also see that there exist two principal low-frequency teleconnection patterns in the Southern Hemisphere, one stretching from Australia to South Africa, and the other is basically round South America. Thus they can be referred to as Australia-South Africa (ASA) pattern and Round-South America (RSA) pattern, respectively.

We have indicated that the EAP low-frequency wave train is connected and interacted with the PNA wave train in the Northern Hemisphere. How about the connection of the 30-60 day low-frequency teleconnection patterns between the Southern and Northern Hemispheres?

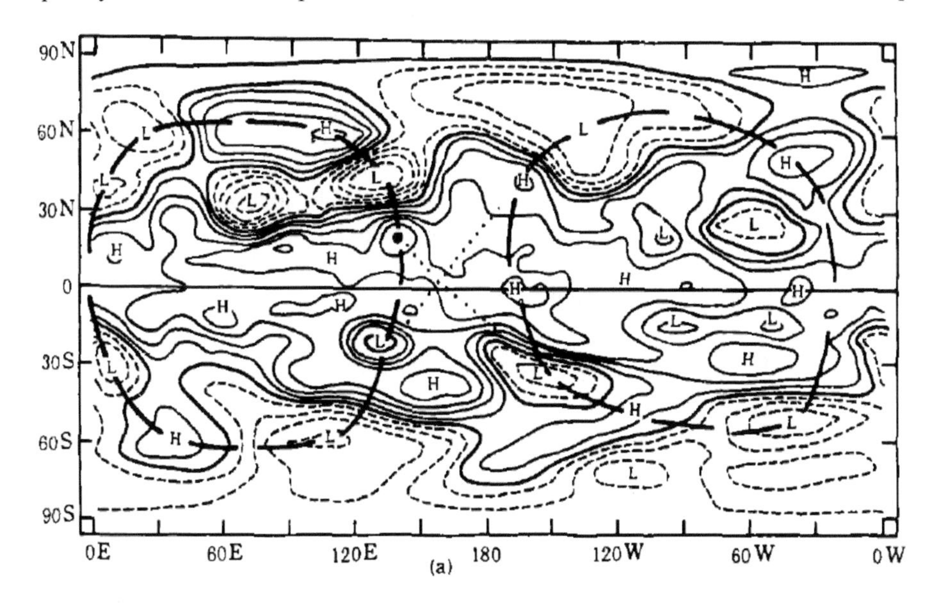

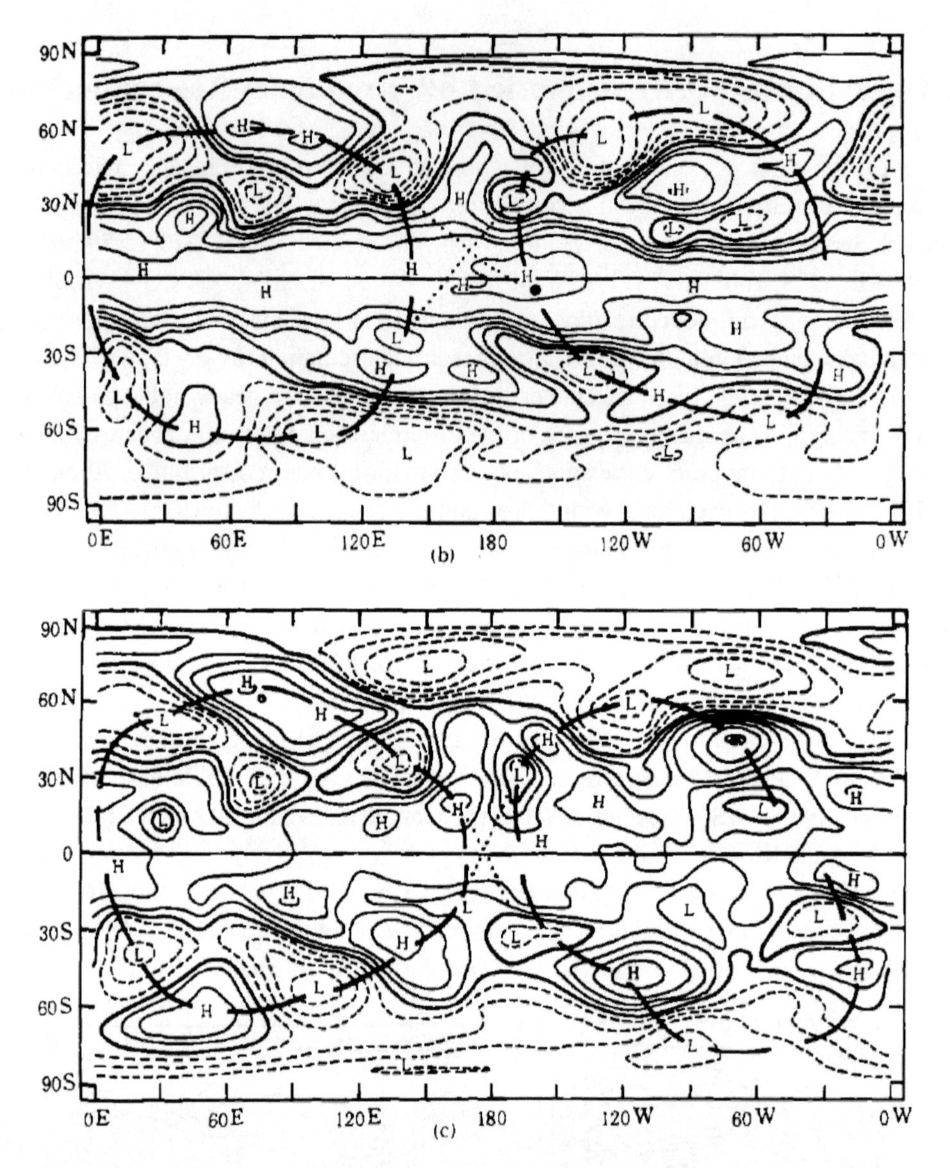

Fig. 2. The global atmospheric low-frequency teleconnection calculated at various base points. The positive (negative) correlation coefficients are denoted by solid (broken) lines. The contour intervals are ±0.2, ±0.4, ±0.6, ±0.8, ±0.9, respectively, (a)The base point (the blackened circle) at 140°E, 20°N; (b)the base point at 160°W, 5°S; (c) the base point at 70°W, 45°N.

The other important feature of the global atmospheric low-frequency teleconnection is also clearly shown in Fig. 2, i.e. the low-frequency wave trains in the Southern and Northern Hemispheres are linked up with each other across the equator. According to the distributions of the correlation coefficients for various base points, it can be found that the connection of low-frequency wave trains between the Southern and Northern Hemispheres has two types. One of them shows that the EAP (PNA) wave train in the Northern Hemisphere links up with the ASA (RSA) wave train in the Southern Hemisphere. And there are two separate closed low-frequency teleconnection patterns and wave trains, i.e. the EAP-ASA and PNA-RSA, in the global atmosphere. The other shows that the EAP (PNA) wave train links up with the RSA (ASA) wave train. And there are two low-frequency teleconnection patterns and wave trains, i.e. the EAP-RSA and ASA-PNA, in the global atmosphere.

The analysis of the year-by-year correlation fields indicates no significant change except minor difference in the two basic wave trains. Fig. 3 presents the global distributions of the contemporary correlation coefficients for the 500 hPa geopotential height data in 1984, 1985, 1986, respectively, with 110°E, 50°N as the fiducial point. It is obvious that for all of the years, the global atmospheric low-frequency teleconnection patterns are similar.

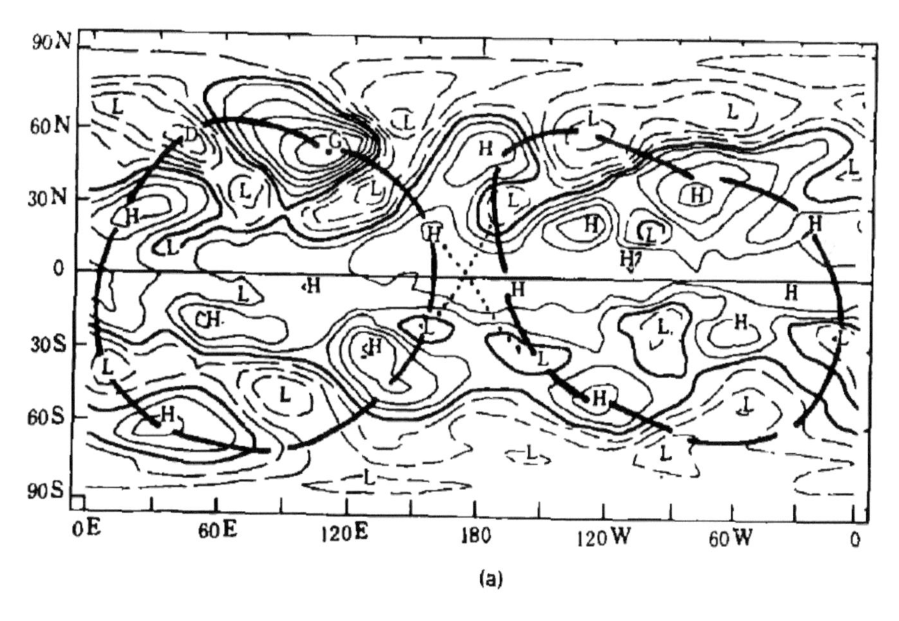

(a)

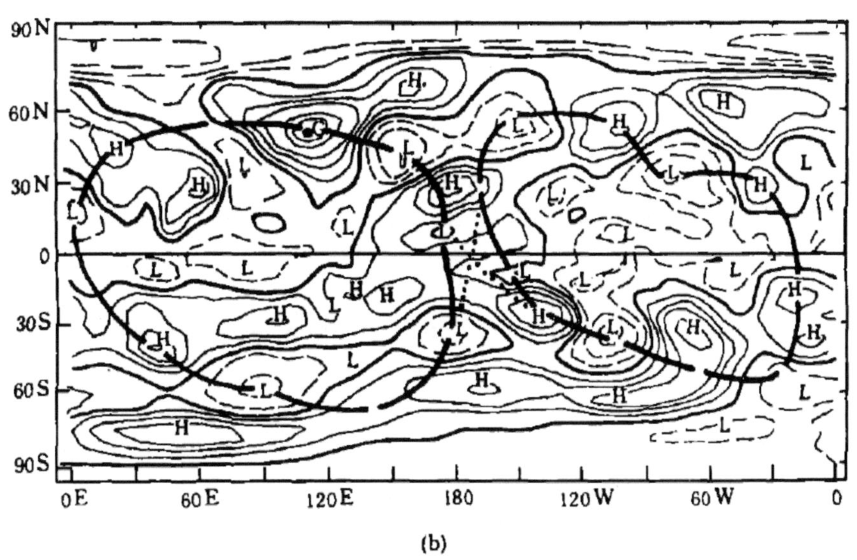

(b)

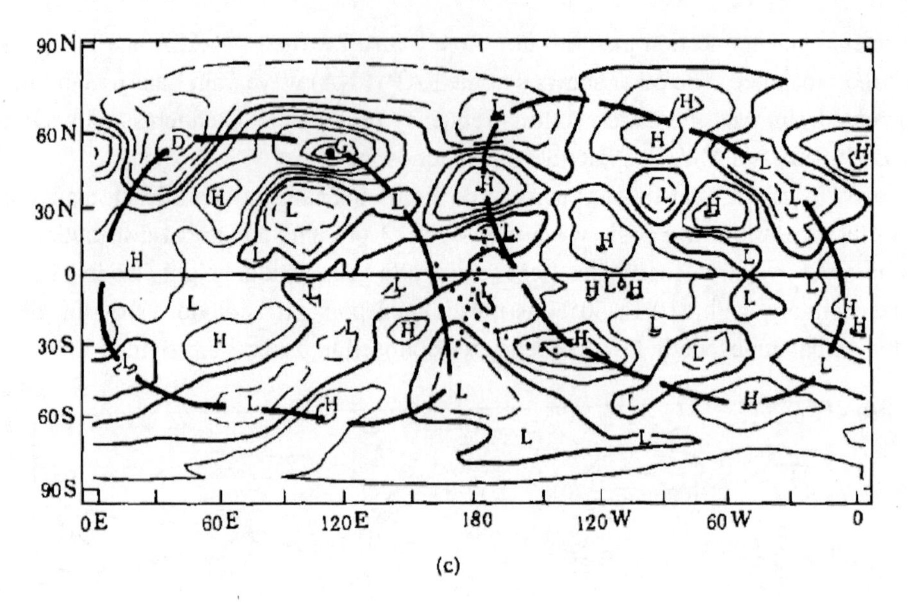

Fig. 3. The global atmospheric low-frequency teleconnection calculated in various years. The positive and negative correlation coefficients are denoted by solid and broken lines, respectively. Contour intervals are 0.2, 0.4, 0.8, 0.9 and −0.1, −0.2, −0.4 and −0.5 with the base point at 110°E, 50°N, (a) 1984, (b) 1985, (c)1986.

In this study, in relation to the connection of low frequency wave trains between the Southern and Northern Hemispheres, two connection types are primarily denoted and the global circular wave trains are affirmed as a major connection. Since the correlation coefficients in the equatorial region are more homogeneous, the results shown in this paper are primary, though much more the correlation charts are analyzed.

Ⅳ. Discussion and Conclusion

A series of studies concerning atmospheric intraseasonal oscillation shows that the 30-60 day oscillation is a universal property of global atmospheric circulations and climatic variations. The present work gives a further indication that the 30-60 day atmospheric oscillation in the global atmosphere has an evident teleconnection feature which is even more significant than that of the general character revealed by Wallace et al. (1981).

Atmospheric low-frequency teleconnections are manifested mainly as EAP and PNA patterns in the Northern Hemisphere, and as ASA and RSA patterns in the Southern Hemisphere. They are connected with each other, constituting thereby an EAP-ASA and a PNA-RSA low-frequency wave train or an EAP-RSA and an ASA-PNA low-frequency wave train, around the globe. Their existence not merely associates the behavior of low-frequency oscillation in the tropical and extratropical atmospheres, but reveals the important passage for interactions between the interhemispheric intraseasonal oscillations.

Our analysis shows that the equatorial central Pacific and central Atlantic areas play an important role in the activities of global atmospheric intraseasonal oscillations. They are the places where the global low-frequency wave trains meet, exerting impacts both on the interactions between the interhemispheric intraseasonal oscillations as a "channel" and on the interactions between atmospheric intraseasonal oscillations in extratropics and tropics as a "transfer stations" for each hemisphere.

References

[1] Bjerknes, J., *Mon. Wea. Rev.*, 97(1969), 162.
[2] Walker, G. T. & Bliss, E.W., *World Weather V.*, Men. Roy. Meteor, Sec., 4(1932), 53.
[3] Wallace, J. M. & Gutzler, D. S., *Mon. Wea. Rev.*, 109(1981), 784.
[4] Hoskins, B. J. & Karoly, D. J, *J. Atmos. Sci.*, 38(1981), 1176.
[5] Murakami, T. et al., *J. Meteor. Sec. Jpn*, 62(1984), 440.
[6] Lau, K. M. & Chan, P. H., *Mon. Wea. Rev.*, 113(1985), 1889.
[7] Anderson, J. R. & Rosen, R. D., *J. Atmos. Sci.*, 40(1983), 1584.
[8] Li Chongyin et al., *Science in China.*, B34(1991), 457.
[9] Li Chongyin, *Chinese J. Atmos. Sci.*, 14(1990), 32.
[10] Lau, K. M. & Phillips, T. J., *J. Atmos. Sci.*, 43(1986), 1164.

The 30-60 Day Oscillations in the Global Atmosphere Excited by Warming in the Equatorial Eastern Pacific

Li Chongyin (李崇银), Xiao Ziniu (肖子牛)

LASG, Institute of Atmospheric Physics, Academia Sinica, Beijing

Abstract 30-60 day oscillations (also called intraseasonal oscillations) have been indicated in the early 1970s[1, 2]. A series of studies in the 1980s not only investigated the 30-60 day oscillations in the tropical atmosphere and revealed their structure characteristics and fundamental moving regularity[3-8], but also exposed the existence of these oscillations in the middle-high latitudes[9, 10]. Then they have been investigated as special properties of the atmospheric motion over the globe[11-13]. Some studies have indicated that the feedback of the cumulus convection is an important mechanism to cause such oscillations in the tropical atmosphere[14], and there is CISK-Kelvin wave as well as CISK-Rossby wave in this theory[15-17]. Based on atmospheric teleconnection research, Wallace regarded external forcing as one of the important mechanisms to cause low-frequency oscillation in the atmosphere[18]. Some numerical simulations have also shown that the anomaly of SST (sea surface temperature) in the equatorial eastern Pacific can excite teleresponse and produce PNA teleconnection pattern[19, 20].

There is a natural question: What is the relationship between the atmospheric teleresponse to the anomalies of SST in the equatorial eastern Pacific and low-frequency oscillation in the atmosphere? In this note, two numerical experiments are completed by using IAP-GCM. And the properties and activities of the teleresponses caused by the anomalies of SST in the equatorial eastern Pacific are discussed based on temporal evolutions of the response fields. It is shown that the anomaly of SST in the equatorial eastern Pacific is an important mechanism to produce 30-60 day oscillations in the atmosphere.

Keywords 30-60 day oscillation in the atmosphere, low-frequency teleresponse.

Ⅰ. Experiments and Data Treatment

The general circulation model, IAP-GCM developed in the Institute of Atmospheric Physics, Academia Sinica, is used in this study. It is a 2-level primitive equation model with

Received January 20, 1991.

horizontal resolution of 5° in longitude and 4° in latitude. All physical processes, such as topographic effect, radiation and condensation, are considered in the model. IAP-GCM is described in detail in Ref. [21].

Two numerical experiments are completed. One is the control run (CE), which is the integration for one year at normal sea surface temperature. In the control run, the sea surface temperature is climatological observation (SSTo) and variable with time. The other is the disturbance experiment (DE), which is the integration for one year at anomalous sea surface temperature. In the disturbance experiment, the sea surface temperature in January is an anomalous SST in the equatorial eastern Pacific (Fig. 1) superposed on SSTo. Thus, during the integration of DE, SST is different from that during CE just in January and the first half of February. The differences between DE and CE (SD = DE − CE) are responses of the atmosphere to the warming in the equatorial eastern Pacific. To analyse the pentad results of the response fields (SD), the properties and activities of the teleresponse in the atmosphere will be clearly expressed.

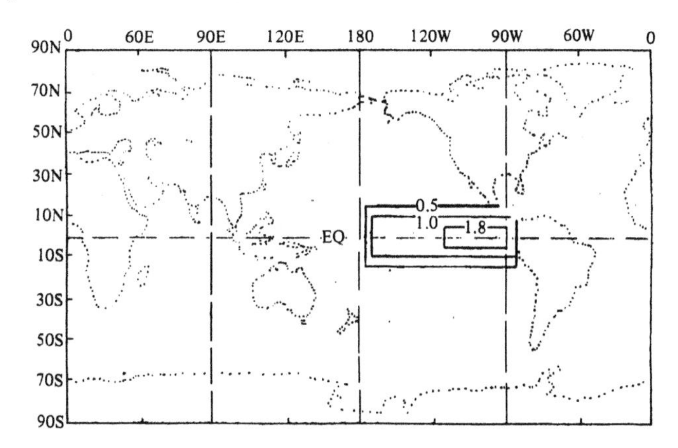

Fig. 1. Anomalous sea surface temperature field.

In order to discuss the properties of SD fields and compare them with 30-60 day oscillations in the atmosphere, the 30-60 day band-pass filtering[22], which is usually used to study the intra-seasonal oscillations in the atmosphere, is used in the analyses of numerical simulation data.

II. Numerical Simulation Results

It has been shown that the anomalies of SST in the equatorial eastern Pacific should excite PNA teleconnection pattern in the Northern Hemisphere atmosphere. In this note, the SD fields either show the PNA pattern or the EAP (Eurasia-Pacific) pattern. In addition, the data analyses in relation to the influences of El Niño on the atmospheric circulation pointed out that the El Niño event should move the major rainfall centre in the tropics from its original location (150°E) eastwards to the date line. The computation in this note shows that a negative zone of the precipitation anomalies during January -March is just over the equatorial 150°E and a positive zone is just over the equatorial date line region. Therefore, some fundamental features of the atmospheric teleresponse to the SST anomalies in the equatorial eastern Pacific have been simulated successfully, showing that the numerical simulation results in this note are dependable.

According to the temporal evolution of the SD fields, the periodical variation of the atmospheric teleresponse is very clear. Taking the responses of geopotential height at 500 hPa as an example, it is obvious that the simulated pattern in the third — fifth pentads is similar to that in the tenth—eleventh pentads, but approximately reverse to that in the seventh-eighth pentads. Thus the evolution of the atmospheric teleresponse has a quasiperiodicity of about 35 days.

In order to show the 30-60 day low-frequency response features in the atmosphere, the time-longitude sections of zonal wind response at 400 hPa along 6°S-6°N latitude zone are given in Fig. 2 for both the 30-60 day band-pass filtered and the unfiltered. The unfiltered results (in Fig. 2a) have clearly shown a quasi-period of 30-60 days. Through the 30-60 day bandpass filtered, the quasi-periodic variation pattern is more obvious and the fundamental circumstance is similar to the unfiltered one. Especially, the maximum easterly and westerly are respectively 11 m/s and 9 m/s for the unfiltered, but 7m/s and 8 m/s for the band-pass filtered. The proportion of 30-60 day oscillation is near 70%, which means that it is the major component in the atmospheric teleresponse in the tropics.

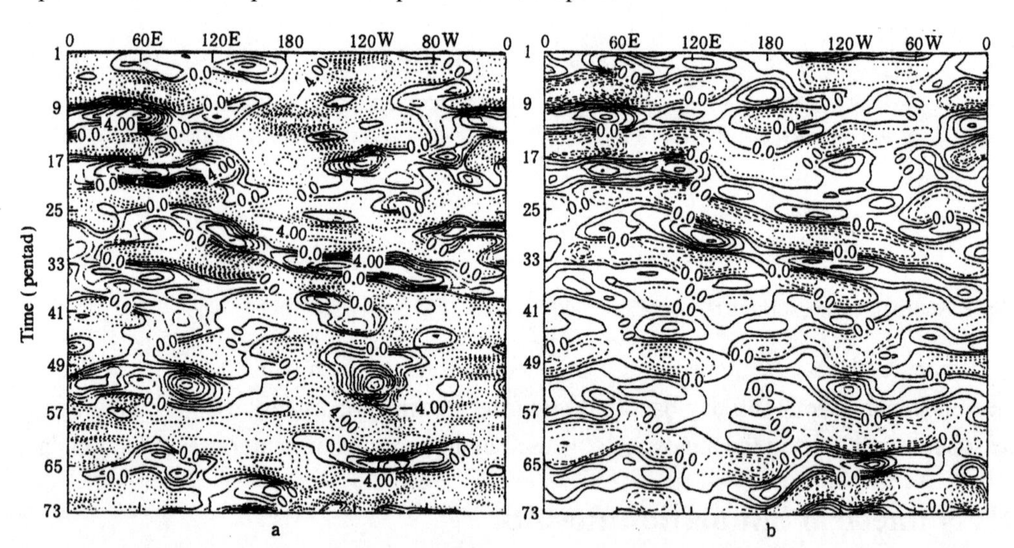

Fig. 2. Time-longitude sections of zonal wind response at 400 hPa along 6°S-6°N latitude zone. The dashed line is easterly.

a. Unfiltered; b. 30-60 day band-pass filtered.

The above mentioned analysis shows that the teleresponse is mainly the 30-60 day low-frequency response. In other words, the SST anomalies in the equatorial eastern Pacific should produce 30-60 day oscillation in the tropical atmosphere. How about that in the middle-high latitudes? Fig. 3 shows the time-longitude sections of geopotential height response at 500 hPa along 66°N latitude for both the unfiltered and the 30-60 day band-pass filtered. It is also clear that the geopotential height responses at 500 hPa have obvious 30-60 day quasi-periodical fluctuation in either case. For the unfiltered case, the maximum geopotential height responses at 500 hPa along 66°N latitude are –210 m and + 150 m. For the 30-60 day band-pass filtered case, the maximum responses are ± 120 m, and the circumstance is similar to that for the unfiltered one. Therefore, the 30-60 day oscillation is also the major component in the atmospheric teleresponse in the middle-high latitudes.

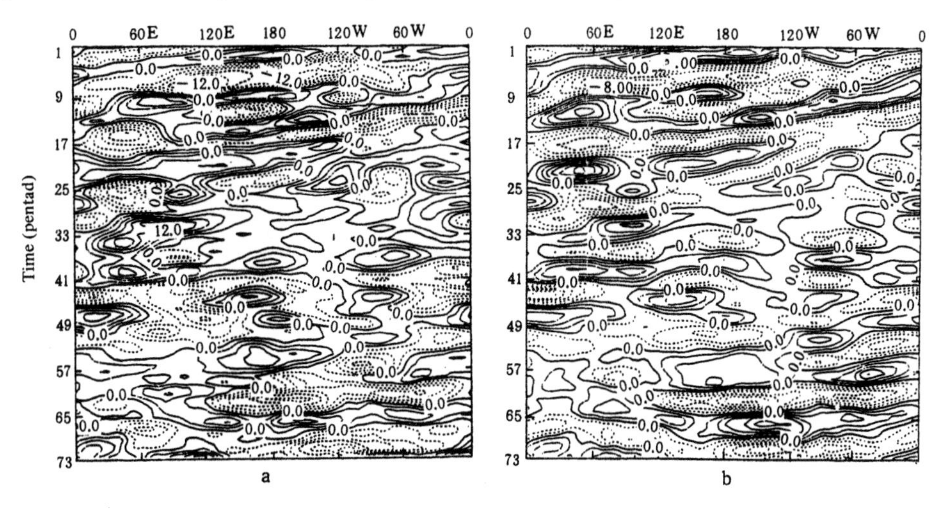

Fig. 3. Time-longitude sections of geopotential height response at 500 hPa along 66°N latitude. The dashed line is negative.
a. Unfiltered; b. 30-60 day band-pass filtered.

According to the above analyses, it is clear that the atmospheric response to the SST anomalies in the equatorial eastern Pacific is mainly the 30-60 day low-frequency teleresponse. In other words, the SST anomalies in the equatorial eastern Pacific should excite 30-60 day oscillations in the global atmosphere.

Ⅲ. Discussion and Conclusion

A lot of studies have pointed out some fundamental properties of 30-60 day oscillation in the atmosphere. In the tropical atmosphere, the slow eastward propagation and baroclinic structure sign change from lower to upper troposphere are the major features of 30-60 day oscillations. But they are propagated westwards and have obvious barotropic structure in the middle-high latitudes. The simulated 30-60 day low- frequency responses in this note have similar properties to the 30-60 day oscillations in the actual atmosphere. In Fig. 2b, it is very clear that the simulated 30-60 day oscillations in the tropical atmosphere are propagated slowly eastwards. According to the evolution of geopotential height at 500 hPa given in Fig. 3b, it is shown that the excited 30-60 day atmospheric oscillations are basically propagated westwards in the middle-high latitudes.

The longitudinal distributions of geopotential height at 500 hPa and the sea surface pressure in January along 6°S and 42°S are shown in Fig. 4 and Fig. 5, respectively. It is very clear in Fig. 4 that the geopotential height and sea surface pressure are obviously in reverse distribution. This means that the excited 30-60 day oscillations in the tropical atmosphere have obvious "baroclinic" structure sign change from lower to upper troposphere. And Fig.5 shows that the excited 30-60 day atmospheric oscillations in the middle-high latitudes are of barotropic structure obviously.

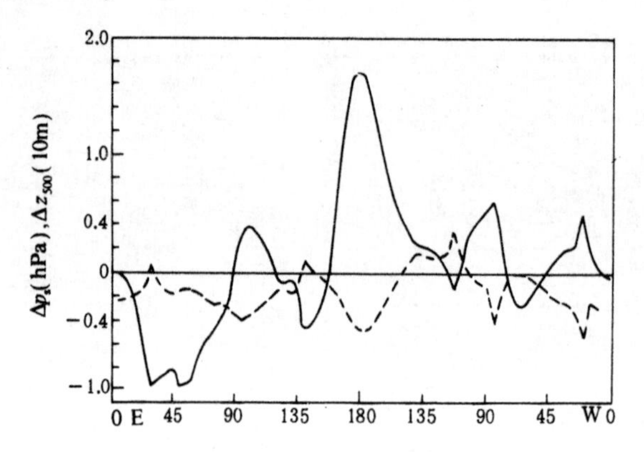

Fig. 4. The longitudinal distribution of geopotential height at 500 hPa (dashed line) and sea surface pressure (solid line) in January along 6°S latitude.

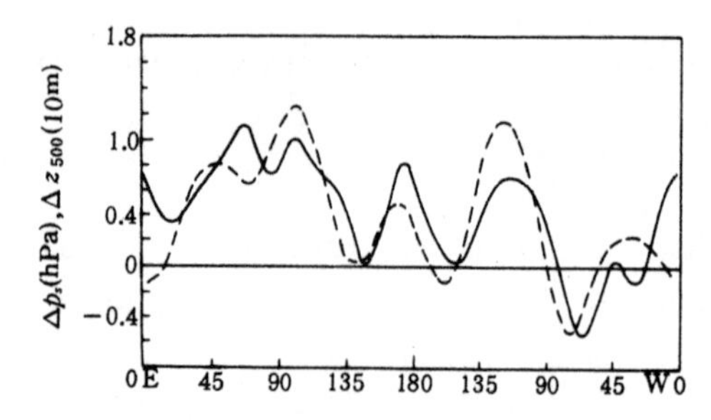

Fig. 5. Same as Fig. 4, but along 42°S latitude.

The simulations and the above analyses show the responses in the global atmosphere to the anomalies of SST in the equatorial eastern Pacific are mainly the 30-60 day low-frequency responses. And the excited 30-60 day oscillations are very similar to 30-60 day oscillations in the actual atmosphere in horizontal propagation and vertical structure. Therefore, the anomalies of SST in the equatorial eastern Pacific can excite 30-60 day oscillations in the global atmosphere. This external forcing is an important mechanism to produce 30-60 day oscillations in the global atmosphere.

References

[1] Madden, R.D. & Julian, P., *J. Atmos. Sci.*, **29**(1972). 1109.
[2] Madden, R.D. & Julian, P., *ibid.*, **28**(1971), 702.
[3] Krishnamurti, T.N.et al., *ibid.*, **39**(1982), 2088.
[4] Murakami, T.et al., *J. Meteor. Soc. Japan*, **62**(1984), 440.
[5] Lau, K.M. et al., *Mon. Wea. Rev.*, **113**(1985), 1889.
[6] Lau, K.M. et al., *ibid.*, **114**(1986), 1354.
[7] Murakami, T. et al., *J. Atmos. Sci.*, **40**(1986), 961.
[8] Lau, K.M. & Lau, N.C., *ibid.*, **43**(1986), 2023.

[9] Anderson, J. R.et al., *ibid.*, **40**(1983), 1584.
[10] 李崇银、肖子牛, 大气科学文集, 科学出版社. 1990, pp. 1-10.
[11] Krishnamurti, T. N. & Gadgil, S., *Tellus* (A), Vol.37, 336-360.
[12] Li Chongyin (李崇银) & Wu Peili (武培立), *Advances Atmos. Sci.*, **7**(1990), 294.
[13] 李崇银, 大气科学, **14**(1990), 33.
[14] 李崇银, 中国科学, B 辑, 1985, 668-675.
[15] Lau, K.M. & Peng, L., *J. Atmos. Sci*, **44**(1987), 950.
[16] Chang, C.P. & Lim, H., *ibid.*, **45**(1988), 1709.
[17] 李崇银, 大气科学, **14**(1990), 83.
[18] Wallace. J. M. et al., *Large Scale Dynamical Processes in the Atmosphere*, *Academic* Press INC (London), 1983, pp. 55-91.
[19] Wallace, J.M. & Shukla, J., *J. Atmos. Sci.*, **40**(1983), 1613.
[20] Keshavarmurty, R.N., *ibid.*, **39**(1982). 1241.
[21] Zeng Qingcun et al., *Climate Change*, *Dynamics and Modellings*, China Meteorological Press. 1990, pp. 303-330.
[22] Murakami, T., *Mon. Wea. Rev.*, **107**(1979), 994.

A Further Inquiry on the Mechanism of 30-60 Day Oscillation in the Tropical Atmosphere

Li Chongyin (李崇银)

LASG, Institute of Atmospheric Physics, Chinese Academy of Sciences, Beijing

Abstract In a simple semi-geostrophic model on the equatorial β-plane, the theoretical analysis on the 30-60 day oscillation in the tropical atmosphere is further discussed based on the wave-CISK mechanism. The convection heating can excite the CISK-Kelvin wave and CISK-Rossby wave in the tropical atmosphere and they are all the low-frequency modes which drive the activities of 30-60 day oscillation in the tropics. The most favorable conditions to excite the CISK-Kelvin wave and CISK-Rossby wave are indicated: There is convection heating but not very strong in the atmosphere and there is weaker disturbance in the lower troposphere.

The influences of vertical shearing of basic flow in the troposphere on the 30-60 day oscillation in the tropics are also discussed.

Ⅰ. Introduction

In recent years, the more attentions have been paid to the research on the 30-60 day oscillation in the atmosphere, because it is considered directly as a cause of the short-term climate anomaly and the occurrence of El Niño event (Lau and Chan, 1986a; Li and Wu, 1990). A series of studies have not only indicated the action regularity and the structure feature of 30-60 day oscillation in the tropical atmosphere (Murakami, et al., 1984; Lau and Chan, 1985, 1986b; Lau and Lau, 1986; Knutson and Weickmann, 1987; and many others), but also exposed the existence and activities of 30-60 day atmospheric oscillation in the middle-high latitudes (Anderson and Rosen, 1983; Li, 1990, 1991; Li and Zhou, 1991).

In relation to the dynamical theory of 30-60 day oscillation in the tropical atmosphere, the feedback of the cumulus convection heating has been considered as an important mechanism. In order to explain theoretically the 30-50 day periodic variation of the monsoon troughs (ridges) in South Asia described by Krishnamurti and Subrahmanyam (1982), the CISK theory was introduced into the investigation on the dynamics of 30-60 day oscillation in the tropical atmosphere and the moving CISK mode was suggested as a mechanism to drive the 30-50 day oscillation of the monsoon troughs (ridges) (Li, 1985a). Lau and Peng (1987) conducted the mobile wave-CISK mechanism to originate the 30-60 day oscillation in the tropical atmosphere and then Takahashi (1987), Chang and Lim (1988) studied this

Received February 13, 1992; revised July 17, 1992.

CISK-Kelvin wave theory further. In which, the slow eastward propagation of 30-60 day oscillation along the equator was explained successfully.

The observations still show some characteristics of 30-60 day atmospheric oscillation in the tropics, such as it also propagates westwards sometimes, especially in the tropics outside the equator (Chen and Xie, 1988), and the 30-60 day oscillation has the wavetrain patterns even though in the tropics (Li, 1991). These properties of 30-60 day oscillation in the tropical atmosphere are hardly explained only by the CISK-Kelvin wave theory. Li (1988) and Liu and Wang (1990) indicated that the CISK-Rossby wave is also a mechanism to drive 30-60 day oscillation in the tropical atmosphere, especially in the tropics outside the equator. This CISK-Rossby wave is able to propagate westwards or eastwards depending on the heating and it has the energy dispersion feature. But, in order to have an analytic solution, Li (1988) took tropical β-plane approximation while Liu and Wang (1990) assumed vertical velocity in the lower troposphere is independent with y-direction in their study. Therefore, those results have a certain limitation.

In the present paper, we will make a further inquiry on the mechanism of 30-60 day oscillation in the tropical atmosphere at the equatorial β-plane and understand the features of the CISK-Kelvin wave and CISK-Rossby wave further. The influences of vertical wind shearing on the CISK-Kelvin wave and CISK-Rossby wave in the tropics are also discussed.

II. Basical Theory

Some studies have indicated that the intraseasonal oscillation has the characteristics of planetary scale motion in the tropical atmosphere. A dynamic study in relation to the atmospheric motion in the tropics has shown that the planetary scale motion in the tropical atmosphere is quasi-geostrophic (Li, 1985b). Therefore, the basical equations with the cumulus heating feedback based on the Boussinesq fluid and the equatorial β-plane can be written as follows:

$$\frac{\partial u}{\partial t}+w\frac{\partial \overline{u}}{\partial z}-\beta yv=-\frac{\partial \varphi}{\partial x} \tag{1}$$

$$\beta yu=-\frac{\partial \varphi}{\partial y} \tag{2}$$

$$\frac{\partial u}{\partial x}+\frac{\partial v}{\partial y}+\frac{\partial w}{\partial z}=0 \tag{3}$$

$$\frac{\partial}{\partial t}\left(\frac{\partial \varphi}{\partial z}\right)+N^2w=N^2\eta w_B \tag{4}$$

There, the influences of the wave-CISK mechanism and the vertical shearing of basic flow are introduced. Generally, the basic flow is smaller in the tropical atmosphere while the vertical wind shearing is quite prominent. So the vertical shearing of basic flow here is just introduced. The Brunt-Vaisala frequency N, Rossby parameter β and vertical wind shearing $\frac{\partial \overline{u}}{\partial z}$ are assumed to be constants for convenience. The term $N^2\eta w_B$ on the right side in Eq.(4) represents the cumulus convection heating (Takahashi, 1987) in which w_B is the vertical velocity at the top of the atmospheric boundary layer and η the nondimensional heating parameter.

Eliminated u from Eqs.(1)-(3), we obtain

$$\frac{\partial}{\partial t}\frac{\partial^2\varphi}{\partial y^2}-\beta y\frac{\partial\bar{u}}{\partial z}\frac{\partial w}{\partial y}-\beta w\frac{\partial\bar{u}}{\partial z}+2\beta^2 yv-\beta^2 y^2\frac{\partial w}{\partial z}=\beta\frac{\partial\varphi}{\partial x}. \tag{5}$$

Then, based on Eqs.(1) and (2), the variable v is represented by φ and we obtain

$$\frac{\partial}{\partial t}\left(\frac{\partial}{\partial y}-\frac{2}{y}\right)\frac{\partial\varphi}{\partial y}+\beta\frac{\partial\varphi}{\partial x}-\beta y\frac{\partial\bar{u}}{\partial z}\frac{\partial\omega}{\partial y}+\beta^2 y^2\frac{\partial w}{\partial z}+\beta w\frac{\partial\bar{u}}{\partial z}=0. \tag{6}$$

Applying normal mode method, the solutions of the Eqs.(4) and (6) can be written as follows:

$$(\varphi,w)=(\Phi,W)e^{i(kx-\sigma t)}. \tag{7}$$

Then, we can obtain

$$-i\sigma\frac{\partial\Phi}{\partial z}+N^2W=N^2\eta W_B \tag{8}$$

$$i\sigma\left(\frac{2}{y}-\frac{\partial}{\partial y}\right)\frac{\partial\Phi}{\partial y}-\beta yU\frac{\partial W}{\partial y}+\beta UW-\beta^2 y^2\frac{\partial W}{\partial z}+ik\beta\Phi=0 \tag{9}$$

respectively, where $U\equiv\dfrac{\partial\bar{u}}{\partial z}$.

Taking simple 2-level model shown in Fig.1, we have the equations:

$$-i\sigma\frac{\Phi_1-\Phi_2}{\Delta}+N^2W_2=N^2\eta_2W_B \tag{10}$$

$$i\sigma\left(\frac{2}{y}-\frac{d}{dy}\right)\frac{d\Phi_1}{dy}-\beta yU\frac{dW_1}{dy}+\beta UW_1+\beta^2y^2\frac{W_2}{\Delta}+ik\beta\Phi_1=0 \tag{11a}$$

$$i\sigma\left(\frac{2}{y}-\frac{d}{dy}\right)\frac{d\Phi_3}{dy}-\beta yU\frac{dW_3}{dy}+\beta UW_3-\beta^2y^2\frac{W_2}{\Delta}+ik\beta\Phi_3=0 \tag{11b}$$

where we have assumed $w=0$ at both the top and bottom of the model, Δ is the height interval and $\Delta=7$km in this 2-level model (Fig. 1).

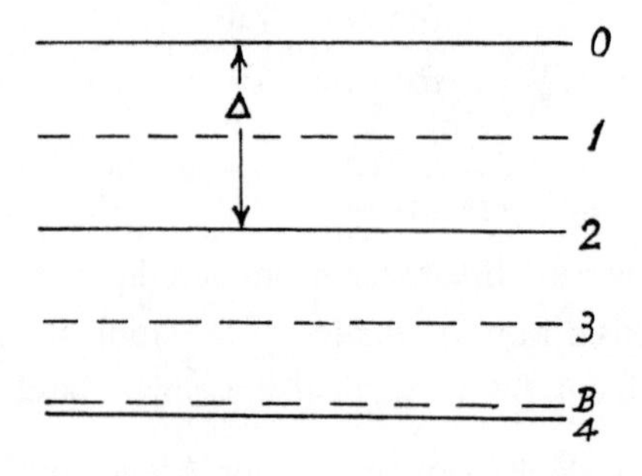

Fig. 1. Schematic illustration of a 2-level model.

From the Eqs.(11 a) and (11b), we obtain

$$i\sigma\left(\frac{2}{y}-\frac{d}{dy}\right)\frac{d(\Phi_1-\Phi_3)}{dy}-\beta yU\frac{d(W_1-W_3)}{dy}+\beta U(W_1-W_3)+2\beta^2y^2\frac{W_2}{\Delta}+ik\beta(\Phi_1-\Phi_3)=0 \quad (12)$$

and the $(\Phi_1-\Phi_3)$ can be represented by using

$$\Phi_1-\Phi_3=-\frac{i\Delta}{\sigma}N^2W_2+\frac{i\Delta}{\sigma}N^2\eta_2W_B. \quad (13)$$

According to the observational distribution of vertical velocity in the tropical atmosphere, the vertical velocity has identical sign and the maximum is at about 300-400hPa for general convection system in the tropics. Therefore, for the convenience and losing no fundamental feature of the tropical atmosphere, we would have approximation $W_1 - W_3 = a_1W_2$, where $a_1\approx$ 0.5-0.8 in general. Identically the vertical velocity at the top of the atmospheric boundary layer can also be simplified as

$$W_B = bW_2, \quad (14)$$

where b describes the intensity of the disturbance (convergence) in the lower troposphere. Thus, from Eqs.(12) and (13), we can obtain an equation in relation to the variable W_2 as follows:

$$y^2\frac{d^2W_2}{dy^2}-\left[2y-\frac{\Delta\beta Ua_1}{c_1^2(1-b\eta_2)}y^3\right]\frac{dW_2}{dy}-\left[\frac{2\beta^2}{c_1^2(1-b\eta_2)}y^4+\left(\frac{\beta k}{\sigma}+\frac{a_1\Delta\beta U}{c_1^2(1-b\eta_2)}\right)y^2\right]W_2=0 \quad (15)$$

where $c_1^2=N^2\Delta^2$, the phase speed of the gravity wave in the atmosphere.

The Eq.(15) is the basic equation in the 2-level model, based on which, the cumulus heating feedback and the effect of vertical wind shearing can be discussed. For the Eq.(I5), the boundary condition in y-direction can be written as follows:

$$W_2\equiv 0,\ \ y\to\pm\infty. \quad (16)$$

Ⅲ. Analytic Solution (1)

At first, we discuss the case without vertical shearing of basic flow, i.e., U = 0 in the Eq.(15). Thus, the Eq.(15) is reduced as follows:

$$y^2\frac{d^2W_2}{dy^2}-2y\frac{dW_2}{dy}-\left[\frac{2\beta^2}{c_1^2(1-b\eta_2)}y^4+\frac{\beta k}{\sigma}y^2\right]W_2=0. \quad (17)$$

For convenience to solve the Eq.(17), it is necessary to alternate the variables. We take

$$\left.\begin{aligned}\zeta&=\frac{\sqrt{2}\beta}{c_2}y^2\\ W_2&=\zeta^{1/4}\omega_2\end{aligned}\right\} \quad (18)$$

where $c_2=\sqrt{(1-b\eta_2)}c_1$.

Since there are some formulations as follows:

$$\left.\begin{aligned} &\frac{dW_2}{dy}=\frac{dW_2}{d\zeta}\left(\frac{2\sqrt{2}\beta}{c_2}y\right) \\ &\frac{d^2W_2}{dy^2}=\frac{4(\sqrt{2}\beta)^2}{c_2^2}y^2\frac{d^2W_2}{d\zeta^2}+\frac{2\sqrt{2}\beta}{c_2}\frac{dW_2}{d\zeta} \end{aligned}\right\} \tag{19}$$

and

$$\left.\begin{aligned} &\frac{dW_2}{d\zeta}=\zeta^{1/4}\frac{d\omega_2}{d\zeta}+\frac{1}{4}\xi^{-3/4}\omega_2 \\ &\frac{d^2W_2}{d\zeta^2}=\zeta^{1/4}\frac{d^2\omega_2}{d\zeta^2}+\frac{1}{2}\xi^{-3/4}\frac{d\omega_2}{d\zeta}-\frac{3}{16}\xi^{-7/4}\omega_2 \end{aligned}\right\}. \tag{20}$$

The Eq.(17) can be reduced as:

$$\frac{d^2\omega_2}{d\zeta^2}-\left(\frac{5}{16}\zeta^{-2}+\frac{kc_2}{4\sqrt{2}\sigma}\zeta^{-1}+\frac{1}{4}\right)\omega_2=0, \tag{21}$$

and the boundary condition (16) can be written as

$$|\omega_2|_{\zeta\to\infty}=0. \tag{22}$$

It can be said that the problem becomes to solve the eigenvalue of the Whittaketer equation (Wang and Kou, 1979; Liu and Wang, 1990):

$$\begin{cases} \dfrac{d^2\omega_2}{d\zeta^2}+\left(-\dfrac{1}{4}+\dfrac{l}{\zeta}+\dfrac{\frac{1}{4}-\mu^2}{\zeta^2}\right)\omega_2=0, \\ \omega_2|_{\zeta\to\infty}=0 \end{cases} \tag{23}$$

where

$$l=-\frac{kc_2}{4\sqrt{2}\sigma},\quad \mu_2=\frac{9}{16}.$$

Making

$$\omega_2=e^{-\zeta/2}\zeta^{\mu+\frac{1}{2}}P. \tag{24}$$

Eq. (23) can be reduced to solve the eigenvalue of the Kummer equation as follows:

$$\begin{cases} \zeta\dfrac{d^2P}{d\zeta^2}+(2\mu+1-\zeta)\dfrac{dP}{d\zeta}-\left(\mu+\dfrac{1}{2}-l\right)P=0 \\ P|_{\zeta\to\infty}=0(\zeta^m) \end{cases} \tag{25}$$

The eigenvalues of Eq. (25) are

$$\mu+\frac{1}{2}-l=-m\quad (m=0,1,2,\cdots), \tag{26}$$

and the corresponding eigenfunctions can be written as

$$P\equiv A_m S_m^{2\mu}(\zeta)=A_m\cdot\frac{(2\mu+1)_m}{m!}K(-m,2\mu+1,\zeta), \tag{27}$$

where A_m is an arbitrary constant and $(2\mu + 1)_m$ is the Gauss symbol which is defined as

$$(2\mu+1)_m = (2\mu+1)(2\mu+2)\cdots(2\mu+m) = \frac{\Gamma(2\mu+m)}{\Gamma(2\mu+1)} \tag{28}$$

and

$$(2\mu+1)_0 = 1;$$

while $S_m^{2\mu}$ and $K(-m, 2\mu+1, \zeta)$ are known as the Sonine polynomial and Kummer function respectively.

Taking $\mu = -\frac{3}{4}$ as general case, from (26), it can be obtained

$$-\frac{1}{4} + \frac{kc_2}{4\sqrt{2}\sigma} = -m \tag{29}$$

and the angular frequency can be represented

$$\sigma = \frac{kc_2}{\sqrt{2}(1-4m)}, \quad (m = 0, 1, 2, \cdots). \tag{30}$$

And from (27) we have

$$P \equiv A_m S_m^{-3/2}(\zeta) = A_m \frac{(-1/2)_m}{m!} K\left(-m, -\frac{1}{2}, \zeta\right), \tag{31}$$

then, according to (18) and (24), it can be obtained

$$W_2^m(y) = A_m \frac{(-1/2)_m}{m!} e^{-y^2\sqrt{2}\beta/2c_2} K\left(-m, -\frac{1}{2}, \frac{\sqrt{2}\beta}{c_2} y^2\right). \tag{32}$$

When $m = 0$, it is clearly shown in (30) that

$$\sigma = \frac{\sqrt{2}}{2} kc_2, \tag{33}$$

and it is the Kelvin wave propagating eastwards. But when $m \neq 0$, then $\sigma < 0$, it represents the westward Rossby wave. Since c_2 has included the cumulus heating, the Kelvin wave and Rossby wave mentioned here are the one with the cumulus convection feedback.

Making $L_0 = (\sqrt{2}c_2 / 2\beta)^{1/2}$, (32) can be written as:

$$W_2^m(y) = A_m \frac{(-1/2)_m}{m!} e^{-\frac{1}{2}(y/L_0)^2} K\left(-m, -\frac{1}{2}, (y/L_0)^2\right). \tag{34}$$

The latitudinal structures for $m = 0$ Kelvin mode and $m = 1$ Rossby mode are shown in Fig. 2. Obviously, the latitudinal structures of these modes with the cumulus convection feedback (the heating is not quite strong) are similar to ordinary modes without the convection heating in the tropical atmosphere, i.e., the Kelvin mode has a maximum at the equator ($y = 0$) and it is symmetrical about y; the Rossby mode has extreme values at $y = 0$ and $y = \pm\sqrt{3/2}L_0$ and it is also symmetrical about y. The convection heating is unable to change the fundamental features of latitudinal structures.

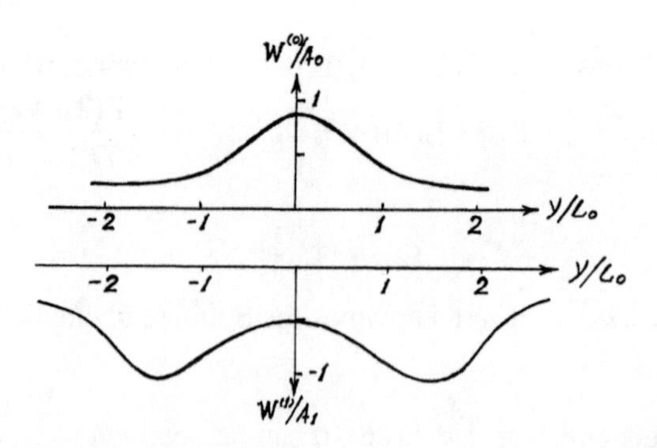

Fig. 2. Meridional structure of $W(y)$ for the CISK-Kelvin mode $m = 0$ (a) and CISK-Rossby mode $m = 1$ (b).

1. CISK-Kelvin Wave

When $m = 0$, the angular frequency (33) can be written as

$$\sigma = \sqrt{(1 - b\eta_2)}kc_1 / \sqrt{2}, \tag{35}$$

this is equatorial Kelvin wave with condensational heating, i.e., the so-called CISK-Kelvin wave. Because when $\eta_2 = 0$ (no heating), (35) becomes $\sigma = kc_1 / \sqrt{2}$, it is the eastward equatorial Kelvin wave with the phase speed:

$$c_{xo} \equiv \frac{\sigma}{k} = \frac{\sqrt{2}}{2}c_1. \tag{36}$$

In general, $N = 10^{-2}\ \mathrm{s}^{-1}$, we then can obtain $c_{xo} = 49.5\mathrm{m/s}$ which is consistent with one in the past.

When the cumulus heating is existent but weaker, then $1 - b\eta_2 > 0$, (35) should indicate $\sigma = \sigma_r > 0$, the CISK-Kelvin wave is stable and propagates eastwards with the phase speed

$$c_x = \frac{\sqrt{2(1 - b\eta_2)}}{2}c_1 < c_{xo}. \tag{37}$$

Taking $b = 0.4$ and $\eta_2 = 2.0$ we then obtain $c_x = 15.7\mathrm{m/s}$. It is consistent with the 30-60 day oscillation in the tropical atmosphere.

The period of the CISK-Kelvin wave can be described as

$$T \equiv \frac{2\pi}{\sigma_r} = \frac{2\pi}{kc_1}\frac{2}{\sqrt{2(1 - b\eta_2)}} = \frac{\sqrt{2}L}{\sqrt{1 - b\eta_2}c_1}, \tag{38}$$

where L is the zonal scale of the CISK-Kelvin wave. Obviously, the period of the CISK-Kelvin wave with a certain zonal scale depends on the intensity of the convection heating (η_2). Taking $L = 3.2 \times 10^7\mathrm{m}$, the periods associated with η_2 can be obtained from (38) and they are shown as in Fig.3. It is clear that the periods of the CISK-Kelvin wave are about 15-70 days for general convection heating $\eta_2 \approx 1.8 - 2.4$ (Hayashi, 1970).

When the cumulus convection heating is strong, then $1 - b\eta_2 < 0$, (35) should indicate $\sigma = i\sigma_i$. The CISK-Kelvin wave is unstable and stationary since $c_x = 0$. These are consistent with the Takahashi's results (1987) even though the model in this study is different from the one used by Takahashi.

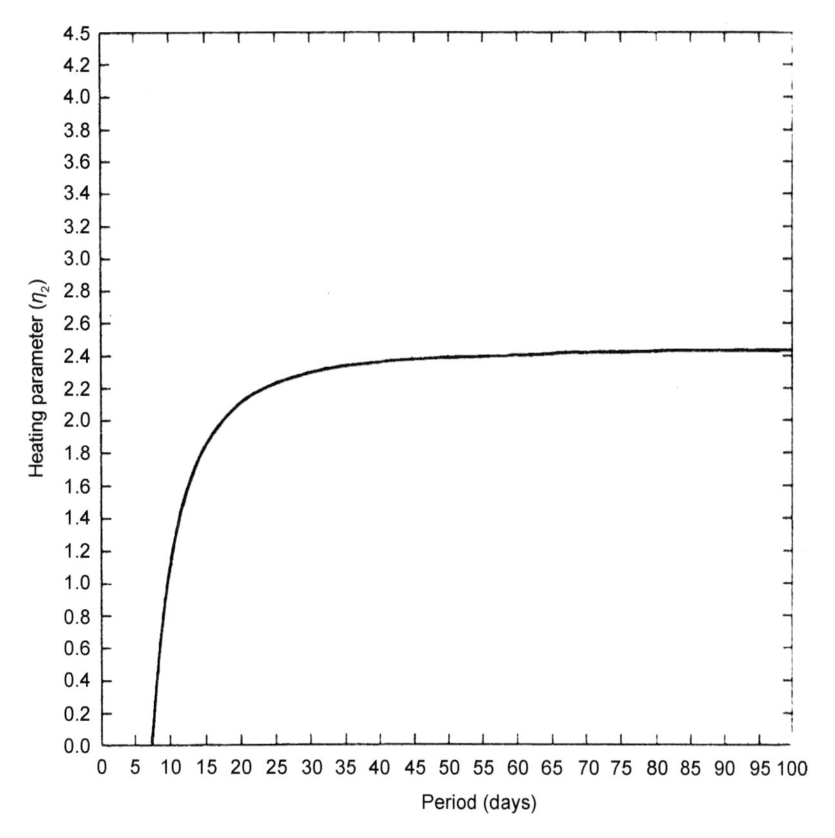

Fig. 3. Variation of the periods of the CISK-Kelvin wave (m = 0) with the heating parameter η_2.

2. CISK-Rossby Wave

If $m \neq 0$, the angular frequency can be written as

$$\sigma = -kc_2 / \sqrt{2}(4m-1) = -\sqrt{1-b\eta_2}\, kc_1 / \sqrt{2}(4m-1), (m = 1, 2, \cdots) \tag{39}$$

When $\eta_2 = 0$, (39) shows a westward equatorial Rossby wave with the phase speed

$$c_{xo} \equiv \frac{\sigma}{k} = -c_1 / \sqrt{2}(4m-1). \tag{40}$$

Therefore, (39) represents the CISK-Rossby wave.

If the cumulus convection heating is not quite strong, then $1 > 1 - b\eta_2 > 0$, we can obtain the phase speed and the period of the CISK-Rossby wave as follows:

$$c_x = \frac{\sigma}{k} = -\sqrt{1-b\eta_2} / \sqrt{2}(4m-1) \tag{41}$$

$$T = \left|\frac{2\pi}{\sigma}\right| = \sqrt{2}(4m-1)L / \sqrt{(1-b\eta_2)}c_1. \tag{42}$$

Eq. (41) shows that the westward phase speed of the CISK-Rossby wave is slower than that of the equatorial Rossby wave. Taking $L = 9.0 \times 10^6$m, for $m = 1$ and $m = 2$, the periods associated with η_2 can be obtained from (42) and they are shown in Fig.4. For general convection heating ($\eta_2 = 1.8 - 2.4$), the periods of the CISK-Rossby wave are about 20-70 days.

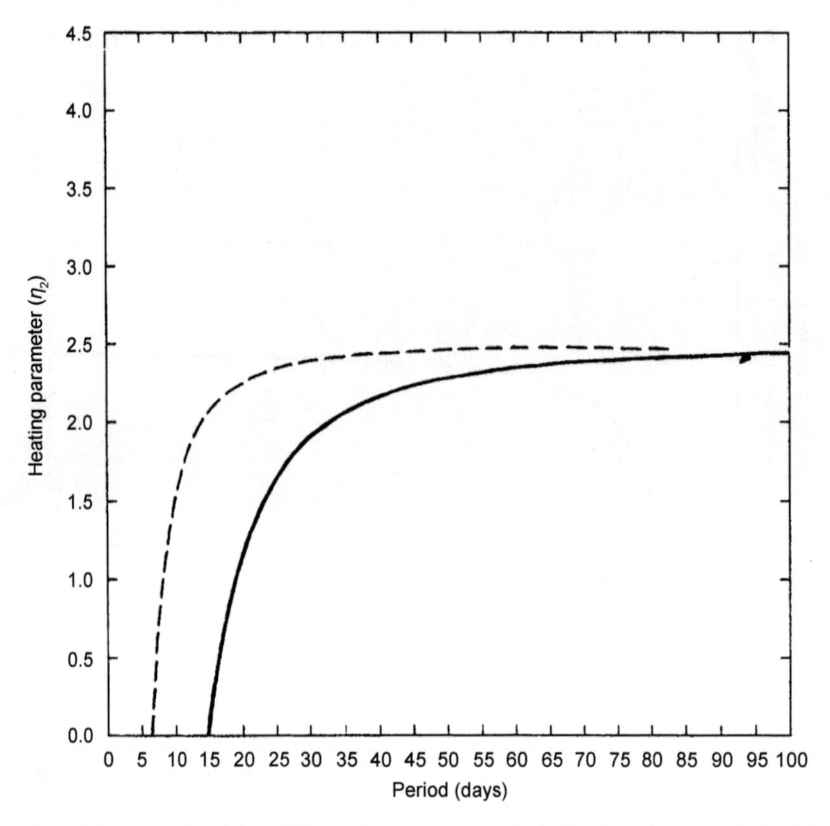

Fig. 4. Variation of the periods of the CISK-Rossby wave $m = 1$ (solid line) and $m = 2$ (dashed line) with the heating parameter η_2.

When the cumulus convection heating is strong, then $1 - b\eta_2 < 0$, it is shown in (39) that the CISK-Rossby wave is reductive and stationary.

Above-mentioned analyses show that there are eastward CISK-Kelvin wave and westward CISK-Rossby wave resulted from the exciting of the cumulus convection heating. They are all the low-frequency modes and the CISK-Rossby wave is not only able to partly explain westward propagation of 30-60 day oscillation in the tropical atmosphere, but also can decrease eastward speed of the CISK-Kelvin wave which is faster than the propagation of 30-60 day oscillation. Therefore, it is obvious that the CISK-Kelvin wave and CISK-Rossby wave are an important mechanism to drive the activity of 30-60 day oscillation in the tropical atmosphere even though the coupled moist Kelvin-Rossby wave (Wang and Rui, 1990) is not obtained in this simple model.

It should be pointed out that the CISK-Kelvin wave is unstable growing but stationary while the CISK-Rossby wave is reductive and stationary when the cumulus convection heating is strong (such as $\eta_2 > 2.5$). This is different from the slower propagation of 30-60 day oscillation and implies that the strong convection heating is unfavourable to excite the 30-60 day oscillation in the tropical atmosphere. The general convection heating, it is not quite strong in the large area, can excite the CISK-Kelvin wave and CISK-Rossby wave which drive the activity of 30-60 day oscillation in the tropics. This result is consistent with the observation, that is, the strong convection heating in the tropical atmosphere will lead to the occurrence of the disturbance like tropical cyclone. Recently, a linear theory study shows that the disturbance caused by wave-CISK in the tropiçal atmosphere has a tendency to generate smaller scale feature as the convection heating increased (Frederiksen and Frederiksen, 1991). It also means that the strong convection heating is unfavourable to produce intraseasonal oscillation which is planetary scale system in the tropical atmosphere.

In this paper, the parameter b is introduced for analysing conveniently. The greater (smaller) value of b represents the stronger (weaker) disturbance in the lower troposphere. An interesting result shows that the existence of stronger disturbance in the lower troposphere is unfavourable to excite the 30-60 day oscillation in the tropical atmosphere for general convection heating but favorable to produce stationary unstable growth of the disturbance. The occurrence of 30-60 day oscillation is more favourable in general convection heating when there is only weaker disturbance in the lower troposphere. Of course, if there is stronger disturbance in the lower troposphere, the 30-60 day oscillation in the tropical atmosphere can be still excited by weaker convection heating. In Fig.5, the relationships between the heating parameter η_2 and the periods of the CISK-Kelvin wave are shown for $b = 0.35$ and $b = 0.25$, respectively. Obviously, the stronger heating is still able to excite the 30-60 day oscillation in the tropical atmosphere as the disturbance is weaker in the lower troposphere.

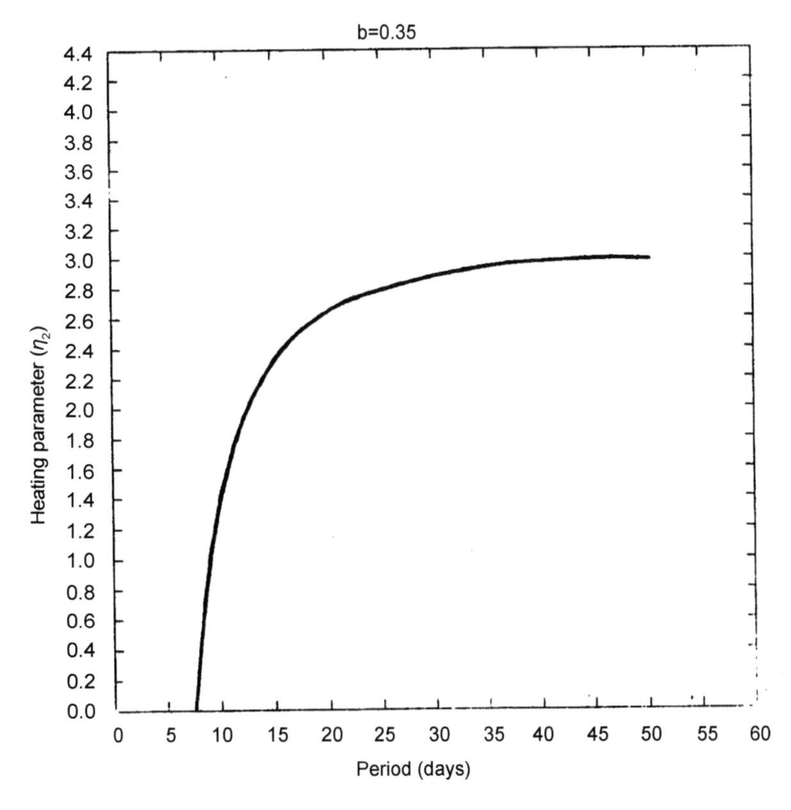

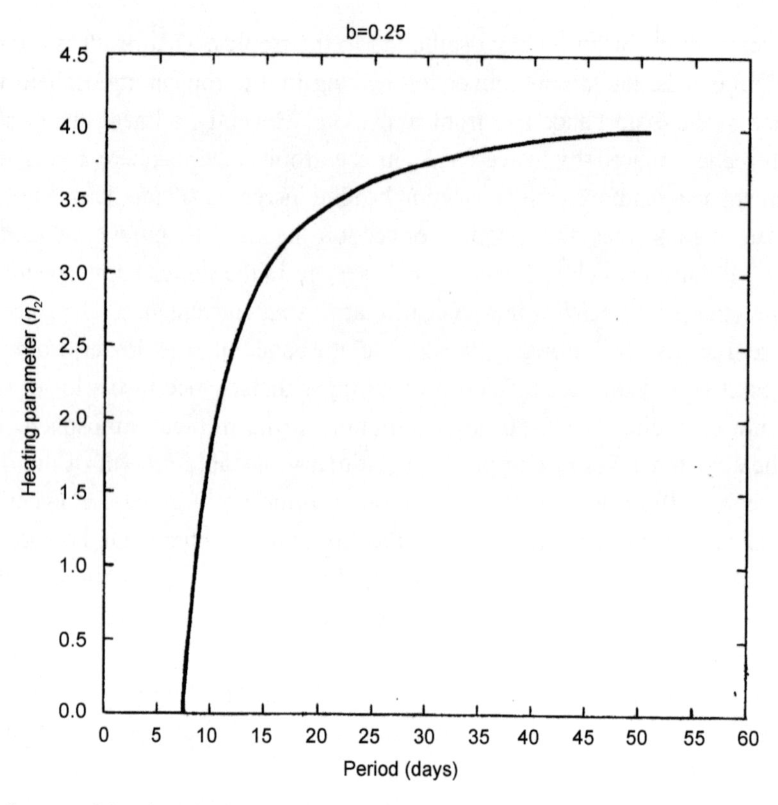

Fig. 5. The influence of the disturbance intensity in the lower troposphere on the CISK-Kelvin wave. Above for $b = 0.35$ and bottom for $b = 0.25$.

Ⅳ. Analytic Solution (2)

The case with vertical shearing of basic flow will be discussed in this section. Since Eq.(15) is complicated and difficult to solve analytically, the simplification is necessary. According to the distribution of the vertical velocity $W(y)$ shown in Fig.2, the following approximate formulation can be used:

$$y\frac{d(W_1-W_3)}{dy} \approx \frac{(W_1-W_3)_{j+1}-(W_1-W_3)_{j-1}}{y_{j+1}-y_{j-1}}y_j \approx -a_2(W_1-W_3), \tag{43}$$

where a_2 is a smaller constant, we can take $a_2 = 0.2$. Thus, Eq.(15) can be written as

$$y^2\frac{d^2W_2}{dy^2}-2y\frac{dW_2}{dy}-\left[\frac{2\beta^2}{(1-b\eta_2)c_1^2}y^4+\left(\frac{\beta k}{\sigma}+\frac{a\beta U\Delta}{(1-b\eta_2)c_1^2}\right)y^2\right]W_2=0, \tag{44}$$

where $a = (1 + a_2)a_1$. If we assume the variation of the difference of vertical velocities at level 1 and level 3 with y being ignored (i.e., $\frac{d(W_1-W_3)}{dy}=0$) to replace the formulation (43), and Eq. (44) is still the same except $a = a_1$.

Using the similar methods to above analyses and taking

$$\zeta = \frac{\sqrt{2}\beta}{c_2} y^2, \quad W_2 = \zeta^{1/4} \omega_2, \tag{45}$$

then Eq.(44) can be written as

$$\frac{d^2 \omega_2}{d\zeta^2} - \left[\frac{5}{16} \zeta^{-2} + \left(\frac{kc_2}{4\sqrt{2}\sigma} + \frac{aU\Delta}{4\sqrt{2}c_2} \right) \zeta^{-1} + \frac{1}{4} \right] \omega_2 = 0, \tag{46}$$

The boundary condition is still

$$\omega_2 \big|_{\zeta \to 0} = 0. \tag{47}$$

There is still the following Whittaker equation

$$\begin{cases} \dfrac{d^2 \omega_2}{d\zeta^2} + \left(-\dfrac{1}{4} + \dfrac{l'}{\zeta} + \dfrac{\frac{1}{4} - \mu^2}{\zeta^2} \right) \omega_2 = 0 \\ \omega_2 \big|_{\zeta \to 0} = 0 \end{cases}, \tag{48}$$

but where

$$\mu^2 = 9/16, \quad l' = -(kc_2 / 4\sqrt{2}\sigma + aU\Delta / 4\sqrt{2}c_2).$$

Corresponding to Eq.(48), the Kummer equation is

$$\begin{cases} \zeta \dfrac{d^2 P}{d\zeta^2} + (2\mu + 1 - \zeta) \dfrac{dP}{d\zeta} - \left(\mu + \dfrac{1}{2} - l' \right) P = 0 \\ P \big|_{\zeta \to \infty} = 0(\zeta^m) \end{cases}. \tag{49}$$

The eigenvalues of Eq. (49) are

$$\mu + \frac{1}{2} - l' = -m, \quad (m = 0, 1, 2, \cdots) \tag{50}$$

Taking $\mu = -\frac{3}{4}$, from (50) we have

$$-\frac{1}{4} + \frac{kc_2}{4\sqrt{2}\sigma} + \frac{aU\Delta}{4\sqrt{2}c_2} = -m, \quad (m = 0, 1, 2, \cdots) \tag{51}$$

Then, the angular frequency can be written as

$$\sigma = \frac{kc_2}{(1 - 4m)\sqrt{2} - aU\Delta / c_2} \tag{52}$$

The corresponding eigenfunction can be still represented by (31) or (32).

In the following, we will discuss the CISK-Kelvin wave and CISK-Rossby wave with vertical shearing of basic flow respectively.

1. The CISK-Kelvin Wave Case

For the CISK-Kelvin wave, $m = 0$, Eq. (52) can be written

$$\sigma = \frac{kc_1^2 (1 - b\eta_2)(aU\Delta + \sqrt{2(1 - b\eta_2)}c_1)}{2(1 - b\eta_2)c_1^2 - a^2 U^2 \Delta^2} \tag{53}$$

It is obvious in (53) that the vertical westerly shearing ($U = \frac{d\bar{u}}{dy} > 0$) will accelerate the eastward propagation of the CISK-Kelvin wave, but the vertical easterly shearing (U<0) will slow down the eastward propagation; And all the vertical shearing should decrease the unstable growth rate of the CISK-Kelvin wave. It is also shown in (53) that the propagation property of the stable CISK-Kelvin wave will be changed when the vertical shearing of basic flow is strong enough and it will propagate westwards.

2. The CISK-Rossby Wave Case

For the CISK-Rossby, $m \neq 0$, taking $m = 1$ to analyze simply. Then, the angular frequency is

$$\sigma = -\frac{k(1-b\eta_2)c_1^2}{3c_1\sqrt{2(1-b\eta_2)} + aU\Delta} \tag{54}$$

We can see from (54) that the westerly (easterly) shearing will slow down (speed up) the westward propagation of the CISK-Rossby wave. When the vertical easterly shearing is very strong (it is difficult to occur), the propagation property of the CISK-Rossby wave will be changed and propagates eastwards.

In general, we can take $a_1 = 0.6$, then $a = 0.72$. Based on Eqs.(53) and (54), the relationships between the vertical shearing of basic flow and the phase speeds of the CISK-Kelvin wave and CISK-Rossby wave for certain convection heating can be obtained. For the CISK-Kelvin wave, the phase speed increases with the decrease of the easterly shearing and the increase of the westerly shearing (Fig.6). But, when the westerly shearing is strong enough, the CISK-Kelvin wave propagates westwards. The very strong easterly or westerly shearing will all lead the CISK-Kelvin wave to quasi-stationary. For the CISK-Rossby wave, the westward phase speed decreases with the decrease of easterly shearing and the increase of the westerly shearing (Fig.7). The easterly and westerly shearing of basic flow will all decrease the reduction of the CISK-Rossby wave.

V. Conclusion Remarks

Throughout this paper, we have discussed further the dynamical mechanism of the 30-60 day oscillation in the tropical atmosphere. Especially, the more favourable conditions to excite the CISK-Kelvin wave and the CISK-Rossby wave which drive the activities of 30-60 day oscillation in the tropical atmosphere and the effects of vertical shearing of basic flow are indicated.

1) The cumulus convection heating, through the CISK mechanism, can excite the CISK-Kelvin wave and CISK-Rossby wave in the tropical atmosphere. These waves are the low-frequency modes and they are able to drive the activities of 30-60 day oscillation in the tropics. The obvious eastward propagation of 30-60 day oscillation near by the equator is dominated by the CISK-Kelvin wave as pointed out in many studies. The obvious westward propagation of 30-60 day oscillation in the tropical atmosphere outside the equator is probably

dominated by the CISK-Rossby wave; and the CISK-Rossby wave is able to decrease eastward speed of the CISK-Kelvin wave which is greater than the observation of 30-60 day oscillation.

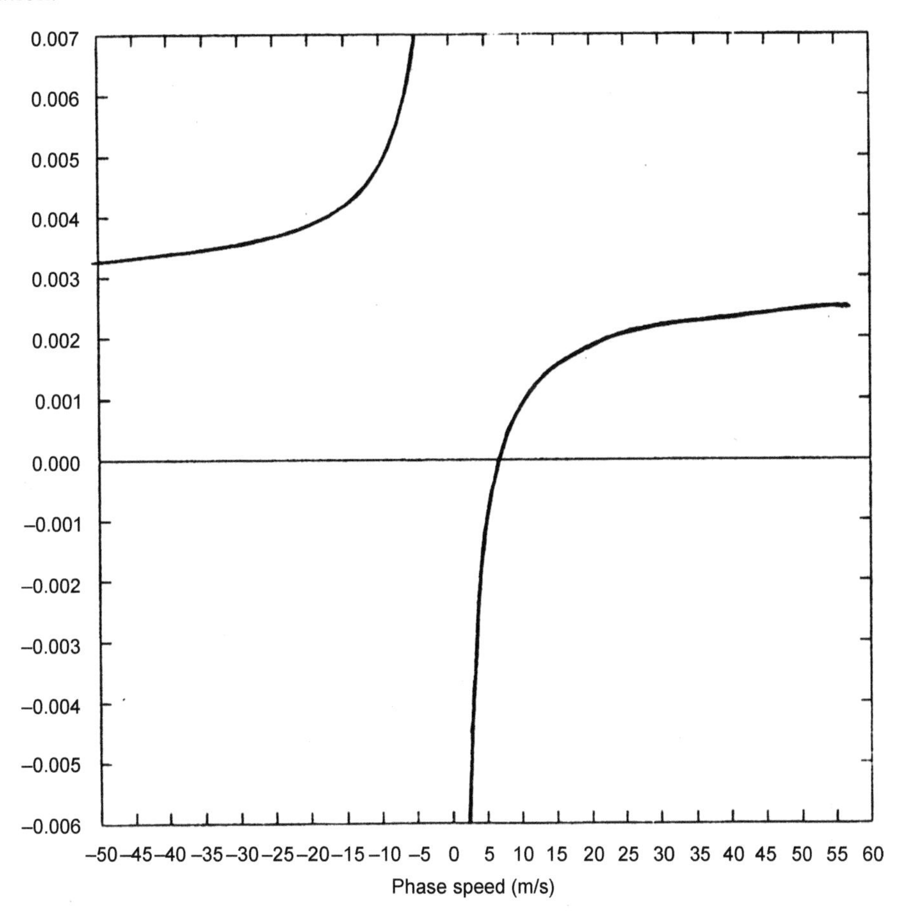

Fig. 6. The influence of vertical shearing of basic flow on the phase speed of the CISK-Kelvin wave.

2) The more favourable conditions to excite the CISK-Kelvin wave and CISK-Rossby wave are:

(a) There is convection heating in the large area but it is not very strong, otherwise the quite strong convection heating is just favourable to produce the disturbance like tropical cyclone.

(b) There is only weaker disturbance in the lower troposphere, otherwise it is unfavourable to excite the 30-60 day oscillation in the tropical atmosphere.

3) The vertical shearing of basic flow in the troposphere is obviously able to affect the propagation of the CISK-Kelvin wave and CISK-Rossby wave. Therefore, the vertical shearing of basic flow will also affect the propagation of 30-60 day oscillation in the tropical atmosphere. The westerly (easterly) shearing will speed up (slow down) the eastward propagation of the CISK-Kelvin wave. But, the CISK-Kelvin wave will propagate westwards when the westerly shearing is strong enough. The westward propagation of the CISK-Rossby

wave will be accelerated with the decrease of the westerly shearing (or the increase of the easterly shearing).

4) The vertical shearing of basic flow will weaken the instability of the CISK-Kelvin wave and the reduction of the CISK-Rossby wave.

This research was completed when the author was visiting in CSIRO Division of Atmospheric Research in Australia. The author is grateful to Dr. J. Frederiksen for reading the manuscript and for his helpful discussions.

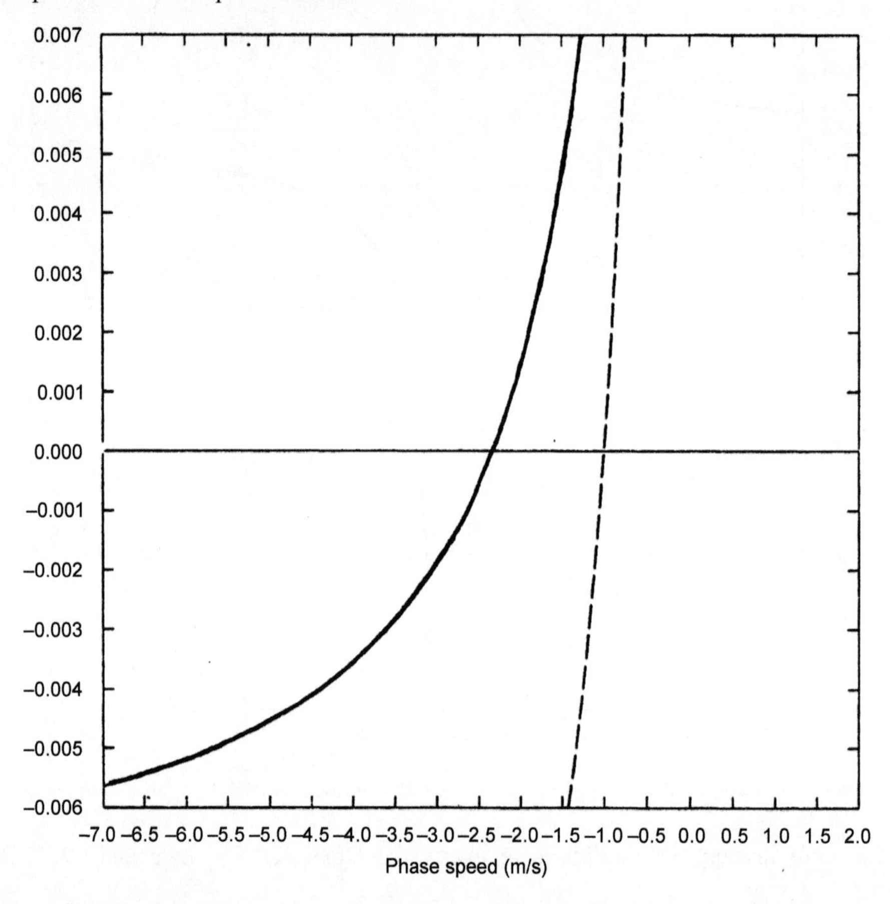

Fig. 7. The influence of vertical shearing of basic flow on the phase speed of the CISK-Rossby wave $m = 1$ (solid line) and $m = 2$ (dashed line).

References

Anderson, J.R., and R.D. Rosen (1983), The latitude-height structure of 40-50 day variations in atmospheric angular momentum, *J. Atmos. Sci.* **40**: 1584-1591.

Chang, C.P., and H. Lim (1988), Kelvin wave-CISK: A possible mechanism for the 30-60 day oscillation, *J. Atmos. Sci.* **45**: 1709-1720.

Chen Longxun and Xie An (1988), Westward propagating low-frequency oscillation and its teleconnections in the Eastern Hemisphere, *Acta Meteorologica Sinica* **2**: 300-312.

Frederiksen, J.S., and C.S. Frederiksen (1991), Monsoon disturbances, intraseasonal oscillations, teleconnection patterns, blocking and storm tracks of the global atmosphere during January 1979: Linear theory, *J. Atmos. Sci.* to be published.

Hayashi, .Y. (1970), A theory of large-scale equatorial wave generated by condensation heat and accelerating the zonal wind, *J. Meteor. Soc. Japan* **48**: 140-160.

Knutson, T.R., and K..M. Weickmann (1987), The 30-50 day atmospheric oscillation: composite life cycles of convection and circulation anomalies, *Mon. Wea. Rev.* **115**: 1407-1436.

Krishnamurti, T.N., and D. Subrahmanyam (1982), The 30-50 day mode at 850 mb during MONEX, *J. Atmos. Sci.* **39**: 2088-2095.

Lau, K.M., and P.H. Chan (1985), Aspects of the 40-50 day oscillation during northern winter as infrared from outgoing longwave radiation, *Mon. Wea. Rev.* **113**: 1889-1909.

——, and ——(1986a), The 40-50 day oscillation and the El Niño / Southern Oscillation: A new perspective, *Bull. Amer. Meteor. Soc.* **67**: 533-534.

Lau, K.M., and P.H. Chan (1986b), Aspects of the 40-50 day oscillation during the northern summer as infrared from outgoing longwave radiation, *Mon. Wea. Rev.* **114**: 1354-1367.

Lau, K.M., and L. Peng (1987), Origin of low-frequency (intraseasonal) oscillation in the tropical atmosphere, Part I: Basic theory, *J. Atmos. Sci.* **44**: 950, 972.

Lau, N.C., and K.M. Lau (1986), The structure and propagation of intraseasonal oscillations appearing in a GFDL general circulation model, *J. Atmos. Sci.* **43**: 2023-2047.

Li Chongyin (1985a), Actions of summer monsoon troughs (ridges) and tropical cyclones over South Asia and the moving CISK. mode, *Scientia Sinica* (*B*) **28**: 1197-1206.

——(1985b), The characteristics of the motion in the tropical atmosphere, *Scientia Atmospherica Sinica* (in Chinese), **9**: 366-376.

——(1988), Intraseasonal (30-50 day) oscillation in the atmosphere, International Summer Colloquium on Large Scale Dynamics of the Atmosphere, Beijing, August 10-20, 1988, 361-393.

——(1990), Intraseasonal oscillation in the atmosphere, *Chinese Journal of Atmos. Sci.* **14**: 35-51.

——(1991), The global characteristics of 30-60 day atmospheric oscillation, *Chinese Journal of Atmos. Sci.* **15**: 130-140.

Li Chongyin and Wu Peili (1990), A further inquiry on the 30-60 day oscillation in the tropical atmosphere, *Acta Meteorologica Sinica* **4**: 525-535.

Li Chongyin and Zhou Yiaping (1991), An observational study of the 30-50 day atmospheric oscillations, Part II: Temporal evolution and the interaction between Southern and Northern Hemisphere, *Advances in Atmos. Sci.* **8**: 399-406.

Liu Shikuo and Wang Jiyong (1990), A baroclinic semi-geostrophic model using the wave-CISK theory and low-frequency oscillation, *Acta Meteorologica Sinica* **4**: 576-585.

Murakami, T., et al. (1984), On the 40-50 day oscillation during the 1979 northern hemisphere summer, Part I: Phase propagation, *J. Meteor. Soc. Japan* **62**: 440-468.

Takahashi, M. (1987), A theory of the slow phase speed of the intraseasonal oscillation using the wave-CISK, *J. Meteor. Soc. Japan* **65**: 43-49.

Wang Bin and H. Rui (1990), Dynamics of the coupled moist Kelvin-Rossby wave on an equatorial β-plane, *J. Atmos. Set* **47**: 397-413.

Wang Zhuxi and Kuo Dengren (1979), Outline of Particular Function (in Chinese), Sciences Press, Beijing.

CISK Kelvin Wave with Evaporation-Wind Feedback and Air-Sea Interaction——A Further Study of Tropical Intraseasonal Oscillation Mechanism

Li Chongyin[1*] (李崇银), Cho Han-Ru[2] (禚汉如), Wang Jough-Tai[2] (王作台)

[1] Institute of Atmospheric Physics, Chinese Academy of Sciences, Beijing

[2] Department of Atmospheric Sciences, National Central University, Taibei

Abstract The wave-CISK (cumulus convection heating feedback), the air-sea interaction and the evaporation-wind feedback are together introduced into a simple theoretical model, in order to understand their effect on driving tropical atmospheric intraseasonal oscillation (ISO). The results showed that among the introduced dynamical processes the wave-CISK plays a major role in reducing phase speed of the wave to be closer to the observed tropical ISO. While the evaporation-wind feedback plays a major role in unstabilizing the wave. The air-sea interaction has certain effect on slowing down the phase speed of the wave. Therefore, the wave-CISK and evaporation-wind feedback can be regarded as fundamental dynamical mechanism of the tropical ISO. This study also shows that since the effects of the evaporation-wind feedback and the air-sea interaction were introduced, the excited wave is zonally dispersive, which can dynamically explain the activity feature of the observed ISO in the tropical atmosphere very well.

Keywords Intraseasonal oscillation, CISK Kelvin wave, Evaporation-wind feedback, Air-sea interaction.

1. Introduction

Since the intraseasonal (30-60 day) oscillation was identified in the early of 1970s (Madden and Julian 1971; 1972), its general characteristics and activity regulations have been investigated in series studies (Krishnamurti and Subrahmanyam 1982; Murakami 1984; Lau and Lau 1986; Knuston and Weickmann 1987; Li 1991; Madden and Julian 1994). So therefore the existence and activity of tropical ISO as an atmospheric system in the tropics have not been suspected wholly.

The cumulus convection heating feedback (wave-CISK) was introduced first in 1985 as the dynamical mechanism of the ISO in the tropical atmosphere (Li 1985). The theoretical analysis showed that the moving CISK mode due to the cumulus convection heating can

Received May 18, 2001; revised October 22, 2001.

*Lcy@lasg.iap.ac.cn

cause the ISO in Asian monsoon region since the convection heating feedback will reduce the phase speed of the excited wave to be closer to the propagation speed of the observed ISO in tropical atmosphere through changing atmospheric stratification. The convection heating feedback as an important mechanism of tropical ISO was also discussed by Lau and Peng (1987) in the study of the origin of low-frequency (intraseasonal) oscillation in the tropical atmosphere. In the meantime, the influence of convection heating profile on phase speed of the excited tropical wave was studied and it was shown that the stronger heating in the middle-lower troposphere will lead phase speed of the excited wave to be closer to that of the observed ISO (Takahashi 1987). The further studies showed that the excited wave will become unstable under the effects of the vertical modes interaction (Chang and Lim 1988) and the moist atmospheric boundary layer (Wang 1988), which can be more advantageous to explain the activities of tropical ISO. In early 1990s, the theoretical studies indicated that the convection heating feedback can excite not only the CISK-Kelvin wave but also the CISK-Rossby wave in tropical atmosphere. These waves play a very important role in driving tropical ISO (Li 1990; 1993). Recently, some studies have emphasized the importance of nonlinearity in the wave-CISK theory (Lim et al. 1990; Cho et al. 1994; Cho and Li 1999).

The evaporation-wind feedback is advanced as another mechanism to drive the ISO in tropical atmosphere (Emanual 1987; Neelin et al. 1987) and still be introduced into the wave-CISK theory (Li and Liu 1993; Kritman and Vernekar 1993; Crum and Dunkerton 1994). However, further studies show that the evaporation-wind feedback alone is difficult to excite the ISO in tropical atmosphere, because the excited wave propagates too fast. But evaporation-wave feedback can lead to unstable wave, so that the combined effect of the cumulus convection heating and the evaporation-wind feedback can be considered as dynamical mechanism of tropical ISO (Li 1995; Zhao and Liu 1996).

Since some studies have shown that intraseasonal oscillation also exists in the ocean, the air-sea interaction was regarded as a mechanism of tropical atmospheric ISO (Lau and Shen 1988; Hirst and Lau 1990; Li and Liao 1996; Wang and Xie 1998). Exceptionally, the eastward propagation of the coupled wave is only favourable for the oceanic ISO but not for the atmospheric ISO due to its slow propagation.

In this paper, the cumulus convection heating (wave-CISK), the air-sea interaction effect and the evaporation-wind feedback are altogether introduced in a theoretical model, in order to understand their role in driving tropical atmospheric ISO. Any process that can lead the phase speed of wave to be closer to the observed tropical ISO or lead to the unstable wave will be regarded as important mechanism of the tropical atmospheric ISO. Since the structure of tropical ISO, particularly the vertical structure, has been described and explained by the wave-CISK theory very well (Lau and Peng 1987; Li 1988), the structure of excited wave will not be discussed in this study. The consideration of physical processes and the numerical model are given in section 2. Dynamical discussion of the phase speed and instability of the excited wave is given in section 3. The section 5 is the general calculation Results are given in section 4. Section 5 is the conclusions.

2. Model and Physics

Generally, the atmospheric governing equations for Kelvin wave on the equatorial beta-plane without basic flow, Rayleigh friction and Newton cooling can be written simply as follows:

$$\frac{\partial U}{\partial t}+U\frac{\partial U}{\partial x}+g\frac{\partial H}{\partial x}=0, \tag{1}$$

$$\frac{\partial H}{\partial t}+U\frac{\partial H}{\partial x}+D\frac{\partial U}{\partial x}=-Q-EU, \tag{2}$$

where the coordinates (t, x) measure time and distance in the eastward direction, U is zonal wind speed, H is the deviation from D (the mean depth of the equivalent atmosphere). In (2), Q is atmospheric diabatic perturbation heating rate; the term EU represents the evaporation-wind feedback, E is a constant.

The atmospheric diabatic perturbation heating Q includes the major internal heating (cumulus convection heating) and the major external heating (the heat flux from the ocean to the atmosphere). The internal heating in the present case can be parameterized as

$$Q_1=-\eta\frac{\partial U}{\partial x}, \tag{3}$$

where η resembles the heating function as a constant. Based on some studies (Philander et al. 1984; Battisti and Hirst 1989), the heat flux from the ocean to the atmosphere can be parameterized as

$$Q_2=\alpha(h-\kappa h^3), \tag{4}$$

where h is the deviation from d (a mean depth of the equivalent ocean); α and κ are coefficients to describe air-sea interaction.

The oceanic model is a one-layer shallow water model driven by the wind and can be written as

$$\frac{\partial u}{\partial t}+u\frac{\partial u}{\partial x}+g\frac{\partial h}{\partial x}=\gamma U, \tag{5}$$

$$\frac{\partial h}{\partial t}+u\frac{\partial h}{\partial x}+(d+h)\frac{\partial u}{\partial x}=0, \tag{6}$$

where u is zonal oceanic current speed, the term γU represents the wind stress with y as stress coefficient. Since the fundamental forcing of the atmosphere on the ocean is wind stress while the fundamental forcing of the ocean on the atmosphere is the heating, only these two key processes are considered in present study for the simplicity.

Obviously, Eqs.(l), (2), (5) and (6) can be regarded as the simplest self-organized system with 4 independent variables. In the positive feedback process, which is defined as atmospheric motion acting on the ocean and the heating from the ocean to the atmosphere, the positive (negative) h will cause atmospheric convergence (divergence) through the effect of term - αh. In the meantime, oceanic convergence (divergence) will further enhance positive (negative) h. In the negative feedback process, which is defined as the gravitational restoring force connecting with the stratification of the ocean and the nonlinear term $\alpha\kappa h^3$, the gravitational restoring force in Eq.(5) produces divergence (convergence) in the maximum (minimum) value region of h. The divergence (convergence) then will decrease (increase) the value of h. The nonlinear term also plays certain role in the negative feedback.

In order to obtain the analytic solution, we only discuss the linear case in the present pa- per. The linearized equations can be written as follows;

$$\frac{\partial U}{\partial t}+g\frac{\partial H}{\partial x}=0, \tag{7}$$

$$\frac{\partial H}{\partial t}+(D-\eta)\frac{\partial U}{\partial x}=-\alpha h-EU, \tag{8}$$

$$\frac{\partial u}{\partial t}+g\frac{\partial h}{\partial x}=\gamma U, \tag{9}$$

$$\frac{\partial h}{\partial t}+d\frac{\partial u}{\partial x}=0, \tag{10}$$

They can be written as

$$\alpha\frac{\partial h}{\partial t}=g(D-\eta)\frac{\partial^2 H}{\partial x^2}+gE\frac{\partial H}{\partial x}-\frac{\partial^2 H}{\partial t^2}, \tag{11}$$

$$gd\frac{\partial^2 h}{\partial x^2}-\frac{\partial^2 h}{\partial t^2}=-\frac{\gamma d}{D-\eta}\left(\frac{\partial H}{\partial t}+\alpha h\right) \tag{12}$$

Adopting the normal mode method, the solutions of Eqs. (11) and (12) are:

$$(H,h)=(H_0,h_0)\mathrm{e}^{\mathrm{i}(kx-\sigma t)} \tag{13}$$

Substituting (13) into Eqs.(11) and (12) yields the following relation:

$$[\alpha C_{\mathrm{o}}^2(D-\eta)+\gamma dC_{\mathrm{a}}^2]\sigma^2=g(D-\eta)dk^2(\gamma C_{\mathrm{a}}^2+\alpha C_{\mathrm{o}}^2)-\gamma d\alpha^2 C_{\mathrm{o}}^2-igEkd\gamma C_{\mathrm{a}}^2, \tag{14}$$

Where $C_{\mathrm{a}}^2=gH_0, C_{\mathrm{o}}^2=gh_0$.

Since $\sigma=\sigma_{\mathrm{r}}+i\sigma_{\mathrm{i}}$, σ_{r} and σ_{i} are respectively the frequency and the growth rate of the perturbation (wave). From the relation (14), σ_{r} and σ_{i} can be expressed as:

$$\sigma_{\mathrm{i}}=\frac{gdEk\gamma C_{\mathrm{a}}^2}{2[(D-\eta)\alpha C_{\mathrm{o}}^2+\gamma dC_{\mathrm{a}}^2]\sigma_{\mathrm{r}}}, \tag{15}$$

$$\sigma_{\mathrm{r}}=k\sqrt{\frac{g(D-\eta)d(\gamma C_{\mathrm{a}}^2+\alpha C_{\mathrm{o}}^2)-\gamma d\alpha^2 C_{\mathrm{o}}^2/k^2+\sqrt{[gd(D-\eta)(\gamma C_{\mathrm{a}}^2+\alpha C_{\mathrm{o}}^2)-\gamma d\alpha^2 C_{\mathrm{o}}^2/k^2]^2+(dgE\gamma C_{\mathrm{a}}^2/k)^2}}{2[\alpha C_{\mathrm{o}}^2(D-\eta)+\gamma dC_{\mathrm{a}}^2]}}, \tag{16}$$

or

$$\sigma_{\mathrm{i}}=\frac{gdE\gamma C_{\mathrm{a}}^2}{2[(D-\eta)\alpha C_{\mathrm{o}}^2+\gamma dC_{\mathrm{a}}^2]C_{\mathrm{x}}}, \tag{17}$$

$$C_{\mathrm{x}}=\sqrt{\frac{g(D-\eta)d(\gamma C_{\mathrm{a}}^2+\alpha C_{\mathrm{o}}^2)-\gamma d\alpha^2 C_{\mathrm{o}}^2/k^2+\sqrt{[gd(D-\eta)(\gamma C_{\mathrm{a}}^2+\alpha C_{\mathrm{o}}^2)-\gamma d\alpha^2 C_{\mathrm{o}}^2/k^2]^2+(dgE\gamma C_{\mathrm{a}}^2/k)^2}}{2[\alpha C_{\mathrm{o}}^2(D-\eta)+\gamma dC_{\mathrm{a}}^2]}}, \tag{18}$$

In expression (16) or (18), only the positive root is selected because the classic atmospheric Kelvin wave should propagate eastwards without the ocean, the heating and the evaporation-wind feedback. It is very clear that the excited wave can be named the CISK-Kelvin like wave, which propagates slower than the classic Kelvin wave due to the convection heating feedback. It is also an unstable and dispersive wave due to the evaporation-wind feedback and the air-sea interaction.

3. Discussion of the Special Solution

Three special cases are considered here to understand the role of various physical processes.

3.1 Evaporation-wind feedback alone

Without the ocean and cumulus convection heating in the atmosphere ($C_o = 0$ and $\eta = 0$), expressions (17) and (18) will give

$$\sigma_i = \frac{gE}{2C_x} \tag{19}$$

and

$$C_x = \sqrt{\frac{gD(1+\sqrt{1+E^2/D^2k^2})}{2}} \equiv C_{XE}, \tag{20}$$

$$C_{gXE} = C_{XE} - \frac{gE^2}{2\sqrt{2}k}\frac{1}{\sqrt{gD(Dk+\sqrt{D^2k^2+E^2})}\sqrt{D^2k^2+E^2}}. \tag{21}$$

Obviously, the excited wave is unstable and dispersive due to the effect of evaporation-wind feedback. The group speed of the excited wave has same propagating direction as the phase speed but slower than the phase speed ($C_{gXE} < C_{XE}$), because of

$$C_{XE} > \frac{gE^2}{2k\sqrt{2gD(Dk+\sqrt{D^2k^2+E^2})}\sqrt{D^2k^2+E^2}}.$$

If the evaporation-wind feedback is eliminated ($E = 0$), the wave will become classic atmospheric Kelvin wave ($\sigma_i = 0$, and $C_{XK} = \sqrt{gD}$). It is a stable and eastward propagating wave with faster speed (about 50-60 m s^{-1}). In general, $C_{XE} = C_{XK}$, the phase speed of wave is very different from that of the observed tropical ISO. Therefore, the evaporation-wind feedback alone is difficult to drive atmospheric intraseasonal oscillation.

Even though the starting equations have some differences, the above mentioned result is consistent with that in previous studies (Kirtman and Vernekar 1993; Li 1995). In this simple dynamical study, the effect of evaporation-wind feedback on instability of the Kelvin wave is shown and its limitation in reducing phase speed of the Kelvin wave is also indicated.

3.2 Air-sea coupling alone

Without the convection heating (η=0) and evaporation-wind feedback (E=0), the air-sea interaction exists alone, expressions (17) and (18) give

$$\sigma_i = 0 \tag{22}$$

and

$$C_x = \sqrt{\frac{Dgd\gamma k^2 C_a^2 + (Dgk^2 - \alpha\gamma)d\alpha C_o^2}{k^2(d\gamma C_a^2 + D\alpha C_o^2)}} \equiv C_{XAS}, \tag{23}$$

$$C_{gXAS} = C_{XAS} + \frac{d\gamma\alpha^2 C_o^2}{k\sqrt{(d\gamma C_a^2 + D\alpha C_o^2)[gDd\gamma k^2 C_a^2 + d\alpha C_o^2(Dgk^2 - \alpha\gamma)]}}. \tag{24}$$

The expression (22) means that the coupled wave is stable. The expression (23) shows that the phase speed is related to the wavenumber. Thus the excited coupled wave is a dispersive wave and the group speed shown in (24) has same direction as the phase speed but faster than the phase speed.

Since $(\gamma dC_a^2 + \alpha DC_o^2) = (\gamma dC_a^2 + \alpha dC_o^2 - \gamma d\alpha^2 C_o^2 / gD)$, the relation $C_{XAS} = C_{XK}$ can be obtained when the air-sea coupling is stronger ($\gamma\alpha$ is larger). This means that the stronger air-sea coupling could favour the tropical ISO.

3.3 Combined effect of cumulus heating and evaporation-wind feedback

Without the effect of the ocean ($C_o = 0$, $\alpha = 0$), expressions (17) and (18) give

$$\sigma_i = \frac{gE}{2C_x}, \tag{25}$$

and

$$C_x = \sqrt{\frac{g(D-\eta)}{2}[1 + \sqrt{1 + E^2/(D-\eta)^2 k^2}]} \equiv C_{XCE}, \tag{26}$$

$$C_{gXCE} = C_{XCE} - \frac{gE^2}{\sqrt{8g(D-\eta)k^2[1 + \sqrt{(D-\eta)^2 k^2 + E^2}]}\sqrt{(D-\eta)^2 k^2 + E^2}}. \tag{27}$$

When the evaporation-wind feedback is omitted (E=0), (26) will represent the atmospheric CISK-Kelvin wave, i.e., $C_{XCK} = \sqrt{g(D-\eta)}$ with σ_i= 0. In general, $C_{XK} = C_{XCK} = C_{XCE}$, the convection heating feedback is very important to reduce the phase speed of excited wave. Therefore, the cumulus convection heating is mainly responsible for the propagating speed of the excited wave and the evaporation-wind feedback for the instability of the excited wave. These results are consistent with that in previous studies (Li and Liu 1993; Kirtman and Vernekar 1993; Li 1995) even though the starting equations have some differences.

It is very clear that the favourable situation for the tropical atmospheric ISO is the generation of an unstable and dispersive wave, particularly with the phase speed C_{XCE}=C_{XK}, through the combined effect of convection heating (wave-CISK) and evaporation-wind feedback.

4. General Calculation Results

The phase speed of the excited wave in various cases can be calculated through using general parameters in the ocean and atmosphere, that is C_a= 40 m s^{-1}; C_o = 1.5 m s^{-1}; D = 250 m;

d=0.28 m; $k=\frac{2\pi}{L}$, where k is wavenumber and L wavelength. The values of parameters E, α, γ and L will be different in different cases.

The result for the case with evaporation-wind feedback alone is shown in Fig. 1. It is obvious that the phase speed (C_{XE}) of excited wave increases very fast with the increase of wavelength. C_{XE} will be greater than 50 m s^{-1} as the wavelength is over 2 × 10^7 m. It is well known that the wavelength of tropical ISO is usually longer than 2 × 10^7 m. Therefore, the wave caused by evaporation-wind feedback alone is not consistent with the intraseasonal wave in the tropical atmosphere. In other words, the evaporation-wind feedback alone is difficult to excite tropical ISO in the atmosphere.

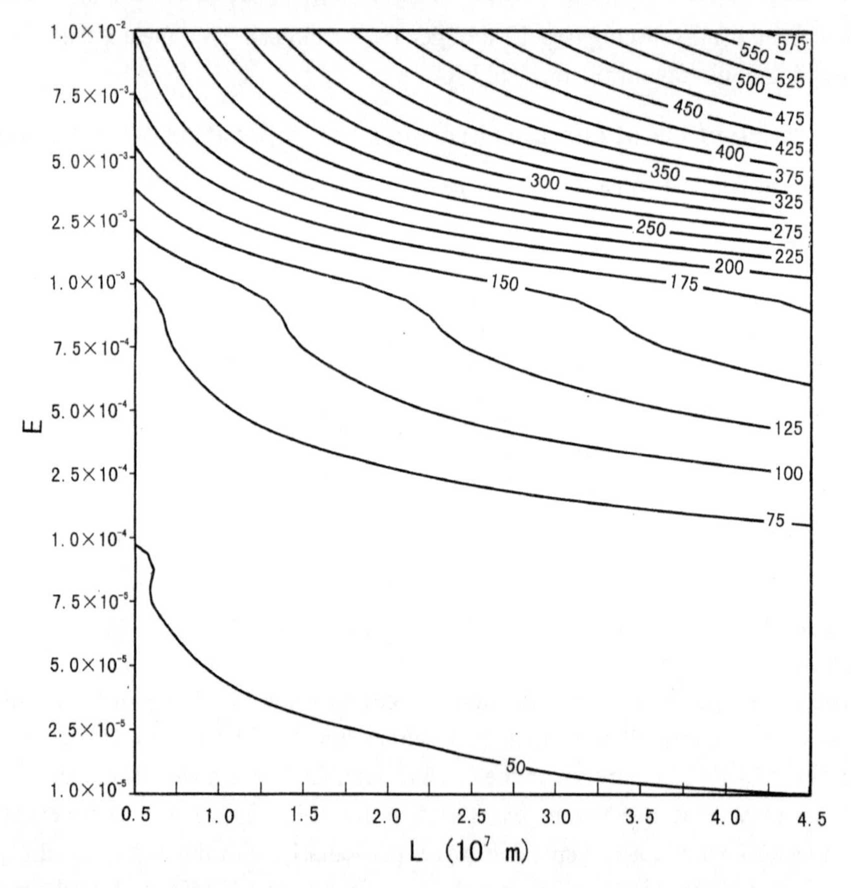

Fig. 1. Relationship between the phase speed of excited wave (C_{XE}) and the parameters (E and L).

The result for the case with combined effect of cumulus heating feedback (wave-CISK) and evaporation-wind feedback is given in Fig. 2, with a moderate intensity of evaporation-wind feedback (fixed E=1.5 × 10^{-5}). Although, we selected the larger wavelength and equivalent atmosphere depth D and weaker heating intensity, the phase speed C_{XCE} is still closer to the propagation speed of observed ISO in tropical atmosphere. Therefore, the convection heating feedback (wave-CISK) is very important in exciting the ISO in tropical atmosphere.

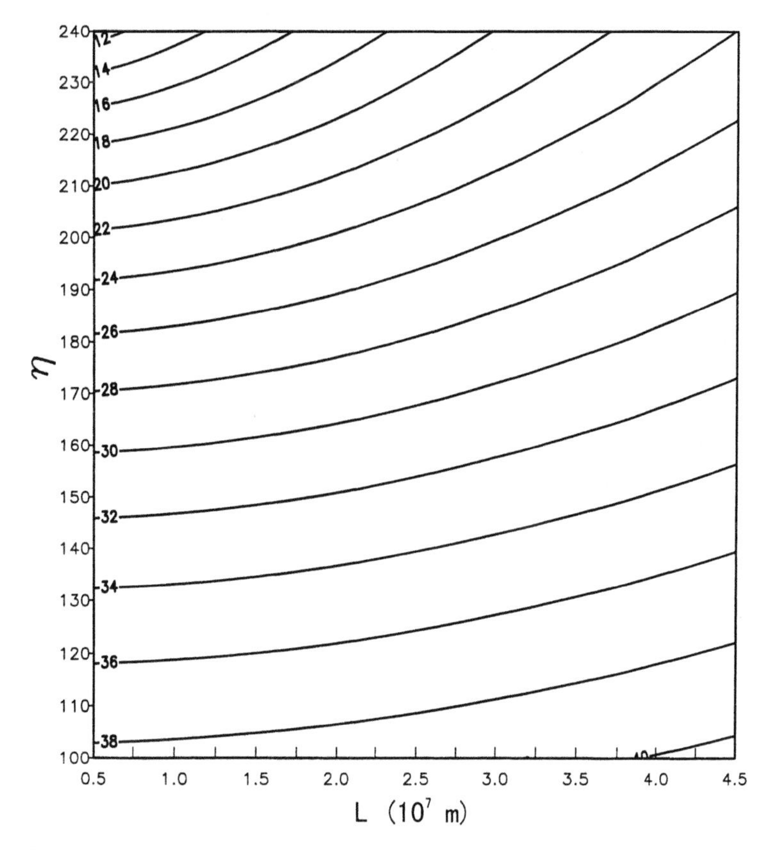

Fig. 2. Relationship between the phase speed of excited wave (C_{XCE}) and the parameters (η and L).

The result for the general case is shown in Fig. 3, with stronger heating (η=150) and moderate evaporation-wind feedback (E=1.5 × 10^{-5}). The comparison between Fig. 3 and Fig. 2 reveals that the phase speed of excited wave is reduced further by air-sea interaction. For example, the C_{XCE} is about 30 m s^{-1} as L = 2.8 × 10^7 m and η=150 in Fig. 2, but C_x in Fig. 3 is smaller than 20 m s^{-1} generally. It is also clear that the air-sea interaction parameter α is quite important to reduce the phase speed of excited tropical low-frequency waves. It is also evident that the larger γ (stronger wind stress) will quicken the phase speed of excited wave.

Although the group speed C_{gX} is not calculated in this study, the dispersive feature of excited wave has been indicated in above analyses. As we know, the OLR or TBB data are usually used to study the tropical ISO, particularly the propagation of the ISO (Lau and Chan 1985; Knuston and Weickmann 1987). An expansion of cloud cluster can be seen during the propagation of tropical ISO according to OLR or TBB data. This expansion can be regarded as a dispersive phenomenon of the ISO and explained by using the dispersive feature of excited wave with the effects of evaporation-wind feedback and air-sea interaction, because the combined effect of evaporation-wind feedback and air-sea interaction will excite a dispersive wave with the phase speed related to the wavenumber. It is shown that the evaporation-wind feedback will decrease the group speed to be slower than the phase speed of the excited wave but the air-sea interaction will increase the group speed to be faster than the phase speed of the excited wave. The expansion of cloud cluster results from the new generation due to the energy dispersion of the ISO.

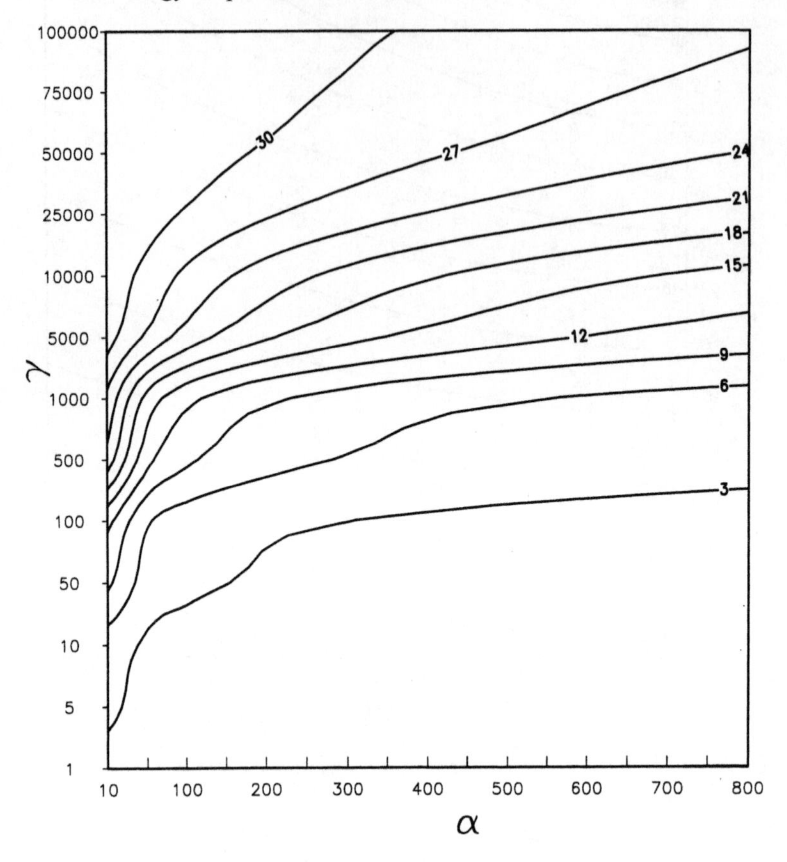

Fig. 3. Relationship between the phase speed of excited wave (C_x) and the parameters (α and γ).

5. Conclusion

The important role of wave-CISK, evaporation-wind feedback and air-sea interaction effect in exciting tropical atmospheric ISO is respectively discussed in this paper, theoretically. The result showed that a CISK-Kelvin like wave is excited in the model. This wave propagates eastward slowly as the observed tropical ISO does due to the convection heating feedback and air-sea interaction. It becomes unstable due to the effect of evaporation-wind feedback. It is also a dispersive wave due to the evaporation-wind feedback and the air-sea interaction.

The convection heating feedback (wave-CISK) is the dominant process in reducing phase speed of the excited wave to be closer to the propagation speed of observed ISO (about 10-15 m s^{-1}) in the tropical atmosphere. The stronger air-sea interaction, particularly the oceanic heating also plays certain role in reducing phase speed of the excited atmospheric wave.

Since the existence of instability is a favourable condition for the ISO activity in the tropical atmosphere, the evaporation-wind feedback cannot be omitted as a mechanism of tropical atmospheric ISO despite this feedback alone is difficult to excite the ISO. When there is the combined effect of cumulus heating feedback (wave-CISK) and evaporation-wind feedback, the phase speed of excited wave is closer to the propagation speed of observed ISO (about 10-15 m s^{-1}) in tropical atmosphere. This means that the wave-CISK plays a key role in reducing the phase speed of excited wave and it can be regarded as a fundamental dynamical mechanism of the ISO in tropical atmosphere.

The evaporation-wind feedback and air-sea coupling interaction will excite a wave with the phase speed related to the wavenumber. Therefore, the excited wave has dispersive feature due to the effect of evaporation-wind feedback and air-sea coupling interaction.

This study is partly supported by National Key Programme for Developing Basic Sciences (G1998040903).

References

Battisti, D. S., and A. C. Hirst, 1989: Interannual variability in a tropical atmosphere-ocean model: influence of the basic state, ocen geometry and nonlinearity. *J, Atmos. Sci.*, **46,** 1687-1712.

Chang, C. P., and H. Lim, 1988: Kelvin wave-CISK: a possible mechanism for 30-50 day oscillations. *J. Atmos. Sci*, **45,** 1709-1720.

Cho, H. R., K. Fraedrich, and J. T. Wang, 1994: Cloud clusters, Kelvin wave-CISK and the Madden Julian Oscillation in the equatorial troposphere. *J. Atmos. Sci.*, **51,** 68-76.

Cho, H. R., and D. Pendlebury, 1997; Wave CISK of equatorial waves and the vertical distribution of cumulus heating. *J. Atmos. Sci.*, **54,** 2429-2440.

Cho, H. R., and C. Y. Li, 1999: Equatorial Kelvin wave and intraseasonal oscillation in the equatorial troposphere. *Chinese J. Atmos. Sci.*, **23,** 237-244.

Crum, F. X., and T. J. Dunkerton, 1994: CISK. and evaporation-wind feedback with conditional heating on an equatorial beta-plane, *J. Meteor. Soc. Japan*, **72**, 11-18.

Emanual, K. A., 1987: An air-sea interaction model of intraseasonal oscillation in the tropics. *J, Atmos. Sci.*, **44,** 2324-2340.

Hirst, A. C., and K. M. Lau, 1990: Intraseasonal and interannual oscillations in coupled ocean-atmosphere models. *J. Climate*, **3,** 713-725.

Kirtman, B., and A. Vernekar, 1993: On wave-CISK and the evaporation-wind feedback for the Madden-Julian oscillation. *J. Atmos. Sci.*, **50**, 2811-2814.

Knuston, T. R., and K. M. Weickmann, 1987: 30-60 day atmospheric oscillation: composite life cyclones of convection and circulation anomalies. *Mon. Wea. Rev.*, **115,** 1407-1436.

Krishnamurti, T. N., and D. Subrahmanyam, 1982: The 30-50 day mode at 850 mb during MONEX. *J. Atoms. Sci*, **39**, 2088-2095.

Lau, N. C., and K. M. Lau, 1986: The structure and propagation of oscillation appearing in a GFDL general circulation model. *J. Atmos. Sci.*, **43,** 2023-2047.

Lau, K. M., and P, H. Chan, 1985: Aspects of the 40-50 day oscillation during the northern winter as inferred from outgoing longwave radiation. *Mon. Wea. Rev.*, **113,** 1889-1909.

Lau, K. M., and L. Peng, 1987: Origin of low-frequency (intraseasonal) oscillation in the tropical atmosphere, Part I: Basic theory. *J. Atmos. Sci.*, **44**, 850-972.

Lau, K. M., and S.Shen, 1988: On the dynamics of intraseasonal oscillation and ENSO. *J. Atmos. Sci.*, **45**, 1781-1791.

Li Chongyin, 1985: Actions of summer monsoon troughs (ridges) and tropical cyclones over South Asia and the moving CISK mode. *Scientia Sinica* (*B*), **28**, 1197-1207.

Li Chongyin, 1988: On the feedback role of tropical convection. *Tropical Rainfall Measurements*, A. Deepak Publishing, New York 141-146.

Li Chongyin, 1990: Dynamical study on 30-60 day oscillation in the tropical atmosphere outside equator. *Chinese J. Atmos. Sci*, **14**, 101-112.

Li Chongyin, 1991: *Low-Frequency Oscillation in the Atmosphere*, China Meteorological Press, Beijing, 276pp (in Chinese).

Li Chongyin, 1993: A further inquiry on the mechanism of 30-60 day oscillation in the tropical atmosphere. *Advances in Atmospheric Science*, **10**, 41-53.

Li Chongyin, 1995: *Introduction to Climate Dynamics*, China Meteorological Press, Beijing, 254-268 (in Chinese).

Li Chongyin, and Q. H. Liao, 1996: Behaviour of coupled modes in a simple nonlinear air-sea interaction model. *Advances in Atmospheric Science*, **13**, 183-195.

Li G. L., and Liu S. K., 1993: Wave-CISK, evaporation-wind feedback and low-frequency oscillation. *Chinese J. Atmos. Sci.*, **17**, 696-702.

Lim, H., T. K. Lim, and C. P. Chang, 1990: Reexamination of wave-CISK theory: Existence and properties of nonlinear wave-CISK modes. *J. Atmos. Sci*, **47**, 3078-3091.

Madden, R. A., and P. R. Julian, 1971: Detection of 40-50 day oscillation in the zonal wind in the tropical Pacific. *J. Atmos. Set*, **28**, 702-708.

Madden, R. A., and P. R. Julian, 1972: Description of global scale circulation cells in the tropics with 40-50 day period. *J. Atmos. Set*, **29**, 1109-1123.

Madden, R. A., and P. R. Julian, 1994: Observations of the 40-50 day tropical oscillation-A review. *Mon. Wea. Rev.*, **112**, 814-837.

Murakami, M., 1984: Analysis of the deep convective activity over the western Pacific and Southeast Asia, Part II: Seasonal and intraseasonal variations during Northern Summer. *J. Meteor. Soc. Japan*, **62**, 88-108.

Neelin, J. D., I. M. Held, and K. H. Cook, 1987: Evaporation-wind feedback and low-frequency variability in the tropical atmosphere. *J. Atmos. Sci.*, **44**, 2341-2348.

Philander, S. G. H., T. Yamagata, and R. C. Pacanowski, 1984: Unstable air-sea interactions in the tropics. *J. Atmos. Sci.*, **41**, 604-613.

Takahashi, M., 1987: A slow phase speed of the intraseasonal oscillation using the wave-CISK. *J. Meteor. Soc. Japan*, **65**, 43-49.

Wang, B., 1988: Dynamics of tropical low-frequency waves: An analysis of the moist Kelvin wave. *J. Atmos. Sci*, **45**, 2051-2065.

Wang, B., and X. Xie, 1998: Coupled modes in the warm pool climate system, Part I: The role of air-sea interaction in maintaining Madden-Julian oscillation. *J. Climate*, **11**, 2116-2135.

Zhao Qiang, and Liu Shikuo, 1996: CISK and Evaporation-wind feedback mechanism and 30-60 day oscillations in the tropical atmosphere. *Acta Meteor, Sinica.*, **54**, 417-426 (in Chinese).

Sensitivity of MJO Simulations to Diabatic Heating Profiles

Chongyin Li[1,2], Xiaolong Jia[1,3], Jian Ling[1,4], Wen Zhou[4], Chidong Zhang[5]*

[1] LASG, Institute of Atmospheric Physics, Chinese Academy of Sciences, Beijing, China
[2] Institute of Meteorology, PLA University of Science and Technology**, Nanjing, China
[3] National Climate Center, China Meteorological Administration, Beijing, China
[4] City U-IAP Laboratory for Atmospheric Sciences, Department of Physics and Materials Science, City University of Hong Kong, Hong Kong, China
[5] University of Miami, 4600 Rickenbacker Causeway, MPO, Miami, FL 33149, USA

Abstract The difficulty for global atmospheric models to reproduce the Madden-Julian oscillation (MJO) is a long-lasting problem. In an attempt to understand this difficulty, simple numerical experiments are conducted using a global climate model. This model, in its full paramterization package (control run), is capable of producing the gross features of the MJO, namely, its planetary-scale, intraseasonal, eastward slow propagation. When latent heating profiles in the model are artificially modified, the characteristics of the simulated MJO changed drastically. Intraseasonal perturbations are dominated by stationary component over the Indian and western Pacific Oceans when heating profiles are top heavy (maximum in the upper troposphere). In contrast, when diabatic heating is bottom heavy (maximum in the lower troposphere), planetary-scale, intraseasonal, eastward propagating perturbations are reproduced with a phase speed similar to that of the MJO. The difference appears to come from surface and low-level moisture convergence, which is much stronger and more coherent in space when the heating profile is bottom heavy than when it is top heavy. These sensitivity experiments, along with other theoretical, numerical, and observational results, have led to a hypothesis that the difficulty for global models to produce the MJO partially is rooted in a lack of sufficient diabatic heating in the lower troposphere, presumably from shallow convection.

1. Introduction

The tropical intraseasonal oscillation is an important element of the atmospheric climate

Received May 3, 2007; accepted August 11, 2008; published online September 2, 2008.
*e-mail: czhang@rsmas.miami.edu
** In 2017, the PLA university of Science and Technology and the PLA Ordnance Engineering College were jointly established as the PLA army engineering university.

system. One of its dominant components is the eastward moving Madden- Julian oscillation (MJO, Madden and Julian 1971, 1972). A related component is the northward moving intraseasonal variation associated with the Asian summer monsoon (e.g., Krishinamurti and Subrahmanyam 1982; Li 1985; Wang and Rui 1990; Li et al. 2001). The importance of the MJO to tropical and global weather and climate has been increasingly appreciated. The MJO influences onset and breaks of the summer monsoons over Asia (Lau and Chan 1986; Mu and Li 2000; Lawrence and Webster 2002), Australia (Hendon and Liebmann 1990; Wheeler and McBride 2005), Americas (Paegle et al. 2000; Higgins and Shi 2001), and Africa (Matthews 2004). It has been suggested that the evolution of the ENSO can be affected by the MJO (e.g., Lau and Shen 1988; Li and Zhou 1994; Kessler et al. 1995; Li and Long 2002; Zhang and Gottschalck 2002). Tropical extreme events, such as cyclogenesis, can be modulated by the MJO (Liebmann et al. 1994; Maloney and Hartmann 2000; Mo 2000; Bessafi and Wheeler 2006; Frank and Roundy 2006). Many weather and climate phenomena outside the tropics are also related to the MJO (e.g., Li and Li 1997; Bond and Vecchi 2003).

There is no doubt that accurate simulations and prediction of the MJO would lead to tremendous societal benefit. However, current numerical prediction skill of the MJO becomes useless after 15 days (e.g., Waliser et al. 2003), which is far shorter than the dominant timescales of the MJO (30-60 days). Most global climate models (GCMs) fail to reproduce the most salient features of the MJO, such as its intraseasonal timescales and eastward propagation (Slingo et al. 1996; Lin et al. 2006). Our inability of explaining the difficulty of simulating the MJO by current state-of-art GCMs in terms of existing MJO theories reflects our lack of understanding of its fundamental dynamics. The purpose of this study is to explore a particular physical process that might be important to the dynamics and numerical simulations of the MJO: the vertical structure of diabatic heating.

Several mechanisms have been proposed for the MJO (see summaries in Wang 2005 and Zhang 2005). Deep convection has always been taken as the central factor for the MJO. Its roles in the MJO may come in play through interactions with boundary-layer moisture convergence (e.g., Li 1985; Lau and Peng 1987; Wang 1988), tropospheric water vapor (e.g., Grabowski and Moncrieff 2005), cloud radiation (Hu and Randall 1994; Raymond 2001), and sea surface temperature (SST) (e.g., Flatau et al. 1997; Waliser et al. 1999). It is commonly thought that deficiencies in parameterizations of deep convection in GCMs are responsible for model incapability of reproducing the MJO (Li and Smith 1995; Slingo et al. 1996). The sensibility of MJO simulations to convective parameterizations has been well demonstrated (Wang and Schlesinger 1999). However, it is uncertain exactly what inadequacy in convective parameterizations contributes most to the deterioration in MJO simulations. This study investigates the dependence of numerical simulations of the MJO on the vertical structure of diabatic heating.

It has been documented that diabatic heating profiles in GCMs are typically deep, with peaks in the mid to upper troposphere (Lin et al. 2004), in contrast to the observed top-heavy profile typical for tropical convective systems dominated by stratiform rain (e.g., Houze 1989). This poses an intriguing question: Is the lack of top-heavy heating profiles in GCMs responsible for their inability of reproducing a realistic MJO? The modeling experiments in this study indicate the opposite: it is bottom-heavy heating profile with its maximum in the lower to mid troposphere that might play a crucial role in simulating a realistic MJO, especially its eastward propagation.

The importance of diabatic heating in the lower troposphere to the MJO was first suggested by Li (1983). He showed that characteristics of tropical unstable modes were sensitively related to the vertical structure of diabatic heating (see more discussion in Sect. 5). The linear theory of the equatorial waves (Matsuno 1966) predicts that the phase speed of the eastward-propagating Kelvin wave is $c = (gh)^{1/2}$, where g is gravity and h the equivalent depth, a measure of the vertical scale. In this theory, h is determined by the vertical scale of diabatic heating, the source of energy for the tropospheric Kelvin wave. A smaller h due to a shallow heating profile leads to a slower phase speed. It has been shown that slow wave-CISK modes can indeed be produced by heating profiles peaking in the lower troposphere, namely, between 500 and 700 hPa (Lau and Peng 1987; Chang and Lim 1988; and Sui and Lau 1989).

Low-level heating may come from shallow precipitating clouds that are abundant in the western equatorial Pacific prior to the deep convective phase of the MJO (e.g., Johnson et al. 1999). It has been hypothesized that these shallow convective clouds may be instrumental to moistening the lower troposphere and set a stage favorable to the following deep convective phase of the MJO (e.g., Slingo et al. 2003). This current study provides new modeling evidence that lower-tropospheric heating is in favor of the MJO and points out a possible role of shallow convective clouds in the MJO through their diabatic heating structure in addition to their moistening effects.

A comparison of three simulations by two GCMs with different cumulus parameterization schemes (Sect. 2) indicates that the one with stronger diabatic heating in the lower troposphere produced a more realistic MJO (Sect. 3). In simple numerical experiments, vertical heating profiles in a GCM were artificially modified to be top heavy (maximum in the upper troposphere) or bottom heavy (maximum in the lower troposphere). Slow eastward propagating intraseasonal perturbations with a deep, first baroclinic mode structure, interpreted as of the MJO, were produced only when diabatic heating is bottom heavy. When diabatic heating is top heavy, the dominant intraseasonal perturbations became stationary (Sect. 4). These results, together with their diagnoses and other numerical, theoretical and observations results (Sect. 5), lay the foundation for a hypothesis that a lack of diabatic heating in the lower troposphere is a reason for many GCMs to fail in reproducing a realistic MJO (Sect. 6).

2. Model and Methodology

2.1 Models

Two global atmospheric models were used in this study. One is the atmospheric component of the Flexible Global Ocean-Atmosphere-Land System model (FGOALS), which was originally from the version of Simmonds (1985) and developed at the Institute of Atmospheric

Physics (IAP) Laboratory for Numerical Modeling for Atmospheric Sciences and Geophysical Fluid Dynamics (LASG) (Wu et al. 1996). This is a global spectral model of R42 (128 × 108 Gaussian grid points, equivalent resolution of 2.8125° longitude × 1.66° latitude) and nine sigma levels with the top level at 17 hPa (R42L9). It employs a unique dynamic core that removes a reference atmosphere to reduce systematic errors due to truncation and topography (Wu 1997). The parameterization package includes the Slingo et al. (1996) scheme for radiation, the Slingo (1980, 1987) scheme for cloud diagnosis, the Holtslag and Boville (1993) scheme for the boundary layer, and the moisture convective adjustment scheme of Manabe et al. (1965). A simplified version of the Simple Biosphere model (SIB) of Sellers et al. (1986) is used for the land surface processes (Xue et al. 1991). A semi-implicit scheme with a time step of 15 min was used. This model will be referred to as SAMIL. Further details on the formulation of SAMIL are available in Wu et al. (1996), Wu (1997) and Wang et al. (2005).

The other model used is the NCAR Community Atmosphere Model (CAM2.0.2) with the Euler dynamic core of T42 (128 x 64 Guassian grid points), 26 vertical levels (top level at 2.917 hPa). The integration scheme is semi-implicit with a time step of 20 min. Standard parameterization package (Collins et al. 2006) was used, except in one simulation where the Zhang and McFarlane (1995) scheme (ZM scheme) in the CAM original package was replaced by the Tiedtke (1989) scheme (T scheme). The original version of CAM with the ZM scheme will be referred to as CAM2+ZM. The modified version using the T scheme will be referred to as CAM2+T.

2.2 Experiment and analysis

Both models were integrated for January 1, 1978-Decem- ber 31, 1989. Initial conditions were based on the NCAR/ NCEP reanalysis (Kalney et al. 1996) for SAMIL and a previous simulation up to January 1, 1987 for CAM2. Time-evolving monthly mean sea surface temperature (SST) and sea ice from Program for Climate Model Diagnostic and Intercomparison (PCMDI) were used in both models. The first year of integration was ignored for diagnostics, thus the model data sets cover 11 years for January 1979-December 1989. The NCAR/NCEP reanalysis and Xie and Arkin (1997) precipitation data (hereafter referred to as XA precipitation) were used to validate the model simulations.

Main diagnostic variables include zonal wind at the 850 hPa level (U850) and 200 hPa level (U200), velocity potential at the 200 hPa level (X200), and precipitation. Anomalous time series were first created by removing the respective 11-year climatology from daily mean data. Then a 30-60 day band-pass filter was applied to isolate the intraseasonal variability. Power spectra, empirical orthogonal function (EOF), and linear regression were used to analyze simulated MJO features.

3. Simulations of the MJO

Detailed descriptions of the mean and MJO simulated by SAMIL and the two CAM models are given by Jia (2006). Here, only the main features of the simulated MJO are discussed. The most prominent feature of the MJO, namely, its slow eastward propagation, is captured by SAMIL. Fig. 1 compares lag regression of band-pass (30-60 day) filtered U200, X200, and U850 based on the NCEP reanalysis (left column) and SAMIL (right). The eastward propagation of the MJO reproduced by SAMIL is evident, although the phase speed is slightly higher than in the reanalysis, especially in the western hemisphere. The ability of SAMIL to reproduce the MJO is further demonstrated by an EOF analysis. The first two leading EOF

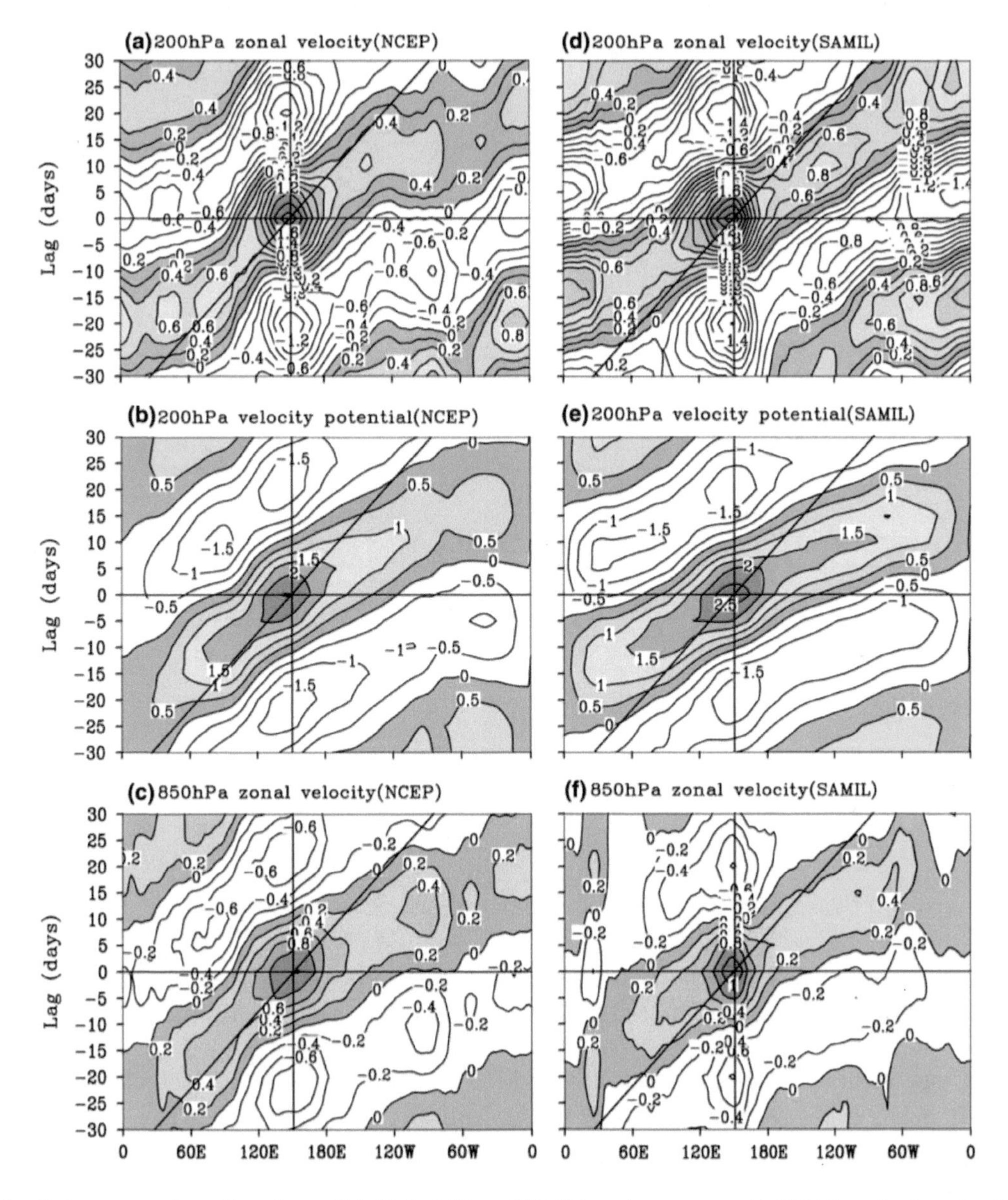

Fig. 1. Lag regression of band-pass (30-60 day) filtered U200 (m s^{-1}, *upper panels*), 200 hPa velocity potential (10^6 m^2 s^{-2}, *middle*), and U850 (m s^{-1}, *lower*) with the reference point at 150°E, all averaged over 10°S-10°N based on the NCEP reanalysis (*left column*) and SAMIL (*right*). Positive values are *shaded*. The *straight lines* indicate 5 m s^{-1} phase speed

modes of band-pass (30-60 day) filtered U850 is in quadrature with each other, as expected for the MJO in the reanalysis (Fig. 2). Their lag correlation reaches the maximum at 9 days, suggesting a period of 36 days. This is shorter than the 44 day period indicated by the maximum lag correlation at 11 days between the first two leading EOF modes in the reanalysis U850. This has been a common problem in MJO simulations for many years (e.g., Slingo et al. 1996). Only few GCMs are able to reproduce the realistic MJO phase speed (e.g., Maloney and Hartmann 1998; Sperber et al. 2005; Zhang et al. 2006).

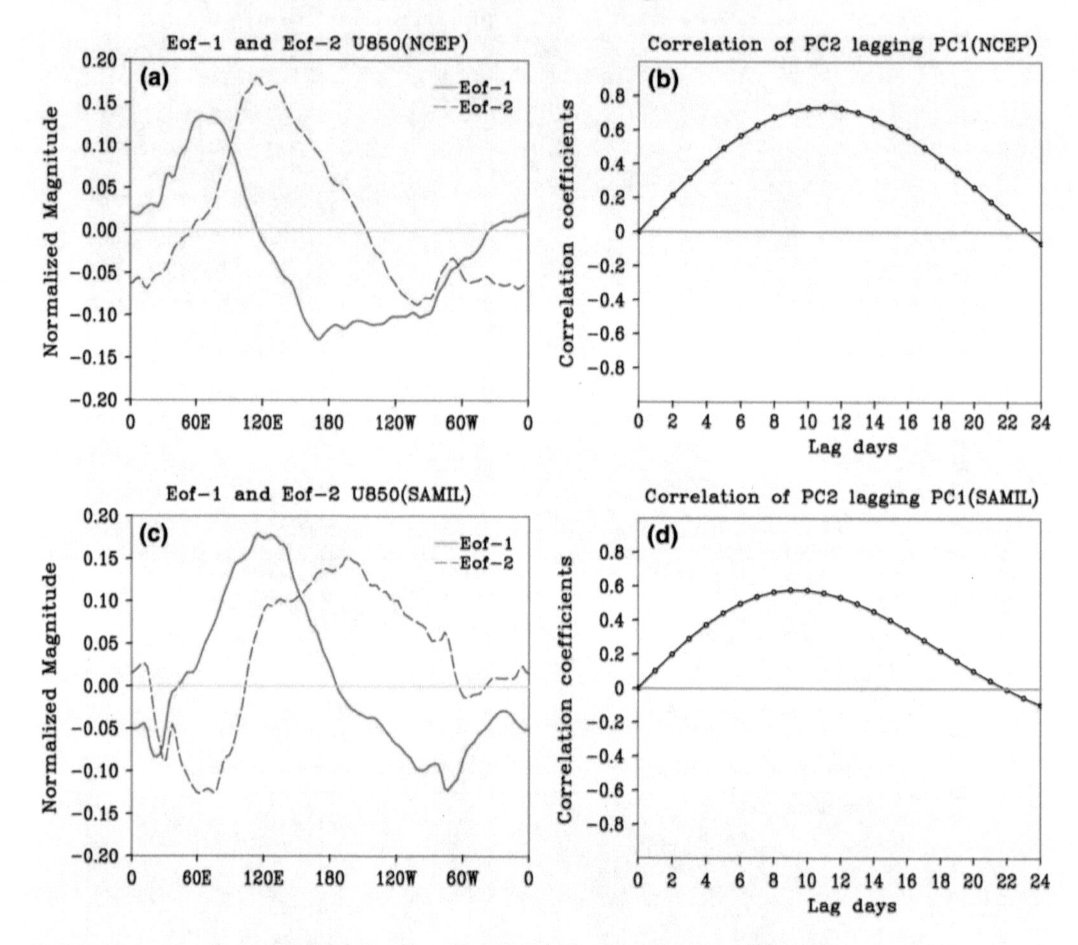

Fig. 2. PCs of the first two leading EOF for band-pass (30-60 day) filtered U850 (*left column*) and their lag correlation (*right column*) based on the NCEP reanalysis (*top row*) and SAMIL (*bottom*)

The agreement and discrepancy between the MJO in SAMIL and the reanalysis can be further seen from their time-space spectra (Fig. 3). An intraseasonal spectral peak near 50-60 days and well separated from the lower frequency power exists clearly for both NCEP U850 and XA precipitation. In the simulation of SAMIL, such spectral peaks can be discerned, if weaker than in the reanalysis and observations. Meanwhile, there are additional spectral peaks near 30 days in SAMIL U850 and precipitation with roughly equal strength as the intraseasonal peaks. This exaggeration of the spectral power near 30 days is another common problem in

GCMs (e.g., Zhang et al. 2006). They are partially responsible for the shorter period and faster eastward phase speed of the simulated MJO. Another problem in the simulation is the lack of eastward moving spectral power at zonal wavenumber two ($k = 2$) in precipitation. In observations, such spectral power exist with the same amplitude as at $k = 1$. These problems in the simulation notwithstanding, the most important feature for the MJO in the spectra of NCEP U850 and XA precipitation is well reproduced by SAMIL: the eastward moving power at the planetary scale ($k = 1$) and intraseasonal period (30-60 days) is much greater that its westward moving counterpart. Most current GCMs fail to reproduce this salient property of the MJO (e.g., Lin et al. 2006). There is hardly any GCM that can produce much more realistic MJO spectrum than SAMIL (Zhang et al. 2006).

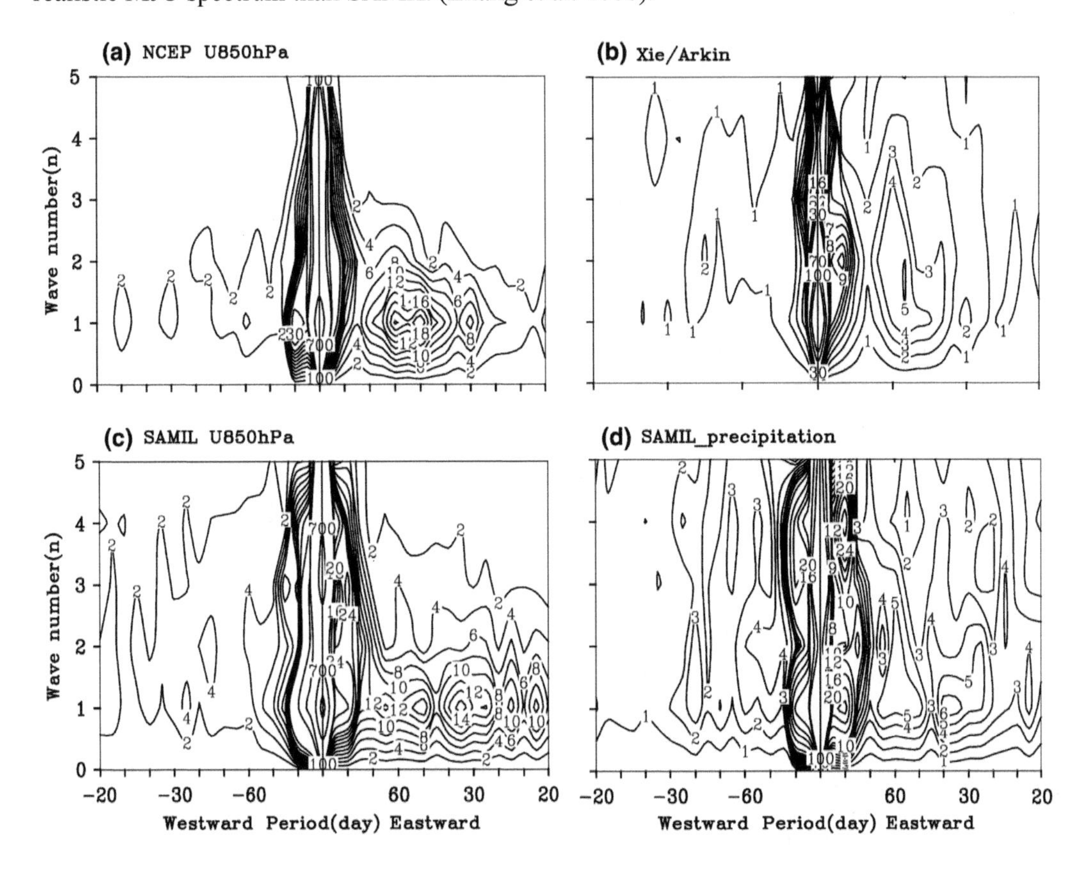

Fig. 3. Time-space spectra of **a** NCEP U850 (m^2 s^{-2}), **b** XA precipitation (mm^2 day^{-2}), **c** U850 from SAMIL (m^2 s^{-2}), and **d** precipitation from SAMIL (mm^2 day^{-2}). The U850 contour interval is 2 m^2 s^{-2} for values smaller than and equal to 20 and contours of 24, 30, 70, 100, 700, 1200 are given for values exceed 20, starting at 2 m^2 s^{-2}. The precipitation contour interval is 1 mm^2 day^{-2} for values smaller than and equal to 10 and, contours of 12, 16, 20, 24, 30, 70, 100, 700 are given for values exceed 10

The ability of CAM+ZM and CAM+T to reproduce observed features of the MJO is much limited than that of SAMIL. While there is a faint hint of eastward propagation in band-pass filtered U200, such hint disappears for X200 and U850 (Fig. 4). There are even strong signs of westward propagation in U850 of CAM+ZM. The time-space spectra of CAM+ZM indeed show stronger westward moving power than eastward moving one for U850 (Fig. 5). Replacing the ZM cumulus scheme by the T scheme apparently increased the eastward moving power. But the resulting spectra remain highly unrealistic. One example is the lack of eastward moving power at $k = 1$ for precipitation. It is interesting to notice that other GCMs with the T scheme are able to produce much more realistic MJOs than CAM+T (Zhang et al. 2006). This suggests that, while cumulus scheme is a critical component of a model for its ability of reproducing the MJO, it alone is not always the determining factor.

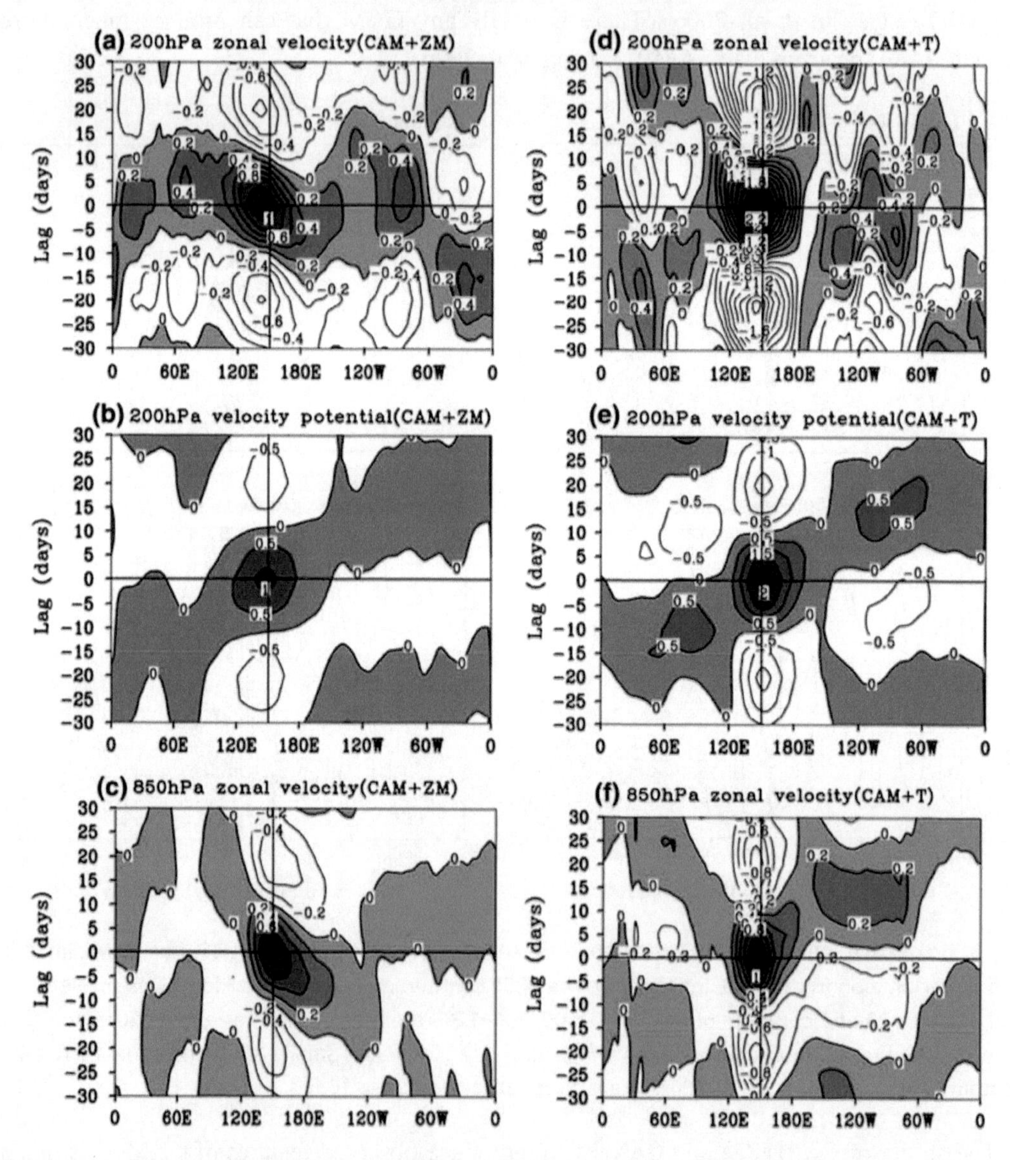

Fig. 4. Lag regression for band-pass (30-60 day) filtered U200 (*upper row*), X200 (*middle*), and U850 (*bottom*), all averaged over 10°S-10°N, from CAM+ZM (*left column*) and CAM+T (*right*). The reference point is at 150°E

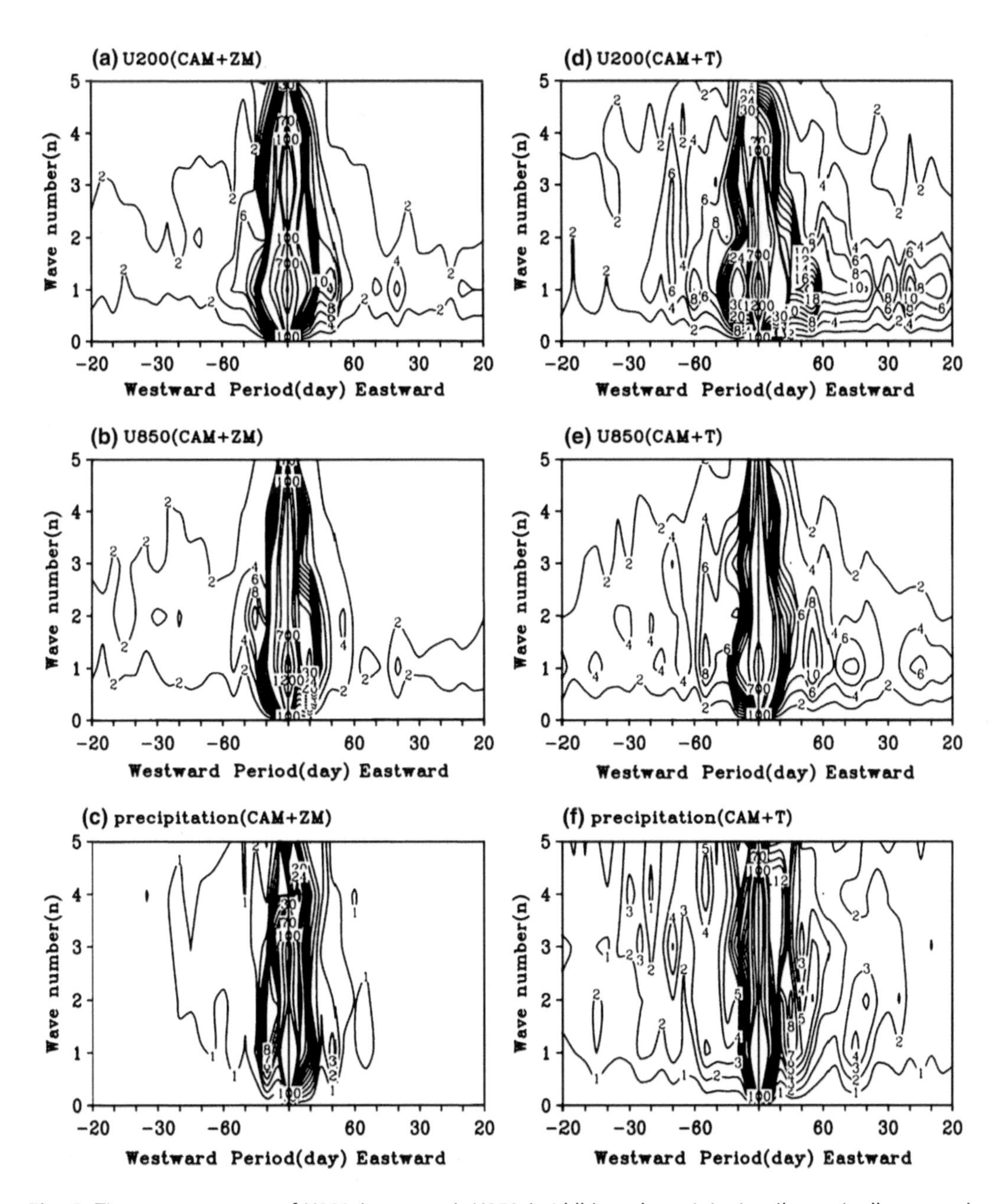

Fig. 5. Time-space spectra of U200 (*upper row*), U850 (*middle*), and precipitation (*bottom*), all averaged over 10°S-10°N, from CAM+ZM (*left column*) and CAM+T (*right*). Contour intervals same as in Fig. 3

These simulations exemplify the intriguing puzzle: What is the most critical factor determining the success and failure of a model to reproduce the MJO and why MJO simulations are so sensitive to cumulus parameterizations? In the next section, we will present simple numerical experiments to demonstrate that the vertical structure of diabatic heating might be one piece of the puzzle.

4. Sensitivity to Heating Profiles

There might be various reasons for the SAMIL and CAM2 simulations to be different. Here, we consider their different vertical structures of diabatic heating. The mean diabatic heating profiles over the western Pacific (140-160°E, 10°N-10°S) during periods with strong precipitation from the three simulations introduced previously are compared in Fig. 6. They indeed exhibit different characteristics. The heating profile had a peak near the 600 hPa level in SAMIL, near the 400-450 hPa levels in CAM2+ZM, and near the 550 hPa level in CAM2+T. The maximum heating is the strongest in CAM2+T (10 K/day) and much weaker in SAMIL and CAM2+ZM (5 K/day). While the heating profiles of CAM2 are typical of GCMs, the heating profile of SAMIL is unusual in that there is a minimum near the 450 hPa level. All the three, meanwhile, miss the top-heavy structure (peak at and above the 200 hPa level) found in observation during the convective peak phase of the MJO (Lin et al. 2004). Another interesting difference among the three heating profiles is that there is a smaller fraction of total heating concentrated in the lower troposphere in CAM2+T and CAM2+ZM than in SAMIL which produced much more realistic MJO signals.

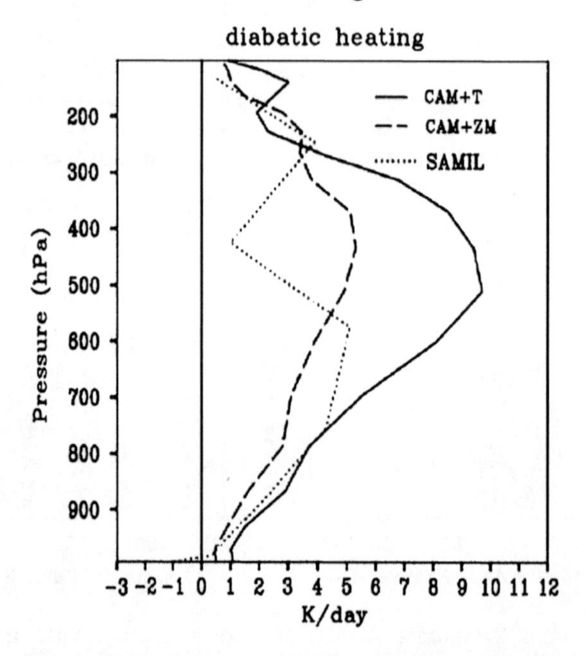

Fig. 6. Mean diabatic heating profile in the western Pacific (140-160°E, 10°S-10°N) during the peak of TIO precipitation from SAMIL (*dotted line*), CAM2+ZM scheme (*dashed*), and CAM2+T (*solid*)

The distinctions among the three heating profiles led to a speculation that, among other things, it was the larger fraction of lower-tropospheric diabatic heating in SAMIL that made its simulation of the MJO more realistic than the other two. This speculation was examined by a set of simulations using SAMIL in which latent heating profiles in the tropics were artificially modified to make them top heavy or bottom heavy. This was done in the following way: to make a given latent heating profile (e.g., solid line in Fig. 7a) top heavy, for example, its amplitude is reduced by 90% at all levels except at levels six and seven, which correspond roughly to 300 and 250 hPa. The modified profile is the dashed line in Fig.7a. Then the amplitude of the new profile is amplified at the two peak levels (dashed line in Fig. 7b) so that the vertically integrated latent heating is the same as the original one. In this procedure, cooling at any level in the original profile remains intact. This was done to all latent heating profiles within 20°S and 20°N. Between 20 and 30 degree latitudes, the reduction in amplitude of original latent heating profiles was tapered from 90% gradually to zero and beyond 30 degree latitudes no latent heating profile was modified. The same procedure was followed to create bottom-heavy latent heating profiles except the peak levels are four and five (600 and 500 hPa). Once a latent heating profile was modified, it was added to other diabative heating terms before total diabatic heating was introduced into the thermodynamic equation. As will be seen below, this procedure effectively changes the vertical structure of total diabatic heating. Notice that when latent heating profiles are modified this way, the moistening effect of clouds is not directly changed.

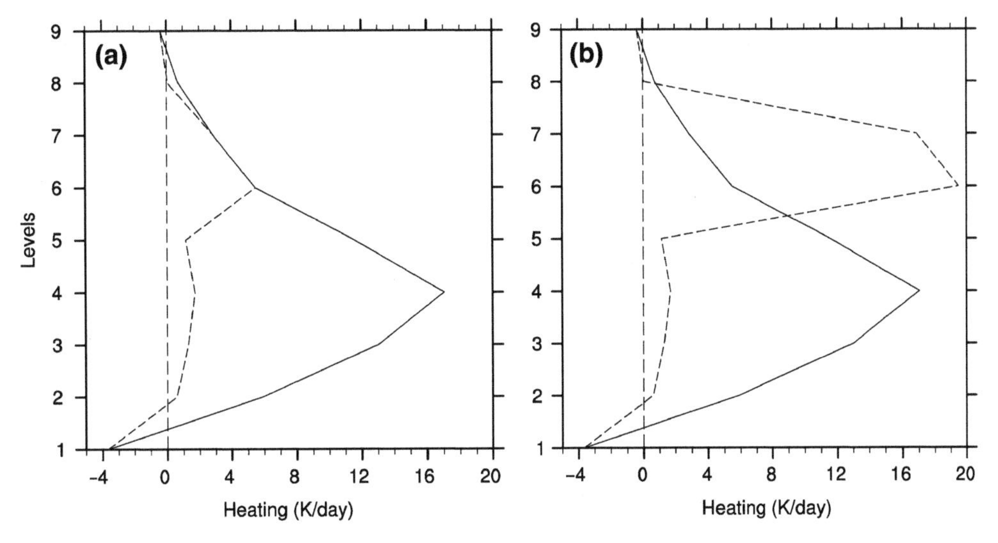

Fig. 7. Example for the procedure of modifying a diabatic heating profile (K/day). **a** An original profile (*solid*) and a new profile (*dashed*) whose amplitude is reduced by 90% except at the peak levels (6 and 7 in this case). **b** The final new profile (*dashed*) whose vertically integrated heating is the same as the original one (*solid*)

Two numerical experiments were made in addition to the control run (CT) in which latent heating profiles generated by the model were not modified. In one experiment, latent heating profiles are modified at each time step and each grid to be top heavy as described above. This run will hereafter be referred to as TH (top heavy). The other experiment is the bottom-heavy counterpart of TH, which will be referred to as BH. Each simulation lasts for 11 years and diagnostics were made for the last 10 years.

The mean tropical (15°N-15°S) total diabatic heating profiles from the three simulations are shown in Fig. 8. The artificial modification of the latent heating profile appears to be highly unrealistic. It may have exaggerated the heating peaks and associated vertical heating gradient for individual profiles (Fig. 7) but not in the mean diabatic heating profiles (Fig. 8). The profiles of CT and BH appear to be similar but one is original without any modification and the other is modified to be bottom heavy in its latent heating component. Realistically, a bottom-heavy heating profile, due to shallow convection, should have a smaller amplitude than a top-heavy profile due to stratiform rain. But this is not necessarily the case in BH, where bottom- heavy profiles can be as strong as top-heavy ones. The mean profile is even stronger in BH than in TH. The modified profiles in TH and BH, however, form a sharp contract between extreme cases in which the role of vertical heating profiles in simulations of the MJO can be cleanly isolated and clearly demonstrated, even if exaggerated.

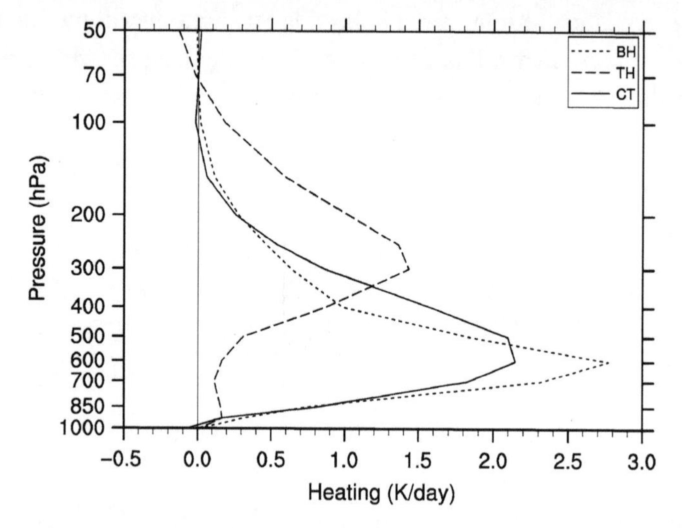

Fig. 8. Mean diabatic heating profiles (K/day) for CT (*solid*), TH (*dashed*) and BH (*dotted*) averaged over 15°N and 15°S

The time evolution of intraseasonal diabatic heating profiles from the three simulations along with their associated zonal-vertical circulations are compared in terms of lag regression in Fig. 9. There is an obvious eastward propagation in diabatic heating and cooling in CT and its associated circulation, with a phase speed about 5.5 m s^{-1}. The associated circulation is of a typical deep, first baraclinic mode extending up to 150 hPa, with mid-level upward motions, low-level convergence and upper-level divergence in region of heating, and mid-level downward motions, low-level divergence and upper-level convergence in regions of cooling. The zonal scale of this deep over-turning circulation is about 13,000 km. The time scale of this apparently convection-circulation coupled pattern is clearly 40 days. In the same lag regression but with the reference point at 90°E, there is a visible westward tilt of the upward motion in the heating region (not shown), which as been reported before (e.g., Lin et al. 2006; Kiladis et al. 2005). There is little doubt that these features represent the MJO.

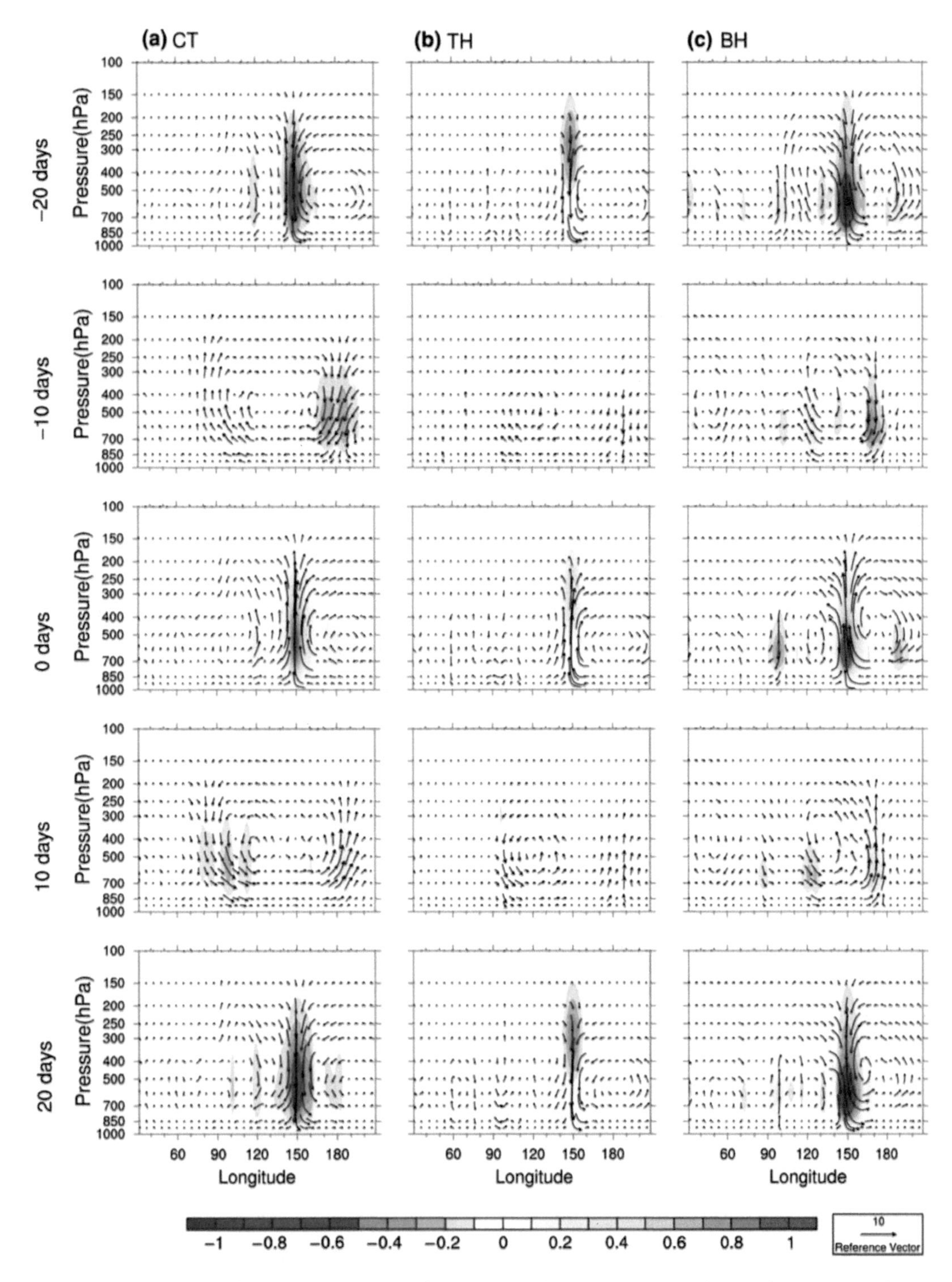

Fig. 9. Lag regression of intraseasonal (30-60 days) diabatic heating (*colors*, K/day) and zonal-vertical wind vectors in the tropics (15°S-15°N) upon maximum heating at 150°E (0 day) from CT (*left column*), TH (*middle*) and BH (*right*)

The manifestation of the MJO in diabatic heating and its associated circulation is completely lost in TH (Fig. 9, middle column). The heating and cooling are weaker and their zonal scales are smaller than in CT. The vertical motion in the mid-troposphere are prominent only in isolated regions. Strong zonal wind is confined to the lower troposphere. The overturning circulation is not only weaker but also shallower than in CT and the large-scale, deep, first mode baroclinic circulation seen in CT is missing here. In a sense, the low-level and upper-level winds are almost decoupled. Furthermore, there is no obvious eastward propagation in either diabatic heating or the circulation. Evidently, no MJO is produced in TH.

Results from BH (Fig. 9, right column) are very similar to those from CT. Especially, the characteristics of the MJO in diabatic heating and wind, namely, the eastward slow propagating large-scale, deep over-turning circulation coupled to diabatic heating, are well reproduced in BH. It is interesting to notice that in BH, even though latent heating profiles in the tropics were modified to be bottom heavy, diabatic heating manages to penetrate into the upper troposphere occasionally, in particular, near the reference point (150°E). This presumably is due to the tendency of the model cumulus parameteriation to produce extraordinarily strong upper-tropospheric latent heating, as seen in CT, that survived somewhat the artificial reduction. The heating maximum is however in the lower troposphere and is much stronger than the maximum in TH. In consequence, strong upward motions are forged and the deep overturning circulation generated. This is in sharp contrast to TH, where latent heating profiles were modified to be top heavy, lower-tropospheric diabatic heating is always very weak. It is therefore justifiable to say that the discrepancies in the circulation and zonal propagation between TH and BH seen in Fig. 9 are mainly due to the difference in lower-tropospheric heating, not in upper-tropospheric heating.

In a more conventional manner, the eastward propagations of intraseasonal U850 along in the three simulations are demonstrated in Fig. 10 in terms of lag regression. In CT, U850 propagates eastward with a phase speed slightly greater than 5m s^{-1} over the Indian Ocean but accelerates east of the dateline (Fig. 10a). The fast propagating (~15 m s^{-1}) perturbations are likely to be free Kelvin waves, which have been seen in the reanalysis wind (Fig. 1c). In TH (Fig. 10b), the slow eastward propagation in U850 over the Indian Ocean seen in CT is almost gone. The intraseasonal perturbations over the western Pacific are more stationary than eastward propagating. The fast eastward propagating Kelvin wave signals east of the dateline remain prominent. In BH (Fig. 10c), the slow eastward propagation over the Indian Ocean is regained with a more realistic phase speed (5 m s^{-1}). Again, the fast propagating Kelvin wave signals east of the dateline are evident. Among the three simulations, BH is the most realistic in comparison to the global reanalysis (Fig. 1c).

The contrast between the intraseasonal eastward propagating perturbations in TH and BH is further demonstrated in Fig. 11, where ratios of eastward versus westward power on intraseasonal (30-60 days) and planetary (zonal wavenumber 1-5) scales are shown for tropical zonal wind and diabatic heating at each level. The large ratios (indicating dominant eastward power and therefore eastward propagation) in zonal wind in the upper (300-200 hPa) and lower (850 hPa) troposphere from CT and BH (Fig. 11a) are obviously associated with the first mode baroclinic structure of the MJO seen in Fig. 9. In TH, eastward power in zonal wind does not dominate westward power at any level. The strong eastward propagating signals in CT and BH is also evident in their diabatic heating fields, but is much stronger in BH than in CT (Fig. 11b). It is interesting that in BH the strongest eastward propagating

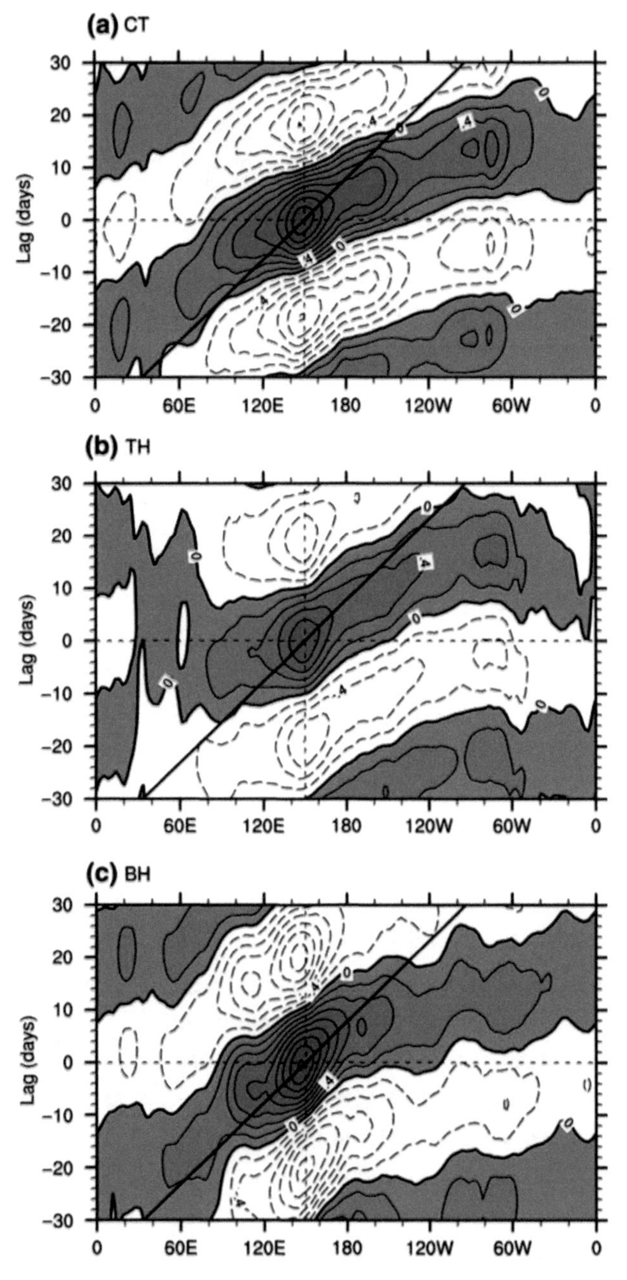

Fig. 10. Lag correlation of band-pass (30-60 day) filtered U850 in the tropics (15°S-15°N) with reference point at 150°E from **a** CT, **b** TH, and **c** BH. *Shading* indicates significance at the 99% confidence level. The *straight lines* indicate 5 m s^{-1} phase speed. Contour interval is 0.2

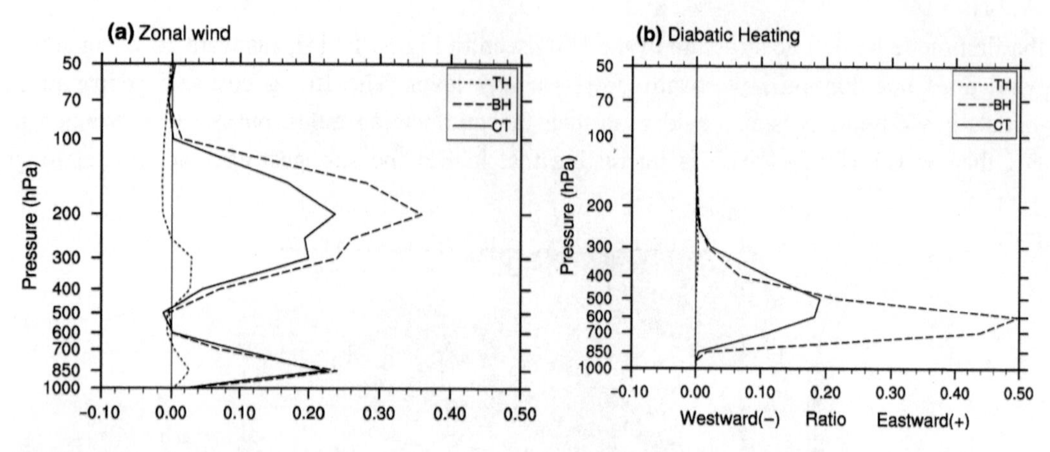

Fig. 11. Ratio of eastward vs. westward power on intraseasonal (30-60 days) and planetary (k = 1-5) scales from CT (*solid*), TH (*dashed*), and BH (*dotted*) for **a** zonal wind and **b** diabatic heating, both averaged over 15°S-15°N

signal in diabatic heating is confined in the lower troposphere because of the modification whereas the strongest eastward propagating signal in zonal wind is in the upper troposphere. Large-scale dynamics, especially the interaction between the circulation and diabatic heating must play critical roles in establishing the deep vertical structure of zonal wind when diabatic heating is strongest in the lower troposphere. It is striking but consistent to Fig. 9b that there is no dominance of eastward power in diabatic heating in TH at all.

Figure 12 shows time-space spectra of for equatorially symmetric and antisymmetric U850 in a broad range, following Wheeler and Kiladis (1999), to examine the signals of the MJO in the context of other equatorial perturbations. For symmetric perturbations, stronger westward power and weaker eastward power on the MJO scales is evident in TH (Fig. 12c) in comparison to CT (Fig. 12a) and BH (Fig. 12e). They contribute to the lack of dominant eastward power (Fig. 11) and therefore the lack of eastward propagation (Figs. 9b, 10b) in TH. The power for the Kelvin waves is weaker too in TH than in CT and BH. For the Rossby waves, in contrast, the power appears to be slightly stronger in TH than in the other two simulations. Differences in antisymmetric power among the three simulations are marginal. It can be said that the modifications of latent heating profiles in TH and BH did not qualitatively change other equatorial perturbations as they did for the MJO.

In summary, these three experiments clearly demonstrate that intraseasonal eastward propagating perturbations, i.e., the MJO, cannot be produced by the model if its latent heating is always top heavy, but can if its latent heating is always bottom heavy or is allowed to vary but peaks more often in the lower troposphere than otherwise. This suggests that if a model fails to produce sufficient lower-tropospheric heating, it may not be able to produce a realistic MJO. Possible reasons for the critical role of bottom-heavy heating profile in the MJO are discussed next.

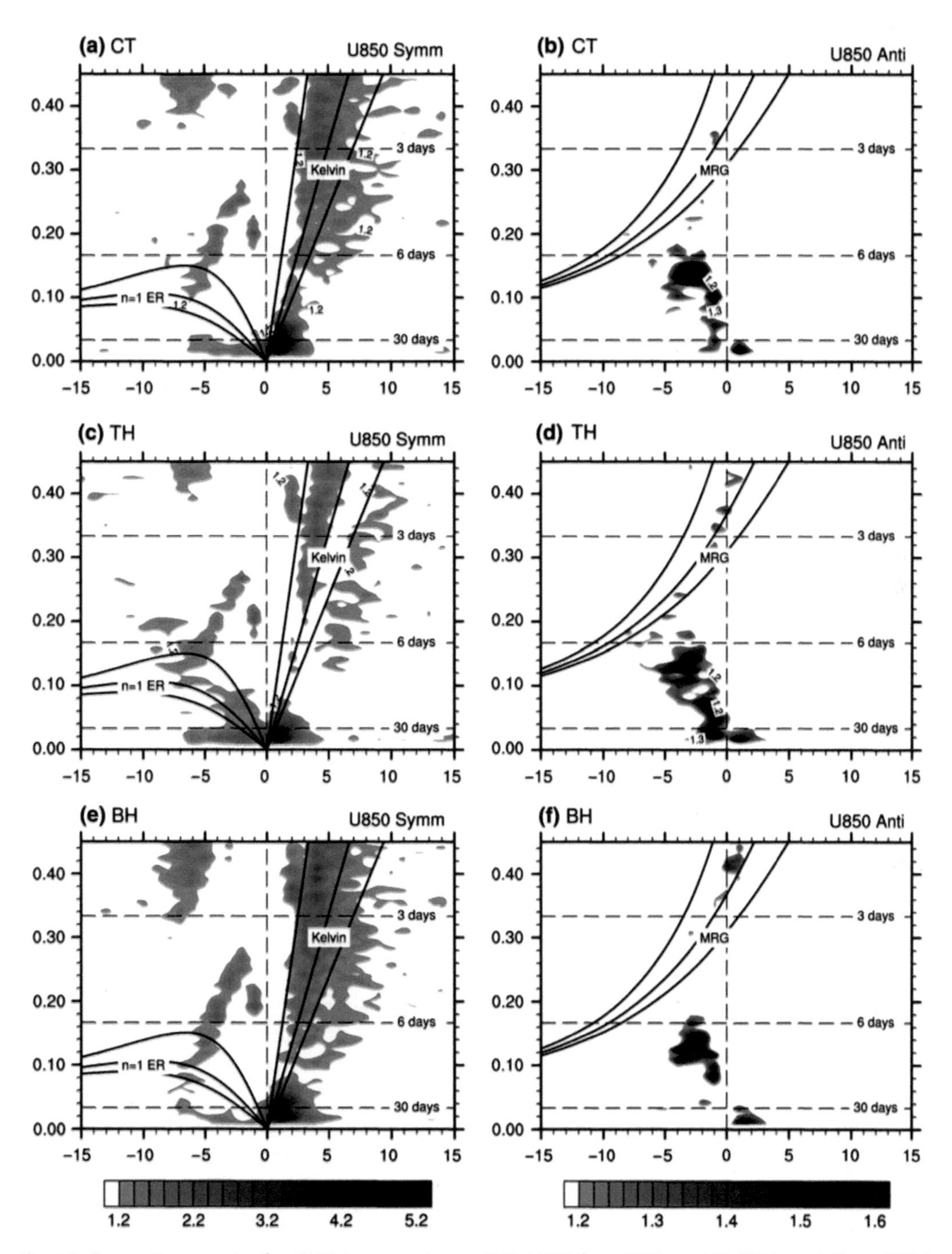

Fig. 12. Space-time spectra for U850 (averaged over 15°S-15°N) from CT (**a** and **b**), TH (**c** and **d**), and BH (**e** and **f**), with the equatorial symmetric components in the left column and asymmetric components in the right column. Dispersion relation curves of equatorial waves were calculated using equivalent depth h = 50, 100, 400 m

5. Discussion

There could be different reasons for the simulated MJO to be sensitive to diabatic heating profiles. But first, we can rule out the possibility that the difference in the intraseasonal eastward propagation in TH and BH is due to changes in the mean state. The important of the mean state in a model to its MJO has been suggested (Slingo et al. 1996; Hendon 2000; Slingo et al. (2003); Sperber et al. 2005; Zhang et al. 2006), even though our understanding of this issue remains qualitative and speculative. Mean U850 and precipitation from TH and BH are shown in Fig. 13. Subtle differences in the tropics indeed exist between the two. The most interesting one is an obvious double ITCZ in mean precipitation over the central Pacific in TH (Fig. 13a) but not in BH (Fig. 13b). More investigations on this are certainly warranted because of the large systematic biases in most climate models in this regard. But this is out of the scope of this study. An erroneous double ITCZ may deteriorate the horizontal structure and distribution of a simulated MJO but may not necessarily affect much its eastward propagation (Zhang et al. 2006). Westerly U850 appears to be stronger over the Bay of Bengal in BH than in TH. But both simulations suffer from the biases of too strong easterlies over the equatorial Indian and western Pacific Ocean, a common problem in GCMs (e.g., Zhang et al. 2006). In short, based on our current knowledge of the possible role of the mean state in MJO simulations, there is no obvious reason to expect the difference in the mean state to cause the distinct zonal propagating behaviors of the intraseasonal perturbations in the two simulations as seen in the previous section.

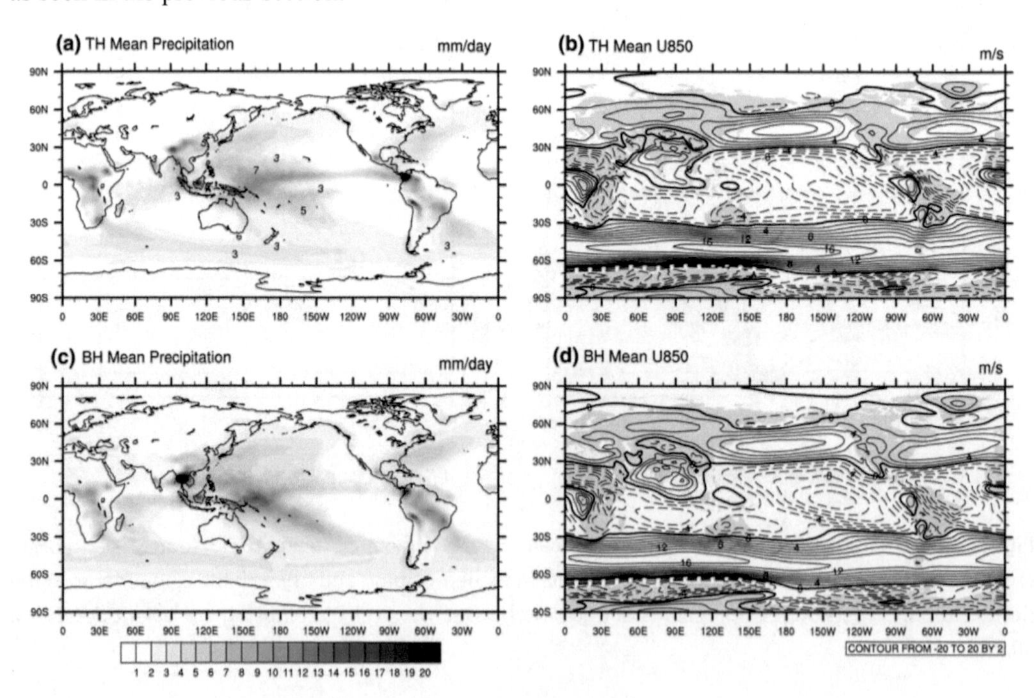

Fig. 13. Mean precipitation from **a** TH and **b** BH and mean U850 from TH (**c**) and BH (**d**)

We suggest that the distinct zonal propagating behaviors are caused by the different diabiatic heating profiles in the two simulations. To support our suggestion, we fist present a simple dynamic argument that may help understand the simulated sensitivity to heating profiles. This argument is based on a study by Li (1983) on tropical wave instability in relation to vertical profiles of convective heating. Results from this study, published in a non-English peer reviewed journal, are unavailable to most readers of this article. The essence of this study is therefore introduced here in the context of the simulated sensitivity described in the previous section.

Without losing generality, tropical wave instability can be studied for a linear atmosphere in an axisymmetric cylinder coordinate system. For the sake of argument, it is assumed that the vertical profile of latent heating rate (Q) can be prescribed, while its amplitude and occurrence are determined by a simple cumulus parameterization, similar to what we have done in our GCM experiments. In this parameterization, Q depends on vertical mass flux due to Ekman pumping and cumulus momentum mixing. When the model atmosphere is divided into a number of pressure layers, an eigenvalue problem can be defined under proper boundary conditions. The solutions to the eigenvalue problem describe the growth rate and vertical structure of corresponding unstable modes.

Four vertical profiles of Q were considered. Three of them have maximum heating rates at certain levels, one is of barotropic structure (Fig. 14). Notice the profiles with maximum heating rates in the upper and lower troposphere are very similar to those in TH and BH of this study (Fig. 8). These vertical profiles of Q lead to distinct characteristics in resulting unstable modes.

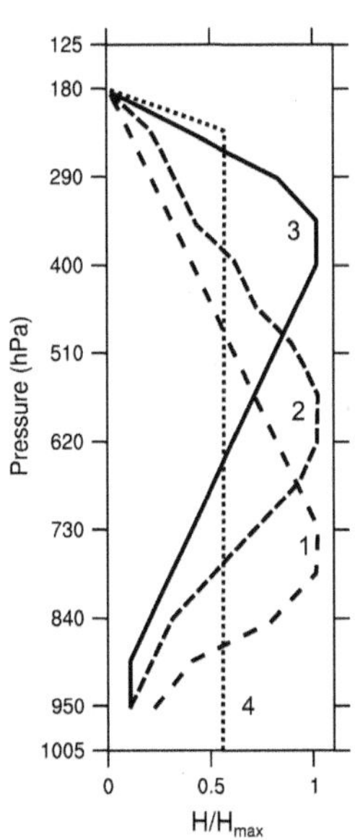

Fig. 14. Heating profiles (normalized by maximum) used in the solution to the eigenvalue problem described in the text (Sect. 5). (From 20)

The growth rate of unstable modes sensitively depends on the vertical profile of Q, as shown in Fig. 15 where dashed curves represent oscillatory unstable modes, which are relevant to the MJO. A maximum of Q in the mid troposphere gives rise to overall the fastest growth rate at almost all horizontal scales (labeled with 2) but only for non-oscillatory modes. Oscillatory unstable modes occur at very large horizontal scales (⩾4,500 km) regardless of the heating profile. But the growth rate is the largest for heating with its maximum in the lower troposphere (dashed curve labeled with "1" with peak at 4,500 km). At intermediate scales (<1,000 km), oscillatory unstable modes are generated only by low-level heating (curve 1) and barotropic heating (curve 4). Most interestingly, however, the instability catastrophe, namely, the maximum growth rate occurring at the smallest scales (Crum and Dunkerton 1992), is avoided only when the peak of Q is in the lower troposphere (labeled with 1). This wavelength selection of instability is due to convective mixing (friction) in this linear model.

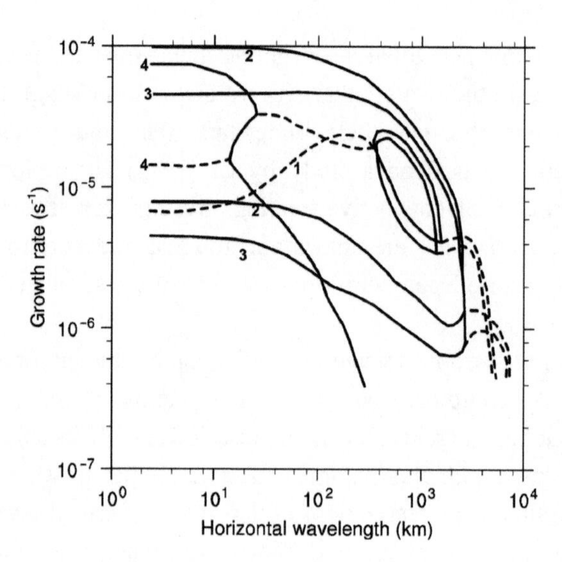

Fig. 15. Growth rates for unstable modes, with labels corresponding to those for the heating profiles in Fig. 11. *Dashed curves* represent oscillatory modes. (From 20)

The vertical structures in the vertical and horizontal motions and in temperature also vary with the vertical profiles of Q. The peaks of Q roughly coincide with levels of maximum ascents, reversal levels of horizontal winds, and warm anomalies (not shown). From an energetic point of view, the maxima of both generation of perturbation available potential energy (G) and conversion from perturbation available potential energy to perturbation kinetic energy (K) are roughly in the same layers of maximum Q, as anticipated (Fig. 16). While K is the largest for Q with its peak in the upper troposphere (Fig. 16a), G is the largest when maximum Q is in the lower troposphere (Fig. 16c).

These results, as all analytical solutions, critically depend on the assumptions and simplifications applied to the linear atmospheric model. In particular, the cumulus parameterization used in this model may play a dictating role in the solutions. The results from this simple model do not provide any information regarding the propagation direction of the unstable modes. Nevertheless, the solutions to the simple linear atmospheric model suggest that the vertical structure of convective heating rate is important factor to large-scale atmospheric disturbances in response to and interacting with diabatic heating.

Several other studies also explored the role of heating profiles in the intraseasonal oscillation using dynamic framework under wave-CISK assumptions. In a simple dynamical model, Lau and Peng (1987) and Sui and Lau (1989) found that a fast eastward propagating (~20 m s^{-1}) wave-CISK mode is produced by heating maximized in the upper troposphere (500-200 hPa), while a slow mode with an eastward propagation (7.6 m s^{-1}) close to the observed MJO phase speed is produced by heating maximized in the lower troposphere (700-500 hPa). Chang and Lim (1988) obtained quite different results. They found that while mid-tropospheric heating generates a fast eastward propagating CISK mode, lower-tropospheric heating generates a

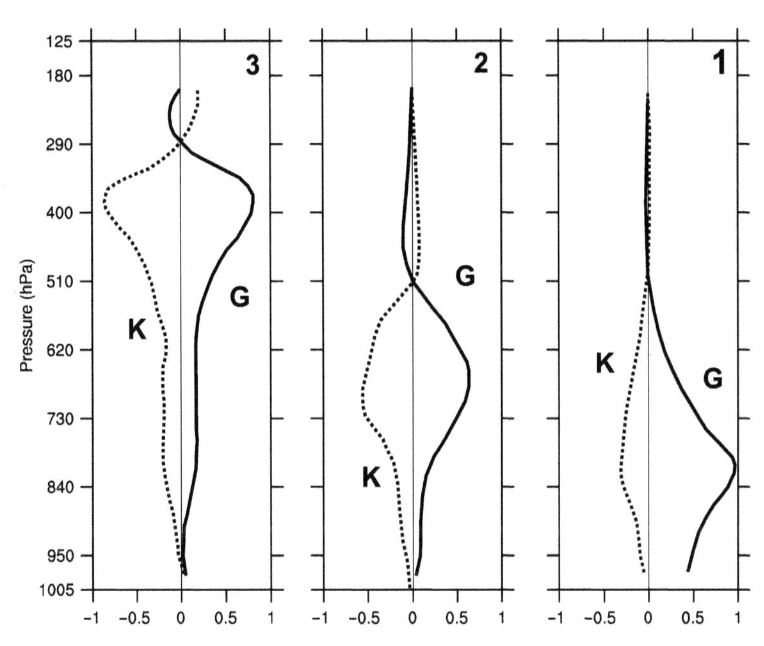

Fig. 16. Generation of perturbation available potential energy (*G*) and conversion from perturbation available potential energy to perturbation kinetic energy (*K*) for unstable modes corresponding to the heating profiles in Fig. 11. The unit for the energy variable is $m^2\ s^{-3}$. (From 20)

stationary instead of a slow eastward propagating mode. They interpreted the slow MJO phase speed as a result of the interaction between the two modes. It is unclear whether these wave-CISK dynamics may explain what observed in our GCM simulations. In our model, when heating profiles are top heavy, there is no fast eastward propagating perturbation as in the three studies using wave-CISK models, and when the heating profile is bottom heavy, there is no stationary perturbations as in Chang and Lim (1988).

In a recent study of GCM simulations, Zhang and Mu (2005) found that an improvement of MJO simulation due to modifications in a cumulus parameterization is apparently related to changes in the behavior of simulated convective heating profiles. In their improved MJO simulation, there is clearly more abundant low-level (shallow) heating during the MJO transition periods that lead to convectively active phases. Their interpretation is that the modified cumulus parameterization makes deep convection more difficult to occur, leaving more chances for shallow convection, which is parameterized by a separated scheme. The improvement of the simulated MJO might have come from the promoted low-level heating. A major difference between their and our simulations is that the overall amount of shallow heating in their simulations remains roughly the same, no matter what cumulus parameterization was used. Shallow heating is only better "organized" into the MJO scale in their improved simulation. Regardless of whether the heating profiles were changed by careful tuning of cumulus parameterizations as in Zhang and Mu (2005) or were brutally forced upon the calculated convective heating term in the model as in this current study, results from both point to the potential role of bottom-heavy (low-level) heating in the dynamics of the MJO and in interpreting the difficulty many GCMs are facing in simulating a realistic MJO.

The importance of bottom-heavy heating to the MJO may come through its effect on low-level moisture convergence. The atmospheric large-scale divergent circulation in the tropics is sensitive to the vertical structure (or more specifically, the vertical gradient) of diabatic heating (e.g., Hartmann et al. 1984). It has been shown that the lower branch of the atmospheric circulation directly responding to the top-heavy heating profile is significant only in the lower troposphere away from the surface (Schneider and Lindzen 1977; Wu et al. 2000, 2001). Only if the heating profile peaks in the lower troposphere, can the atmospheric circulation respond with significant amplitude near the surface (Bergman and Hendon 2000; Schumacker et al. 2004). Such a low-level heating maximum is needed for the MJO if the feedback from its surface winds (through either its moisture convergence or evaporation) is crucial (Wu 2003).

Low-level large-scale responses in the circulation to diabatic heating profiles indeed appear to be a reason for the difference between TH and BH. When bottom-heavy heating dominates in CT and BH, low-level moisture convergence is strong, deep coherent relative to heating regions, and exhibits large zonal scales (Fig. 17a, c). In contrast, when top-heavy heating dominates in TH, low-level moisture convergence becomes weak, shallow, and fragmental in its zonal scale (Fig. 17b). There is a perceptible sign of weak and shallow low-level moisture convergence immediately east of the main heating centers in CT and BH. This, consistent with one MJO theory in terms of boundary-layer frictional convergence (Wang 1988; Wang and Rui 1990), might be the reason for the eastward propagation promoted by bottom-heavy heating profiles.

If bottom-heavy heating profiles are crucial to MJO simulations, an inevitable question would be whether they exist in the real world. It is known that tropical convective heating is dominated by top-heavy profiles due to stratiform precipitation (Houze 1989). The early studies on the role of heating profile in the MJO struggled to justify the low-level heating maximum because of a lack of observed evidence for it. In fact, bottom-heavy heating profiles indeed exist in the tropics. Over the western Pacific, for example, apparent heating profiles (Q1, Yanai et al. 1973) derived from sounding observations are dominated by the top-heavy structure as manifested by the leading EOF mode (Fig. 18). But, the second leading EOF mode is bottom heavy. Even though this second, bottom-heavy EOF mode only explains 10% of the total variance of the Q1 time series, in comparison to 74% variance explained by the first, top-heavy mode, this EOF analysis clearly shows that bottom-heavy heating profiles indeed exist and their importance might not be negligible. It is ensuring that such second, bottom-heavy EOF mode can be derived also from sounding observations taken in other tropical regions. A full report on this will be given separately. In short, the existence of bottom-heavy diabatic heating in the tropics is beyond doubt.

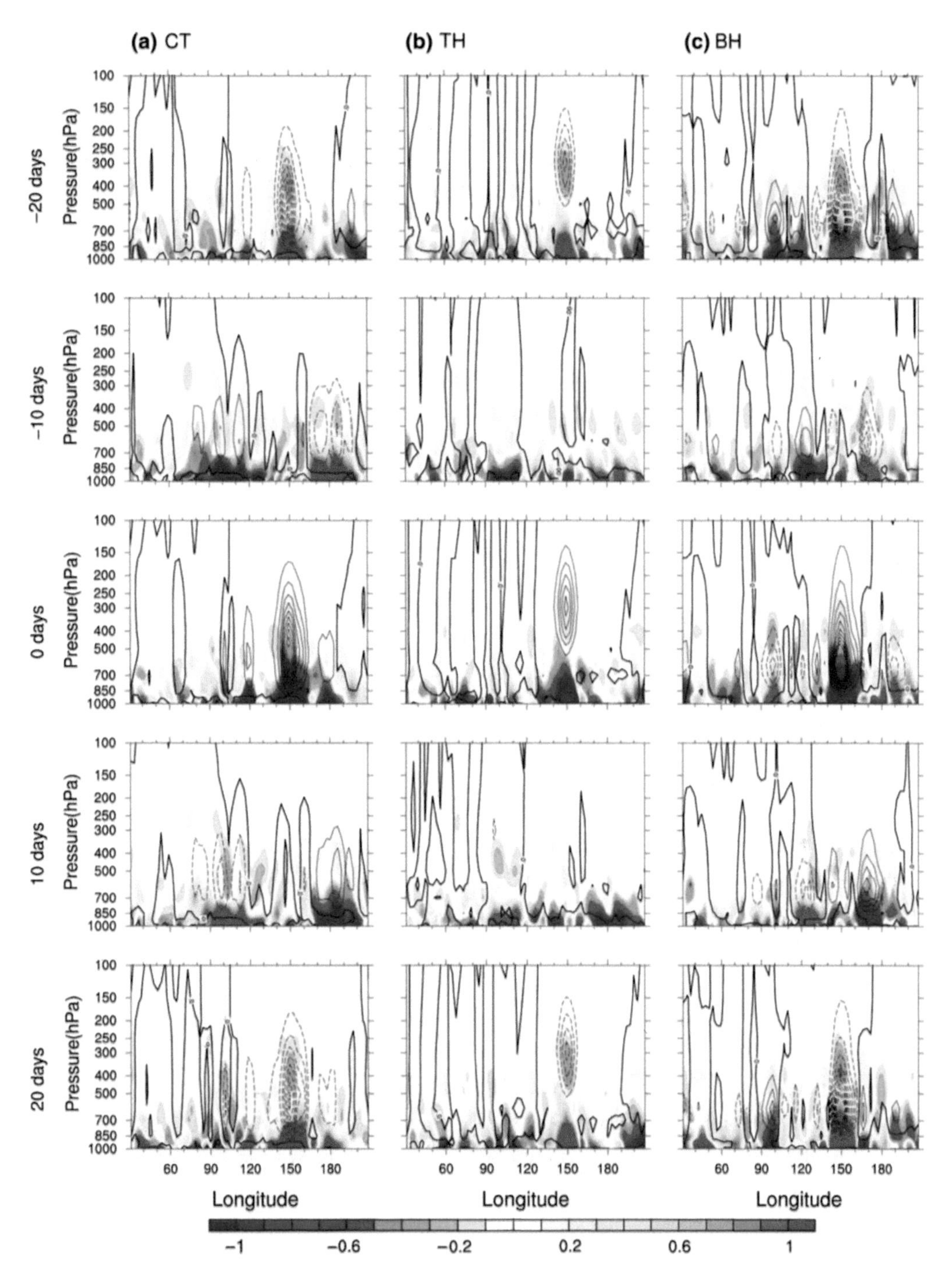

Fig. 17. Lag regression of intraseasonal (30-60 days) diabatic heating (contours, interval of 0.2 K/day) and moisture convergence (*color*, 10^{-6} g kg^{-1} s^{-1}) upon maximum heating at 150°E (0 day) from CT (*left column*), TH (*middle*) and BH (*right*)

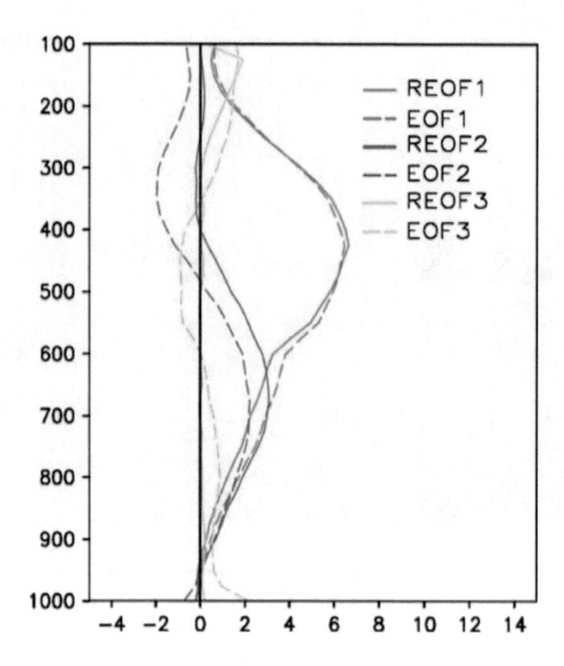

Fig. 18. First three leading modes of EOF (*dashed lines*) and rotated EOF (*solid*) analyses for apparent heating (Ql) profiles of TOGA-COARE (30)

6. Summary

The most intriguing result from this study is that slow eastward propagation of intraseasonal perturbations resembling the MJO in a GCM can be produced only when there is sufficient bottom heavy (maximum in the lower troposphere) diabatic heating; when diabatic heating is top heavy (maximum in the upper troposphere) only, a stationary component dominated intraseasonal perturbations over the Indian and western Pacific Oceans. This result provides new modeling evidence for the importance of lower-tropospheric heating to the MJO, supporting similar results from previous studies (Li 1983; Lau and Peng 1987; Chang and Lim 1988; Sui and Lau 1989). But there are important distinctions between this and the previous studies using wave-CISK framework. First, there is no wave-CISK assumption made in our study. Second, the stationary intraseasonal perturbations are produced by top-heavy heating profile only in our model, not in any of the wave-CISK models. It would be desirable, therefore, to reproduce and interpret our results using simple dynamical framework without any wave-CISK assumption. Third, the previous studies mainly emphasized the dependence of the phase speed on the vertical scale in their interpretation of the sensitivity to the heating profile; an alternative interpretation has been presented in this study in terms of surface and low-level moisture convergence that is enhanced by bottom-heavy heating but weakened by topheavy heating. Forth, we provide observational evidence for the bottom-heavy heating profile that is missing in the previous studies.

Possible reasons for bottom-heavy diabatic heating to play a critical role in our simulations to produce the MJO appear to come from its efficiency in promoting low-level moisture convergence, which in turn feeds back to deep convection. Such feedback has been suggested crucial to the MJO in theories (e.g., Wang 1988; Wang and Rui 1990). In consequence, a large-scale, deep, first-mode baroclinic circulation emerges, which is one of the fundamental signatures of the MJO. Low-level moisture convergence might have also been instrumental to the slow eastward propagation of the MJO. When heating profiles are top heavy, low-level moisture convergence is much weaker, moist convection also weaker and less organized on the large scale, the low- and upper-level circulations decoupled, and the signature of the MJO in the first-mode baroclinic structure lost. Some of these details might have been consequences of the particular way by which latent heating profiles were artificially modified or have been related to the particular cumulus parameterization used. Numerical experiments with less stringent test of the role of diabatic heating profiles in the MJO are needed to confirm the results from this study. The biggest remaining uncertainty is why bottom-heavy heating profile would promote eastward propagation in our simulation. This uncertainty is inherently related to the lack of understanding of why the MJO propagates eastward at all. Based on the only observationally validated theory that provides an explanation for this zonal asymmetry in the MJO, namely, the frictional convergence theory (Wang 1988; Wang and Rui 1990), the promotion of the eastward propagation of the MJO in our simulation is through a stronger boundary-layer easterly wind response when diabatic heating is bottom heavy.

Based on the results from this study and discussion in Sect. 5, we propose the following hypothesis: The failure of global models to produce a realistic MJO is partially rooted in their insufficient diabatic heating in the lower troposphere. This hypothesis by no means dismisses the importance of deep convective heating, including its observed top-heavy structure, to the MJO (Lin et al. 2004). For example, the observed vertical structure of the MJO (e.g., Sperber 2003) is likely to be closely related to deep convective heating. A typical top-heavy heating profile would generate strong convergence in the mid troposphere (Bergman and Hendon 2000; Wu 2003), which tends to bring relatively dry air into convective region of the MJO. Dry air into convection regions is known to be detrimental to deep convection (e.g., Tompkins 2001). Taking convectively inactive and active phases as equally important components of the MJO, one can argue that top-heavy heating during active phases of the MJO might be necessary to the transition from active to inactive phases. The evolution from bottom-heavy to top-heavy heating profiles through the life cycle of the MJO in observations (e.g., Lin et al. 2006) and model simulations (Zhang and Mu 2005) might be an essential ingredient in the MJO dynamics. Such an evolution in the heating profile must have been distorted by the brute-force method used in our simulations. Nonetheless, the sensitivity demonstrated by the simulations with modified heating profile has exposed bluntly the possible role of bottom-heavy heating in the MJO. To completely understand the sensitivity of the simulated MJO to the vertical structure of convective heating, we need to consider more carefully designed numerical experiments and more advanced observations.

Acknowledgments The authors thank Brian Mapes, Eric Maloney, Paul Roundy, Jun-Ichi

Yano and two anonymous reviewers for their comments on an earlier version of the manuscript. This study was support by the National Nature Science Foundation of China under grant no. 40575027 (Li, Jia, and Ling), by a grant from City University of Hong Kong under grant no. 7002329 (Zhou), and by US National Science Foundation under grant ATM0739402 (Zhang). Chidong Zhang thanks the State Key Laboratory of Numerical Modeling for Atmospheric Sciences and Geophysical Fluid Dynamics (LASG), Institute of Atmospheric Physics, Chinese Academy of Sciences for hosting his visits in 2006 and 2007, during which he collaborated with LASG scientists on this study.

References

Bergman JW, Hendon HH (2000) Cloud radiative forcing of the low latitude tropospheric circulation: linear calculations. J Atmos Sci 57: 2225-2245.

Bessafi M, Wheeler MC (2006) Modulation of south Indian Ocean tropical cyclones by the Madden-Julian Oscillation and convectively coupled equatorial waves. Mon Weather Rev 134: 638656. doi: 10.1175/ MWR3087.1.

Bond NA, Vecchi GA (2003) The influence of the Madden-Julian oscillation on precipitation in Oregon and Washington. Weather Forecast 18: 600-613. doi: 10.1175/1520-0434(2003)018<0600: TIOTMO>2.0.CO; 2.

Chang C-P, Lim H (1988) Kelvin wave-CISK: a possible mechanism for the 30-50 day oscillation. J Atmos Sci 45: 1709-1720.

Collins WD et al (2006) The community climate system model, version 3 (CCSM3). J Clim 19: 2122-2143.

Crum FX, Dunkerton TJ (1992) Analytic and numerical models of wave-CISK with conditional heating. J Atmos Sci 49: 1693-1708.

Edwards JM, Slingo A (1996) Studies with a flexible new radiation code. I: choosing a configuration for a large-scale model. Q J R Meteorol Soc 122: 689-719. doi: 10.1002/ qj.49712253107.

Flatau M, Flatau PJ, Phoebus P, Niiler PP (1997) The feedback between equatorial convection and local radiative and evaporative processes: the implications for intraseasonal oscillations. J Atmos Sci 54: 2373-2386. doi: 10.1175/1520-0469(1997)054<2373: TFBECA>2.0.CO; 2.

Frank WM, Roundy PE (2006) The role of tropical waves in tropical cyclogenesis. Mon Weather Rev 134: 2397-2417. doi: 10.1175/ MWR3204.1.

Grabowski WW, Moncrieff MW (2005) Moisture-convection feedback in the tropics. Q J R Metab Soc 130: 3081-3104. doi: 10.1256/qj.03.135.

Hartmann DL, Hendon HH, Houze RA Jr (1984) Some implications of the mesoscale circulations in tropical cloud clusters for large scale dynamics and climate. J Atmos Sci 41: 113-121.

Hendon HH, Liebmann B (1990) The intraseasonal (30-50 day) oscillation of the Australian summer monsoon. J Atmos Sci 47: 2909-2923.

Hendon HH (2000) Impact of air-sea coupling on the Madden-Julian oscillation in a general circulation model. J Atmos Sci 57: 3939-3952.

Higgins RW, Shi W (2001) Intercomparison of the principal modes of interannual and intraseasonal variability of the North American monsoon system. J Clim 14: 403-417. doi: 10.1175/1520-0442 (2001) 014<0403: IOTPMO> 2.0.CO; 2.

Holtslag AAM, Boville B (1993) Local versus nonlocal boundary layer diffusion in a global climate model. J Clim 6: 1825-1842. doi: 10.1175/1520-0442(1993)006<1825: LVNBLD>2.0.CO; 2.

Houze RA Jr (1989) Observed structure of mesoscale convective systems and implications for large-scale heating. Q J R Meteorol Soc 115: 425-461. doi: 10.1002/qj.49711548702.

Hu Q, Randall DA (1994) Low-frequency oscillations in radiative- convective systems. J Atmos Sci 51: 1089-1099. doi: 10.1175/ 1520-0469(1994)051<1089: LFOIRC>2.0.CO; 2.

Jia xiaolong, *Numerical Simulations of the Tropical Intraseasonal Oscillation*, Doctor's thesis, Institute of Atmospheric Physics, Chinese Academy of Sciences, 2006.

Johnson RH, Rickenbach TM, Rutledge SA, Ciesielski PE, Schubert WH (1999) Trimodal characteristics of tropical convection. J Clim 12: 2397-2418. doi: 10.1175/1520-0442(1999)012<2397: TCOTC>2.0.CO; 2.

Kalnay E, Kanamitsu M, Kistler R, Collins W, Deaven D, Gaudin L, Iredell M, Saha S, White G, Woollen J, Zhu Y, Chelliah M, Ebisuzaki W, Higgins W, Janowiak J, Mo K, Ropelewski C, Wang J, Leetrnaa A, Reynolds R, Jenne R, Joseph D (1996) NCEP/NCAR 40-year reanalysis project. Bull Am Meteor Soc 77: 437-471.

Kessler WS, McPhaden MJ, Weickmann KM (1995) Forcing of intraseasonal Kelvin waves in the equatorial Pacific. J Geophys Res 100: 10613-10631. doi: 10.1029/95JC00382.

Kiladis GN, Straub KH, Haertel PT (2005) Zonal and vertical structure of the Madden-Julian oscillation. J Atmos Sci 62: 2809-2890. doi: 10.1175/JAS3520.1.

Krishinamurti TN, Subrahmann D (1982) The 30-50 day mode at 850mb during MONEX. J Atmos Sci 39: 2088-2095. doi: 10. 1175/1520-0469(1982)039<2088: TDMAMD>2.0.CO; 2.

Lau KM, Chan PH (1986) Aspects of the 40-50 day oscillation during the northern summer as inferred from outgoing longwave radiation. Mon Weather Rev 114: 1354-1367. doi: 10.1175/1520-0493(1986)114< 1354: AOTDOD>2.0.CO; 2.

Lau KM, Peng L (1987) Origin of low-frequency (intraseasonal) oscillation in the tropical atmosphere, Part I: basic theory. J Atmos Sci 45: 1781-1791. doi: 10.1175/1520-0469(1988)045< 1781: OTDOIO>2, O.CO; 2.

Lau K-M, Shen S (1988) On the dynamics of intraseasonal oscillations and ENSO. J Atmos Sci 45: 1781-1797. doi: 10.1175/1520-0469(1988)045<1781: OTDOIO>2.0.CO; 2.

Lawrence DM, Webster PJ (2002) The boreal summer intraseasonal oscillation: relationship between northward and eastward movement of convection. J Atmos Sci 59: 1593-1606. doi: 10.1175/1520-0469 (2002)059< 1593: TBSIOR>2.0.CO; 2.

Li C (1983) Convection condensation heating and unstable modes in the atmosphere. Chin J Atmos Sci 7: 260-268 in Chinese.

Li C (1985) Actions of summer monsoon troughs (ridges) and tropical cyclone over South Asia and the moving CISK mode. Scientia Sin B 28: 1197-1206.

Li C, Zhou Y (1994) Relationship between intraseasonal oscillation in the tropical atmosphere and ENSO. Chin J Geophys 37: 213-223 in Chinese.

Li C, Smith I (1995) Numerical simulation of the tropical intraseasonal oscillation and the effect of warm SSTs. Acta Meteor Sin 9: 1-12.

Li C, Li G (1997) Evolution of intraseasonal oscillation over the tropical western Pacific/South China Sea and its effect to the summer precipitation in Southern China. Adv Atmos Sci 14: 246-254. doi: 10.1007/ s00376-997- 0053-6.

Li C, Long Z, Zhang Q (2001) Strong/weak summer monsoon activity over the South China Sea and atmospheric intraseasonal oscillation. Adv Atmos Sci 18: 1146-1160. doi: 10.1007/s00376-001-0029-x.

Li C, Long Z (2002) Intraseasonal oscillation anomalies in the tropical atmosphere and El Niño events. Exchanges 7(2): 12-15.

Liebmann B, Hendon HH, Glick JD (1994) The relationship between tropical cyclones of the western Pacific and Indian Oceans and the Madden-Julian oscillation. J Meteorol Soc Jpn 72: 401-411.

Lin J, Mapes BE, Zhang M, Newman M (2004) Stratiform precipitation, vertical heating profiles, and the

Madden-Julian Oscillation. J Atmos Sci 61: 296-309. doi: 10.1175/1520- 0469(2004)061<0296: SPVHPA> 2.0.CO; 2.

Lin J-L, Kiladis GN, Mapes BE, Weickmann KM, Sperber KR, Lin W et al (2006) Tropical intraseasonal variability in 14 IPCC AR4 climate models Part I: convective signals. J Clim 19: 2665-2690. doi: 10.1175/JCLI3735.1.

Lin X, Johnson RH (1996) Heating, moistening and rainfall over the western Pacific warm pool during TOGA COARE. J Atmos Sci 53: 3367-3383. doi: 10.1175/1520-0469(1996)053<3367: HMAROT> 2.0.CO; 2.

Maloney ED, Hartmann DL (1998) Frictional moisture convergence in a composite life cycle of the Madden-Julian Oscillation. J Clim 11: 2387-2403. doi: 10.1175/1520-0442(1998)011<2387: FMCIAC> 2.0.CO; 2.

Maloney ED, Hartmann DL (2000) Modulation of eastern North Pacific hurricanes by the Madden-Julian oscillation. J Clim 13: 1451-1460. doi: 10.1175/1520-0442(2000)013<1451: MOENPH>2.0. CO; 2.

Maloney ED, Hartmann DL (2001) The sensitive of intraseasonal variability in the NCAR CCM3 to changes in convection parameteriazation. J Clim 14: 2015-2034. doi: 10.1175/1520-0442(2001)014<2015: TSOIVI> 2.0.CO; 2.

Manabe S, Smagorinsky J, Strickler RF (1965) Simulated climatology of general circulation model with a hydrologic cycle. Mon Weather Rev 93: 769-798. doi: 10.1175/1520-0493(1965)093< 0769: SCOAGC> 2.3.CO; 2.

Matsuno T (1966) Quasi-geostrophic motions in the equatorial area. J Meteorol Soc Jpn 44: 25-43.

Matthews AJ (2004) Intraseasonal variability over tropical Africa during northern summer. J Clim 17: 2427-2440. doi: 10.1175/1520-0442(2004)017<2427: IVOTAD>2.0.CO; 2.

Mo KC (2000) The association between intraseasonal oscillations and tropical storms in the Atlantic basin. Mon Weather Rev 128: 4097-4107. doi: 10.1175/1520-0493(2000)129<4097: TABIOA>2.0. CO; 2.

Mu M, Li C (2000) The onset of summer monsoon over the South China Sea in 1998 and action of atmospheric intraseasonal oscillation. Clim Environ Reserch 5: 375-387 in Chinese.

Paegle JN, Byerle LA, Mo KC (2000) Intraseasonal modulation of South American summer precipitation. Mon Weather Rev 128: 837-850. doi: 10.1175/1520-0493(2000)128<0837: IMOSAS>2.0.CO; 2.

Raymond DJ (2001) A new model of the Madden-Julian oscillation. J Atmos Sci 58: 2807-2819. doi: 10.1175/1520-0469(2001)058 <2807: ANMOTM>2.0.CO; 2.

Schneider EK, Lindzen RS (1977) Axially symmetric steady-state models of the basic state for instability and climate studies. Part I. Linearized calculations. J Atmos Sci 34: 263-279.

Sellers PJ, Min Y, Sud YC, Dalcher A (1986) A simple biosphere model (SIB) for use within general circulation models. J Atmos Sci 43: 505-531. doi: 10.1175/1520-0469(1986)043<0505: ASBMFU> 2.0.CO; 2.

Simmonds I (1985) Analysis of the "spinning" of a global circulation model. J Geophys Res 90: 5637-5660. doi: 10.1029/ JD090iD03p05637.

Simpson J, Kummerow C, Tao W-K, Adler R (1996) On the tropical ainfall measuring mission (TRMM). Meteorol Atmos Phys 60: 19-36. doi: 10.1007/BF01029783.

Schumacher C, Houze RA Jr, Kraucunas I (2004) The tropical dynamical response to latent heating estimates derived from the TRMM precipitation radar. J Atmos Sci 61: 1341-1358. doi: 10. 1175/1520-0469(2004) 061<1341: TTDRTL>2.0.CO; 2.

Slingo A (1980) A cloud parameterization scheme derived from GATE data for use with a numerical model. Q J R Meteorol Soc 106: 747-770. doi: 10.1002/qj.49710645008.

Slingo A (1987) The development and verification of a cloud prediction scheme for the ECMWF model. Q J

R Meteorol Soc 113: 899-927. doi: 10.1256/smsqj.47708.

Slingo, J. M. and Coauthers, Intraseasonal oscillations in 15 atmospheric general circulation models: Results from an AMIP diagnostic subproject. Climate Dyn., 1996, 13: 325-357. doi: 10.1007/BF00231106.

Slingo JM, Inness P, Neale R, Woolnough S, Yang G-Y (2003) Scale interactions on diurnal to seasonal timescales and their relevance to model systematic errors. Ann Geophys 46: 139-155.

Sperber KR (2003) Propagation and the vertical structure of the Madden-Julian Oscillation. Mon Weather Rev 131: 3018-3037. doi: 10.1175/1520-0493(2003)131<3018: PATVSO>2.0.CO; 2.

Sperber KR (2004) Madden-Julian variability in NCAR CAM 20 and CCSM2.0. Clim Dyn 23: 259-278. doi: 10.1007/s00382-004-0447-4.

Sperber KR, Gualdi S, Legutke S, Gayler V (2005) The Madden-Julian oscillation in ECHAM4 coupled and uncoupled GCMs. Clim Dyn 25: 117-140.

Sui C-H, Lau K-M (1989) Origin of low-frequency (intraseasonal) oscillations in the tropical atmosphere. Part. II: Structure and propagation of mobile wave-CISK modes and their modification by lower boundary forcings. J Atmos Sci 46: 37-56.

Tiedtke M (1989) A comprehensive mass flux scheme for cumulus parameterization in large-scale models. Mon Wea Rev 117: 1779-1800.

Tompkins AM (2001) Organization of tropical convection in low vertical wind shears: the role of water vapor. J Atmos Sci 58: 529-545. doi: 10.1175/1520-0469(2001)058<0529: OOTCIL>2.0. CO; 2.

Waliser DE, Lau KM, Kim JH (1999) The influence of coupled sea surface temperatures on the Madden-Julian oscillation: a model perturbation experiment J Atmos Sci 56: 333-358. doi: 10.1175/1520-0469 (1999)056< 0333: TIOCSS>2.0.CO; 2.

Waliser DE, Lau KM, Stern W, Jones C (2003) Potential predictability of the Madden-Julian Oscillation. Bull Am Meteorol Soc 84: 33-50. doi: 10.1175/BAMS-84-l-33.

Wang B (1988) Dynamics of tropical low-frequency waves: an analysis of the moist Kelvin wave. J Atmos Sci 45: 2051-2065. doi: 10.1175/1520-0469(1988)045<2051: DOTLFW>2.0.CO; 2.

Wang B (2005) Theory, 307-360. In: Lau WKM, Waliser DE (eds) Intraseasonal variability of the atmosphere-ocean climate system. Praxis, Chichester, pp 436.

Wang B, Rui H (1990) Synoptic climatology of transient tropical intraseasonal convective anomalies: 1975-1985. Meteorol Atmos Phys 44: 43-61. doi: 10.1007/BF01026810.

Wang W, Schlesinger ME (1999) The dependence on convective parameterization of the tropical intraseasonal oscillation simulated by the UIUC 11-layer atmospheric GCM. J Clim 12: 1423-1457. doi: 10.1175/1520-0442(1999)012<1423: TDOCPO>2.0. CO; 2.

Wang Z-Z, Wu G-X, Liu P, Wu T-W (2005) The development of Goals/LASG AGCM and its global climatological features in climate simulation I-influence of horizontal resulotion. J Trop Meteorol 21(3): 225-237 In Chinese.

Wheeler M, Kiladis GN (1999) Convectively coupled equatorial waves: analysis of clouds and temperature in the wavenumber- frequency domain. J Atmos Sci 56: 374-399.

Wheeler MC, Hendon HH (2004) An all-season real-time multivariate MJO index: development of an index for monitoring and prediction. Mon Weather Rev 132: 1917-1932. doi: 10.1175/ 1520-0493(2004)132< 1917: AARMMI>2.0.CO; 2.

Wheeler MC, McBride JL (2005) Intraseasonal variability in the atmosphere-ocean climate system. In: Lau WKM, Waliser DE (eds) Praxis, Chichester, pp 125-173.

Wu G-X, Liu H, Zhao Y-C and Liw-p, (1996) A nine-layer atmospheric general circulation model and its performance. Adv Atmos Sci 13(1): 1-18. doi: 10.1007/BF02657024.

Wu G, Zhang X, Liu H, Yu Y, Jin X, Guo Y, Sun S, Li W, Wang B, Shi G, 1997 Global ocean-atmosphere-land system model of LASG (GOALS/LASG) and its performance in simulation study, *Quart. J. Appl. Meteor.*, Supplement Issue, 15-28. (In Chinese).

Wu Z (2003) A shallow CISK, deep equilibrium mechanism for the interaction between large-scale convection and large-scale circulations in the tropics. J Atmos Sci 60: 377-392. doi: 10.1175/ 1520-0469 (2003)060<0377: ASCDEM>2.0.CO; 2.

Wu Z, Sarachik ES, Battisti DS (2000) Vertical structure of convective heating and the three-dimensional structure of the forced circulation on an equatorial beta plane. J Atmos Sci 57: 2169-2187. doi: 10.1175/1520-0469 (2000)057<2169: VSOCHA>2.0. CO; 2.

Wu Z, Sarachik ES, Battisti DS (2001) Thermally driven tropical circulations under Rayleigh friction and Newtonian cooling: analytic solutions. J Atmos Sci 58: 724-741. doi: 10.1175/1520-0469(2001)058< 0724: TDTCUR>2.0.CO; 2.

Xie P, Arkin PA (1997) Global precipitation: a 17-year monthly analysis based on gauge observations, satellite estimates, and numerical model outputs. Bull Am Meteor Soc 78: 2539-2558.

Xue Y, Sellers PJ, Linter JL, Shukla J (1991) A simplified biosphere model for global climate studies. J Clim 4: 345-364. doi: 10.1175/ 1520-0442(1991)004<0345: ASBMFG>2.0.CO; 2.

Hui Yang, Chongyin Li (2003) The relation between atmospheric intraseasonal oscillation and summer severe flood and drought in the Changjiang-Huaihe basin. Adv Atmos Sci 20: 540-553. doi: 10.1007/BF02915497.

Yanai M, Esbensen S, Chu J-H (1973) Determination of bulk properties of tropical cloud clusters from large-scale heat and moisture budgets. J Atmos Sci 30: 611-627.

Zhang C (2005) 2005: Madden-Julian Oscillation. Rev Geophys 43: RG2003. doi: 10.1029/2004RG000158.

Zhang GJ, McFarlane NA (1995) Sensitivity of climate simulations to the parameterization of cumulus convection in the CCC-GCM. Atmos Ocean 3: 407-446.

Zhang C, Gottschalck J (2002) SST anomalies of ENSO and the Madden-Julian Oscillation in the equatorial Pacific. J Clim 15: 2429-2445. doi: 10.1175/1520-0442(2002)015<2429: SAOEAT>2.0. CO; 2.

Zhang C, Dong M, Gualdi S, Hendon HH, Maloney ED, Marshall A, et al (2006) Simulations of the Madden-Julian oscillation in four pairs of coupled and uncoupled global models. Clim Dyn 27: 573-592. doi: 10.1007/s00382-006-0148-2.

Zhang G, Mu M (2005) Simulation of the Madden-Julian oscillation in the NCAR CCM3 using a revise Zhang-McFarlane convection parameterization scheme. J Clim 18: 4046-4064. doi: 10. 1175/ JCLI3508.1.

Impact of East Asian Winter Monsoon on MJO over the Equatorial Western Pacific

Xiong Chen[1]*, Chongyin Li[1,2], Jian Ling[2]**, Yanke Tan[1]

[1] College of Meteorology and Oceanography, PLA University of Science and Technology, Nanjing, China
[2] LASG, Institute of Atmospheric Physics, Chinese Academy of Sciences, Beijing, China

Abstract This paper investigates the processes and mechanisms by which the East Asian winter monsoon (EAWM) affects the Madden-Julian oscillation (MJO) over the equatorial western Pacific in boreal winter (November-April). The results show that both the EAWM and MJO over the equatorial western Pacific have prominent interannual and interdecadal variabilities, and they are closely related, especially on the interannual timescales. The EAWM influences MJO via the feedback effect of convective heating, because the strong northerlies of EAWM can enhance the ascending motion and lead the convection to be strengthened over the equatorial western Pacific by reinforcing the convergence in the lower troposphere. Daily composite analysis in the phase 4 of MJO (i.e., strong MJO convection over the Maritime Continent and equatorial western Pacific) shows that the kinetic energy, outgoing longwave radiation (OLR), moisture flux, vertical velocity, zonal wind, moist static energy, and atmospheric stability differ greatly between strong and weak EAWM processes over the western Pacific. The strong EAWM causes the intensity of MJO to increase, and the eastward propagation of MJO to become more persistent. MJO activities over the equatorial western Pacific have different modes. Furthermore, these modes have differing relationships with the EAWM, and other factors can also affect the activities of MJO; consequently, the relationship between the MJO and EAWM shows both interannual and interdecadal variabilities.

1. Introduction

The East Asian winter monsoon (EAWM) Huang Huang is one of the most active atmospheric circulation systems in the boreal winter. It not only affects the weather and climate over the East Asia but can also modify the variations of tropical atmosphere and ocean. The EAWM also plays an important role in the global weather and climate anomalies (Chang et al. 1979; Chan and Li 2004; Huang et al. 2007a; Yan et al. 2009; Zhou 2011). Consequently,

Received January 6, 2015;accepted September 24, 2015.

*cx3212007753@163.com
**lingjian@lasg.iap.ac.cn

the EAWM is one of the most important issues facing researchers, especially Chinese scholars, and a good understanding of its activities and variability has been developed. In recent years, an increasing attention has focused on the anomalies of EAWM, such as the interaction between the EAWM and ENSO (Li and Mu 2000; Zhou et al. 2007), the characteristics of EAWM variability and its mechanism (Huang et al. 2003; Wang and Ding 2006), and the relationship between the EAWM and the activities of quasi-stationary planetary wave (Chen et al. 2005; Huang et al. 2007b; Wang et al. 2009).

Madden-Julian Oscillation (MJO) first discovered by Madden and Julian (1971, 1972) is characterized with a period about 30-60 days. Most of the researches have shown that MJO has a significant impact on the global weather and climate and is closely related to the anomalous weather and climate in many areas (Madden and Julian 1994; Zhang 2005, 2013; Li et al. 2014). For example, MJO activities in boreal summer are directly related to the outbreak and interruption of the East Asian summer monsoon and its precipitation (Yang and Li 2003; Zhou and Chan 2005; Lü et al. 2012). MJO activities are also closely related to the ENSO cycle. The MJO over the equatorial western Pacific strengthens significantly in the winter and spring prior to El Niño occurrence but decreases rapidly during the mature stage of El Niño (Li and Zhou 1994; Zhang and Gottschalck 2002; Hendon et al. 2007). Furthermore, the MJO also influences the formation and activities of the typhoon over the western Pacific (Zhu et al. 2004; Pan et al. 2010; Li et al. 2012, Li and Zhou 2013a, 2013b). The importance of extratropical influences on the initiation and maintenance of the MJO has been argued using reanalysis data and numerical analysis for a long time (Liebmann and Hartmann 1984; Murakami 1988; Hsu et al. 1990; Ray et al. 2009, Ray and Zhang 2010; Ling et al. 2013, 2014). Wang et al. (2012) studied a case in which a northerly surge reinforced MJO over the Indian Ocean during its initiation phase and found that such subtropical cold surges are likely to strengthen and accelerate the buildup of deep MJO convection during the initiation phase of MJO. The latitudinal transport of momentum from the extratropics is crucial to the generation of westerly associated with the MJO initiation in the lower troposphere in the tropics (Ray et al. 2009; Ray and Zhang 2010).

Many studies have shown that the strong EAWM can enhance the convection over the South China Sea and tropical western Pacific and lead to an anomalous cyclonic circulation to the east of the Philippines (Chang et al. 1979; Ji et al. 1997). However, the feedback effect of convective heating is one of the main mechanisms for the initiation and maintenance of MJO (Li 1985; Lau and Peng 1987; Li et al. 2009), which means that the EAWM is closely related to the variation of the MJO over the equatorial western Pacific. In the late 1980s, Li (1989) pointed out that the persistent and strong northerly of EAWM can promote the occurrence of El Niño event, as it can strengthen the anomalous westerly and the 30-60 days oscillation in the atmosphere over the western Pacific by reinforcing convection there. However, the processes and mechanisms by which the EAWM affects the MJO over the equatorial western Pacific remain unclear. Therefore, this study aims to further investigate these processes and mechanisms using reanalysis data.

The remainder of this paper is organized as follows. The data and methods are described briefly in section 2. The relationship between the EAWM and MJO and the processes by which the EAWM impacts the MJO are analyzed in section 3. A further study of these influences and mechanisms in the phase 4 of MJO is provided in section 4 using composite analysis. Finally, the summary and discussion are provided in section 5.

2. Data and Methods

The daily mean atmospheric data are from the National Centers for Environmental Prediction/ National Center for Atmospheric Research (NCEP/NCAR) (Kalnay et al. 1996), including wind, relative humidity covering the period from 1 January 1948 to 31 December 2011 with a horizontal resolution of 2.5° × 2.5°. The daily mean outgoing longwave radiation (OLR) data with a horizontal resolution of 2.5° × 2.5° are from the National Oceanic and Atmospheric Administration (NOAA) (Liebmann and Smith 1996) from 1 January 1975 to 31 December 2011. The real-time multivariate MJO (RMM) index developed by Wheeler and Hendon (2004) is from the Australian Bureau of Meteorology.

The main methods used in this study were composite and correlation analyses. The anomalies were generated by removing the annual cycle. The MJO signal was obtained by using the 30-60 days Lanczos bandpass filter with a 200-point window (Duchon 1979).

3. Relationship Between the MJO and EAWM

3.1 Correlation of the MJO and EAWM

The strength of the EAWM can be represented by the EAWM index. The EAWM index defined by meridional wind in the lower troposphere can better reflect the impact of the EAWM on the tropical atmosphere and ocean (Wang and Chen 2010). Here, the EAWM index is defined as the meridional wind averaged over 10°-30° N, 115°-130° E at 1000 hPa (Ji et al. 1997). Therefore, a higher/lower index value indicates a weaker/stronger EAWM. The intensity of MJO is represented by the MJO kinetic energy at 850 hPa (Li and Zhou 1994), and the MJO intensity index is defined as the MJO kinetic energy at 850 hPa averaged over the equatorial western Pacific (10°S-10°N, 120°-150°E).

The normalized EAWM and MJO indices in boreal winter from 1948 to 2010 (Fig. 1) clearly show the pronounced interannual and interdecadal variabilities of them, as well as the obviously out of phase variation between them. Their correlation coefficient is –0.417 (exceeding the 99% confidence level), which indicates that a strong EAWM corresponds to a strong MJO over the equatorial western Pacific. The interannual and interdecadal timescale components of the EAWM and MJO indices can be obtained using a 9-point Gaussian low-pass filter. Their correlation coefficients are –0.466 (exceeding the 99% confidence level) and –0.242 (exceeding the 90% confidence level) on interannual and interdecadal timescales, respectively. These results suggest that the relationship between the MJO over the equatorial western Pacific and EAWM on interannual timescales is much stronger than that on interdecadal timescales.

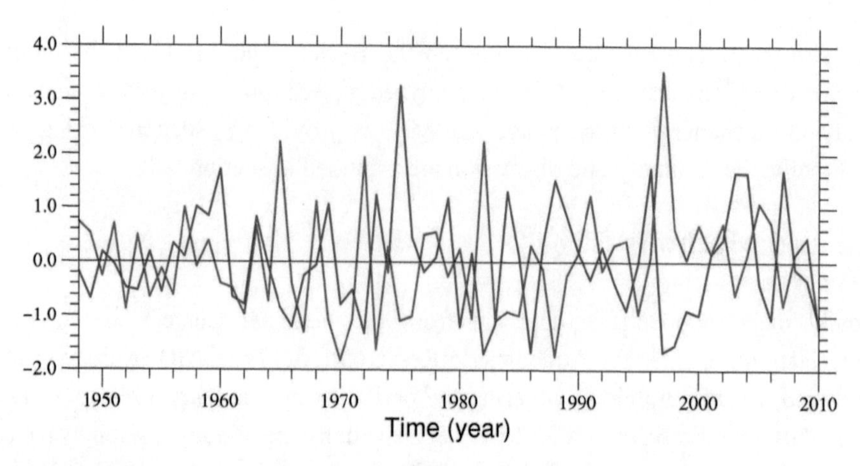

Fig. 1. The normalized MJO (*blue*) and EAWM (*red*) indices in boreal winter

The distribution of correlation coefficients of the 850 hPa MJO kinetic energy with EAWM index is shown in Fig. 2. Areas of negative correlation are concentrated in the equatorial western Pacific west of 160° E with high values over 10° S-10° N, 110°-140° E. However, the MJO near the dateline is positively correlated with the EAWM index, which means the MJO activities there will be weakened in a strong EAWM winter. In addition, the areas with significant correlation are more prominent in the northern hemisphere, whereas the strongest MJO activities are mainly centered in the south of the equator during boreal winter (not shown). A probable explanation is that the MJO over western Pacific has a variety of activity modes, and each has a distinct relationship with the EAWM. Therefore, the influences of the EAWM on each mode differ (discussed below). On the other hand, the mean intertropical convergence zone (ITCZ), South Pacific convergence zone (SPCZ), and ascending branch of Hadley circulation in the boreal winter are located in the southern hemisphere. Thus, although a strong EAWM can enhance the ascending motion and convection north of the equator, the mean background circulation also plays an important role in the southern hemisphere, which causes the main activities center of MJO in the southern hemisphere.

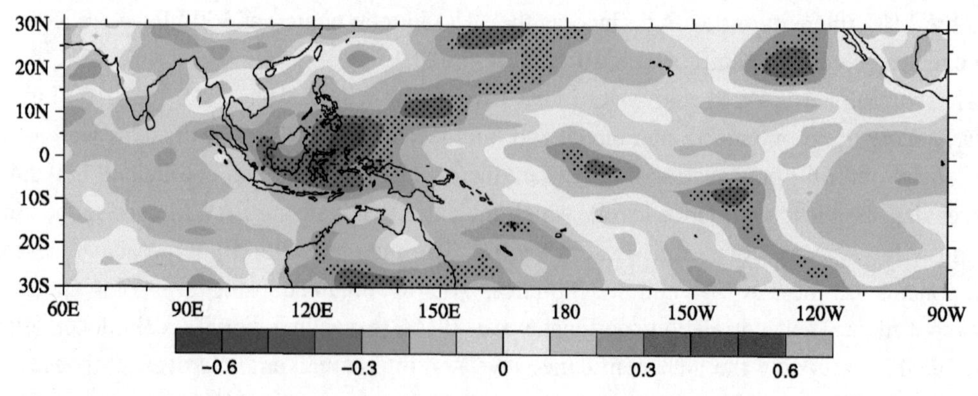

Fig. 2. Distribution of the correlation coefficients between MJO kinetic energy at 850 hPa and EAWM index. Results passing the significant test at the 90% confidence level are stippled

The empirical orthogonal function (EOF) decomposition on the MJO kinetic energy at 850 hPa over western Pacific (15° S-15° N, 115°-165° E) revealed five modes passing the North significant test (North et al. 1982). The first three modes account for 56.88% of the total variance, and each mode has its own characteristics (Fig. 3). The first mode shows the consistent variation of the MJO over the equatorial western Pacific. The second mode shows the dipole structure of MJO activities in the meridional direction. The third mode primarily shows the tripole structure of MJO activities over the western Pacific.

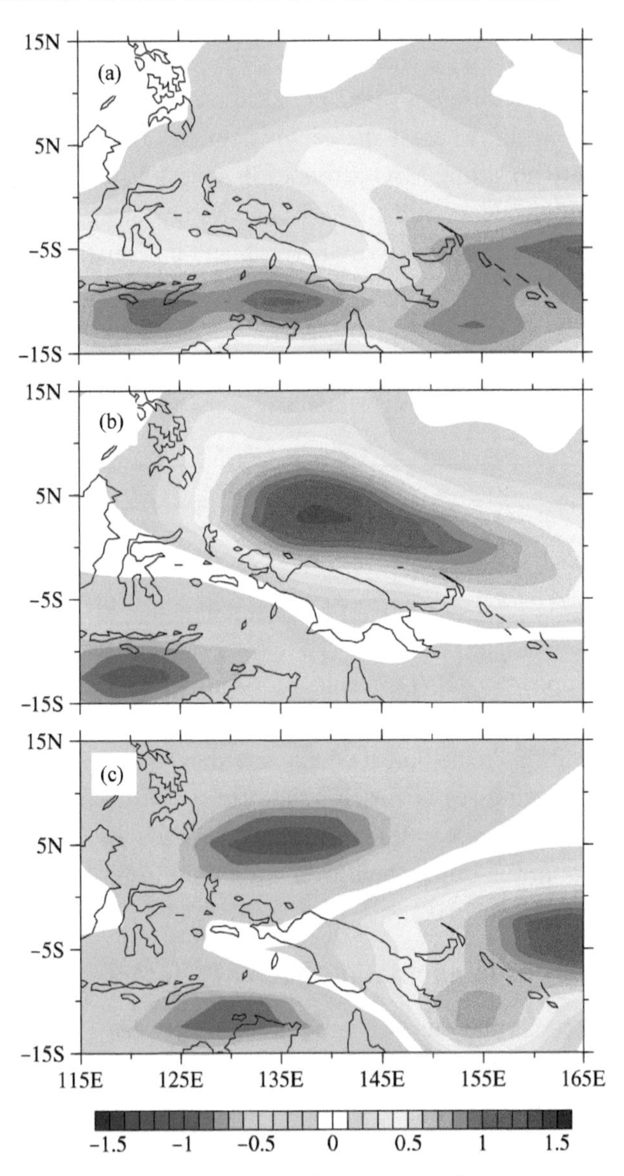

Fig. 3. Distribution of the **a** first, **b** second, and **c** third EOF modes of MJO kinetic energy ($m^2\ s^{-2}$) at 850 hPa over the tropical western Pacific (15° S-15°N, 115°-165°E)

Table 1 lists the correlation coefficients between the EAWM index and the time series of these three modes on different timescales. The first and third modes are closely related to the EAWM index on raw (unfiltered) and interannual timescales. The first and third modes reflect the consistent and tripole variations of MJO over the western Pacific, respectively, which may explain why the maximum correlation region appears in the area of10° S-10° N, 110°-140° E (Fig. 2). The first mode is also closely related to the EAWM on interdecadal timescales. Thus, the relationships between the various modes of MJO and EAWM are different, which increases the complexity of the relationships between the EAWM and MJO. As each mode has its own activities and variability characteristics, the relationship between the EAWM and MJO may also be different in different years. For example, the EAWM was strong in 1962/63, 1970/71, and 1983/84 winters, whereas the activities of MJO over the western Pacific were weak. However, there are no strong MJO activities in a weak EAWM winter. That is, there is indeed a certain relationship between the EAWM and MJO, but the impact of the EAWM on the MJO is asymmetric for strong and weak EAWM, and this relationship shows interannual and interdecadal variabilities.

Table 1. Correlation coefficients between the EAWM index and the time series of different EOF modes on different timescales. (Results passing the significant test at the 90, 95, and 99% confidence level are marked with *, **, and ***, respectively)

	First	Second	Third
Raw data	0.244*	–0.102	0.272**
Interannual timescales	0.231*	–0.145	0.355***
Interdecadal timescales	0.271**	0.060	–0.059

Singular value decomposition (SVD) results (not shown) of the MJO kinetic energy over the western Pacific (20° S-20° N, 110°E-180°) and meridional wind at 850 hPa over the East Asia(10°-40°N, 100°-150° E) in the boreal winter also show that the meridional wind over the East Asian and its neighboring sea is closely related to the MJO over the equatorial western Pacific. These features also suggest a close relationship between the EAWM and MJO, because the strong northerlies at lower troposphere over East Asia in boreal winter can represent a strong EAWM (Wang and Chen 2010).

3.2 Composite analysis in anomalous EAWM winters

The above analysis points the out-of-phase variation between EAWM and MJO indices, from which some questions arise. Firstly, is there a prominent difference in the activities of MJO between strong and weak EAWM winters? Furthermore, what is the mechanism of the EAWM influencing the MJO? To address these questions and discuss the influence of anomalous EAWM on the activities of MJO, composite analysis was performed on strong and weak EAWM winters (defined as the amplitude of EAWM index exceeding 0.7 standard deviation). The 18 identified strong EAWM winters are 1952/53, 1962/63, 1964/65, 1967/68,

1969/70, 1970/71, 1971/72, 1973/74, 1975/76, 1976/77, 1981/82, 1983/84, 1984/85, 1985/86, 1988/89, 1995/96, 2007/08, and 2010/11; the 12 identified weak winters are 1948/49, 1951/52, 1957/58, 1965/66, 1968/69, 1972/73, 1982/83, 1991/92, 1997/98, 2002/03, 2005/06, and 2006/2007.

The horizontal distribution of composite MJO kinetic energy and its anomalies at 850 hPa in strong and weak EAWM winters are shown in Fig. 4. In strong EAWM winters, there are two major MJO activity centers over the western Pacific: one is located at 0°-10° N, 120°-150° E and the other is in the southern hemisphere (5°-15°S, 110°-150° E) with their maximum strength exceeding 3.0 and 4.0 $m^2 s^{-2}$, respectively. The anomalous MJO kinetic energy clearly indicates that the MJO over the western Pacific west of 140° E is enhanced in the strong EAWM winters, whereas it is weakened over the central Pacific (Fig. 4a). However, the activities of MJO have almost opposite characteristics in the weak EAWM winters, featuring reduced/strengthened over the western/central Pacific (Fig. 4b). The activity center in the northern hemisphere is weakened significantly and that in the southern hemisphere is also reduced but not as prominently as the northern center. Moreover, there is a new activity center west of the dateline (10° S, 175° E), with its maximum strength exceeding 3.0 $m^2 s^{-2}$. The difference of MJO between strong and weak EAWM winters shows that the activities of MJO over the equatorial western Pacific and near the dateline differ significantly between strong and weak EAWM winters, which indicate that EAWM activities have a significant effect on the activities of MJO over these areas.

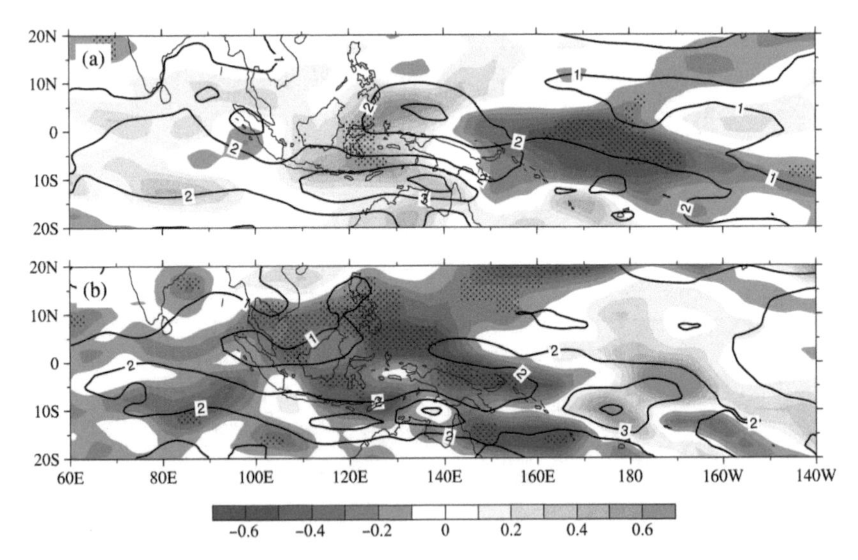

Fig. 4. Distribution of MJO kinetic energy at 850 hPa (*contours*, $m^2 s^{-2}$) and its anomalies (*color*, $m^2 s^{-2}$) in **a** strong and **b** weak EAWM winters. Results passing the significant test at the 90 % confidence level are stippled for anomalies

The activities of MJO are significantly different in strong and weak EAWM winters, but how does the EAWM affect them? The difference of wind and divergence at 850 hPa between strong and weak EAWM winters shows that anomalous westerlies are over the equatorial

Indian Ocean and equatorial Pacific west of 150° E, anomalous easterlies are over the equatorial Pacific east of150° E, a strong northerlies are over the eastern China and the South China Sea, and an anomalous cyclonic circulation is present to the east of the Philippines in strong EAWM winters (Fig. 5a). Therefore, the convergence over the equatorial western/ eastern Pacific is strengthened/ weakened. In the upper troposphere (Fig. 5b), the anomalous circulation and divergence patterns are almost opposite to those at 850 hPa. The equatorial

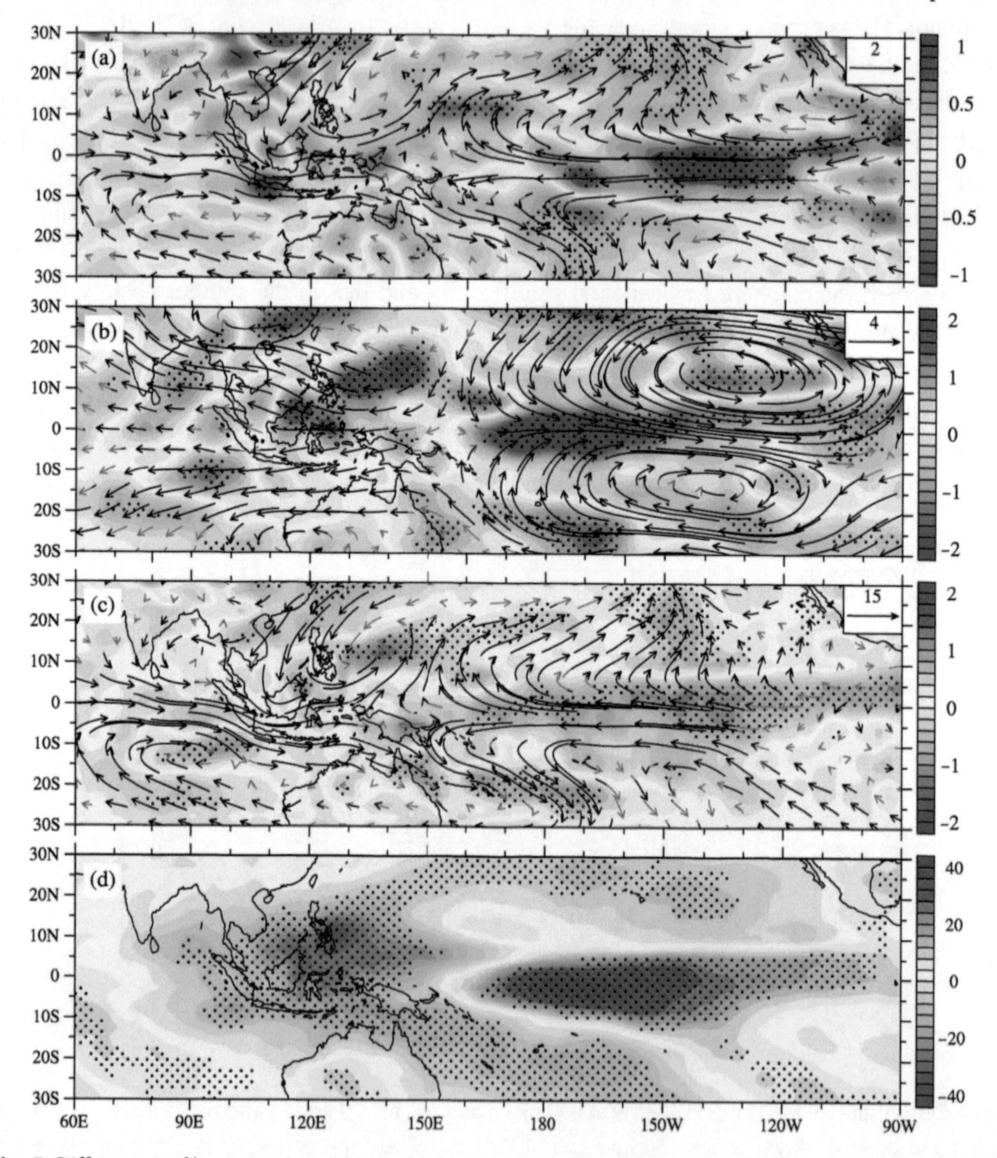

Fig. 5. Differences of horizontal circulation (*vector*, m s^{-1}) and divergence (*color*, $10^{-6}s^{-1}$) at **a** 850 hPa and **b** 200 hPa, **c** moisture flux (*vector*, g kg^{-1} m s^{-1}) and its convergence (color, 10^{-5}g kg s^{-1}) averaged from 1000 to 700 hPa, and **d** OLR (W m^{-2}) between the strong and weak EAWM winters. Results passing the significant test at the 90 % confidence level are stippled for divergence, moisture flux convergence, and OLR and marked by *black* for wind and moisture flux

Indian Ocean and equatorial Pacific region west of160°Eis dominated by easterlies anomalies, the region east of 160° E is dominated by westerlies anomalies, and anomalous cyclonic circulation is present in both hemispheres over the central-eastern Pacific. The equatorial western/eastern Pacific shows anomalous divergence/ convergence. As the strengthening of convergence, the moisture flux convergence at lower troposphere over the western Pacific is also enhanced in the strong EAWM winters (Fig. 5c). The moisture flux convergence at lower troposphere strengthens significantly over the western Pacific, especially in the northern hemisphere, while it is weakened over the eastern Pacific (Fig. 5c). The anomalous convergence/divergence in the lower/upper troposphere leads to an anomalous ascending motion over the equatorial eastern Indian Ocean and western Pacific from 80° E to 160° E (Fig. 6). The convection over the equatorial western Pacific is also strengthened during the strong EAWM winters, as reflected by the difference of OLR (Fig. 5d). Therefore, the activities of the MJO are enhanced over the equatorial western Pacific, because the feedback effect of convective heating is an important mechanism for the initiation and maintenance of MJO (Li 1985; Lau and Peng 1987; Li et al. 2009).

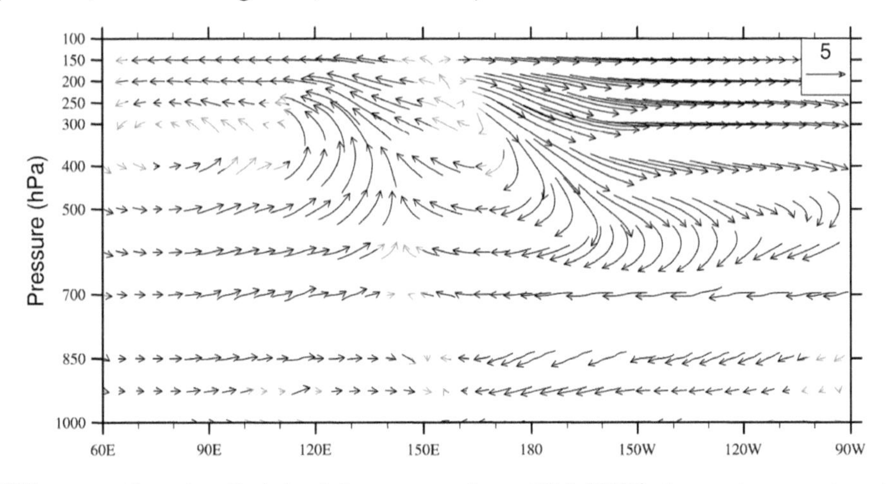

Fig. 6. Differences of zonal-vertical circulation averaged over 10° S-10° N between strong and weak EAWM winters. Results passing the significant test at the 90 % confidence level are marked by *black*. The vertical velocity is timed by 500

Based on the above analysis, the EAWM influences the MJO over the equatorial western Pacific mainly through the following processes. First, the northerly surge of the strong EAWM reaches into tropics and strengthens the convection there. Subsequently, the feedback effect of convective heating inspires strong CISK-Rossby waves and CISK-Kelvin waves, which strengthen the activities of MJO (Li 1985; Lau and Peng 1987).

4. Impact of EAWM on MJO in Its Phase 4

The northerlies associated with the strong EAWM can enhance the convection over the equatorial western Pacific and affect the activities of MJO. However, the process of EAWM is discrete pulse-like event, and the MJO activities also have an eastward propagation feature. Wheeler and Hendon (2004) divided the life cycle of the MJO into eight phases according to the location of the active convective center of MJO during its eastward propagation. Therefore, to further study the impact of EAWM on the MJO, we examined strong/weak EAWM processes occurring in the phase 4 of the MJO, in which the convection center of MJO almost locates in the Maritime Continent and the equatorial western Pacific.

A 3-day running mean was firstly applied to the daily EAWM index. The weak EAWM process is defined as the daily EAWM index exceeding 0.8 standard deviation at least for 3 consecutive days, whereas the strong EAWM process is identified if daily EAWM index is smaller than –0.8 standard deviation at least for 3 consecutive days. The first day that exceeds 0.8 standard deviation (or smaller than –0.8 standard deviation) is regarded as day 0 of the process. The strong/weak EAWM processes whose onset date is between November 10 and April 20 were used for the composite, and 19 strong and 11 weak EAWM processes were identified and shown in Table 2.

Table 2. Start and end time of strong and weak EAWM processes

Strong EAWM	Weak EAWM
1982.01.14-1982.01.18	1984.12.13-1984.12.15
1986.02.19-1986.03.04	1992.02.13-1992.02.17
1987.01.30-1987.02.03	1994.12.05-1994.12.11
1989.04.08-1989.04.10	1995.01.19-1995.01.21
1990.03.03-1990.03.07	2000.03.02-2000.03.04
1991.04.18-1991.04.21	2003.02.06-2003.02.08
1992.02.20-1992.02.24	2003.02.26-2003.03.04
1993.04.04-1993.04.12	2003.03.29-2003.04.01
1994.03.22-1994.03.28	2003.11.15-2003.11.19
1996.01.31-1996.02.09	2004.12.01-2004.12.03
1996.04.01-1996.04.03	2005.04.05-2005.04.08
1997.02.16-1997.02.21	
2000.02.24-2000.02.26	
2001.04.09-2001.04.11	
2001.11.12-2001.11.24	
2002.11.21-2002.11.24	
2004.03.04-2004.03.07	
2008.02.06-2008.02.18	
2009.11.16-2009.11.20	

The evolution of latitudinal distribution of composite anomalous meridional wind at 850 hPa averaged over 110°-130° E in strong and weak EAWM processes is shown in Fig. 7. During the strong EAWM process, East Asia north of 30°N features anomalous northerlies on day –8. On day 0, the strong northerlies of the EAWM reach into the tropics and persist for 6 days.

During the weak EAWM process, the anomalous northerlies are active over East Asia before day 0, whereas the strong anomalous southerlies dominate the110°-130° E region in the tropics on day 0 and persist for 5 days. OLR can reflect the activities of convection in the tropics and widely be used in the studies of MJO (Wheeler and Hendon 2004; Wang et al. 2012). Negative OLR anomalies in the phase 4 of MJO are located mainly in 20° S-20° N, 70°-150°E (not shown), and the large values (<−12 W m^{-2}) are concentrated mainly in 10°S-10°N, 80°-140°E. Figure 8a, b shows the daily evolution of the equatorially averaged OLR anomalies over 10° S-10° N under strong and weak EAWM processes. There are no prominent differences in the OLR between strong and weak EAWM processes before day 0. However, the OLR is prominently different over Maritime Continent and western Pacific (especially over the western Pacific) after day 0. On the one hand, the convection is suppressed after day 0 under the weak EAWM process; in contrast, it becomes more active under the strong EAWM process. These differences are more prominent over the western Pacific, which means that the EAWM has a stronger impact on the convection there. The convection reaches its maximum near day 5 in the strong EAWM process, which is consistent with the disappearance of strong EAWM on day 5 (Fig. 7a). On the other hand, the eastward propagation of OLR is more persistent and prominent under the strong EAWM process; in contrast, it is discontinued under the weak EAWM process. The zonal wind and kinetic energy of MJO at 850 hPa also show similar characteristics (not shown), which suggests that the EAWM not only influence the intensity of MJO but also impact its eastward propagation over the equatorial Pacific.

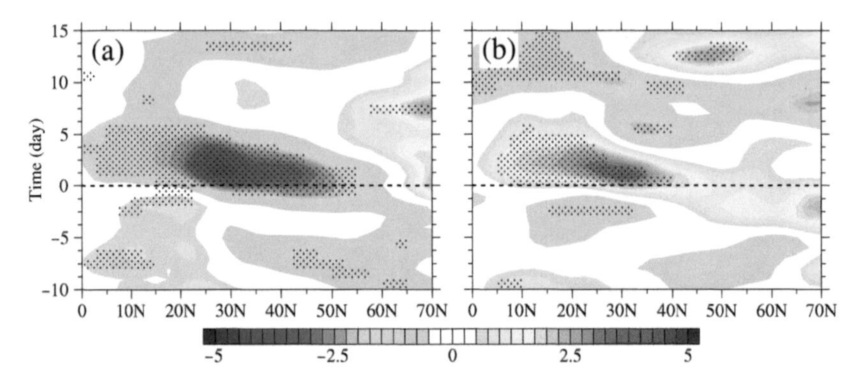

Fig. 7. Evolution of latitudinal distributions of anomalous meridional wind (m s^{-1}) at 850 hPa averaged over 110°-130° E in **a** strong and **b** weak EAWM processes. Negative (positive) in time coordinate indicates before (after) the onset of EAWM. Results passing the significant test at the 90 % confidence level are stippled

The composite results of the vertically averaged (1000-700 hPa) moisture flux convergence averaged over 10°S-10°N (Fig. 8c, d) clearly show that under the strong EAWM process, the moisture flux convergence is strengthened after day 0 over the Maritime Continent and western Pacific. In contrast, the moisture flux convergence is weakened after day 0 under the weak EAWM process there. Further exploration shows that the strengthening of moisture flux convergence over the western Pacific after day 0 under strong EAWM process (Fig. 8c) mainly due to the meridional component of moisture flux convergence (Fig. 8g), whereas the meridional component (Fig. 8h) reduces the moisture flux convergence under weak EAWM process (Fig. 8d). The zonal component of moisture flux convergence has a contribution to the moisture flux convergence around 120° E after day 0 under weak EAWM process (Fig. 8f), but the negative anomalies of meridional component of moisture flux convergence are stronger (Fig. 8h), which leads to the decrease of horizontal moisture flux convergence under weak EAWM process (Fig. 8d).

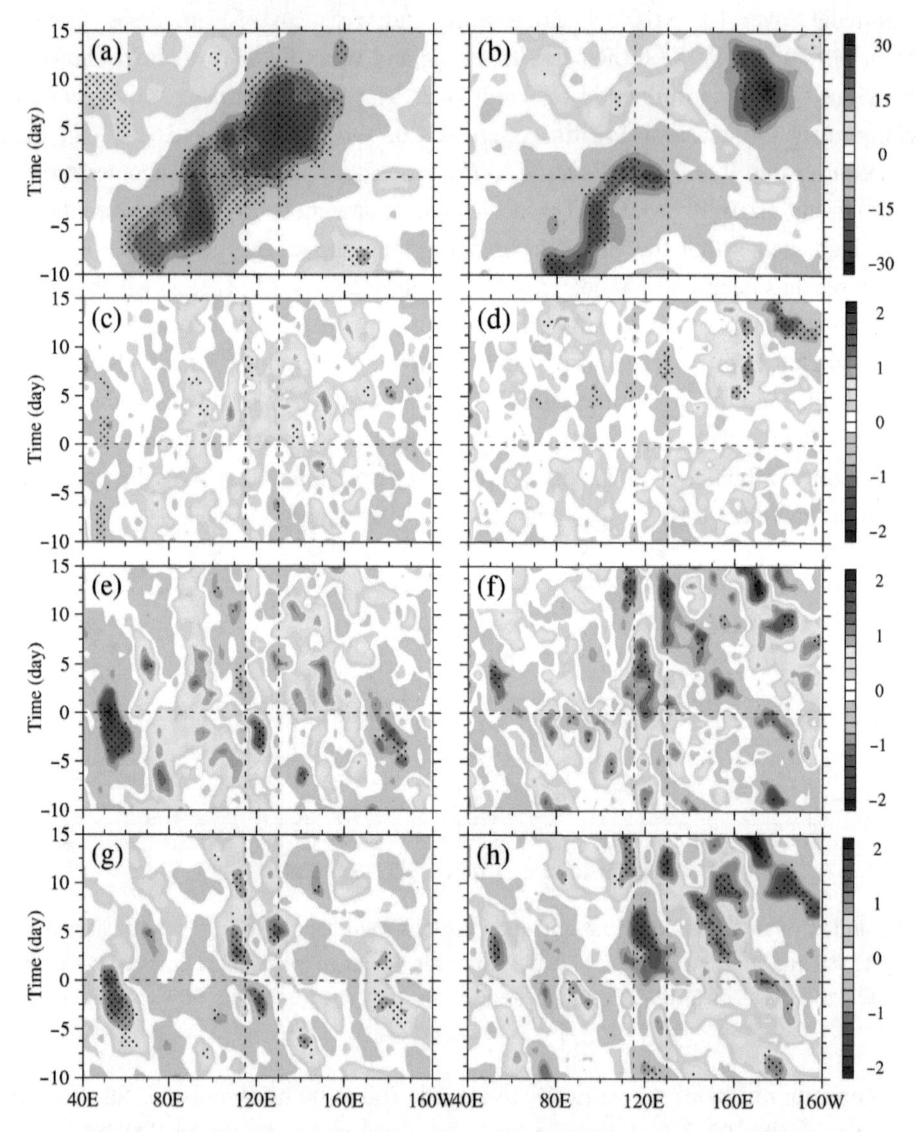

Fig. 8. a-h Evolution of longitudinal distribution of anomalous (from *top* to *bottom*) OLR (W m^{-2}), horizontal moisture flux convergence (10^{-5} g kg^{-1}m s^{-1}) averaged from 1000 to 700 hPa and its zonal and meridional components averaged over 10° S-10° N in (*left column*) strong and (*right*) weak EAWM processes. Results passing the significant test at the 90 % confidence level are stippled. The two *vertical dotted lines* represent the west and east boundaries of the area used to define the EAWM index

Convection is always accompanied by the ascending motion. The evolution of averaged (10°S-10°N, 120°-150°E) anomalous vertical velocity (Fig. 9a, b) clearly shows that the anomalous vertical velocity over the equatorial western Pacific is dominated by descending motion before day -5, which gradually converts into ascending motion after day -5 under strong EAWM process. This strong anomalous ascending motion can persist to day 10. However, under weak EAWM process (Fig. 9b), the anomalous ascending motion is weakened rapidly after day 0, although it is stronger before day 0.

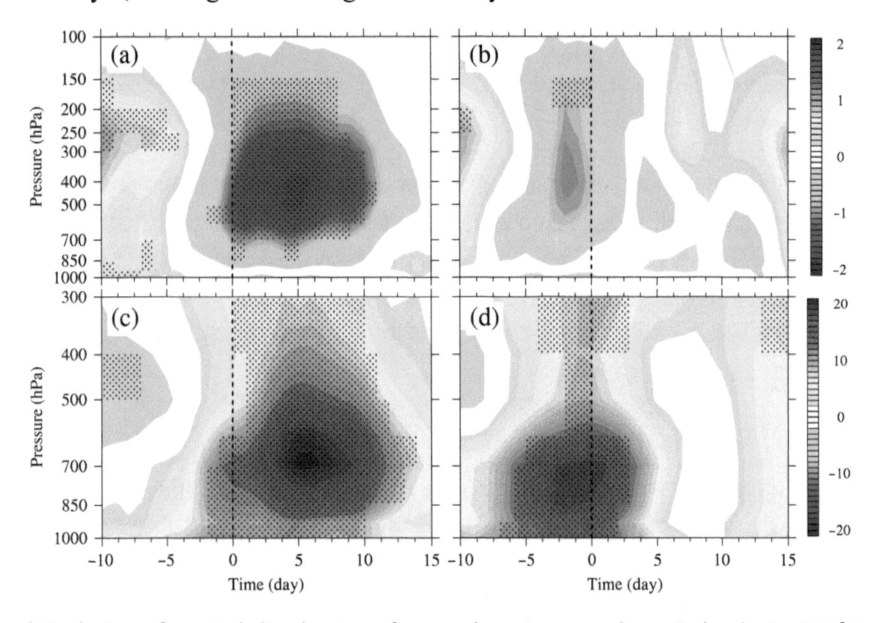

Fig. 9. a-d Evolution of vertical distribution of anomalous (*top panel*) vertical velocity (10^{-2}Pa s^{-1}) and (*bottom*) moist static energy (10^3 J kg^{-1}) averaged over 10° S-10° N, 120°-150° E in (left *column*) strong and (*right*) weak EAWM processes. Results passing the significant test at the 90 % confidence level are stippled

Moist static energy is a good indicator of MJO activities and has been widely used in the studies of MJO (Kemball- Cook and Weare 2001; Wang et al. 2012). The evolution of the averaged (10° S-10° N, 120°-150° E) anomalous moist static energy (Fig. 9c, d) clearly shows that it gradually increases after day -3 and reaches its maximum near day 5 under strong EAWM process, which is consisted with the maximum of anomalous northerly and OLR (Figs. 7a and 8a). Under weak EAWM process, however, the moist static energy gradually decreases after day 0, although it is very strong between day -5 and day 0. A strong EAWM can clearly enhance the moist static energy over the equatorial western Pacific, which also reflects the strong activities of MJO under strong EAWM process. The EAWM can also influence the stability of the atmosphere over the western Pacific. The atmospheric instability index (Fig. 10) defined by Kemball-Cook and Weare (2001) as the difference of moist static energy between 1000 and 500 hPa shows that the atmosphere is unstable from day 2 to day 10 under strong EAWM process, but it is almost stable under the weak EAWM process. The instability of atmosphere is favor to the activities of convection and also to the activities of MJO.

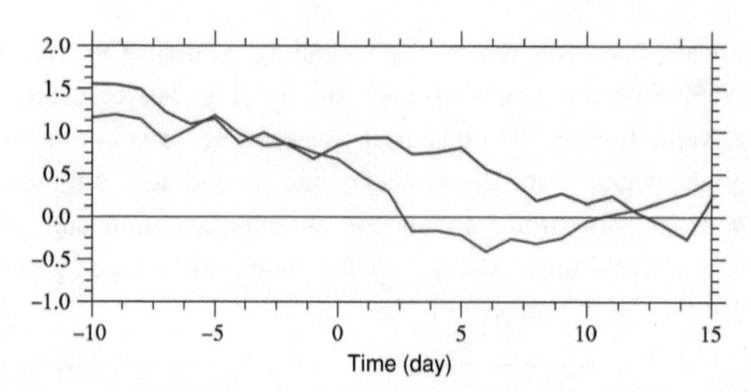

Fig. 10. The atmospheric instability index (10^3 J kg^{-1}) in strong (*red*) and weak (*blue*) EAWM processes averaged over 10° S-10° N, 120°-150° E

Above analysis shows that the physical quantities, such as MJO kinetic energy, convection, moisture flux convergence, moist static energy, and atmospheric stability, differ significantly between strong and weak EAWM processes. And these differences can lead to the anomalies of MJO intensity and propagation.

5. Summary and Discussion

The activities of MJO and their anomalies are the frontier scientific issues receiving broad attentions. The equatorial western Pacific is one of the primary activities areas of MJO. This study has investigated the relationship between the EAWM and MJO over the equatorial western Pacific using reanalysis data, and the main conclusions are as follows:

Both the EAWM and MJO over the equatorial western Pacific have significant interannual and interdecadal variabilities, and they are related significantly, especially on the interannual timescales. The processes by which the EAWM reinforces the MJO over the western Pacific are due to the northerlies associated with strong EAWM intruding into equatorial regions and leading to a strengthened convection over the equatorial western Pacific. Based on feedback effect of convective heating, the strengthened convection can generate a stronger MJO. The relationship between the MJO and EAWM is unstable and shows interannual and interdecadal variabilities. One possible explanation for these variabilities is that the MJO activities over the western Pacific have different modes, and the relationship between each mode and the EAWM differ.

The composite analysis in the phase 4 of the MJO shows that the MJO kinetic energy, convection, moisture flux convergence, vertical velocity, zonal wind at 850 hPa, moist static energy, atmospheric stability, and other physical quantities over the Maritime Continent and western Pacific all differ significantly between strong and weak EAWM processes. The generation and evolution of these differences are directly linked to the intrusion of strong northerlies associated with the strong EAWM into the tropics. The strong EAWM not only enhances the intensity of MJO over the equatorial western Pacific but can also lead to the eastward propagation of MJO more persistent over the equatorial Pacific.

The EAWM has a reinforcing effect on the MJO over the Maritime Continent and equatorial western Pacific, but this effect can also be modified by other factors, such as the influence of the southern hemisphere systems, the distribution of anomalous convective heating profiles caused by the EAWM and so on. Therefore, the EAWM is an important factor that affects the activities of MJO over the Maritime Continent and western Pacific, but it is not the only factor, which also leads to the interannual and interdecadal variabilities in the relationship between the EAWM and MJO.

Some simple numerical experiments about the EAWM exciting the activities of convection over tropical western Pacific has been done, but more numerical simulation studies for the results revealed by this paper using a high-resolution mode will be the following work.

Acknowledgments This research is sponsored by the National Basic Research Program of China (Grant No. 2015CB453200, 2013CB956200) and National Nature Science Foundation of China (Grant No., 41475070, 41575062, 41520104008).

References

Chan JCL, Li CY (2004) The East Asian winter monsoon, East Asian Monsoon, CP Chang Ed., World Scientific Publisher, Singapore 54-106.

Chang CP, Erickson JE, Lau KM (1979) Northeasterly cold surges and near-equatorial disturbances over the winter MONEX area during December 1974. Part I: synoptic aspects. Mon Weather Rev 107: 812-829.

Chen W, Yang S, Huang RH (2005) Relationship between stationary planetary wave activity and the East Asian winter monsoon. J Geophys Res 110: D14. doi: 10.1029/2004JD005669.

Duchon CE (1979) Lanczos filtering in one and two dimensions. J Appl Meteorol 18: 1016-1022.

Hendon HH, Wheeler MC, Zhang C (2007) Seasonal dependence of the MJO-ENSO relationship. J Clim 20: 531-543.

Hsu HH, Hoskins BJ, Jin FF (1990) The 1985/86 intraseasonal oscillation and the role of the extratropics. J Atmos Sci 47: 823-839.

Huang RH, Zhou LT, Chen W (2003) The progresses of recent studies on the variabilities of the East Asian monsoon and their causes. Adv Atmos Sci 20: 55-69.

Huang RH, Chen JL, Huang G (2007a) Characteristics and variations of the East Asian monsoon system and its impact on climate disasters in China. Adv Atmos Sci 24(6): 993-1023. doi: 10.1007/s00376-007-0993-x.

Huang RH, Wei K, Chen JL, Chen W (2007b) The East Asian winter monsoon anomalies in the winters of 2005 and 2006 and their relations to the quasi-stationary planetary wave activity in the Northern Hemisphere (in Chinese). Chinese J Atmos Sci 31(6): 1033-1048.

Ji LR, Sun SQ, Arpe K, Bengisson L (1997) Model study on the interannual variability of Asian winter monsoon and its influence. Adv Atmos Sci14(1): 1-22.

Kalnay E, Kanamitsu M, Kistler R, Collins W, Deaven D, Gandin L, Iredell M, Saha S, White G, Woollen J, Zhu Y, Chelliah M, Ebisuzaki W, Higgins W, Janowiak J, Mo KC, Ropelewski C, Wang J, Leetma A, Reynolds R, Jenne R, Joseph D (1996) The NCEP/NCAR 40-year reanalysis project. Bull Am Met Soc 77: 437-471.

Kemball-Cook SR, Weare BC (2001) The onset of convection in the Madden-Julian oscillation. J Clim 14: 780-793.

Lau KM, Peng L (1987) Origin of low-frequency (intraseasonal) oscillations in the tropical atmosphere. Part I: basic theory. J Atmos Sci 44: 950-972.

Li CY (1985) Actions of summer monsoon troughs (ridges) and tropical cyclone over South Asia and the moving CISK mode. Scientia China (B)28: 1197-1206.

Li CY (1989) Frequent activities of stronger aerotroughs in East Asia in wintertime and the occurrence of the El Niño event. Sci China (B) 32: 976-985.

Li CY, Mu MQ (2000) Relationship between East-Asian winter monsoon, warm pool situation and ENSO cycle (in Chinese). Chin Sci Bull 45: 1448-1455.

Li CY, Zhou YP (1994) Relationship between intraseasonal oscillation in the tropical atmosphere and ENSO (in Chinese). Chinese J Geophysics 37: 213-223.

Li RCY, Zhou W (2013a) Modulation of western north Pacific tropical cyclone activity by the ISO. Part I: genesis and intensity. J Clim 26: 2904-2918.

Li RCY, Zhou W (2013b) Modulation of western north Pacific tropical cyclone activity by the ISO. Part II: tracks and landfalls. J Clim 26: 2919-2930.

Li CY, Jia XL, Ling J, Zhou W, Zhang CD (2009) Sensitivity of MJO simulations to diabatic heating profiles. Climate Dyn 32: 167-187.

Li RCY, Zhou W, Chan JCL (2012) Asymmetric modulation of western north Pacific cyclogenesis by the Madden-Julian oscillation under ENSO conditions. J Clim 25: 5374-5385.

Li CY, Ling J, Song J, Pan J, Tian H, Chen X(2014) Research progress in China on the tropical atmospheric intraseasonal oscillation. J Meteorol Res 28: 671-692.

Liebmann B, Hartmann DL (1984) An observational study of tropical-midlatitude interaction on intraseasonal time scales during winter. J Atmos Sci 41(23): 3333-3350.

Liebmann B, Smith CA (1996) Description of a complete (interpolated) outgoing longwave radiation dataset. Bull Am Met Soc 77: 1275-1277.

Ling J, Zhang C, Bechtold P (2013) Large-scale distinctions between MJO and Non-MJO convective initiation over the tropical Indian Ocean. J Atmos Sci 70: 2696-2712.

Ling J, Li CY, Zhou W, Jia XL (2014) To begin or not to begin? A case study on the MJO initiation problem. The or Appl Climatol 115: 231-241.

Lü JM, Ju JH, Ren JZ, Gan WW (2012) The influence of the Madden- Julian oscillation activity anomalies on Yunnan's extreme drought of 2009-2010. Sci China Earth Sci 55: 98-112. doi: 10.1007/s11430-011-4348-1.

Madden RA, Julian PR (1971) Detection of a 40-50 day oscillation in the zonal wind in the tropical Pacific. J Atmos Sci 28: 702-708.

Madden RA, Julian PR (1972) Description of global scale circulation cells in the tropics with 40-50 day period. J Atmos Sci 29: 1109-1123.

Madden RA, Julian PR (1994) Observations of the 40-50-day tropical oscillation—A review. Mon Weather Rev 122: 814-837.

Murakami T (1988) Intraseasonal atmospheric teleconnection patterns during the Northern Hemisphere winter. J Clim 1(2): 117-131.

North GR, Bell T, Cahalan R, Moeng FJ (1982) Sampling errors in the estimation of empirical orthogonal function. Mon Weather Rev 110: 699-706.

Pan J, Li CY, Song J (2010) The modulation of Madden-Julian oscillation on typhoons in the northwestern Pacific Ocean (in Chinese). Chinese J Atmos Sci 34(6): 1059-1070.

Ray P, Zhang C (2010) A case study of the mechanics of extratropical influence on the initiation of the Madden-Julian oscillation. J Atmos Sci 67: 515-528. doi: 10.1175/2009JAS3059.1.

Ray P, Zhang C, Dudhia J, Chen SS (2009) A numerical case study on the initiation of the Madden-Julian oscillation. J Atmos Sci 66: 310-331. doi: 10.1175/2008JAS2701.1.
Wang L, Chen W (2010) How well do existing indices measure the strength of the East Asian winter monsoon? Adv Atmos Sci 27: 855-870.
Wang ZY, Ding YH (2006) Climate change of the cold wave frequency of China in the last 53 years and the possible reasons (in Chinese). Chinese J Atmos Sci 30(6): 1068-1076.
Wang L, Huang RH, Gu L, Chen W, Kang LH (2009) Interdecadal variations of the East Asian winter monsoon and their association with quasi-stationary planetary wave activity. J Clim 22: 4860-4872.
Wang L, Kodera K, Chen W (2012) Observed triggering of tropical convection by a cold surge: implications for MJO initiation. Quart J R Meteorol Soc 138(668): 1740-1750. doi: 10.1002/qj.1905.
Wheeler MC, Hendon HH (2004) An all-season real-time multivariate MJO index: development of an index for monitoring and prediction. Mon Weaker Rev 132(8): 1917-1932.
Yan HM, Zhou W, Yang H, Cai Y (2009) Definition of an East Asian winter monsoon index and its variation characteristics (in Chinese). Trans Atmos Sci 32(3): 367-376.
Yang H, Li CY (2003) The relation between atmospheric intraseasonal oscillation and summer severe flood and drought in the Changjiang- Huaihe river basin. Adv Atmos Sci 20(4): 540-553.
Zhang CD (2005) Madden-Julian oscillation. Rev Geophy 43: RG2003. doi: 10.1029/2004RG000158.
Zhang CD (2013) Madden-Julian oscillation: bridging weather and climate. Bull Am Meteorol Soc 94: 1849-1870. doi: 10.1175/BAMS- D-12-00026.1.
Zhang CD, Gottschalck J (2002) SST anomalies of ENSO and the Madden-Julian oscillation in the equatorial Pacific. J Clim 15: 2429-2445.
Zhou LT (2011) Impact of East Asian winter monsoon on rainfall over southeastern China and its dynamical process. Int J Climatol 31(5): 677-686. doi: 10.1002/joc.2101.
Zhou W, Chan JCL (2005) Intraseasonal oscillations and the South China Sea summer monsoon onset. Int J Climatol 25(12): 1585-1609.
Zhou W, Wang X, Zhou TJ, Li C, Chan JCL (2007) Interdecadal variability of the relationship between the East Asian winter monsoon and ENSO. Meteor Atmos Phys 98: 283-229.
Zhu CW, Nakazawa T, Li JP (2004) Modulation of tropical depression/ cyclone over the Indian-western Pacific oceans by Madden-Julian oscillation (in Chinese). Acta Meteorologica Sinca 62(1): 42-50.

Evolution of the Madden-Julian Oscillation in Two Types of El Niño

Xiong Chen[1], Jian Ling[2*], Chongyin Li[3]

[1] College of Meteorology and Oceanography, PLA University of Science and Technology, Nanjing, China
[2] LASG, Institute of Atmospheric Physics, Chinese Academy of Sciences, Beijing, China
[3] College of Meteorology and Oceanography, PLA University of Science and Technology, Nanjing, and LASG, Institute of Atmospheric Physics, Chinese Academy of Sciences, Beijing, China

Abstract Evolution characteristics of the Madden-Julian oscillation (MJO) during the eastern Pacific (EP) and central Pacific (CP) types of El Niño have been investigated. MJO activities are strengthened over the western Pacific during the predeveloping and developing phases of EP El Niño, but suppressed during the mature and decaying phases. In contrast, MJO activities do not show a clear relationship with CP El Niño before their occurrence over the western Pacific, but they increase over the central Pacific during the mature and decaying phases of CP El Niño. Lag correlation analyses further confirm that MJO activities over the western Pacific in boreal spring and early summer are closely related to EP El Niño up to 2-11 months later, but not for CP El Niño. EP El Niño tends to weaken the MJO and lead to a much shorter range of its eastward propagation. Anomalous descending motions over the Maritime Continent and western Pacific related to El Niño can suppress convection and moisture flux convergence there and weaken MJO activities over these regions during the mature phase of both types of El Niño. MJO activities over the western Pacific are much weaker in EP El Niño due to the stronger anomalous descending motions. Furthermore, the MJO propagates more continuously and farther eastward during CP El Niño because of robust moisture convergence over the central Pacific, which provides adequate moisture for the development of MJO convection.

1. Introduction

The Madden-Julian oscillation (MJO; Madden and Julian 1971, 1972), a dominant component of the tropical atmosphere on the intraseasonal time scale, has important influences on the weather and climate around the world (Zhang 2005, 2013; Li et al. 2014). El Niño was first discovered to be characterized as an anomalous rising of the sea surface temperature (SST) over the equatorial eastern Pacific caused by large-scale atmosphere-ocean

Received July 16, 2015; in final form December 4, 2015.
*E-mail: lingjian@lasg.iap.ac.cn

interaction (Bjerknes 1966; Rasmusson and Carpenter 1982). It is one of the pivotal factors for the interannual variation of the climate system. It has evident and wide influences on weather and climate globally (Zhou et al. 2010; Yuan and Yang 2012; Yuan et al. 2012). The relationship between El Niño and weather and climate has been a hot topic since the 1980s.

Lau and Chen (1986) first showed that the MJO could excite the El Niño event through energy transfer. Many following studies suggested that the enhanced MJO activities over the western Pacific in boreal spring and early summer favor the occurrence and development of El Niño events as a stochastic forcing (Hendon et al. 2007; Marshall et al. 2009). Joint effects of stronger MJO activities and anomalous oceanic Kelvin waves over the equatorial central-western Pacific could lead to the occurrence of El Niño (Li and Liao 1998; McPhaden et al. 2006). The MJO influences ENSO mainly through the interannual variability of its intensity (Li and Liao 1998; Li et al. 2003; McPhaden et al. 2006; Zavala-Garay et al. 2005). The MJO energy is a small fraction of the total energy in the atmosphere, but it might have a pivotal effect on promoting the development of El Niño (Zavala-Garay et al. 2005; Kapur et al. 2012). Some numerical model results show that the MJO signal, especially its amplitude, might closely relate to model prediction skills of El Niño (McPhaden 1999; Kapur and Zhang 2012). For example, the simulated El Niño of 1997/98 became stronger by 50% when MJO forcing was included (Kessler and Kleeman 2000). However, other studies did not find a clear simultaneous relationship between the MJO and El Niño (Hendon et al. 1999; Slingo et al.1999). Indeed, anomalous MJO activities over the western Pacific lead El Niño events by several months (Li and Liao 1998; Zhang and Gottschalck 2002; Li et al. 2003; Tang and Yu 2008). Hendon et al. (2007) indicated that the MJO over the western Pacific in boreal spring and early summer is closely related to El Niño. The occurrence of El Niño may not be a result of any individual MJO event, but rather the cumulative effect of a sequence of several MJO events (Zhang and Gottschalck 2002; Zavala-Garay et al. 2005). Furthermore, the relationship between the MJO and El Niño has a decadal variation (Zhang and Gottschalck 2002; Tang and Yu 2008). So the relationship between them may have seasonal, regional, and decadal dependence.

The variability of the MJO also can be influenced by El Niño (Li and Smith 1995; Roundy and Kravitz 2009; Gushchina and Dewitte 2012; Kapur and Zhang 2012; Chen et al. 2015). The horizontal distribution of SST may be important to certain aspects of the MJO, such as its initiation, intensity, propagation, and prediction (Pegion and Kirtman 2008a, b; Ray et al. 2009; Ray and Zhang 2010; Kim et al. 2010; Wang et al. 2015). The eastward propagation of MJO is enhanced during an El Niño developing summer, while it is weakened during the decaying summer (Lin and Li 2008). In an El Niño winter, the MJO over the Pacific (especially over the western Pacific) is weakened, discontinuous in its eastward propagation, and inclines to a "barotropic" vertical structure (Li and Smith 1995; Chen et al. 2015). Certain changes have been found in MJO characteristics during El Niño based on observations and numerical simulations. They include enhanced (weakened) low-level zonal winds over the central (western) Pacific, slower eastward propagation, and an eastward extension of MJO convection (McPhaden 1999; Tam and Lau 2005). The feedback effect of SST to the MJO in a coupled model is also very important in the relationship between the MJO and ENSO (Kapur and Zhang 2012). The interaction between the MJO and El Niño and their relationship needs further exploration.

Recently, a new type of El Niño, in which warm SST anomalies are in the equatorial central Pacific near the date line, has drawn more research attention (Ashok et al. 2007; Kao and Yu 2009, Kug et al. 2009; Ren and Jin 2011). In many recent studies, El Niño events are classified into two types: eastern Pacific (EP) El Niño and central Pacific (CP) El Niño. These two types of El Niño have different impacts, such as tropical cyclones over the Atlantic and Pacific (Kim et al. 2009; Chen and Tam 2010) and precipitation over East Asia, the United States, and Australia (Wang and Hendon 2007; Weng et al. 2007; Yuan and Yang 2012; Yuan et al. 2012). Their impacts are also different in their different phases (Yuan and Yang 2012; Yuan et al. 2012).

However, most previous studies on the relationship between the MJO and El Niño did not distinguish these two types of El Niño events. Recent studies that do consider the different relationships between the MJO and the two types of El Niño have yielded controversial results. Feng et al. (2015) showed that the MJO could promote the development of CP El Niño but not EP El Niño. The activities of the MJO are enhanced during the predeveloping, mature, and decaying phases of CP El Niño, whereas there are no significantly intensified MJO activities prior to EP El Niño. However, Gushchina and Dewitte (2012) showed that enhanced MJO activities appeared in boreal spring and summer prior to EP El Niño, but also appeared in the mature and decaying phases of CP El Niño. They suggested that the MJO might contribute to the triggering of EP El Niño, and to the persistence of positive SST anomalies of CP El Niño. Yuan et al. (2015) showed that enhanced MJO activities over the western Pacific and its eastward propagation are important to the onset for both types of El Niño. Therefore, the relationship between MJO and these two types of El Niño events needs further investigation.

This study documents the evolution features of the MJO during the life cycle of these two types of El Niño and proposes possible mechanisms. Data and methods are described in section 2. Characteristics of SST and OLR related to these two types of El Niño are briefly shown in section 3. Different relationships between the MJO and two types of El Niño are discussed in section 4, and their possible reasons are explored in section 5. A summary and discussion are given in section 6.

2. Data and Methods

Daily mean atmospheric data of wind and specific humidity are from the National Centers for Environmental Prediction (NCEP)-National Center for Atmospheric Research (NCAR) reanalysis (Kalnay et al. 1996), with a horizontal resolution of 2.5° × 2.5°. Daily mean outgoing longwave radiation (OLR) data with a 2.5° × 2.5° horizontal resolution are from the National Oceanic and Atmospheric Administration (NOAA) (Liebmann and Smith 1996). Monthly SST data are from the Hadley Centre Sea Ice and Sea Surface Temperature datasets

(HadISST1) of the Met Office (Rayner et al. 2003) with a horizontal resolution of 1° × 1°. The quality and precision of the reanalysis data improved dramatically after the 1970s due to the inclusion of the satellite observations into the assimilation (Liu et al. 2014). Therefore, to be consistent with the period of OLR, all results presented in this study are based on the period from 1 January 1975 to 31 December 2011.

An anomalous time series of a given variable was generated by removing its annual climatology and its long-time linear trend. The MJO signal was obtained using a 30-90-day Lanczos bandpass filter with 200 days of smoothing (Duchon 1979; Cai et al. 2013). Filtered signals thus obtained may contain other intraseasonal signals, but the MJO should be the dominant one. The method proposed by the Wheeler and Kiladis (1999) using the eastward spectral to reconstruct the MJO signal may overestimate the MJO strength (Ling et al. 2013), because all eastward spectral power does not belong to eastward-propagating perturbations; only that which is incoherent with its westward counterparts does (Hayashi 1982).

The El Niño index used in this study for EP and CP El Niño follows the definition of Ren and Jin (2011):

$$\begin{cases} N_{\mathrm{EP}} = N_3 - \alpha N_4 \\ N_{\mathrm{CP}} = N_4 - \alpha N_3 \end{cases} \quad \text{and} \quad \alpha = \begin{cases} 2/5 & N_3 N_4 > 0 \\ 0 & \text{otherwise} \end{cases}, \tag{1}$$

where N_{EP} and N_{CP} represent the EP and CP El Niño index, respectively; N_3 and N_4 denote the Niño3 (5°S-5°N, 150°-90°W) and Niño-4 (5°S-5°N, 160°E-150°W) index, respectively. These two indices have less simultaneous correlation, and they can clearly capture the main features of these two types of El Niño (Ren and Jin 2011). A 3-month running mean was first applied to the indices. If the amplitudes of both types of El Niño indices averaged in boreal winter (November-March) are less than a half standard deviation, a normal year was then identified. There are seven normal years (1980/81, 1981/82, 1989/90, 1992/93, 1993/94, 2001/02, and 2003/04) from 1975 to 2011. If the values of either index exceed one standard deviation for five consecutive months including at least one month in boreal winter, then a mature phase of an El Niño event is identified. According to this definition, three EP El Niño events (1976/77, 1982/83, and 1997/98) and six CP El Niño events (1977/78, 1990/91, 1994/95, 2002/03, 2004/05, and 2009/10) are identified.

The methods used in this paper include composite analyses, partial correlation and multivariate linear regression. Because N_{EP} and N_{CP} have certain simultaneous correlation, the partial correlation was used to calculate the correlation between one of them and a third party time series in order to minimize the influence of such simultaneous correlation. For example, the partial correlation between the time series of N_{EP} and MJO OLR is calculated, based on Anderson (2003, 136-144), as

$$\mathrm{rp}_{\mathrm{EP,OLR}} = \frac{r_{\mathrm{EP,OLR}} - r_{\mathrm{EP,CP}} r_{\mathrm{CP,OLR}}}{\sqrt{(1 - r_{\mathrm{EP,CP}}^2)(1 - r_{\mathrm{CP,OLR}}^2)}} \tag{2}$$

where $\mathrm{rp}_{\mathrm{EP,\,OLR}}$ represents the partial correlation coefficient between N_{EP} and MJO OLR with minimized the influence of the time series of N_{CP}, and r represents the correlation coefficient between two time series indicated in the subscript. The t test was used as the significance test.

Because of one more variable used in the calculation, the degree of freedom should be $n-3$, where n is the total number of time series.

Multivariate linear regression, also known as partial regression, was used in this study to reduce the impacts of simultaneous correlation between N_{EP} and N_{CP}. The regression coefficient represents the contribution of one independent variable to the dependent variable and minimizes the influence of other variables. For example, the multivariate regression of the indices of the two types of El Niño to MJO OLR is calculated, also based on Anderson (2003), as

$$\hat{OLR}=a_0+b_{EP}N_{EP}+b_{CP}N_{CP}, \tag{3}$$

where $\hat{OLR}$ represents the regressed time series of MJO OLR, and b_{EP} (b_{CP}) represents the directly contribution of EP (CP) El Niño to the regressed MJO OLR while partly eliminating the influence of CP (EP) El Niño. They are calculated as follows:

$$b_{EP}=\frac{S_{CP}S_{EP,OLR}-S_{EP,CP}S_{CP,OLR}}{S_{EP}S_{CP}-S_{EP,CP}^2} \text{ and } b_{CP}=\frac{S_{EP}S_{CP,OLR}-S_{CP,EP}S_{EP,OLR}}{S_{CP}S_{EP}-S_{CP,EP}^2}, \tag{4}$$

where S represents the variance of the variable or the covariance between the two variables indicated in the subscript. The F test was used for statistical significance test of b_{EP} and b_{EP}.

3. Characteristics of SST and OLR in the Two Types of El Niño

Monthly mean SST anomalies averaged over 10°S-10°N partially regressed onto the two types of El Niño indices (averaged over December-February) clearly show their differences in the distribution and intensity of SST anomalies (Fig. 1, color). Positive SST anomalies in EP El Niño are mainly to the east of date line (Fig. 1a), in contrast, they are concentrated between 160°E and 130°W with the maximum center located in central Pacific near the date line in CP El Niño (Fig. 1b). The intensity of SST anomalies is also stronger in EP El Niño. Meanwhile, the intensity and range of negative SST anomalies over the western Pacific are also more prominent, and positive SST anomalies appear over the eastern Indian Ocean after the occurrences of EP El Niño (Fig. 1a).

SST anomalies can directly affect the activities of atmosphere through its thermal effects, such as the ascending motion and precipitation (Wang and Sobel 2011). OLR is well in describing the convective activities in tropics. Monthly mean OLR anomalies averaged over 10°S-10°N and partially regressed onto the two types of El Niño indices (Fig. 1, contour) show that enhanced convective activities exist over the western Pacific before the occurrences and gradually move eastward after the onset for both types of El Niño. The OLR anomalies are stronger and wider in EP El Niño than in CP El Niño. The eastward movement of the anomalous convection center along with the development of El Niño is more prominent in EP than in CP El Niño. During the mature phase of El Niño, the anomalous convection center is around 150°W in EP El Niño and around date line in CP El Niño. The locations of anomalous convection centers are coincident with the locations of positive SST anomalies in both types of El Niño. Meanwhile, suppressed convection appears over the eastern Indian Ocean and, especially, the western Pacific after the onset of these two types of El Niño, but it emerges earlier and more strongly in EP El Niño.

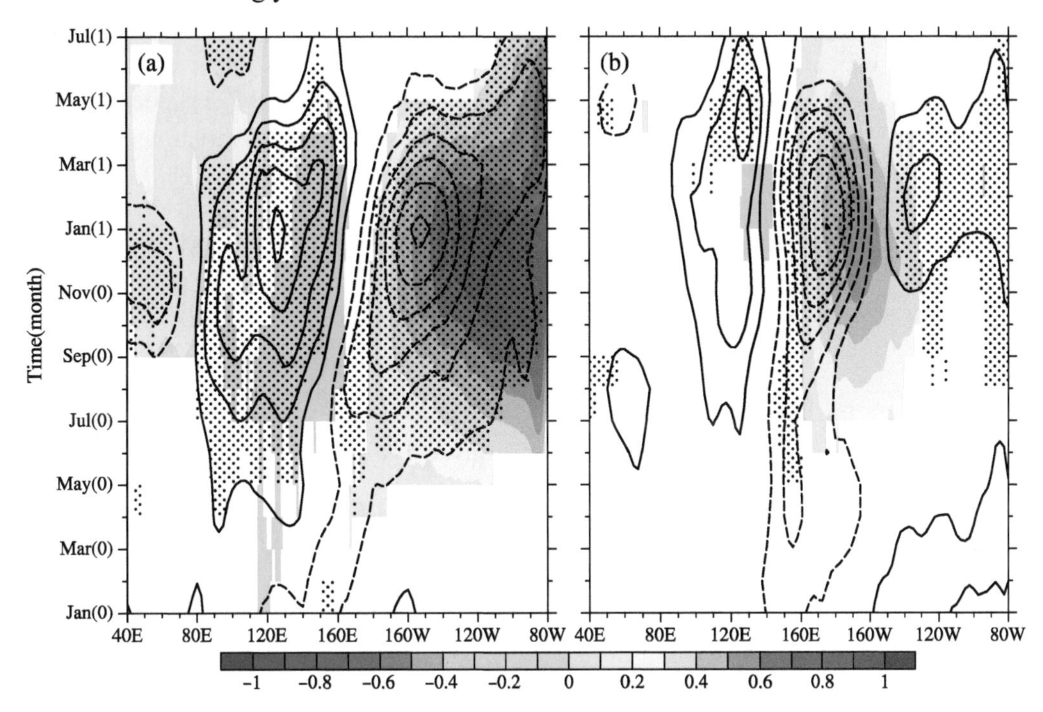

Fig. 1. Evolution of longitudinal distribution of monthly mean anomalous SST (colors, K) and OLR (contours; interval = 2W m^{-2}) averaged over 10°S-10°N partially regressed onto (a) EP and (b) CP El Niño indices averaged from December to February. Dashed contours are for negative values, and zero contours are omitted. Results passing the significant test at 90% confidence level are presented for anomalous SST and stippled for anomalous OLR; 0 (1) represents El Niño onset (next) year.

The anomalous activities of convection reflect the impacts of El Niño events on the atmosphere and further illustrate that these impacts are different for these two types of El Niño. Previous studies have shown the important role of feedback from convective latent heating in the dynamics of the MJO (Lau and Peng 1987; Li et al. 2002). The significant difference of convective activities in these two types of El Niño could lead to different activities of the MJO.

4. Relationships Between the MJO and the Two Types of El Niño

Figure 2 shows composite of longitudinal evolution of the MJO in terms of the monthly mean amplitude and anomalies of zonal wind at 850 hPa and OLR averaged over 10°S-10°N in two types of El Niño events. The monthly mean amplitude is calculated by taking the square root of the variance of MJO daily zonal wind and OLR within a 3-month moving window following Hendon et al. (2007).

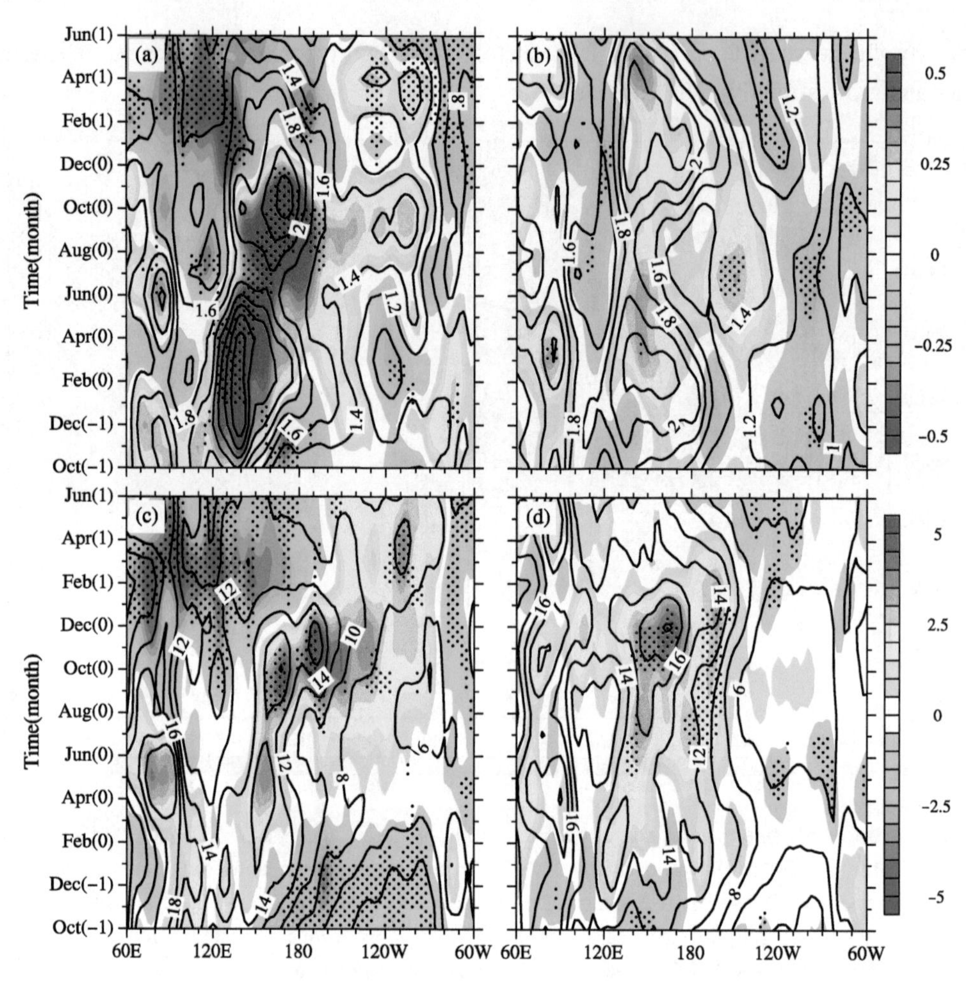

Fig. 2. Evolution of longitudinal distribution of monthly mean amplitude of MJO (a), (b) zonal wind at 850 hPa (contours, m s^{-1}) and (c), (d) OLR (contours, W m^{-2}) and their corresponding anomalies (colors) averaged over 10°S-10°N composited in the (left) EP and (right) CP El Niño. Results passing the significant test at 90% confidence level are stippled for anomalies; 0, −1, and 1 indicate the El Niño onset, previous, and subsequent year.

The amplitude of MJO zonal wind at 850 hPa is enhanced over the equatorial western Pacific during the predeveloping and developing stages of EP El Niño, but it is weakened over the eastern Indian Ocean and western Pacific during the mature phase (Fig. 2a). Such enhanced MJO amplitude over the western Pacific occurs prior to CP El Niño (Fig. 2b), but it also appears at the mature and decaying phases without eastward movement. The composite results of the MJO amplitude coincide with those of MJO kinetic energy in Yuan et al. (2015). Furthermore, the occurrence of the strengthened MJO amplitude over the western Pacific is earlier and farther west in EP than in CP El Niño. The composite results in the amplitude of MJO OLR (Figs. 2c, d) show almost the same characteristics. The enhanced MJO amplitude is mainly over the western Pacific, and the weakened MJO amplitude is near the date line in boreal winter before the occurrence of EP El Niño (Figs. 2a, c). After EP El Niño onset, the

MJO over the eastern Pacific is strengthened significantly. But it does not occur in CP El Niño, which shows a different relationship between the MJO and SST during boreal winter. Some differences in Fig. 2 do not pass the significance test, which may be due to the limited sample size of El Niño events. These differences in the MJO amplitude in the two types of El Niño suggest their different impacts on the MJO, especially over the equatorial Indian Ocean and western Pacific.

The different features of the MJO in the two types of El Niño raise a question as whether these differences are physically related to El Niño. Figure 3 shows the monthly mean amplitude of MJO zonal wind at 850 hPa and OLR averaged over 10°S-10°N partially regressed onto the two types of El Niño indices (averaged over December-February). It clearly illustrates evolutions of MJO activities under these two types of El Niño. Significantly enhanced MJO activities over the western Pacific emerge as early as the previous boreal

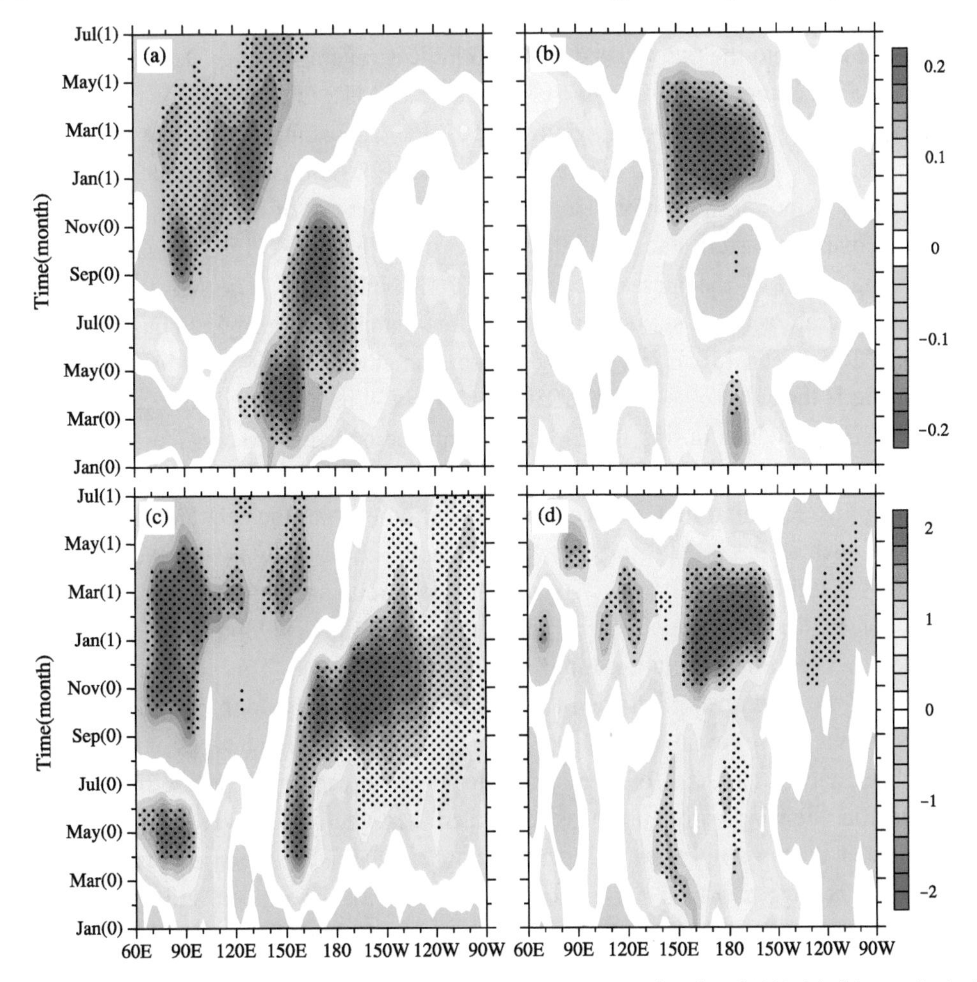

Fig. 3. Evolution of longitudinal distribution of monthly mean amplitude of MJO (a), (b) zonal wind at 850 hPa (m s^{-1}) and (c), (d) OLR (Wm^{-2}) averagedover10°S-10°N partially regressed on to (left) EP and (right) CP El Niño indices averaged from December to February. Results passing the significant test at 90% confidence level are stippled; 0 (1) represent El Niño onset (next) year.

winter before the onset of EP El Niño, and then the center of the enhanced MJO gradually moves eastward along with the development of EP El Niño. Notable suppressed MJO activities appear over the eastern Indian Ocean after the onset of EP El Niño and expand eastward gradually into the western Pacific (Fig. 3a). However, there are no significantly enhanced MJO activities before the onset of CP El Niño. The MJO amplitude over the eastern Indian Ocean and Maritime Continent does not show significant changes after the onset of CP El Niño, but it is significantly increased over the central Pacific (Fig. 3b). The enhanced MJO activities over the western Pacific may play an important role in the onset of EP El Niño, but they do not occur before the onset of CP El Niño. Furthermore, the MJO amplitude over the central Pacific is increased during the mature phase of CP El Niño, but it is decreased over the western Pacific in EP El Niño. These suggest different roles of the MJO in the initiation of the two types of El Niño and different impacts of the two types of El Niño on the MJO.

The evolutions of the monthly mean amplitude of MJO OLR averaged over 10°S-10°N partially regressed onto the two types of El Niño indices (averaged over December-February) are shown in Figs. 3c and 3d. Different characteristics of MJO OLR under the two types of El Niño are similar to the MJO zonal wind at 850 hPa. The amplitude of the MJO OLR is enhanced over the western Pacific before the onset of EP El Niño, although not as noticeably as MJO zonal wind at 850 hPa. However, the amplitude of MJO OLR is increased significantly over the eastern Pacific after the onset of EP El Niño (Fig. 3c), which may relate to the heating effect of the increased SST in EP El Niño there. The MJO OLR over the western and central Pacific is strengthened during the mature and decaying phases of CP Niño. The MJO OLR between 130°E and 170°W is also strengthened during the developing phases of CP Niño, as is the MJO zonal wind at 850 hPa. This will be discussed later.

Only the MJO activities during the El Niño events were taken into consideration in Fig. 2, whereas all MJO activities during both cold and warm ENSO phases were used in the regression in Fig. 3, which may contain information other than the relationship of the MJO and El Niño, such as the relationship between the MJO and La Niña. However, both results agree well with each other and clearly suggest the close relationships between MJO activities over the western Pacific in boreal spring and summer and the development of EP El Niño in the following winter. Such a relationship is absent for CP El Niño. This confirms the seasonal dependence of the relationship between the MJO and El Niño (Hendon et al. 2007) and also demonstrates that this seasonal dependence is not the same for the two types of El Niño.

To further analyze the sensitivity of El Niño to the preceding MJO activities over the western Pacific, their lag correlation was calculated. MJO activities over the western Pacific are represented by the amplitude of its zonal wind at 850 hPa averaged over 10°S-10°N, 120°E-180°. The lagged partial correlation between the El Niño index and MJO amplitude over the western Pacific as a function of starting month is shown in Fig. 4 for both types of El Niño. MJO activities over the western Pacific show evidently simultaneous correlation with the EP El Niño index from July through October (Fig. 4a). There is a significant lagged correlation between the MJO amplitude over the western Pacific in boreal spring and summer and the EP El Niño index up to 2-11 months later. However, a strong simultaneous correlation between the MJO amplitude over the western Pacific and CP El Niño index only occurs from January to June (Fig. 4b). The MJO over the western Pacific in boreal spring and summer does not show significant lagged correlation with the CP El Niño index in the following fall or winter.

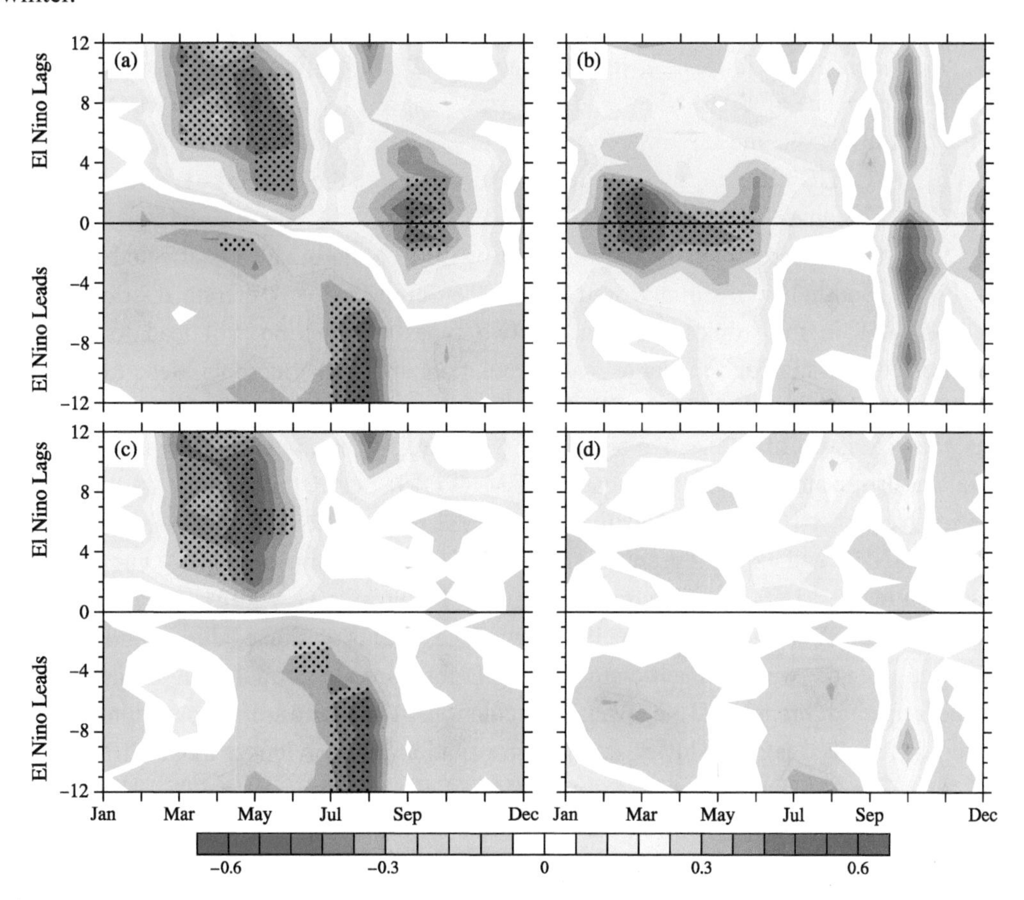

Fig. 4. Lagged partial correlation of the monthly (a) EP and (b) CP El Niño indices with respect to MJO intensity index over western Pacific as a function of start month. The abscissa indicates the start month for MJO intensity index. (c), (d) As in (a), (b), but the MJO intensity index linear regressed onto El Niño index at zero lag was first removed. Results passing the significant test at 90% confidence level are stippled.

Hendon et al. (2007) pointed out that this significant relation of MJO activities over western Pacific in boreal spring and summer with the EP El Niño may partly stem from significant self-correlation of SST. To exclude this possibility, the MJO amplitude linear regressed onto the El Niño indices at zero lag was first removed before calculating the lagged correlation. Results (Fig. 4c) clearly show that the correlation of the MJO over the western Pacific in boreal spring and early summer with the EP El Niño index up to 2-11 months later is still significant, which indicates that the strengthened MJO activities over the western Pacific in boreal spring and early summer may play an important role in the development of EP El Niño in the following autumn and winter. No such significant lag correlation appears in

CP El Niño (Fig. 4d). It further demonstrates that MJO activities over the western Pacific may have no noticeable interaction with the development of CP El Niño. The results of MJO OLR have the similar characteristics (not shown). The summer MJO has a significant negative relationship with the EP El Niño index, leading it by 2-12 months (Figs. 3a, c), indicating that EP El Niño may suppress MJO activities in its decaying summer (Lin and Li 2008).

As indicated in Figs. 3b and 3d, there are strong MJO activities over the central Pacific before the occurrence of CP El Niño. The lag correlation relationship between MJO activities over the central Pacific and CP El Niño has been evaluated. The results (not shown) suggest that such strong MJO activities over the central Pacific before the onset of CP El Niño mainly stem from significant self-correlation of SST.

The MJO not only promotes the development of EP El Niño but also is affected by the El Niño (Li and Smith 1995; Tam and Lau 2005; Chen et al. 2015). Different distributions of anomalous SST in the tropics between the two types of El Niño will lead to different distributions of moisture and the large-scale circulation, and furthermore will lead to different impacts on the characteristic of the MJO, such as its intensity and propagation.

The characteristics of convective signals of MJO propagation over the Indian Ocean and Pacific are also examined. The propagation of MJO OLR that regressed onto its time series averaged over 10°S-10°N, 120°-150°E during boreal winter (November-March) of the two types of El Niño and normal years is shown in the left column of Fig. 5. The eastward propagation of the MJO, one of its most prominent features, is very different between these two types of El Niño (Figs. 5a, c). The intensity of the MJO is weakened during both types of EP El Niño over the western Pacific around 130°E compared to normal years because of the weakened upward branch of the Walker circulation. The eastward propagation is more continuous, its speed is little slower, and the propagation range is longer in CP El Niño than EP El Niño. The MJO intensity over the central Pacific is also stronger in CP El Niño than in normal years and its propagation is more continuous (Figs. 5c, e). The propagation of the MJO was also evaluated using the zonal wind at 850 hPa (Figs. 5b, d, f) and the MJO index averaged over different longitude ranges (150°E-180°, 150°-160°E, and 160°E-180°). The results are almost the same.

5. Causes for the Different Impacts of the Two Types of El Niño on the MJO

To explain the different characteristics of MJO propagation and intensity under the two types of El Niño, composites of the anomalous large-scale background circulation during boreal winter of the two types of El Niño are investigated. The composite anomalous zonal-vertical circulation (Fig. 6) shows that anomalous convergence (divergence) occurs over the equatorial eastern (western) Pacific in the lower troposphere during EP El Niño. Anomalous westerlies are over the equatorial Pacific from 150°E to 90°W. The Indian Ocean is dominated by easterly anomalies (Fig. 6a). Anomalous ascending (descending) motions occur over the eastern (western) Pacific in EP El Niño. For CP El Niño, a similar anomalous circulation exists but is weaker, and the anomalous convergence and ascending motion are mainly over the central Pacific (Fig. 6b). The western Pacific becomes dry in both types of El Niño, and the eastern Pacific becomes moist in EP El Niño while the central Pacific becomes moist in CP El Niño. The anomalies of moisture are related to the anomalous circulation and SST (Fig. 6, blue line) of the two types of El Niño. The anomalous circulation and moisture lead to insufficient moisture flux convergence in the lower troposphere over the western Pacific, especially in EP El Niño (Fig. 6, red line), which is a reason for the weakened intensity of the MJO over the Maritime Continent and western Pacific during El Niño (Fig. 5). The anomalous large-scale convergence as well as the positive specific humidity anomaly in the

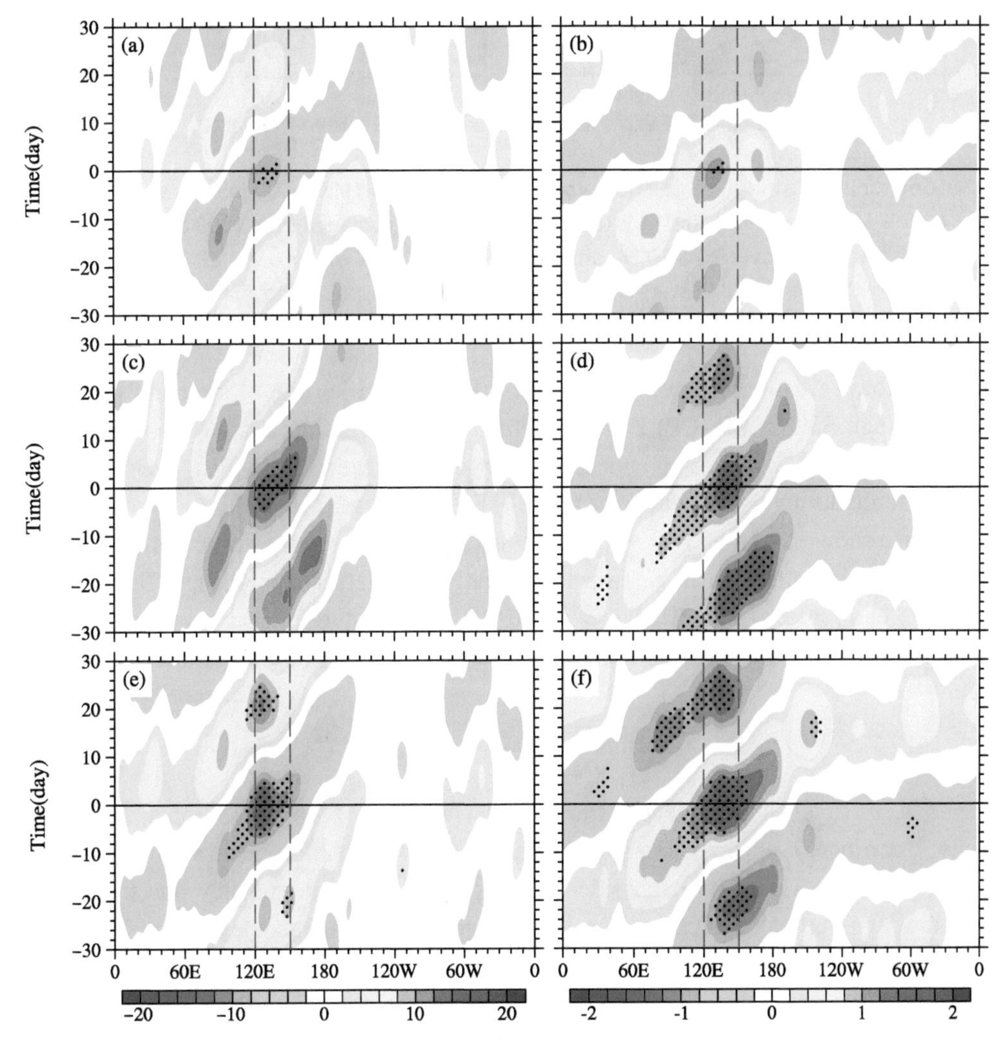

Fig. 5. The eastward propagation of (a), (c), (e) MJO OLR (W m^{-2}) and (b), (d), (f) MJO zonal wind at 850 hPa (m s^{-1}) averaged over 10°S-10°N lag regressed onto their time series averaged over 10°S-10°N, 120°-150°E during (top)-(bottom) EP and CP El Niño and normal boreal winters. Results passing the significant test at 90% confidence level are stippled. The two vertical dashed red lines represent the west and east boundaries of the area used to define the MJO time series.

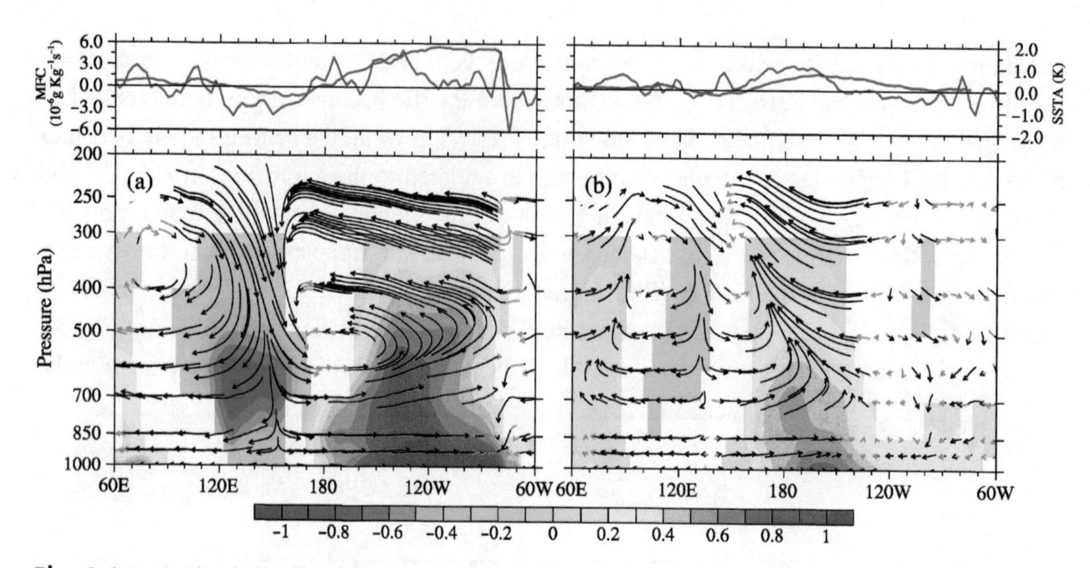

Fig. 6. Longitudinal distribution of anomalous zonal-vertical circulation (vectors), specific humidity (colors; g kg^{-1}), SST (blue line; K), and moisture flux convergence averaged from 1000 to 700 hPa (red line; 10^{-6} g kg^{-1} s^{-1}) averaged over 10°S-10°N composited in the boreal winter of (a) EP and (b) CP El Niño. Results passing the significant test at 90% confidence level are presented for specific humidity and marked black for vectors.

middle and lower troposphere over the eastern (central) Pacific (Fig. 6) are mainly due to the underneath high SST of EP (CP) El Niño, which favors initiation and development of deep convections. This is the reasons for the enhanced convection activities over the eastern (central) Pacific during boreal winter of EP (CP) El Niño (Figs. 1 and 2).

It has been shown that the low-level tropospheric moistening to the east of MJO convection center might be important for its eastward propagation (Andersen and Kuang 2012; Ling et al. 2013; Zhao et al. 2013; Li 2014; Hsu et al. 2014). The moisture budget equation at pressure level is used to diagnose the tropospheric moistening in this study as follows (Hsu and Li 2012):

$$\frac{\partial q}{\partial t} = -u\frac{\partial q}{\partial x} - v\frac{\partial q}{\partial y} - \omega\frac{\partial q}{\partial p} - \frac{Q_2}{L}, \tag{5}$$

where q, u, v, ω, Q_2, and L denote the specific humidity; zonal, meridional, and vertical velocity; atmospheric apparent moisture sink; and latent heat of condensation, respectively. The first three terms on right-hand side of Eq. (5) are moisture tendency due to the zonal, meridional, and vertical advection. They mainly reflect the moisture exchange between the MJO convection and the environment. The moisture tendency due to the atmospheric apparent moisture sink, which is the fourth term on the right-hand side of Eq. (5), is mainly dominated by the inner physical processes in the subgrid scale, including the cumulus convection, large-scale condensation, and evaporation (Chikira 2014).

The time evolution of regressed large-scale MJO moisture tendency due to advection, which is the sum of the first three terms on the right-hand side of Eq. (5), and the corresponding zonal-vertical circulation averaged over 10°S-10°N are further compared between the two types of El Niño during boreal winter in Fig. 7. Both the low-level moisture advection and zonal-vertical circulation of MJO are stronger and deeper, and their zonal scale is larger in CP El Niño. The low-level positive moisture advection is to the east of the MJO convection center as reflected by the regressed MJO OLR. The large-scale circulation and positive moisture advection have a notable eastward propagation feature in both types of El Niño. Such MJO zonal-vertical circulation and low-level moistening are totally lost in EP El Niño when active convection of the MJO moves to around 170°E at day 12 (Fig. 7g), whereas they are still very strong in CP El Niño (Fig. 7h). This may be the reason for the different intensity and propagation features of MJO in the two types of El Niño (Fig. 5).

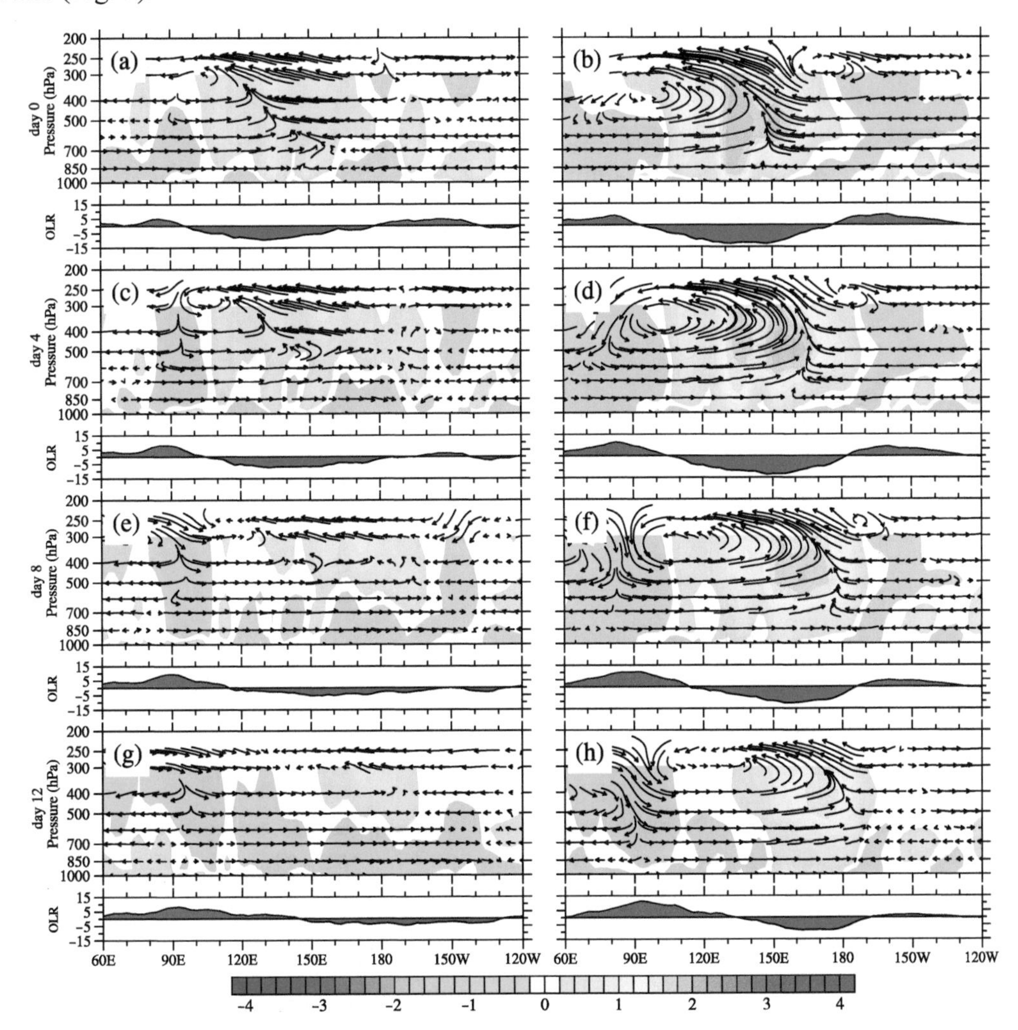

Fig. 7. The 3-day mean MJO OLR (curve; W m^{-2}), MJO zonal-vertical circulation (vectors), and MJO moisture tendency due to the advection (colors; 10^{-6} g kg^{-1} s^{-1}) averaged over 10°S–10°N lag regressed onto the reference MJO OLR time series averaged over 10°S–10°N, 120°–150°E in the boreal winter of (a), (c), (e), (g) EP and (b), (d), (f), (h) CP El Niño.

The time evolution of moisture tendency due to the atmospheric apparent moisture sink is almost the same as that due to the advection except for the opposite sign (not shown). To further explore the relative importance for each individual term on the right-hand side of Eq. (5), they are first regressed onto the reference MJO OLR time series and then averaged over 1000-500 hPa and 170°E-170°W, which is at the low-level troposphere to the east of the MJO convection center at day 8, and the results are shown in Fig. 8a. It is clearly shown that the value of local moisture change ($\partial q / \partial t$) is very small compared to the moisture advection and atmospheric apparent moisture sink in both EP and CP El Niño. In EP El Niño, meridional and vertical advection of moisture lead a negative moisture tendency, while the zonal advection and atmospheric apparent moisture sink lead a positive one. The total effect of moisture advection is to reduce the moisture to the east of MJO convection center. The positive atmospheric apparent moisture sink to the east of the MJO convection center corresponds to the negative latent heating release (evaporation) there. Even if the evaporation to the east of MJO convection contributes a positive moisture tendency, it is almost canceled out by the moisture advection there; furthermore, such negative latent heating corresponding to the positive atmospheric apparent moisture sink will further induce the downward motion and lead the atmosphere to be more stable there. Therefore, the eastward propagation signal of MJO almost disappears after day 8 in the central Pacific Ocean in EP El Niño (Fig. 7g).

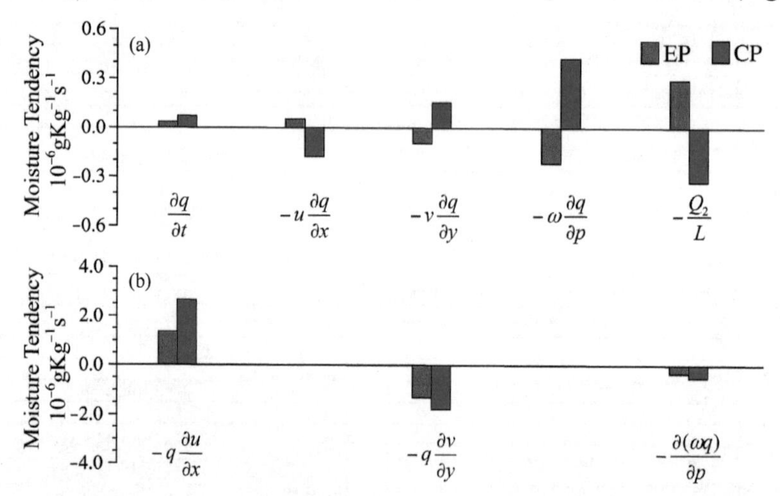

Fig. 8. Vertically averaged (1000-500 hPa) individual terms of Eq. (5) averaged over 10°S-10°N, 170°E-170°W regressed onto the reference MJO OLR time series over 10°S-10°N, 120°-150°E at day 8 in the boreal winter of EP (red) and CP El Niño (blue) (10^{-6} g kg^{-1} s^{-1}).

However, in CP El Niño, the effort for each term is almost opposite to that of EP El Niño. The zonal advection and atmospheric apparent moisture sink have negative effects to the moisture tendency, while the meridional and vertical advection lead to a positive moisture tendency. The upward motion delivers the moisture from the bottom to the upper level in the troposphere and enhances condensation, which corresponds to the negative effort of the atmospheric apparent moisture sink. Such positive latent heating released from the condensation will further induce upward motion and lead to the moisture convergence there. The importance of diabatic heating in the lower level in MJO simulation has been already indicated by Li et al. (2009). The adequate moisture and strong upward motion in the central Pacific to the east of MJO convection center will lead to the MJO signal continuing to propagate eastward after day 8 in EP El Niño (Fig. 7h).

Chikira (2014) regarded the net effect of moisture tendency due to the vertical advection and cloud processes as the "column process." The atmospheric apparent moisture sink is almost dominated by the cloud processes. It is reasonable and convenient to use the column process to diagnose the moisture tendency; however, it also conceals many aspects of the moisture change. Such a column process works as a moistening factor both in CP and EP El Niño to the east of the MJO convection center and day 8 (Fig. 8a) even though the value is much smaller in EP El Niño. In EP El Niño, the downward motion delivers the moisture out of the convection and works as a drying factor, while the evaporation works as a moistening factor. The net effect of the column process leads the atmosphere more stable. In CP El Niño, the vertical advection works as a moistening factor while the condensation works as a drying factor to the atmosphere. The net effect leads the atmosphere to become more unstable. When analyzing the low-level moistening to the east of MJO convection, we need to consider not only the tendency of moisture, but also the source of the moisture and the stability of atmosphere (the release of latent heating), which is important in maintaining the eastward propagation of MJO and may be neglected using analysis of the column process. Therefore, both the different of vertical advection of moisture, which is the dominant component of the moisture advection term, and the corresponding atmospheric apparent moisture sink are the most important factors accounting for the different eastward propagation characteristics of MJO between CP and EP El Niño in the central Pacific.

Based on the mass continuity, the vertical advection term can be further decomposed into the zonal $q(\partial u / \partial x)$ and meridional $q(\partial v / \partial y)$ moisture convergence and the vertical moisture flux convergence $(\partial \omega q / \partial p)$ following Hsu and Li (2012). The vertical averages (1000-500 hPa) of these three terms to the east of the MJO convection center at day 8 are shown in Fig. 8b. It is clearly shown that the zonal and meridional moisture convergences have somehow opposite effects with each other, and the vertical moisture flux convergence is relative small compared to them. Figure 8b clearly shows that the strong positive vertical advection of moisture $(\omega \partial q / \partial p)$ is mainly attributed to the zonal moisture convergence $(q \partial u / \partial x)$.

To identify the relative contribution of eddy-eddy and eddy-mean flow interactions following Hsu and Li (2012), the specific humidity and zonal and vertical velocity can be further decomposed into three components, the low-frequency background state (LFBS; with a period longer than 90 days), the MJO (30-90 day) component, and the high-frequency (with a period shorter than 30 days) component following Zhao et al. (2013):

$$q = \overline{q} + q' + q^*; u = \overline{u} + u' + u^*; \omega = \overline{\omega} + \omega' + \omega^* \tag{6}$$

where the overbar, prime, and asterisk denote the LFBS, MJO, and high-frequency components, respectively. Therefore, the terms $q(\partial u / \partial x)$ and $\omega(\partial q / \partial p)$ can be further divided into nine terms as follows:

$$q\frac{\partial u}{\partial x}=\bar{q}\frac{\partial \bar{u}}{\partial x}+q'\frac{\partial \bar{u}}{\partial x}+q^*\frac{\partial \bar{u}}{\partial x}+\bar{q}\frac{\partial u'}{\partial x}+q'\frac{\partial u'}{\partial x}+q^*\frac{\partial u'}{\partial x}+\bar{q}\frac{\partial u^*}{\partial x}+q'\frac{\partial u^*}{\partial x}+q^*\frac{\partial u^*}{\partial x} \quad \text{and} \tag{7}$$

$$\omega\frac{\partial q}{\partial p}=\bar{\omega}\frac{\partial \bar{q}}{\partial p}+\omega'\frac{\partial \bar{q}}{\partial p}+\omega^*\frac{\partial \bar{q}}{\partial p}+\bar{\omega}\frac{\partial q'}{\partial p}+\omega'\frac{\partial q'}{\partial p}+\omega^*\frac{\partial q'}{\partial p}+\bar{\omega}\frac{\partial q^*}{\partial p}+\omega'\frac{\partial q^*}{\partial p}+\omega^*\frac{\partial q^*}{\partial p}. \tag{8}$$

The vertically integrated (from 1000 to 500 hPa) terms from the right-hand side of Eqs. (7) and (8) averaged over 10°S-10°N, 170°E-170°W at day 8 are shown in Fig. 9. It indicates that the zonal convergence of moisture is mainly attributed to the convergence of LFBS specific humidity induced by the MJO zonal wind (Fig. 9a). This term is much stronger in CP than EP El Niño. The other terms from the right-hand side of Eq. (7) contribute little to the zonal convergence of moisture. The vertical advection of moisture is mainly attributed to the advection of LFBS specific by the MJO vertical velocity (Fig. 9b). This term is positive in CP El Niño and negative in EP El Niño. The combined efforts of LFBS specific humidity and MJO wind play an important role in both zonal moisture convergence and vertical moisture advection. They are may be the key reason for the different propagation characteristics of MJO between these two types of El Niño.

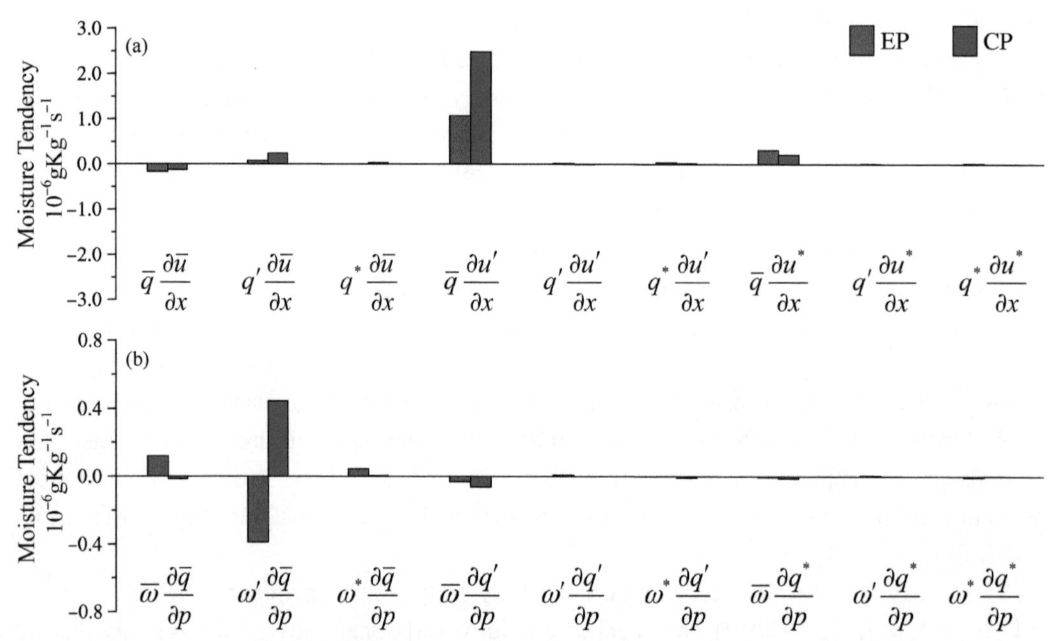

Fig. 9. Vertically averaged (1000-500 hPa) individual terms of (a) $\omega(\partial q/\partial p)$ and (b) $q(\partial u/\partial x)$ averaged over 10°S-10°N, 170°E-170°W regressed onto the reference MJO OLR time series over 10°S-10°N, 120°-150°E at day 8 in the boreal winter of EP (red) and El Niño (blue) (10^{-6} g kg^{-1} s^{-1}).

The composite results of anomalous LFBS specific humidity (Fig. 6) and the regressed MJO zonal-vertical circulation at day 8 (Figs. 7e, f) show that both anomalous LFBS specific humidity and MJO circulation differ significantly between the two types of El Niño. They are weaker in EP than in CP El Niño to the east of MJO convection center. The strong anomalous moistening over the eastern Pacific due to the warming sea surface there in EP El Niño is much too far east of the convection center and the MJO easterly is mainly to the east of 150°W. Such a distribution of specific humidity and zonal wind cannot lead to low-level tropospheric moistening near the date line and support the initiation of the convection to the east of an existing convection center of the MJO. In contrast, maximum anomalous specific humidity is located just east of the date line and the anomalous easterly of the MJO in CP El Niño prevails. The anomalous easterlies lead to robust moistening to the east of an existing convective center and set a favorable state for the generation of deep convection there. The MJO vertical velocity is prominently upward in CP El Niño in the lower troposphere to the east of its convection center at day 8. This is not so in EP El Niño. Such a distribution of the vertical velocity in CP El Niño can transport moisture from the planetary boundary layer to the free atmosphere to help the generation of deep convection. Therefore, the different distributions of LFBS moisture and MJO circulation lead to the different characteristics of MJO eastward propagation in the two types of El Niño.

These results suggest that anomalous descending motions over the eastern Indian Ocean and western Pacific can lead to the negative anomalies of moisture flux convergence there, which are unfavorable to the generation and maintenance of deep convection. All would lead to weaker MJO activities over the western Pacific during the mature phase of El Niño, especially EP El Niño. The different distributions of positive SST anomalies of the two types of El Niño lead to the different distributions of anomalous ascending motions and the associated anomalous moisture advection over the central and eastern Pacific. As a result, low-level moisture moistening is totally lost to the east of the active convection center of the MJO in EP El Niño when the MJO convection center moves to 170°E (Fig. 7g), which leads to the short eastward propagation range of the MJO in EP El Niño.

6. Summary and Discussion

El Niño events are classified into the eastern Pacific (EP) type and the central Pacific (CP) type according to the locations of their maximum positive SST anomalies. The differences in the intensities and locations of SST anomalies between these two types of El Niño lead to their different impacts on the tropical atmosphere, especially on convection. Enhanced OLR over the western Pacific appears in boreal spring and early summer before the onset of CP and EP El Niño, and the enhanced activity center moves eastward along with the development of CP and EP El Niño. During the mature stages, OLR is strengthened over the eastern (central) Pacific in EP (CP) El Niño, and weakened over the western Pacific.

Further analyses show that MJO activities over the western Pacific in boreal spring and early summer are closely related to EP El Niño 2-11 months later. In contrast, there is no such significant lagged relation between the MJO and CP El Niño. MJO activities are enhanced over the Indian Ocean and western Pacific prior to EP El Niño and reduced during its mature and decaying stages. There is no strengthened MJO over the Indian Ocean or western Pacific closely related to CP El Niño before its occurrence. But MJO activities near the date line are strengthened during the mature and decaying phases of CP El Niño, which may be due to positive SST anomalies there.

El Niño impacts the intensity and propagation of the MJO mainly through the anomalous circulation related to El Niño. Anomalous divergence and descending motions over the western Pacific during El Niño lead to the weakened convection and insufficient moisture flux convergence there, which result in reduced MJO activities through the feedback effect of convective heating. The anomalous circulation related to EP El Niño is stronger, so its impacts on the MJO are also more prominent. The increased intraseasonal (30-90 day) oscillation over the eastern Pacific during the mature phase of EP El Niño may be due to increased SST there and may not directly relate to the MJO over the Indian and Pacific Oceans. The eastward propagation of the MJO is more continuous and farther eastward during CP El Niño, because strong and deep low-level atmospheric moistening to the east of the MJO convection center can still be maintained even if the MJO passes the date line. Such low-level moistening totally disappears during EP El Niño when the MJO reaches around 170°E. The robust low-level moistening in CP El Niño is primarily due to the LFBS specific humidity advection by the MJO upward motion, which is induced by the convergence of LFBS specific humidity by the zonal gradient of MJO zonal winds. Both LFBS specific humidity and MJO circulation are stronger to the east of the MJO convection center in CP than EP El Niño, which leads to stronger low-level moistening and more continuous eastward propagation of the MJO. Furthermore, the different characteristics of latent heating release corresponding to the atmosphere's apparent moisture sink are also important in leading to different eastward propagation of MJO in the two types of El Niño.

Detailed results in this study may be influenced by the limited sample size of El Niño events used in the diagnosis, but the main evolution characteristics of the MJO in the two types of El Niño are robust. There are only three EP El Niño events after 1975. However, four more EP El Niño events (1951/52, 1963/64, 1965/66, and 1972/73) were further identified during the period from 1948 to 1975, which are only covered by the NCEP reanalysis and SST data. The composite results using all the seven events (not shown) are almost the same as those shown in this study. Regression results using a longer period (from 1948 to 2011) also show the same results. The relationship between the MJO and the two types of El Niño investigated in this study is only the linear relationship between them. Their nonlinear relation needs further investigation. A comprehensive understanding of the interaction between the MJO and the two types of El Niño may need further studies using numerical simulations.

Acknowledgments Three anonymous reviewers provided careful comments on the submitted manuscript, which helped improve this article. This research is sponsored by the National Basic Research Program of China (Grants 2015CB453200, 2013CB956200) and National Nature Science Foundation of China (Grants 41475070, 41575062, 41520104008).

References

Andersen, J. A., and Z. Kuang, 2012: Moisture static energy budget of MJO-like disturbances in the atmosphere of a zonally symmetric aquaplanet. *J. Climate*, **25**, 2782-2804, doi: 10.1175/ JCLI-D-11-00168.1.

Anderson, T. W., 2003: *An Introduction to Multivariate Statistical Analysis*. 3rd ed. John Wiley & Sons, 752 pp.

Ashok, K., S. K. Behera, S. A. Rao, H. Weng, and T. Yamagata, 2007: El Niño Modoki and its possible teleconnection. *J. Geophys. Res.*, **112**, C11007, doi: 10.1029/2006JC003798.

Bjerknes, J., 1966: A Possible response of the atmospheric Hadley circulation to equatorial anomalies of ocean temperature. *Tellus*, **18**, 820-829, doi: 10.1111/j.2153-3490.1966.tb00303.x.

Cai, Q., G. J. Zhang, and T. Zhou, 2013: Impacts of shallow convection on MJO simulation: A moist static energy and moisture budget analysis. *J. Climate*, **26**, 2417-2431, doi: 10.1175/ JCLI-D-12-00127.1.

Chen, G., and C.-Y. Tam, 2010: Different impacts of two kinds of Pacific Ocean warming on tropical cyclone frequency over the western North Pacific. *Geophys. Res. Lett.*, **37**, L01803, doi: 10.1029/2009GL041708.

Chen, X., C. Li, and Y. Tan, 2015: The influences of El Niño on MJO over the equatorial Pacific. *J. Ocean Univ. China*, **14**, 1-18, doi: 10.1007/s11802-015-2381-y.

Chikira, M., 2014: Eastward-propagation intraseasonal oscillation represented by Chikira-Sugiyama cumulus parameterization. Part II: Understanding moisture variation under weak temperature gradient balance. *J. Atmos. Sci.*, **71**, 615-639, doi: 10.1175/JAS-D-13-038.1.

Duchon, C. E., 1979: Lanczos filtering in one and two dimensions. *J. Appl. Meteor.*, **18**, 1016-1022, doi: 10.1175/ 1520-0450(1979)018<1016: LFI0AT>2.0.C0; 2.

Feng, J., P. Liu, W. Chen, and X. C. Wang, 2015: Contrasting Madden-Julian oscillation activity during various stages of EP and CP. *Atmos. Sci. Lett.*, **16**, 32-37, doi: 10.1002/asl2.516.

Gushchina, D., and B. Dewitte, 2012: Intraseasonal tropical atmospheric variability associated with the two flavors of El Niño. *Mon. Wea. Rev.*, **140**, 3669-3681, doi: 10.1175/ MWR-D-11-00267.1.

Hayashi, Y., 1982: Space-time spectral analysis and its applications to atmospheric waves. *J. Meteor. Soc. Japan*, **60**, 156-171.

Hendon, H. H., C. Zhang, and J. Glick, 1999: Interannual variation of the Madden-Julian oscillation during austral summer. *J. Climate*, **12**, 2538-2550, doi: 10.1175/1520-0442(1999)012<2538: IVOTMJ>2.0.CO; 2.

——, M. Wheeler, and C. Zhang, 2007: Seasonal dependence of the MJO-ENSO relationship. J. *Climate*, **20**, 531-543, doi: 10.1175/ JCLI4003.1.

Hsu, P.-C., and T. Li, 2012: Role of the boundary layer moisture asymmetry in causing the eastward propagation of the Madden-Julian oscillation. J. *Climate*, **25**, 4914-4931, doi: 10.1175/JCLI-D-11-00310.1.

——, ——, and H. Murakami, 2014: Moisture asymmetry and MJO eastward propagation in an aqua-planet general circulation model. J. *Climate*, **27**, 8747-8760, doi: 10.1175/ JCLI-D-14-00148.1.

Kalnay, E., and Coauthors, 1996: The NCEP/NCAR 40-Year Reanalysis Project. *Bull. Amer. Meteor. Soc.*, **77**, 437-471, doi: 10.1175/1520-0477(1996)077<0437: TNYRP>2.0.CO; 2.

Kao, H. Y., and J. Y. Yu, 2009: Contrasting eastern-Pacific and central-Pacific types of El Niño. *J. Climate*, **22**, 615-632, doi: 10.1175/2008JCLI2309.1.

Kapur, A., and C. Zhang, 2012: Multiplicative MJO forcing of ENSO. *J.* Climate, **25**, 8132-8147, doi: 10.1175/ JCLI-D-11-00609.1.

——, ——, J. Zavala-Garay, and H. H. Hendon, 2012: Role of stochastic forcing in ENSO in observations and a coupled GCM. *Climate Dyn.*, **38**, 87-107, doi: 10.1007/s00382-011-1070-9.

Kessler, W., and R. Kleeman, 2000: Rectification of the Madden- Julian oscillation into the ENSO cycle. *J. Climate*, **13**, 3560-3575, doi: 10.1175/1520-0442(2000)013<3560: ROTMJO>2.0.CO; 2.

Kim, H.-M., P. J. Webster, and J. A. Curry, 2009: Impact of shifting patterns of Pacific Ocean warming on North Atlantic tropical cyclones. *Science*, **325**, 77-80, doi: 10.1126/science.1174062.

——, D. Hoyos, J. Webster, and I.-S. Kang, 2010: Ocean-atmosphere coupling and boreal winter MJO. *Climate Dyn.*, **35**, 771-784, doi: 10.1007/s00382-009-0612-x.

Kug, J.-S., F.-F. Jin, and S.-I. An, 2009: Two types of El Niño events: Cold tongue El Niño and warm pool El Niño. *J. Climate*, **22**, 1499-1515, doi: 10.1175/2008JCLI2624.1.

Lau, K.-M., and P. H. Chan, 1986: The 40-50 day oscillation and the El Niño/Southern Oscillation: A new perspective. *Bull. Amer. Meteor. Soc.*, **67**, 533-534, doi: 10.1175/1520-0477(1986)067<0533: TDOATE> 2.0.CO; 2.

——, and L. Peng, 1987: Origin of low-frequency (intraseasonal) oscillations in the tropical atmosphere. I: Basic theory. *J. Atmos. Sci.*, **44**, 950-972, doi: 10.1175/1520-0469(1987)044<0950: OOLFOI> 2.0.CO; 2.

Li, C., and I. Smith, 1995: Numerical simulation of the tropical intraseasonal oscillation and the effect of warm SSTS. *Acta Meteor. Sin.*, **9**, 1-12.

——, and Q. Liao, 1998: The exciting mechanism of tropical intraseasonal oscillation to El Niño event. *J. Trop. Meteor.*, **14**(4), 113-121.

——, H.-R. Cho, and J.-T. Wang, 2002: CISK Kelvin wave with evaporation-wind feedback and air-sea interaction—A further study of tropical intraseasonal oscillation mechanism. *Adv. Atmos. Sci.*, **19**, 379-390, doi: 10.1007/s00376-002-0073-1.

——, Z. Long, and M. Mu, 2003: Atmospheric intraseasonal oscillation and its important effect (in Chinese). *Chin. J. Atmos. Sci.*, **27**, 518-535.

——, X. Jia, J. Ling, W. Zhou, and C. Zhang, 2009: Sensitivity of MJO simulations to diabatic heating profiles. *Climate Dyn.*, **32**, 167-187, doi: 10.1007/s00382-008-0455-x.

——, J. Ling, J. Song, J. Pang, H. Tian, and X. Chen, 2014: Research progress in China on the tropical atmospheric intraseasonal oscillation. *J. Meteor. Res.*, **28**, 671-692, doi: 10.1007/s13351-014-4015-5.

Li, T., 2014: Recent advance in understanding the dynamics of the Madden-Julian oscillation. *J. Meteor. Res.*, **28**, 1-33, doi: 10.1007/s13351-014-3087-6.

Liebmann, B., and C. A. Smith, 1996: Description of a complete (interpolated) outgoing longwave radiation dataset. *Bull. Amer. Meteor. Soc.*, **77**, 1275-1277.

Lin, A., and T. Li, 2008: Energy spectrum characteristics of boreal summer intraseasonal oscillations: climatology and variations during the ENSO developing and decaying phases. *J. Climate*, **21**, 6304-6320, doi: 10.1175/ 2008JCLI2331.1.

Ling, J., C. D. Zhang, and P. Bechtold, 2013: Large-scale distinctions between MJO and non-MJO convective initiation over the tropical Indian Ocean. *J. Atmos. Sci.*, **70**, 2696-2712, doi: 10.1175/JAS-D-13-029.1.

Liu, C., X. Ren, and X. Yang, 2014: Mean flow-storm track relationship and Rossby wave breaking in two types of El Niño. *Adv. Atmos. Sci.*, **31**, 197-210, doi: 10.1007/s00376-013-2297-7.

Madden, R. A., and P. R. Julian, 1971: Detection of a 40-50 day oscillation in the zonal wind in the tropical Pacific. *J. Atmos. Sci.*, **28**, 702-708, doi: 10.1175/1520-0469(1971)028<0702: DOADOI>2.0.CO; 2.

——, and——, 1972: Description of global-scale circulation cells in the tropics with 40-50-day period. *J. Atmos. Sci.*, **29**, 1109-1123, doi: 10.1175/1520-0469(1972)029<1109: DOGSCC>2.0.CO; 2.

Marshall, A. G., O. Alves, and H. H. Hendon, 2009: A coupled GCM analysis of MJO activity at the onset of El Niño. *J. Atmos. Sci.*, **66**, 966-983, doi: 10.1175/2008JAS2855.1.

McPhaden, M. J., 1999: Genesis and evolution of the 1997-98 El Niño. *Science*, **283**, 950-954, doi: 10.1126/

science.283.5404.950.

——, X. Zhang, H. H. Hendon, and M. C. Wheeler, 2006: Large scale dynamics and MJO forcing ENSO variability. *Geophys. Res. Lett.*, **33**, L16702, doi: 10.1029/2006GL026786.

Pegion, K., and B. P. Kirtman, 2008a: The impact of air-sea interactions on the simulation of tropical intraseasonal variability. *J. Climate*, **21**, 6616-6635, doi: 10.1175/2008JCLI2180.1.

——, and ——, 2008b: The impact of air-sea interactions on the predictability of the tropical intraseasonal oscillation. *J. Climate*, **21**, 5870-5886, doi: 10.1175/2008JCLI2209.1.

Rasmusson, E. M., and T. H. Carpenter, 1982: Variations in tropical sea surface temperature and surface wind fields associated with the Southern Oscillation/El Niño. *Mon. Wea. Rev.*, **110**, 354-384, doi: 10.1175/ 1520-0493 (1982)110<0354: VITSST>2.0.CO; 2.

Ray, P., and C. Zhang, 2010: A case study of the mechanics of extratropical influence on the initiation of the Madden-Julian oscillation. *J. Atmos. Sci.*, **67**, 515-528, doi: 10.1175/ 2009JAS3059.1.

——, ——, J. Dudhia, and S. S. Chen, 2009: A numerical case study on the initiation of the Madden-Julian Oscillation. *J. Atmos. Sci.*, **66**, 310-331, doi: 10.1175/2008JAS2701.1.

Rayner, N. A., D. E. Parker, E. B. Horton, C. K. Folland, L. V. Alexander, D. P. Rowell, E. C. Kent, and A. Kaplan, 2003: Global analyses of sea surface temperature, sea ice, and night marine air temperature since the late nineteenth century. *J. Geophys. Res.*, **108**(D14), 4407, doi: 10.1029/2002JD002670.

Ren, H.-L., and F.-F. Jin, 2011: Niño indices for two types of ENSO. *Geophys. Res. Lett.*, **38**, L04704, doi: 10.1029/ 2010GL046031.

Roundy, P. E., and J. R. Kravitz, 2009: The association of the evolution of intraseasonal oscillations to ENSO phase. *J. Climate*, **22**, 381-395, doi: 10.1175/2008JCLI2389.1.

Slingo, J. M., D. P. Rowel, and K. R. Sperber, 1999: On the predictability of the interannual behavior of the Madden-Julian oscillation and its relationship with El Niño. *Quart. J. Roy. Meteor. Soc.*, **125**, 583-609, doi: 10.1002/qj.49712555411.

Tam, C. Y., and N. C. Lau, 2005: Modulation of the Madden-Julian oscillation by ENSO: Inferences from observations and GCM simulations. *J. Meteor. Soc. Japan*, **83**, 727-743, doi: 10.2151/ jmsj.83.727.

Tang, Y., and B. Yu, 2008: MJO and its relationship to ENSO. *J. Geophys. Res.*, **113**, D14106, doi: 10.1029/ 2007JD009230.

Wang, G., and H. H. Hendon, 2007: Sensitivity of Australian rainfall to inter-El Niño variations. *J. Climate*, **20**, 42114226, doi: 10.1175/JCLI4228.1.

Wang, S., andA. H. Sobel, 2011: Responseofconvection to relative sea surface temperature: Cloud-resolving simulations in two and three dimensions. *J. Geophys. Res.*, **116**, D11119, doi: 10.1029/2010JD015347.

——, ——, F. Zhang, Q. Sun, Y. Yue, and L. Zhou, 2015: Regional simulation of the October and November MJO events observed during the CINDY/DYNAMO field campaign at gray zone resolution. *J. Climate*, **28**, 2097-2119, doi: 10.1175/ JCLI-D-14-00294.1.

Weng, H., K. Ashok, S. K. Behera, and S. A. Rao, 2007: Impacts of recent El Niño Modoki on dry/wet conditions in the Pacific Rim during boreal summer. *Climate Dyn.*, **29**, 113-129, doi: 10.1007/s00382-007-0234-0.

Wheeler, M., and G. N. Kiladis, 1999: Convectively coupled equatorial waves: Analysis of clouds and temperature in the wavenumber-frequency domain. *J. Atmos. Sci.*, **56**, 374-399, doi: 10.1175/1520-0469 (1999)056<0374: CCEWA0>2.0.C0; 2.

Yuan, Y., and S. Yang, 2012: Impacts of different types of El Niño on the East Asian Climate: Focus on ENSO cycles. *J. Climate*, **25**, 7702-7722, doi: 10.1175/JCLI-D-11-00576.1.

——, ——, and Z. Zhang, 2012: Different evolutions of the Philippine Sea anticyclone between eastern and central Pacific El Niño: Possible effect of Indian Ocean SST. *J. Climate*, **25**, 7867-7883, doi: 10.1175/

JCLI-D- 12-00004.1.

——, C. Y. Li, and J. Ling, 2015: Different MJO activities between EP El Niño and CP El Niño (in Chinese). *Sci. Sin. Terr.*, **45**, 318-334.

Zavala-Garay, J. Z., C. D. Zhang, A. M. Moore, and R. Kleeman, 2005: The linear response of ENSO to the Madden-Julian oscillation. *J. Climate*, **18**, 2441-2459, doi: 10.1175/JCLI3408.1.

Zhang, C., 2005: Madden-Julian oscillation. *Rev. Geophys.*, **43**, RG2003, 10.1029/2004RG000158.

——, 2013: Madden-Julian oscillation: Bridging weather and climate. *Bull. Amer. Meteor. Soc.*, **94**, 1849-1870, doi: 10.1175/ BAMS-D-12-00026.1.

——, and J. Gottschalck, 2002: SST anomalies of ENSO and the Madden-Julian oscillation in the equatorial Pacific. *J. Climate*, **15**, 2429-2445, doi: 10.1175/1520-0442(2002)015<2429: SAOEAT> 2.0.CO; 2.

Zhao, C. B., T. Li, and T. J. Zhou, 2013: Precursor signals and processes associated with MJO initiation over the tropical Indian Ocean. *J. Climate*, **26**, 291-307, doi: 10.1175/ JCLI-D-12-00113.1.

Zhou, L.-T., C.-Y. Tam, W. Zhou, and J. C. L. Chan, 2010: Influence of South China Sea SST and the ENSO on winter rainfall over South China. *Adv. Atmos. Sci.*, **27**, 832-844, doi: 10.1007/ s00376-009-9102-7.

Challenges and Opportunities in MJO Studies

Jian Ling[1*], Chongyin Li[1], Tim Li[2], Xiaolong Jia[3], Boualem Khouider[4], Eric Maloney[5], Frederic Vitart[6], Ziniu Xiao[1], Chidong Zhang[7]

[1] LASG, Institute of Atmospheric Physics, Chinese Academy of Sciences, Beijing, China
[2] University of Hawaii at Mānoa, Honolulu, Hawaii, and Nanjing University of Information Science & Technology, Nanjing, China
[3] National Climate Center, China Meteorological Administration, Beijing, China
[4] University of Victoria, Victoria, British Columbia, Canada
[5] Colorado State University, Fort Collins, Colorado
[6] European Centre for Medium-Range Weather Forecasts, Reading, United Kingdom
[7] NOAA Pacific Marine Environmental Laboratory, Seattle, Washington and Rosenstiel School of Marine and Atmospheric Science, University of Miami, Miami, Florida

Abstract The Madden-Julian oscillation (MJO) plays an essential role in connecting weather and climate. During the past decade, great progress has been made in observations, modeling, and theory of the MJO. There is a pressing need to synthesize our current understanding of the MJO, its impact on global weather and climate, and its prediction capabilities to gauge the progress in its study to be made in the coming decade. The International Workshop on the Madden-Julian Oscillation was held to meet this need.

The workshop included 128 participants from 39 institutes and universities from seven countries. There were 37 oral and 29 poster presentations. The workshop covered various scientific issues related to the MJO and included in-depth discussions. Recent progress in the MJO study was presented; gaps in our knowledge of understanding its global impact, dynamics, and predictability were identified; and future research targets were proposed. Highlights from the workshop are briefly summarized below. Presentation slides are posted at the workshop website (www.lasg.ac.cn/mjo-workshop/).

1. Main conclusions. Interaction with other weather-climate phenomena

Progress. The MJO is connected to many weather-climate systems globally. Its structure is crucial to its influence on the onset of the East Asian monsoon and tropical cyclone genesis in the western North Pacific and other basins. The MJO influences phenomena in remote areas, such as extreme precipitation in China and the western Himalayas, and the North Atlantic Oscillation (NAO) through its modulation on the western Pacific subtropical high, monsoon circulations, Rossby wave trains, subtropical jets, upper-level troughs, potential vorticity

In final form November 4, 2016.
*Corresponding author: E-mail: Jian Ling, lingjian@lasg.iap.ac.cn

anomalies, and moisture and temperature advection. Its dipole heating structure produces a stronger extratropical response than monopole heating. Most MJO effects on weather systems also depend on the El Niño-Southern Oscillation (ENSO) cycle and other factors. Even when MJO teleconnections are well reproduced in state-of-the-art numerical models, its impact on the Euro-Atlantic sector is still too weak.

ENSO modulation of the MJO is complex because of its diversities and interaction with the seasonal cycle. Interaction between El Niño and state-dependent MJO enhances El Niño's amplitude and skewness, but limits El Niño predictability.

One of the most exciting discoveries in recent MJO studies is that MJO activities can be modulated by the quasi-biennial oscillation (QBO). MJO events with stronger amplitude, slower and more persistent eastward propagation, and longer periods tend to occur in the easterly phase of the QBO in boreal winter in both observations and global models. MJO prediction skill is usually higher during easterly phases of the QBO in boreal winter, especially when MJO convection is located over the Maritime Continent (MC).

Research recommendations

- The statistical significance of MJO effects on global weather-climate phenomena, especially for the extreme events, needs to be rigorously tested.
- Physical mechanisms for the MJO effects on NAO need to be identified.
- The degree to which the MJO impacts on the Euro-Atlantic sector are reproduced by numerical models needs to be explored.
- The potential impacts of El Niño on the predictability of the MJO should be further investigated.
- The possible physical connection between QBO and the barrier effect of the MC on MJO propagation should be explored.

2. Initiation and propagation

Progress. MJO initiation is still an unsolved problem. It may involve a number of processes. They include moisture advection by anomalous equatorial easterlies, anomalously high sea surface temperatures (SSTs) over the thermocline dome of the southern Indian Ocean, and anomalous ascending motion induced by anomalously warm advection, all associated with a preceding suppressed phase of MJO. Convergence of Rossby wave activity flux originating from midlatitudes, especially the Southern Hemisphere, and internal stochasticity of organized convection may also play some roles in MJO initiation.

Eastward propagation of simulated MJOs depends on horizontal transport of lower-tropospheric mean moist static energy by its circulation, which is mediated by the seasonal mean low-level tropospheric moisture pattern over the MC and western equatorial Indian Ocean.

The MC drew special attention at the workshop because of its barrier effect on eastward propagation of the MJO and the coming international project Years of the Maritime Continent (YMC) that will take place in 2017-19. A YMC pilot study (9 November to 25 December

2015) revealed a shift of maximum rainfall from land to the ocean during the active phase of the MJO. Propagation patterns of the MJO over the MC can be affected by the ENSO cycle, QBO, and the tropical Pacific-Indian Ocean combined mode. Furthermore, the percent of MJO events that do not cross the MC is higher in models than in reanalysis.

Research recommendations.

- The barrier effect of the MC on MJO propagation should be more extensively documented in the observations and in model simulations.
- Physical mechanisms for the MC barrier effect and its exaggeration in numerical models and the dynamics that determine MJO behavior over the MC need to be further explored.
- The capability of general circulation models with stochastic parameterization of convection to simulate MJO convection may provide an important research opportunity.
- Special attention should be paid to initiation of the primary MJO events where stochasticity may play a central role.

3. Theories

Progress. This was a rare opportunity for four distinct MJO theories to be presented, compared, discussed, and evaluated by their leading proponents at the workshop. Moisture is explicitly included and the MJO is considered as a large-scale mode in the first three theories listed below.

New "general" framework frictional convergence theory. This theory is an expansion of traditional frictional convergence theory in that it includes prognostic moisture and a simplified Betts-Miller cumulus parameterization scheme. The instability (growth), scale selection (planetary, slow eastward propagation), and vertical tilt of the MJO come from a three-way interaction among Rossby-Kelvin wave coupling, boundary layer frictional convergence, and moisture. The frictional convergence feedback plays an essential role in coupling Kelvin and Rossby waves with convective heating and selecting a preferred eastward propagation. The moisture feedback can enhance the Rossby wave component, thereby substantially slowing down eastward propagation.

"Moisture mode" MJO theory. This theory assumes that the free troposphere is regulated by weak tropical temperature gradients and that a strong coupling between MJO convection and free tropospheric water vapor regulates the dynamics of the disturbance. In this theory, prognostic moisture is advected by the steady-state Gill model response to convection heating. The MJO instability arises from cloud-radiation interaction and its impacts on moisture. The MJO propagates eastward because of the moisture advection induced by the wind anomalies. The most recent iteration of this theory uniquely produces an MJO that is dispersive with a westward group velocity.

Skeleton MJO model theory. The MJO is presented as a neutrally unstable mode in a planetary-scale dynamical system forced by synoptic-scale convective activity. The model presented at the meeting is extended with inclusion of stochasticity of synoptic and convective activities to reproduce the intermittency, growth and decay, seasonal variation, and vertical tilt of the MJO through the inclusion of a passive second baroclinic mode which responds to congestus and stratiform heating. The important features of this theory include a slow eastward propagation; a quadrupole vortex structure in the horizontal; a vertical tilt; and the realistic interaction between the planetary dynamics, the planetary moisture field, and the planetary envelope of synoptic/convective activity, leading to intermittent statistical behavior of MJO events.

Gravity-wave theory. In contrast to the above theories that consider the MJO as a large-scale mode, this theory based on gravity wave dynamics treats the MJO as a large-scale envelope of high-frequency, small-scale gravity waves; their zonal asymmetry determines the eastward propagation of the MJO, and a scaling of convective strength and gravity wave speed selects its zonal scale.

Discussions covered different opinions on the roles of the second baroclinic mode, the MJO's eastward propagation mechanism, the most salient structural elements needed to explain its dynamics, and the representation of the MJO dispersion relationship. The basic requirements that must be met by a successful MJO theory were also debated.

Research recommendations.

- A review article summarizing and comparing the existing MJO theories would facilitate their general appreciation and further advancement of theoretical understanding of the MJO.
- Assumptions and parameters in MJO theories should be validated against observations in different seasonal mean backgrounds.
- Thought should be given to how to test all the theories against the observations and model outputs using the same diagnostic process.
- The observed irregularity of the MJO should be explained by successful MJO theories.

4. Modeling and prediction

Progress. What makes a model capable of simulating a realistic MJO has long been puzzling. It was reported that most global models capable of reproducing the MJO should exhibit realistic sensitivity of their parameterized convection to environmental moisture; a realistic convective adjustment time scale in response to departures from the moisture "quasi-equilibrium" state; and realistic vertical profiles in temperature, humidity, and diabatic heating. Different models disagree on the role of ocean coupling, surface flux feedbacks, and large-scale radiative-convective feedbacks to the MJO. While tuning model parameterizations can lead to improved MJO simulations, it is often at the expense of a deterioration elsewhere, including the mean state.

The MJO is a major source of predictability on subseasonal time scales (3-6 weeks). The predictability of precipitation over the United States can be enhanced by the MJO with its error growth slower than that on other scales. However, the current forecast skill of the MJO is good only around 3 weeks, in contrast to its intrinsic predictability of about 4-5 weeks. Several methods were reported to be useful for improving MJO prediction skill by several days. Coupling to an ocean model did not improve the MJO forecast skill significantly in the Beijing Climate Center's model but did improve considerably in the NCEP forecast systems if accurate SST is produced. Stochastic perturbation ensembles such as SKEBS (Stochastic Kinetic Backscatter Scheme) and SPPT (Stochastically Perturbed Parameterization Tendencies) can improve the model mean state and variability, especially over the MC, and are valuable for exploring predictability and model predictive skill and uncertainty. To avoid ambiguity in prediction skill-translation from the anomaly correlation coefficient and root-mean-square error, several new methods of evaluating MJO prediction skill are proposed.

Research recommendations

- The possible sensitivity of MJO diagnostics, for example, horizontal and vertical advection of moisture static energy, to analysis domain size needs to be explored.
- The importance of radiation and surface flux feedbacks to the MJO should be quantified through cross-model comparisons.
- The possible dependence of the barrier effect of the MC on MJO propagation on model resolution should be explored.
- MJO forecast skill should be measured using new methods presented at the workshop.
- Possible relationships between forecast spread and skill among different models participating in the Subseasonal-to-Seasonal (S2S) Prediction Project should be identified.

5. New data

Progress. In situ observations play an invaluable role in validating global reanalysis and satellite data and in providing physical insights uncontaminated by deficiencies in numerical models and remote sensing retrievals. A large volume of in situ observations (atmospheric and oceanic profiles) have been collected from the South China Sea using GPS soundings, conductivity-temperature-depth (CDT) casts, and acoustic Doppler current profiler (ADCP) from ship cruises, automated weather stations, moored buoys, an air-sea boundary flux tower, a wind profiler, and other ground-based remote sensing instruments. Against these mooring and tower data, most biases in gridded latent heat flux products, which are relatively large in coastal regions, mainly come from errors in near-surface specific humidity and wind speed. Networks of 170 surface stations, 34 weather radar sites, and 22 radiosonde sites have been established in Indonesia, augmented by measurements from ship cruises. These networks and the special field observations will constitute the main body of the YMC data archive.

The model database of the World Weather Research Programme/World Climate Research Programme S2S Prediction Project, already opened to the research community (www.s2sprediction.net), provides a rare opportunity to quantify prediction skill of the MJO up to 2 months ahead as well as its global impacts.

Acknowledgments This workshop was financially supported by State Key Laboratory of Numerical Modelling for Atmospheric Sciences and Geophysical Fluid Dynamics, Chengdu University of Information Technology, Nanjing University of Information Science Technology, State Key Laboratory of Tropical Oceanography, the Major International (Regional) Joint Research Project of National Science Foundation of China through Grant 41520104008, the National Nature Science Foundation of China through Grant 41575062 and 41575090, the National Basic Research Program of China through Grant 2015CB453200, and the Key Research Program of Frontier Sciences at the Chinese Academy of Sciences through Grant QYZDB-SSW-DQC017. Logistic support provided by the Chinese Meteorological Administration's Institute of Plateau Meteorology is gratefully acknowledged.

2

The ENSO, IOD and PIOAM (Pacific–Indian Ocean Associated Mode)

Frequent Activities of Stronger Aero-Troughs in East Asia in Wintertime and the Occurrence of the El Niño Event

Li Chongyin (李崇银)

Institute of Atmospheric Physics, Academia Sinica, Beijing

Abstract In this paper, some data analyses and theoretical discussion show that there are frequent activities of stronger aerotroughs (cold surges) in East Asia during the wintertime prior to the occurrence of El Niño, and the disturbance energies acompanied with the stronger cold surges in East Asia are frequently dispersed southeastward from Siberia to the central western Pacific. The trade winds will be weakened and the convections will be enhanced over the equatorial central western Pacific by the strong cold surges. Then the El Niño event will occur. A possible important process or mechanism to cause the El Niño event is suggested.

Keywords aerotroughs in East Asia, cold surges, El Niño event.

Ⅰ. Introduction

El Niño is an anomalous phenomenon occurring in the equatorial eastern Pacific. At present, El Niño is expressed by using the mean sea surface temperature (SST) in (0°—10°S, 180°—90°W) region. An El Niño event is confirmed when the positive anomalies of SST appear continuously in this region, and this year is called an El Niño year. When there are continuous negative anomalies of SST in the above region, that year is called an inverse El Niño year. In general, the positive anomalies of SST start during March—April in an El Niño year, and they can continue for one year or more. The maximum positive anomaly which is more than 1℃ usually occurs in November—December.

El Niño has held the public attention, since it not only influences the circulation and weather in the tropics directly, but also the general circulation and the weather/climate changes globally through the teleconnection. This has been confirmed by lots of studies[1-4]. But the cause of the El Niño event is still an important question for study. Since the atmosphere and ocean are the coupling geophysical fluid and the wind stress is a fundamental effect of the atmosphere to the ocean, in some studies the anomalies of the trade wind over the equatorial Pacific have been regarded as an important mechanism to cause El Niño. Analyzing the low-level winds data obtained by the satellite in 1975—1978, Sadler found that trade wind

Received June 15, 1987; revised March 1, 1988.

over the eastern Pacific exhibits obvious abnormality in the El Niño year. The trade winds were obviously on the weak side in 1976 with an El Niño event, but in 1975 (the inverse El Niño year), the trade winds were 1—4 m/s stronger than those in 1976[5]. The sea surface height in the equatorial eastern Pacific is obviously raised in an El Niño year, and the occurrence of this raising is earlier than the positive anomaly of SST[6]. No doubt, this anomaly of the sea surface height is concerned with the variation of the trade wind. Therefore, it is suggested that a certain reason leads to the anomalies of the trade wind over the equatorial Pacific and then causes the El Niño event.

The results of a numerical simulation experiment completed by O'Brien et al. are convincing[1]. They introduced the winds over the equatorial Pacific period of 1961—1978 into a numerical model. The raising of the sea surface height as observed in El Niño years was obtained with computation. The simulation also showed that the sea surface height is obviously descendant in the equatorial western Pacific in an El Niño year and this descent began in winter or the next spring. Therefore, the precedence sign of El Niño is most probably in the central western Pacific. It is necessary to study the wind over the central western Pacific prior to the El Niño event.

In another paper on the interactions between the atmospheric circulation over East Asia and El Niño, we have indicated that the anomalies of the atmospheric circulation over East Asia that happen during the wintertime prior to El Niño event and the frequent activities of the stronger cold surges in East Asia are the fundamental features[8]. Analyzing the 1982—1983 El Niño event, Erickson also indicated that there are anomalous frequent aerotroughs over East Asia in winter of 1981[9]. Therefore, to study the anomalies of the atmospheric circulation over East Asia prior to an El Niño event and investigate the relationship between the cold surges and the anomalies of atmospheric circulation over the equatorial Pacific and El Niño are probably an important way to find the mechanism of the El Niño event.

Ⅱ. Activities of Cold Surges in East Asia Prior to El Niño

Through analyzing meteorological data over the period of 1950—1979, we have found that frequent and stronger cold surges are an important feature of the atmospheric circulation anomalies in East Asia prior to El Niño. In order to further confirm the relationship between the occurrence of El Niño and frequent cold surges in East Asia, the statistical analyses of the temperatures in Beijing, Shanghai and Qingdao for 75 years (1910—1984) are completed. There are 19 El Niño years in this span of 75 years. The temperature anomalies in the wintertime (October—April) prior to the 7 strongest El Niño events and the mean anomaly for all the 19 El Niño events are listed in Table 1. The negative temperature anomalies in Table 1 show that there are frequent and stronger cold surges in East Asia prior to the occurrence of an El Niño event.

Table 1. Anomalies of Surface Temperatures (℃) During October—April Prior to the Occurrence of El Niño at Beijing, Shanghai and Qingdao

	1917—1918	1929—1930	1950—1951	1956—1957	1968—1969	1971—1972	1981—1982	Mean for 19 El Niño Events
Beijing	0.0	−1.0	−3.6	−14.0	−7.7	0.4	−2.4[a)]	−1.58
Shanghai	−8.2	− 5.3	−3.0	−8.4	0.5	−3.6	−5.6	−2.41
Qingdao	−8.1	−1.2	− 3.6	−13.3	−6.6	−2.8	−2.1	−2.33

a) Average of Beijing and Taiyuan.

Table 2. Anomalies of Surface Temperatures (℃) During October—April Prior to the Occurrence of Inverse El Niño at Beiiing, Shanghai and Qingdao

	1915—1916	1936—1937	1948—1949	1953—1954	1974—1975	Mean for 14 Inverse El Niño Events
Beijing	2.8	1.3	5.0	4.7	1.7	1.15
Shanghai	5.9	1.5	5.3	6.3	3.0	2.83
Qingdao	3.9	5.7	4.8	3.3	1.7	1.92

In order to stick out the relationship between El Niño and the cold surges in East Asia, the temperature anomalies prior to 5 strongest inverse El Niño phenomena and the mean anomaly for all 14 inverse El Niño phenomena at Beijing, Shanghai and Qingdao in the wintertime are given in Table 2. It is quite different from that prior to El Niño. The temperatures in Beijing, Shanghai and Qingdao during the wintertime prior to the inverse El Niño are higher than normal. These show that the activities of the cold surges in East Asia are weaker during the wintertime prior to the inverse El Niño. The statistical test to the temperature anomalies shows that there should be different temperature anomalies prior to El Niño and prior to the inverse El Niño for the notability 0.01.

Variations of the monthly temperature anomlies averaged for 19 El Niño events (solid line) and 14 inverse El Niño events (dashed line) at Shanghai and Qingdao are given in Fig. 1. It is clear that there are negative temperature anomalies before April for El Niño and positive temperature anomalies before April for the inverse El Niño.

Therefore, we can believe that the occurrence of El Niño is closely related to the frequent cold surges in East Asia.

The rates (R) between the meridional circulation index and zonal circulation index at 500 hPa in midlatitude over Eurasia (0—170°E) in winter (December—February) for a span of 31 years (1950—1980) are investigated. The results show that $R = 0.52$ on the average for 7 El Niño events and that it is greater than normal (0.49) and that (0.45) averaged for 7 inverse El Niño phenomena. In other words, the stronger meridional circulation patterns are rife over Eurasia in the winter prior to El Niño. And those are favourable for the frequent cold surges in East Asia.

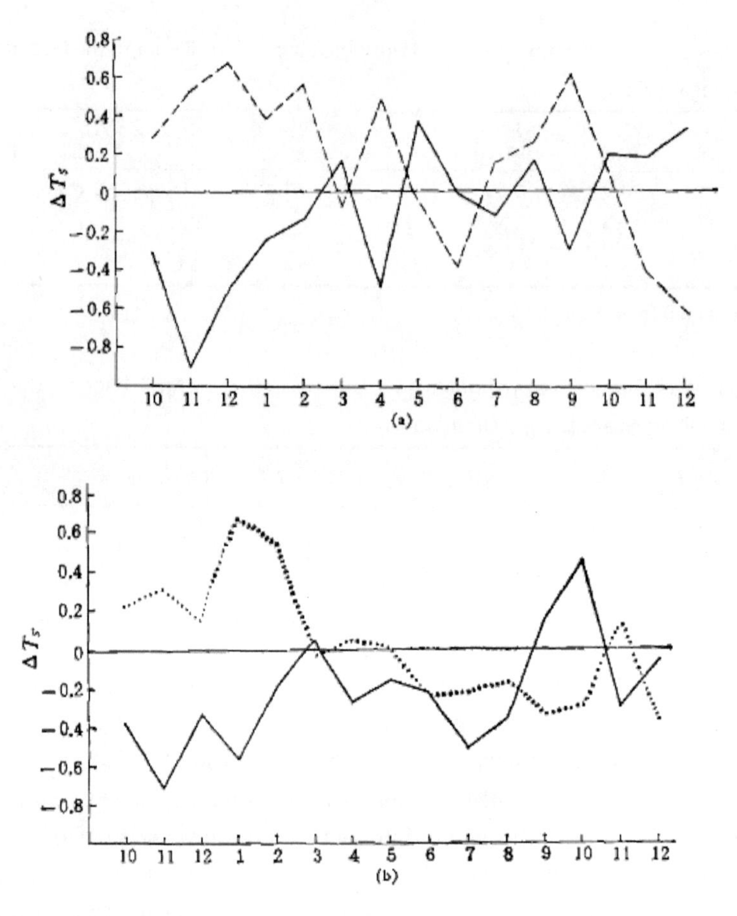

Fig. 1. The variations of the monthly temperature anomalies (℃) at Shanghai (a) and Qingdao (b).

Ⅲ. The Surface Wind over the Tropical Central Western Pacific

In general, there is a steady trade wind zone over the Pacific near the equator and the northeast trade wind is rife at the sea surface. The numerical simulation has indicated that the sea surface height in the equatorial central western Pacific was descendant during the wintertime prior to El Niño. This descent results from the anomalies of the trade wind. In this section, the variations of the trade wind over the central western Pacific during the wintertime prior to El Niño will be investigated.

Based on the statistical analyses in general circulation by Oort[10], the anomalies of the zonal wind at the equator and 25°N over the central western Pacific during December—April for 4 El Niño events are given in Table 3. It is clear that whether for every El Niño event or for the average of 4 El Niño events, there are obvious positive anomalies of the zonal wind over the equator at 150°E and 180°E and there are the most of positive anomalies of the zonal wind at 25°N. The occurrences of these positive anomalies show that the trade winds over the equatorial central western Pacific have been weakened during the wintertime prior to the El Niño event.

Table 3. Monthly Averaged Anomalies of Zonal Surface Wind (m/s) Over Tropical Central Western Pacific Prior to the Occurrence of El Niño

Time \ Lat. \ Lon.	150°E		180°E	
	25°N	Equator	25°N	Equator
December 1962—April 1963	1.16	0.86	3.02	0.50
December 1964—April 1965	0.24	0.36	1.88	0.20
December 1968—April 1969	−0.04	0.20	−0.60	1.16
December 1971—April 1972	−0.66	0.54	0.40	1.62
Mean for 4 wintertimes	0.18	0.49	1.18	0.87

For comparison some relevant statistical results for 4 inverse El Niño phenomena are given in Table 4. They are different from those prior to El Niño, the zonal winds are negative anomalies except during the 1969—1970 winter (1969 is an El Niño year). In other words, the easterlies over the equatorial central western Pacific during the wintertime prior to the inverse El Niño phenomena are on the stronger side.

Table 4. Monthly Averaged Anomalies of Zonal Surface Wind (m/s) Over Tropical Central Western Pacific Prior to the Occurrence of the Inverse El Niño

Time \ Lat. \ Lon.	150°E		180°E	
	25°N	Equator	25°N	Equator
December 1963—April 1964	−0.24	−0.38	−0.06	0.00
December 1966—April 1967	−0.18	−0.02	−0.38	−0.58
December 1969—April 1970	−0.28	0.20	0.40	0.26
December 1972—April 1973	−0.06	−1.74	−0.54	−0.12
Mean for 4 wintertimes	−0.14	−0.49	−0.15	−0.24

We can therefore suggest that the trade winds over the equatorial central western Pacific have been weakened during the wintertime prior to an El Niño event. According to Yamagata's study①, the anomaly of the trade wind in this region is favourable for the abnormality of the Kelvin wave. And the weakened trade wind over the equatorial central western Pacific will lead to the falling of the sea level in the equatorial central western Pacific and the rising in the equatorial eastern Pacific. Then an El Niño event will occur as observed.

Ⅳ. Propagation of the Geopotential Height Anomalies at 500 hPa Toward the Equatorial Central Western Pacific

In Section Ⅱ, we have indicated that there are frequent and stronger cold surges in East Asia during the wintertime prior to El Niño. It is also shown in Section III that the trade winds

① Presented at a seminar in the Institute of Atmospheric Physics, Academia Sinica, 1986.

over the equatorial central western Pacific during the wintertime prior to El Niño are weakened continuously. The question then arises: will the activities of the cold surges in East Asia influence the atmospheric circulation over the equatorial central western Pacific directly?

The influences of the cold surges in East Asia on the atmospheric circulation over the equatorial area have been investigated by Chang and Lau[11-14]. They indicated that the cold surges are an important way affecting the midlatitude disturbance to the tropical circulation. And the cold surges in East Asia will enhance the convective activities and local Hadly cell and two Walker circulation branches in the maritime continent. Then the entire tropical atmosphere will be affected. In order to illustrate the direct influences of the stronger cold surges in East Asia on the atmospheric motions over the tropical central western Pacific, the variations of the height anomalies at 500 hPa averaged for 5 days along a section from (80°N, 85°E) to (20°N, 140°E) are analyzed. Fig. 2(a) shows the situations prior to the occurrence of the 1972 El Niño event. We can see that there are 5 negative height anomalies (corresponding to 5 aerotroughs) propagated toward the tropical central western Pacific from North Asia in the 1971—1972 winter and causing obvious negative anomalies of the height in the area (20°N, 140°E). But in the wintertime prior to the occurrence of the 1975 inverse El Niño, just one negative height anomaly at 500 hPa propagates into the tropical central western Pacific area (Fig. 2(b)) and all other negative anomalies do not reach the south of 30°N.

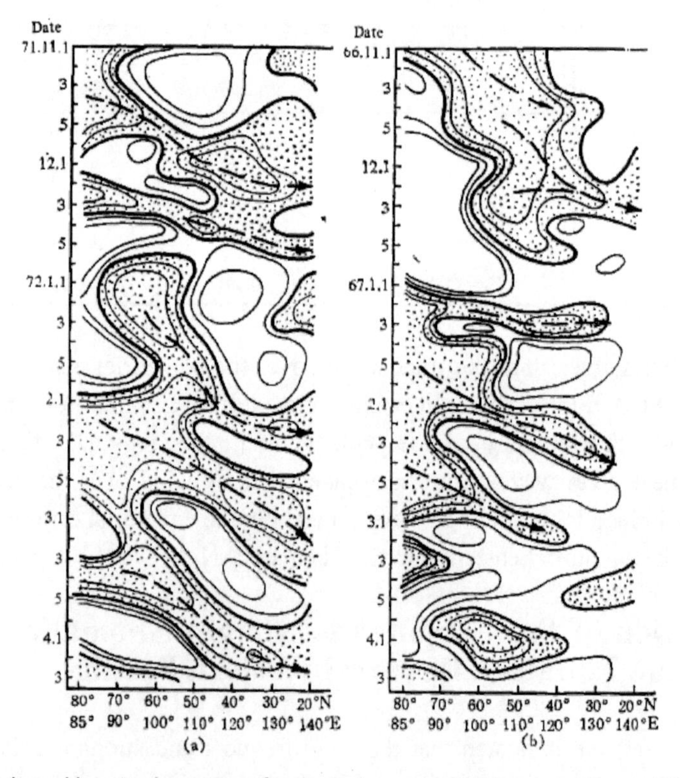

Fig. 2. Time-latitude and longitude sections for the 5-day mean height anomalies at 500 hPa (here 40 and 80 metres lines are given and the negative anomalies are shaded). (a) Situations prior to the 1972 El Niño events, (b) situations prior to the 1967 inverse El Niño.

These analyses for positive and opposite examples show that the frequent activities of the stronger troughs in East Asia during the wintertime prior to the El Niño event will affect the atmospheric motions in the equatorial central western Pacific area. Thus the trade wind will be weakened and the convection will be enhanced over the equatorial central western Pacific. But the troughs are weaker during wintertime prior to the inverse El Niño and they cannot frequently affect the atmospheric motions in the tropical central western Pacific area.

Ⅴ. The Energy Dispersion of the East Asia Trough

In this section, the energy propagation of the East Asia trough will be investigated in dynamics. In spherical coordinate system, the linearized barotropic vorticity equation can be written as

$$\frac{\partial q'}{\partial t}+\frac{\partial \overline{q}}{a\partial \varphi}v'+\overline{u}\frac{\partial q'}{a\cos\varphi\partial\lambda}=0, \tag{1}$$

where $\overline{u}$ is the mean zonal wind speed, v' is the perturbance wind component in meridional direction, q' is tbe disturbance vorticity as

$$q'=\frac{1}{a\cos\varphi}\left[\frac{\partial v'}{\partial\lambda}-\frac{\partial}{\partial\varphi}(u'\cos\varphi)\right], \tag{2}$$

and $\overline{q}$ is the vorticity for the basic flow, in spherical coordinates, we can obtain the formula

$$\frac{\partial\overline{q}}{a\partial\varphi}=\frac{1}{a}\left\{2\Omega\cos\varphi-\frac{\partial}{\partial\varphi}\left[\frac{1}{a\cos\varphi}\frac{\partial}{\partial\varphi}(\overline{u}\cos\varphi)\right]\right\}. \tag{3}$$

For the large-scale motion, with the geopotential ϕ' introduced, Eq. (1) will become

$$\left(\frac{\partial}{\partial t}+\overline{U}\frac{\partial}{\partial\lambda}\right)\left\{\frac{1}{f}\left[\frac{f^2}{\cos\varphi}\frac{\partial}{a\partial\varphi}\left(\frac{\cos\varphi}{f^2}\frac{\partial\phi'}{a\partial\varphi}\right)+\frac{1}{a^2\cos^2\varphi}\frac{\partial^2\phi'}{\partial\lambda^2}\right]\right\}+\frac{\partial\overline{q}}{a\partial\varphi}\times\frac{1}{f}\frac{\partial\phi'}{a\cos\varphi\partial\lambda}, \tag{4}$$

where

$$\overline{U}=\overline{u}\,/\,a\cos\varphi. \tag{5}$$

Since the coefficients in Eq. (4) are independent of λ, we can assume

$$\phi'(\lambda,\varphi,t)=\mathrm{Re}\sum_{k=1}^{n}\Phi_k(\varphi,t)e^{ik\lambda}. \tag{6}$$

Substituting Eq. (6) into Eq. (4), we obtain

$$\left(\frac{\partial}{\partial t}+\overline{U}\frac{\partial}{\partial\lambda}\right)\left\{\frac{1}{f}\left[\frac{f^2}{\cos\varphi}\frac{\partial}{a\partial\varphi}\left(\frac{\cos\varphi}{f^2}\frac{\partial\Phi_k}{a\partial\varphi}\right)-\frac{k^2}{a^2\cos^2\varphi}\Phi_k\right]\right\}+\frac{ik}{af\cos\varphi}\frac{\partial\overline{q}}{a\partial\varphi}\Phi_k=0. \tag{7}$$

Defining $dy=(af^2/\cos\varphi)d\varphi$, the relation $\overline{q}_y=\frac{\cos\varphi}{af^2}\frac{\partial\overline{q}}{\partial\varphi}$ is existential. Taking $\eta=f^2/\cos\varphi$, Eq.(7) can become

$$\left(\frac{\partial}{\partial t}+\overline{U}\frac{\partial}{\partial\lambda}\right)\left[\eta^2\frac{\partial^2\Phi_k}{\partial y^2}-\frac{k^2}{(a\cos\varphi)^2}\Phi_k\right]+\frac{ik\eta}{a\cos\varphi}\overline{q}_y\Phi_k=0. \tag{8}$$

The variation of the geopotential is slower for the planetary waves, we can therefore assume Φ_k is a relaxing function to y and t. Thus WKBJ method can be used to solve Eq. (8). Introducing the slow coordinates X, Y and T which are respectively defined as

$$X=\varepsilon\lambda, Y=\varepsilon y, T=\varepsilon t, \tag{9}$$

where ε is a small parameter over zero ($0<\varepsilon\ll 1$), Eq. (8) can be written as

$$\varepsilon\left(\frac{\partial}{\partial T}+\bar{U}\frac{\partial}{\partial X}\right)\left[\eta^2\varepsilon^2\frac{\partial\Phi_k}{\partial Y^2}-\frac{k^2}{(a\cos\varphi)^2}\Phi_k\right]+\frac{ik\eta}{a\cos\varphi}\bar{q}_y\Phi_k=0. \tag{10}$$

Assuming the wave-packet solution of Eq. (10) as

$$\Phi_k=\hat{\Phi}_k(Y,T)e^{i\theta(Y,T)/\varepsilon}, \tag{11}$$

where θ is a phase function and

$$\frac{\partial\theta}{\partial Y}=m(Y,T), \frac{\partial\theta}{\partial T}=-\omega(Y,T), \tag{12}$$

and taking the expansions of $\hat{\Phi}_k(Y,T)$ in the parameter ε, i. e.

$$\hat{\Phi}_k(Y,T)=\hat{\Phi}_0(Y,T)+\varepsilon\hat{\Phi}_1(Y,T)+\varepsilon^2\hat{\Phi}_2(Y,T)+\cdots, \tag{13}$$

a frequency equation can be obtained as follows from Eq. (10) by assuming constant k and m and taking the zeroth ε approximation

$$\omega=\bar{U}k-\frac{af^2k\bar{q}_y}{a^2f^4m^2+k^2}. \tag{14}$$

Then, the group velocity components can be obtained

$$C_{gX}=\frac{\partial\omega}{\partial k}=\bar{U}+\frac{af^2(k^2-m^2a^2f^4)\bar{q}_y}{(k^2+m^2a^2f^4)^2}, \tag{15}$$

$$C_{gY}=\frac{\partial\omega}{\partial m}=\frac{2kma^3f^6\bar{q}_y}{(k^2+m^2a^2f^4)^2}. \tag{16}$$

In general, the distribution of the basic flow could satisfy $\bar{q}_y>0$. Therefore, we can get $C_{gY}>0$ for the leading waves (troughs) since $km>0$ while $k>0$ in general. The sign of C_{gX} depends on $k^2 - m^2a^2f^4$. If $k^2>m^2a^2f^4$ then $C_{gX}>0$, and the perturbation energies propagate northeastward. If $k^2<m^2a^2f^4$ and their difference is much greater, C_{gX} will be negative and the energies will propagate northwestward. Usually, C_{gX} is positive except for the planetary wave with a very short meridional scale. Therefore, the leading planetary waves will disperse the energies northeastward. For the trailing waves (troughs), since $km<0$, C_{gY} is negative and C_{gX} also depends on $k^2 - m^2a^2f^4$. Therefore, C_{gX} is positive except for the planetary wave with a very short meridional scale. In other words, the trailing waves will disperse the energies southeastward. This is identical with the result obtained by Zeng from Rossby-wave packet theory[15].

The studies on the East Asia cold surges showed that the stronger cold surges are closely linked with the deeper aerotroughs in the trailing type[16, 17]. Therefore, we can suggest that the East Asia troughs corresponding with the stronger cold surges will directly disperse the energies southeastward. If there are frequent and stronger aerotroughs over East Asia, the disturbance energies will continuously be propagated to the equatorial central western Pacific area from the midlatitude in Asia. Then the anomaly of the atmospheric circulation will occur there and it will lead to the anomalous motions of the tropical atmosphere and ocean.

Ⅵ. Conclusions

According to the data analyses and the theoretical investigation completed in this study, the relationship between the occurrence of El Niño and the frequent activities of the stronger aerotroughs in East Asia can lead to the following results:

1) There are frequent and stronger cold surges in East Asia during the wintertime prior to the El Niño event. But the stronger cold surges are fewer in East Asia during the wintertime prior to the inverse El Niño phenomenon.

2) The trade winds have been weakened over the equatorial central western Pacific during the wintertime prior to the El Niño event. But there are enhanced trade winds over the equatorial central western Pacific in the wintertime prior to the inverse El Niño.

3) The negative height anomalies at 500 hPa propagate to the tropical central western Pacific area frequently from midlatitude in East Asia during the wintertime prior to the El Niño event. But the midlatitude disturbances in East Asia which can influence the atmospheric motions over the tropical central western Pacific are fewer during the wintertime prior to the inverse El Niño,

4) The theoretical analysis shows that the trailing planetary waves can disperse the energies southeastward. However, the East Asia aerotroughs acompanied with the stronger cold surges are usually in the trailing type. Therefore, the aerotroughs which are intensely developed in East Asia can propagate the energies to the tropical central western Pacific area and lead to the abnormalities of the atmospheric circulation in the tropics.

5) Based on the above results, we can conclude that frequent activities of the stronger aerotroughs in East Asia will propagate the energies into the tropical central western Pacific area continually. The trade winds will be weakened and the convective actions will be enhanced in this region because of the effects of cold surges. Then, continual anomalies of the atmospheric circulation over the tropical central western Pacific and their expansion will lead to the occurrence of the El Niño event. Therefore, the frequent activities of the stronger aerotroughs and cold surges in East Asia during wintertime can be believed as an important mechanism to cause the El Niño event.

References

[1] 陈烈庭, 大气科学, **1**(1977), 1-12.
[2] Wallace, J. M. & Gutzler, D. S., *Mon. Wea. Rev.*, **109**(1981), 785-812.
[3] Rowntree, P. R., *Dyn. Atmos. Oceans.*, **3**(1979), 373-390.
[4] Li Chongyin, *Kexue Tongbao*, **31**(1986), 538-542.
[5] Sadler, J. C., *Proceedings of the Fourth Annual Climate Diagnostic Workshop*, 1979, pp. 307-313.
[6] Quinn, W. H., *ibid.*, 1979, 233-239.

[7] O'Brien, J. J. & Busalacchi, A. J., *WCRP Publications Series*, No. 1, 1983, 111-122.
[8] 李崇银、胡季, 大气科学, **11**(1987), 359-364.
[9] Erickson, C. O. & Livezey, R. E., *Mon. Wea. Rev.*, **110**(1982), 46-54, 152-158.
[10] Oort, A. H., *Global Atmospheric Circulation Statistics* (1958—1973), NOAA, 1983.
[11] Chang, C. P. et al., *Mon. Wea. Rev*., **107**(1979), 812-829.
[12] Chang, C. P. & Lau, K. M., *ibid*., **108**(1980), 298-312.
[13] ——, *ibid*., **110**(1982), 933-946
[14] Lau, N. C. & Lau, K. M., *ibid*., **112**(1984), 1309-1327 .
[15] Zeng Q. C., *J. Atmos. Sci*., **40**(1983), 73-84.
[16] 陶诗言, 气象学报, **30**(1959), 227-230.
[17] 北京大学地球物理系, 天气分析与预报, 科学出版社, 北京, 1976.

Interaction Between Anomalous Winter Monsoon in East Asia and El Niño Events

Li Chongyin(李崇银)

LASG, Institute of Atmospheric Physics, Academia Sinica, Beijing

Abstract Based on a series of data analyses, the intimate relations between anomalous winter monsoon in East Asia and El Niño are studied in this paper.

Anomalistic circulation in the Northern Hemisphere caused by El Niño event can lead to enhancing the Ferrel cell and the westerlies in the mid-latitudes as the Hadley cell and result in the location of the front zone in East Asia to the north. These are unfavourable for the cold wave breaking out southward in East Asia. Therefore, there are warmer weather and weaker winter monsoon in East Asia in El Niño winter.

There are stronger and frequent cold waves in East Asia during the wintertime prior to the occurrence of El Niño event. They will induce stronger winter monsoon in East Asia. Thus, the weakened trade wind and enhanced cumulus convection in the equatorial middle-western Pacific area caused by the stronger winter monsoon will play an important role in the occurrence of El Niño event. Therefore, the anomalously strong winter monsoon in East Asia during wintertime might be an important mechanism to cause El Niño event.

Ⅰ. Introduction

Some data analyses and numerical simulations have shown that El Niño event should induce the anomalies of atmospheric circulation and the climate in the whole global. The studies in relation to the influences of El Niño event on the climate anomalies in China showed that the anomalous climate is closely related to El Niño event. However, most of the studies are the analyses in relation to anomalistic weather and climate in the summertime of El Niño year, for example, the influences of El Niño on the actions of typhoon over western Pacific (Li, 1986; Li, 1988), the relation between El Niño event and the lower temperature in northeastern China in summer (Zhang et al., 1982; Zeng et al., 1987) and the influences of El Niño on the rainfall in the mid-lower reaches of Yangtze River in the flood season (Chen, 1977; Li, M. et al., 1987). These studies show from various angles that El Niño event is an important factor to cause the anomalies of the weather and climate in China in the summertime. In El Niño year, the typhoon frequency is lower than normal and the landing typhoon on the continent of

Received March 6, 1989.

China is fewer, there is often the lower temperature in northeastern China in summer, and the rainfall in the flood season is on the less side in the mid-lower reaches of Yangtze River. It has been indicated by a lot of researches that El Niño event makes summer monsoon weaken and induces continuous drought in India (Pant, et al., 1981; Rasumusson, et al., 1983; Bhalme et al., 1984).

There is a natural question: Are there any relationships between the El Niño and the climate anomaly in China in winter? In other words, what is the influence of El Niño event on the winter monsoon in East Asia? In this paper, the relationship between El Niño and the winter monsoon in East Asia is exposed by data analyses at first, and then, the causes of anomalous winter monsoon through anomalistic variation of the atmospheric circulation are discussed.

The occurring mechanism of El Niño event has not been understood clearly as yet. Some studies indicate that the occurrence of El Niño event is closely related to the anomalous weakening of the trade wind over the equatorial Pacific (Sadler, 1979; Eagger, 1981). The numerical simulation by using ocean model shows that the sea level in the equatorial western Pacific subsides obviously in El Niño year. And this subsiding of the sea level began in the previous winter or that spring prior to the rise of SST and the sea level in the equatorial eastern Pacific (O'Brien, et al., 1983). Therefore, the precedence sign of El Niño event may be in the equatorial middle-western Pacific area.

According to the data analyses and theoretical study, we have pointed out that the frequent activities of stronger East-Asia deep trough (cold wave) are closely related to the occurrence of El Niño event (Li, et al, 1987; Li, 1988). In wintertime, the frequent activities of stronger East-Asia deep trough can disperse energy southeastwards to the equatorial middle-western Pacific area and lead to the weakening of the trade wind and the enhancement of the cumulus convection there. Thus, El Niño event would be caused through the anomalous Kelvin waves and 30-50 day oscillation. In this paper, the data analyses will further show that the strong winter monsoon is very important for weakening trade wind and enhancing convection over the equatorial middle-western Pacific.

II. Data

El Niño event and inverse El Niño phenomenon will be involved in this paper. They are defined by using Angell's (1981) and Rasmusson's (1982) data before 1950. We define El Niño event and inverse El Niño phenomenon, which occurred after 1950, according to the anomalies of the sea surface temperature in (0-10°S, 180-90°W) region directly. Thus, the El Niño events and inverse El Niño phenomena since 1910 can be shown in Table 1. They are similar to that defined by others. We would merely like to say that: 1969 used to be called El Niño year. But obvious positive anomalies of SST in the equatorial eastern Pacific began in autumn, 1968. It should be called 1968-69 El Niño event.

Table 1. The El Niño Events and Inverse El Niño Phenomena since 1910

El Niño events (years)	1911, 1913, 1918, 1923, 1925, 1930, 1935, 1940, 1944, 1948, 1951, 1953, 1957, 1963, 1965, 1968—1969, 1972, 1976, 1982—1983, 1986—1987
Inverse El Niño phenomena (years)	1912, 1916, 1921, 1924, 1937, 1942, 1949, 1954—1955, 1964, 1967, 1970, 1973, 1975

As we know, the winter monsoon in East Asia is closely related to cold waves. Frequent and stronger (weaker) cold wave in East Asia can induce stronger (weaker) winter monsoon. Some studies in relation to the cold waves in East Asia have shown that the surface temperature in East Asia and the surface high pressure in Siberia are important parameters to represent the activities of cold wave in East Asia. In this paper, the surface temperature and pressure in East Asia will be used to study the action of winter monsoon and its relationship with ENSO. Some of them are from "Monthly Climate Data for the World".

Ⅲ. Weaker Winter Monsoon in East Asia and El Niño Event

Most of SST anomalies in the equatorial eastern Pacific, which correspond to El Niño event, begin in March-April. A few of them begin in summer. In general, there is the greatest anomaly of SST in November or December. Therefore, in order to study the influence of El Niño event on the temperature in eastern China in winter, the anomalies of the surface temperature in Shenyang, Beijing, Qingdao, Wuhan, Shanghai, Fuzhou and Guangzhou from November in El Niño year to next February are computed for 20 El Niño events in the period of 1910—1988. Statistical results are given in Table 2. Since there is the greatest positive anomaly of SST in the equatorial eastern Pacific in November, we regard November-February as wintertime in this paper. This is a little difference from that in general winter (December-February). It can be seen from Table 2 that 14 El Niño events led to higher temperature in eastern China out of 19 El Niño events (since the observation is absent, 1944 is not considered), just 5 El Niño winters (1918—1919, 1930—1931, 1935—1936, 1963—1964, 1976—1977) occur negative anomaly of the temperature at more than 4 stations. Therefore, we can suggest that the surface temperature in eastern China in El Niño winter is mostly higher than normal. About 70% of El Niño events will produce warmer winters in eastern China.

In order to compare with the above results, the cases in relation to inverse El Niño are analyzed. For inverse El Niño phenomenon, there are negative anomalies of SST in the equatorial eastern Pacific, which is inverse type to that in El Niño event. The temperature anomalies at above 7 stations in 13 inverse El Niño winters (November- next February) in the period of 1910—1980 are given in Table 3. Based on 10 inverse El Niño winters in Table 3, the observations are complete for them, we can see that in most inverse El Niño winters (70%), there appears lower temperature in eastern China.

In most of El Niño winters, there exists higher surface temperature in eastern China. However, there exists lower surface temperature in most of inverse El Niño witners.

Table 2. Anomalies of the Surface Temperature (℃) in El Niño Winter (Nov.-Feb.)

year	Shenyang	Beijing	Qingdao	Hankou	Shanghai	Fuzhou	Guangzhou
1911-12	7.5	2.9*	−0.6	1.0	0.6	2.5	−0.6*
1913-14	4.7	4.3*	2.2	4.5	2.2	−1.8	1.5*
1918-19	−0.6	−4.7*	−2.6	−1.2	0.7	1.9	4.8
1923-24	0.5	1.6	1.1	1.9	1.4	5.4	4.4
1925-26	7.1	3.0	3.0	3.6	0.7	1.7	4.0
1930-31	−3.3	0.5	−1.4	−1.3	−1.3	4.0	1.3
1935-36	−12.1	−10.5	−8.4	−4.7	−0.5	0.3	1.8
1940-41	(6.9)	0.8	6.5	(3.4)	7.2	1.6	1.7
1944-45	/	−7.4	−4.1	/	/	/	/
1948-49	(5.0)	5.2	3.5	1.1	(3.9)	5.7**	2.2
1951-52	3.0	4.0	2.2	2.3	2.6	3.0	5.8
1953-54	3.4	3: 4	2.0	1.5	5.3	4.9	3.8
1957-58	0.8	2.1	2.5	4.4	0.0	−0.4	−1.0
1963-64	−3.8	−0.8	−0.7	−2.9	0.6	0.1	−0.5
1965-66	1.7	1.6	4.0	2.9	2.8	4.7	5.4
1968-69	0.6	−3.8	−1.8	−4.5	2.4	4.1	1.9
1972-73	7.5	0.4	2.9	2.4	1.4	3.7	3.8
1976-77	−6.8	−4.8	−5.8	−6.4	−7.5	−7.0	−7.6
1982-83	11.0	4.0	3.0	2.0	4.0	0.0	−3.0
1986-87	3.0	4.1	5.9	2.4	4.6	8.3	6.9

( ) — data are not complete, * — data from Tianjing
** —data from Shaxian, + —data from Hongkong

Table 3. Anomalies of the Surface Temperature (℃) in Inverse El Niño Winter (Nov.-Feb.)

year	Shenyang	Beijing	Qingdao	Hankou	Shanghai	Fuzhou	Guangzhou
1912-13	−8.7	−6.4*	−6.3	−2.9	−1.4	−3.7	−0.7*
1916-17	−4.4	−3.2*	−6.1	−3.6	−3.8	−3.5	−2.5
1921-22	−1.5	−1.4*	−0.9	1.2	2.4	7.7	4.1
1924-25	0.3	−4.2*	−3.0	−0.9	−4.8	−5.1	−5.1
1937-38	1.1	/	/	−0.5	0.1	−0.3	0.7
1942-43	/	1.6	1.4	/	−0.2	−4.1	/
1949-50	/	1.4	2.8	/	/	/	/
1954-55	−6.4	−0.4	−0.2	−2.7	−0.8	−0.8	−0.9
1964-65	−4.1	2.5	4.1	3.5	1.5	1.4	2.5
1967-68	0.5	−8.9	−8.4	−7.1	−8.5	−6.0	−5.4
1970-71	1.3	−1.8	−1.4	0.2	−2.8	0.0	−1.5
1973-74	2.9	1.9	−1.0	−0.6	0.0	−5.1	−4.1
1975-76	9.0	4.6	2.3	−0.6	0.1	−2.6	−3.4

This elucidates from both the right side and the reverse side that the influence of El Niño on the temperature in eastern China in winter is very important. In other words, the winter monsoon in East Asia is also closely related to El Niño; El Niño event will weaken the winter monsoon in East Asia.

It has been indicated that the breaking of cold air (cold wave) in East Asia can enhance the convection and rainfall over the equatorial western Pacific area (Lau, 1982; Lau et al., 1983). Therefore, we will analyze the variation of precipitation over the equatorial western Pacific to study the variaton of winter monsoon in East Asia. The first line in Table 4 shows the anomalies of the precipitation (ΔR) at Truk, Caroline Islands (7°28′N, 151°51′E) in El Niño winter (Nov.- Apr.). It is clear that there are obvious negative anomalies of the precipitation during the wintertime for every El Niño event except 1986-87. This result indicates that El Niño event can lead to obvious weakening of the witner monsoon in East Asia.

Table 4. The Precipitation Anomalies (ΔR) in Truk, Caroline Islands (7°28′N, 151°51′E), the Surface Temperature Anomalies (ΔT_M and ΔT_N) at Minamidaitojima (25°50′N, 131°14′E) and Nada (26°14′N, 127°41′E) and the Monthly Mean Meridional Wind Anomalies ($\Delta \bar{V}$) at 850 hPa at Minamidaitojima during the Wintertime (Nov-Apr.) in El Niño Year

	1963.11-1964, 4	1965.11-1966.4	1968.11-1969.4	1972.11-1973.4	1976.11-1977.4	1982.11-1983.4	1986.11-1987.4
ΔR (mm)	−412	−497	−285	−610	−289	−457	139
ΔT_M (℃)	1.5	3.4	2.0	5.4	0.7	3.1	2.1
ΔT_N (℃)	5.8	6.3	5.8	5.9	−2.5	4.2	2.9
$\Delta \bar{V}$ (m/s)	1.3	1, 0	0.8	0.8	−0.1	1.6	0.2

The second line and third line in Table 4 give respectively the temperature anomalies at Minamidaitojima (25°50′N, 131°14′E) and Nada (26°14′N, 127°41′N) in El Niño wintertime (Nov-Apr.). Obviously, the temperatures in El Niño winter are all higher than normal. The fourth line in Table 4 shows the anomalies of the monthly meridional wind component at 850 hPa over Minamidaitojima in El Niño winter. There are obvious southerlies anomalies (positive anomalies) in El Niño winter. According to the anomalies of surface temperature and meridional wind, it can be suggested that the cold waves (winter monsoon) in East Asia are weaker in El Niño winter.

A series of above analyses show that El Niño event plays an important role in the variations of East-Asia winter monsoon. El Niño event should lead to obvious weakening of the winter monsoon in East Asia.

Ⅳ. The Analyses of General Circulation Anomalies in El Niño Winter

The occurrence of the anomalous climate associated with weaker winter monsoon in East Asia in El Niño winter results from the general circulation anomalies caused by El Niño event.

Previous studies have shown: The westerlies in lower troposphere at middle latitude will be enhanced in El Niño winter due to the effect of positive anomaly of SST in the equatorial eastern Pacific (Bjerlenes, 1969); A numerical simulation shows that the continued positive anomaly of SST in the equatorial eastern Pacific would enhance the Hadley circulation exceedingly (Wu, et al., 1987). Recently, an analysis on the general circulation in December 1982 compared with December 1980 shows that the westerlies in whole troposphere are enhanced obviously and the temperature in lower troposphere (including the surface temperature) at middle latitude is also on the warmer side. These anomalies result not only from the enhanced Hadley circulation but also from the enhanced Ferrel circulation caused by the positive anomaly of SST in the equatorial eastern Pacific. The deviations of zonal mean meridional wind component between in December 1982 (El Niño winter) and in December 1980 (normal winter) are given in Fig.1. Fig.2 shows the deviations of zonal mean meridional temperature transportation (TV) between in December 1982 and in December 1980. Obviously, there are obvious southerly anomaly and anomalous transportation of the temperature northwards from 35° through 65°N in El Niño winter, especially in the lower troposphere. The enhanced zonal westerly in middle latitudes and the stronger southerly in the lower troposphere are unfavourable to the intrusion of cold wave southwards. And they are important circulation background producing weaker winter monsoon in East Asia.

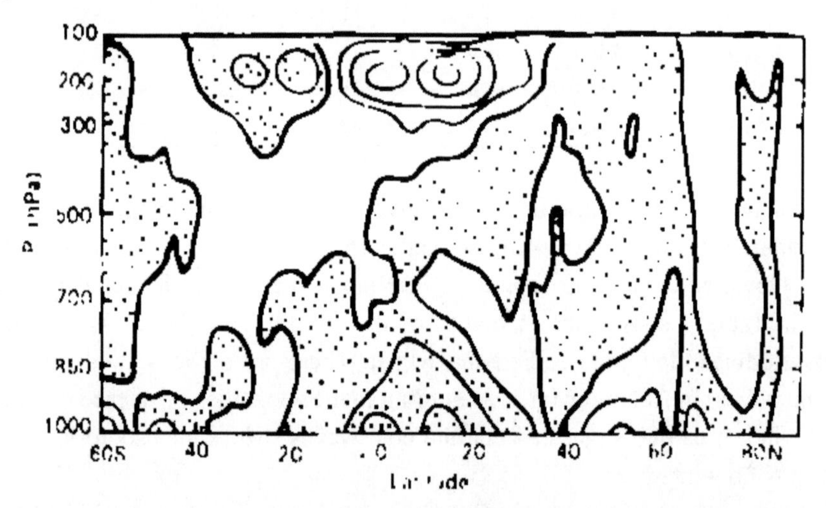

Fig. 1. Deviations of zonal mean meridional wind component between in December 1982 and in December 1980. The contour intervals are 0.3 m/s, shaded region indicates northerly anomalies.

The above discussion is in relation to zonal mean anomalies of the general circulation in El Niño winter. It represents the averaged situation in the Northern Hemisphere. The discussion in relation to that in East Asia is as follows. We take the anomaly of geopotential height at 500 hPa in November 1972-February 1973 as a representation of the circulation anomaly in El Niño winter (Fig.3a), while the anomaly in November 1954-February 1955 as a representation of the circulation anomaly in inverse El Niño winter (Fig.3b). It is clear in Fig.3a that there are negative anomalies to the north of 50°N in East Asia / northwestern Pacific area, but positive anomalies to the south of 50°N. This illustrates that the westerly is enhanced at 50°N region in El Niño winter and it is similar to the anomaly of zonal mean westerly. Simultaneously, the position of the polar front in East Asia is on the north side in El Niño winter and there is anomalous high ridge over eastern China and Japan area. Thus, the activity of cold wave is weaker there, the surface temperature is on the high side in eastern China and the winter monsoon in East Asia is on the weak side. During 1954-55 winter, the westerly is weaker at 40°-55°N in East Asia and stronger at 35°N region; The position of the polar front is on the south side and the negative anomalies of geopotential height maintain over northern China. These represent that the activity of cold wave is frequent in East Asia, the winter monsoon is on the strong side and there is often colder winter in eastern China in inverse El Niño winter.

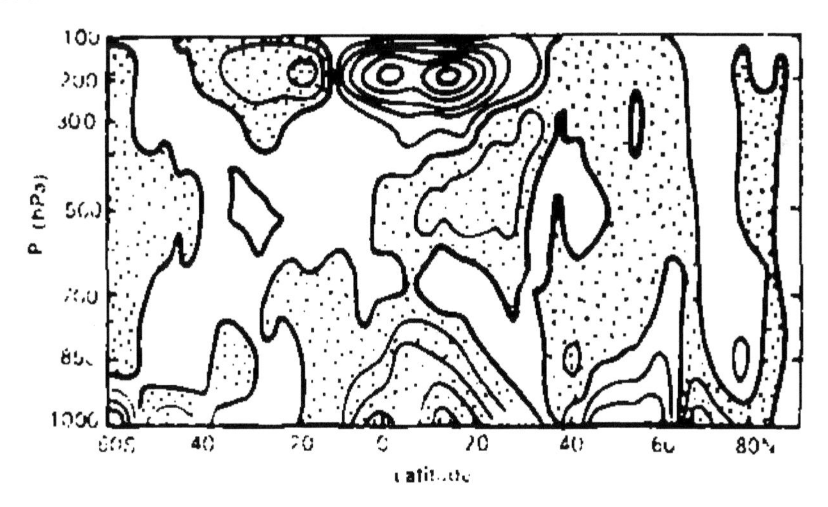

Fig. 2. Deviations of zonal mean meridional transportation of the temperature between in December 1982 and in December 1980. The contour intervals are 100 K • m/s, shaded region indicates anomalous transportation southwards.

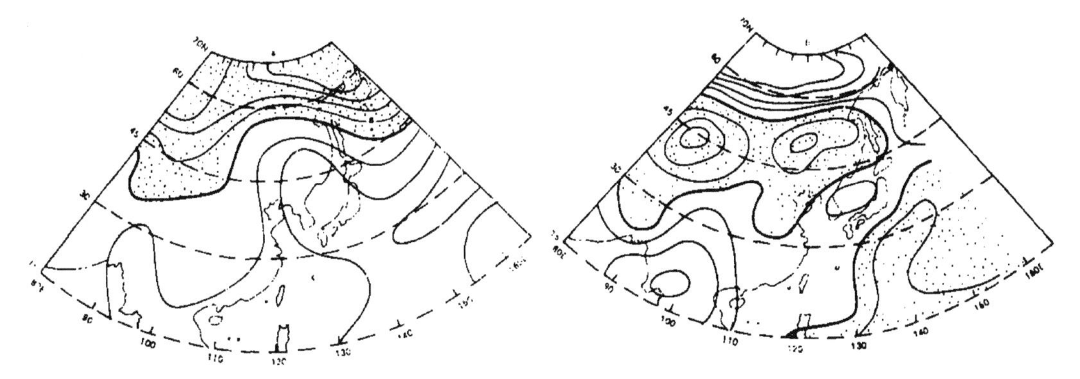

Fig. 3. Anomalies of geopotential height at 500 hPa over East Asia / northwestern Pacific (contour intervals are 40m. Shaded region indicates negative), a) Nov. 1972 - Feb. 1973, b) Nov. 1954 - Feb. 1955.

V. The Occurrence of El Niño Event and Strong Winter Monsoon in East Asia

The occurrence mechanism of El Niño event has drawn public attention since some influences of El Niño event on general circulation and the weather situation were understood. But now it is not quite clear. According to some previous studies, the weakening of trade wind in the equatorial Pacific might regard as an important factor to cause El Niño event and the precedence sign of El Niño event seems to appear in the equatorial middle-western Pacific area.

Based on the data analyses and theoretic study, we have obtained an initiatory conclusion: In wintertime, the frequent activity of stronger East-Asian troughs (cold waves) would weaken the trade wind and enhance the convection over the equatorial middle-western Pacific area, and then El Niño event might be excited. In this study, we will further indicate the importance of strong winter monsoon in East Asia to cause El Niño event.

In Table 5, we list the anomalies of the surface pressure in central region (105°-120°E, 40°-60°N) of cold high in Asia and the anomalies of the surface temperature at 15 stations in eastern China during wintertime (Nov.-April) prior to every El Niño event period 1950-1980. It is obvious that the Mongolia cold high is on the strong side and the surface temperature in eastern China is on the low side during wintertime prior to every El Niño event except 1965. Especially, the strong cold waves are more prominent before the occurrence of strong El Niño event (such as in 1951, 1957, 1968-1969 and 1972). Therefore, it might be confirmed that there is strong winter monsoon in East Asia during wintertime prior to El Niño event.

In order to show the influence of strong winter monsoon in East Asia on the weakening trade wind and the enhancement of convection over the equatorial middle-western Pacific and the relationship between strong winter monsoon in East Asia and El Niño event, the variations of some elements with time are given in Fig.4. They are averaged surface temperature anomalies (ΔT_s) at Ishigaki (24°20′N, 124°10′E) and Minamidaitojima (25°50′N, 131°14′E) during wintertime, the precipitation anomalies (ΔR) at Truk, Caroline Island during wintertime, the monthly mean anomalies of zonal wind at 850 hPa (ΔU) at Truk, Caroline Island during wintertime and the anomalies of SST in the equatorial eastern Pacific (SSTA). We can easily find two important phenomena. The first, for the most of strong winter monsoon (ΔT_s is negative), AR and ΔU are positive. That means there are strong convection and weak trade wind over the equatorial middle-western Pacific corresponding to strong winter monsoon in East Asia. The second, corresponding to every El Niño event (marked with dark for SSTA), there are negative anomalies ΔT_s (marked with C). That means there are strong winter monsoon in East Asia during wintertime prior to El Niño event. According to the curve ΔT_s, the wintertime 1969-1970, 1973-1974 and 1983-1984 seems to be strong winter monsoon in East Asia, but no El Niño event appears after those. This is a challenge to the relationship between strong winter monsoon in East Asia and the occurrence of El Niño event. Comparing curves ΔT_s, ΔR and ΔU in Fig.4, we can find that ΔR and ΔU are negative during 1983-1984 and 1969-1970 winter though ΔT_s is negative. The winter monsoon in East Asia might be not strong during 1969-1970 and 1983-1984 winter. But in 1973-1974 winter, ΔR and ΔU are positive associated with negative ΔT_s, the winter monsoon in East Asia is stronger. Although El Niño does not present in 1974, SST in the equatorial eastern Pacific is on the rise while 1973 inverse El Niño is on the drop. Therefore, we suggest that the occurrence of El Niño event is closely related to the strong winter monsoon in East Asia.

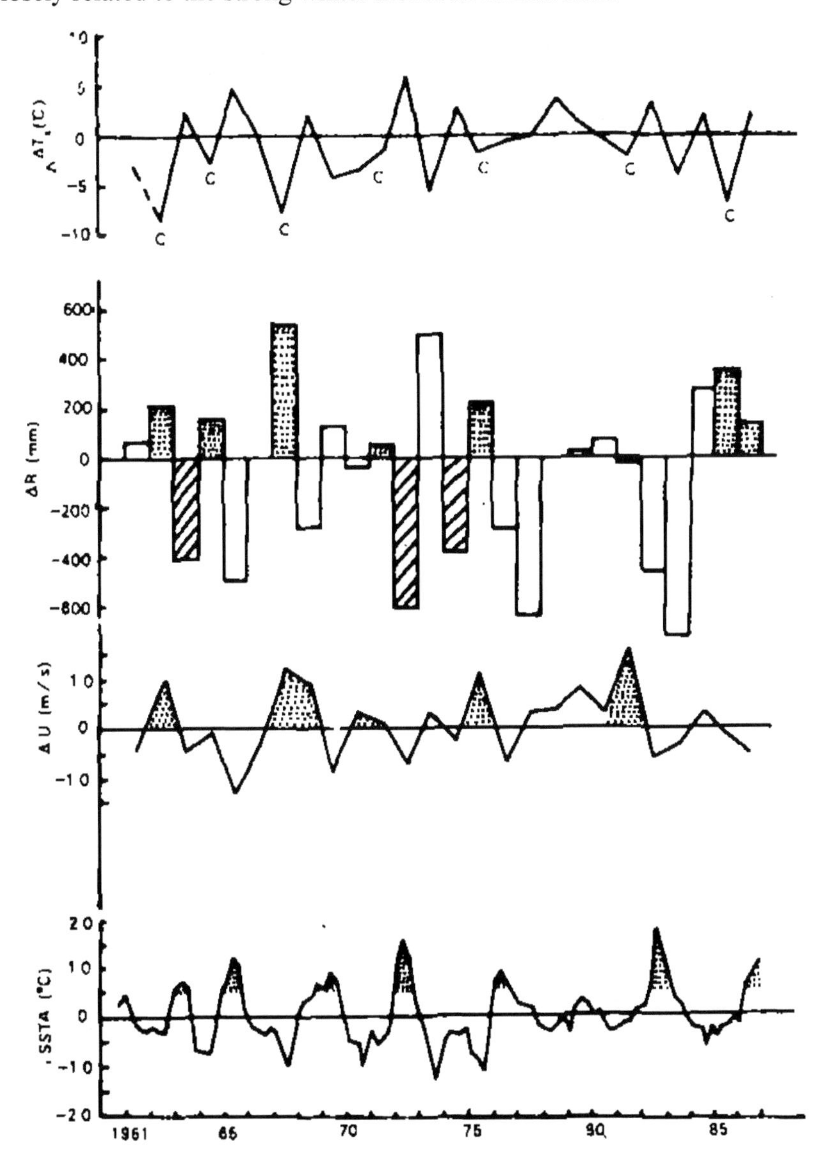

Fig. 4. The Interannual variations in relation to the averaged temperature anomalies (ΔT_s) at Ishigakijima and Minamidaitojima in wintertime (Nov.- Apr.), the precipitation anomalies (ΔR) at Truk, Caroline Island in wintertime, the monthly mean anomalies of zonal wind at 850 hPa ($\Delta \overline{U}$) and the anomalies of SST in the equatorial eastern Pacific (SSTA).

Table. 5. The Surface Pressure Anomalies in the Region (105°-120°E, 40°-60°N) and the Temperature Anomalies in Eastern China during Wintertime (Nov.- Apr.) and the Occurrence of El Niño Event

		1950.11-1951.4	1952.11-1953.4	1956.11-1957.4	1962.11-1963.4	1964.11-1965.4	1967.11 1968.4	1971.11 1972.4	1975.11-1976.4
ΔP(hPa)		4.3	12.0	21.4	2.3	−2.4	12.0	6.8	5.8
ΔT (℃)	Beijing	−2.8	−2.5	−13.2	6.3	0.5	−6.8	0.3	2.3
	Taiyuan	−8.1	−6.2	−8.2	0	2.0	−0.1	−2.2	−0.9
	Shijia zhuang	−5.1	−5.2	−13.0	3.7	2.2	−3.4	2.7	2.9
	Jinan	−5.1	−4.0	−15.6	−2.3	1.4	−6.9	−2.2	1.2
	Qingdao	−2.3	−2.9	−12.8	−1.9	3.6	−8.3	−1.7	2.1
	Zhengzhou	−3.3	**−0.0**	−8.9	−0.9	1.7	−4.3	−0.7	−1.4
	Xuzhou	/	/	/	−1.6	1.4	−4.8	−3.0	−0.8
	Hefei	/	0.4	−10.1	0.8	2.0	−5.1	−4.7	−0.6
	Shanghai	−3.2	2.6	−7.7	−3.1	−1.0	−8.1	−2.0	−0.7
	Wuhan	−5.0	1.6	−7.1	−1.6	1.5	−4.1	−5.0	−2.6
	Nanchang	−6.1	−0.6	−8.2	3.3	3.5	−5.8	−2.8	−3.1
	Ganzhou	−6.3	−1.6	−7.1	−0.6	2.2	−4.7	−1.7	−5.7
	Huangyan–Haimen	0.6	1.6	−4.1	−2.9	−0.6	−7.5	−2.6	−3.7
	Fuzhou	−2.9	0.2	−5.9	−0.3	−0.2	−5.3	−2.3	−5.1
	Guangzhou	−2.9	−0.8	−4.6	−2.8	1.9	−7.1	−2.0	−6.4
Mean ΔT		−4.1	−1.2	−9.0	−0.3	1.5	−6.0	−2.5	−1.5

If we set out ΔT_s, ΔR and $\Delta \bar{U}$, which are prior to every El Niño event (in Table 6), the anomalies of general circulation over western Pacific are very clear. In the wintertime prior to the occurrence of El Niño event, the winter monsoon in East Asia is on the strong side, the trade wind is on the weak side and the cumulus convection is on the strong side over the equatorial middle-western Pacific.

Table 6. The Surface Temperature Anomalies (ΔT_s) Averaged in Ishigakijima (24°20′N, 124°10′E) and Minamidaitojima (25°50′N, 131°14′E), the Precipitation Anomalies (ΔR) and Monthly Zonal Mean Wind Anomalies ($\Delta \bar{U}$) at 850 hPa in Truk, Caroline Islands in North Pacific (7°28′N, 151°51′E) during Wintertime (Nov.-Apr.) Prior to the Occurrence of El Niño Events

	1962.11-1963.4	1964.11-1965.4	1967.11-1968.4	1971.11-1972.4	1975.11-1976.4	1981.11-1982.4	1985.11-1986.4
ΔT(℃)	−8.3	−2.7	−7.7	−1.5	−1.5	−2.2	−4.4
ΔR (mm)	209	150	544	51	223	−34	347
$\Delta \bar{U}$ (m/s)	0.9	−0.3	1.2	0.1	1.1	1.6	−0.1

Analysing the influence of individual cold wave on the tropical atmosphere, it can be also got that stronger cold wave (winter monsoon surge) in East Asia will weaken the trade wind over the equatorial middle-western Pacific. Fig.5 shows the activities of cold waves in

1975-76 winter. The pentad-averaged pressure anomalies (solid line) and temperature anomalies (dashed line) at Shanghai and temperature anomalies at Zhengzhou (dotted line) are shown in Fig. 5a. Several major processes of cold air breaking are also shown (marked with C). The variation of averaged zonal wind with time in the (5°-20°N, 140°-150°N) area in the same period is given in Fig.5b. It is very clear that the trade wind over the equatorial middle-western Pacific will be obviously weakened after each enhancement of the winter monsoon in East Asia (or each stronger cold wave in East Asia).

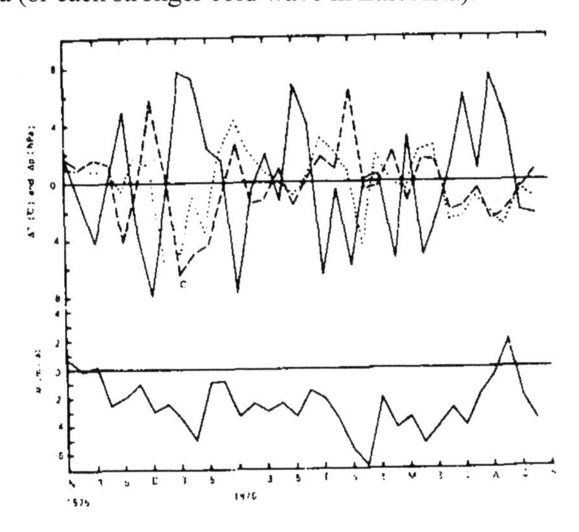

Fig. 5. The enhancement of the winter monsoon in East Asia and the weakening of the trade wind over the equatorial middle-western Pacific, a. Temporal variations of the pressure anomalies (solid line) and temperature anomalies (dashed line) at Shanghai and temperature anomalies at Zhengzhou, b. Temporal variation of the averaged zonal wind in (5°-20°N, 140°-150°N) area.

Above materials further indicate that the occurrence of El Niño event is closely related to anomalous winter monsoon in East Asia. During the wintertime prior to the occurrence of El Niño event, there are frequent activities of stronger cold waves and which enhance the winter monsoon in East Asia. The continuous strong winter monsoon in East Asia should lead to obvious enhancement of the cumulus convection and obvious weakening of the trade wind over the equatorial middle-western Pacific. The 30-50 day low-frequency oscillations caused by stronger convection and the anomalous Kelvin waves in the equatorial Pacific caused by reducing trade wind are important factors to produce El Niño event (Lau, et al., 1986, 1988; Wyrtki, 1975; Yamagata, 1986). Therefore, the continuous strong winter monsoon in East Asia during wintertime might be an important mechanism to cause El Niño event.

Ⅵ. Conclusion

A series of analyses in this study show that the relationship between the anomalies of the winter monsoon in East Asia and El Niño event is very close.

At first, the anomalous atmospheric circulation in winter caused by El Niño event, such as the enhancements of the Hadley cell and Ferrel cell, the strengthening westerlies and the polar front to the north in East Asia, are all unfavourable to the intrusion of cold wave southwards. Therefore, there are higher surface temperature and weaker winter monsoon in East Asia.

During the wintertime prior to the occurrence of El Niño event, there are frequent and stronger cold waves in East Asia. They should lead to the continuous strong winter monsoon in East Asia. And then, the trade wind is weakened and the cumulus convection and rainfall are enhanced over the equatorial middle-western Pacific. The stronger 30-50 day oscillations caused by the strong convection over the equatorial Pacific and the anomalous Kelvin waves caused by the weakened trade wind in the equatorial Pacific play an important role in the occurrence of El Niño event. Therefore, we can suggest that the continuous and stronger winter monsoon in East Asia is an important mechanism to cause El Niño event.

The research is partly supported by State Natural Science Foundation No. 9488009.

References

Angell, J. K. (1981), Comparison of variation in atmospheric quantities with sea surface temperature variations in equatorial eastern Pacific, *Mon. Wea.*, **109**: 230-243.

Bhalme, H. N. and S. K. Jadhav (1984), The southern oscillation and its relation to the monsoon rainfall, *J. Climat*, **4**: 509-520.

Bjerlenes, J. (1969), Atmospheric teleconnections from the equatorial Pacific, *Mon. Wev. Rev.*, **97**: 163-172.

Chen Lieting (1977), The effects of the anomalous sea-surface temperature of the equatorial eastern Pacific Ocean on the tropical circulation and rainfall during the rainy period, *Scientia Atmospherica Sinica*, **1**: 1-12(in Chienese).

Eagger. J., et al. (1981), Pressure, wind and cloudiness in the tropical Pacific relted to the Southern Oscillation, *Mon. Wea. Rev.*, **109**: 39-1149.

Lau, K.-M. (1982), Equatorial responses to northeasterly cold surges as inferred from cloud satellite imagery, *Mon.Wea. Rev.*, **110**: 1306-1313.

Lau, K.-M., et al. (1983), Short term planetary—scale interactions over the tropics and midlatitudes, Part II: winter MONEX period, *Mon.Wea. Rev.*, 111: 1372-1388.

Lau. K. M., and P. H. Chan (1986), The 40-50 day oscillation and the El Niño/Southern Oscillation: A new perspective, *Bull. Amer. Meteor. Soc.*, **67**: 533-534.

Lau. *K.* M., and P. H. Chan (1988), Interannual and intraseasonal variations of tropical convection, A possible link between the 30-50 day oscillation and ENSO, *J. Atmos. Sci.*, **44**: 506-521.

Li Chongyin (1986), El Niño and typhoon action over the western Pacific, *Kexue Tongbao*, **31**: 538-542.

Li Chongyin (1988), Actions of typhoon over the western Pacific (including the South China Sea) and El Niño, *Advances in Atmospheric Sciences*, **5**: 107-116.

Li Chongyin, et al. (1987), A study on interaction between the East-Asia atmospheric circulation and El Niño, *Chinese J. Atmos. Sci.*, **11**: 411-420.

Li Chongyin (1989), The frequent activities of stronger aerotroughs in East Asia in wintertime and the occurrence of the El Niño event, *Scientia Sinica*, series B, **32**: 976-985.

Li Maicun, et al. (1987), The relationspip between the monsoon rainfall over eastern China and the eastern equatorial Pacific sea surface temperature, *Scientia Aimospherica Sinica*, **11**: 372-380 (in Chinese).

O'Brien, J. J. et al. (1983), The Pacific Ocean response to El Niño conditions, WCRP Publications Series, **1**:

111-122.

Pant, G. B. and B. Parthasarsthy (1981), Some aspects of an association between the southern oscillation and Indian summer monsoon, *Arch, Met. Geoph. Biokl.*, Ser. B, **29**: 245-252.

Rasmusson, E. M. and T. H. Carpenter (1983), The relationship between eastern equatorial Pacific sea surface temperatures and rainfall over India and SriLanka. *Mon. Wea. Rev.*, **111**: 517-528.

Rasmusson, E. M. et al. (1982), Variations in tropical sea surface temperature and surface wind fields associated with the southern oscillation/El Niño, *Mon. Wea. Rev.*, **110**: 354-384.

Sadler, J. C. (1979), Trade wind anomalies using satellite cloud motion vectors, *Proceedings of the Fourth Annual Climate Diagnostics Workshop*, P. 233.

Wu Guoxiong and U. Cubasch (1987), The impact of the El Niño anomaly on the mean meridional circulation and atmospheric transformations, *Scientia Sinica*, series B, **30**: 535-546.

Wyrtki, K. (1975), EL Niño the dynamic response of the equatorial Pacific Ocean to atmospheric forcing, *J. Phys. Oceartogr.*, **5**: 572-584.

Yamagata, T. (1986), On the recent development of simple coupled ocean-atmosphere models of ENSO, *J. Oceanogr. Soc. Japan*, **42**: 299-307.

Zeng Zhaomei and Zhang Mingli (1987), Relationship between the key region SST of the tropical eastern Pacific and air temperature of Northeast China, *Scientia Atmospherica Sinica*, **11**: 382-389 (in Chinese).

Zhang Mingli, et al. (1982), A study of global surface temperature field in 1970's (1), *Scientia Atmospherica Sinica*, **6**: 229-236.

Relationship Between East Asian Winter Monsoon, Warm Pool Situation and ENSO Cycle

Li Chongyin, Mu Mingquan

LASG, Institute of Atmospheric Physics, Chinese Academy of Sciences, Beijing, China

Abstract Based on the observational data analyses and numerical simulations with the air-sea coupled model (CGCM), a new perspective on the occurrence mechanism of ENSO is advanced in this paper. The continuous strong (weak) East Asian winter monsoon will lead to continuous westerly (easterly) wind anomalies over the equatorial western Pacific region. The anomalous equatorial westerly (easterly) winds can cause eastward propagation of the subsurface ocean temperature anomalies (SOTA) in the warm pool region, the positive (negative) SOTA have been in the warm pool region for quite a long time. The eastward propagating of positive (negative) SOTA along the thermocline will lead to positive (negative) SSTA in the equatorial eastern Pacific and the occurrence of El Niño (La Niña) event. After the occurrence of ENSO, the winter monsoon in East Asia will be weak (strong) due to the influence of El Niño (La Niña).

Keywords East Asian winter monsoon, warm pool in the western Pacific, subsurface ocean temperature (SOT), ENSOcycle.

ENSO has been paid much attention in the world because it can always cause serious climate anomalies and disasters in vast areas. Previous studies have attributed ENSO to the interaction between the atmosphere and ocean in the tropical Pacific region[1, 2]. In relation to the mechanism of El Niño (ENSO), even though the oceanic relaxation theory[3], the unstable oceanic waves theory[4-6] and the delayed action oscillation theory[7-9], have been advanced, the origin of ENSO has not been understood very well.

Moreover, putting stress on atmospheric anomaly in the air-sea interaction, our research indicated that the abnormal strong winter monsoon in East Asia plays an important role in exciting the El Niño event[10, 11], The data analyses and numerical simulations in a CGCM have shown that strong (weak) winter monsoon in East Asia can excite El Niño (La Niña) through the produced westerly (easterly) wind anomalies and strong (weak) convection in the equatorial western Pacific region, as important physical processes to excite ENSO, the former will lead to warm (cold) oceanic Kelvin waves and the latter will lead to strong (weak) intraseasonal oscillation[12, 13].

In this paper, we will indicate that there are clear interactions between East Asian Winter Monsoon (EWM), Warm Pool Situation (WPS) and ENSO, the EWM-WPS-ENSO seems to

Received January 20, 2000.

be a large climate system. According to this system and its evolution, the ENSO mechanism will be understood better.

The NCEP/NCAR reanalysis data, the Joint Environmental Data Analysis Center (JEDAC) data and the COADS data were mainly used in this study.

1. Interaction between Abnormal East Asian Winter Monsoon and ENSO

The relationship of ENSO with the anomalies of East Asian winter monsoon has been indicated in a series of studies and their interactions are also evident[10, 11, 14]. In this paper, we just give some composite results for the El Niño and La Niña events in order to show the interactions between ENSO and abnormal East Asian winter monsoon.

As we know, the activity of East Asian winter monsoon can be represented by using some meteorological elements in the East Asia region, such as the sea level pressure, surface air temperature, surface wind and geopotential height at 500 hPa. For continuous strong East Asian winter monsoon, there are usually strong surface cold high system in the Siberia-Mongolia region, deepening trough at 500 hPa in East Asia, strong northerly wind and low surface air temperature in eastern China and the northwestern Pacific region. Conversely, the above systems will be in opposite situations for continuous weak East Asian winter monsoon. Fig. 1 shows the interaction between abnormal East Asian winter monsoon and El Niño. Fig. 1(a), (d) can represent the activity of winter monsoon in East Asia, fig. 1(e), (f) represent the zonal wind anomaly over the equatorial western Pacific and the El Niño event (SSTA in Niño 3). It is very evident that the El Niño (composited) event outbreaks in spring, the westerly wind anomalies over the equatorial western Pacific are about 2—3 months earlier than El Niño outbreak, and there was strong winter monsoon in East Asia during the wintertime prior to the El Niño outbreak. This means that strong winter monsoon plays an important role in exciting the El Niño event through producing westerly wind anomalies over the equatorial western Pacific. It is also clear that there is weak winter monsoon in East Asia during the wintertime after the El Niño outbreak. This suggested that the El Niño event should reduce the winter monsoon in East Asia.

The relationship between abnormal East Asian winter monsoon and La Niña is similar to that in fig. 1 (figure omitted). It is also evident that the La Niña event outbreaks in spring, the easterly wind anomalies over the equatorial western Pacific are about 2—3 months earlier than La Niña outbreak, and there was weak winter monsoon in East Asia during the wintertime prior to the La Niña outbreak. This means that weak winter monsoon plays an important role in exciting the La Niña event through producing easterly wind anomalies over the equatorial western Pacific. Moreover, during the wintertime after the La Niña outbreak, there is strong winter monsoon in East Asia. This can suggest that the La Niña event should enhance winter monsoon in East Asia.

In order to show the exciting effect of abnormal East Asian winter monsoon on occurrence of ENSO, some numerical simulations are completed with a CGCM, which was developed in

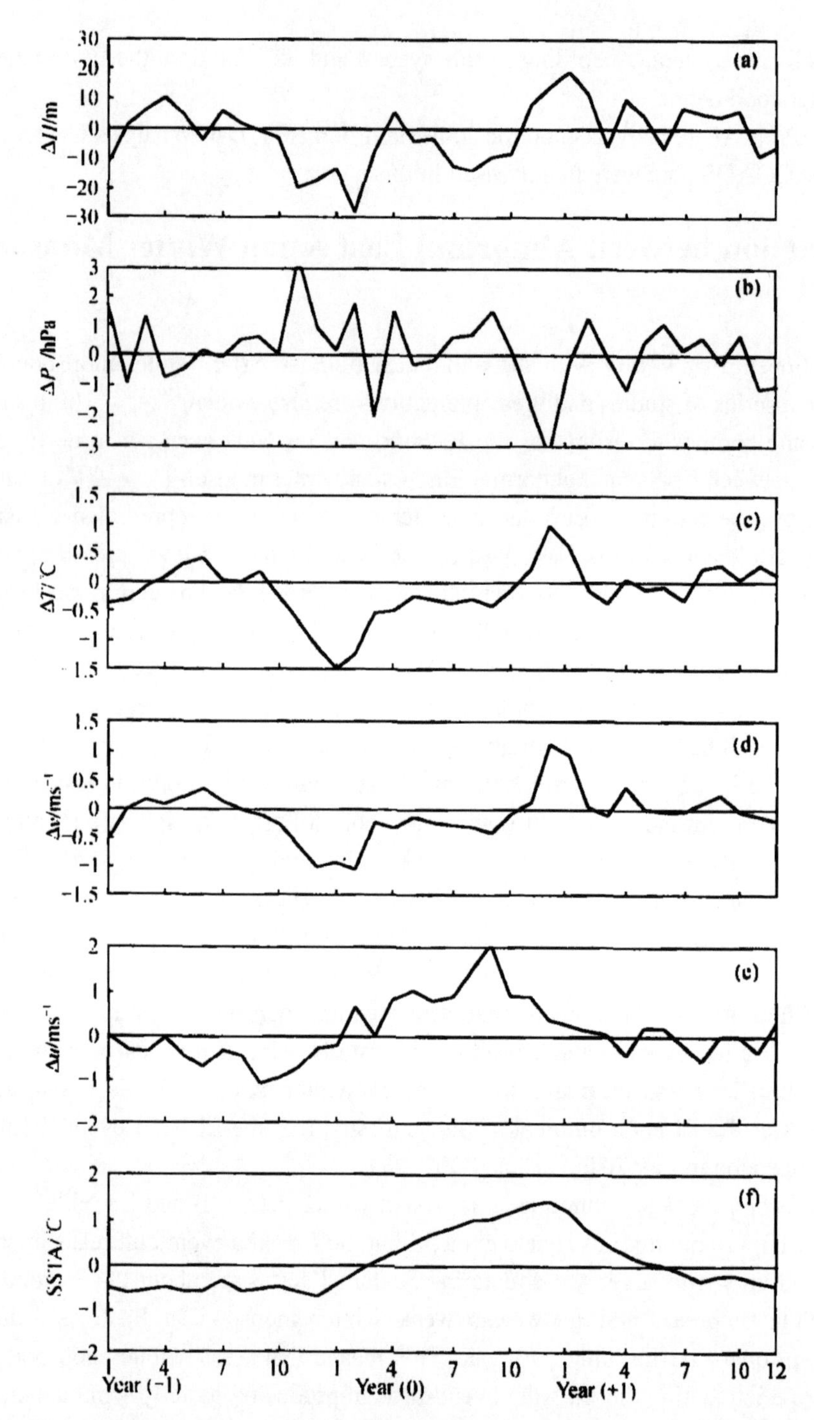

Fig. 1. The relationship between abnormal East Asian winter monsoon and El Niño. Geopotential high anomalies at 500 hPa (gm) in (30°—40°N, 100°—130°E) region (a), sea level pressure anomalies (hPa) in (35°—50°N, 80°—110°E) region (b), surface air temperature anomalies (℃) in (30°—40°N, 120°—130°E) region (c), meridional wind anomalies (m/s) in (25°—35°N, 120°—130°E) region (d). zonal wind anomalies (m/s) over the equatorial western Pacific (e) and the SSTA in Niño 3 region (f) for composite El Niño case.

the Institute of Atmospheric Physics, Chinese Academy of Sciences. The atmospheric component is two-level GCM formulated in the σ-coordinate with the resolution of 4° in latitude and 5° in longitude[15]. The oceanic component is a free surface tropical Pacific Ocean general circulation model with resolution of 1° in latitude and 2° in longitude and 14 unequal layers in the vertical with flat-bottom (4 000 m), of which the domain is 121°E—69°W and 30°S—30°N[16]. In order to control the "climate drift" in the coupled system, the linear statistical correlation is used in the model integration[17].

The numerical simulations in the CGCM showed some similar results to those in the data analyses, abnormal strong (weak) East Asian winter monsoon in the wintertime will lead to positive (negative) SSTA in the equatorial eastern Pacific as El Niño (La Niña) like pattern. Since the space is limited, the simulated SSTA in Niño 1+2 and in Niño 3 caused by abnormal strong East Asian winter monsoon, which is represented by using positive anomalies of the surface pressure (maximum +14 hPa) and negative anomalies of surface air temperature (maximum – 4℃) in the Siberia-Mongolia region in the period of November—April, are just given in this paper. It is clearly shown in fig. 2 that abnormal strong East Asia winter monsoon can excite positive SSTA in the equatorial eastern Pacific like El Niño, through air-sea coupled interaction. Moreover, similar to the observation results, it can be shown that the westerly wind anomalies and enhanced intraseasonal oscillation over the equatorial central-western Pacific caused by strong East Asian winter monsoon play an important role in occurrence of El Niño (figure omitted). In the above simulations, the anomalies of East Asian

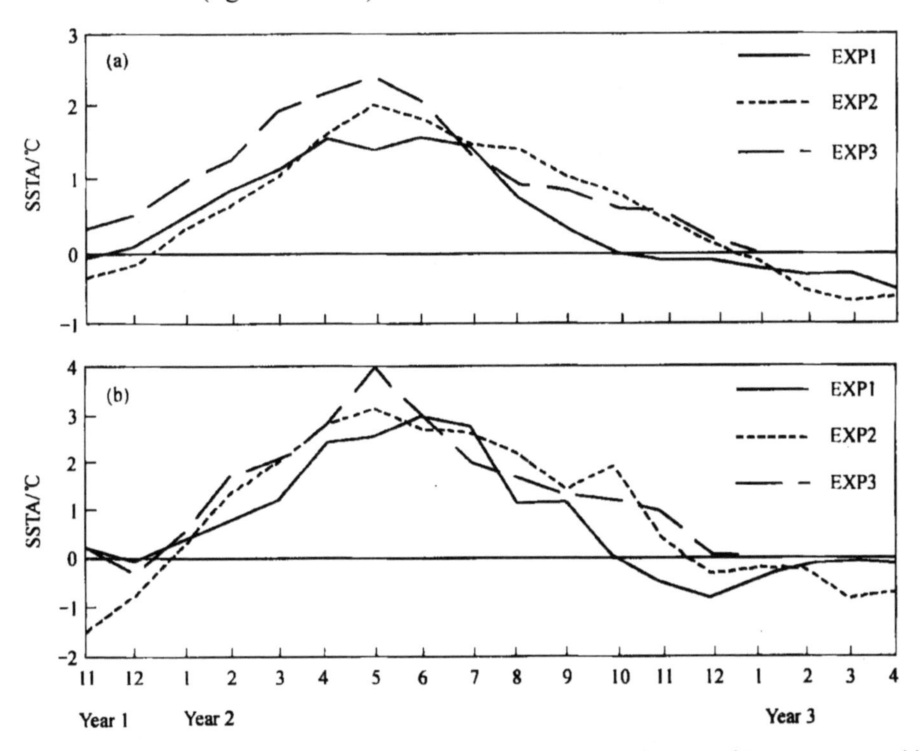

Fig. 2. Temporal variations of the simulated SSTA in Niño 1+2 (a) and Niño 3 (b) regions caused by the anomalous strong winter monsoon in East Asia. Three lines represent the results for different initial fields.

winter monsoon were just in the period of November—April, which were shorter than those observed. If the anomalies continued into summer as shown in fig. 1, the simulated result will be much similar to the observation, which showed the maximum of SSTA in November.

2. Occurrence of ENSO and Subsurface Ocean Temperature Anomaly in the Warm Pool Region

The strongest El Niño event in this century outbroke in summer of 1997. But the observation showed that there were obviously positive anomalies of subsurface ocean temperature (SOT) in the warm pool region before the occurrence of El Niño, the anomaly of SOT began in the autumn of 1996. The eastward propagation of positive anomalies of SOT from the equatorial western Pacific to the equatorial eastern Pacific and expanding to the sea surface can be regarded as the origin of the El Niño event. When the El Niño event was onset, there were negative anomalies of SOT in the warm pool, then, the negative anomalies of SOT propagated eastwards from the warm pool region to the equatorial eastern Pacific along the thermocline and excited the La Niña in 1998.

In relation to the evolution of SOT anomalies in the equatorial Pacific in the 1997—1998 ENSO period, most of us have found it in *Climate Diagnostics Bulletin* published in USA, it is not necessary to show that again. But it can be suggested that the 1997—1998 ENSO is closely related to the anomalies of SOT in the warm pool region and the eastward propagation of the anomalous SOT.

In fact, the important role of the subsurface ocean temperature anomalies in the warm pool region in the occurrence of ENSO was not only shown in the 1997—1998 ENSO, but also in most of the historical events. As we know, the SSTA in Niño 3 can represent the ENSO process very well, and the anomaly of SOT during 100—200 m layer in (10°S —10°N, 140°E —180°) region can represent thermal regime of subsurface in the warm pool. In order to investigate the relationship between ENSO and anomaly of SOT in the warm pool region, the temporal variations of the SSTA in Niño 3 and the SOTA in warm pool arc given in fig. 3, in which the SOTA and SSTA are computed by using the Joint Environmental Data Analysis Center (JEDAC) data and the COADS data, respectively. It is very clear that the positive anomalies of the SOT always exist in the warm pool region prior to the occurrence of El Niño event, but after the onset of El Niño, the SOTA in the warm pool region are out-of-phase to the SSTA in Niño 3. Moreover, it is also shown that the subsurface warning in the warm pool region leads to the onset of El Niño event for half to two years.

By analyzing each El Niño event, it can be shown that the eastward propagation of positive anomalies of the SOT is directly related to the occurrence of El Niño event, when positive anomalies of the SOT propagated into the equatorial eastern Pacific (from the warm pool region), the positive anomalies of SST will occur since the thermocline is gradually raised in the equatorial eastern Pacific. In other words, the positive SOTA in the warm pool region and its eastward propagation into the equatorial eastern Pacific can be regarded as an important origin of the El Niño occurrence. In fig. 4, the time-longitude sections of SOTA in the

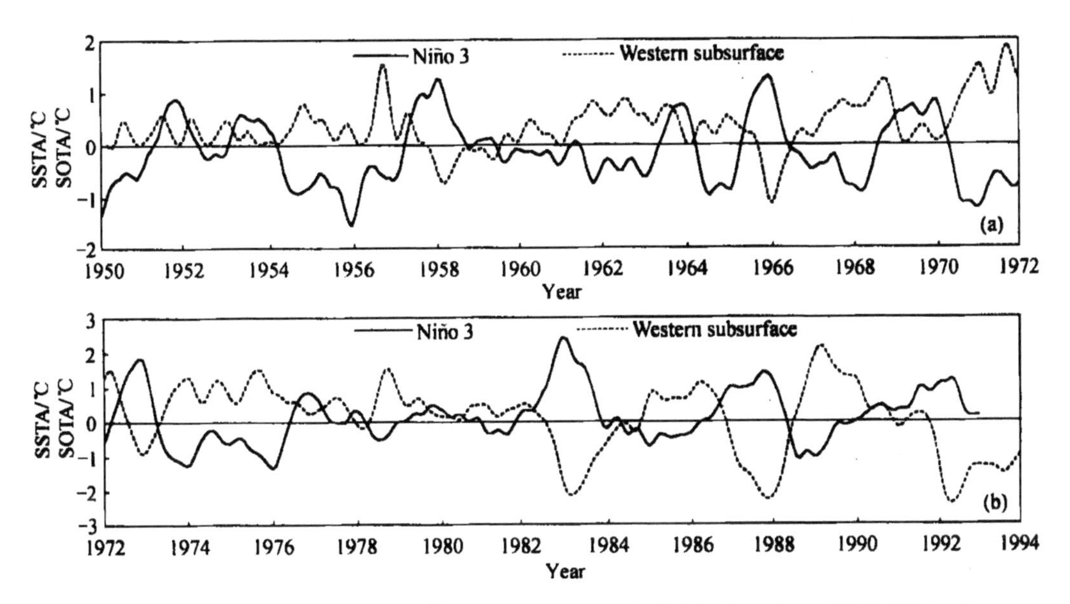

Fig. 3. Temporal variations of the SSTA (℃) in Niño 3 region (solid line) and the SOTA (℃) in the warm pool region (dashed line).

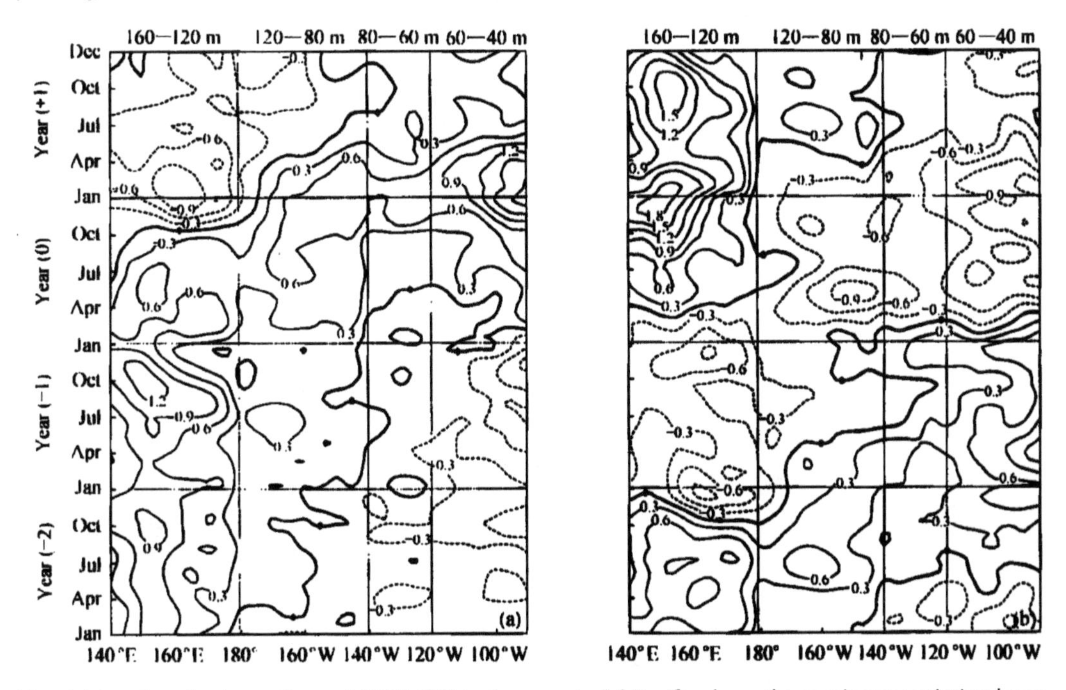

Fig. 4. Time-longitude sections of SOTA (℃) in the equatorial Pacific along the maximum variation layer (thermocline) of ocean temperalure for composite El Niño case (a) and La Niña case (b).

equatorial Pacific are given for composited El Niño cases and La Niña cases, respectively. The data of SOTA are respectively adopted in different depths, such as 160—120 m in the equatorial western Pacific, 120—60 m in the equatorial centre Pacific and 60—40 m in the equatorial eastern Pacific, because the thermocline is deeper and thicker in the equatorial western Pacific than that in the equatorial eastern Pacific. It is also evident that there were obviously positive (negative) anomalies of SOT in the warm pool region for quite a long time prior to the El Niño (La Niña) event, when the positive (negative) SOTA propagated eastwards into the equatorial eastern Pacific (east of 160°—150°W) along the thermocline, the El Niño (La Niña) will outbreak.

In order to expose further the important role of SOTA in the warm pool region in exciting ENSO, the numerical simulation data in the CGCM (which has been introduced in the first section) are analyzed for the 61st —100th years integrations. During the 40 years (the 61st—100th), 13 simulated warm (El Niño) events and 12 simulated cold (La Niña) events are clearly shown, even though the intensity of SSTA is weaker than the observation. The temporal variations of the simulated SSTA in Niño 3 region and the SOTA in the warm pool region (6°S—6°N, 140°E—180°) are shown in fig. 5. It can be seen that there are obviously positive anomalies of SOT in the warm pool region prior to most of the warm events and the SOTA in the warm pool region become negative after the appearance of warm events. On the contrary, the cold events are always corresponding to negative SOTA in the warm pool region in the earlier stage.

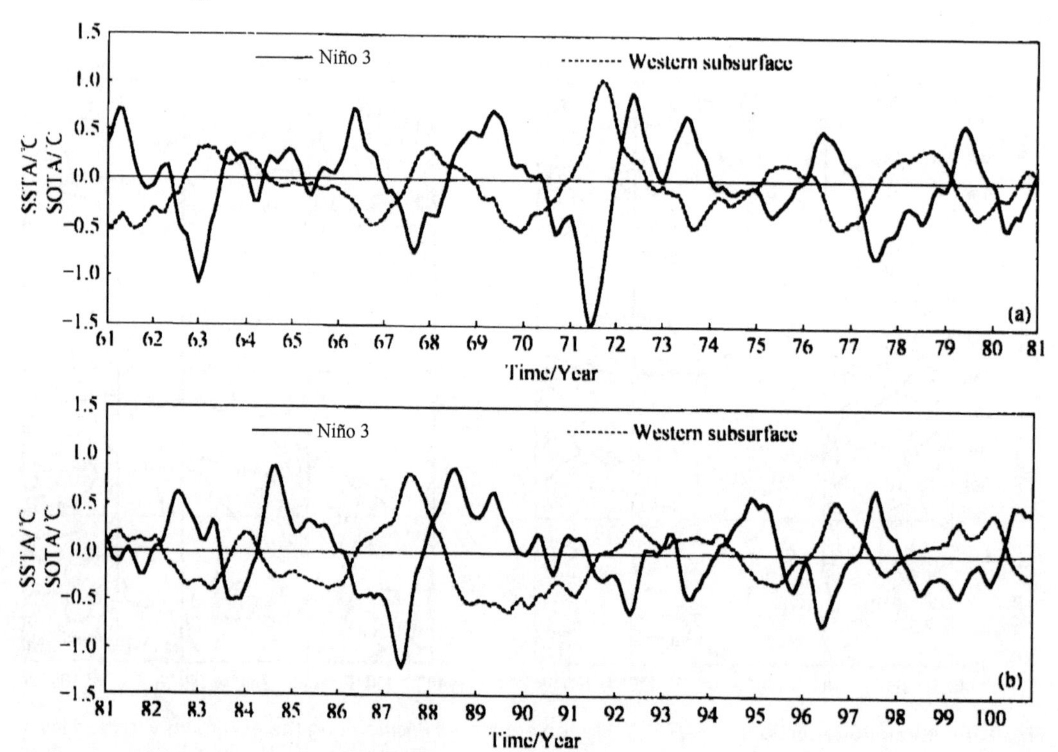

Fig. 5. Temporal variations of the simulated SSTA in Niño 3 region (solid line) and SOTA in the warm pool (6°S—6°S, 140°E—180°) region (dashed line).

The above-mentioned discussions have shown that the data analyses and numerical simulations in the CGCM indicated the same results: the occurrence of ENSO is closely related to the anomaly of SOT in the warm pool region, there are positive (negative) SOTA in the warm pool region prior to the El Niño (La Niña) event; the onset of ENSO is directly related with the eastward propagation of SOTA in the warm pool region, the eastward propagating of positive (negative) SOTA into the equatorial eastern Pacific and expanding to the sea surface will lead to the anomalies of SST in the equatorial eastern Pacific and the onset of El Niño (La Niña) event. After the occurrence of ENSO, the SSTA in Niño 3 and the SOTA in the warm pool region are in the opposite phase.

3. Important Role of the Westerly Wind Anomaly over the Equatorial Western Pacific

The above discussion clearly shows that the occurrence of ENSO is directly related to eastward propagation of the SOTA in the warm pool region. What is the factor to cause the eastward propagation of the SOTA? In fig. 6, the temporal variations of the composite SOTA in the warm pool region, westerly wind anomalies over the equatorial western Pacific

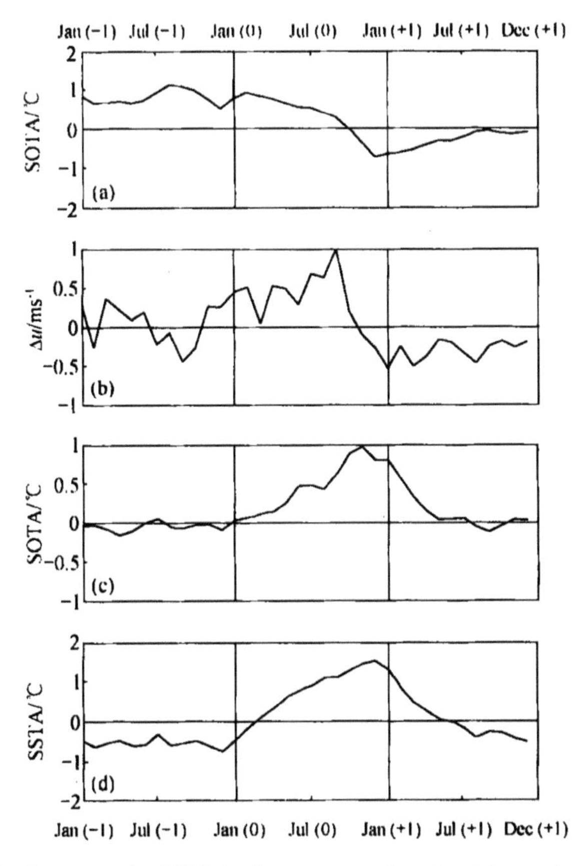

Fig. 6. The relationships between the SOTA in the warm pool region (a), zonal wind anomalies over the equatorial western Pacific (b), the SOTA in (5°S—5°N, 170°—130°W) region (c), which can represent the eastward propagation of SOTA from the warm pool region, and the SSTA in Niño 3 region (d) for the composite El Niño case.

(10°S—10°N, 120°—160°E), SOTA in the equatorial eastern Pacific (5°S—5°N, 170°—130°W) region and SSTA in Niño 3 region (5°S—5°N, 150°—90°W) are respectively shown for El Niño cases. Fig. 7(d) clearly shows the evolution of SSTA in Niño 3 region for the El Niño event; before the occurrence of positive SSTA in Niño 3, positive SOTA have propagated into the equatorial eastern Pacific as shown in fig. 6(c); but the westerly wind anomalies over the equatorial western Pacific occurred earlier (fig. 6(b)) and it can be regarded as a mechanism to cause eastward propagation of SOTA in the warm pool region; fig. 6(a) shows that before the occurrence of El Niño event (about 1 year), there were obviously positive SOTA in the warm pool region.

The similar evolution for La Niña cases is shown in fig. 7, negative SSTA in Niño 3 corresponds to early negative SOTA and the easterly wind anomalies over the equatorial western Pacific. In other words, the easterly wind anomalies over the equatorial western Pacific will lead to the eastward propagation of negative SOTA in the warm pool region, when negative SOTA propagated into the equatorial eastern Pacific, negative SSTA will appear in the equatorial eastern Pacific and La Niña will occur.

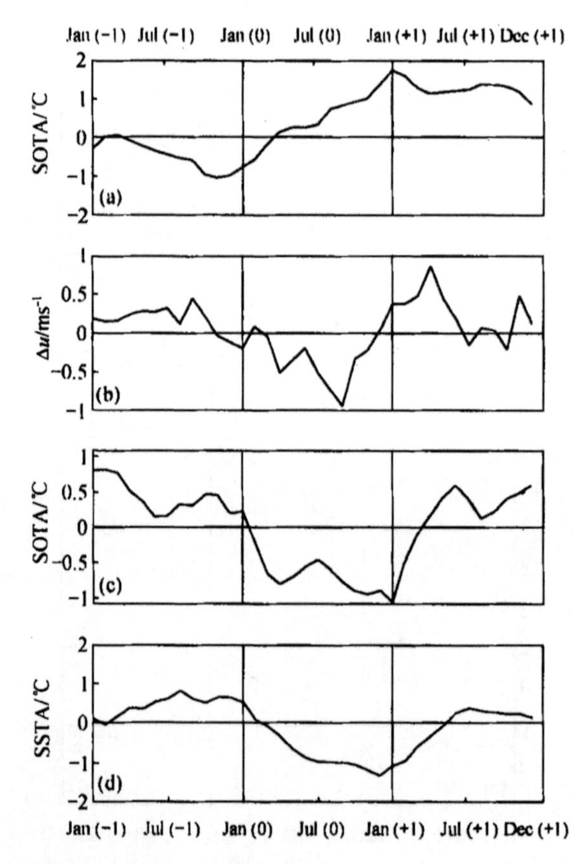

Fig. 7. Same as in fig. 6, but for composite La Niña case.

Based on figs. 6 and 7, it can be suggested that the westerly wind anomalies over the equatorial western Pacific play an important role in the eastward propagation of SOTA in the warm pool region and the occurrence of ENSO. The oceanic Kelvin wave may be a fundamental mechanism causing the eastward propagation of the SOTA in the warm pool region, because some studies have indicated that the westerly (easterly) wind anomalies over the equatorial western Pacific can excite anomalous oceanic warm (cold) Kelvin wave[18, 19].

Figs. 1 and 2 clearly show that continued westerly (easterly) wind anomalies over the equatorial western Pacific are closely related to the strong (weak) winter monsoon in East Asia. Therefore, we can still say that the anomaly of zonal wind over the equatorial western Pacific, which is mainly caused by anomalous winter monsoon in East Asia, is an important mechanism to drive the eastward propagation of SOTA in the warm pool region and excite the ENSO.

4. Conclusion

Through the data analyses and numerical simulations with a CGCM, the new perspective associated to the occurrence of ENSO is advanced in this paper. It can be represented as follows:

The continued strong (weak) winter monsoon in East Asia will lead to continued westerly (easterly) wind anomalies over the equatorial western Pacific; then anomalous westerly (easterly) wind causes the eastward propagation of positive (negative) anomalies of SOT in the warm pool region, which have been there for quite a long time; the eastward propagating positive (negative) SOTA along the thermocline will lead to positive (negative) SSTA in the equatorial eastern Pacific and the occurrence of El Niño (La Niña) event. After the occurrence of ENSO, the winter monsoon in East Asia will be weaker (stronger) due to the influence of the El Niño (La Niña) event, as shown in previous studies.

Therefore, ENSO cycle is closely related to the SOTA in the warm pool region and the anomaly of winter monsoon in East Asia. There is a cycle evolution which showed the interactions between the East Asian winter monsoon, SOTA in the warm pool region and ENSO. Of course, two questions need to be investigated further: what is the mechanism to cause the anomaly of SOT in the warm pool region? How about the physical process, through which anomalous zonal wind over the equatorial western Pacific can lead to the eastward propagation of the anomalous SOT in the warm pool region?

Acknowledgements We would like to thank Wang Xuan for her typing this paper. This work was supported by the National Key Basic Science Program in China (Grant No. 1998040903) and the National Natural Science Foundation of China (Grant Nos. 49823002 and 49635180).

References

[1] Bjerknes, J., Atmospheric teleconnections from the equatorial Pacific, Mon. Wea. Rev., 1969, 97: 163.
[2] Rasmusson, E. M., Wallace, J. M., Meteorological aspects of El Niño/Southern Oscillation, Science, 1983,

222: 1195.
[3] Wyrtki, K., El Niño —— the dynamic response of the equatorial Pacific ocean to atmospheric forcing, J. Phys. Oceanogr., 1975, 5: 572.
[4] Philander, S. G., Yamaguta, T., Pacanowski, R. C., Unstable air-sea interactions in the tropics, J. Atmos. Sci., 1984, 41: 604.
[5] Cane, M. A., Zebiak, S. E,, A theory for El Niño and Southern Oscillation, Science, 1985, 228: 1085.
[6] Hirst, A. C., Unstable and damped equatorial modes in simple coupled ocean-atmosphere model, J. Atmos. Sci., 1986, 43: 606.
[7] Suarez, M. J., Schopf, P., A delayed action oscillator for ENSO, J. Atmos. Sci., 1988, 45: 3283.
[8] Battisit, D. S., The dynamics and thermodynamics of a warm event in a coupled ocean-atmosphere model, J. Atmos. Sci., 1988, 45: 2889.
[9] Neelin. J. D., The slow sea surface temperature mode and the fast-wave limit: Analytic theory for tropical interannual oscillations and experiments in a hybrid coupled model, J. Atmos. Sci., 1991, 48: 584.
[10] Li. C., The frequent activities of stronger aerotroughs in East Asia in wintertime and the occurrence of the El Niño event, Scientia Sinica, Ser. B, 1989, 32: 976.
[11] Li. C., Interaction between anomalous winter monsoon in East Asia and El Niño events, Advances in Atmospheric Sciences, 1990, 7: 36.
[12] Li, C., Mu, M., ENSO cycle and anomalies of winter monsoon in East Asia, in East Asia and Western Pacific Meteorology and Climate, Singapore World Scientific, 1998, 60-73.
[13] Li, C., Mu, M., Numerical simulations on anomalous winter monsoon in East Asia exciting ENSO, Chinese J. Atmos. Sci., 1998, 22: 481.
[14] Li, C., ENSO cycle and anomalies of winter monsoon in East Asia, Workshop on El Niño, Southern Oscillation and Monsoon, ICTP, SMR/930-18, Trieste, 15—26 July 1996.
[15] Zeng, Q., Zhang, X., Liang, X. et al,, Documentation of IAP Two-Level AGCM, TR044, DOE/ER/603I4-HI, U.S. DOE, 1989, 383.
[16] Zhang, R., Endoh, M., A free surface general circulation model for the tropical Pacific Ocean, J. Geophys. Res., 1992, 97(c7): 11237.
[17] Zhou, G., Li, C., Simulation of the relation between the subsurface temperature anomaly in western Pacific and ENSO by using CGCM, Climatic and Environmental Research, 1999, 4: 346.
[18] Yamagata, T., On the recent development of simple coupled ocean-atmosphere models of ENSO, J. Oceanogr. Soc. Japan, 1986, 42: 299.
[19] Huang, R. H., Fu, Y. K, Zhan, X. Y., Asian monsoon and ENSO cycle interaction. Climate and Environmental Research.1996, 1: 38.

A Further Study of the Essence of ENSO*

Li Chongyin (李崇银), Mu Mingquan (穆明权)

LASG, Institute of Atmospheric Physics, Chinese Academy of Sciences, Beijing

Abstract Based on scientific theory — the ENSO results from air—sea interaction in the tropical Pacific, the analyses in this study show that the precedence of ENSO is in the warm pool of equatorial western Pacific and the occurrence of ENSO is directly related to the subsurface ocean temperature anomalies (SOTA) in the equatorial western Pacific and its eastward propagation. The SOTA shows an interannual cycle very clearly, the eastward propagation of positive (negative) SOTA along the equator always associated to the westward propagation of negative (positive) SOTA along 10°N and 10°S latitudes. The analyses still show that the cycle of SOTA in the tropical Pacific is driven by zonal wind anomalies over the equatorial center-western Pacific, which is mainly caused by anomalous East-Asian winter monsoon. So that the data analyses and numerical simulations can show that the El Niño(La Niña) is closely related to anomalous strong (weak) winter monsoon in East Asia. Therefore summarily, it can be suggested that ENSO is exactly the cycle of SOTA in the tropical Pacific driven by zonal wind anomalies over the equatorial center-western Pacific, which mainly results from anomalous East-Asian winter monsoon.

Keywords ENSO cycle, East-Asian monsoon anomaly, zonal wind anomaly, subsurface ocean temperature anomaly.

1. Introduction

ENSO cycle is a strongest signal of interannual climate variation, its occurrence usually cause to produce climate anomalies and disasters (especially the flood and drought) within great scopes in the world, so that it has been paid maximum attention and studied extensively in the world (Chen, 1977; Rasmusson and Wallace, 1983; Zhang and Wang 1984). In order to explain the reasons for ENSO occurrence, the trade wind tension-relaxation theory (Wyrtki, 1975), instable oceanic waves theory (Philander et al., 1984; Cane and Zebiak, 1985) and the delay-oscillator theory (Hirst, 1986; Suarez and Schopf, 1988) were advanced in succession. Chinese scientists also have a few studies in this field and indicated that equatorial oceanic Rossby wave also pays an important role in occurrence of the El Niño event (Neelin, 1991). Above studies and theories can partly explain the occurrence and evolution features of the

Received February 20, 2002; revised October 8, 2002.
*Project supported by the National Natural Science Foundation of China under Grant No. 40233033, the Chinese Academy of Sciences under Grant No.ZKCX2-SW-210, and the National Key Program (No.2001BA603B-04).

ENSO, but they still exist individual deficiency. The occurrence mechanism of ENSO has not been understood really, the prediction for the ENSO is still in experiment stage.

In above theories, the oceanic effect is particularly emphasized, it seems to be contrary to Bjerknes' viewpoint: ENSO is the result of air-sea interaction in the equatorial Pacific (Bjerknes, 1969). The important impact of atmospheric circulation anomalies on the occurrence of ENSO should be attached. It has been indicated in our previous studies that the continued strong (weak) anomaly of East-Asian winter monsoon will lead to continued westerly (easterly) wind anomalies and continued strong (weak) anomaly of atmospheric intraseasonal oscillation over the equatorial western Pacific, then the El Niño (La Niña) event can be excited through the air-sea interaction (Li, 1989; 1990). A numerical simulation with the CGCM also fully prove that the atmospheric anomalies in the tropics caused by strong (weak) anomaly of East-Asian winter monsoon, through the air—sea interaction, can lead to positive (negative) SSTA in the equatorial eastern Pacific or the El Niño (La Niña) event during spring—autumn (Li and Mu, 1999). Therefore, it is clear that the occurrence of ENSO is closely related to zonal wind anomalies over the western Pacific caused by anomalous East-Asian winter monsoon; but we don't know how does zonal wind anomaly excite the ENSO?

Recently, the analyses with Joint Environmental Data Analysis Center (JEDAC) ocean temperature data and National Center for Environmental Prediction/National Center for Atmospheric Research (NCEP/NCAR)reanalysis data showed that the occurrence of ENSO is closely related to positive (negative) SOTA (subsurface ocean temperature anomalies) in the western Pacific warm pool and its eastward propagation along the equator (Zhou and Li, 1999). The numerical simulation in a CGCM also has the same results. Linking with our previous studies on the interaction between anomalous East-Asian winter monsoon and the ENSO cycle, a new conception on the interaction of East—Asian winter monsoon, thermal situation of the warm pool and ENSO cycle was advanced in a further study (Li and Mu, 2000).

In this study, the evolution regularity of the SOTA in tropical Pacific will be investigated further to reveal its relation with ENSO cycle and the anomalies of zonal wind over the equatorial western Pacific and East-Asian winter monsoon. So that a new viewpoint on the ENSO cycle will be advanced: The ENSO is actually cycle of the SOTA in the tropical Pacific driven by anomalous zonal wind over the equatorial western Pacific, which resulted mainly from anomalous East-Asian winter monsoon.

2. ENSO and Winter Monsoon Anomaly in East Asia

The interaction between the ENSO and East-Asian winter monsoon anomaly has been indicated in our previous studies. Some major results will be shown here, because we need to consider them for putting forward a new result.

A series of analyses with observation data showed that the East-Asian winter monsoon was strong prior to the occurrence of El Niño event, and the winter monsoon is weak in the outbreak year of El Niño. Generally, strong (weak) winter monsoon in East Asia can be represented by deep (shallow) 500 hPa trough over East Asia, strong (weak) Siberia surface high, low (high) temperature in Eastern China and the northerly (southerly) wind anomalies over the northwestern Pacific/East Asia region. In Fig. 1, the composite time-longitude section of 500 hPa height anomalies for El Niño and La Niña are respectively shown to represent the relationship between the ENSO and East-Asian winter monsoon anomaly. It is very clear that there were negative anomalies of the height over East Asia before the occurrence of El Niño, but positive anomalies after the outbreak of El Niño (Fig. la). This means strong winter monsoon in East Asia can probably excite the occurrence of El Niño and the El Niño event will reduce winter monsoon in East Asia. For the composite pattern of La Niña (Fig. lb), the result is similar, i.e., weak winter monsoon in East Asia can probably excite the occurrence of La Niña and the occurrence of La Niña event will enhance winter monsoon in East Asia. The same result can be got by using other meteorological elements mentioned the above. Therefore, the interaction between the ENSO and winter monsoon anomaly in East Asia is an irrefutable fact.

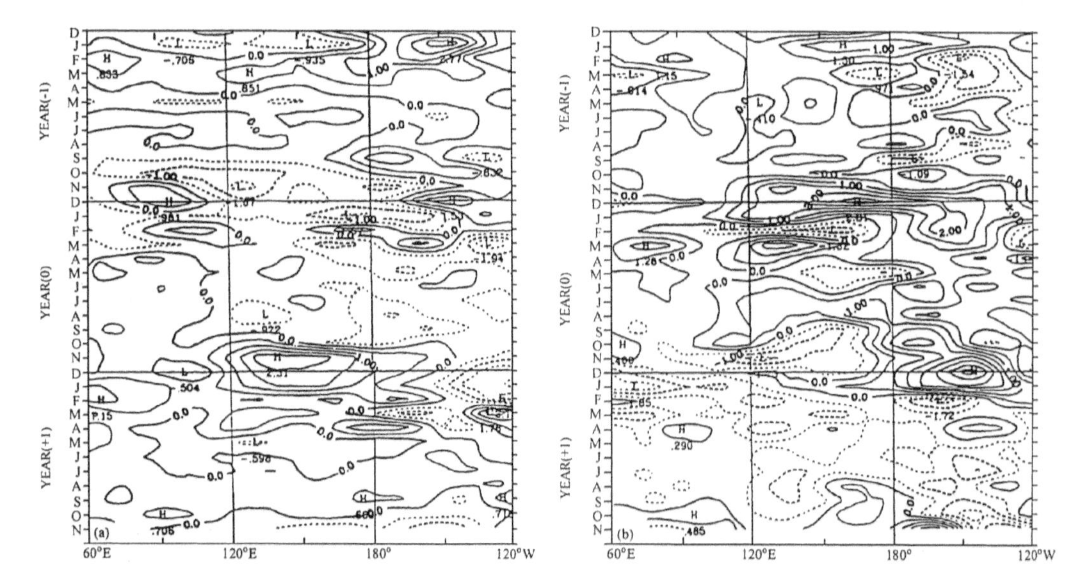

Fig. 1. Time-longitude sections of the composite height anomalies at 500 hPa in (20-45°N) latitudes for El Niño cases (A) and La Niña cases (B). The dashed lines represent negative anomalies.

To use a coupled atmosphere-ocean model, the impact of anomalous winter monsoon in East Asia on the oceanic temperature in the tropical Pacific was simulated. The difference between anomalous experiment and control run can be regarded as the response of tropical pacific to winter monsoon forcing. Here the excited SSTA in Niño l+2 region and Niño3 region due to strong winter monsoon forcing (November-April) are shown in Fig. 2, respectively. It can be seen clearly that the strong winter monsoon excited obvious positive SSTA in the equatorial eastern-center Pacific in the next year. This means that strong winter monsoon in

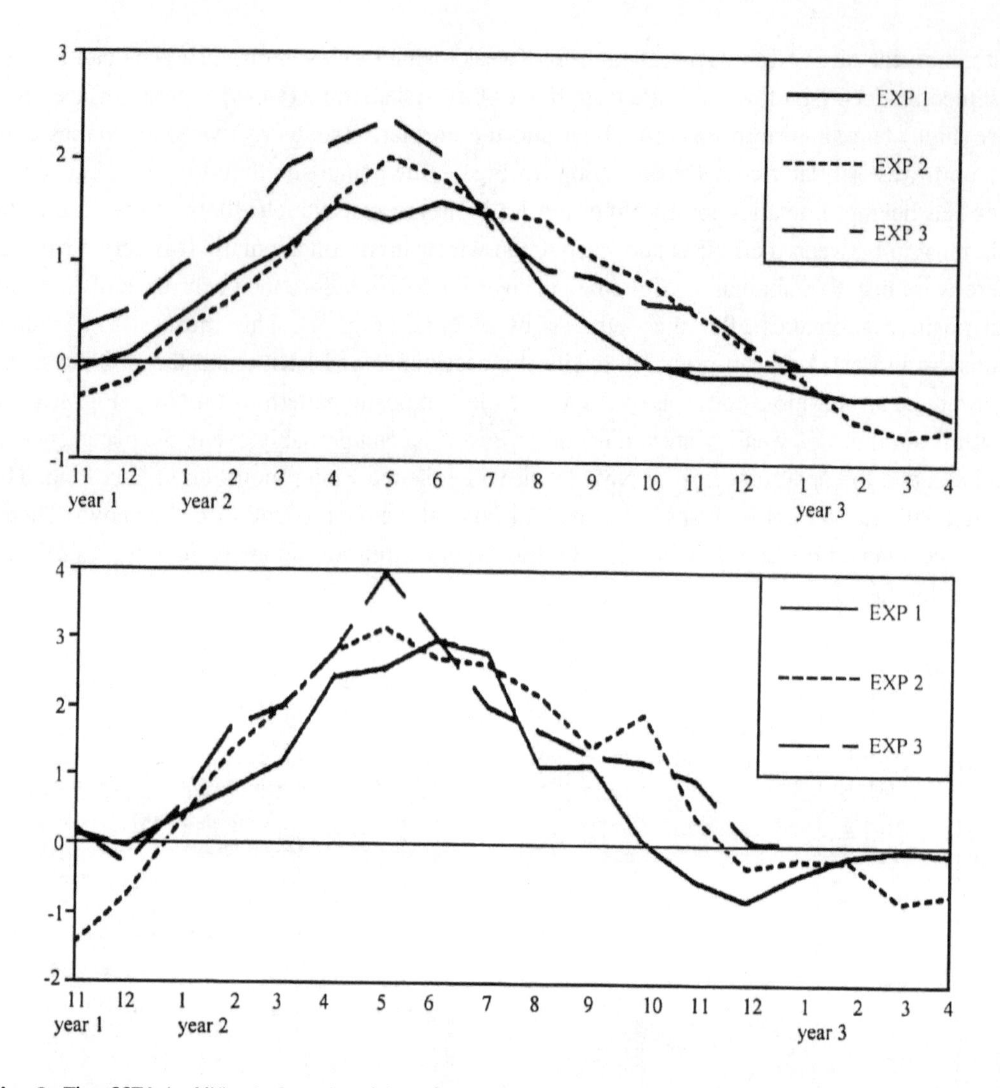

Fig. 2. The SSTA in Niño 1+2 region (a) and in Niño3 region (b) excited by strong East-Asian winter monsoon in the CGCM for 3 deferent initial conditions.

East Asia play an important role in the occurrence of El Niño event. The simulation also showed that the weak winter monsoon in East Asia will be favourable to excite La Niña (figure omitted).

3. The Cycle of Sota and ENSO

The strongest El Niño event in the 20-century was occurred in early summer 1997. But before the occurrence of El Niño event, positive SOTA has been in the equatorial western Pacific warm pool for a long time. The eastward propagation of positive SOTA from warm pool to the equatorial eastern Pacific and upward extending to the surface was direct reason of the El Niño occurrence. After the El Niño occurrence, the SOTA in warm pool became negative, then this negative SOTA propagated eastwards and extended to the surface, the La Niña also occurred in 1998 (Zhou and Li, 1999; Li and Mu, 2000). A interesting phenomenon can be still shown, when positive SOTA propagated eastwards along the equator, there were westward propagations of negative SOTA along 10°N and 10°N latitudes (the propagation along 10°N is even more evident), then warm pool will be controlled by the negative SOTA. Similarly, when negative SOTA in the warm pool propagated eastwards along the equatorial thermocline and the La Niña took shape, the positive SOTA propagated westwards along 10°N and 10°S latitudes from the equatorial eastern Pacific, then warm pool will be controlled by the positive SOTA. In other words, the ENSO (El Niño-La Niña) cycle can be regarded as the cycle of SOTA along the equator and 10°N/10°S two latitudes in the tropical Pacific thermocline.

What is the reason to cause this cycle of the SOTA? In our previous studies (Zhou and Li, 1999; Li and Mu, 2000), it has been indicated that the direct important reason is zonal wind anomalies over the equatorial western Pacific caused by the continued anomaly of East—Asian winter monsoon. For the El Niño occurrence in 1997, it was also shown that the anomaly of atmospheric circulation (actually East—Asian winter monsoon anomaly) in mid-latitude of East-Asian region played an important role in the occurrence of El Niño event (Yu and Rienecker, 1998). Obviously, the exciting effect of anomalous East-Asian winter monsoon on occurrence of the El Niño is not the isolated case.

In order to reveal the circulatory process of SOTA in the tropical Pacific during 1997-98 ENSO and its relationship with westerly wind anomalies over the equatorial western Pacific and East-Asian winter monsoon anomaly, the monthly anomalous patterns of SST in the tropical Pacific and the atmospheric circulation over that region are analyzed. Since the space is limited, in Fig. 3, we only showed the 850 hPa meridional wind anomalies (VA) over East Asia and tropical Pacific region, 850 hPa zonal wind anomalies (UA) over the tropical Pacific, SOTA and SSTA in the tropical Pacific. Here the value of SOTA is selected along the thermocline, it is the value at 160-120 m depth in the warm pool, at 120-80 m depth in the center—eastern Pacific and at 80—40 m depth in the eastern Pacific. It can be shown in November 1996 that there were some old scraps of negative SSTA in the equatorial eastern Pacific, but there were positive SOTA in the warm pool and negative SOTA in the equatorial eastern Pacific (Fig. 3a). Then positive SOTA in the warm pool were enhanced and expanded eastwards; in February 1997 there were still negative SSTA in the equatorial eastern Pacific, but positive SOTA have been propagated into the equatorial eastern Pacific; when positive SOTA propagated eastwards from the warm pool, it is also evident that the negative SOTA propagated westwards from eastern Pacific along 10°N and 10°S, respectively (Fig. 3b). In June 1997, the El Niño event has occurred and there were strong positive SSTA in the equatorial eastern Pacific, positive SOTA were also in the equatorial eastern Pacific but there were obvious negative SOTA in 10°N region of the western Pacific (Fig. 3c). In September 1997 (Fig. 3d), the El Niño was quite strong and the SSTA reached 4℃and more, the positive SOTA in the equatorial eastern Pacific became even more and there were obvious negative

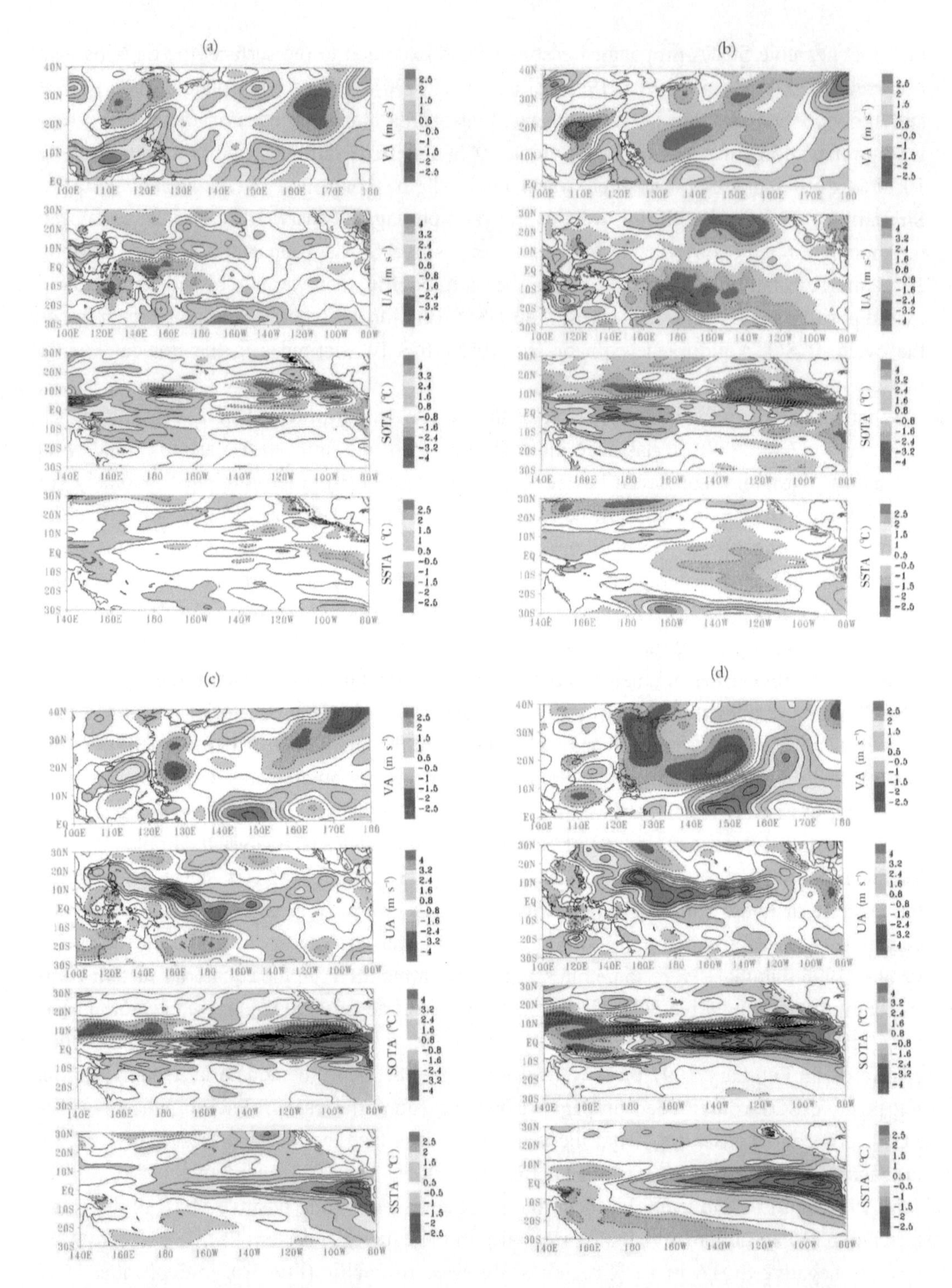

Fig. 3. Monthly anomalous patterns in November 1996 (a), February 1997 (b), June 1997 (c), and September 1997 (d) for 850 hPa meridional wind (VA), 850 hPa zonal wind (UA), the subsurface ocean temperature (SOTA), and sea surface temperature (SSTA).

SOTA (–4℃ and more) in the tropical western Pacific; along 10°N, there were negative SOTA, which may resulted from that the westward propagation was slower. Corresponding to the existence and evolution of SOTA and SSTA in the tropical Pacific, the atmospheric circulation anomalies were also remarkable and began early than ocean temperature. There were anomalous northeasterly wind over eastern China and the northwestern Pacific in November 1996 (Fig. 3a), anomalous easterly wind still dominated the equatorial western Pacific. In February 1997, the northeasterly wind had expanded into tropical western Pacific region, there had been anomalous westerly wind over the equatorial western Pacific and anomalous easterly wind contracted into the equatorial eastern Pacific (Fig. 3b). Then anomalous northeasterly wind slightly moved eastwards and anomalous westerly wind over the equatorial western Pacific strengthened and propagated eastwards. Up to June 1997, anomalous northeasterly wind had disappeared but anomalous westerly wind had expanded into the equatorial eastern Pacific (Fig. 3c). In September 1997, there were anomalous southerly wind over eastern China and anomalous easterly wind over the South China Sea and equatorial western Pacific to the west of the 130°E.

During the generating and developing stage of El Niño, above analyses show clearly that positive SOTA exist in the warm pool and the eastward propagation of this positive SOTA and expanding to the surface of the equatorial eastern Pacific will cause positive SSTA in the equatorial eastern Pacific and El Niño event; the eastward propagation of positive SOTA along the equatorial was directly related to anomalous westerly wind over the equatorial western Pacific and there have been anomalous northeasterly wind over eastern China/northwestern Pacific region (it means the East—Asian winter monsoon were strong continuously) prior to anomalous westerly wind over the equatorial western Pacific. Another phenomenon is that negative SOTA will propagate westwards along 10°N and 10°S latitudes as positive SOTA propagated eastwards along the equator, then a negative SOTA will form in the warm pool and replace original positive SOTA. In other words, accompanying with the occurrence of El Niño, positive SOTA propagated into the equatorial eastern Pacific along the equator from warm pool, negative SOTA in the equatorial eastern Pacific will propagate westwards along the 10°N and 10°S latitudes. Above propagations of the SOTA have formed a cycle, but it is only a half circle for the ENSO cycle because it just showed the propagation of positive SOTA in the tropical Pacific. And anyhow, this half circle of the SOTA is directly related to the westerly wind anomalies over the equatorial western Pacific and continued strong winter monsoon in East Asia.

Using the same method, the patterns of monthly anomalies in December (1997), February 1998, May 1998 and August 1998 are given in Fig. 4, they can show the situation of the transformation from El Niño into La Niña. It is clear that negative SOTA propagated into the equatorial eastern Pacific from warm pool and expanded to the sea surface, then the negative SST and La Niña occurred in the equatorial eastern Pacific. The eastward propagation of negative SOTA along the equator was directly related to the easterly wind anomalies over the equatorial western Pacific and there have been anomalous southerly wind over eastern China/northwestern Pacific region (continued weak East-Asian winter monsoon) prior to anomalous easterly wind over the equatorial western Pacific. While propagating eastwards of negative SOTA along the equator, positive SOTA will propagate westwards along the 10°N and 10°S latitudes. Then, positive SOTA will form in the warm pool, it will provide essential condition to occur next El Niño event. Obviously, Fig. 4 showed another half circle of the SOTA, i.e., negative SOTA propagated eastwards along the equator from warm pool and positive SOTA propagated eastwards along the 10°N and 10°S latitudes from the equatorial eastern Pacific. Thus, Fig. 3 and Fig. 4 together showed a full cycle of SOTA in the tropical Pacific, i.e. the ENSO cycle. For the SOTA cycle—positive (negative) SOTA propagating eastward along the equator and negative (positive) SOTA propagating westwards along the 10°N/ 10°S latitudes, the anomalous zonal wind over the equatorial western Pacific played an important role. Therefore, it can be suggested that the eastward propagation of positive (negative) SOTA in warm pool along a the equator is driven by anomalous westerly (easterly) wind over the equatorial western Pacific, which is mainly caused by anomalous winter monsoon in East Asia.

The analyses on all previous ENSO cycles since 1950 showed similar evolving situation to the 1997-98 ENSO cycle. As an example, Fig. 5 showed the evolving situation of La Niña in 1988. It is clearly shown that negative SOTA in warm pool region propagated eastwards along the equator into the equatorial eastern Pacific and expanded to the sea surface and led to occurrence of La Niña, while positive SOTA propagated westwards along 10°N and 10°S latitudes from equatorial eastern Pacific to western Pacific; The eastward propagation of negative SOTA in warm pool was related to easterly wind anomalies over the equatorial western Pacific and the continued weak winter monsoon in East Asia. Therefore, the cycles of SOTA in the tropical Pacific driven by anomalous zonal wind over the equatorial western Pacific caused mainly by anomalous winter monsoon in East Asia is a reliable interannual variation phenomenon of the climate system.

Even though some studies have shown that the eastward propagation of anomalous sea temperature along the equator mainly results from anomalous Kelvin wave excited by zonal wind anomalies over the equatorial western Pacific; while the westward propagation of anomalous sea temperature along 10°N and 10°S latitudes is caused by oceanic Rossby wave and the equatorial sea currents. Of course, the concrete process and dynamics are required to study further.

4. Temporal Section of the Sota Cycle

We have been shown that the ENSO cycle is actually a cycle of SOTA in the tropical Pacific, while this cycle is closely related to the continued zonal wind anomalies over the equatorial western Pacific and anomalous East-Asian monsoon (mainly the winter monsoon). In order to reveal the cycle feature of SOTA in the tropical Pacific, the time-longitude section of SOTA along the equator (left C) and along 10°N latitude (left A), and the time—latitude section of SOTA in the tropical western Pacific (left B) during 1979-1993 are given in Fig. 6 corresponding to SSTA in Niño3 (right). For convenience, we begin the discussion since 1981 at

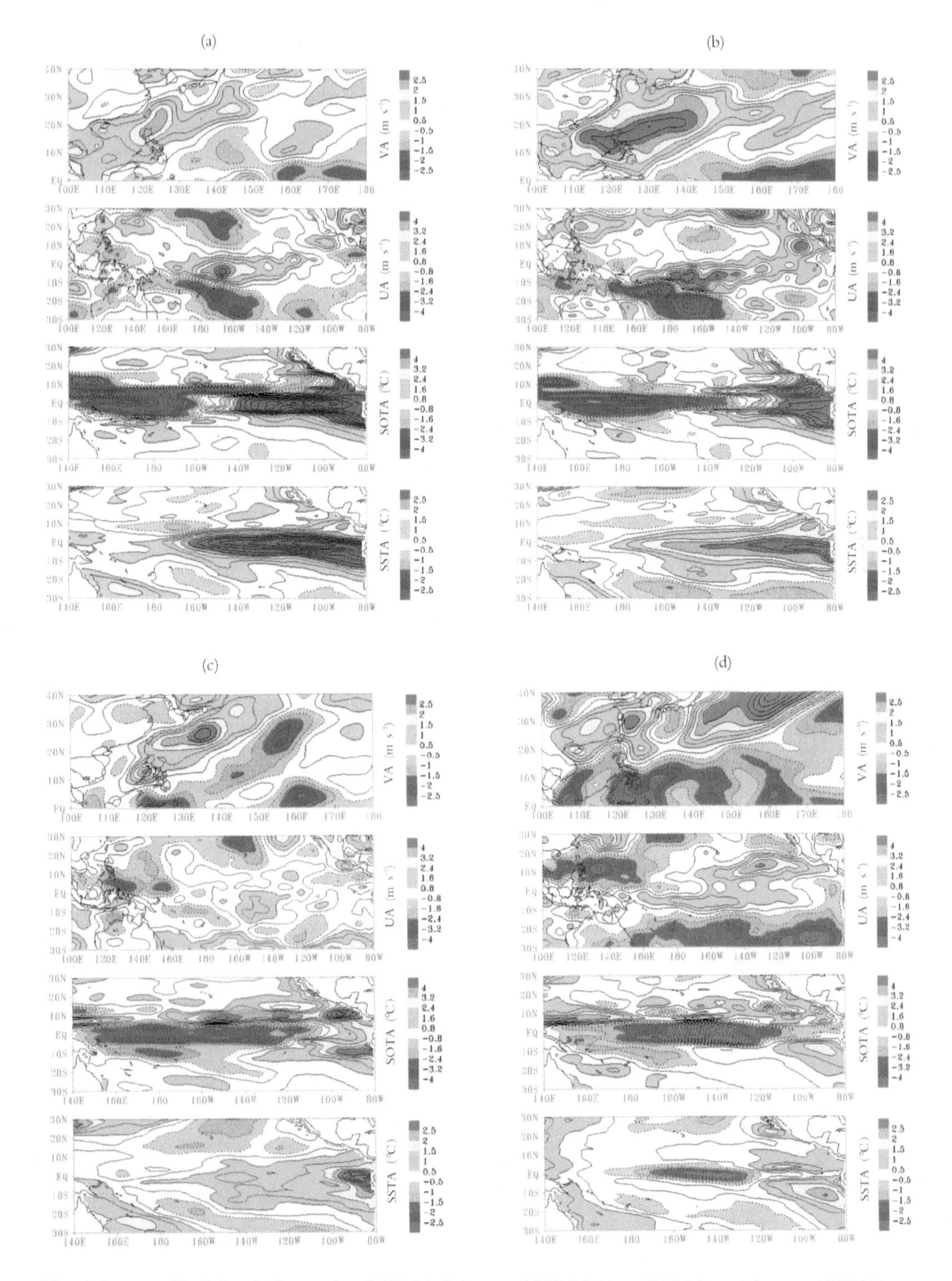

Fig. 4. Same as Fig.3, but in December 1997 (a), February 1998 (b), May 1998 (c), and August 1998 (d).

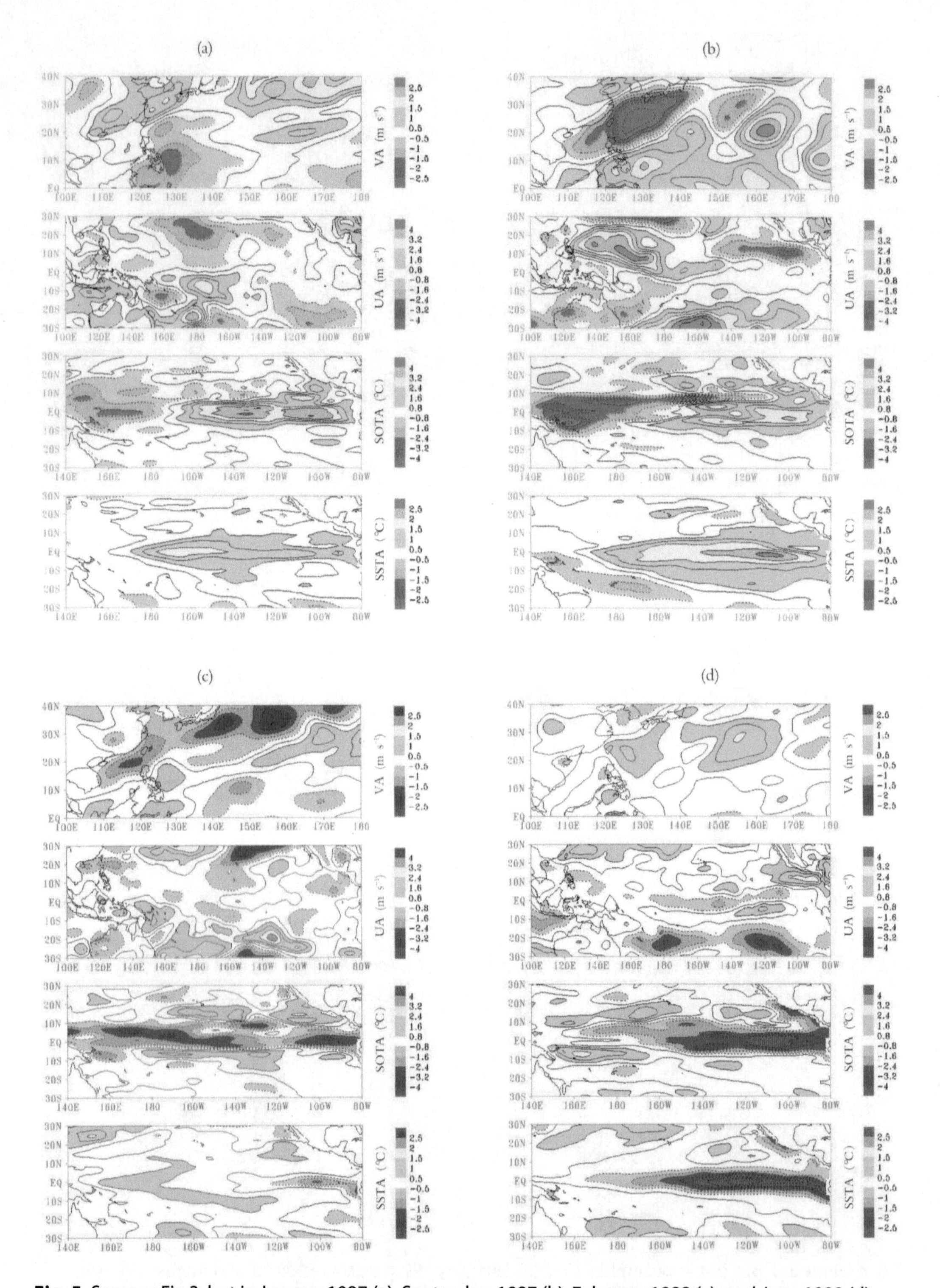

Fig. 5. Same as Fig.3, but in January 1987 (a), September 1987 (h), February 1988 (c), and June 1988 (d).

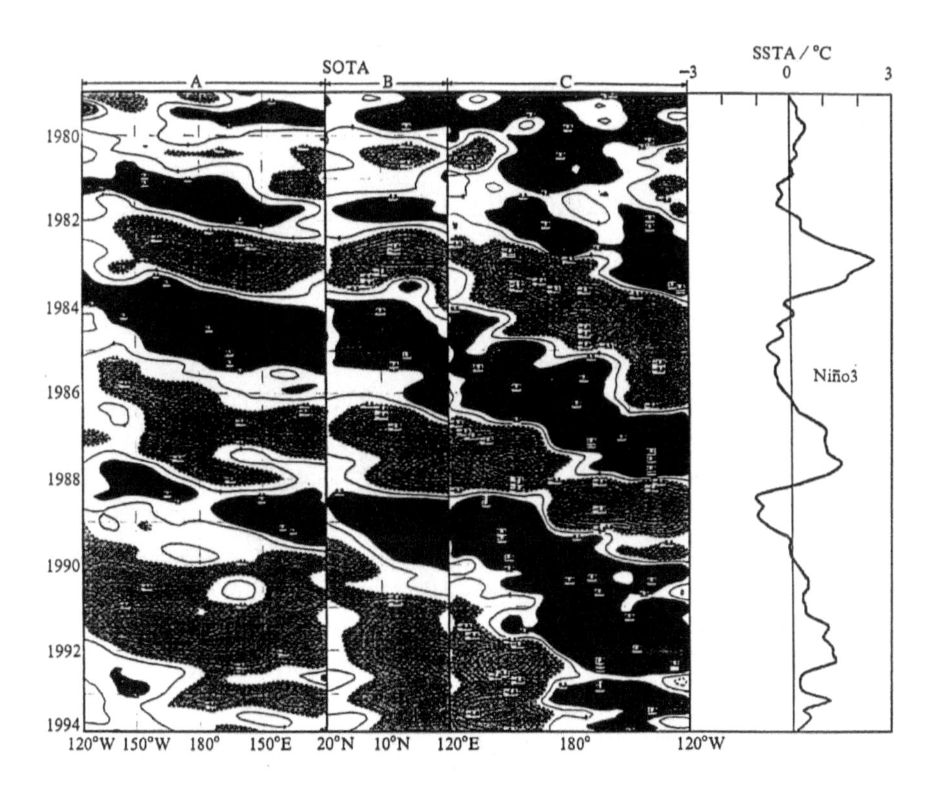

Fig. 6. Time—longitude sections of SOTA in the tropical Pacific along the equator (6°N-6°N, Left part C)and 10°N (Left part A), time—latitude section of SOTA in the western Pacific (120-160°E, Left part B) and the temporal variation of SSTA in the Niño3 (Right).

the left part C in Fig. 6. It is clear that there was eastward propagation of positive SOTA from (140—180°E) region, positive SOTA reached the equatorial eastern Pacific and caused strong SSTA there. Then, positive SOTA propagated westwards along the 10°N latitude (left A) and reached the western Pacific in 1984, and reached the equatorial western Pacific through southward propagation (left B). After that, positive SOTA in the warm pool region propagated eastwards again in 1985 (left C) and formed positive SOTA in the equatorial eastern Pacific in summer 1986, the El Niño event also occurred in summer 1986, ······. Similarly, negative SOTA went the same cycle processes and formed the La Niña in 1984 and in 1988.

There only is the SOTA propagations along the equator and 10°N latitude in Fig. 6, the result very similar with Fig. 6 will be shown if we analyze the SOTA propagations along the equator and 10°S latitude. Therefore, it will be said again that the ENSO (El Niño—La Niña) cycle is actually a reflection in the equatorial eastern Pacific of interannual cycle of the SOTA in the tropical Pacific along the equator and 10°N/ 10°S latitudes. The strongest SOTA is in warm pool of the western Pacific, but there the thermocline is deeper (150—200 m), so that SSTA in the equatorial western pacific is not large. Since the thermocline in the equatorial eastern pacific is shallow, there the SOTA is a little short of same to the SSTA. Thus the strongest SSTA is in the equatorial eastern Pacific, it links directly with the ENSO. Of course, so far as talking on the anomaly of sea surface temperature, it is natural to define the ENSO cycle by using SSTA in the equatorial eastern Pacific. But genuine origin of the ENSO is in warm pool of the western Pacific, because the SSTA in the equatorial eastern Pacific is caused by the eastward propagation of the SOTA in warm pool.

5. Conception Model of the Sota Cycle in the Equatorial Pacific Driven by Anomalous East-Asian Winter Monsoon

In order to illustrate the driving effect of zonal wind anomalies over the equatorial western Pacific caused by anomalous East—Asian winter monsoon to the SOTA cycle in the tropical Pacific, some composite results (for all El Niño during 1950—1955) on the anomaly of East—Asian winter monsoon, zonal wind anomalies over the equatorial western pacific and the eastward propagation of positive SOTA along the equator are shown in Fig. 7. It is clearly shown that positive SOTA has propagated eastwards into the equatorial eastern Pacific (Fig. 7d) before the occurrence of El Niño event (Fig. 7e). But more early, the westerly wind anomalies have occurred over the equatorial western Pacific (Fig. 7c) and anomalous northerly wind (strong East-Asian winter monsoon) has begun in the wintertime (Fig. 7b). The continued positive SOTA has been in warm pool of the equatorial western Pacific for longer time (Fig. 7a). The composite analyses showed that the westerly wind anomalies over the equatorial western Pacific caused by strong East-Asian winter monsoon play an important role in eastward propagation of positive SOTA in the warm pool and the occurrence of El Niño event.

The correlations of the eastward propagation of the SOTA in the equatorial Pacific with zonal wind anomalies over the equatorial western Pacific and with meridional wind anomalies over the northwestern Pacific (120°—140°E, 25°—35°N) are also analyzed. Temporal variations of their time-lag correlation coefficients are given in Fig. 8, the abscissa expresses the lag months, negative value means the SOTA eastward propagation along the equator is later than the anomalies of zonal wind over the equatorial western Pacific and meridional wind over the northwestern Pacific. It is very clear that the eastward propagation of the SOTA from the western Pacific to eastern Pacific along the equator has positive correlation with zonal wind anomalies over the equatorial western Pacific and negative correlation with meridional wind anomalies over the northwestern Pacific. There is the largest correlation coefficient when zonal wind anomaly takes the lead in 2—3 months and meridional wind anomaly takes the lead in 4—5 months (Fig. 8).

Based on the above case analysis and section analysis, the composite analyses for every El Niño and La Niña since 1950 showed that the evolution features of the ENSO cycle, the SOTA cycle in the equatorial Pacific, zonal wind anomalies over the equatorial Pacific western Pacific and meridional wind anomalies over the northwestern Pacific are all similar with the patterns shown in Fig. 3, Fig. 4 and Fig. 5. Since the limitation of space, it is difficult to show the real composite results (monthly patterns in the last year of ENSO, in the ENSO year and in the next year of ENSO). But the sketch diagram is given in Fig. 9 corresponding to the composite analyzing results, it can reflect completely the cycle feature of SOTA in the tropical Pacific driven by anomalous zonal wind caused mainly by the East-Asian winter monsoon anomaly, i.e., the ENSO cycle feature.

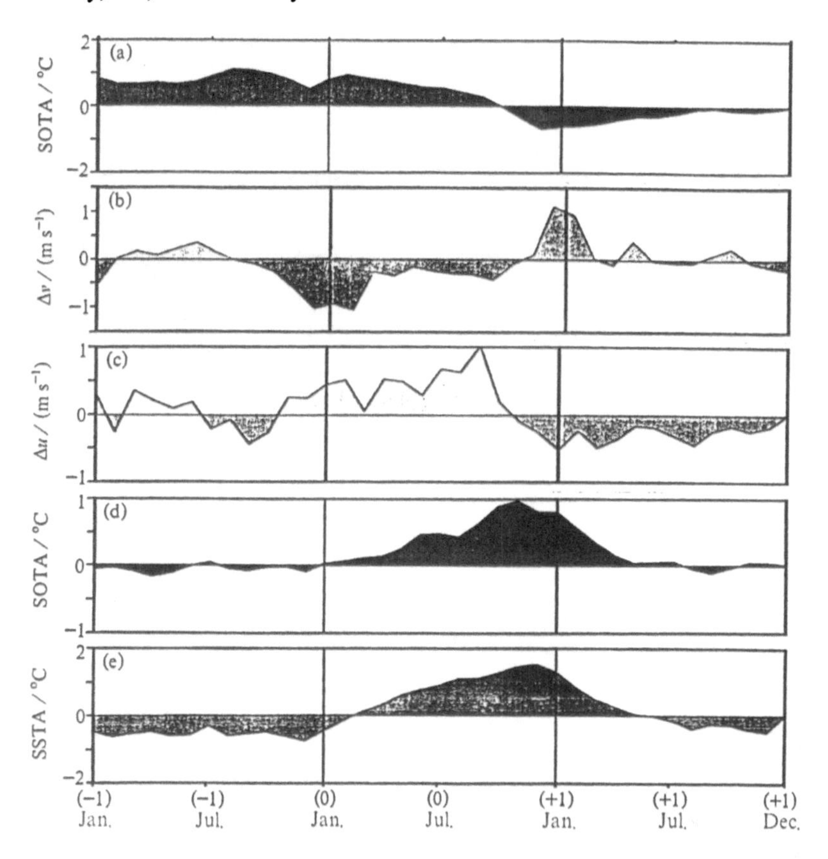

Fig.7. Composite analyses on the El Niño occurrence, (a) The SOTA in warm pool in the western Pacific; (b) Meridional wind anomalies over the region (25°-35°N, 120°-140°E); (c)Zonal wind anomalies over the equatorial western Pacific; (d) The SOTA in the equatorial eastern Pacific (5°S-5°N, 170-130°E); (e) The SSTA in Niño3 region.

There are 4 typical and important patterns in Fig. 9. Figure 9a shows positive SOTA in the warm pool but negative SOTA in the equatorial eastern Pacific, this is corresponding to the situation before occurrence of the El Niño event or the transforming period from the La Niña to the El Niño (in winter—spring generally). If the East—Asian winter monsoon is strong continuously in this period (continued northerly wind anomalies), there will be the westerly wind anomalies over the equatorial western Pacific and the positive SOTA in the warm pool will begin to propagate eastwards. The anomalous westerly wind will occur over the whole equatorial center—western Pacific along with strong winter monsoon effect; the positive SOTA propagates eastwards into the equatorial eastern Pacific and the positive SSTA/El Niño event will be excited (Fig. 9b). While propagating eastwards of positive SOTA along the equator, negative SOTA propagates westwards along 10°N and 10°S latitudes, then the main negative SOTA will be formed in the equatorial western Pacific. Figure 9b is corresponding to the developing and mature period of the El Niño (in autumn—winter generally). Figure 9c showed the transforming period from El Niño to La Niña (in winter—spring generally), there is negative SOTA in the equatorial western Pacific warm pool, but positive SOTA in the equatorial eastern Pacific. If the East—Asian winter monsoon is weak (continued southerly wind anomalies) at this time, anomalous easterly wind will be excited over the equatorial western Pacific, and negative SOTA will begin to propagate eastwards. Under the impact of the continued weak winter monsoon, anomalous easterly wind will be over all equatorial centre-western Pacific, negative SOTA propagates into the equatorial eastern Pacific and negative SSTA/La Niña will occur (Fig. 9d). While eastwards propagating of negative SOTA along the equator, positive SOTA propagates westwards along 10°N and 10°S latitudes, then the major positive SOTA will be formed in the equatorial western Pacific. Figure 9d is corresponding to the developing and mature period of the La Niña (in autumn—winter generally).

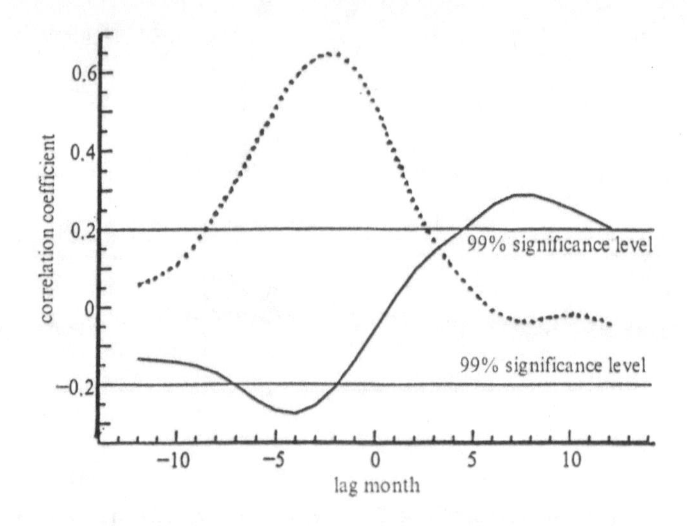

Fig. 8. Temporal variations of the time-lag correlation of the eastward propagation of the SOTA along the equatorial with zonal wind anomalies over the equatorial western Pacific (dashed line) and with meridional wind anomalies over the northwestern Pacific (solid line).

6. Conclusion and Discussion

Based on the previous studies, through the analyses and study in the JEDAC data and NCEP reanalysis data, some important new results can be summed as follows:

(1) The strongest signal of SOTA is in the western Pacific warm pool and the propagation of the SOTA in warm pool plays a key role in the occurrence of ENSO, therefore, the real origin of the ENSO cycle is in the subsurface of warm pool.

(2) The ENSO (El Niño-La Niña) cycle, in point of fact, is the expression in the equatorial eastern Pacific of interannual cycle of the SOTA along the equator and along 10°N/ 10°S latitudes.

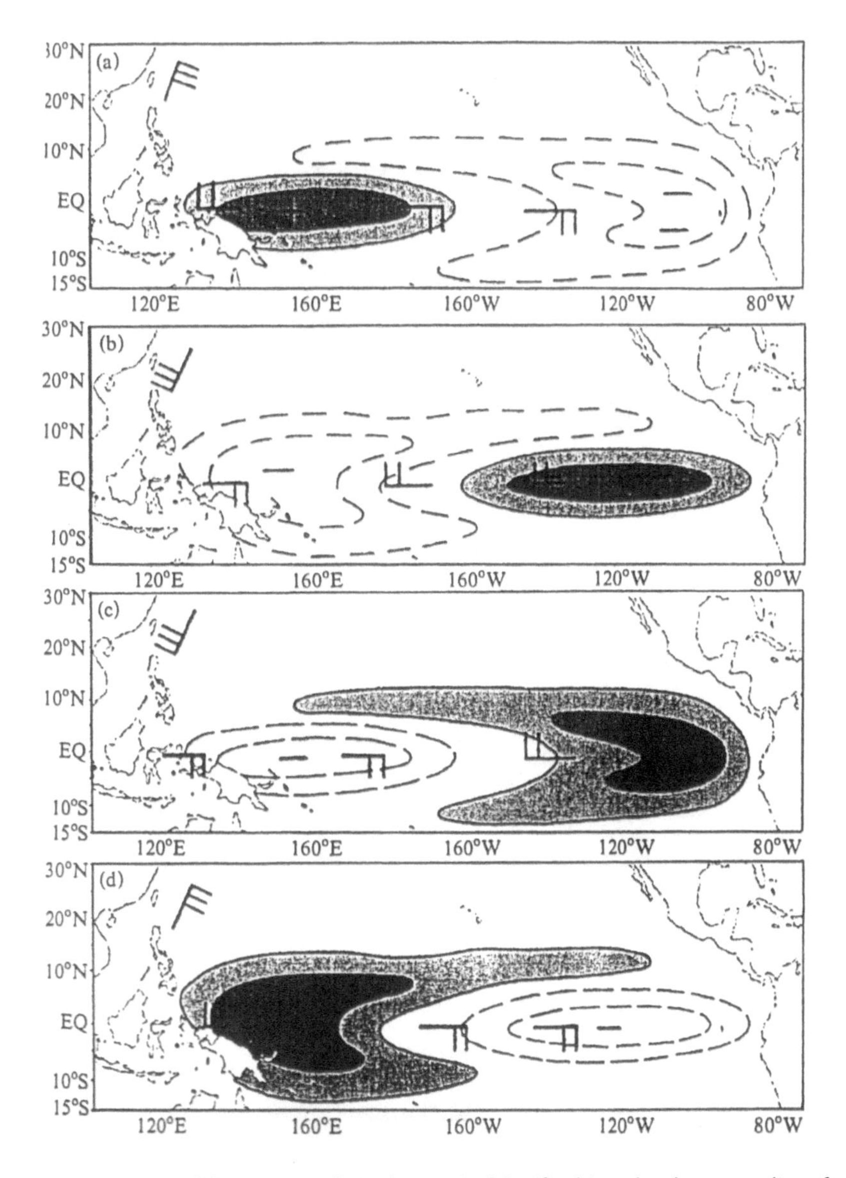

Fig. 9. The sketch diagram of the SOTA cycle in the tropical Pacific driven by the anomalies of zonal wind over the equatorial western Pacific and East-Asian winter monsoon.

(3) The eastward propagation of the SOTA in warm pool along the equatorial thermocline and the systematic westward propagation from the equatorial eastern Pacific result from zonal wind anomalies over the equatorial western Pacific. So that zonal wind anomalies over the equatorial western Pacific can be regarded as the diver of SOTA cycle (ENSO cycle).

(4) The analyses and numerical simulations showed that the westerly (easterly) wind anomalies over the equatorial western Pacific result mainly from continued strong (weak) anomaly of the East-Asian winter monsoon and play an important role in the occurrence of ENSO.

(5) A new theory on the ENSO cycle can be advanced as follows: ENSO is in fact the cycle of SOTA in the tropical Pacific driven by zonal wind anomalies over the equatorial western Pacific, which result mainly from anomalous East-Asian winter monsoon.

(6) Although the dynamics of SOTA propagation along the thermocline is not studied in this paper, the analyses show that the oceanic Kelvin waves play a major role in eastward propagation of SOTA along the equator and the oceanic Rossby waves and currents are common factors to cause westward propagation of SOTA along the 10°N and 10°S latitudes.

It should be indicated that we emphasized the important role of anomalous East—Asian winter monsoon to cause zonal wind anomalies over the equatorial western Pacific, but anomalous Australian winter monsoon will play a certain role. In addition, some studies have shown that the anomaly of East—Asian winter monsoon has the continued influence, it is not only in the wintertime. After strong (weak) East—Asian winter monsoon, the summer monsoon in East Asia is often weak (strong). Therefore, sometimes the "monsoon anomaly in East Asia" is used in this paper, it in fact is consistent with the "East—Asian winter monsoon anomaly" in previous studies.

Finally, it was known that the delay response of the ocean to local wind forcing was not only introduced in the delay oscillator theory, but also the weak coupling in the equatorial western Pacific and boundary reflection were assumed, in order to maintain a continual oscillation of the system. These assumptions are not very reliable in fact, so that the delay oscillator theory still faces some problems to explain the ENSO cycle. In this paper, we emphasize the cycle of subsurface temperature anomalies (thermocline variations), which can maintain continual oscillation under the driving of anomalous zonal wind over the equatorial western Pacific. Moreover, in this cycle of the subsurface temperature anomalies, the boundary reflection couldn't be considered certainly and the air—sea coupling interaction in the equatorial western Pacific region plays an important role in the occurrence of ENSO. These are exactly fundamental features of the atmosphere—ocean system in the equatorial Pacific region.

References

Bjerknes, J., 1969: Atmospheric teleconnections from the equatorial Pacific, *Mon. Wea. Rev.*, 97, 163-172.

Cane, M. A., and S. E. Zebiak, 1985: A theory for El Niño and Southern Oscillation, *Science*, 228, 1085-1087.

Chen Lieting, 1977: Influence of sea surface temperature anomaly in the equatorial eastern Pacific on atmospheric circulation in the tropics and rainfall in China, *Chinese J. Atmos. Sci.*, No.1, 1-12. (in Chinese)

Hirst, A. C., 1986: Unstable and damped equatorial modes in simple coupled ocean-atmosphere model, *J. Atmos. Sci.*, 43, 606-630.

Huang Ronghui, Fu Yunfei, and ZangXiaoyun, 1996: Asian monsoon and ENSO cycle interaction, *Climatic and Environmental Research*, 1, 38-50. (in Chinese)

Li Chongyin, 1985: Westerly anomalies over the equatorial western Pacific and Asian winter monsoon, Proceeding of International Scientific Conference on the TOGA Programme, WCRP-91-WMO/ TP, No.

717, 557-561.

Li Chongyin, 1989: The frequent activities of stronger aerotroughs in East Asia in the wintertime and the occurrence of El Niño event, *Scientia Sinica*(Series B), 32, 976-985.

Li Chongyin, 1990: Interaction between anomalous winter monsoon in East Asia and El Niño events, *Adv. Atmos. Sci.*, 7, 36-46.

Li Chongyin, and Mu Mingquan, 1998: ENSO cycle and anomalies of winter monsoon in East Asia. In: *East Asia and Western Pacific Meteorology and Climate*, Edited by C. P. Chang, J. C. L. Chan and J. T. Wang, Word Scientific, 60-73.

Li Chongyin, and Mu Mingquan, 1999: ENSO occurrence and sub—surface ocean temperature anomalies in the equatorial warm pool, *Chinese J. Atmos., Sci.*, 23, 217-225.

Li Chongyin, and Mu Mingquan, 2000: Relationship between East Asian winter monsoon, warm pool situation and ENSO cycle, *Chinese Science Bulletin*, 45, 1448-1455.

Neelin, J. D., 1991: The slow sea surface temperature mode and the fast-wave limit: Analytic theory for tropical interannual oscillations and experiments in a hybrid coupled model, *J. Atmos. Sci.*, 48, 584-606.

Philander, S.G., T. Yamagata, and R. C. Pacanowski, 1984: Unstable air-sea interactions in the tropics, *J. Atmos. Sci.*, 41, 604-613.

Rasmusson, E.M., and J. M. Wallace, 1983: Meteorological aspects of El Niño/Southern Oscillation, *Science*, 222, 1195-1202.

Suarez, M. J., and P. Schopf, 1988: A delayed action oscillator for ENSO, *J. Atmos. Sci.*, 45, 3283-3287.

Wyrtki, K., 1975: El Niño-the dynamic response of equatorial Pacific ocean to atmospheric forcing, *J. Phy. Oceanogr.*, 5, 572-583.

Yamagata, T., 1986: On the recent development of simple coupled ocean-atmosphere models of ENSO, *J. Oceanogr. Soc. Japan*, 42, 299-307.

Yu, L., and M. Rienecker, 1998: Evidence of an extratropical atmospheric influence during the onset of the 1997-98 El Niño, *Geuphys. Res. Lett.*, 25, 3537-3540.

Zang Hengfan, and Wang Shaowu, 1984: The influence of sea temperature in the equatorial eastern Pacific on atmospheric circulation in the low latitudes, *Acta Ocean. Sin.*, 6, 16-24. (in Chinese)

Zhou Guangqing, and Li Chongyin, 1999: Simulation of the relation between the subsurface temperature anomaly in western Pacific and ENSO by using CGCM, *Climatic and Environmental Research*, 4, 346-353. (in Chinese)

Indian Ocean Dipole and Its Relationship with ENSO Mode*

Mu Mingquan(穆明权), Li Chongyin (李崇银)

LASG, Institute of Atmospheric Physics, Chinese Academy of Sciences. Beijing

Abstract Near 100-year observed data sets are analyzed, and the results show that the variation of sea surface temperature (SST) in the equatorial Indian Ocean has a feature as a dipole oscillation. The situation of the dipole oscillation mainly shows the positive phase pattern (higher SST in the west and lower SST in the east than normal) and the negative phase pattern (higher SST in the east and lower SST in the west). The amplitude of the positive phase is larger than that of the negative phase. The dipole is stronger in September—November and weaker in January — April than in other months. It principally shows obviously inter-annual (4—5 year period) and interdecadal variation (25—30 year period). Although the Indian Ocean dipole in the individual year seems to be independent of ENSO in the equatorial Pacific Ocean, in general, the Indian Ocean dipole has obviously negative correlation with the Pacific Ocean "dipole" (similar to the inverse phase of ENSO) . The atmospheric zonal (Walker) circulation is fundamental for relating the two dipoles to each other.

Keywords the Indian Ocean dipole, the Pacific Ocean dipole (the inverse ENSO mode), Walker circulation.

Ⅰ. Introduction

A strong El Niño event occurred in early summer 1997 (McPhaden 1999). It caused seriously large-scale climate disasters nearly all over the world, such as drought and forest fire in Indonesia, flood in the northern region of South America, and so on. According to studies based on long-term observed data analyses (Rasmusson and Carpenter 1983; Ropelweski and Halpert 1987), there should be weaker summer monsoon and drought in Indian Peninsula in El Niño year. But this situation was reverse in 1997, the mean summer precipitation was approximate to or slightly more than normal only in the partial region of India (Bell and Halpert 1998), and more precipitation occurred in East Africa (Birkett et al. 1999). Why did these situations occur? Some diagnostic studies have shown that much stronger SSTA (sea surface temperature anomaly) than normal occurred in the equatorial Indian Ocean during the

Received June 18, 2001; revised May 15, 2002

*This work was supported jointly by National Key Programme for Developing Basic Sciences (G1998040900—part 1) in China and Chinese Academy of Sciences under Grant KZCX 2-SW-210.

period of 1997—1998 El Niño event. The maximum SSTA in the equatorial western Indian Ocean was over 2℃, which was seldom seen in the history. Thus, much more attentions have been paid to studies on the reason for the large SSTA in the equatorial Indian Ocean and its impacts.

Through analyzing SSTA in the Indian Ocean, a dipole oscillation was indicated, which represents the reverse situation of SSTA in two regions (10°S—10°N, 50°—70°E) and (10°S —EQ, 90°—110°E). Webster et al. (1999) suggested that the Indian Ocean dipole in 1997—1998 was independent of ENSO and was excited by the strong atmosphere- land-sea interaction. Studies by Sagi et al. (1999) also revealed that this dipole just occupies 12% of the total variance and is not always related to ENSO. Usually, the dipole appears not only in the variation of SST, but also in the variation of subsurface ocean temperature (SOT) in the Indian Ocean (Anderson 1999).

In previous studies, the occurrence of the 1997—1998 event was quite emphasized, particularly the dipole was regarded as an independent event of ENSO. In fact, some features of the Indian Ocean dipole still indicated to be related to the El Niño event. Therefore, we should analyze and study the composite feature of the Indian Ocean dipole and further explore its general characteristics.

In fact, the importance of SSTA in the Indian Ocean and its influence on summer rainfall in the middle and lower reaches of the Yangtze River have been indicated early (Luo et al. 1998: Chen 1991). But the Indian Ocean SSTA as a dipole needs to be studied further, particularly the influence of the dipole on the Asian monsoon circulation and climate.

The SST data from 1900 to 1997 in this study were taken from the Hadley Center in UK. The spatial-temporal features of the Indian Ocean dipole and its relationship with the Pacific Ocean dipole (ENSO) are studied in this paper. At the same time, by using NCEP/NCAR reanalysis data, wė further study the relationship of the Indian Ocean dipole with atmospheric circulation, particularly with the anomalous equatorial zonal wind in the tropics.

Ⅱ. Spatial-Temporal Features of Indian Ocean Dipole

The observational data analyses exhibited that the SST variation in the northwest of the equatorial Indian Ocean has opposite feature to that in the southeast. In order to quantify this dipole oscillation, we define the difference of averaged SSTA in the region (5°S—10°N, 50°—65°E) from that in another region (10°S—5°N, 85°—100°E) as the dipole index. This is better than that by Saji et al. (1999). because some islands and the Laut Jawa sea have been ruled out the definition domain here. The temporal variation of the dipole index from 1900 to 1997 and its power spectrum are shown in Fig. 1. It is very clear that the dipole index has the clearly inter-annual variation (4—5 year period and 2 year period) and the interdecadal (25—30 year period) variation. Besides these, there is another interesting phenomenon in Fig. 1. The dipole index before 1961 was mainly the stronger negative value, after then it became positive and meanwhile most of the stronger signals were positive. In other words, the

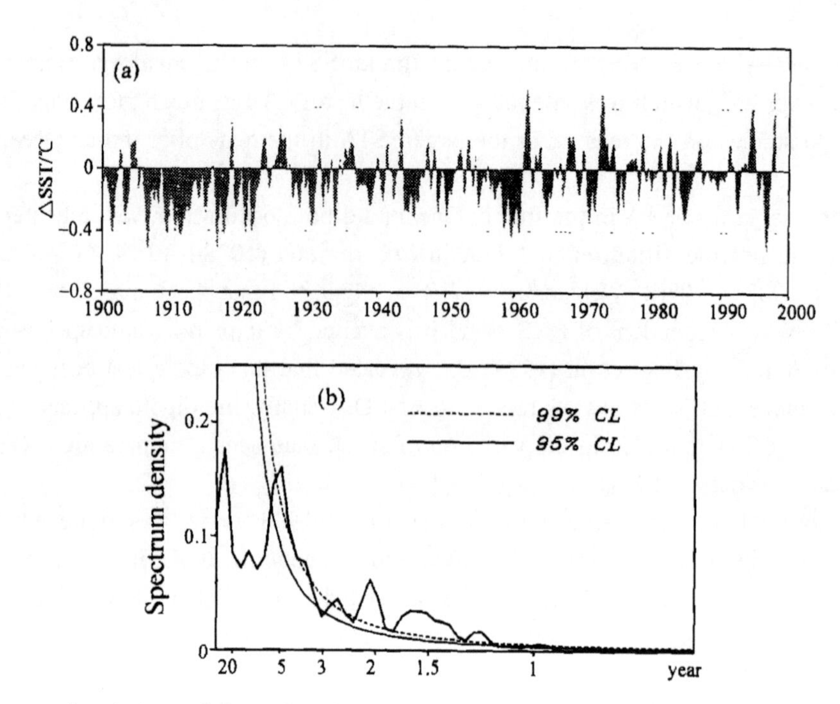

Fig. 1. Temporal variations of the Indian Ocean dipole (a) and its power spectrum (b). The solid and dashed lines represent 95% and 99% confidence level.

fundamental characteristic of SST in the equatorial Indian Ocean is higher in the eastern region and lower in the western region before 1961, but it was the reverse variation after 1961. This means that the clearly inter- decadal variation also exists in the Indian Ocean.

The temporal variation of the dipole index showed that the Indian Ocean dipole has obviously interdecadal variations besides inter-annual variations. Therefore, for the study of the interdecadal climate variability (WMO. ICSU and UNESCO 1995) in the Pacific- Indian Ocean region (CLIVAR/DecCen-D3). the Indian Ocean dipole should be regarded as an important phenomenon and studied further in another paper.

In order to abstract the general feature of the Indian Ocean dipole, we separately take 4 positive cases (1961. 1972, 1994 and 1997) with larger positive index and 5 negative cases (1958, 1959, 1960, 1970 and 1996) with larger negative index to make composite analyses. The horizontal distributions of the composite SSTA for the positive and negative phase (case) of the Indian Ocean dipole in the region (30°S—50°N. 30°E — 80°W) are shown in Fig. 2. Figures 2a and 2b indicate the results of the positive phase in July and December, which can represent the situations in summer and in winter, respectively, even though the strongest dipole is in October, not in July and December. Analogously, Figs. 2c and 2d are for the negative phase in July and December. The basic feature of the Indian Ocean dipole is exhibited clearly in Fig. 2. The SSTA is positive in the equatorial western Indian Ocean and negative in the eastern Indian Ocean for the positive phase, but the SSTA for the negative phase has reverse distribution compared to that for the positive phase. The dipole shown in Fig. 2a and in Fig. 2b is stronger than that in Fig. 2c and Fig. 2d, This means that the dipole is stronger in the positive phase generally. Comparing the SSTA distributions in the equatorial Pacific Ocean with that in the equatorial Indian Ocean, their relationship is obviously noticed. When the Indian Ocean dipole is in the positive phase. SSTA is positive in the equatorial eastern Pacific Ocean and negative in the western Pacific Ocean. When the Indian Ocean dipole is in the negative phase, SSTA in the equatorial western Pacific Ocean is positive and negative in the eastern Pacific Ocean. How about the connection between the Indian Ocean dipole and the SSTA in the equatorial Pacific, we will discuss below.

Comparing Fig. 2a with Fig. 2b and Fig. 2c with Fig. 2d, it can be seen that the intensity of the Indian Ocean dipole is different in different months. In Fig. 3, the annual variations of the dipole intensity are shown for the positive and negative phases respectively. The seasonal variation of the dipole intensity is prominent. The Indian Ocean dipole is stronger during the July—December and weaker during January—May, the strongest is in October and the weakest in February. Comparing the two curves in Fig. 3. it can be also shown that the dipole intensity is stronger in the positive phase than in the negative phase generally.

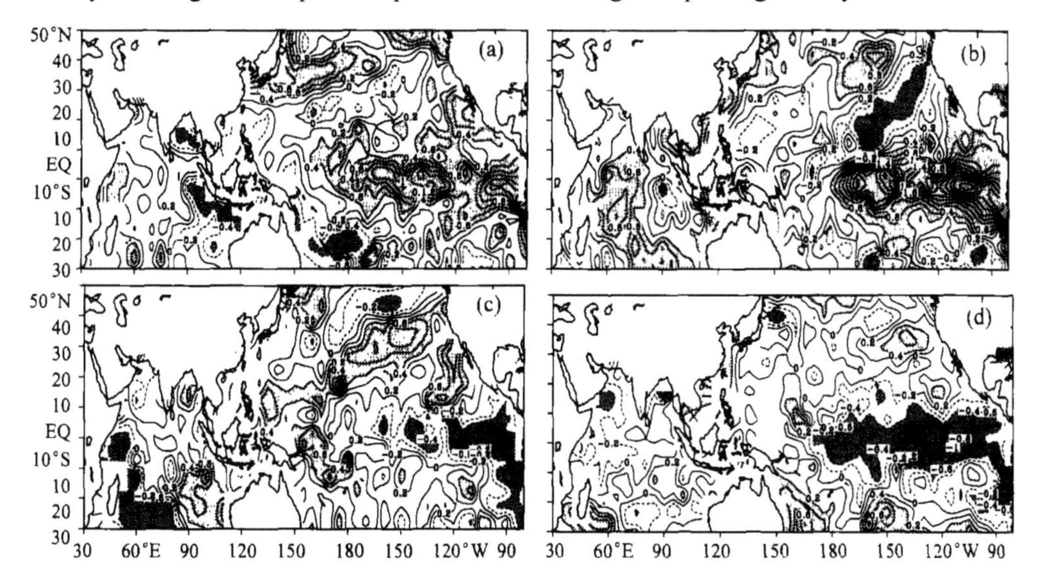

Fig. 2. Composite SSTA for the positive phase (1961, 1972. 1994 and 1997) in July (a) and December (b) and for the negative phase (1958, 1959, 1960, 1970 and 1996) in July (c) and December (d).

Ⅲ. Relationship Between Indian Ocean Dipole and Pacific ENSO Mode

We have discussed the associations of the Indian Ocean dipole with SSTA in the equatorial Pacific Ocean before. For the positive (negative) phase of the Indian Ocean dipole, SST in the equatorial Pacific Ocean is higher (lower) in the east and lower (higher) in the west than normal. In order to further determine their relationship, the correlation coefficients of the Indian Ocean dipole index with SSTA in the global ocean are calculated. The horizontal

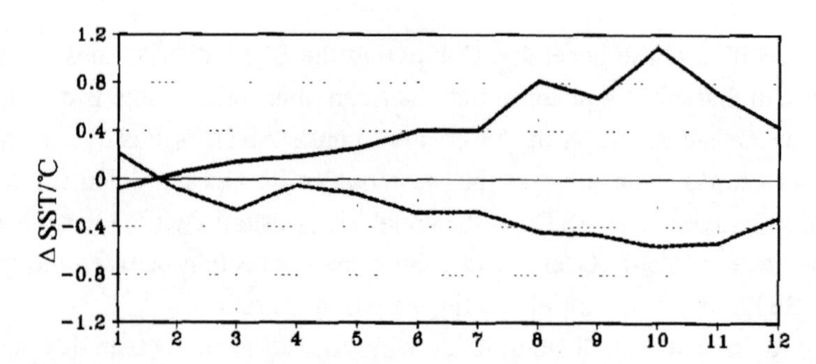

Fig. 3. Annual variation of intensity of the Indian Ocean dipole. The dark and light curves represent the composite results for the positive and negative phases, respectively.

distribution of their correlation coefficients is given in Fig. 4. In the figure, the shaded area represents the coefficient over the statistical significance test. Evidently, there is not only a dipole pattern in the Indian Ocean, but also an inverse pattern in the equatorial Pacific Ocean where the correlation coefficients is positive in the east and negative in the west and its distribution is very similar to the ENSO mode. According to the distribution of SSTA in the equatorial Pacific Ocean and for convenience, it can be suggested that another ocean dipole exists in the equatorial Pacific Ocean although the amplitude of SSTA in the equatorial eastern and western Pacific is not same. This kind of dipole is not mathematical dipole, it is just an oscillating pattern and can be asymmetric pattern (positive and negative directions have different values). Thus, we can define the Pacific Ocean dipole index as. similar to the Indian Ocean dipole index, the difference of averaged SSTA in the region (5°S —10°N, 140°E—180°) from that in the region (10°S—5°N. 90—130°W). Apparently, the Pacific Ocean dipole, defined as the above, is basically consistent with the ENSO mode, but their phase is reverse to each other.

The analyses with the SSTA in Niño 3 region have shown that there are two major spectrum peaks (Li 1998; Mu and Li 1999); one is about 4-year period and the other is about 2-year period, which are similar to the results shown in Fig. 1b. In a certain sense, this also represented the relationship between the Indian Ocean dipole and the Pacific Ocean dipole.

In order to further examine the relationship between the Indian Ocean dipole and the Pacific Ocean dipole, temporal variations of their correlation coefficient are shown in Fig. 5. Obviously, these two dipoles have clearly negative correlation. This means that the positive (negative) phase of the Indian Ocean dipole is corresponding to the negative (positive) phase of the Pacific Ocean dipole. In general, there is the same mark of SSTA in the equatorial eastern Indian Ocean as that in the equatorial western Pacific; or SSTA in the equatorial western Indian Ocean and in the equatorial eastern Pacific has a consistent sign. Figure 5 also indicates that the two dipoles have basically contemporary negative correlation, But in minor years, the Pacific Ocean dipole may lead ahead of the Indian Ocean dipole. The 1997—1998 event can be an example, during the period October—June 1998, the SSTA and SSHA (Webster's Fig. 3 1999) showed that the Indian Ocean dipole was also negative correlation with the Pacific dipole; the different relationship was only in the period May — September

1997. This means that the Indian Ocean dipole had lag negative correlation with the Pacific Ocean dipole (inverse ENSO mode) in the 1997—1998 case.

What are reasons for the SST variation in the equatorial eastern Indian Ocean consistent with that in the equatorial western Pacific Ocean so that above-mentioned two dipoles have a good negative correlation? Besides that the Indonesian throughflow links up the Pacific Ocean and the Indian Ocean (Potemra et al. 1997: Meyers 1996; Masumoto and Yamagata 1996), the wind field in the equatorial lower troposphere plays an important role too. In Fig. 6, anomalous wind fields at 1000 hPa averaged in September —November are shown for the positive phase and negative phase of the Indian Ocean dipole, respectively. It is very evident that the flow pattern in the positive phase of the Indian Ocean dipole is not only opposite to that in the negative phase over the equatorial Indian Ocean (easterly for the positive phase, but westerly for the negative phase) but also over the equatorial Pacific Ocean, particularly over the equatorial western Pacific Ocean (westerly for the positive phase, but easterly for the negative phase).

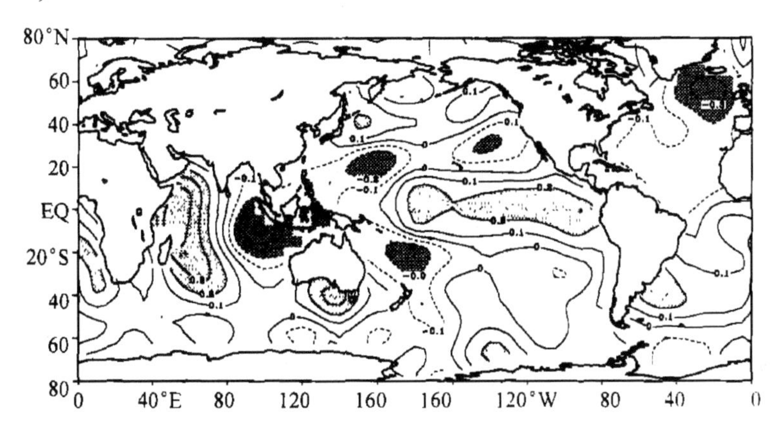

Fig. 4. Horizontal distribution of correlation coefficient of the Indian Ocean dipole index with SSTA in the global ocean. The shaded area represents the correlation coefficient beyond 99. 9% statistical significance test.

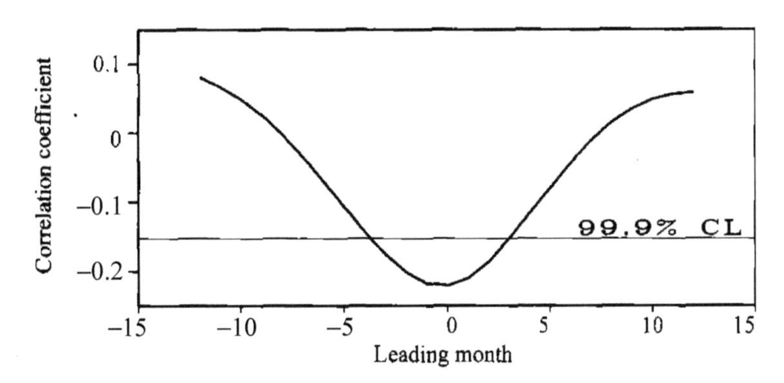

Fig. 5. Temporal variation of correlation coefficient between the Indian Ocean dipole index and the Pacific Ocean dipole index. The horizontal axis represents the leading time. The negative month means the Indian Ocean dipole leading the Pacific Ocean dipole.

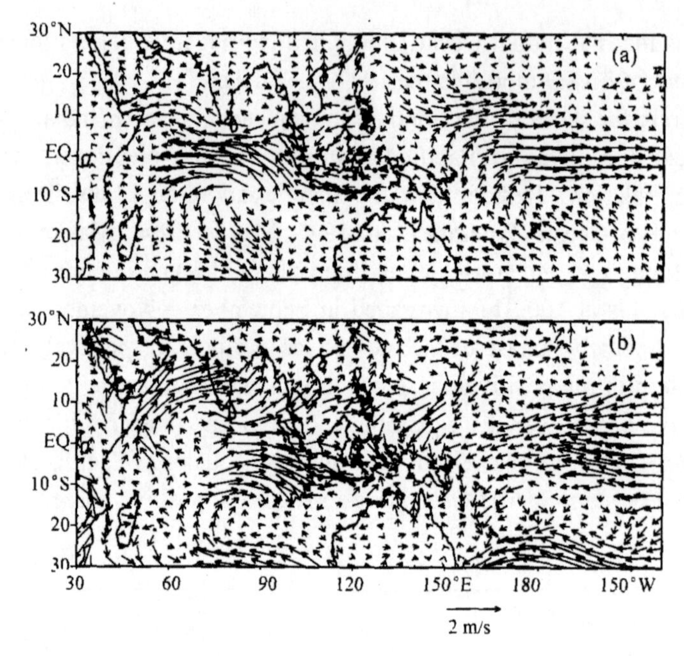

Fig.6. Anomalous wind fields at 1000 hPa averaged in September-November for the positive phase (a) and the negative phase (b) of the Indian Ocean dipole.

In fact, the above-mentioned flow pattern for the stronger Indian Ocean dipole exists not only in September— November but also in other months (not shown).

In the upper troposphere (200 hPa), the anomalous flow pattern over the equatorial region for the positive phase and the negative phase of the Indian Ocean dipole also has the opposite feature (Fig. 7). That means there is anomalous westerly (easterly) wind over the equatorial Indian Ocean and anomalous easterly (westerly) wind over the equatorial Pacific Ocean.

Corresponding to the positive and negative phases of the Indian Ocean dipole, the anomalous zonal-vertical (Walker) circulation patterns in the equatorial atmosphere are given in Fig. 8, respectively. It is very clear that there is an inverse anomalous Walker circulation in the atmosphere for the positive and negative phases of the Indian Ocean dipole, in particular over the equatorial Indian and Pacific Oceans. Anomalous easterly (westerly) wind in the lower troposphere over the equatorial Indian Ocean is favorable to form and maintain the positive (negative) phase of the Indian Ocean dipole. At the same time, through the Walker circulation, the anomalous westerly (easterly) wind in the lower troposphere over the equatorial Pacific Ocean will be excited, and further leads the negative (positive) phase of the Pacific Ocean dipole to form and maintain. Of course, we can also use the same theory to explain how dose the occurrence of the Pacific Ocean dipole lead the Indian Ocean dipole to form and maintain. Therefore, anomalous atmospheric zonal-vertical (Walker) circulation can be regarded as an important physical factor to relate the Indian Ocean dipole and the Pacific Ocean dipole to each other, and leads to the remarkable mutual-correlation of two dipoles.

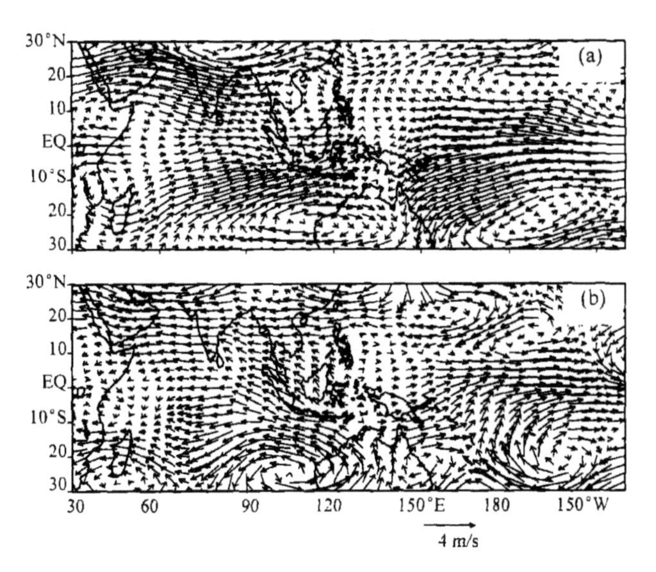

Fig. 7. As in Fig. 6, but for 200 hPa.

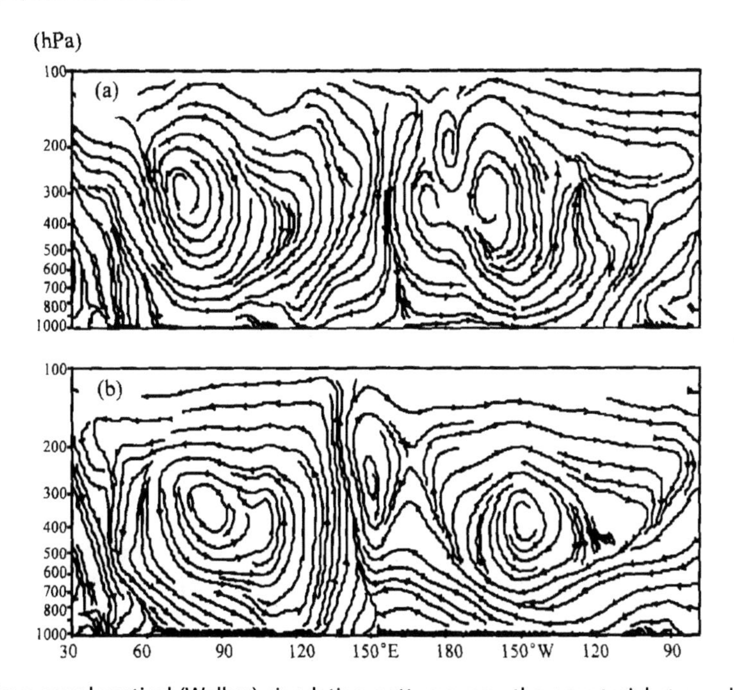

Fig. 8. Anomalous zonal-vertical (Walker) circulation patterns over the equatorial atmosphere in October for the positive phase case (a) and the negative phase case (b) of the Indian Ocean dipole, respectively.

Ⅳ. Conclusions

Based on above observational analyses, some interesting results on the Indian Ocean dipole are obtained, the major points can be summarized as follows:

(1) The SST or SSTA in the equatorial Indian Ocean has really a zonal oscillation feature similar to a dipole. This Indian Ocean dipole has distinctly inter-annual (4 —5 year period) and interdecadal (25 — 30 year period) variations. Its seasonal variation is also clear, stronger in September— November and weaker in January— April.

(2) Two major patterns exist in the equatorial Indian Ocean. One is the higher SST center in the west and the lower SST in the east. The other is the reverse situation, higher SST in the east and lower SST in the west. They can be regarded as the positive and negative phases of the Indian Ocean dipole, respectively. In general, the positive phase of the dipole is stronger than the negative phase.

(3) Although the Indian Ocean dipole in the individual year seems to be independent of ENSO, generally, the Indian Ocean dipole has obviously positive correlation with ENSO mode. If we regard the positive anomaly of SSTA in the west (east) and the negative anomaly in the east (west) as the positive (negative) phase of the Pacific Ocean dipole, the positive and negative phase patterns are analogous to La Nina and El Niño, respectively. Thus, the positive (negative) phase of the Indian Ocean dipole is corresponding to the negative (positive) phase of the Pacific Ocean dipole (inverse ENSO mode) and means the negative correlation of these two dipoles is very obvious. But in some years, this correlation has the time lag with months.

(4) Anomalous atmospheric zonal-vertical (Walker) circulation is the main link pass to relate the Indian Ocean dipole to the Pacific Ocean dipole. The change of the Walker circulation, through the air-sea interaction, can lead the Indian Ocean dipole or the Pacific Ocean dipole to occur. Of course, even though the Indonesian throughflow is also important to link up the Pacific and Indian Oceans.

References

Anderson, D. (1999). Extremes in the Indian Ocean. *Nature*, 401: 337-339.

Bell, G., and Halpert. M. (1998), Climate assessment for 1997. *Bull. Am. Meteor. Soc.*, 79: 810-850.

Birkett, C., Murtugdde, R. and Allan T. (1999). Indian Ocean climate event brings floods to East Africa's lakes and the Sudd Marsh, *Geophys. Res. Lett.*, 26: 1031-1034.

Chen Lieting (1991), Influence of zonal difference of the SSTA from Arabian Sea to South China Sea on the precipitation in the middle-lower reaches of the Yangtze River, *Chinese J. Atmos. Sci.*, 15: 33-42 (in Chinese).

Li Chongyin (1998). The qusi-decadal oscillation of air-sea system in the northwestern Pacific region. *Adv. Atmos. Sci.*, 15: 31-40.

Luo Shaohua. Jin Zhuhui and Chen Lieting (1985), Correlation analyses of the SST in the Indian Ocean/ the Southe China Sea and the precipitation in the middle-lower reaches of the Yangtze River, *Chinese J. Atmos. Sci.*, 9: 336-342.

Masumoto, Y. and Yamagata, T. (1996). Seasonal variations of Indonesian throughflow in a general circulation model, *J. Geophy. Res.*, 101: 12287-12293.

McPhaden, M. J. (1999), Climate oscillations-genesis and evolution of the 1997-1998 El Niño. *Science*, 283: 930-940.

Meyers. G., (1996) Variation of Indonesian throughflow and El Niño/Southern Oscillation, *J. Geophy. Res.*,

101: 12255-12264.

Mu Mingquan and Li Chongyin (1999), ENSO signals in interannnual variability of East Asian winter monsoon, Part I: Observed data anlayses. *Chinese J. Atmos. Sci.*, 23: 134-143.

Potemra, J. T., Lukas, R. and Mitchum, G. T, (1997). Large-scale estimation of transport from the Pacific to the Indian Ocean, *J. Geophy. Res.*, 102: 27795-27812.

Rasmusson. E. M., and Carpenter. T. H. (1983). The relationship between eastern equatorial Pacific sea surface temperatures and rainfall over India and Sri Lanka, *Mon. Wea. Rev.*, 111: 517-528.

Ropelewski, C. F., and Halpert, M. S, (1987), Global and regional scale precipitation patterns associated with the El Niño/southern Oscillation, *Mon. Wea. Rev.*, 115: 1606-1626.

Saji, N. H., Goswami, B. N., Viayachandrom. P. N. and Yomagada, T. (1999), A dipole mode in the tropical Indian Ocean. *Nature*. 401: 360-363.

Webster. P. T., Moore. A. M. Loschning, J. P. and Leben, R. R. (1999), Coupled ocean-atmosphere dynamics in the Indian Ocean during 1997-1998. *Nature*, 401: 356-360.

WMO, TCSU, and UNESCO (1995). CLIVAR-A study of climate Variability and Predictability, Science Plan, WMD/TD, No. 690. WCRP-89, Geneva.

The Tropical Pacific-Indian Ocean Temperature Anomaly Mode and Its Effect

Yang Hui[1], Jia Xiaolong[1,2], Li Chongyin[1,3]

1. LASG, Institute of Atmospheric Physics, Chinese Academy of Sciences, Beijing, China
2. Graduate University of Chinese Academy of Sciences, Beijing, China
3. Meteorological College, PLA University of Science and Technology, Nanjing, China

Abstract Temperature anomaly in the Indian Ocean is closely related to that in the Pacific Ocean because of the Walker circulation and the Indonesian throughflow. So only the El Niño/Southern Oscillation (ENSO) in the Pacific cannot entirely explain the influence of sea surface temperature anomaly (SSTA) on climate variation. The tropical Pacific-Indian Ocean temperature anomaly mode (PIM) is presented based on the comprehensive research on the pattern and feature of SSTA in both Indian Ocean and Pacific Ocean. The features of PIM and ENSO mode and their influences on the climate in China and the rainfall in India are further compared. For proving the observation results, numerical experiments of the global atmospheric general circulation model are conducted. The results of observation and sensitivity experiments show that presenting PIM and studying its influence are very important for short-range climate prediction.
Keywords tropical Pacific-Indian Ocean temperature anomaly, composite mode, influence on climate.

The impacts of anomalous sea temperature on general circulation and climate have attracted researchers' attention. The strong sea surface temperature anomaly in the equatorial eastern Pacific El Niño/Southern Oscillation (ENSO) event especially causes serious flood or drought disasters in many regions and countries and is an important subject of many studies for many years in the world[1-4]. Although ENSO is the strongest signal in the interannual climate variation, it is not the only cause of the anomalous climate. For example, the summer precipitations were more than normal in some regions of India[5, 6] and eastern China[7]. On the other side, the sea surface temperature anomaly (SSTA) in the Indian Ocean has been paid attention to for a long time[8]. Moreover, the Indian Ocean temperature dipole was discovered a few years ago[9, 10]. The further work discussed the important influences of the dipole including the influence on the climate in China[11-14]. Generally, the Indian Ocean temperature dipole was significantly related to the ENSO in the Pacific[15]. The equatorial western Pacific and the eastern Indian Ocean are together called warm pool[16]. The Indian through-flow[17, 18] and anomalous atmospheric zonal-vertical (Walker) circulation[15] are the main cause for

Received February 14, 2006; accepted March 30, 2006.
Correspondence should be addressed to Yang Hui (email: yanghui@mail.iap.ac.cn)

relating the SSTA in Indian Ocean to that in the Pacific Ocean. Generally, some climate phenomena including SSTA in the middle-high latitude ocean all lag SSTA in the eastern equatorial Pacific[19, 20]. But observation research has revealed that the Indian Ocean dipole is almost simultaneous with the Pacific Ocean dipole (namely ENSO) without significant lag[15]. The equatorial ocean sea surface temperature (SST) change (response) occurring almost simultaneously is attributed mainly to the close interaction of the two anomalous atmospheric Walker circulations. On that account, the tropical Pacific-Indian Ocean temperature anomaly mode (PIM) has a clear physical meaning: the SSTA pattern and change in both the equatorial Pacific and the Indian Ocean are related to the two anomalous Walker circulations, that is, the two anomalous Walker circulations result in the SSTA patterns in the two oceans, respectively. Apparently, those can be regarded as dipoles. In order to reveal the influence rule of the tropical ocean SSTA on climate, the PIM will be pre-sented through analyzing its SSTA patterns and features. And the influences of every SSTA pattern on Chinese climates are studied.

The monthly SST data of the Hadley Centre in UK on a 1.0°×1.0° grid from 1900—1999 were used. The monthly rainfall and temperature data at 160 stations of China compiled by the China Meteorological Administration and the global land monthly precipitation data (PREC/L) on 2.5°×2.5°grids (1948—2001) were also used.

1. Definition of the PIM

El Niño and La Niña are defined by the SSTA in the equatorial eastern Pacific. Actually, when there is the positive (negative) SSTA in the eastern equatorial Pacific, the negative (positive) SSTA in the western equatorial Pacific occurs. On the other hand, the so-called Indian Ocean dipole is defined by the difference of the SSTA in the western equatorial Indian Ocean from that in the eastern equatorial Indian Ocean, indicating zonal heat contrast of the Indian Ocean SSTA. Although it is named dipole, actually it is not related to the mathematic meaning (SSTA distribution with positive (negative) in the west and opposite in the east)[15]. Considering the close relationship between the Pacific ENSO mode and the Indian Ocean dipole, the index of the PIM can be defined as the respectively normalized east-west SSTA differences of the equatorial areas in the two oceans. As to the SSTA, the SSTA of ENSO is stronger than that in the equatorial Indian Ocean because of the bigger Pacific basin. However, as to the influence of the SSTA on East Asia, a series of numerical experiments clearly indicate that the effect of SSTA forcing of the Indian Ocean is stronger than that of the eastern equatorial Pacific[21-23]. So the composite index will be defined based on the normalized dipoles in the Pacific and the Indian Ocean respectively. In this way, the normalized dipoles are comparable and the composite index has mathematical and physical base. The composite PIM index I_{com} is defined as

$$I_{\text{com}} = \nabla T_i + \nabla T_p,$$
$$\nabla T_i = T_1 - T_2,$$

$$\nabla T_p = T_3 - T_4,$$

where T_1, T_2, T_3 and T_4 means the averaged SSTA in the region (50°E—65°E, 5°S—10°N), (85°E—100°E, 10°S—5°N), (130°W—80°W, 5°S—5°N) and (140°E—160°E, 5°S—10°N), respectively; ∇T_i and ∇T_p are normalized.

Fig. 1 gives the time series of the I_{com} and SSTA in Niño3.4 (also called ENSO index). It can be found that although I_{com} is closely related to ENSO index, the difference between them is large. In most cases the Indian Ocean SSTA makes the I_{com} stronger than the normalized ENSO index, and partly changes the phase of ENSO mode. This indicates that studying composite mode and its index I_{com} are very important.

In order to further explain the difference between the composite mode and the pure Pacific ENSO mode, the two types of positive phase and quasi-normal phase are compared. Based on the PIM index, the positive phase (the years of 1951, 1965, 1972, 1982, 1983, 1987 and 1997) is selected with the composite index more than or equal to 3.8 and the quasi-normal phase (the years of 1952, 1956, I960, 1967, 1968, 1979, 1980, 1981 and 1990) is picked with the index near zero. For comparing with the pure ENSO mode, the pure El Niño years are chosen when the west-east difference of SSTA in the Indian Ocean (the Indian Ocean dipole index) is small: the years of 1951, 1953, 1957, 1963, 1965, 1969, 1976, 1986 and 1991; and the quasi-normal years of the Pacific are also picked when the Niño3.4 SSTA is about zero: the years of 1959, 1960, 1962, 1980, 1981 and 1990.

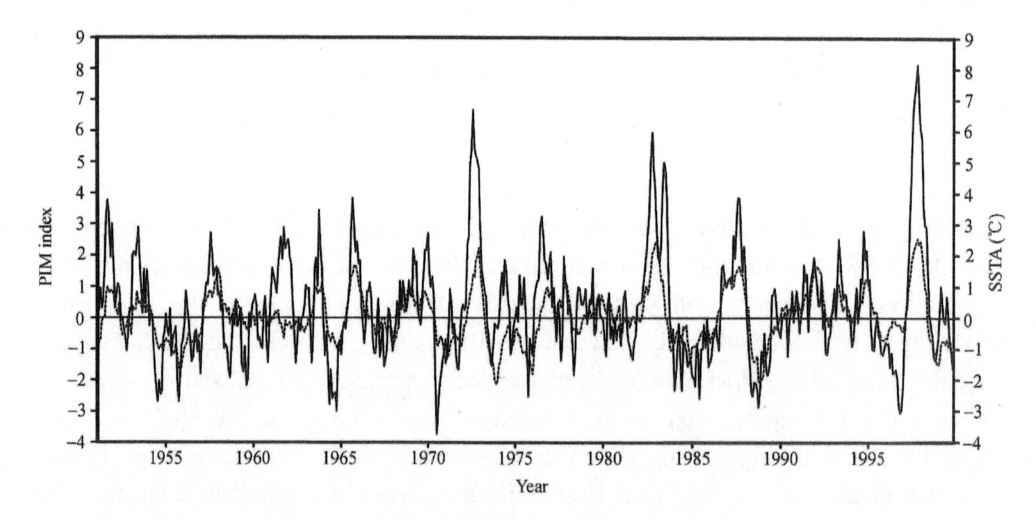

Fig. 1. Time series of the PIM index (solid line) and SSTA in Niño3.4 (dash line).

The composite summer SSTA patters of the above mentioned types are shown in Fig. 2 respectively. The quasi-normal year of the composite index reflects the quasi-normal feature of SST better than the ENSO index does. The SSTA in equatorial sea areas is very small in the quasi-normal year of the composite index. Moreover, there are a lot of differences between the positive phase year of the composite index and that of the ENSO index. The composite

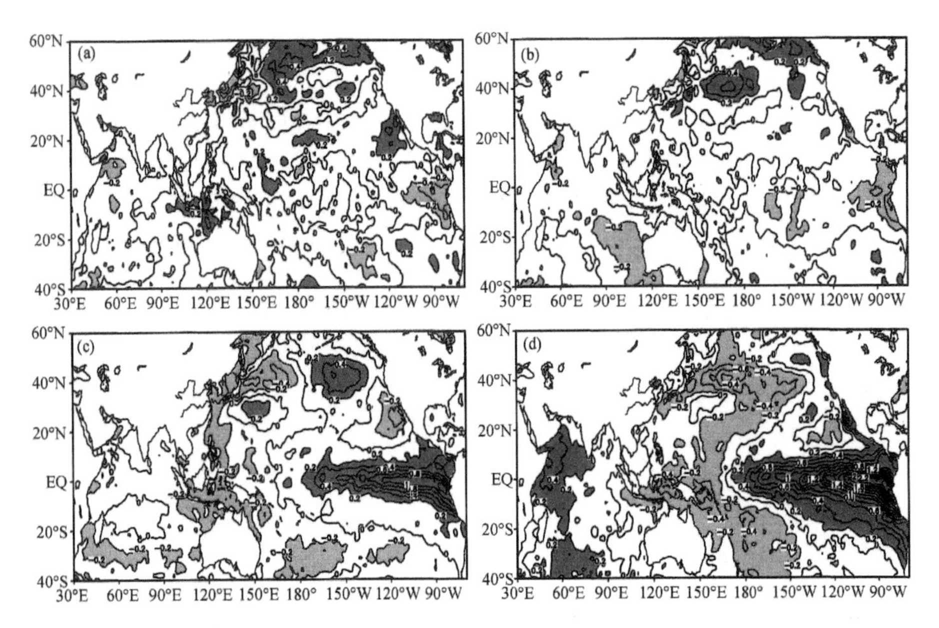

Fig. 2. Composite chart of SSTA in summer (June-July-August, i.e. JJA) in (a) the quasi-normal phase of ENSO mode, (b) the quasi-normal phase of the PIM, (c) the positive phase (El Niño) of the ENSO mode, and (d) the positive phase of the PIM (the same below). Unit: ℃. The dark shaded areas and light shaded areas represent the anomalies greater than 0.2 and those less than −0.2, respectively.

index displays the west-east difference of the tropical ocean SSTA, not only in the Pacific, but also in the Indian Ocean. These comparisons once again indicate that the PIM considerably differs from the Pacific ENSO mode, and proposing and studying the composite index is very significant.

2. Influence of the Composite PIM Index on Summer Climate in China

Based on the corresponding Chinese climates for the four types of Fig. 2, the influences of all SSTA patterns on climate are quite different, which can be easily found. Because of long duration of SSTA, the analysis results here can provide some base for Chinese summer climate prediction. Owing to the finite time length of observation data, the number of composite analysis cases is small. However, to a certain extent they explain the considerable effects of SSTA patterns on East China rainfall. Fig. 3 shows the summer rainfall anomalies in East China corresponding to the four types of SSTA distribution in Fig. 2 respectively. Corresponding to the quasi-normal phase of the Pacific SSTA, more rainfall can be measured in the coastal regions of South and Southeast China, the Yangtze-Huaihe River Basin and the southern part of North China, but less rainfall is observed from south of the Yangtze River to Yunnan Province (Fig. 3(a)). Corresponding to the quasi-normal phase of the PIM, the areas to the south of the Yangtze River, North China and the southern part of Northeast China receive less rainfall, and more rainfall is recorded in the areas from the Yangtze-Huaihe River

Basin to Sichuan Province (Fig. 3(b)). In the pure El Niño year, there is more rainfall in the Yangtze-Huaihe River Basin and Northeast China, less rainfall occurs to the areas to the south of the Yangtze River (Fig. 3(c)). For the positive phase of the PIM, There is more rainfall in Southeast China coastal region and from the Yangtze-Huaihe River Basin to Sichuan Province, and less rainfall in North China and Northeast China (Fig. 3(d)). Comparing Fig. 3(a) with Fig. 3(b), and Fig. 3(c) with Fig. 3(d), respectively, it clearly indicates that the impacts of the pure Pacific SSTA on the summer rainfall in China are very different from the composite impacts of the Pacific-Indian Ocean SSTA.

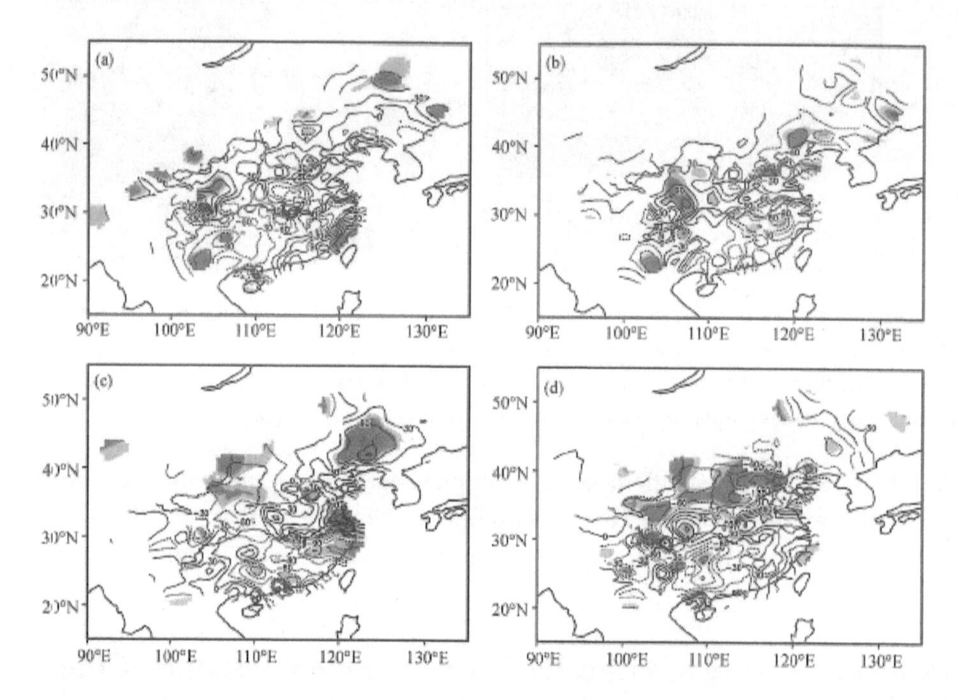

Fig. 3. Composite plot of summer (JJA) total precipitation anomaly in China. Unit: mm. The dark shaded areas and light shaded areas represent the *t*-test significance at 0.05 and 0.1 levels, respectively. The four phases in Fig. 3(a)—(d) are the same with Fig. 2(a)—(d), respectively.

In a similar way, Fig. 4 respectively illustrates the surface air temperature anomalies in East China corresponding to the four types of SSTA distributions in Fig. 2. The features include that: noticeable positive temperature anomaly occurs in the Yangtze-Huaihe River Basin (Fig. 4(a)) in the quasi-normal year of the Pacific SSTA; there is remarkable positive temperature anomaly in South China (Fig. 4(b)) in the quasi-normal year of the Pacific-Indian Ocean SSTA; in the pure El Niño year, negative temperature anomaly takes place in Northeast China; but corresponding to the positive phase of the PIM, negative temperature anomaly occurs in middle-low reaches of the Yangtze River, and there is positive temperature anomaly in North China and the southern part of Northeast China (Fig. 4(d)). Evidently, the results indicate that the impacts of the pure Pacific SSTA on the summer temperature in East China are also different from those of the Pacific-Indian Ocean SSTA composite mode. In the summer of

2003, persistent severe hot weather attacked large areas of South China. The main reason was that the Pacific-Indian Ocean SST remained a quasi-normal state[24].

Previous researches of the influence of El Niño on summer climates in China already showed that during El Niño year more rainfall above average occurred in the Yangtze-Huaihe River Basin[25, 26], and low temperature and more rainfall were observed in Northeast China[27, 28]. Our results involving the pure El Niño are almost the same as previous research results. Consequently, we can conclude that the analysis in this section is credible to a certain degree.

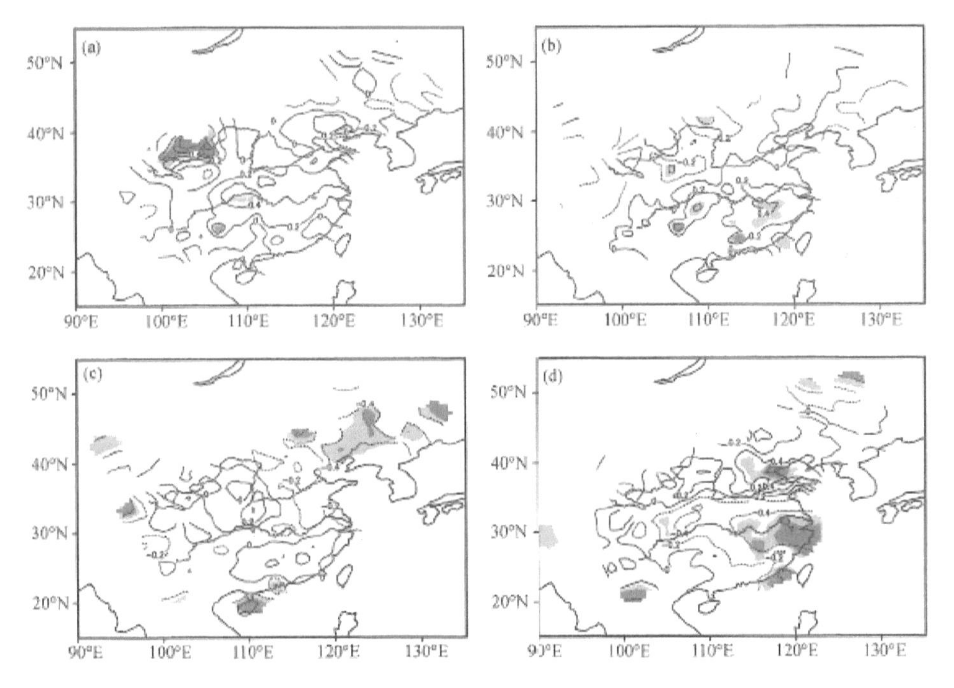

Fig. 4. The distribution of surface air temperature anomaly in China. Unit: ℃. The four phases in Fig. 4 (a)—(d) are the same with Fig. 2(a)—(d), respectively.

3. Summer Rainfall in India

Although the relationship between the Indian summer precipitation and ENSO in recent years is somewhat weakened, on the whole, the summer anomalous rainfall in India is still a typical example of the ENSO impacts. One of the causes of the weakened relationship between the Indian summer rainfall and ENSO is considered the influence of the Indian Ocean SSTA. If the composite effect of the SSTA in both the Pacific Ocean and the Indian Ocean is taken into account, it may make the relationship between our presented composite mode and the Indian summer rainfall become relatively good. Fig. 5 shows the Indian summer rainfall corresponding to the pure El Niño year and the positive phase of the PIM, respectively. More rainfall than average occurs in the middle part of India, less rainfall happens in the northern part and the southwestern part of India in the pure El Niño year (Fig. 5(a)). The whole India receives substantially less rainfall than average for the positive phase of the PEM (Fig. 5(b)). Compar-ing Fig. 5(a) to Fig. 5(b), it can be found that the difference between the influence of the pure El Niño and that of the positive phase of the PIM is quite large. The influence of the positive phase of the PIM can increase the drought in India. Although the mean summer rain- falls in 1983 (figure omitted) and 1997 are near average in India and slightly more than average in some regions of India[29], other 5 years among the 7 cases receive less rainfalls. It indicates that less rainfall in India is a main feature during the positive phase year of the PIM. Therefore, it is worth discussing that the occurring of the Indian Ocean temperature dipole causes the Indian rainfall in El Niño year being not negative anomaly. The relationship between ENSO and Indian rainfall is weakened recently much because of other factors such as the role of Atlantic circulations[30].

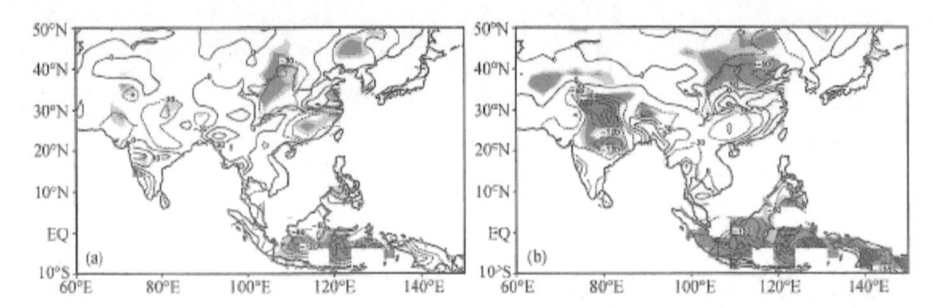

Fig. 5. Composites of summer (JJA) total precipitation anomalies from the precipitation reconstruction over land (PREC/L) in summer in positive phase (El Niño) of the ENSO mode in (a), and positive phase of the PIM in (b). Unit: mm. The dark shaded areas and light shaded areas represent the *t*-test significance at the 0.05 and 0.1 levels, respectively.

4. Numerical Simulation

The above observation analysis suggests that the PIM and its role are significantly different from those of the Pacific ENSO. Numerical simulation is a very effective method of research, which can isolate the influence of SSTA on climate change from other factors. Ensemble simulation is used for further investigating the climate effect of the PIM. The R42L9 numerical model in this paper is a global generation spectral model developed by State Key Laboratory of Numerical Modeling for Atmospheric Sciences and Geophysical Fluid Dynamics (LASG), Institute of Atmospheric Physics, Chinese Academy of Sciences. The horizontal resolution is roughly 2.8125 (lon) × 1.66(lat) with a rhomboidal truncation at wave number 42. The vertical adopts the terrain-following Sigma coordinate with 9 levels. The model uses a reference atmosphere, and semi-implicit time-integration scheme with 15-minute time step. This model was successful in modeling climate characteristics[31].

Control experiments and sensitivity experiments were performed with the time integration from January 1 to September 30. The monthly mean SST was used in the control experiment. In the sensitivity experiment, the forcing SST used the monthly mean SST plus the SSTA in the forcing region of 15°S—15°N, 40°E—75°W which was twice that of the positive phase of the PIM (Fig. 2(d)) after April 16. In order to decrease influences of initial error and other errors, the control experiments and sensitivity experiments applied 7 ensemble members starting from some different initial states. The mean value of 7 ensemble members was used. Based on the difference between the control experiment and the sensitivity experiment, the anomalous response to the SSTA forcing could be obtained.

Fig. 6 gives the precipitation anomaly forced by the SSTA of positive phase of the PIM. It can be seen that substantial negative anomaly occurs in Indian Peninsula, Indochina and Indonesia, which is quite consistent with Fig. 5(b). But it differs from Fig. 5(a) of ENSO. The numerical results agree well with the observation. The positive rainfall anomaly occurs in the areas from Southwest China to Northeast China with the center in Sichuan Province and Guizhou Province of China and North Korea as well in numerical simulation. The Yangtze-Huaihe River Basin and large part of South China receive more rainfall. These indicate that the numerical results in the major parts of China are the same as those observed. But in the minor parts of China the numerical results do not agree with those observed. The mechanism of summer precipitation in China is complex and middle-high latitudes are not all influenced by SSTA. Generally, the numerical results are consistent with those observed. The numerical simulation further indicates that the PIM has important impact on Asian climate which differs from that of ENSO mode.

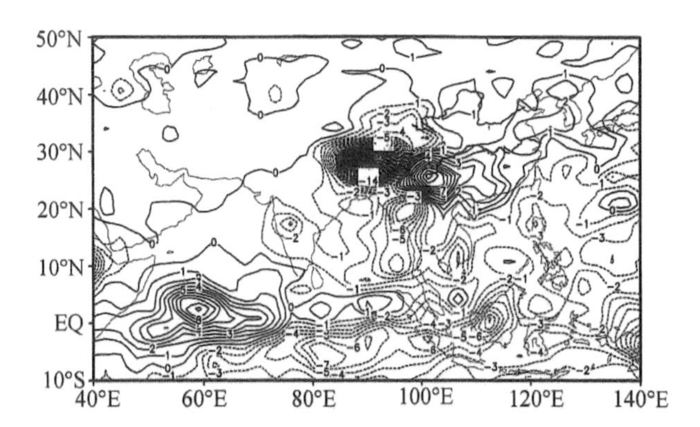

Fig. 6. Precipitation anomalies in summer (JJA) by the SSTA forcing of positive phase of the PIM. Unit: mm/d.

5. Conclusions

Based on the anomalous pattern and features of the whole Pacific-Indian Ocean sea temperature, the idea of the PIM has been put forward. Moreover, the characteristics of the PIM and ENSO mode and their influences on climates in China and summer precipitation in India have been compared. The conclusions can be drawn as follows:

(1) The quasi-normal year of the PIM can better reveal the quasi-normal feature of SST compared with that of ENSO mode. The SSTAs are all near zero around the equator. In the

positive phase year of the PIM it presents a west-east difference of the tropical SSTA not only in the Pacific but also in the Indian Ocean, which is distinct from that in the positive phase year of ENSO mode.

(2) There are many differences between the influence of the PIM and that of the ENSO mode on the summer precipitation and temperature in China. Thus analyzing and forecasting the Asian climate change must consider the SSTA pattern in both the Pacific Ocean and the Indian Ocean, namely the PIM.

(3) Considering the effect of SSTA on Indian rainfall alone, the PIM induces a more significant decrease in Indian rainfall than pure El Niño does. It may not be true that the tropical Indian Ocean dipole is the cause of the weakening of the relationship between ENSO and Indian precipitation in recent years. Other factors may be more important.

(4) The numerical simulation obtains almost the same results as the observational study. It proves that the PIM has important effect on Asian climate, which differs from ENSO mode.

Therefore, to provide better scientific explanation for short-term climate prediction, the PIM and its influence should be considered and investigated.

Acknowledgements This work was supported by the National Natural Science Foundation of China (Grant No. 40233033), the National Key Basic Research and Development Project of China (Grant No. 2004CB18300), and the Innovation Key Program of the Chinese Academy of Sciences (Grant No. ZKCX2-SW-226).

References

[1] Rasmusson E M, Carpenter T H. The relationship between eastern equatorial Pacific sea surface temperatures and rainfall over Indian and Sri Lanka. Mon Wea Rev, 1983, 111: 517-528.

[2] Ropelewsk C F, Halpert M S. Global and regional scale precipitation patterns associated with the El Niño/Southem Oscillation. Mon WeaRev, 1987, 115: 1606-1626.

[3] Huang R H, Wu Y F. The influence of ENSO on the summer climate change in China and its mechanism. Adv Atmos Sci, 1989, 6: 21-32.

[4] Li C Y. Introduction to Climate Dynamics (in Chinese). 2nd ed. Beijing: Meteorology Press, 2000.

[5] Kumar K K, Rajagopalan B, Cane M A. On the weakening relationship between the Indian monsoon and ENSO. Science, 1999, 284: 2156-2159.

[6] Torrence C, Webster P J. Interdecadal changes in ENSO-monsoon system. J Clim, 1999, 12: 2679-2690.

[7] Ye D Z, Huang R H, eds. Study on the Law and Causes of Drought/Flood in Yangtse River and Yellow River Valleys (in Chinese). Jinan: Shandong Scientific and Technological Press, 1996.

[8] Luo S H, Jin Z H, Chen L T. Correlation analyses of the SST in the middle-lower reaches of the Yangtze River. Chin J Atmos Sci (in Chinese), 1985, 9(3): 336-342.

[9] Saji N H, Goswami B N, Viayachandrom P N, et al. A dipole mode in the tropical Indian Ocean. Nature, 1999, 401: 360-363.

[10] Webster P T, Moore A M, Loschning J P, et al. Coupled ocean-atmosphere dynamics in the Indian Ocean during 1997-98. Nature, 1999, 401: 356-360.

[11] Wang D X, Wu G X, Xu J J. Interdecadal variability in the tropical Indian Ocean and its dynamic explanation. Chin Sci Bull, 1999, 44(17): 1620-1626.

[12] Li C Y, Mu M Q. Influence of the Indian Ocean dipole on atmospheric circulation and climate. Adv Atmos Sci, 2001, 18: 831-843.

[13] Xiao Z N, Yan H M, Li C Y. The relationship between Indian Ocean SSTA dipole index and the precipitation and temperature over China. J Tropical Meteor (in Chinese), 2002, 18(4): 335-344.

[14] He J H, Zhang R H, Tan Y K, et al. The features of the interannual variation of sea surface temperature anomalies in the tropical Indian Ocean. In: Chao J P, Li C Y, Chen Y Y, et al, eds. Study on the Mechanism and Prediction of ENSO Cycle (in Chinese). Beijing: Meteorology Press, 2003. 279-293.

[15] Li C Y, Mu M Q, Pan J. Indian Ocean temperature dipole and SSTA in the equatorial Pacific Ocean. Chin Sci Bull, 2002, 47: 236-239.

[16] Niiler P, Stevenson J. The heat budget of tropical ocean warm-water pools. J Mar Res, 1982, 40(Suppl): 465-480.

[17] Potemra, J T, Lukas R, Mitchum G T. Large-scale estimation of transport from the Pacific to the Indian Ocean. J Geophy Res, 1997, 102: 27795-27812.

[18] Meyers G Variation of Indonesian throughflow and the El Niño-Southem Oscillation. J Geophy Res, 1996, 101(C5): 12255-12263.

[19] Pan Y H, Oort A H. Correlation analyses between sea surface temperature anomalies in the eastern equatorial Pacific and the world ocean. Clim Dyn, 1990, 4: 191-205.

[20] Alexander M A, Bladé I, Newman M, et al. The Atmospheric bridge: the influence of ENSO teleconnections on air-sea interaction over the global oceans. J Clim, 2002, 15(16): 2205-2231.

[21] Shen X S, Kimoto M, Sumi A, Numaguti A, et al. Simulation of the 1998 East Asian Summer Monsoon by the CCSR/NIES AGCM. J Meteor Soc Japan, 2001, 79(3): 741-757.

[22] Guo Y F, Zhao Y, Wang J. Numerical simulation of the relationships between the 1998 Yangtze River valley floods and SST anomalies. Adv Atmos Sci, 2002, 19(3): 391-404.

[23] Guo Y F, Wang J, Zhao Y. Numerical simulation of the 1999 Yangtze River valley heavy rainfall including sensitivity experiments with different anomalies. Adv Atmos Sci, 2004, 19(3): 391-404.

[24] Yang H, Li C Y. Diagnostic study of serious high temperature over South China in 2003 summer. Clim Environ Res (in Chinese), 2005, 10(1): 90-95.

[25] Fu C B, Teng X L. The relationship between the climate anomaly in China and El Niño/Southern Oscillation phenomenon. Chin J Atmos Sci (in Chinese), 1988, special issue: 133-141.

[26] Gong D Y, Wang S W. Impacts of ENSO on the seasonal rainfall in China. J Nat Disasters (in Chinese), 1998, 7: 44-52.

[27] Liu Y S, Zhi J H, Zhou Z H. The rule of period change of temperature and group occurring of low temperature in summer in Northeast China. In: Edit Group of Long-range Forecast of Summer Low Temperature in Northeast China, ed. Collected Papers of Long- range Forecast of Summer Low Temperature in Northeast China (in Chinese). Beijing: Meteorology Press, 1983. 17-21.

[28] Li C Y. El Niño event and the temperature anomalies in the eastern China. J Tropical Meteor (in Chinese), 1989, 5: 210-219.

[29] Bell G, Halpert M. Climate assessment for 1997. Bull Am Meteor Soc, 1998, 79: S1-S50.

[30] Chang C P, Harr P, Ju J. Possible role of Atlantic circulations on the weakening Indian monsoon rainfall-ENSO relationship. J Clim, 2001, 14: 2376-2380.

[31] Wu T W, Liu P, Wang Z Z, et al. The Performance of Atmospheric Component Model R42L9 of GOALS/LASG. Adv Atmos Sci, 2003, 20(5): 726-742.

Seasonal Evolution of Dominant Modes in South Pacific SST and Relationship with ENSO

Li Gang*[1] (李刚), Li Chongyin[1, 2] (李崇银), Tan Yanke[1] (谭言科), Bai Tao[1, 3] (白涛)

[1] Institute of Meteorology and Oceanography, PLA University of Science and Technology, Nanjing
[2] LASG, Institute of Atmospheric Physics, Chinese Academy of Sciences, Beijing
[3] No. 94162 Troops of PLA, Xi'an

Abstract A season-reliant empirical orthogonal function (S-EOF) analysis was applied to the seasonal mean SST anomalies (SSTAs) based on the HadISST1 dataset with linear trend removed at every grid point in the South Pacific (60.5°—19.5°S, 139.5°E—60.5°W) during the period 1979—2009. The spatiotemporal characteristics of the dominant modes and their relationships with ENSO were analyzed. The results show that there are two seasonally evolving dominant modes of SSTAs in the South Pacific with interannual and interdecadal variations; they account for nearly 40% of the total variance. Although the seasonal evolution of spatial patterns of the first S-EOF mode (S-EOF1) did not show remarkable propagation, it decays with season remarkably. The second S-EOF mode (S-EOF2) showed significant seasonal evolution and intensified with season, with distinct characteristics of eastward propagation of the negative SSTAs in southern New Zealand and positive SSTAs southeast of Australia. Both of these two modes have significant relationships with ENSO. These two modes correspond to the post-ENSO and ENSO turnabout years, respectively. The S- EOF1 mode associated with the decay of the eastern Pacific (EP) and the central Pacific (CP) types of ENSO exhibited a more significant relationship with the EP/CP type of El Niño than that with the EP/CP type of La Niña. The S-EOF2 mode contacted with the EP type of El Niño changing into the EP/CP type of La Niña showed a more significant connection with the EP/CP type of La Niña.

Keywords South Pacific, sea surface temperature anomalies, season-reliant empirical orthogonal function (S-EOF), El Niño-Southern Oscillation (ENSO).

1. Introduction

The Southern Hemisphere oceans comprise a vast domain, and they play important roles in the global climate system. Many studies have investigated the SST variations in these region

Received October 12, 2011; revised February 14, 2012
*Corresponding author: LI Gang, Liang. 1983@163.com

and their impacts on the global climate, including the South Indian Ocean (Behera and Yamagata, 2001; Jia and Li, 2005; Yan et al., 2009), the South Atlantic Ocean (Venegas et al., 1997, 1998; Weijer et al., 2002; Nnamchi and Li, 2011), the southeastern Pacific (Shaffer et al., 2000; Falvey and Garreaud, 2009), and the southwestern Pacific (Holbrook and Bindoff, 1999; Holbrook et al., 2005). The SST anomalies in these regions exhibit significant interannual and decadal variations and have close relationships with El Niño-Southern Oscillation (ENSO).

The South Pacific region covers nearly half of the Southern Hemisphere oceans, spanning from the equator to 60°S and from ~140°E to 70°W. However, relative to the North Pacific, fewer ocean observations take place in the South Pacific because of not only its inhospitable environment but also the paucity of observation stations (Reynolds and Smith, 1994; Linsley et al., 2000; Smith et al., 2008). Therefore, the ocean reanalyzed datasets based on assimilation models and interpolated methods have to be used to study South Pacific ocean-atmosphere interactions. In fact, the lack of high-quality sea temperature data hampers the study of the South Pacific to some extent.

Despite the lack of high-quality ocean observations in the South Pacific, ocean temperature datasets based on the satellite are available dating to the late 1970s; they provide a basis for study the ocean-atmosphere interactions in the South Pacific. Many studies have been conducted on the South Pacific; they are mainly focused on the sea temperature (surface and subsurface temperatures) variability on the interannual and decadal time scales (Wang and Liu, 2000; Luo and Yamagata, 2001; Giese et al., 2002; Kidson and Ren- wick, 2002; Luo et al., 2003; Yu and Boer, 2004; Wang et al., 2007; Yang et al., 2007; Shakun and Shaman, 2009) and its impact on global climate variability (Barros and Silvestri, 2002; Hsu and Chen, 2011). These studies have not only suggested that both the surface and subsurface temperature anomalies in South Pacific can propagate to the tropical Pacific, but they have also reported that ENSO has an important impact on the South Pacific SST anomalies. In addition, the SST anomalies in the South Pacific feature significant decadal and interdecadal variations.

However, what are the dominant modes of SST anomalies in the entire South Pacific on an interannual time scale? The answer to this question is not straightforward because the SST variations in this region seem to have a closer association with the evolution of ENSO than in the North Pacific (Wang et al., 2003a, b; Shakun and Shaman, 2009). When conventional EOF analysis and singular value decomposition (SVD) analysis are applied to the SST anomalies in the South Pacific, the first mode obtained has a close relationship with ENSO in the mature phase, which reflects the response of SST variations in the South Pacific to ENSO. Previous studies have applied EOF analysis to either boreal summer SST anomalies or boreal winter SST anomalies in the South Pacific (Hsu and Chen, 2011; Li et al., 2011). The resultant modes of SST anomalies are closely associated with the evolution of ENSO. Therefore, the SST anomalies in the South Pacific may show significant seasonal evolution characteristics. In this study, we used a season-reliant EOF (S-EOF) analysis to extract the major modes of SST anomalies in the South Pacific.

The goals of this study were to identify the spatiotemporal characteristic of the leading seasonally evolving modes of the South Pacific and to determine their relationships with ENSO during recent decades (i.e., after the 1976/1977 climate shift).

This paper is organized as follows: Section 2 briefly describes the datasets and analysis methods used in present study. In section 3, we identify the main seasonally evolving modes of the South Pacific. Section 4 describes the relationships between these patterns of variability and ENSO. The conclusions and some discussions are summarized in section 5.

2. Data and Analysis Methods

2.1 Data

The principal data used in this study were sea surface temperature (SST) data (horizontal resolution 1°×1°) from the Hadley Centre Sea Ice and Sea Surface Temperature datasets (HadISST1) of the UK Meteorological Office (Rayner et al., 2003). Although this dataset is available from 1870 to the present, our analyses only considered decades following the 1976/1977 climate shift. We chose to analyze the period 1979—2009 (Fig. 1).

Anomalous quantities were computed by removing the monthly mean climatology and the linear trend based on a linear least squares fit. Given that array $y(n)$ is the SST data and array $x(n)$ is the time index of SST data $y(n)$, the linear trend was fitted as $y(n) = a \times x(n) + b$. Then we removed this linear trend from the SST.

We used the Niño3.4 index (available online at http: //www.esrl.noaa.gov/psd/gcos_wgsp/ Timeseries/) as an indicator of the ENSO variability. The Niño3.4 index was calculated from the HadISST1 using the averaged SST over the study region (5°S—5°N, 170°E—120°W).

2.2 Analysis methods

Season-reliant empirical orthogonal function (S-EOF) was the primary method used in this study. Although conventional EOF analysis can depict the spatiotemporal variations of a physical field, it cannot produce coherently seasonal evolution patterns. S-EOF analysis is similar to the extended EOF (EEOF; Weare and Nasstrom, 1982) analysis to some extent. The idea of S-EOF has been discussed in detail by Wang and An (2005). In this study we applied the S-EOF analysis to SST anomalies in a seasonal sequence beginning from the winter of a year denoted as D(–1)JF(0) to the following fall, denoted as SON(0). We treated the anomalies for D(–1)JF(0), MAM(0), JJA(0), and SON(0) as a "yearly block", where –1 denotes the year before year 0. When the conventional EOF analysis was conducted, the yearly block was divided into four consecutive seasonal anomalies to obtain a seasonally evolving mode of the SST anomalies. In our analysis, we also used the Morlet wavelet to identify the periodicity of the seasonally evolving modes (Torrence and Compo, 1998). In addition, we used regression and correlation methods; statistical significance of correlation coefficients was assessed using the Student's *t*-test.

We defined the South Pacific to be the area between 60.5°—19.5°S and 139.5°E—60.5°W. The choice of 19.5°S as the northern boundary excluded the impact of the farthest ENSO.

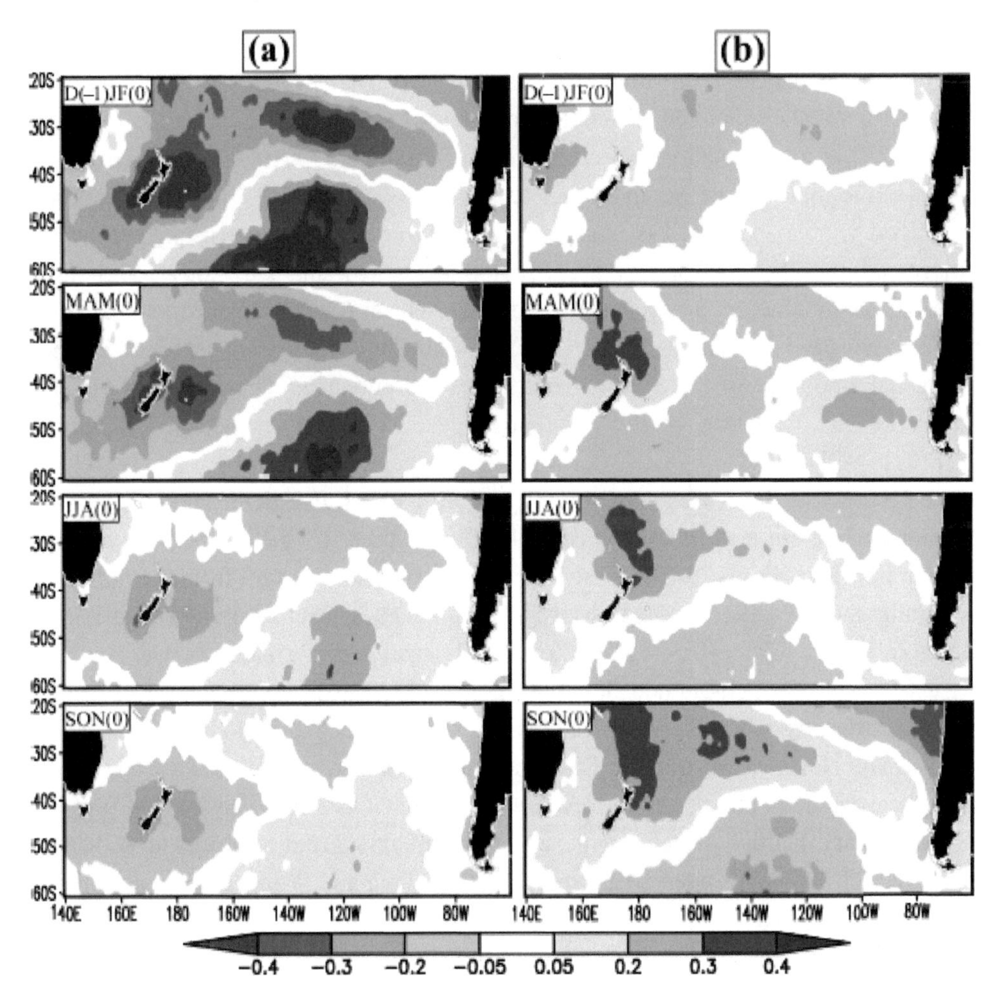

Fig. 1. (a) Seasonally evolving spatial patterns of S-EOF1 mode derived from the HadISST1 SST anomalies over the South Pacific during the period 1979—2009 (℃). (b) The same as in panel (a), but for S-EOF2 mode.

3. Dominant Seasonally Evolving SST Anomaly Patterns

We first applied S-EOF analysis to identify the leading modes of SST anomalies in the South Pacific. The leading two modes accounted for 25.7% and 14.1% of the total variance, respectively. According to the rule of North et al. (1982), we found that these two modes were well distinguished from each other and from the remaining modes. Therefore, these two modes are considered statistically distinguishable and significant. We explored the distinctive seasonal evolution characteristics of these two dominant modes, and we report our findings the following discussion.

Figure 1a shows the seasonally evolving spatial patterns of S-EOF1 mode. Notably, the patterns are derived using regression of the South Pacific SST anomalies from boreal winter

[D(−1)JF (0)] to fall [SON(0)], with the first normalized principal component (PC1). In general, we found that S-EOF1 mode is characterized by a tripolar configuration, with a positive SST anomalies centered in the high latitudes of the South Pacific region (60°—50°S, 160°—120°W), surrounded by two negative SST anomalies centered over eastern New Zealand and the mid-latitude South Pacific region (35°—25°S, 140°—100°W), respectively. The strongest tripolar pattern was observed in D(−1)JF(0); it was consistent with the SST pattern forced by ENSO (Wang et al., 2007; Shakun and Shaman, 2009). Although the positive SST anomalies of S-EOF1 show a slight eastward shift, the positions of both the positive and negative SST anomalies of S-EOF1 do not show significant seasonal dependence. However, the amplitude of the S-EOF1 mode shows pronounced characteristics of seasonal evolution. In fact, both the positive and negative SST anomalies decayed rapidly from D(−1)JF(0) to SON(0).

Figure 1b presents the seasonally evolving spatial patterns of S-EOF2 mode. The S-EOF1 and S-EOF2 mode are clearly distinguished from each other by their seasonal evolution. In sharp contrast with S-EOF1, S-EOF2 mode features a wavelike pattern from the northwestern to the southeastern Pacific, with positive SST anomalies in southeastern Australia and the southeastern Pacific (60°—40°S, 130°—80°W) and negative SST anomalies over southeastern New Zealand and the mid-latitude South Pacific. In MAM(0), the positive SST anomalies intensify significantly while the negative SST anomalies over the mid-latitude South Pacific begin to decay. Notably, the positive SST anomalies center over the southeastern Australia shift northeastward to the northern New Zealand from D(−1)JF(0) to MAM(0) and then become stationary, but they develop with season remarkably. From MAM(0) to SON(0), both the positive SST anomalies over the northern New Zealand and negative SST anomalies over the southern New Zealand extend eastward and intensify with season. However, the positive SST anomalies over the southeastern South Pacific decay from MAM(0) to SON(0). The eastward propagation of the positive SST anomalies in the eastern Australia is consistent with the results of Kidson and Renwick (2002), who found that large-scale South Pacific SST variations on the interannual time scale are primarily driven by ENSO.

Compared with conventional EOF analysis, S-EOF analysis can depict the seasonal evolution of a physical field. However, the pattern of S-EOF1 mode in D(−1)JF(0) over the South Pacific (Fig. 1a) resembles the pattern obtained from conventional EOF analysis of SST anomalies in D(−1)JF (0), not only using the same dataset (Wang et al., 2007; Shakun and Shaman, 2009; Terray, 2011) but also using different datasets (Li et al., 2011), even though both the time period and the spatial coverage of the data used in the analyses were different. In addition, the pattern of S-EOF2 mode in D(−1)JF(0) (Fig. 1b) also resembles that based on conventional EOF analysis of SST anomalies in D(−1)JF(0), only with the reverse sign (Li et al., 2011). This resemblance indicates that both S-EOF1 and S-EOF2 mode of the South Pacific are stable and credible. Notably, the resemblance can be observed not only between the patterns of S- EOF1 mode in JJA(0) (Fig. 1a) and S-EOF2 mode in D(−1)JF(0) (Fig. 1b) but also between the patterns of S-EOF1 mode in D(−1)JF(0) (Fig. 1a) and S-EOF2 mode in SON(0). This resemblance between S-EOF1 and S-EOF2 mode reveals that both of these modes have a connection with some important signals (e.g., ENSO; Kidson and Renwick, 2002), but with different lead-lag correlations, which are discussed hereafter.

Although the combination of these two dominant S-EOF modes accounted for nearly 40% of the total variance, the fractional variance explained by the two S-EOF modes varied with season and location remarkably. The spatial and seasonal geographic distributions of the fractional variance for these two modes are shown in Fig. 2. For S-EOF1 mode (Fig.2a), the foremost prominent feature is that the largest fractional variance occurs in D(–1)JF(0) over the mid-latitude South Pacific along 30°S, over the southwestern Pacific that encompasses the New Zealand, and over the high-latitude South Pacific, where the accounted variance exceeds 60%. The second most notable feature in Fig. 2a is that the fractional variance over the three regions decreases gradually from D(–1)JF(0) to SON(0). For S-EOF2 (Fig.2b), the prominent feature is that the maximum variance appears in SON(0) over the regions that resemble the first mode in D(–1)JF(0), where the fractional variance exceeds 40%. Notably, the fractional variance carried by the second mode over the northern New Zealand and high-latitude South Pacific features a significant eastward propagation and becomes large gradually with season appreciably. We also noted that the regions with the larger variance are associated with significant SST anomalies.

Figure 3 presents the time series of principal components of S-EOF1 (Fig.3a) and S-EOF2 (Fig.3b). The black thick line is the normalized 7-year low-pass filtered time series. The first notable feature in Fig. 3 is that both of these modes have not only prominent interannual variations but also significant interdecadal variations. The second feature is that the amplitude of these two principal components decayed rapidly after 2000. The third prominent feature is the reversal of interdecadal variations (the black thick line in Fig. 3) of the two S-EOF modes.

To investigate the periodicities of the two modes in detail, the local wavelet power spectrum analysis based on Morelet wavelet was applied to their principal components. Figure 4 presents the local wavelet power spectrum of the normalized PC1 (Fig. 4a) and PC2 (Fig.4b). PC1 shows high power in the 3-5.5- year and 2-3- year bands for the periods 1980—1995 and 1995—2000 on the interannual time scale, respectively. However, PC2 has significantly high power in the 2-4-year band for the period 1993-2000 on the interannual time scale. Both of the two modes have pronounced high power in the 8-15-year band on the interdecadal time scale. In addition, the high power of the two modes on the interannual and interdecadal time scales decayed significantly after the year 2000.

4. Relationships Between the Dominant Modes and ENSO

ENSO is the largest well-known signal in the climate system on the interannual time scale. Analysis of the dominant, seasonally evolving SST patterns of the South Pacific show that they may have an association with ENSO. To examine the relationships between the two S-EOF modes and ENSO, we calculated the lead-lag correlations between the PC1 and PC2 and ENSO.

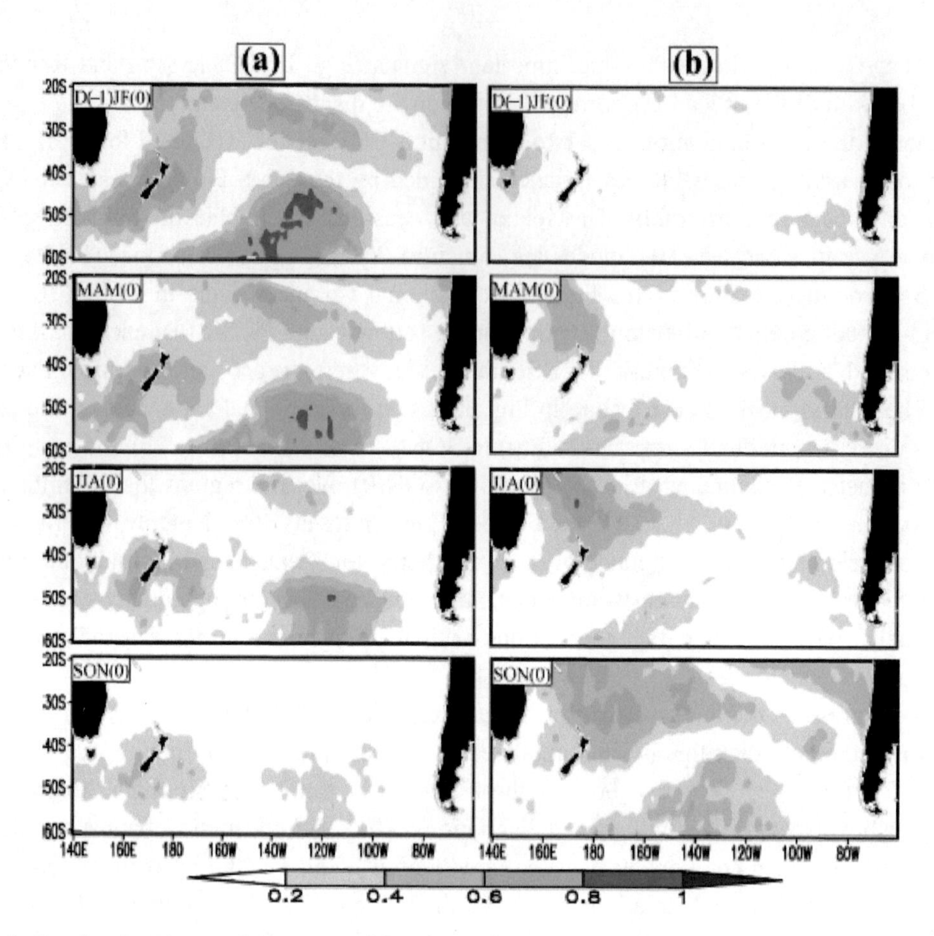

Fig. 2. Fractional variance of the seasonal South Pacific SST anomalies accounted by (a) S-EOF1 and (b) S-EOF2.

The lead-lag correlations of PC1 with Niño3.4 index are shown in Fig.5 (blue line). Here the year (–1) denotes that Niño3.4 index leads the PC1 and the year (1) denotes that Niño3.4 index lags the PC1. Notably, the PC1 was positively correlated with the Niño3.4 index from JJA(–1) to MAM(0), with the correlation coefficient passing the threshold of significance in the 95% Studenfs *t*-test. The PC1 shows a maximum positive correlation coefficient that exceeds 0.8 (statistically significant at the 95% Student's *t*-test significance) in D(–1)JF(0). After the correlation coefficient reaches its peak in D(–1)JF(0), the correlation between them decays rapidly and becomes negative (not passing the 5% Student's t-test significance) that persists after SON (0).

The lead-lag correlation between PC2 and the Niño3.4 index is also shown in Fig.5 (red line). Notably, the correlation is positive and weak (not passing Student's *t*-test at 5% significance) from MAM(–1) to D(–1)JF(0). However, the correlation becomes negative and rapidly develops from MAM (0) to SON (0). A maximum negative correlation coefficient occurs in SON(0) that passes the threshold of the 5% Student's *t*-test significance. After the correlation reaches its peak in SON(0), it decays significantly after D(0)JF(1).

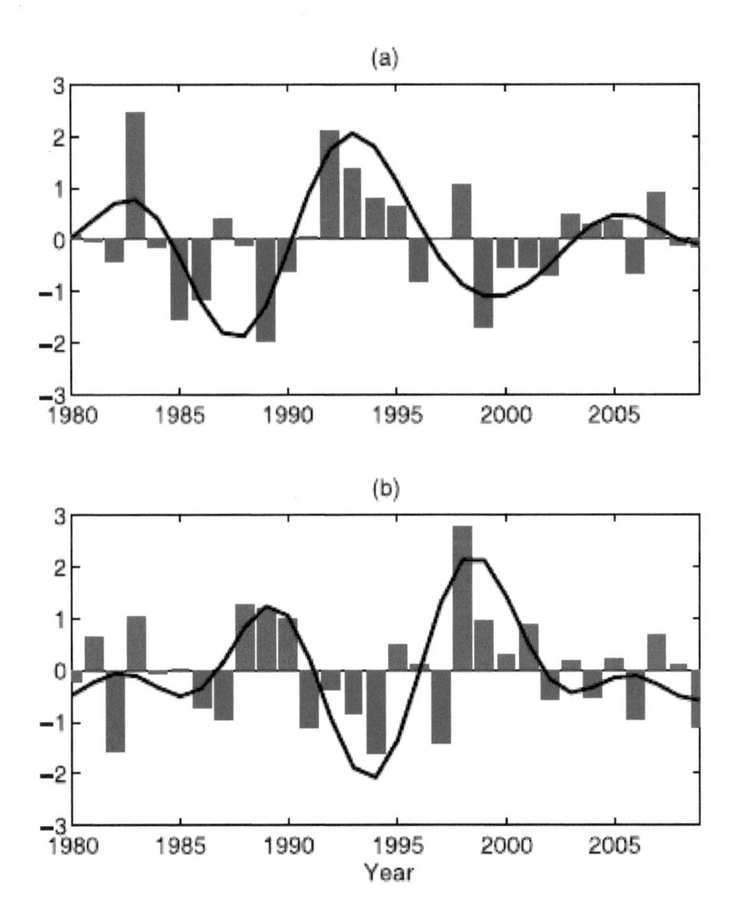

Fig. 3. Normalized principal components of (a) S-EOF1 mode and (b) S-EOF2 mode. The superimposed black thick line is the normalized 7 year low-pass filtered time series.

As we know, ENSO features significant phase-locking and seasonal cycle behavior, starting in the boreal spring, normally maturing in the boreal winter, and decaying in the next spring (Rasmussen and Carpenter, 1982; Neelin et al., 2000). Based on the correlations between the two modes and ENSO, we can conclude that S-EOF1 mode occurs following the peak of the ENSO, and it may be viewed as responses to ENSO (Shakun and Shaman, 2009). However, S-EOF2 mode is primarily associated with the development of ENSO, and it may impact ENSO (Zhang et al., 2005) and thus may make a significant contribution to ENSO prediction methods.

To further demonstrate the relationship between season-evolving variability of South Pacific SST and ENSO, the spatial patterns of seasonally evolving SST in the tropical Pacific associated with the two S-EOF modes are shown in Figs. 6 and 7, respectively. Here the SST anomalies in the tropical Pacific are regressed with PC1 and PC2. To show the evolution of ENSO, the figures from the last year [year (–1)] to the next year [year (1)] are presented.

Figure 6 shows the seasonal evolution of spatial patterns of tropical Pacific SST anomalies associated with S-EOF1. In D(–2)JF(–1), the positive SST anomalies are primarily located over the central tropical Pacific and extend northeastward to North America. The negative

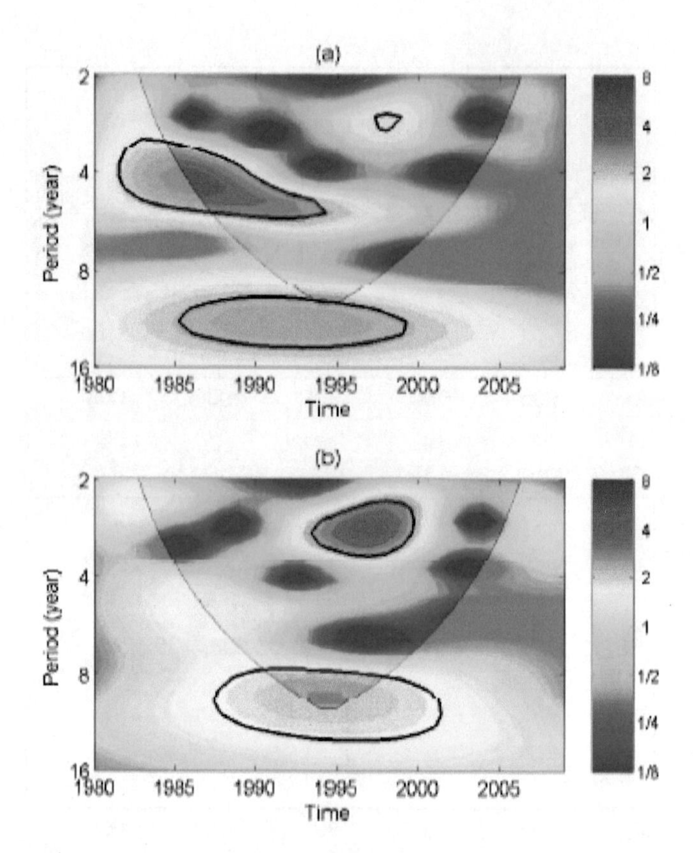

Fig. 4. The local wavelet power spectrum (using the Morlet wavelet) of (a) S-EOF1 and (b) S-EOF2. The thick black contour is the 10% significance level against the red noise. The cone of influence (COI) where edge effects might distort the picture is show as a lighter shade.

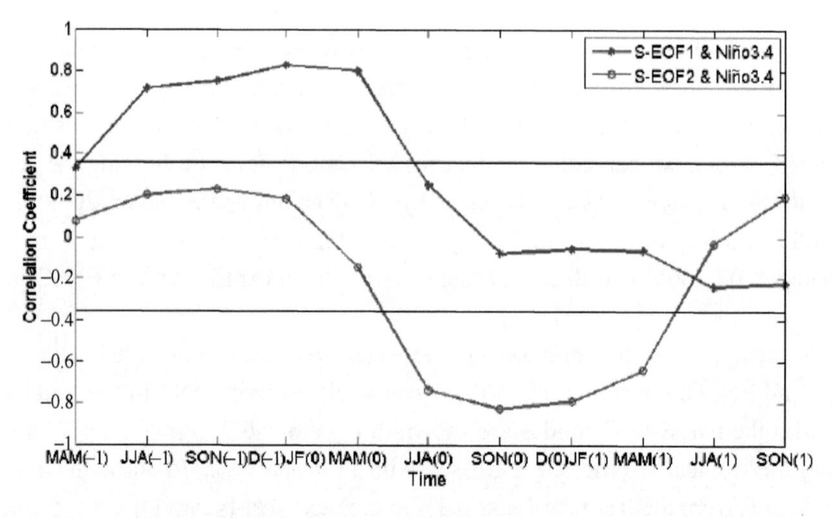

Fig. 5. (a) Lead-lag correlations between the first S- EOF principal component and Niño 3.4 index. (b) The same as in panel (a), but for the second mode. The dashed lines indicate the Student's *t*-test at 5% significance.

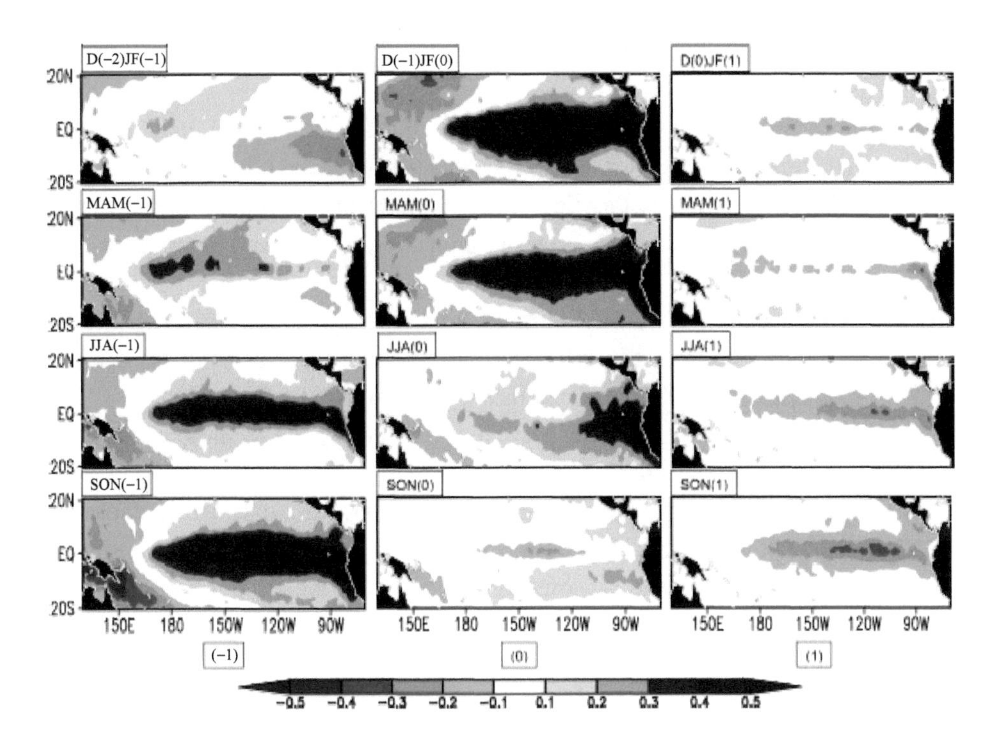

Fig. 6. Seasonally evolving patterns of tropical Pacific SST anomalies regressed with the principal component of S-EOF1 (Units: ℃).

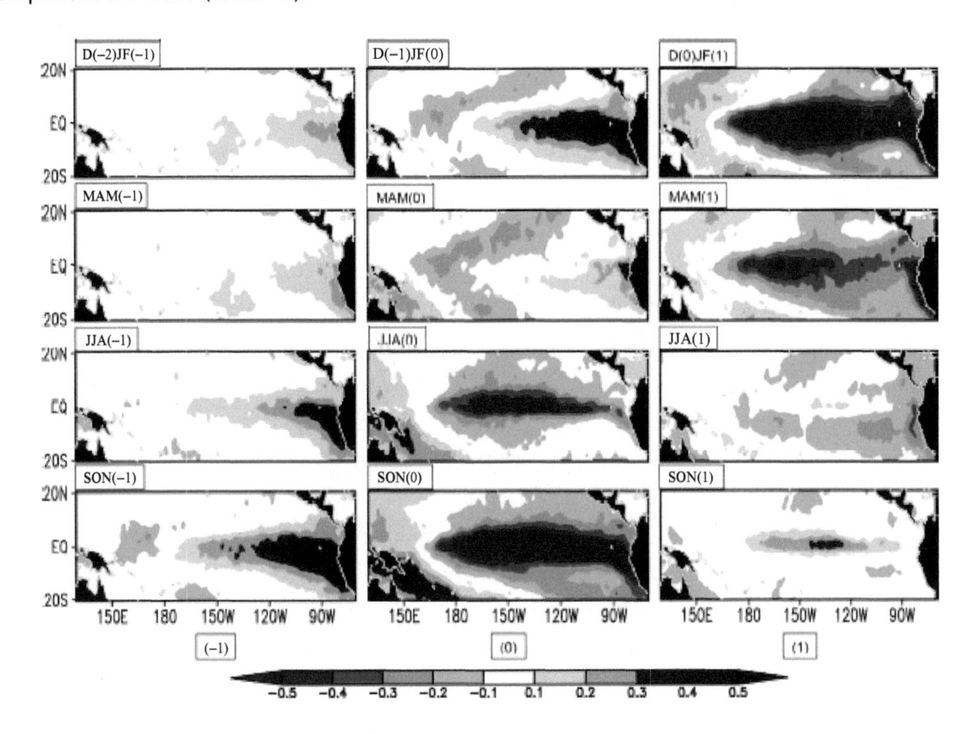

Fig. 7. The same as in Fig. 6, but for S-EOF2 mode.

SST anomalies are mainly found over the southeastern tropical Pacific centered near the western Peruvian coast.

After MAM(–1), the positive SST anomalies in the central tropical Pacific strengthen significantly and propagate eastward along the equator. A weak positive SST anomaly is noticeable in the eastern tropical Pacific and develops rapidly with season. Notably, the amplitude of the eastern tropical Pacific SST anomalies is weaker than that of the SST anomalies in the tropical central Pacific. However, the negative SST anomalies in the tropical southeastern Pacific decay rapidly, and the meridional extent of the negative SST anomalies increases toward North Pacific. This indicates that El Niño develops with the season. The positive SST anomalies in the tropical central-eastern Pacific reach maximum (~1.1℃) in D(–1)JF(0). The negative SST anomalies are mainly located in the tropical western Pacific. This indicates that El Niño reached the mature phase.

Notably, the positive SST anomalies in the central-eastern tropical Pacific decay rapidly after MAM(0), which indicates that El Niño began to decay. It is necessary to mention that the decay in the central tropical Pacific is more rapid than that in the eastern tropical Pacific. In SON(0), the pattern is nearly a reversal of SST anomalies in D(–2)JF(–1), which indicates that El Niño died out.

From D(0)JF(1) to SON(1), the weak negative SST anomalies in the central tropical Pacific developed with season. The other weak negative SST anomalies in the eastern tropical Pacific emerged near the Peruvian coast, which indicates that a weak La Niña developed and then strengthened with season.

For S-EOF1 mode, ENSO first occurred in the central tropical Pacific and then occurred in the eastern tropical Pacific, which started in the boreal spring and reached the mature phase in boreal winter. In addition, S-EOF1 mode showed a more remarkable correlation with the warm ENSO events than the cold ENSO events. Notably, ENSO-associated S-EOF1 mode had a period of quasi-quadrennial (~4 years), which resembled the low-frequency (LF) component of ENSO identified by Barnett (1991).

Figure 7 shows the seasonal evolution of spatial patterns of the tropical Pacific SST anomalies associated with S-EOF2. Notably, warming first appears near the Peruvian coast in D(–2)JF(–1) and develops rapidly with season. This warming features a significant westward propagation and reaches its peak (~0.7℃) in SON–1), which indicates that El Niño reached mature phase. The warming decays significantly from D(–1)JF(0) to MAM(0), and the negative SST anomalies in the central tropical Pacific develop into the well-known "horseshoe" pattern with season.

In JJA(0), the other negative SST anomalies emerged in the eastern tropical Pacific. The negative SST anomalies in the central-eastern tropical Pacific showed remarkable development from JJA(0) to SON(0), and the positive SST anomalies in the western tropical Pacific also developed rapidly with the season. This indicates the development of La Niña, which reached its peak in D(0)JF(1). Notably, the negative SST anomalies in the central-eastern tropical Pacific begin to decay in MAM(1), while the positive SST anomalies in the western tropical Pacific begin to decay in D(0)JF(1), which indicate the decline of La

Niña. In addition, the negative SST anomalies in the central tropical Pacific show a more rapid decay than that in the eastern tropical Pacific.

On the basis of these results, we can infer that the warm phase of ENSO (El Niño) associated with S-EOF2 mode is mainly located in the eastern tropical Pacific. However, the cold phase of ENSO (La Niña) associated with S-EOF2 mode first appears in the central tropical Pacific and then emerges in the eastern tropical Pacific. S-EOF2 mode has a more remarkable relation with La Niña than that with the S-EOF1 mode. In addition, ENSO events normally begin to develop in boreal summer and reach the mature phase in boreal winter. The ENSO events associated with S-EOF2 have a period of quasi-biennial, which is consistent with the quasi-biennial (QB) component of ENSO (Barnett, 1991).

Rasmusson and Wallace (1983) portrayed two kinds of tropical SST variations: (1) The SST is confined to the eastern third of tropical Pacific and spreads westward. (2) The SST firstly appears near the dateline and shows a eastward propagation. Kao and Yu (2009) refer to these two kinds of SST variations as the eastern Pacific (EP) and the central Pacific (CP) types of ENSO, respectively. Yu et al. (2011) further point out that most ENSO events during 1958-2001 are a combination of both the EP and CP types of ENSO, and they refer to these ENSO events as a EP/CP type of ENSO.

Therefore, based on our analyses and on previous studies, we can conclude that S-EOF1 mode has a significant relation with the EP/CP type of ENSO, whereas S-EOF2 mode is associated with not only the EP type of ENSO warm events (EP, El Niño) but also the EP/CP type of ENSO cold events (EP/CP, La Niña).

5. Discussion and Conclusion

Based on the season-reliant EOF analysis, we revealed two dominant modes of South Pacific SST anomalies and their relationships with ENSO in this study. In this section we first summarize the major seasonal-evolution characteristics of the South Pacific SST anomalies and then discuss their relationships with ENSO.

First, we studied the seasonally evolving spatial patterns of the two dominant modes of the South Pacific SSTAs, which accounted for nearly 40% of the total variance. The results of the S-EOF1 featured a tripolar pattern with positive SST anomalies centered over the high-latitude South Pacific region (60°-50°S, 160°-120°W) and two negative SST anomalies centered east of New Zealand and the mid-latitude South Pacific region (35°-25°S, 140°-100°W), respectively. The configuration of the tripolar pattern changed little with season. But the amplitude of this pattern decayed significantly from D(−1)JF(0) to SON(0). S- EOF2 showed a wavelike pattern from northwest to southeast of South Pacific. Both of the positive and negative SST anomalies showed significant eastward propagation, and their amplitude developed remarkably close to season.

Secondly, we noted that the two modes had not only interannual variations but also significant interdecadal variations. S-EOF1 showed high power in the 3-5.5- and 2-3-year band for the period 1980-1995 and 1995-2000 on the interannual time scale, respectively. However, S-EOF2 had significantly high power in the 2-4-year band for the period 1993-2000 on the interannual time scale. Both modes had prominent high power in the 8-15-year band on the interdecadal time scale. In addition, the amplitude of the two principal components decayed rapidly after 2000. The two modes showed reversed variations on the interdecadal time scale.

Finally, we studied the relationships between the two modes and ENSO. The results show that there are significant correlations between the two modes and ENSO. Based on the lead-lag correlations, it is noted that S-EOF1 can be viewed as a response to ENSO and S-EOF2 is primarily associated with the development of ENSO. Moreover, S-EOF2 may make a significant contribution to ENSO prediction. Based on further study, we noted that results of S-EOF1 had a significant relation with EP/CP type of ENSO and the results of S-EOF2 were associated with both the EP type of ENSO warm events (EP-El Niño) and the EP/CP type of ENSO cold events (EP/CP-La Niña).

In this study, some connections between South Pacific SST variability and ENSO were illustrated using data analyses, but our explanations remain somewhat hypothetical. At first, the SST anomalies in both the northern and southern Pacific appear to be a response to ENSO, which is the strongest signal of climate variation; however, the S-EOF1 mode should be mainly a response of SST in the South Pacific to ENSO. ENSO may influence South Pacific SST variations by changing the surface heat fluxes via an "atmospheric bridge" (Alexander et al., 2002; Li et al., 2006). This hypothesis will be further validated using observation analysis and model experiments. Secondly, Fig.1b shows that SST anomalies propagate to the equatorial western Pacific from southeastern Australia and southwestern South America. The propagation of SST anomalies to the equatorial western Pacific is favorable to the occurrence of ENSO, because the SST anomalies in the western tropical Pacific (particularly in the subsurface) and its eastward propagation are a fundamental condition of the occurrence of ENSO (Li and Mu, 2002). Rossby waves play an important role in this propagation (Jacobs et al., 1994; Wang et al., 2007). Therefore, the S-EOF2 is even more closely associated with the occurrence of ENSO than is shown by this study.

In this study we confirmed the relationships between the two S-EOF modes and ENSO. In future work, the EP and CP types of ENSO will be investigated: they may have different impacts on the seasonal evolution of the South Pacific SST, and the influence mechanism remains to be clarified. We will use both the observational analyses and model experiments in these studies. In addition, our analysis is based on data covering 31 years (1979-2009), which may not be long enough to determine the impacts of different types of ENSO on South Pacific SST. Although these results need to be verified using longer datasets, the outputs of coupled climate models may be used to help us understand the results.

Acknowledgements The authors thank the editor and the anonymous reviewers for their suggestions and comments to significantly improve our manuscript.

References

Alexander, M. A., I. Bladé, M. Newman, J. R. Lanzante, N.-C. Lau, and J. D. Scott, 2002: The atmospheric bridge: The influence of ENSO teleconnections on air-sea interaction over the global oceans. *J. Climate*, **15**, 2205-2231.

Barnett, T. P., 1991: The interaction of multiple time scales in the tropical climate system. *J. Climate*, 4, 269-285.

Barros, V. R., and G. E. Silvestri, 2002: The relationship between sea surface temperature at the subtropical south-central Pacific and precipitation in southeastern South America. *J. Climate*, 15, 251-267.

Behera, S. K., and T. Yamagata, 2001: Subtropical SST dipole events in the southern Indian Ocean. *Geophys. Res. Lett.*, 28(2), 327-330.

Falvey, M., and R. D. Garreaud, 2009: Regional cooling in a warming world: Recent temperature trends in the southeast Pacific and along the west coast of subtropical South America (1979-2006). *Geophys. Res. Lett.*, 114, D04102, doi: 10.1029/2008JD010519.

Giese, B. S., S. C. Urizar, and N. S. Fuckar, 2002: Southern hemisphere origins of 1976 climate shift. *Geophys. Res. Lett.*, 29(10), 2215-2231.

Holbrook, N. J., and N. L. Bindoff, 1999: Seasonal temperature variability in the upper southwest Pacific Ocean. *J. Phys. Oceanogr.*, 29, 366-381.

Holbrook, N. J., P. S. L. Chan, and S. A. Venegas, 2005: Oscillatory and propagating modes of temperature variability at the 3-3.5 and 4-4.5 time scales in the upper southwest Pacific Ocean. *J. Climate*, 18, 719-736.

Hsu, H. H., and Y. L. Chen, 2011: Decadal to bi-decadal rainfall variation in the western Pacific: A footprint of South Pacific decadal variability? *Geophys. Res. Lett.*, 38, L03703, doi: 10.1029/2010GL046278.

Jacobs, G. A., H. E. Hurlburt, J. C. Kindle, E. J. Metzger, J. L. Mitchell, W. J. Teague, and A. J. Wallcraft, 1994: Decadal-scale trans-Pacific propagation and warming effects on an El Niño anomaly. *Nature*, 370, 360-363, doi: 10.1126/science.281.5374.240.

Jia, X. L., and C. Y. Li, 2005: Dipole oscillation in the Southern Indian Ocean and its impacts on climate. *Chinese J. Geophys.*, 48(6), 1238-1249. (in Chinese)

Kao, H. Y., and J. Y. Yu, 2009: Contrasting Eastern- Pacific and Central-Pacific types of El Niño. *J. Climate*, 22, 615-632.

Kidson, J. W., and J. A. Renwick, 2002: The Southern Hemisphere evolution of ENSO during 1981-99. *J. Climate*, 15, 847-863.

Li, C., and M. Q. Mu, 2002: A further inquiry on essence of the ENSO cycle. *Advance in Earth Sciences*, 17(5), 631-638.

Li, C. H., D. Wang, J. Liang, D. Gu, and Y. Liu, 2006: A local positive feedback of the tropical Pacific ocean-atmosphere system on interdecadal timescales. *Chinese Science Bulletin*, 51(5), 601-606.

Li, G., C. Y. Li, Y. K. Tan, and T. Bai, 2011: Principal modes of the winter SST anomaly in South Pacific and their relationships with ENSO. *Acta Oceanologica Sinica*, 34(2), 48-56. (in Chinese)

Linsley, B. K., G. M. Wellington, and D. P. Schrag, 2000: Decadal sea surface temperature variability in the subtropical South Pacific from 1726 to 1997 AD. *Science*, 290(5494), 1145-1148.

Luo, J. J., and T. Yamagata, 2001: Long-term El Niño-Southern Oscillation (ENSO)-like variation with special emphasis on the South Pacific. *J. Geophys. Res.*, 106(C10), 22211-22227.

Luo, J. J., S. Masson, S. Behera, P. Delecluse, S. Gualdi, A. Navarra, and T. Yamagata, 2003: South Pacific origin of the decadal ENSO-like variation as simulated by a coupled GCM. *Geophys. Res. Lett.*, 30(24), 2250-2259.

Nnamchi, H. C., and J. P. Li, 2011: Influence of the South Atlantic Ocean Dipole on west African summer precipitation. *J. Climate*, 24, 1184-1197.

Neelin, J. D., F. F. Jin, and H. H. Syu, 2000: Variations in ENSO phase locking. *J. Climate*, 13, 2570-2590.

North, G. R., T. L. Bell, R. F. Cahalan, and F. J. Moeng, 1982: Sampling errors in the estimation of empirical orthogonal functions. *Mon. Wea. Rev.*, 110, 699-706.

Rasmusson, E. M., and T. H. Carpenter, 1982: Variations in tropical sea surface temperature and surface wind fields associated with the Southern Oscillation El Niño. *Mon. Wea. Rev.*, 110, 354-384.

Rasmusson, E. M., and J. M. Wallace, 1983: Meteorological aspects of the El Niño /Southern Oscillation. *Science*, 222(4629), 1195-1202.

Rayner, N. A., D. E. Parker, E. B. Horton, C. K. Folland, L. V. Alexander, D. P. Rowell, E. C. Kent, and A. Kaplan, 2003: Global analyses of sea surface temperature, sea ice, and night marine air temperature since the late nineteenth century. *J. Geophys. Res.*, 108(D14), 4407, doi: 10.1029/2002JD002670.

Reynolds, R. W., and T. M. Smith, 1994: Improved global sea surface temperature analyses using optimum interpolation. *J. Climate*, 7, 929-948.

Shaffer, G., O. Leth, O. Ulloa, J. Bendtsen, G. Daneri, V. Dellarossa, S. Hormazaball, and P. I. Sehlstedt, 2000: Warming and circulation change in the eastern South Pacific Ocean. *Geophys. Res. Lett.*, 27(9), 1247-1250.

Shakun, J. D., and J. Shaman, 2009: Tropical origins of North and South Pacific decadal variability. *Geophys. Res. Lett.*, 36, L19711, doi: 10.1029/2009GL040313.

Smith, T. M., R. W. Reynolds, T. C. Peterson, and J. Lawrimore, 2008: Improvements to NOAA's historical merged land-ocean surface temperature analysis (1880-2006). *J. Climate*, 21, 2283-2296.

Terray, P., 2011: Southern Hemisphere extra-tropical forcing: A new paradigm for El Niño-Southern Oscillation. *Climate Dyn.*, 36, 2171-2199.

Torrence, C., and G. P. Compo, 1998: A practical guide to wavelet analysis. *Bull. Amer. Meteor. Soc.*, 79(1), 61-78.

Venegas, S. A., L. A. Mysak, and D. N. Straub, 1997: Atmosphere-ocean coupled variability in the South Atlantic. *J. Climate*, 10, 2904-2920.

Venegas, S. A., L. A. Mysak, and D. N. Straub, 1998: An interdecadal climate cycle in the South Atlantic and its links to other ocean basins. *J. Geophys. Res.*, 103C, 24723-24736.

Wang, B., and S. L. An, 2005: A method for detecting season-dependent modes of climate variability: S-EOF analysis. *Geophys. Res. Lett.*, 32, L15710, doi: 10.1029/2005GL022709.

Wang, D., and Z. Liu, 2000: The pathway of the interdecadal variability in the Pacific Ocean. *Chinese Science Bulletin*, 45(17), 1555-1561. (in Chinese)

Wang, D., J. Wang, L. Wu, and Z. Liu, 2003a: Regime shifts in the North Pacific simulated by a COADSdriven isopycnal model. *Adv. Atmos. Sci.*, 20(5), 743-754.

Wang, D., J. Wang, L. Wu, and Z. Liu, 2003b: Relative importance of wind and buoyancy forcing for interdecadal regime shifts in the Pacific Ocean. *Science in China (D)*, 46(5), 417-427.

Wang, X., C. Li, and W. Zhou, 2007: Interdecadal mode and its propagating characteristics of SSTA in the South Pacific. *Meteor. Atmos. Phys.*, 98, 115-124.

Weare, B. C., and J. S. Nasstrom, 1982: Examples of extended empirical orthogonal function analyses. *Mon. Wea. Rev.*, 110, 481-485.

Weijer, W., W. P. M. De Ruijter, A. Sterl, and S. S. Drijfhout, 2002: Response of the Atlantic overturning circulation to South Atlantic sources of buoyancy. *Global and Planetary Change*, 34, 293-311.

Yan, H. M., C. Y. Li, and W. Zhou, 2009: Influence of subtropical dipole pattern in southern Indian Ocean on ENSO event. *Chinese J. Geophys.*, 52(10), 2436-2449. (in Chinese)

Yang, X., Y., R. X. Huang, and D. Wang, 2007: Decadal change of wind stress over the Southern Ocean associated with Antarctic ozone depletion. *J. Climate*, 20, 3395-3410.

Yu, B., and G. J. Boer, 2004: The role of the western Pacific in decadal variability. *Geophys. Res. Lett.*, 31, L02204, doi: 10.1029/2003GL018471.

Yu, J. Y., H. Y. Kao, T. Lee, and S. T. Kim, 2011: Subsurface ocean temperature indices for Central-Pacific and Eastern-Pacific types of El Niño and La Niña events. *Theor. Appl. Climatol.*, 103, 337-344.

Zhang, Q., H. Wang, Y. Zhong, and D. Wang, 2005: An idealized study of the impact of extratropical climate change on El Niño-Southern Oscillation. *Climate Dyn.*, 25, 869-880.

The Tropical Pacific-Indian Ocean Associated Mode Simulated by LICOM2.0

Xin Li*[1,2], Chongyin Li[1,2]

[1] Institute of Meteorology and Oceanography, National University of Defense Technology, Nanjing, China

[2] LASG, Institute of Atmospheric Physics, Chinese Academy of Sciences, Beijing, China

Abstract Oceanic general circulation models have become an important tool for the study of marine status and change. This paper reports a numerical simulation carried out using LICOM2.0 and the forcing field from CORE. When compared with SODA reanalysis data and ERSST.v3b data, the patterns and variability of the tropical Pacific-Indian Ocean associated mode (PIOAM) are reproduced very well in this experiment. This indicates that, when the tropical central-western Indian Ocean and central-eastern Pacific are abnormally warmer/colder, the tropical eastern Indian Ocean and western Pacific are correspondingly colder/warmer. This further confirms that the tropical PIOAM is an important mode that is not only significant in the SST anomaly field, but also more obviously in the subsurface ocean temperature anomaly field. The surface associated mode index (SAMI) and the thermocline (i.e., subsurface) associated mode index (TAMI) calculated using the model output data are both consistent with the values of these indices derived from observation and reanalysis data. However, the model SAMI and TAMI are more closely and synchronously related to each other.

Keywords ocean general circulation model, numerical simulation, tropical Pacific-Indian Ocean associated mode, subsurface ocean temperature anomaly.

1. Introduction

El Niño-Southern Oscillation (ENSO) is an important interannual climate system. Since Bjerknes (1969) noted that ENSO is a result of air-sea interaction, it has been the subject of much detailed research that has revealed its physical mechanisms and global-scale climate impacts. However, Saji et al. (1999) found that a similar interannual zonal oscillation of SST anomalies (SSTAs) also exists in the equatorial Indian Ocean and named it the Indian Ocean dipole (IOD). Since then, many papers have discussed its mechanisms and impacts. Originally, the IOD was thought to be the result of ocean-atmosphere interaction in the Indian Ocean itself (Saji et al., 1999; Webster et al., 1999). Gradually, however, research has focused

Received January 5, 2017; revised June 7, 2017; accepted June 16, 2017.

*Corresponding author: Xin LI Emial: indo_pacific@sina.cn

on the relationship between the IOD and ENSO. For example, the asymmetric SSTA between the eastern and western Indian Ocean in 1997/98 was probably triggered by the strong El Niño of that year (Yu and Rienecker, 1999; Ueda and Matsumoto, 2000), as ENSO can affect the sea surface wind field through the anti-Walker circulation over the equator (Yu and Rienecker, 1999). Based on statistical analysis, Li and Mu (2001) reported a close relationship between the SSTA dipole mode in the equatorial Indian Ocean and ENSO (which can also be treated as a dipole) in the Pacific. By analyzing the SSTA features of the Indian Ocean in the warm and cold phases of the ENSO cycle, Yan et al. (2001) showed that a significant dipole oscillation phenomenon develops in the Indian Ocean during ENSO events. In addition, SST variations in the Indian Ocean commonly play an important role in the development of El Niño events (Annamalai et al., 2005; Izumo et al., 2010; Yuan et al., 2011, 2013). Thus, the IOD can also influence ENSO (Annamalai et al., 2005; Yuan, 2005; Yuan et al., 2011). Taken together, the above studies indicate that there are strong interactions between the Indian and Pacific oceans, and hence we should consider the SSTA field of the tropical Pacific and Indian oceans as a whole.

Based on this idea, the tropical Pacific-Indian Ocean SSTA was analyzed by Ju et al. (2004) using EOF decomposition. The first mode shows that the SSTA in the equatorial central-western Indian Ocean and central-eastern Pacific is opposite to that in the equatorial western Pacific and eastern Indian Ocean. As this mode contributes over 50% of the total variance, it is referred to as the tropical Pacific-Indian Ocean temperature anomaly mode or the Pacific-Indian Ocean associated mode (PIOAM), and this has been further researched in other studies (e.g., Wu et al., 2005; Yang and Li, 2005). This mode is meaningful because it better reflects the SSTA differences between the east and west in the tropical basin, and has an important influence on the South Asian high and even on the Asian climate (Yang and Li, 2005; Yang et al., 2006). It is also closely associated with the evolution and propagation of the subsurface ocean temperature anomaly (SOTA) (Wu and Li, 2009).

The above studies were based mainly on SST data. However, the ocean temperature anomaly in the subsurface zone, especially in the thermocline, is much stronger than that at the surface (Qian et al., 2004; Chao et al., 2005). The temperature anomalies usually appear first in the thermocline and then propagate along this layer. For example, the SOTA in the equatorial western Pacific and its eastward propagation are an important driver of El Niño (Li and Mu, 1999). In fact, the variation in the SOTA in the western Pacific warm pool is closely associated with the entire ENSO cycle, and they interact with each other (Li and Mu, 2000). Moreover, the IOD is also more prominent in the subsurface zone than at the surface, and it appears as a dipole in a real physical sense (Chao et al., 2005). Thus, Li et al. (2013) further studied the tropical Pacific-Indian Ocean temperature anomaly mode in the thermocline, analyzing the monthly thermocline temperature anomaly (TOTA) over the period 1958-2007 and the weekly sea surface height (SSH) anomaly between 1992 and 2011 in the tropical Pacific-Indian Ocean using the EOF method. Both of the first two modes show coupled variations between the tropical Indian Ocean and the Pacific. That is, when the subsurface temperature in the tropical central-western Indian Ocean and central-eastern Pacific is abnormally warmer/colder, the subsurface temperature in the tropical eastern Indian Ocean and western Pacific is abnormally colder/warmer. This is seen as a major tripole pattern and is referred to as the tropical Pacific-Indian Ocean thermocline temperature anomaly associated mode. This mode shows a good correlation with both ENSO and the IOD, and its evolution is closely related to the propagation of the TOTA. This mode also has a high positive correlation with that defined by the SSTA, but can better represent the spatial distribution and temporal variation of the associated features between temperature anomalies in the tropical Indian Ocean and Pacific. Considering the spatial patterns of SST and SSH in the first modes of multi-variable EOF of SST, SSH and surface wind stress both resemble a tripole, and Lian et al. (2014) named the dominant mode in the tropical Pacific-Indian Ocean the Indo-Pacific tripole. For simplicity, these modes are referred to collectively as the PIOAM in this paper.

As observational data are sparse in the ocean, especially in the subsurface zone, extended simulations produced by ocean general circulation models (OGCMs) are needed for further exploration of the characteristics and dynamics of the PIOAM in the tropical Pacific-Indian Ocean. Using an OGCM developed by Jin et al. (1999) and Li (2005), Wu and Li (2009) investigated the three-dimensional thermal and dynamic structures of the associated mode and its mechanism of evolution. Their subsequent numerical experiments indicated that the Indonesian Throughflow plays an important role in the formation of the PIOAM, especially in the subsurface (Wu et al., 2010). However, the spatial resolution of their OGCM was relatively low and so could not identify the narrow channels in the Indonesian Sea. Thus, a relatively high-resolution OGCM is required to further improve the simulation of the PIOAM (Wu et al., 2010). In the present study, a state-of-the-art eddy-resolving OGCM, LICOM2.0, is used to examine the properties of the surface and subsurface PIOAM, with an emphasis on the evolution of the subsurface PIOAM and its relationship to the surface PIOAM.

2. Model and Data

LICOM2.0 is the newest version of the fourth-generation OGCM developed by the State Key Laboratory of Numerical Modeling for Atmospheric Science and Geophysical Fluid Dynamics at the Institute of Atmospheric Physics, Chinese Academy of Sciences. It is also the oceanic component of FGOALS2.0, which participated in CMIP5. The model was forced by the normal forcing data derived from CORE (Griffies et al., 2009), as detailed in Table 1. More information on LICOM2.0 can be found in Liu et al. (2012).

With respect to the standard edition of LICOM2.0 (Liu et al., 2012), for this study we made some improvements to the model according to Yu et al. (2012), as follows: (1) The horizontal resolution was increased to an eddy-resolving 1/10° and the number of vertical layers to 55. In the upper 300 m, there were 36 uneven layers with a mean thickness of less than 10 m, and the depth of the first layer was 5 m. (2) To exclude the Arctic Ocean, the model domain was set to 66°N-79°S. (3) Biharmonic viscosity and diffusivity schemes were adopted in the horizontal direction in the momentum and thermohaline equations, respectively, while the parameterization of mesoscale eddies was turned off in the thermohaline equations. (4) The methods of barotropic and baroclinic decomposition were improved.

Table 1. The forcing data of LICOM2.0.

Variable	Dataset	Period	Resolution	Time interval
Air temperature at 2 m	CORE v2	1948-2007	T62	Daily
Relative humidity at 2m	CORE v2	1948-2007	T62	Daily
Sea level pressure	CORE v2	1948-2007	T62	Daily
Wind at 10m	CORE v2	1948-2007	T62	Daily
Downward shortwave radiation	CORE v2	1948-2007	T62	Daily
Downward longwave radiation	CORE v2	1948-2007	T62	Daily
Precipitation	CORE v2	1948-2007	T62	Daily
Runoff	CORE v2	1948-2007	1°	Annual
Ice density	NSIDC	1979-2006	1°	Monthly
SST and SSS	WOA05	Before 2005	1°	Monthly

In our simulation, LICOM2.0 was spun-up for 500 years from zero velocity and initialized from the observed temperature and salinity obtained from WOA05, repeating the daily-corrected Normal Year Forcing data from Large and Yeager (2004) as the forcing condition. Then, according to the parameterization schemes of Liu et al. (2014a, 2014b), the model was forced using the forcing data in Table 1 and integrated for 60 years. The last 50 years of output, from January 1958 to December 2007, were used for the analysis and discussion below.

To validate the model results, we used two datasets— one observational (ERSST.v3b) and one reanalysis (SODA v2.2.4). The ERSST.v3b data were obtained from the National Oceanic and Atmospheric Administration with a horizontal resolution of 2° × 2° and were constructed using ICOADS SST data and improved statistical methods (Smith et al., 2008). The SODA v2.2.4 ocean reanalysis data (Carton and Giese, 2008) were provided by the University of Maryland with a horizontal resolution of 0.5°× 0.5° and 40 levels in the vertical direction.

3. Model Verification

Firstly, the error associated with simulated variables, such as SST, thermocline depth and heat content, was analyzed to examine the results of the numerical simulation. Here, the "error" is the difference between the model data and the ERSST or SODA data, as these two datasets are relatively realistic in their representation of ocean conditions (Smith et al., 2008; Carton and Giese, 2008).

3.1. SST

Figures 1a and b show the January climatological SST simulated by LICOM2.0 and that obtained from ERSST, respectively. The results show that they are essentially consistent in their spatial pattern, although the warm pool SST in the simulation is significantly higher than that in the observations, as is the SST at the East African coast. On the other hand, although the spatial pattern of the SST standard deviation in the simulation (Fig. 1c) is similar to that in the ERSST data (Fig. 1d), the values of the former are generally larger than those of the latter. One possible cause of this is that the relatively high resolution of the model may contain some dynamic and thermodynamic processes that are undetectable in the observational data. Another possibility is that the model systematically overestimates the SST variability in the tropical Pacific-Indian Ocean.

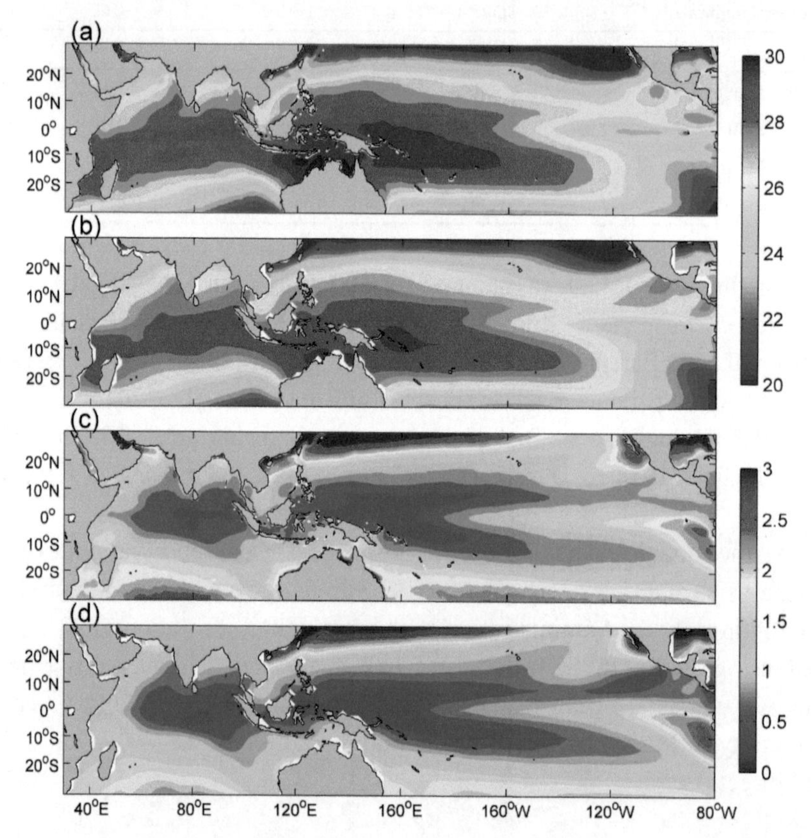

Fig. 1. The (a, b) January climatological SST and (c, d) SST standard deviation (units: ℃) in (a, c) LICOM2.0 and (b, d) ERSST in the tropical Pacific-Indian Ocean.

We also examined the SST trend in the LICOM2.0 simulation, as shown in Fig. 2a. Compared with the result based on the ERSST data (Fig. 2b), the simulated SST trend is obviously higher in the central and eastern Pacific but markedly lower in the warm pool region and Indian Ocean. Thus, the Niño3 region (5°S-5°N, 150°-90°W) and tropical Indian Ocean basin region (20°S-20°N, 40°-110°E) were selected to further compare the SST trend

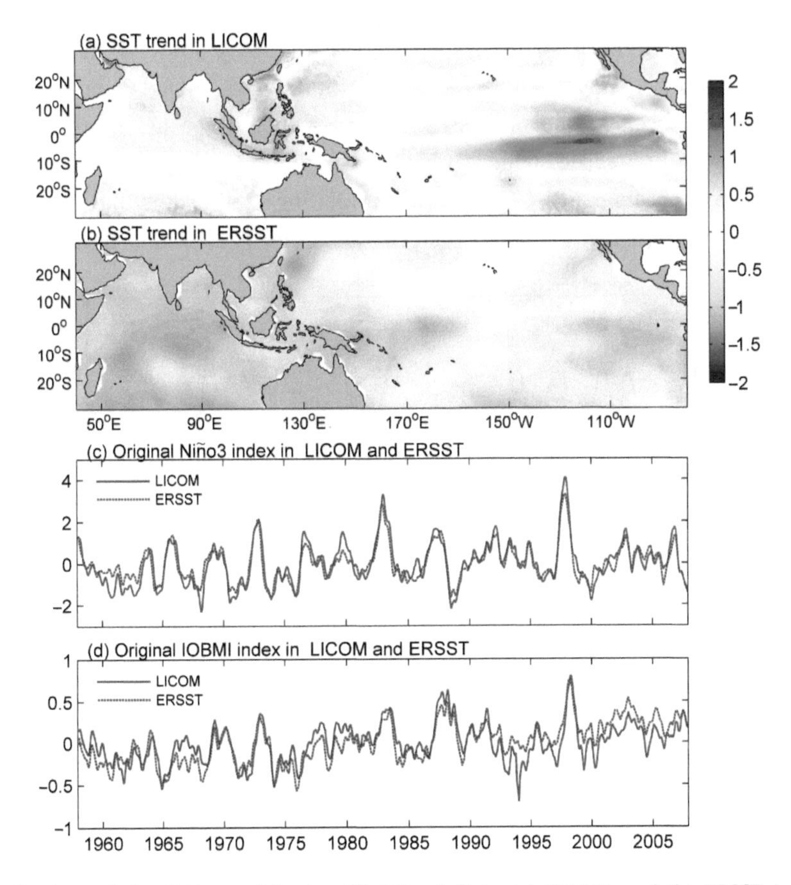

Fig. 2. Distribution of the SST trend [units: ℃ (50 yr) $^{-1}$] in (a) LICOM and (b) ERSST in the tropical Pacific-Indian Ocean, and the time series of (c) the Niño3 index and (d) the IOBMI calculated using the original LICOM2.0 (solid line) and ERSST (dotted line) data.

in the simulation and observational data. Figure 2c shows the time series of the original Niño3 index in the LICOM2.0 simulation and the ERSST data, respectively. In LICOM2.0, the index shows a warming trend [up to 0.92℃ (50 yr)$^{-1}$], whereas in ERSST this warming trend is relatively weaker [only 0.51℃ (50 yr)$^{-1}$]. However, the warming trend of the Indian Ocean basin mode index (IOBMI) in the simulation is not as obvious as it is in the observational data [0.27℃ (50 yr)$^{-1}$ versus 0.55℃ (50yr)$^{-1}$]. These results again indicate that LICOM2.0 overestimates the warming trend in the central and eastern Pacific but underestimates the warming trend in the tropical Indian Ocean. These deficiencies may stem from systematic biases in the model. Thus, we removed all the trends in the model and observational data hereinafter.

Figure 3 shows the Niño3.4 index and Dipole Mode Index (DMI) calculated separately using the simulated SST and that in the ERSST observations. The correlation coefficient between the LICOM2.0 Niño3.4 index and the ERSST Niño3.4 index reaches 0.94 (exceeding the 99% significance level; Fig. 3a), indicating that LICOM2.0 can simulate the ENSO variability very well. Meanwhile, the correlation coefficient between the two DMI time series is about 0.71 (exceeding the 99% significance level; Fig. 3b), showing that the model also performs well in simulating the main SST variability in the Indian Ocean, but not as well as in Pacific. This may be because the processes of air-sea interaction are more complex in the Indian Ocean than in the Pacific, which is hard for a single OGCM to simulate. Regardless, considering the performance ofLICOM2.0 in simulating ENSO and the IOD, we had no reason to doubt that the surface PIOAM could be satisfactorily reproduced in this model.

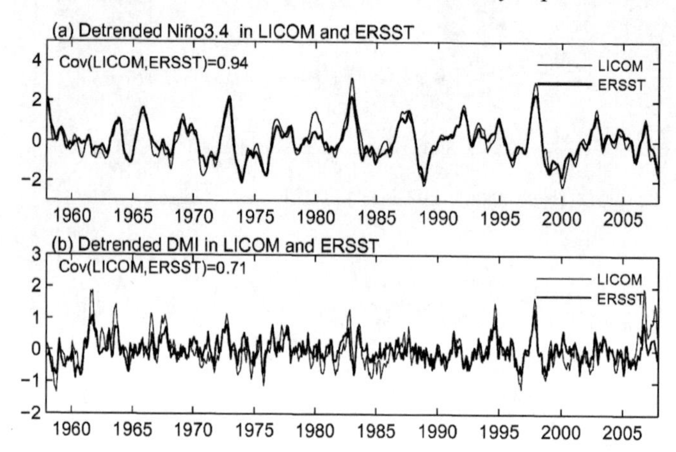

Fig. 3. Time series of (a) the Niño3.4 index and (b) the DMI calculated using the detrended LICOM2.0 (solid line) and ERSST (dotted line) data.

3.2. Thermocline Depth

The thermocline depth was calculated using the vertical gradient method by separately applying the LICOM2.0 and SODA data. The results showed that the climatology of the thermocline depth in LICOM2.0 (Fig. 4a) is consistent with that of SODA (Fig. 4b) in most regions, but is much deeper than the latter in the equatorial northwestern Pacific and in the ITCZ domain (5°-15°N, 150°-110°W), leading to the thermocline ridge along 10° N in the northeastern Pacific being poorly defined in the model (Fig. 4a). By comparing the amplitude (standard deviation) of thermocline depth in these two datasets, we found that the spatial pattern in the model (Fig. 4c) is generally similar to that in the reanalysis data (Fig. 4d). The major differences are that the oscillation of the thermocline in the equatorial eastern Pacific and eastern Indian Ocean are larger in LICOM2.0 than they are in SODA, whereas the reverse is true for the tropical northeastern Pacific, equatorial western Pacific, and western Indian Ocean (Figs. 4c and d). This may indicate that the oceanic long waves in LICOM2.0 are stronger than those observed in the eastern part of the ocean basin, but weaker in the western part.

The simulated upper ocean heat content is also consistent with the observations (figures not shown). These results indicate that LICOM2.0 is also capable of accurately simulating the tropical Pacific-Indian Ocean variability in the subsurface zone. Thus, we were confident that the model could be used to examine the PIOAM both at the sea surface and in the thermocline, and in considering their relationship with each other. These points will be discussed in the next section.

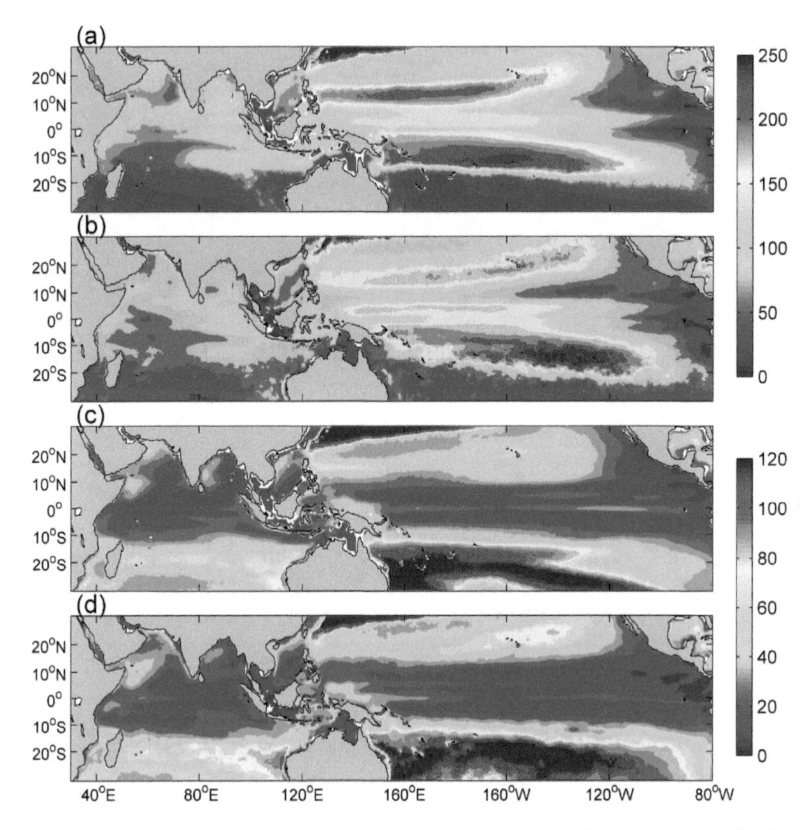

Fig. 4. (a, b) January climatological thermocline depth (units: m) in (a) LICOM2.0 and (b) SODA, and (c, d) thermocline depth standard deviation (units: m) in (c) LICOM2.0 and (d) ERSST, in the tropical Pacific-Indian Ocean.

4. Characteristics of the PIOAM in LI COM2.0

We began by analyzing the SSTA in LICOM2.0 using the EOF method and comparing it with the same in the ERSST data. The spatial pattern of EOF1 presents as a tripole mode; that is, the SSTAs are positive across most of the tropical Indian Ocean and central eastern Pacific, whereas they are negative across the equatorial southeastern Indian Ocean and in the tropical western Pacific (Fig. 5a), which is consistent with the pattern seen in the ERSST data (Fig. 5b). The first mode in the model and in the ERSST data explains 51.4% and 49.8% of the variance of the SSTA, respectively, indicating that this SSTA mode is typical. The correlation coefficients between the corresponding PC1 in LICOM2.0 and in the ERSST data reaches 0.94 (exceeding the 99% significance level). Thus, the major mode of SSTA variation in the tropical Pacific-Indian Ocean, the surface PIOAM, is well represented in the LICOM2.0 simulation.

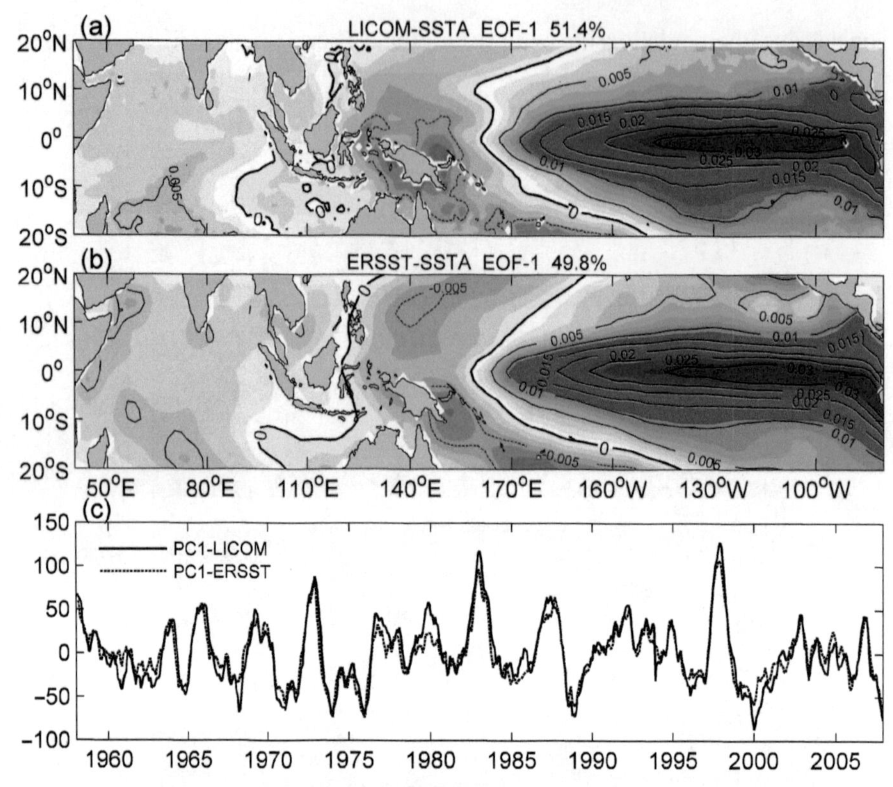

Fig. 5. EOF-1 of tropical Pacific-Indian Ocean SSTA derived from (a) LICOM2.0 and (b) ERSST; (c) temporal variation of the corresponding PC1 in LICOM2.0 (solid line) and ERSST (dotted line).

Following the definition proposed by Yang and Li (2005), the SSTA associated mode index (SAMI) was calculated from the LICOM2.0 and from the ERSST data. Their time series can be seen in Fig. 6a. Generally, the SAMI in the model is consistent with that in ERSST (with correlation coefficients above 0.8 and exceeding the 99% significance level), although it is obviously higher or lower than observed in several years. Positive SAMI years were selected when the SAMI was larger than +1.0 standard deviation in three consecutive months, and vice versa. The composite analysis from these data (Fig. 6b) shows that positive SAMI in the model tends to develop rapidly in late spring and summer, peaks in autumn, and decays in the following spring. In the year prior to positive SAMI years, weak negative SAMI often occurs. However, the negative PIOAM in the model seems to grow slowly from the previous autumn, peaks in the early part of the concurrent autumn, and can then last through to the following spring. Notably, the evolution of positive SAMI in the model is very close to that in the observational data (Fig 6b), whereas negative SAMI in the model develops 6-9 months earlier and decays 2-3 months later than observed. Moreover, the amplitude of negative SAMI in the model is larger than that in the observational data. In short, the simulated negative PIOAM is more durable and stronger than observed. This may indicate that there are some problems related to the ability of LICOM2.0 to simulate the negative PIOAM and the amplitude asymmetry of PIOAM.

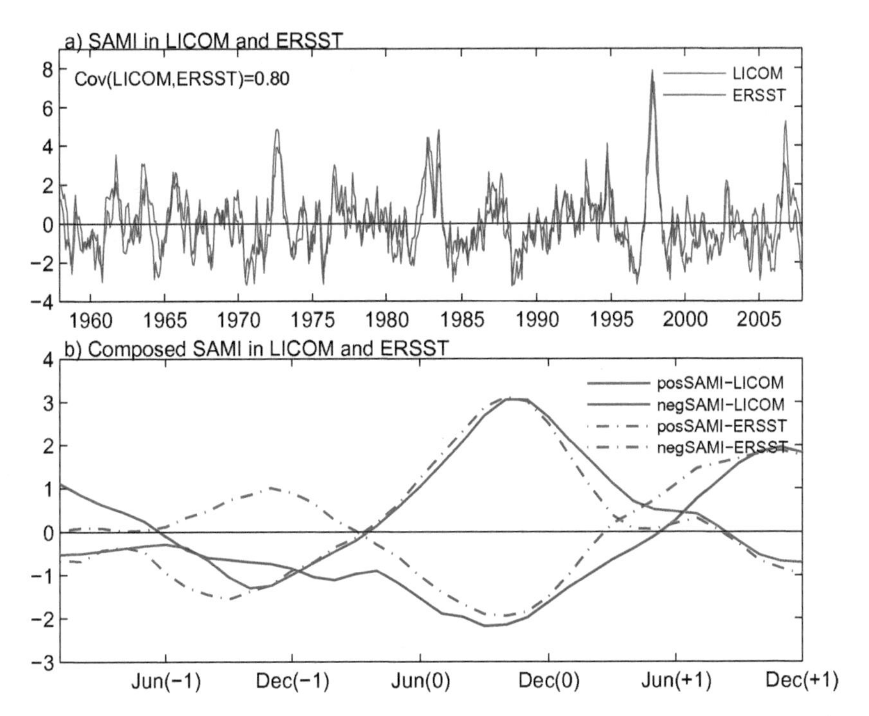

Fig. 6. Time series of the SAMI (a) and the composed SAMI (b) in LICOM and ERSST, respectively.

The next step was to investigate the performance of LICOM2.0 in reproducing the subsurface PIOAM. In section 3, we showed that there are systematic "errors" in the climatology and standard deviation of the thermocline depth in LICOM2.0, indicating the temperature variation in the simulated thermocline does not accurately represent the upper ocean thermal anomaly seen in the observations. However, the climatology and standard deviation of the upper ocean heat content anomaly (HCA) are close to those in the reanalysis data; consequently, we used the HCA of the upper 300 m (HCA300) to examine the subsurface ocean variability in LICOM2.0. The EOF1 of HCA300 in LICOM2.0 explains 32.6% of the variance. In this mode, the HCA300 in both the Pacific and Indian oceans shows a significant dipole pattern, while the HCA300 in the eastern Indian Ocean has the same sign as that of the western Pacific (Fig. 7a). Compared to the surface PIOAM (Fig. 5a), the pattern exhibits clearer physical significance and more associated features between the Pacific and Indian oceans. Thus, it forms an HCA300 associated mode in the tropical Pacific-Indian Ocean that is similar to the TOTA mode referred to by Li et al. (2013). The PC1 of the simulated HCA300 (Fig. 7b) shows stronger amplitude asymmetry than that of the simulated SSTA (Fig. 5c). These results indicate that the subsurface PIOAM is also significant in LICOM2.0. However, there are still some differences between the HCA300 associated mode in LICOM2.0 and that in the SODA data; e.g., the variability in the model is often stronger than that seen in the reanalysis data, especially in the Niño3 region and eastern Indian Ocean. This may indicate a need to further improve the representation of the dynamic and thermodynamic processes in these regions.

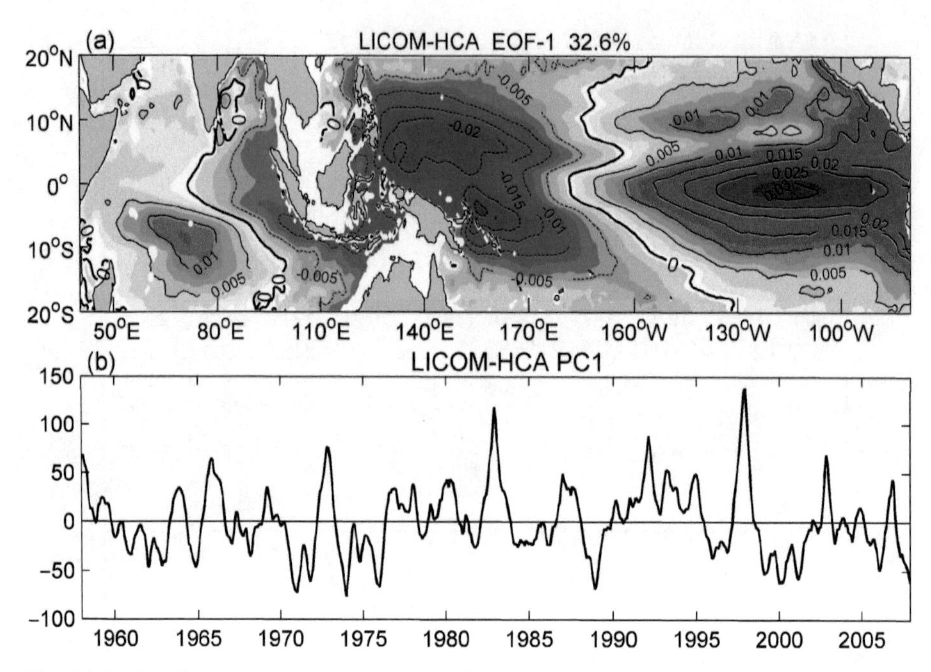

Fig. 7. The (a) EOF-1 of HCA300 in the tropical Pacific-Indian Ocean derived from LICOM 2.0 and (b) the temporal variation of the corresponding PC1.

The subsurface associated mode index, referred to as TAMI by Li et al. (2013), was calculated using the HCA300 from LICOM2.0. From the simulated TAMI series (Fig. 8a), it is apparent that the amplitude of positive SAMI is clearly larger than that of negative SAMI, indicating the subsurface PIOAM also has a significant feature of amplitude asymmetry. Wavelet analysis shows that the TAMI mainly has 2-4-yr interannual variability and a 10-14-yr interdecadal periodicity (Figs. 8b and c). Composite analysis shows that the subsurface PIOAM usually develops in summer, peaks in late autumn, and ends in early spring, and the amplitude of its warm phase is larger than that of the cold phase (figure not shown). These characteristics are similar to the findings of Li et al. (2013), again indicating that the PIOAM is a robust mode in the subsurface of the Pacific-Indian Ocean and that LICOM2.0 can accurately describe the covariations of temperature in the tropical Pacific and Indian oceans.

Figure 9 shows that the maximum correlation coefficient between TAMI and SAMI in LICOM2.0 is 0.75 (exceeding the 99% significance level), at a lag of 0 months. This indicates that there is close relationship between the temperature variations at the surface and in the subsurface zone of the Pacific–Indian Ocean. The zero lag in the maximum correlation between TAMI and SAMI may imply that the exchange of momentum and heat between the surface and subsurface of the Pacific–Indian Ocean occurs rapidly, and the progress of both is likely to be forced by the wind anomaly.

Using ocean reanalysis data, Li et al. (2013) revealed that the propagation of the SOTA results in the development and transition of the subsurface PIOAM. It can also affect the surface, and leads to the evolution of the surface PIOAM. We further explored whether

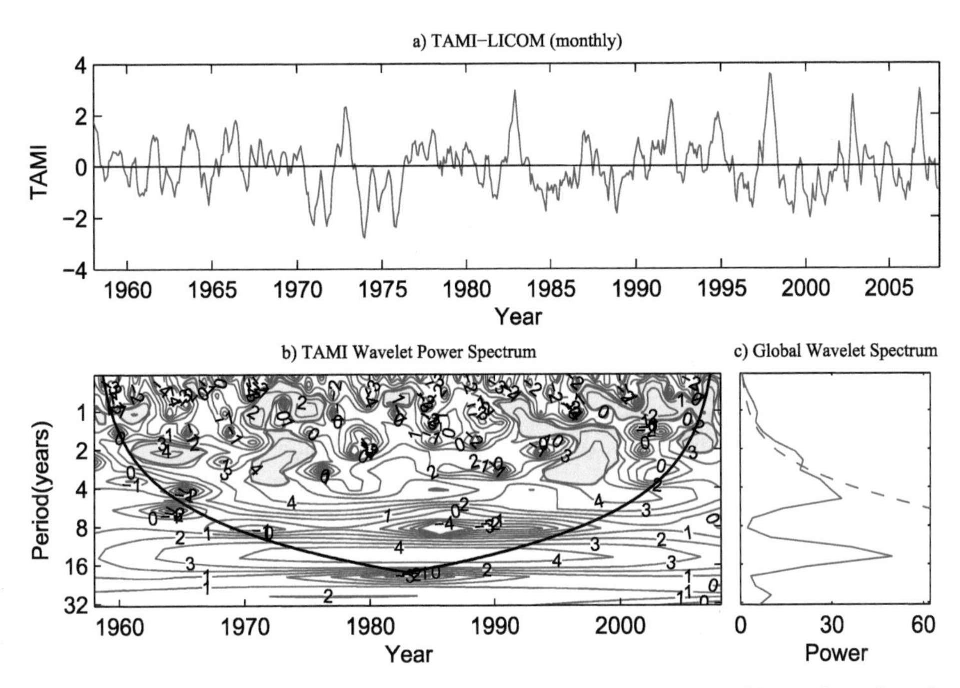

Fig. 8. (a) Time series of the TAMI in LICOM2.0. (b) Wavelet power spectrum of TAMI, the yellow shading denotes where the value exceed 95% confidence level. (c) Global wavelet spectrum of TAMI, the dotted line represent the 95% confidence level.

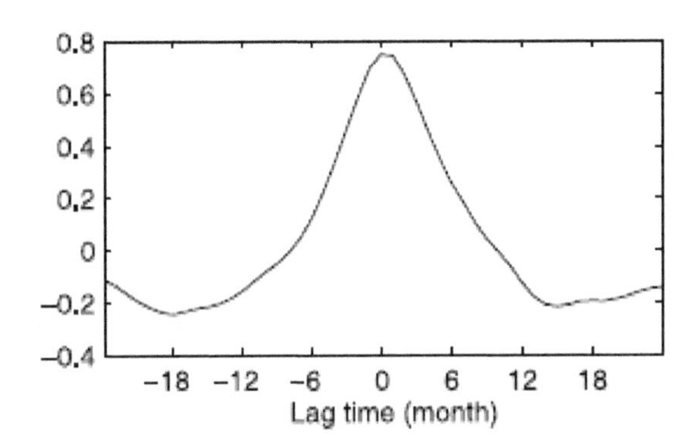

Fig. 9. Time-lag correlation between TAMI and SAMI in LI COM2.0.

LICOM2.0 can reproduce these processes and relationships. The evolution of TAMI (Fig. 10a), combined with the longitude-time profile of HCA300 along key latitude belts (Fig. 10b-d), shows that the upper ocean heat content in the equatorial western Pacific and eastern Indian Ocean is significantly warmer than normal before the PIOAM develops into its warm phase. Then, the warming signals propagate along the equator (Fig. 10b) and westwards along the zonal belt of 10°-15°S in the southern Indian Ocean (Fig. 10d). Thus, the equatorial eastern Pacific and western Indian Ocean become regions of warming, thereby contributing to the development of a positive PIOAM. After the positive PIOAM reaches its peak, the warming signals in the eastern Pacific begin to propagate westwards along 10°-15°N (Fig. 10c). Meanwhile, the

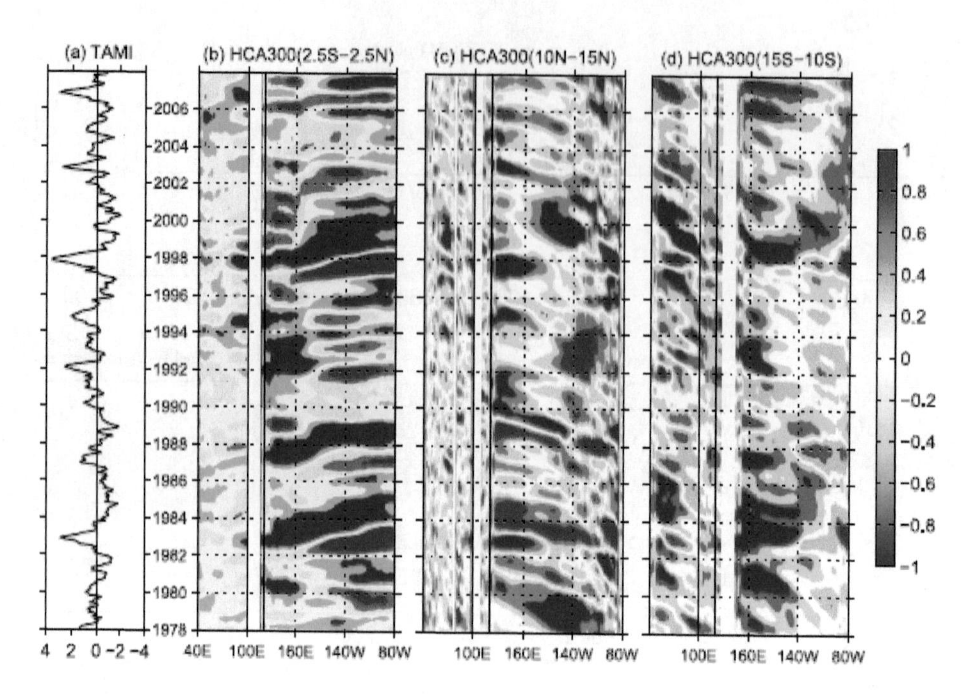

Fig. 10. Time series of (a) TAMI, and Hovmöller diagrams of HCA300 (color scale bar; units: ℃) propagating along (b) the equatorial zone (2.5°S-2.5°N) in the Pacific-Indian Ocean, (c) the zonal belt off the equator in the Northern Hemisphere (10°-15°N), and (d) the zonal belt off the equator in the Southern Hemisphere (10°-15°S). Positive/negative values denote an anomalously warm/cold upper ocean.

equatorial western Pacific and eastern Indian Ocean have been gradually occupied by anomalously cold water, due to upwelling Kelvin and Rossby waves, respectively. The anomalous cold signals in the equatorial western Pacific then propagate eastwards along the equator, while some of the warming signals propagate eastwards along the equator in the Indian Ocean (Fig. 10b), leading to decay of the PIOAM. When the warming signals return to the western Pacific and eastern Indian oceans, the PIOAM transitions to a negative phase. In the development of a negative PIOAM, these processes are opposite to those during a positive PIOAM. Clearly, the evolution of the PIOAM is closely related to the propagation of the ocean HCA (i.e., the SOTA) in LICOM2.0, with a specific passage of the SOTA signals in the Pacific and in the Indian Ocean. These results support the findings of Li et al. (2013) and further confirm the important role of the SOTA in the evolution of the PIOAM. Meanwhile, there is a suggestion that LICOM2.0 can describe the subsurface dynamic and thermodynamic processes well, such as oceanic long waves.

5. Discussion and Conclusions

Following improvements including the vertical resolution, horizontal range, parameterization schemes, and barotropic and baroclinic split, LICOM2.0 was integrated with forcing data from CORE. The outputted monthly ocean temperature data for the period

1958-2007 were analyzed and compared with those of the SODA and ERSST datasets. The results showed that the climatology and variability of temperature in the tropical Pacific-Indian Ocean are reproduced well by LICOM2.0, as are the basic features and evolution of the PIOAM.

EOF analysis of the surface and subsurface temperature anomalies showed that the co-variations of the Indian and Pacific oceans (i.e., the associated modes) are captured by LICOM2.0. The surface and subsurface PIOTM indexes (SAMI and TAMI, respectively) calculated from the model data are consistent with those based on the SODA data, and the high correlation between these two associated modes is also reproduced by LICOM2.0. This indicated that the model can reliably simulate the upper ocean dynamic and thermodynamic processes in the tropical Pacific-Indian Ocean.

The zonal propagations of the HCA300 in LICOM2.0 showed that the evolution of the PIOAM is closely related to the propagations of HCA300 along the equator and in the off-equatorial zonal belt. In the Pacific, the anomaly signals propagate mainly eastwards along the equator and westwards along 10°-15°N, but the westward propagation is not present, or is very weak, in the South Pacific. However, in the Indian Ocean the signals propagate mainly westwards along 10°-15°S and then partially return to the east along the equator, while the westward propagation in the North Indian Ocean is rather weak. These features are similar to the propagation signals of the SOTA based on SODA data, as reported by Li et al. (2013).

In summary, the results obtained from the LICOM2.0 model further confirm that the PIOAM is the principle pattern of the tropical Pacific-Indian Ocean temperature anomaly in both the surface and subsurface zones. As seen in the observational data, the evolution of the PIOAM is related to the development and propagation of the SOTA, and there are also close relationships between the variations of the surface and subsurface PIOAM.

Based on the SODA ocean reanalysis data and NCEP atmospheric circulation reanalysis data, the physical mechanisms of the PIOAM were initially inferred by Li (2015). The general framework is that the anomalous atmospheric circulation (often the Walker circulation) incites the surface wind stress and sea level anomalies, exciting anomalous currents, oceanic long waves, and upwelling. This can further lead to surface and subsurface ocean temperature anomalies through a series of thermodynamic-dynamic processes. These anomalous heat conditions of the ocean then change the atmospheric circulation and SST through evaporation, precipitation, sensible heat fluxes, and latent heat fluxes, which can eventually lead to the system returning to normal or even reversing. In these processes, the coupled Walker circulation above the Pacific and Indian oceans plays an important bridging role, while evaporation, precipitation, shortwave radiation and the reflection of oceanic long waves at the ocean boundary play a role in regulation and feedback.

However, this framework is based only on data analysis and deductions, combined with an OGCM simulation, which limits its capacity to reveal the mechanisms involved in the evolution of the PIOAM in more depth. Thus, an ocean-atmosphere coupled model is needed to explore the key factors and physical processes associated with the development and evolution of the PIOAM, supported by a series of sensitivity experiments; e.g., by turning the

Indonesian Through-flow on or off or by modulating the parameters of air-sea coupling.

Acknowledgements Two anonymous reviewers provided careful comments on the submitted manuscript, which helped improve the article, and for which we are grateful. The authors also sincerely thank Professor Hailong LIU for his help with LICOM2.0. This work was supported by the National Basic Research Program of China (Grant No. 2013CB956203), the National Natural Science Foundation of China (Grant Nos. 41490642 and 41575062), and the Open Fund of LASG.

References

Annamalai, H., S. P. Xie, J. P. McCreary, and R. Murtugudde, 2005: Impact of Indian ocean sea surface temperature on developing El Niño. *J. Climate*, 18, 302-319, doi: 10.1175/JCLI-3268.1.

Bjerknes, J., 1969: Atmospheric teleconnections from the equatorial Pacific. *Mon. Wea. Rev.*, 97, 163-172, doi: 10.1175/1520- 0493(1969)097<0163: ATFTEP>2.3.CO;2.

Carton, J. A., and B. S. Giese, 2008: A reanalysis of ocean climate using simple ocean data assimilation (SODA). *Mon. Wea. Rev.*, 136, 2999-3017, doi: 10.1175/2007MWR1978.1.

Chao, J.P., Q.C. Chao, and L. Liu, 2005: The ENSO events in the tropical Pacific and Dipole events in the Indian Ocean. *Acta Meteologica Silica*, 63, 594-602, doi: 10.3321/j.issn: 0577- 6619.2005.05.005. (in Chinese with English abstract)

Griffies, S. M., and Coauthors, 2009: Coordinated ocean-ice reference experiments (COREs). *Ocean Modelling*, 26, 1-46, doi: 10.1016/j.ocemod.2008.08.007.

Izumo, T., and Coauthors, 2010: Influence of the state of the Indian Ocean Dipole on the following year's El Niño. *Nature Geoscience*, 3, 168-172, doi: 10.1038/ngeo760.

Jin, X. Z., X. H. Zhang, and T. J. Zhou, 1999: Fundamental framework and experiments of the third generation of IAP/LASG world ocean general circulation model. *Adv. Atoms. Sci.*, 16, 197-215, doi: 10.1007/BF02973082.

Ju, J. H., L. L. Chen, and C. Y. Li, 2004: The preliminary research of Pacific-Indian Ocean sea surface temperature anomaly mode and the definition of its index. *Journal of Tropical Meteorology*, 20, 617-624, doi: 10.3969/j.issn.1004- 4965.2004.06.001. (in Chinese with English abstract)

Large, W. G., and S. Yeager, 2004: Diurnal to decadal global forcing for ocean and sea-ice models: The data sets and flux climatologies. NCAR/TN-460+STR, 1-105, doi: 10.5065/D6KK98Q6.

Li, C. Y., and M. Q. Mu, 1999: El Niño occurrence and sub-surface ocean temperature anomalies in the pacific warm pool. *Chinese Journal of Atmospheric Sciences*, 23, 513-521, doi: 10.3878/j.issn.1006-9895.1999. 05.01. (in Chinese with English abstract)

Li, C. Y., and M. Q. Mu, 2000: Relationship between East Asian winter monsoon, warm pool situation and ENSO cycle. *Chinese Science Bulletin*, 45, 1448-1455, doi: 10.1007/BF02898885.

Li, C. Y., and M. Q. Mu, 2001: The influence of the Indian Ocean dipole on atmospheric circulation and climate. *Adv. Atmos. Sci.*, 18, 831-843.

Li, D. H., 2005: Establishment and application of oceanic general circulation model. PhD. dissertation, PLA University of Science and Technology, 200 pp. (in Chinese with English abstract).

Li, X., 2015: The joint evolution of subsurface ocean temperature in tropical Indo-Pacific and its climate impacts. PhD dissertation, PLA University of Science and Technology, 165 pp. (in Chinese with English abstract).

Li, X., C. Y. Li, Y. K. Tan, R. Zhang, and G. Li, 2013: Tropical Pacific-Indian Ocean thermocline temperature associated anomaly mode and its evolvement. *Chinese Journal of Geophysics*, 56, 3270-3284, doi: 10.6038/cjg20131005. (in Chinese with English abstract).

Lian, T., D. K. Chen, Y. M. Tang, and B. G. Jin, 2014: A theoretical investigation of the tropical Indo-Pacific tripole mode. *Science China Earth Sciences*, 57, 174-188, doi: 10.1007/s11430-013-4762-7.

Liu, H. L., P. F. Lin, Y. Q. Yu, and X. H. Zhang, 2012: The baseline evaluation of LASG/IAP Climate system Ocean Model (LICOM) version 2. *Acta Meteologica Sinica*, 26, 318-329, doi: 10.1007/s13351-012-0305-y.

Liu, H. L., Y. Q. Yu, P. F Lin., and F. C. Wang, 2014a: High-resolution LICOM. *Flexible Global Ocean-Atmosphere-Land System Model*, T. Zhou et al., Eds., Springer-Verlag, Berlin Heidelberg, 321-331.

Liu, H. L, P. F. Lin, Y. Q. Yu, F. C. Wang, X. Y. Liu, and X. H. Zhang, 2014b: LASG/IAP climate system ocean model version 2: LICOM2. *Flexible Global Ocean-Atmosphere-Land System Model*, T. Zhou et al., Eds., Springer-Verlag, Berlin Heidelberg, 15-26.

Qian, W. H., Y. F. Zhu, and J. Y. Liang, 2004: Potential contribution of maximum subsurface temperature anomalies to the climate variability. *International Journal of Climatology*, 24, 193-212, doi: 10.1002/joc.986.

Saji, N. H., B. N. Goswami, P. N. Vinayachandran, and T. Ya- magata, 1999: A dipole mode in the tropical Indian Ocean. *Nature*, 401, 360-363.

Smith, T. M., R. W. Reynolds, T. C. Peterson, and J. Lawrimore, 2008: Improvements to NOAA's historical merged land-ocean surface temperature analysis (1880-2006). *J. Climate*, 21, 2283-2296, doi: 10.1175/2007JCLI2100.1.

Ueda, H., and J. Matsumoto, 2000: A possible triggering process of east-west asymmetric anomalies over the Indian Ocean in relation to 1997/98 El Niño. *J. Meteor. Soc. Japan*, 78, 803- 818, doi: 10.2151/jmsj1965.78.6_803.

Webster, P. J., A. M. Moore, J. P. Loschnigg, and R. R. Leben, 1999: Coupled ocean-atmosphere dynamics in the Indian Ocean during 1997-98. *Nature*, 401, 356-360, doi: 10.1038/43848.

Wu, H. Y., and C. Y. Li, 2009: Numerical simulation of the tropical Pacific-Indian Ocean associated temperature anomaly mode. *Climatic and Environmental Research*, 14(6), 567-586, doi: 10.3878/j.issn.1006-9585.2009.06.01. (in Chinese with English abstract)

Wu, H. Y., C. Y. Li, and M. Zhang, 2010: The preliminary numerical research of effects of ITF on tropical Pacific-Indian Ocean associated temperature anomaly mode. *Journal of Tropical Meteorology*, 26(5), 513-520, doi: 10.3969/j.issn.1004- 4965.2010.05.001. (in Chinese with English abstract)

Wu, S., Q. Y. Liu, and R. J. Hu, 2005: The main coupled mode of SSW and SST in the tropical Pacific South China Sea-Tropical Indian Ocean on interannual time scale. *Periodical of Ocean University of China*, **35**, 521-526, doi: 10.3969/j.issn.1672-5174.2005.04.001. (in Chinese with English abstract)

Yan, H. M., J. H. Ju, and Z. N. Xiao, 2001: The variable characteristics analysis of SSTA over the Indian ocean during the two phases of ENSO cycle. *Journal of Nanjing Institute of Meteorology*, 24, 242-249, doi: 10.3969/j.issn.1674- 7097.2001.02.014. (in Chinese with English abstract)

Yang, H., and C. Y. Li, 2005: Effect of the tropical Pacific-Indian ocean temperature anomaly mode on the south Asia high. *Chinese Journal of Atmospheric Sciences*, 29, 99-110, doi: 10.3878/j.issn.1006-9895.2005.01.12. (in Chinese with English abstract)

Yang, H., X. L. Jia, and C. Y. Li, 2006: The tropical Pacific-Indian Ocean temperature anomaly mode and its effect. *Chinese Science Bulletin*, 51, 2878-1884, doi: 10.1007/s11434- 006-2199-5.

Yu, L. S., and M. M. Rienecker, 1999: Mechanisms for the Indian Ocean warming during the 1997-98 El Niño. *Geophys. Res. Lett.*, 26, 735-738, doi: 10.1029/1999GL900072.

Yu, Y. Q., H. L. Liu, and P. F. Lin, 2012: A quasi-global 1/10° eddy-resolving ocean general circulation model and its preliminary results. *Chinese Science Bulletin*, 57, 3908-3916, doi: 10.1007/s11434-012-5234-8.

Yuan, D. L., 2005: Role of the Kelvin and Rossby waves in the seasonal cycle of the equatorial Pacific Ocean circulation. *J. Geophys. Res.*, 110, C04004, doi: 10.1029/2004JC002344.

Yuan, D. L., and Coauthors, 2011: Forcing of the Indian Ocean Dipole on the interannual variations of the tropical Pacific Ocean: Roles of the Indonesian Throughflow. *J. Climate*, 24, 3593-3608, doi: 10.1175/2011JCLI3649.1.

Yuan, D. L., H. Zhou, and X. Zhao, 2013: Interannual climate variability over the tropical Pacific Ocean induced by the Indian Ocean Dipole through the Indonesian Throughflow. *J. Climate*, 26, 2845-2861, doi: 10.1175/JCLI-D-12-00117.1.

3

Monsoon and Tropical Cyclone

Actions of Typhoons Over the Western Pacific (Including the South China Sea) and El Niño

Li Chongyin (李崇银)

Institute of Atmospheric Physics, Academia Sinica, Beijing

Abstract According to the time cross-section of SST in the equatorial eastern Pacific and the historical data on typhoon actions over the western Pacific (including the South China Sea), a composite analysis of the actions of typhoon over the western Pacific in El Niño year (SST in the equatorial eastern Pacific are continuously higher than normal) and in the inverse El Niño year (there are continuative negative anomalies of SST in the equatorial eastern Pacific) is carried out. The results show that the actions of typhoon are in close relation with El Niño: The annual average number of typhoons over the western Pacific and South China Sea is less than normal in El Niño year and more in the inverse El Niño year; The annual average number of the landing typhoon on the continent of China bears the same relationship with El Niño; The anomalies of typhoon actions mainly occur during July-November and their starting are behind the anomaly of SST in the equatorial eastern Pacific.

Based on the generation and development conditions of typhoons, the circulations and state of the tropical atmosphere and SST in the western Pacific are respectively analysed in El Niño year and in the inverse El Niño year. Then some possible influence mechanisms of El Niño on the actions of typhoons are discussed.

Ⅰ. Introduction

In El Niño year, there is a continued anomalous warming phenomenon of sea-water in the equatorial eastern Pacific. It is regarded as a sign of El Niño occurrence that the average sea surface temperature (SST) in the region (0°-10°S, 180°-90°W) is continuously higher than normal. In general, the positive anomaly of SST begins in March-April and continues for about one year or more. If a negative anomaly of SST in the equatorial eastern Pacific takes place continuously, the year is called an inverse El Niño year.

A number of studies in recent years have shown that the appearance of El Niño influences not only the circulation and weather in the tropical atmosphere but also the circulation and short-range climate variation over the global area (Chen 1977, Rowntree 1979, Wallace et al. 1981). For example, in El Niño years, the Pacific-North American (PNA) pattern can be produced and leads to special weather in most parts of the United States; The anomaly of SST in

Received August 30, 1986

the equatorial eastern Pacific is closely correlated with the precipitation in China during summer. One may then ask what relation exists between the action of typhoon over the western Pacific and El Niño? Studies on their relationship have resulted in different conclusions (Pan 1982, Ramage et al. 1981).

In order to understand exactly the relationship between El Niño and the actions of typhoon over the western Pacific, some statistical analyses are completed, using the typhoon data over the western Pacific and the SST data in the equatorial eastern Pacific during the period 1900-1979 in this paper. Based on the environmental conditions of the generation and development of typhoon, some possible influence mechanisms of El Niño on the actions of typhoon are discussed.

II. Data Used

The time cross-section of SST in 1806-1979 provided by Angell (1981) is basic data for analysis. But in order to pledge the reliability of the analyzed results, the SST data (1921-1938, 1950-1976) provided by Rasumusson et al. (1982) is also used for comparison. According to these data, the basic characteristics of SST variation with time are consistent. For the typhoon actions, the statistical results (1884-1955) compiled by Gao Youxi et al. (1957) and some typhoon Almanacs (1951-1979) are used together.

Based on the above-mentioned SST data, 24 El Niño years and 16 inverse El Niño years over a span of 80 years (1900-1979) are defined at first. They are given in Table 1. The typhoon frequency is based on the number presented in every month. If a typhoon was present in two months, it was counted into the month in which the typhoon was formed. The landing typhoon is only counted when it lands on continent, while it is not counted landing on islands.

Table 1. El Niño Years and Inverse El Niño Years during 1900-1979

El Niño	1902	1904	1905	1911	1913	1918	1919	1923
Years	1925	1930	1935	1940	1941	1944	1945	1948
	1951	1953	1957	1963	1965	1969	1972	1976
Inverse El Niño	1907	1909	1912	1917	1921	1924	1937	1942
Years	1949	1954	1955	1964	1967	1970	1973	1975

III. Analyzed Results

In order to expose the influence on typhoons caused by El Niño, composite analyses of the typhoon actions in total El Niño years and in total inverse El Niño years were carried out. Some statistical results are given in Table 2. There the first line is the average number in a span of 80 years (1900-1979), while the other lines are the average number in a span of 30 years (1950-1979). The results show that an average of 24.3 typhoons occurred every year over the western Pacific and South China Sea. The average occurrences of typhoons in El Niño were only 21.4, less by 3 than those in the normal and by 5 than those in the inverse El Niño year. In the inverse El Niño year, more typhoons occurred than in the normal. Just the same, the average number of typhoons entered the South China Sea from the western Pacific, the average number of typhoons generated over the South China Sea and the frequencies of typhoons landing on the continent of China in El Niño year are less than those in the normal. But in the inverse El Niño year all these are more than those in the normal. The statistical tests of them using T-test indicate that the above-mentioned statistics are believable. It shows that the occurrence of typhoons over the western Pacific and South China Sea is closely related with El Niño.

Table 2. Average Number of Typhoons Generated over Western Pacific and South China Sea and Landing on the Continent of China

	Multi-Year Average (Normal)	El Niño Year Average	Inverse El Niño Year Average
Occurrence of Typhoons over Western Pacific and South China Sea	24.3	21.4	26.2
Number of Western Pacific Typhoons Entering in South China Sea	6.9	4.9	8.7
Number of Typhoons Generated over South China Sea	3.4	2.0	4.1
Frequencies of Typhoons Landing on the Continent of China	6.2	5.2	7.3

To understand further the relationship between the El Niño and the typhoon actions over the western Pacific, we have made a composite analysis of the monthly average occurrences of typhoons. The result is given in Fig. 1. It reflects the fundamental situation in different months. The solid line is the average for 80 years, the dashed line for the inverse El Niño years and the dotted line for the El Niño years. It can be seen that the departure from the normal occurrence of typhoon over the western Pacific appears mainly in July-November. In general, the El Niño begins in March-April, while the anomaly of typhoon begins in July, lagging behind the occurrence of the El Niño. This indicates that the El Niño might exert an effect on the tropical atmosphere circulation at first and the typhoon action over the western Pacific afterwards.

Fig. 2 shows the percentage distribution of the monthly average anomaly of typhoon occurrence over the South China Sea in El Niño year and in the inverse El Niño year. The difference between that in El Niño year and that in the inverse El Niño year is very clear. Moreover, the anomaly of typhoons over the South China Sea mainly occurs in October and November.

The above-mentioned analysis shows that the occurrence of typhoons over the western Pacific and South China Sea is in close relation with El Niño and the anomaly of typhoons occurs behind the El Niño event. Therefore, we may conclude that El Niño is an important cause leading to the anomaly of typhoons.

Ⅳ. Environment Conditions Causing Typhoon Anomaly

As we know, El Niño events can cause anomalistic variation of the atmospheric circulation in vast areas, especially in the tropics. These anomalistic variations of the

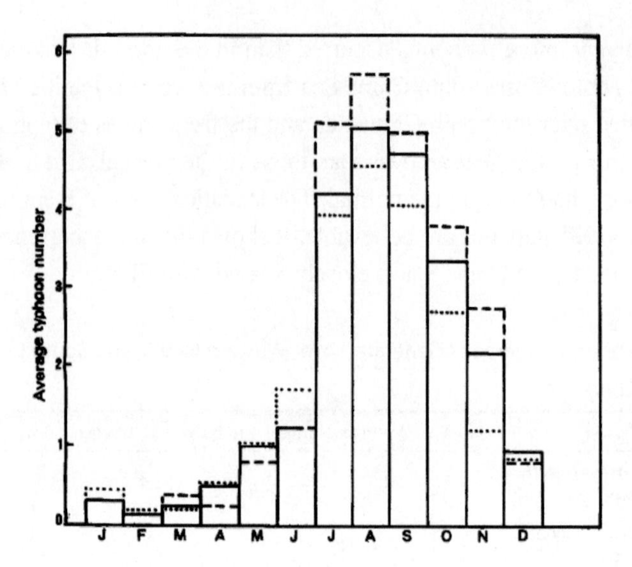

Fig. 1. Monthly average occurrences of typhoons over the western Pacific (period 1900-1979).

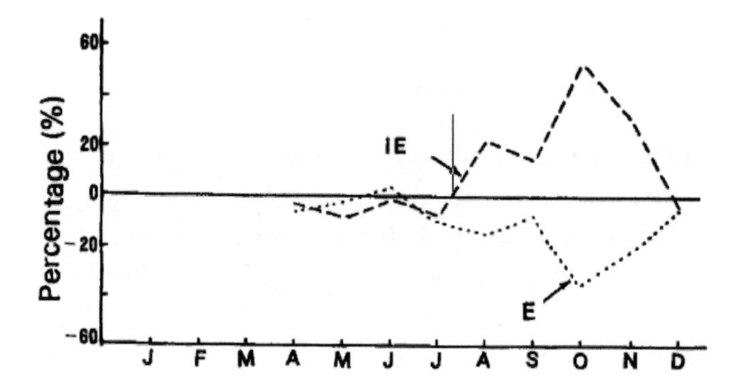

Fig. 2. Percentage distribution of the monthly average anomaly of typhoon occurrence over South China Sea (period 1950-1979).

atmospheric circulation are sure to influence the occurrence of typhoons. Therefore, according to the generating conditions of typhoons, different variations of the atmospheric circulation and SST in the western Pacific during summer in El Niño years and inverse El Niño years will be analyzed respectively. Then, some qualitative explanations about the influence of El Niño on the typhoon over the western Pacific can be obtained.

The surface pressure in the western Pacific and East Asia shows that an identical positive anomaly appears in the vast areas to the south of 35°N latitude during the summer in El Niño year (Fig. 3). The average surface pressure differences during the summer between an El Niño year and an inverse El Niño year show that the surface pressure in El Niño year is higher than that in the inverse El Niño year to the south of 30°N latitude and it is lower in the middle-latitude area in El Niño year. The annual variations of the surface pressure anomalies in El Niño year (dotted line) and in the inverse El Niño year (dashed line) at Guangzhou and Haikou are shown in Fig. 4. It is obvious that basically there is a positive anomaly in El Niño year and a negative anomaly in the inverse El Niño year after March in those tropical regions. The above-mentioned distribution and variation of the surface pressure show that there is an anomalous action of the high pressure system in the tropical western Pacific and South China Sea in El Niño year. Then ITCZ in this area may be weaker and its position is on the southern side.

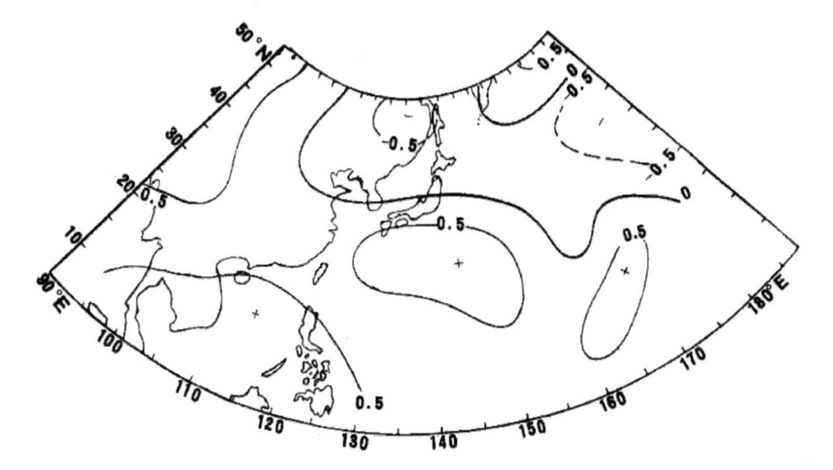

Fig. 3. Surface pressure anomalies (hPa) during July-August in El Niño year (period 1950-1979).

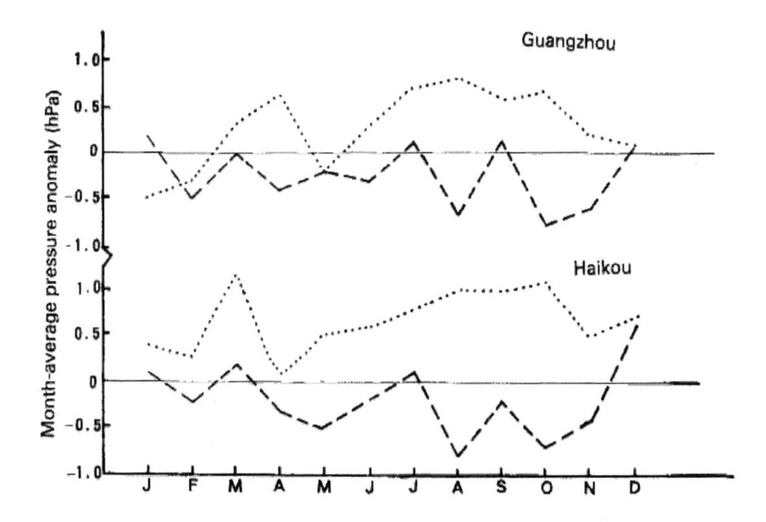

Fig. 4. Annual variation of the surface pressure anomaly at Guangzhou and Haikou in China in El Niño year (dotted line) and in the inverse El Niño year (dashed line) for the period 1950-1979.

El Niño events also influence the height field at 500 hPa. The average height difference at 500 hPa between El Niño year and inverse El Niño year during summer (Fig. 5) has a similar horizontal distribution to the surface pressure difference. There is a lower geopotential height in the middle-latitude region in El Niño year and a higher geopotential height to the south of 30°N latitude. Then, the position of subtropical high over the western Pacific is obviously on the southern side in El Niño summer.

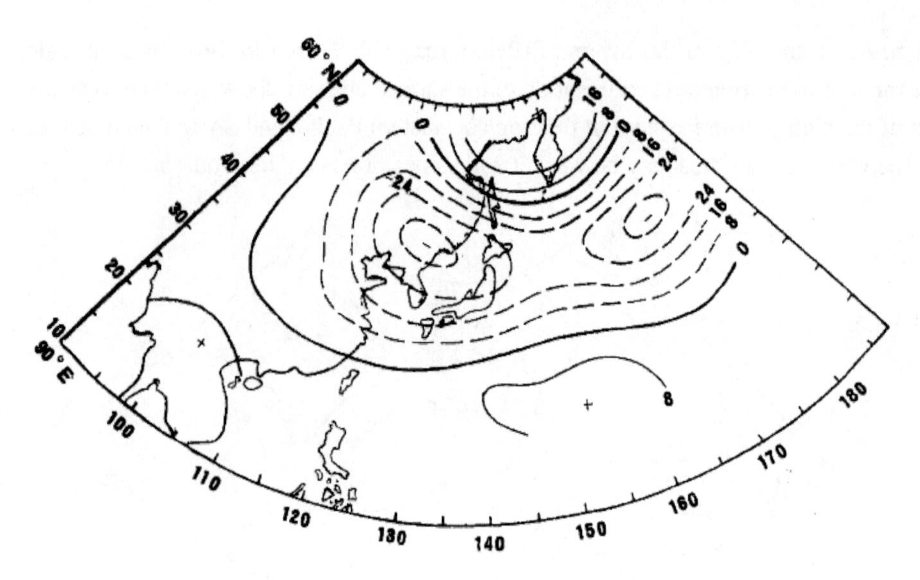

Fig. 5. Average height difference at 500 hPa during July-September between El Niño year and the inverse El Niño year for the period 1950-1979.

In El Niño year, the above-mentioned abnormities of the geopotential field and the surface pressure field inevitably make the position of ITCZ to be on the southern side. This can be seen from an example analyzed. The average position variations of ITCZ with time at 700 hPa in the 130°-150°E area during summer in 1976 (dotted line) and 1967 (dashed line) are counted and shown in Fig. 6. It is obvious that the latitude position of ITCZ over the western Pacific in El Niño year is lower than that in the inverse El Niño year.

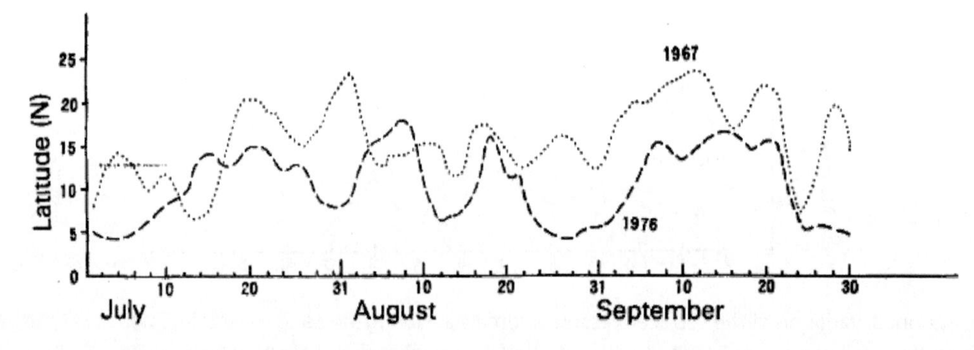

Fig. 6. The average position variations of ITCZ with time at 700 hPa in 130°-150° E area during summer.

The above-mentioned abnormities of the geopotential field and the surface pressure field in El Niño year also cause larger vertical wind shearing in the source region of typhoons. Since the latitude position of ITCZ is lower and the vertical wind shearing is larger in the source region of typhoons during El Niño summer, it is unfavourable to generating typhoons in El Niño year. Therefore, the occurrence of typhoons in El Niño year is less than normal.

The variation of Walker circulation caused by El Niño is well known. In El Niño year there is an anomalous rising motion in the atmosphere over the equatorial eastern Pacific and an anomalous falling motion region is formed near 150°-160°E (Julian et al. 1978). Because of this abnormity of Walker circulation in El Niño year, the anomalous falling motion will be formed in the source region of typhoons. The convective actions will suffer restraint and the occurrence of typhoons will certainly be less. Reversely, the rising motion in the source region of typhoons is enhanced in the inverse El Niño year and is favourable to generating typhoons.

The El Niño event can directly influence the tropical atmospheric circulation over the western Pacific. Simultaneously, the influence of an El Niño event still has a propagating phenomenon westward. In El Niño year, the variation of positive anomalies of height at 500 hPa over the western Pacific is shown in Fig. 7. The expansion of positive anomalies of height at 500 hPa westward is very obvious in El Niño year. This might explain why the anomaly of typhoons appears behind the occurrence of the El Niño.

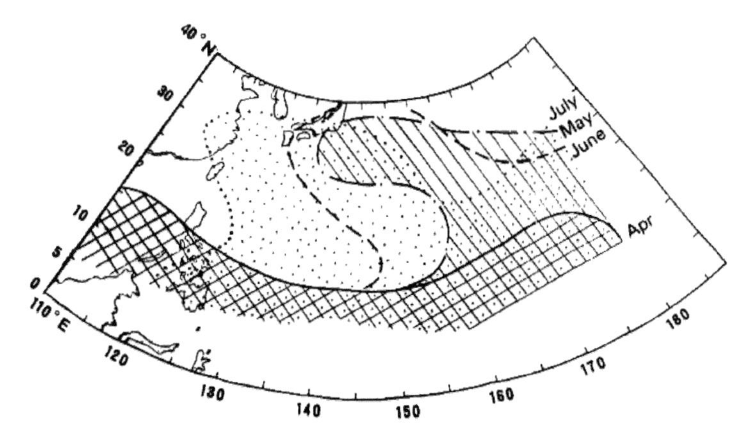

Fig. 7. The variation of positive anomalies of height at 500 hPa over the western Pacific in El Niño year (period 1950-1979).

SST in typhoon source region also plays an important role in the generation of typhoons. In general, it is difficult to generate a typhoon if the SST is lower than 28℃. The distributions of average SST anomaly during June-September in 1976 (El Niño year) and in 1967 (the inverse El Niño year) show that the SST in the typhoon source region is anomalously lower than normal in El Niño year. Therefore, the occurrence of typhoons in El Niño year is less than normal. But in the inverse El Niño year, the SST in the typhoon source region is on the high side so that the occurrence of typhoons is not normal.

CISK is looked upon as a main mechanism for the generation and development of typhoons. And the stability parameter has an important influence on the unstable growth of depression in the CISK theory. In order to compare the atmospheric state in the typhoon source region in El Niño year and in the inverse El Niño year, as an example, the computation of stability parameter between 850-500 hPa in the region nearby Guam Island in July 1976 and 1967 has been completed. The computed static stability σ was respectively 2.28×10^{-2} m^2. hPa^{-2}. s^{-2} in 1976 and 1.94×10^{-2} m^2 hPa^{-2}. s^{-2} in 1967. It is obvious that the atmospheric stability in the typhoon source region in El Niño year is greater than that in the inverse El Niño year. This could also explain why the occurrence of typhoons is greater in 1967 and less in 1976.

V. Conclusion

The average occurrence of typhoons over the western Pacific and South China Sea area is closely correlative with El Niño. There are fewer typhoons in El Niño years than in normal years but more typhoons in the inverse El Niño years.

The anomaly of typhoons over the western Pacific and South China Sea in El Niño year or in the inverse El Niño year occurs mainly during July-November. It lags behind the anomaly of SST in equatorial eastern Pacific. This lag of typhoon anomaly may be correlative with the propagating of the circulation anomaly caused by an El Niño event progressively westward.

The appearance of an El Niño event will cause anomalous variations of atmospheric circulation and state in the tropics, such as: the latitude position of ITCZ is on the southern side; the vertical wind shearing and static stability in the typhoon source region become greater; the convective action decreases in the source region of the forming typhoon. It also causes abnormity (decrease) of SST in the equatorial western Pacific. The above-mentioned anomalous variations are unfavorable for the generation and development of typhoons. So there are fewer typhoons in El Niño year. Contrarily, the environment conditions are favorable for generation and development of typhoons in the inverse El Niño year. There are therefore more typhoons in the inverse El Niño year.

By investigating the different features of every El Niño event and the relevant actions of typhoons further, we should get some concrete targets on the long-range forecast of typhoon action.

References

Angell, J.K. (1981), Comparison of variations in atmospheric quantities with sea surface temperature variations in the equatorial eastern Pacific, *Mon. Wea. Rev.*, 109: 230-243.

Chen Lieh-ting (1977), The effects of the anomalous sea-surface temperature of the equatorial eastern Pacific Ocean on the tropical circulation and rainfall during the rainy period, *Scientia Atmospherica Sinica*, 1: 1-12 (in Chinese with English abstract).

Gao Youxi, et al. (1957), Route Diagram and Statistic of Typhoons, Science Press, Beijing (in Chinese).

Julian, P.R., and Chervin R.M. (1978), A study of the southern oscillation and Walker circulation phenomenon. *Mon. Wea. Rev.*, 106: 1433-1451.

Pan Yi-hang (1982), The effect of the thermal state of equatorial eastern Pacific on the frequency of typhoons over western Pacific, *Acta Meteorologica Sinica*, 40: 24-34 (in Chinese with English abstract).

Ramage, C.S., and Hori A.M. (1981), Meteorological aspects of El Niño, *Mon. Wea. Rev.*, 109: 1827-1835.

Rasmusson, E.M., et al. (1982), Variations in tropical sea surface temperature and surface wind fields associated with the southern oscillational/El Niño, *Mon. Wea. Rev.*, 110: 354-384.

Rowntree, P.R. (1979), The effects of changes in ocean temperature on the atmosphere, *Dyn. Atmos. Oceans*, 3: 373-390.

Wallace, J.M., et al. (1981), Teleconnections in the geopotential height field during the Northern Hemisphere winter, *Mon, Wea. Rev.*, 109: 785-812.

On the Onset of the South China Sea Summer Monsoon in 1998①

Li Chongyin (李崇银), Wu Jingbo (吴静波)

LASG, Institute of Atmospheric Physics, Chinese Academy of Sciences, Beijing

Abstract Through analyzing the NCEP/NCAR reanalysis data, the satellite observational data and the ATLAS-2 mooring buoy observational data, it is shown that May 21 is the onset date of the South China Sea summer monsoon in 1998. There were abrupt variations in the general circulation pattern at the lower troposphere and the upper troposphere, in upper jet stream location and in the convection and rainfall over the South China Sea region corresponding to the outbreak of the South China Sea summer monsoon. It is also indicated that there was rainfall in the southern China coastal region before onset of summer monsoon, but it resulted from the (cold) front activity and cannot be regarded as the sign of summer monsoon outbreak in the South China Sea.

Keywords Onset, South China Sea summer monsoon, General circulation pattern, Jet stream, Convection.

1. Introduction

Chinese scientists have pointed out since the 1980s that the Asian summer monsoon is composed of the South Asian (Indian) monsoon system and the East Asian monsoon system, which have their particular characteristics respectively but also interact on each other; and the Asian summer monsoon breaks out in the South China Sea (SCS) region at first, then spreads northwestward and northward respectively, finally the South Asian summer monsoon and the East Asian summer monsoon are set up (Tao and Chen, 1987; Jin and Chen, 1985; Zhu et al., 1986).

Much attention has been paid to the South China Sea summer monsoon from different aspects since the 1990s, especially more about the onset of the South China Sea summer monsoon (Xie and Zhang, 1994; He et al., 1996; Matsumoto, 1997; Yan, 1997), but the results were different. Since Webster's monsoon index fits only for the Indian monsoon but not for the East Asian (South China Sea) monsoon activity, we offered a new monsoon index (the differentiation of divergence between the upper troposphere and the lower troposphere), which is advantageous for representing East Asian monsoon activity, and studied the charac-

---

Received May 7, 1999; revised November 15, 1999.

① This work was supported by the State Key Project for Research—"The South China Sea Monsoon Experiment", CAS (KZ951-B1-408) and CNSF (49823002).

teristics of the South China Sea summer monsoon activity and its impacts (Li and Zhang, 1999). We also analyzed the circulation characteristics in association with the onset of the South China Sea summer monsoon and the spreading of onset dates. It has been suggested that May 16 is the mean onset date of the South China Sea summer monsoon system. Before the establishment, the southwesterly wind in the southern China coastal region is not real summer monsoon, even though this kind of southwesterly wind and rainfall play an important role in the onset of the South China Sea summer monsoon, especially in the withdrawal of the subtropical high ridge (Li and Qu, 1999).

This work analyzed the onset and its characteristics of the South Chine Sea summer monsoon in 1998 and suggested further that it is not appropriate to divide the onset of South China Sea summer monsoon into two parts—north part (early) and south part (late) . The NCEP/NCAR reanalysis data, the satellite observational data and the ATLAS buoy observational data are used in this work.

2. The Wind Field at 850 hPa

The wind field at 850 hPa during May 15-May 24, 1998 is shown in Fig.1. It is clearly shown that although there had appeared southwesterly wind in the southern China coastal region from May 15 on, the principal winds over the large area of the South China Sea were southeasterly and southerly until May 20. A clockwise subtropical high ridge controlled the South China Sea region from the start. From May 21 on, the southwesterly wind and southerly wind were principal winds controlling the South China Sea region and the subtropical high ridge had been out of the South China Sea completely. Therefore, the evolution of the circulation pattern at 850 hPa indicates that the onset date of the South China Sea summer monsoon in 1998 should be May 21, which was later than normal.

It is also clear in Fig.1 that the southwesterly wind in the southern China coastal region before May 21 was caused by strong northwesterly flow over the Indian peninsula, but on just May 21 and after it, the southwesterly wind in the southern China coastal region resulted from the strong westerly wind over the equatorial Indian Ocean, which is caused by the cross-equatorial Somali flow. The property of flow in the southern China coastal region is different between the days before and after the onset of SCS summer monsoon (May 21). Therefore, the local southwesterly wind is not enough to define the onset of SCS summer monsoon, the origin of the flow is also important.

Fig. 2 showed the temporal evolution of the average zonal wind at 850 hPa in the South China Sea region (5-20°N, 105-120°E). It is clearly indicated that during May 19-21 not only the zonal wind speed increased abruptly, but also the wind direction changed steadily to westerly. On May 19 it was easterly wind with about 0.5 m/ s, but westerly wind with speed speed about 4 m/s on May 21. It can also be suggested that May 21 is the onset date of the South China Sea summer monsoon. During May 12-18, there was westerly wind, but it was very weak and not steady, they cannot be regarded as the signal of the summer monsoon onset.

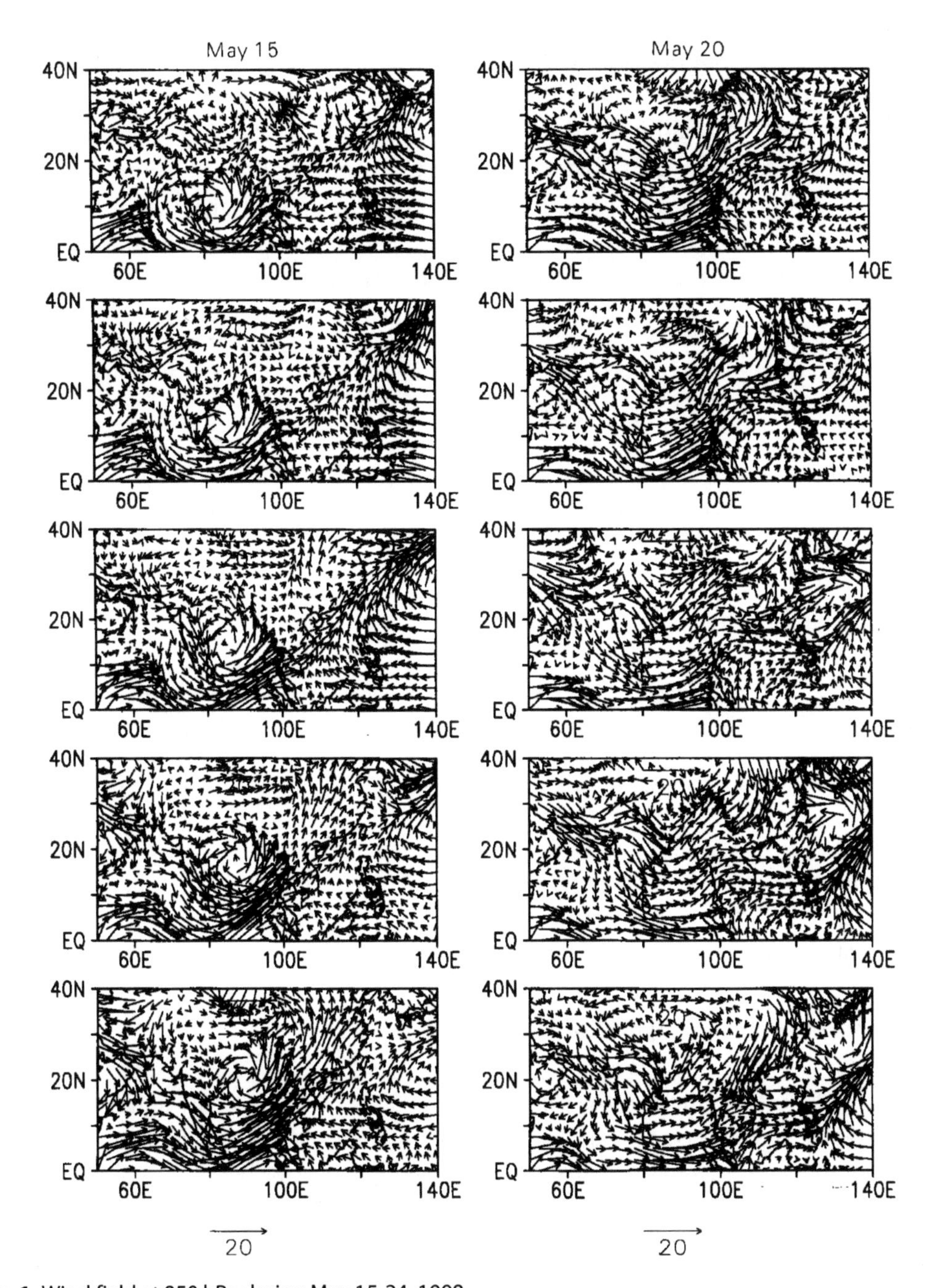

Fig. 1. Wind field at 850 hPa during May 15-24, 1998.

3. The Upper Tropospheric Circulation

The temporal evolution of the geo potential height field at 200 hPa (figure not shown) shows that a barometric high center had the characteristic of propagation from the neighborhood of (10°N, 105°E) northwestward to the neighborhood of (15°N, 95°E) before May 20 and its center intensity increased obviously on May 20 and May 21. The temporal evolution of the geopotential height field at 100 hPa (Fig.3) has an evident characteristic of a barometric high center moving eastward from North Arabian Sea to the Tibetan Plateau before May 20 and its center intensity increased obviously on May 21 and May 22. Such temporal evolution characteristics of the upper tropospheric circulation clearly indicated that the thermodynamic state variations of the Plateau and the circulation variability caused by the thermodynamic variation of the Plateau reached the biggest value on May 21.

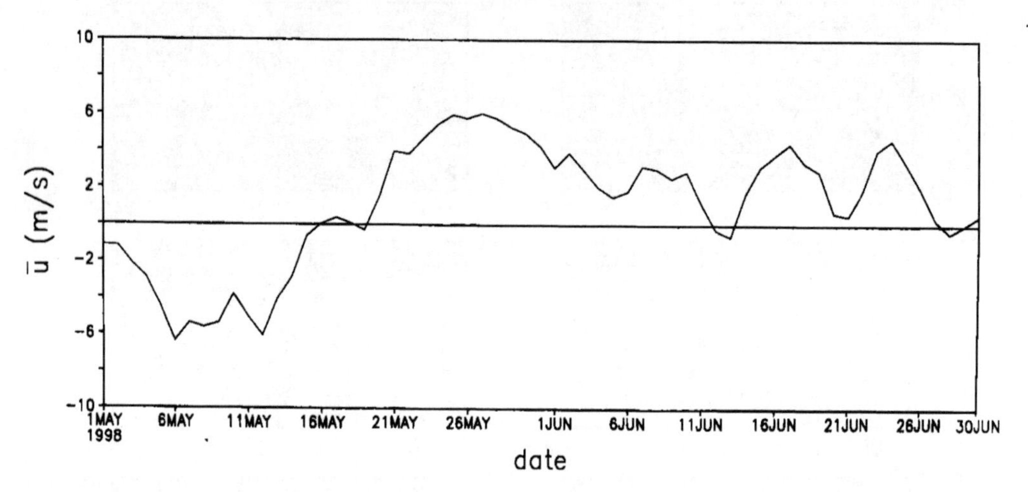

Fig. 2. Temporal evolution of mean zonal wind in the South China Sea region (5-20°N, 105-120°E) at 850 hPa.

Therefore, the upper tropospheric circulation was also changed corresponding to the outbreak of the SCS summer monsoon, and the greatest variation occurred on May 21—the onset date of summer monsoon.

4. The Variation of Upper Jet Stream

Early study pointed out that the upper jet stream axis jumps suddenly northward in June in association with the seasonal variation of northern hemisphere general circulation pattern from winter to summer (Ye et al., 1959). How about the upper jet stream activity in association with the onset of the South China Sea summer monsoon? The altitude-latitude section of the zonal wind in the South China Sea region (105-120°E) during May 12-30, 1998 is shown in Fig. 4. The most interesting thing is that the jet center was located at about 25-28°N before May 20, but it suddenly jumped northward to 35-40°N during May 20-21. The variation of latitude location reached 10 degrees only within two days. It suggested that not only the location of jet in East Asia had an obvious northward jump in association with the establishment of the South China Sea summer monsoon, but also the onset date of the South China Sea summer monsoon was on May 21 in 1998.

The above result shows that the sudden northward jump of the upper jet stream axis exists

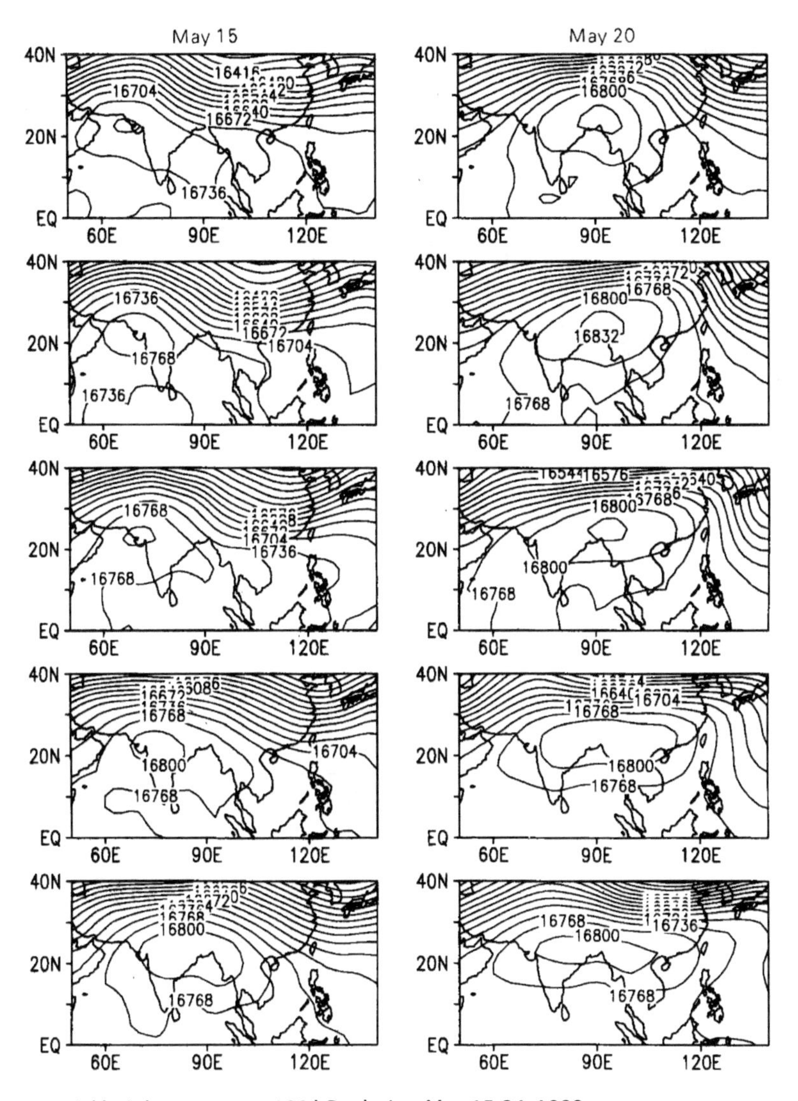

Fig. 3. Geopotential height pattern at 100 hPa during May 15-24, 1998.

not only in June corresponding to the seasonal variation of northern hemisphere circulation pattern from winter to summer, but also in May in association with the onset of SCS summer monsoon, though it is shown in the East Asian region. It can be suggested that the seasonal variation of atmospheric circulation is completed in different stages and the onset of SCS summer monsoon is an important variation stage from winter to summer.

5. Convection Activity in the South China Sea Region

The convection activity and rainfall are important characteristics of the outbreak of summer monsoon. The satellite observational data and the ATLAS buoy observational data indicated

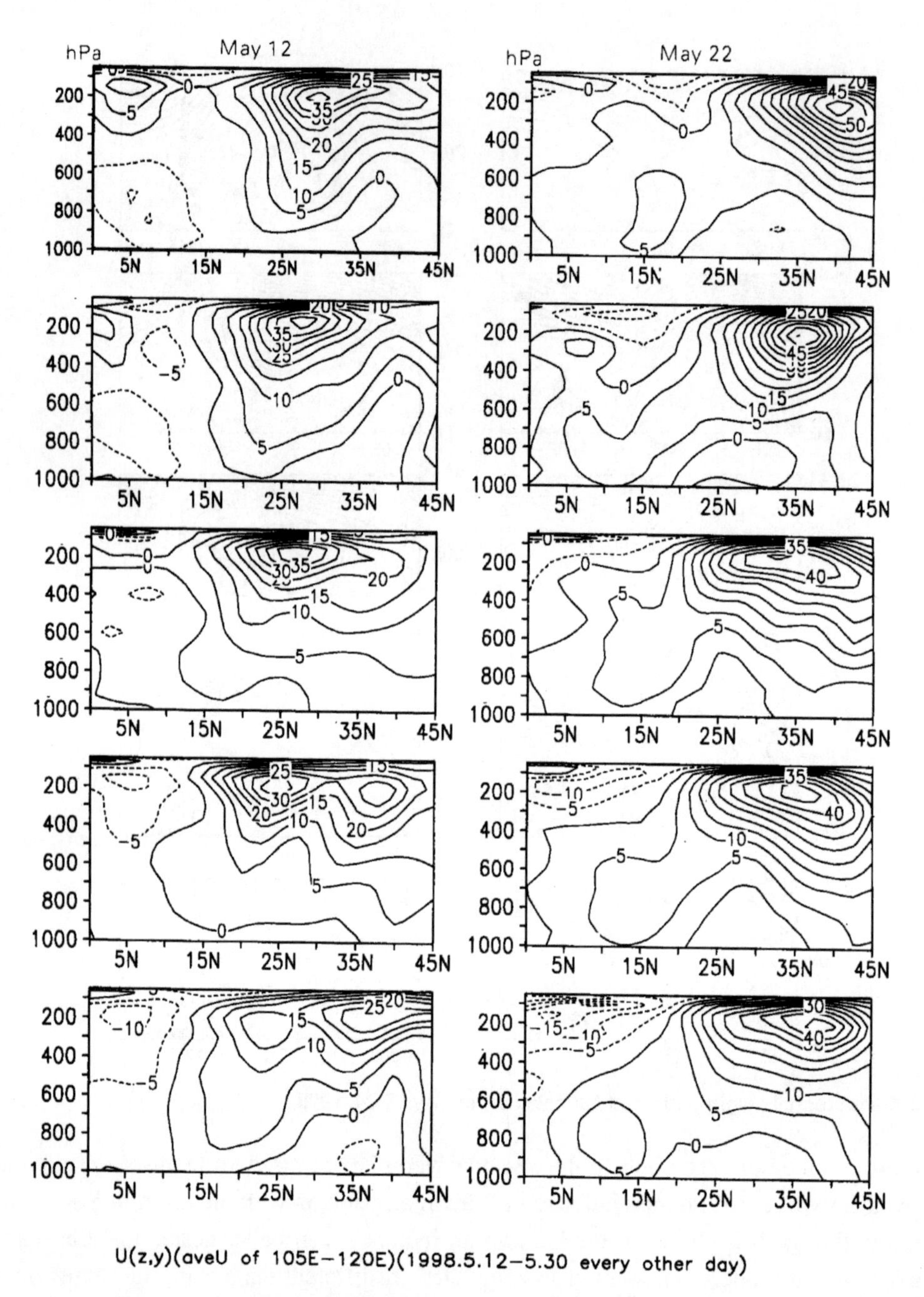

Fig. 4. Temporal evolution of time-latitude section of the zonal wind in the South China Sea region (105-120°E) during May 12-30 (every other day), 1998.

clearly that obvious variations occurred in the convection in the South China Sea region before and after May 21. The time-latitude section of the mean TBB in region 105-120°E is shown in Fig. 5 (black area: TBB>280K, no convection; greyish area: strong convection). It shows that there is no strong convection before May 21 but after that day the convection becomes stronger in the South China Sea region.

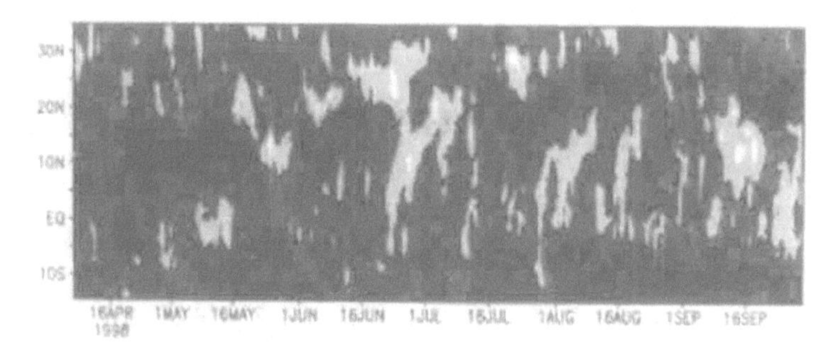

Fig. 5. The time-latitude section of the mean TBB in region (105-120°E)

The ATLAS-2 buoy observational data at (12°N, 114°E) are given in Fig. 6. Both mean daily rain ratio (a) and mean radiation (b) indicated that the atmospheric state changed obviously before and after May 21 in the South China Sea region, the rainfall increased obviously and radiation intensity decreased from 550 W/m^2 to 200 W/m^2.

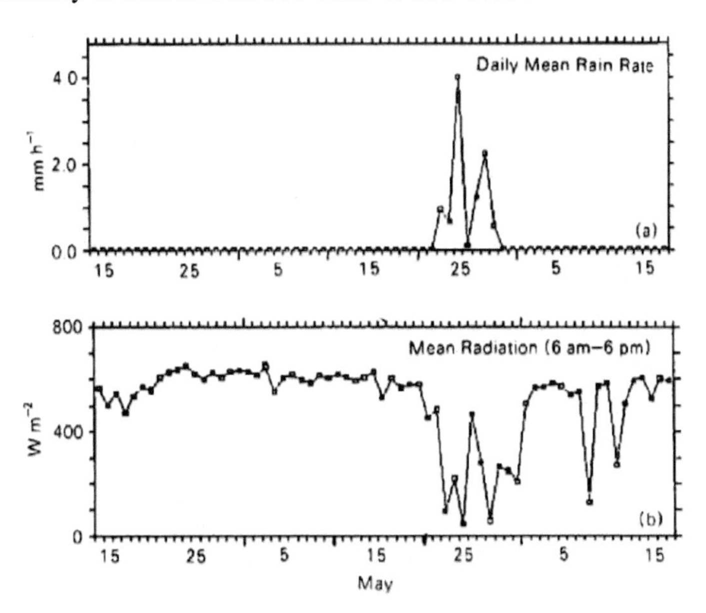

Fig. 6. ATLAS-2 observational data at (12°N, 114°E). (a) Mean daily rain ratio. (b) Mean radiation.

Thus, it can be suggested from the variations of the convection activities in the South China Sea region that May 21 is the onset date of the South China Sea summer monsoon in 1998.

6. The Southwesterly Wind and Rainfall in the Northern South China Sea Prior to the Onset of SCS Summer Monsoon

Sometimes there are southwesterly wind and rainfall in the northern South China Sea region prior to the onset of SCS summer monsoon, so that the onset of SCS summer monsoon is considered over the northern part at first in some studies. In 1998. the onset of SCS summer monsoon is also suggested on May 17 over the northern part in the discussion, but it is not true. The detailed analyses are shown in this section.

We have indicated that there was southwesterly wind in the southern China coastal region before May 21, but it could not be the sign of SCS summer monsoon onset because this southwesterly wind resulted from strong northwesterly flow over the Indian peninsula. The regional southwesterly wind is not enough to define the onset of SCS summer monsoon, and the flow origin should be considered.

Secondly, the satellite photographs showed that there were frontal activities over the northern South China Sea region from May 15 to May 20. The frontal activities not only caused the rainfall over the southern China coastal region, but also caused the southwesterly wind to the south of the front. In Fig. 7, some satellite photographs on May 15, 17, 19 and 20 are shown respectively, two frontal processes are clear over the northern South China Sea region. In general, the frontal rainfall cannot be regarded as the summer monsoon precipitation. In order to expose the rainfall feature in the northern South China Sea region during May 1-July 31, the temporal variations of daily precipitation at Yangjiang (59663), Qionghai (59855), Sanya (59948) and Xisa (59981) are shown in Fig.8. It is evident that the systematic

Fig. 7. Satellite photographs on May 15 (a), May 16(b), May 19(c) and May 20 (d), 1998, respectively.

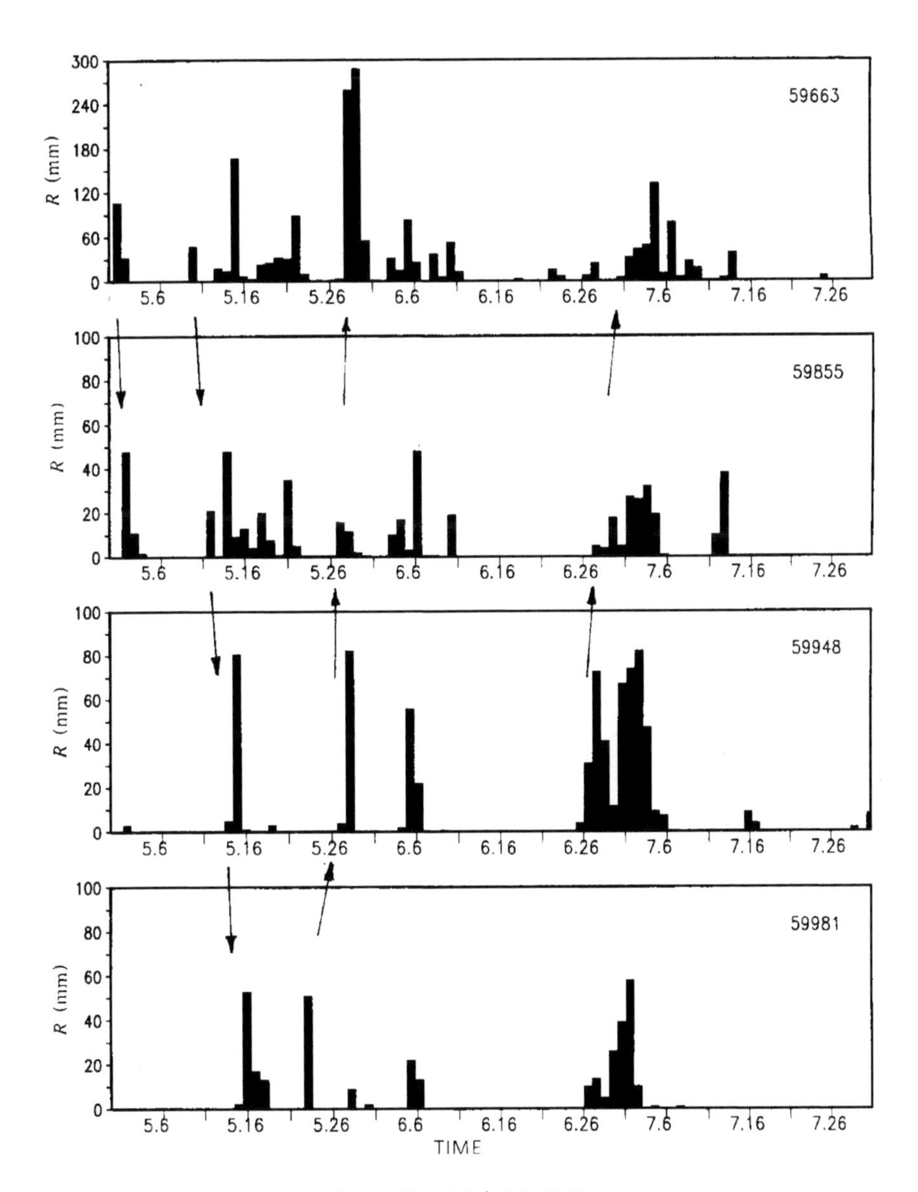

Fig. 8. Daily precipitation at 4 stations during May 1-July 31, 1998.

rainfall propagated from north to south before the onset of SCS summer monsoon (May 21) corresponding to the southward movement of the front. But after May 21, the northward propagation of the systematic rainfall is clearly shown. Therefore, the rainfall in the northern South China Sea region was the frontal feature just before May 21, but the summer monsoon feature after that day.

7. Conclusions

Through analyzing and comparing the NCEP/NCAR reanalysis data, the TBB satellite observational data and the ATLAS buoy observational data, it is clear that May 21 is the onset date of the South China Sea summer monsoon in 1998.

There were abrupt variations of the troposphere general circulation, especially that of the lower troposphere stream in association with the onset of the South China Sea summer monsoon. The subtropical high ridge moving out of the South China Sea, southwesterly wind controlling the South China Sea and westerly jet axis jumping from infra 30°N to ultra 35°N are most important characteristics of the summer monsoon onset in the South China Sea.

Before the onset of the South China Sea summer monsoon, there were rainfall and the southwesterly wind in southern China coastal region, but they did not mean that the summer monsoon first broke out in the northern South China Sea region. The rainfall there was caused mainly by the front and the southwesterly wind came mainly from strong northwesterly wind in Indian peninsula.

The onset of SCS summer monsoon is an important stage and component of the seasonal variation of atmospheric circulation, which occurs in the East Asian monsoon region. Studying on the onset of SCS summer monsoon should investigate the features of large-scale circulation and their changes.

References

He Jinhai, Zhu Qian'gen, and M. Murakami, 1996: The characteristics of seasonal variation and establishment of summer monsoon in Asia-Australia monsoon region revealed by TBB data. *J. Trop. Meteor.*, 12, 34-42. (in Chinese).

Jin Zuhui, and Chen Longxun, 1985: Medium time scale variation of East Asian summer monsoon system and its interaction with Indian monsoon system in summer. *Proceedings of the Symposium on the Summer Monsoon In South East Asia China.* People's Press of Yunnan Province, Kunming 204-217.

Li Chongyin, and Zhang Liping, 1999: The South China Sea summer monsoon activity and its impacts. *Chinese J. Atmos. Sci.*, 23, 111-120.

Li Chongyin and Qu Xin, 2000: The temporal evolution of the large scale general circulation in association with the onset of the South China Sea summer monsoon. *Chinese J. Atmos. Sci.*, 24, 1-14 (in Chinese).

Matsumoto, J., 1997: Seasonal transition of summer rain season over Indochina and Adjacent monsoon region. *Advances in Atmospheric Sciences*, 14, 231-245.

Tao, S. Y., and L. Chen, 1987: A review of recent research on the East Asian summer monsoon in China. *Monsoon Meteorology*, Oxford University Press, 60-92.

Xie An and Zhang Zhenghai, 1994: The spread of the South China Sea summer monsoon. *Acta Meteorologica Sinica*, 52, 374-378 (in Chinese).

Yan Junyue, 1997: Observational study on the onset of the South China Sea southwest monsoon. *Advances in Atmospheric Sciences*, **14,** 275-287.

Ye D, S. Tao, and M. Li, 1959: The abrupt change of circulation over the Northern Hemisphere during June and October. *The Atmosphere and the Sea in Motion*, The Rockefeller Institute Press and Oxford University Press, 249-267.

Zhu Qian'gen, He Jinhai, and Wang Panxing, 1986: A study of circulation differences between East Asian and Indian summer monsoons with their interaction. *Advances in Atmospheric Sciences*, 3, 466-477.

The Influence of the Indian Ocean Dipole on Atmospheric Circulation and Climate①

Li Chongyin, Mu Mingquan (穆明权)

LASG, Institute of Atmospheric Physics. Chinese Academy of Sciences, Beijing

Abstract The SST variation in the equatorial Indian Ocean is studied with special interest in analyzing its dipole oscillation feature. The dipole oscillation appears to be stronger in September-November and weaker in January-April with higher SST in the west region and lower SST in the east region as the positive phase and higher SST in the east region and lower SST in the west region as the negative phase. Generally, the amplitude of the positive phase is larger than the negative phase. The interannual variation (4-5 year period) and the interdecadal variation (25-30 year period) also exist in the dipole. The analyses also showed the significant impact of the Indian Ocean dipole on the Asian monsoon activity, because the lower tropospheric wind fields over the Southern Asia, the Tibetan high in the upper troposphere and the subtropical high over the northwestern Pacific are all related to the Indian Ocean dipole. On the other, the Indian Ocean dipole still has significant impact on atmospheric circulation and climate in North America and the southern Indian Ocean region (including Australia and South Africa).

Keywords Indian Ocean dipole, Sea surface temperature anomaly (SSTA), Asian summer monsoon, Climate impact.

1. Introduction

A strong El Niño event occurred in the early summer 1997 (McPhaden, 1999), it caused serious climate disasters in the world wide, such as the drought and forest fire in Indonesia, the flood in the northern region of South America. The previous studies in observation data (Rasmusson and Carpenter, 1983; Ropeleweski and Halpert, 1987) showed that Asian summer monsoon should be weaker with drought in Indian Peninsula in the El Niño year. But it was exactly reverse in 1997 as the mean summer precipitation in India was normal with even more rainfall in the partial region (Bell and Halpert, 1998), and there was also more precipitation in East Africa (Birkett et al., 1999). To investigate the abnormal situations, some analyses have shown that large SSTA in the equatorial Indian Ocean during the 1997-98 El Niño event, the

Received September 1, 2000.

① This work was supported by the National Key Basic Science Program in China (Grant No. 1998040903) and Chinese NSF (Grant No. 49823002). We would like to thank Wang Xuan for her typing this paper.

maximum SSTA was over 2℃, can be identified as the origin. Thus, more attentions have been paid to studying the anomalies of SST in the equatorial Indian Ocean and its impacts.

Dipole oscillation of SSTA in the equatorial Indian Ocean has been studied (Saji et al., 1999). Their study shows that mean SST in (10°S-10°N, 50°-70°E) region represents reverse variation feature with one in (10°S-EQ, 90°-110°E) region, the dipole is just 12% in the total variability of the Indian Ocean SST and it is not always related to the ENSO. Webster et al. (1999) also suggested that the Indian Ocean dipole in 1997-1998 was independent of the ENSO and caused by strong atmosphere-land-sea interaction. At same time, the study (Anderson, 1999) indicated that the dipole is not only shown in the variation of SST, but also in the variation of subsurface ocean temperature (SOT) of the Indian Ocean.

In fact, the important effect of SSTA in the Indian Ocean on summer rainfall in the middle and lower reaches of the Yangtze River have been indicated earlier by Chinese scientists (Chen et al., 1991; Luo et al., 1985). For better understanding the influence of the Indian Ocean SSTA on the weather and climate, the dipole pattern of the Indian Ocean SSTA need to be studied further, particularly the influence of the dipole on atmospheric circulation and climate.

2. Data

The data used for this study are the monthly 5° × 5° SST data (1900-1997) provided by the Hadley center in UK. The quality of the data set is approvable within meteorological research (Parker et al., 1994; Smith et al., 1998). Beside, the NCEP-NCAR reanalysis data and other data are also used to the statistical analysis in the present study.

3. Spatial-Temporal Features of Indian Ocean Dipole

According to the observation data, the SST variation in northwestern region is opposite to that in southeastern region in the equatorial Indian Ocean. In order to identify this dipole oscillation, a dipole index is defined, which is the difference between the averaged SSTA in the (5°S-10°N, 50°-65°E) region and the (10°S-5°N, 85°-100°E) region. In our definition, differ from that in Saji's study (Saji et al., 1999), some Islands and the Laut Jawa sea have been ruled out the domain. The temporal variation of dipole index in 1900-1997 and its power spectrum are shown in Fig. 1, It is very clear that the dipole index appears interannual variation (4-5 year period) and interdecadal variation (25-30 year period). Another interesting character of interdecadal variability in Fig. 1 is that the dipole index was mainly larger negative value before, 1961, but it was mainly positive value since 1961.

In order to reveal the feature of Indian Ocean dipole, the composite analyses are completed, respectively taking 4 years (1961, 1972, 1994 and 1997) with larger positive index and 5 years (1958, 1959, 1960, 1970 and 1996) with larger negative index to engage in. The horizontal distributions of the composite SSTA in the (30°S-50°N, 30°E-80°W) region for the positive and negative phases of the Indian Ocean dipole are respectively shown in Fig. 2.

Figures 2a and 2b (Figs. 2c and 2d) display the patterns of dipole for positive phase (negative phase) in July and in December, respectively. The basic feature of Indian Ocean dipole can be clearly seen in Fig. 2, the SSTA is positive (negative) in the equatorial western Indian Ocean and negative (positive) in the equatorial eastern Indian Ocean for the positive (negative) phase. The dipole patterns shown in Figs. 2a and 2b are stronger than that shown in Figs. 2c and 2d, because the dipole index in positive phase is larger than that in negative phase.

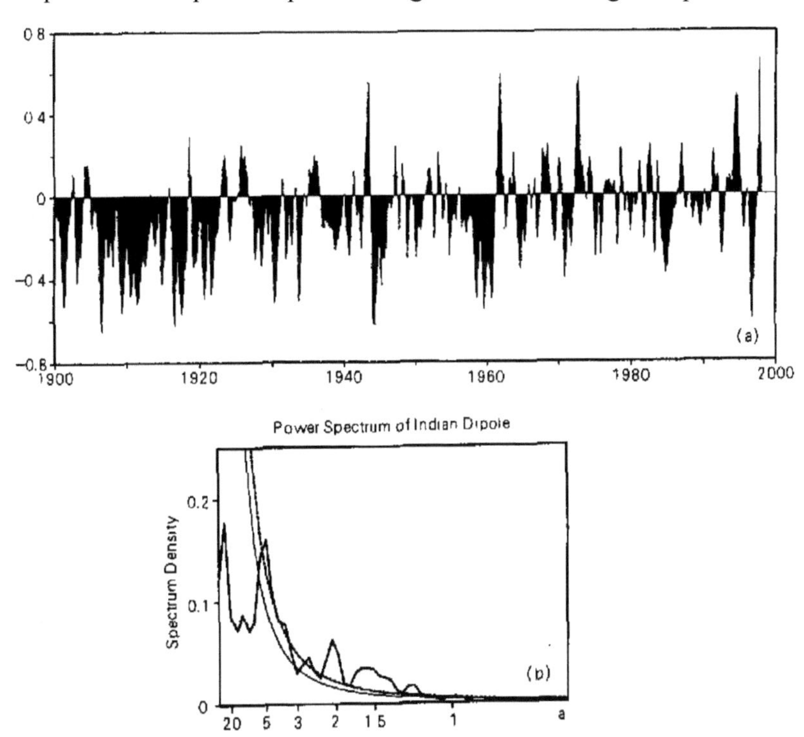

Fig.1. Temporal variation of monthly Indian Ocean dipole index (a) and its power spectrum (b). The solid and dashed lines represent significance level of 95% and 99% respectively.

Comparing Fig. 2a and Fig. 2b, or Fig. 2c and Fig. 2d, it can be seen that the intensity of Indian Ocean dipole is different in various months. An analysis of seasonal variations of the dipole intensity shows that the seasonal variation of the dipole intensity is very clear, the Indian Ocean dipole is stronger during the July-December and weaker during January-May with the strongest in October and the weakest in February (figure omitted).

The distribution of SSTA in the tropical Pacific is also shown in Fig. 2. The comparison of the SSTA distributions in the equatorial Indian Ocean and in the equatorial Pacific shows that when the Indian Ocean dipole is in the positive phase, the SSTA will be positive in the equatorial eastern Pacific and negative in the equatorial western Pacific. When the Indian Ocean dipole is in the negative phase, the SSTA will be positive in the equatorial western Pacific and negative in the equatorial eastern Pacific. How about the connection between the Indian Ocean dipole and the SSTA in the equatorial Pacific will be discussed in another paper.

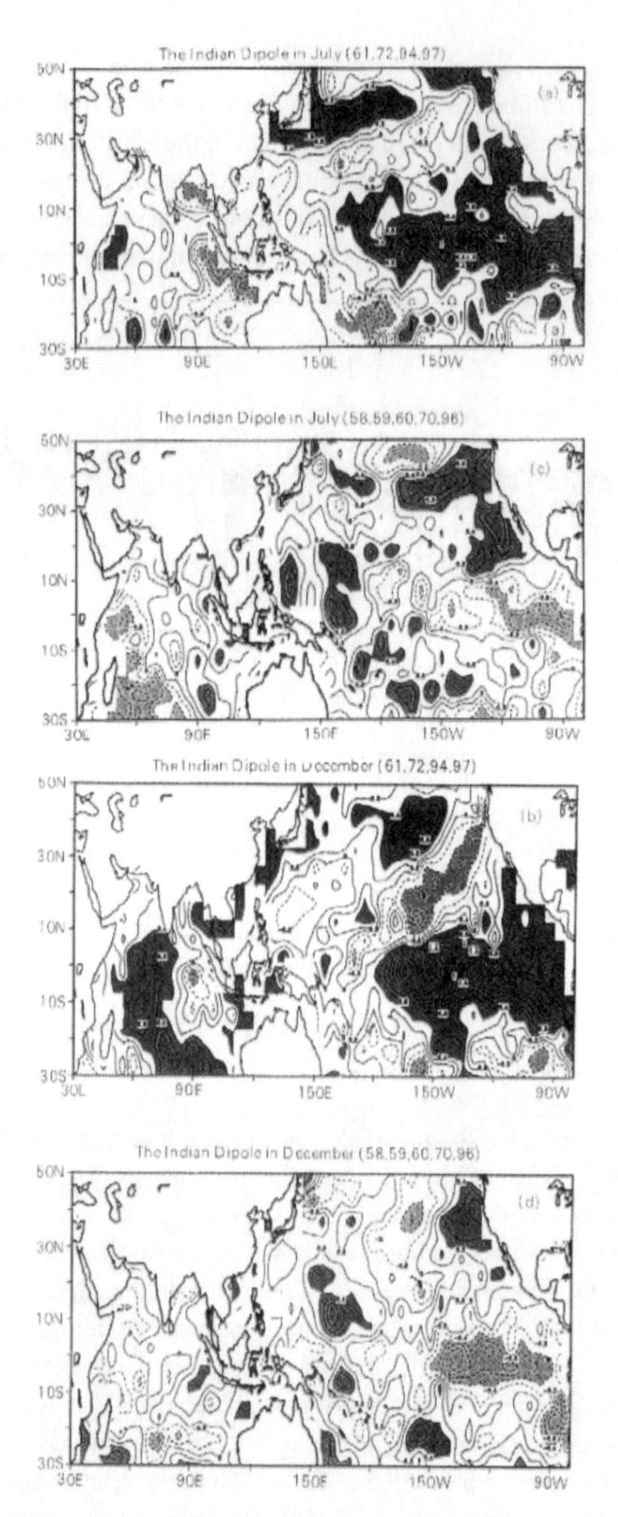

Fig. 2 Composite patterns of SSTA for positive phase (1961, 1972, 1994 and 1997) and negative phase (1958, 1959, 1960, 1970 and 1996) of Indian Ocean dipole. (a) July positive phase, (b) December positive phase, (c) July negative phase, (d) December negative phase

4. Influences of the Indian Ocean Dipole on Atmospheric Circulation and Climate

The ENSO has been regarded as important signal and factor of interannual climate variation (Namias and Cayan, 1981; Rasmusson and Wallace, 1983; Li, 1995) and it may be related to the Indian Ocean dipole (Li and Mu, 2001). In the following analyses, some important impacts of the Indian Ocean dipole on the atmospheric circulation and climate will be revealed without the consideration of ENSO.

4.1 The influences on Asian summer monsoon

Since the Asian monsoon, particularly Asian summer monsoon has important impact on the life and economy in this area, our attention will focus on the influence of Indian Ocean dipole on the Asian monsoon system first. The anomalous circulation patterns at 850 hPa in summer (June-August) for the positive phase and negative phase of the Indian Ocean dipole (Fig. 3) showed the difference of the Asian monsoon activity in different phase of the dipole. Corresponding to the positive phase of the Indian Ocean dipole, there are southeasterly wind anomalies over the equatorial Indian Ocean, anomalous westerly wind over Indian Peninsula and anomalous westerly wind over the region from the Bay of Bengal to the South China Sea. This means that the summer monsoon over the South China Sea and Indian Peninsula are stronger in the positive phase of the Indian Ocean dipole. Corresponding to the negative phase, the summer monsoon is weaker over the South China Sea but stronger over the southern India. Because there are weaker southerly wind anomalies over the equatorial western Indian Ocean, weaker northwesterly wind anomalies over the equatorial eastern Indian Ocean, westerly wind anomalies over the southern Indian Peninsula and easterly wind anomalies over the region from the South China Sea to the Bay of Bengal. Therefore, from the synoptic point of view, the Asian summer monsoon will be directly related to the Indian Ocean dipole.

It is known that the anticyclone in upper troposphere over the Qinghai-Xizang Plateau (called Tibetan high or South Asia high) is an important component of the Asian summer monsoon system, the intensity of South Asia high can partly represent the activity of the Asian summer monsoon. In order to show further the influence of the Indian Ocean dipole on the Asian summer monsoon, the correlation coefficients of the Indian Ocean dipole index with the geopotential height at 200 hPa calculated by using the NCEP reanalysis data (1958—1997) are given in Fig. 4. It shows clearly that the dipole index has negative correlation with the intensity of the South Asia high and a strong negative correlation center is over the Tibetan Plateau. In other words, the South Asian high is weaker (stronger) in the positive (negative) phase of the Indian Ocean dipole. The influences of South Asia high on the Asian summer monsoon, especially the East Asian summer monsoon, have been investigated in some studies (Tao and Zhu, 1964; Luo et al., 1982). Therefore, the influence of the Indian Ocean dipole on the Asian summer monsoon is in the affirmative, especially on the East Asian summer monsoon.

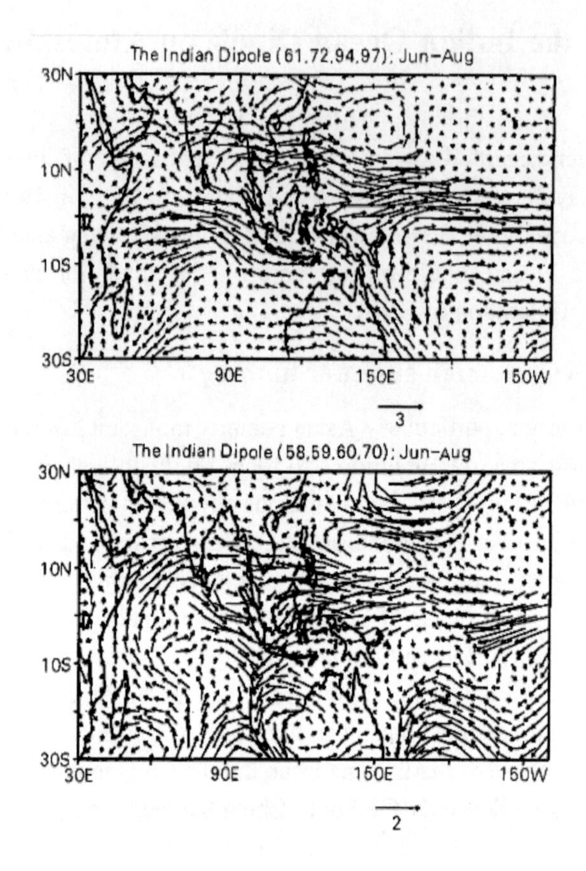

Fig. 3. The circulation patterns at 850 hPa in summer (June-August) over South Asia in the positive phase (a) and negative phase (b) of the Indian Ocean dipole, respectively.

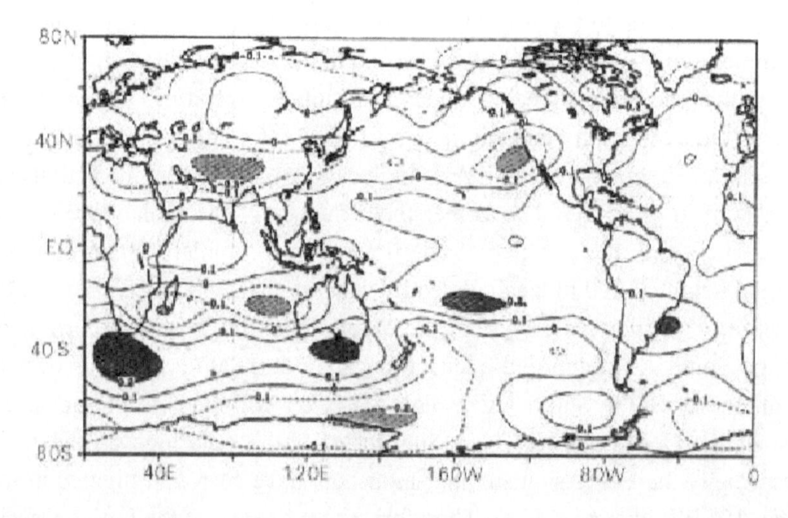

Fig. 4. Distribution of correlation coefficient of the Indian Ocean dipole index with the geopotential height at 200 hPa over the globe. The shadow represents the area, in which the correlation coefficient is more than statistical significance test.

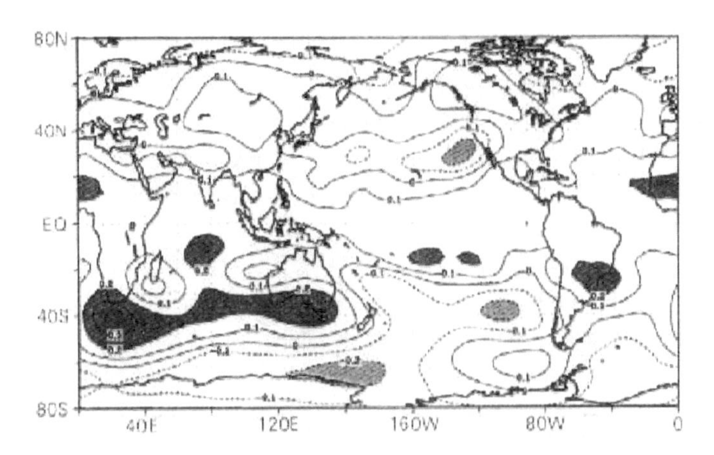

Fig.5. Same as Fig.4, but for the geopotential height at 500 hPa.

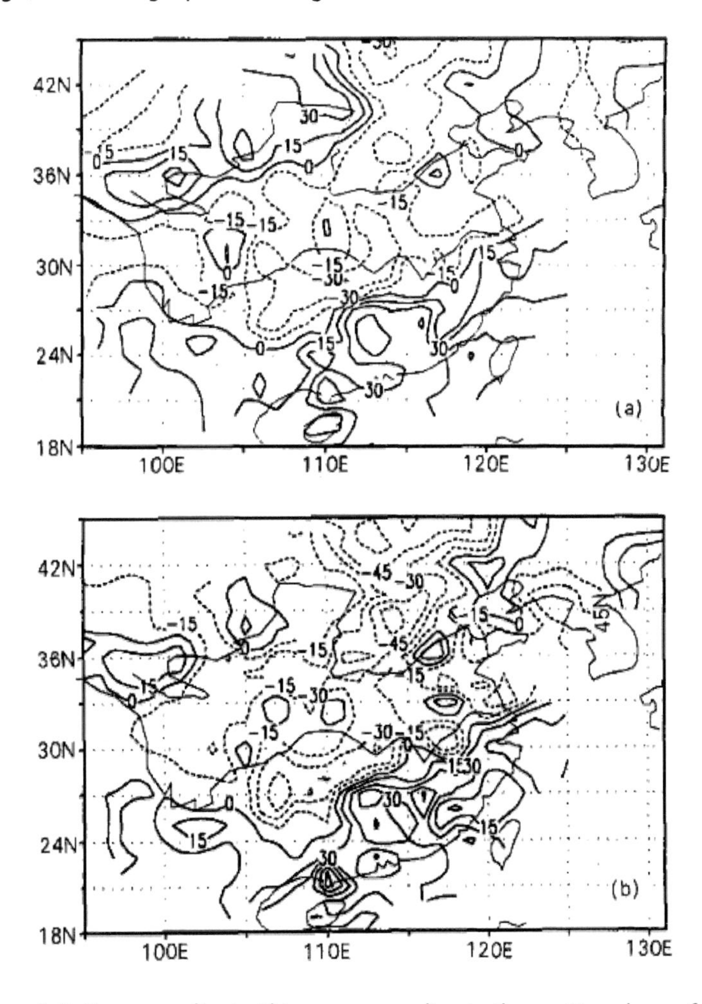

Fig. 6. Summer precipitation anomalies in China corresponding to the positive phase of the Indian Ocean dipole (a) and the differences of summer precipitation anomalies between the positive phase and negative phase of the Indian Ocean dipole (b).

The subtropical high over the northwestern Pacific is also an important component of the East-Asian summer monsoon system, and it is an important climate system to cause climate anomalies in East Asia (Huang and Yu, 1972; Tao et al., 1998). It is also clearly shown in Fig.5 that the subtropical high over the North western Pacific is related to the Indian Ocean dipole, in which the distribution of correlation coefficient of the Indian Ocean dipole index with the geopotential height at 500 hPa over the globe is given. There is a stronger negative correlation zone in 25°-40°N latitudes over the North Pacific and means that the subtropical high is weaker (stronger) in the positive (negative) phase of the Indian Ocean dipole. Over the East Asian continent, the positive correlation is also shown in Fig.5 and means that there exists an anomalous ridge (trough) at 500 hPa over the East Asian continent corresponding to the positive (negative) phase of the Indian Ocean dipole.

The precipitation anomalies in Eastern China during summer can also reveal the influence of the Indian Ocean dipole on Asian summer monsoon, particularly on the East Asian summer monsoon. The composite precipitation anomalies during summer (June - August) in China corresponding to 4 strong positive phase years and their difference from 5 strong negative phase years are shown in Fig. 6, respectively. Although the Indian Ocean dipole is weaker in summer as we indicated, the figure still showed that the positive (negative) phase of the Indian Ocean dipole is advantageous (not advantageous) to the summer rainfall in southeastern China, southern China, Yunnan and Qinghai regions but not advantageous (advantageous) to the summer rainfall in the other regions of China.

4.2 The influences on atmospheric circulation and climate in other regions

Strong correlation zone in 40°-50°S latitudes over the southern Indian Ocean and in the western coast of North America can be also shown in Fig. 4 and Fig. 5. This means that the variations of atmospheric circulation and climate in those regions are related to the Indian Ocean dipole.

The correlation coefficients between the Indian Ocean dipole index and the geopotential heights at 200 hPa and at 500 hPa all have a strong negative center over the northeastern Pacific/western coast of North America region. Therefore, there is an anomalous trough (ridge) at upper troposphere over the western coast region of North America corresponding to the positive (negative) phase of the Indian Ocean dipole. Thus, the weather and climate in the USA, particularly in the western region, will be impacted prominently.

Figure 7 shows the composite wind field at 850 hPa in November-January corresponding to the positive phase and negative phase of the Indian Ocean dipole, respectively. Obviously, systemic wind anomalies appear in Australia region and South Africa region. During the positive phase of the Indian Ocean dipole, there are easterly wind anomalies over northeastern Australia and the cyclonic circulation over southwestern Australia and western South Africa. During the negative phase of the dipole, there are easterly wind anomalies over southeastern Australia and an anticyclonic circulation over South Africa. These different circulation patterns will lead to different climate variation in the above mentioned regions. Therefore, the Indian Ocean dipole will also impact obviously the atmospheric circulation and climate in the Southern Indian Ocean region including south Africa and Australia.

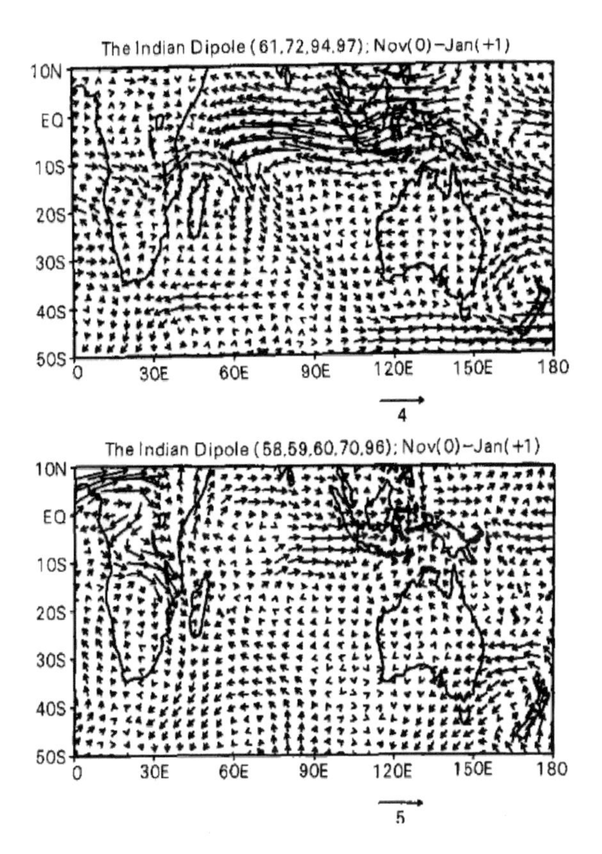

Fig. 7. Composite 850 hPa wind fields for November-January over the South Indian Ocean and nearby regions, corresponding to the positive phase of the Indian Ocean dipole (a) and the negative phase of the Indian Ocean dipole (b).

5. Omen Significance of the Indian Ocean Dipole

Since the variations of atmospheric circulation and climate in some regions have been shown to lag the variation of the dipole in the present study, the Indian Ocean dipole index can be one of the indicators for coming climate variation in some regions.

Figure 8 shows the distributions of correlation coefficients of the dipole index five months ahead of global geopotential height at 200 hPa and 700 hPa (similar to that at 500 hPa), respectively. It is shown clearly that there is an obvious PNA pattern over the eastern Pacific / the North America region and a negative value center of lag correlation coefficient over the Tibetan Plateau. Therefore, the positive (negative) phase of the Indian Ocean dipole can be regarded as a factor to predict the appearance of inverse (direct) PNA pattern and weak (strong) Tibetan high after 5 months. Although the geopotential height is not the best to use in the tropics, Fig. 8 still indicated that the easterly wind will be enhanced (weakened) in all tropical troposphere after the occurrence of positive (negative) phase of the Indian Ocean dipole for 5 months.

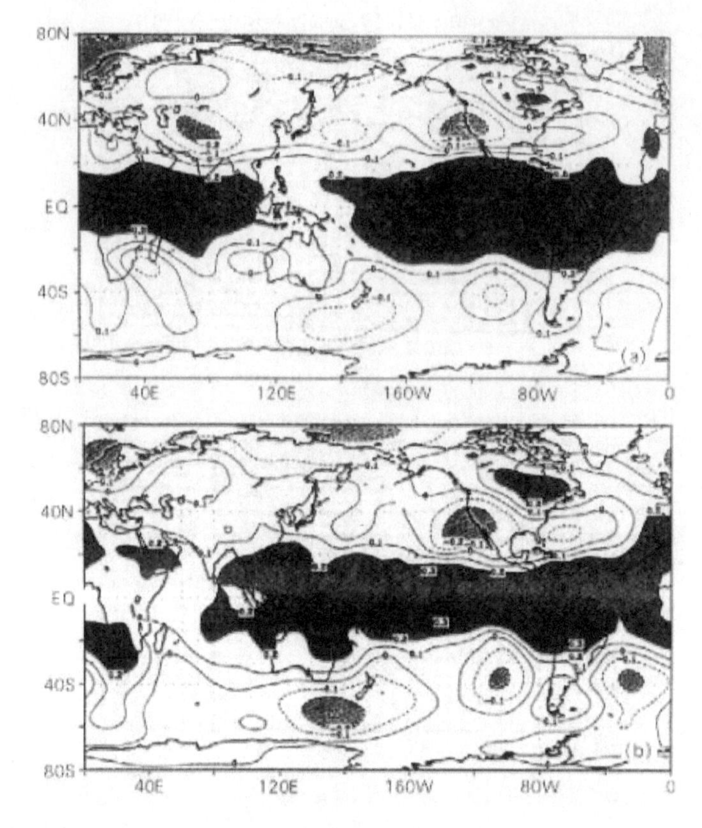

Fig.8. Global distribution of correlation coefficient between the Indian Ocean dipole index and geopotential height at 200 hPa (a) and 700 hPa (b) with the dipole index five months ahead of the geopotential height.

6. Conclusion

Based on above data analyses, some interesting results on the Indian Ocean dipole can be sum up as follows:

(1) A zonal oscillation as the dipole exists in the SST or SSTA in the equatorial Indian Ocean. This Indian Ocean temperature dipole is of interannual variation (4-5 year period) and interdecadal variation (25-30 year period); and its seasonal variation is also clear, stronger in September-November but weaker in January-April.

(2) Two major patterns of the dipole are: higher SST in the west and lower SST in the east of the equatorial Indian Ocean; higher SST in the east and lower SST in the west. The former is defined as the positive phase and the latter the negative phase of the Indian Ocean dipole, respectively. In general, the dipole is stronger in the positive phase than that in the negative phase.

(3) The Indian Ocean dipole will directly affect the Asian summer monsoon through impacting the lower tropospheric wind field. Corresponding to the positive phase of the Indian Ocean dipole, stronger summer monsoon can be identified over India and South China

Sea. But summer monsoon is weaker over the South China Sea and stronger over southern part of Indian Peninsula corresponding to the negative phase of the Indian Ocean dipole.

(4) The Tibetan high as well as subtropical high are getting weaker (stronger) due to the occurrence of the positive (negative) phase of the Indian Ocean dipole. Since the Tibetan high and subtropical high over the northwestern Pacific are important components of the Asian-summer monsoon system, the Indian Ocean dipole can also affect indirectly the Asian-summer monsoon (particularly the East Asian summer monsoon) through affecting the Tibetan high and the subtropical high.

(5) Corresponding to positive (negative) phase of the Indian Ocean dipole, there is an anomalous upper trough (ridge) over the western coast of North America. So the weather and climate will be impacted prominently in the USA, particularly in the western region of USA. The Indian Ocean dipole can also lead to the change of atmospheric circulation and climate in Australia and South Africa regions.

(6) The positive (negative) phase of the Indian Ocean dipole can be regarded as one of the factors used to predict the occurrence of inverse (direct) PNA pattern and the weak (strong) Tibetan high in five months.

References

Anderson, D., 1999: Extremes in the Indian Ocean. *Nature*, 401, 337-339.

Bell, G., and M. Halpert, 1998: Climate assessment for 1997. *Bull. Amer. Meteor. Soc.*, 79(5), S1-S50.

Birkett, C., R, Murtugdde, and T. Allan, 1999: Indian Ocean climate event brings floods to East Africa's lakes and the Sudd Marsh. *Geophys. Res. Lett.*, 26, 1031-1034.

Chen Lieting, 1991: Influence of zonal difference of the SSTA from Arabian Sea to South China Sea on the precipitation in the middle-lower reaches of the Yangtze River. *Chinese J. Atmos. Sci.*, 15, 33-42.

Huang Sisong, and Yu Zhihao, 1972: On the structure of the subtropical high and some associated aspects of the general circulation of atmosphere. *Acta Meteor. Sinica*, 31, 339-359 (in Chinese).

Li Chongyin, 1995: *Introduction to Climate Dynamics*, China Meteorological Press, Beijing 461 pp (in Chinese).

Li Chongyin, and Mu Mingquan, 2001: The dipole in the equatorial Indian Ocean and its impacts on climate. *Chinese J. Atmos. Sci, 25*, 433-443 (in Chinese).

Luo Saohua, Jin Zhuhui and Chen Lieting, 1985: Correlation analyses of the SST in the Indian Ocean / the South China Sea and the precipitation in the middle-lower reaches of the Yantze River, *Chinese J. Atmos. Sci.*, 9, 336-342.

Luo Siwei, Qian Zhengan, and Wang Qianqian, 1982: The climate and synoptical study about the relation between the Qinghai-Xizang high pressure on the 100mb surface and the flood and drought in East China in summer, *Plateau Meteorology*, 1(2), 1-10 (in Chinese).

Masumoto, Y., and T. Yamagata, 1996: Seasonal variations of Indonesian throughflow in a general circulation model *J. Geophys. Res.*, 101, 12287-12293.

McPhaden, M. J., 1999: Climate Oscilltions-Generis and evolution of the 1997-98 El Niño, *Science*, 283, 930-940.

Meyers, G., 1996: Variation of Indonesian throughflow and El Niño / Southern Oscillation. *J. Geophys. Res.*, 101, 12255-12264.

Namias, J., and D. R, Cayan, 1981: Large-scale air-sea interactions and short-period climate fluctuations. *Science*, 214, 868-876.

Parker, D. E., P. D. Jooes, C. K. Folland, and A. Bevan, 1994: Interdecadal changes of surface temperature since the late nineteenth century. *J. Geophys. Res*., 99, 14373-14399.

Potemra, J. T., R. Lukas, and G. T. Mitchum, 1997: Large-scale estimation of transport from the Pacific to the In- dian Ocean, *J. Geophys. Res*., 102, 27795-27812.

Rasmusson, E. M., and J. M. Wallace, 1983: Meteorological aspects of El Niño/ Southern Oscillation. *Science*, *222*, 1195-1202.

Rasmusson, E. M., and T. H. Carpenter, 1983: The relationship between eastern equatorial Pacific sea surface tem- peratures and rainfall over India and Sri Lanka. *Mon. Wea. Rev*., 111, 517-528.

Ropelewski, C. F., and M. S. Halpert, 1987: Global and regional scale precipitation patterns associated with the El Niño / southern Oscillation. *Mon. Wea. Rev*., 115, 1606-1626.

Saji, N. H., B. N. Goswami, P. N. Viayachandrom, and T. Yomagada, 1999: A dipole mode in the tropical Indian Ocean, *Nature*, 401, 360-363.

Smith, M. S., R. E. Livezey, and S. S. Shen, 1998: An improved method for analyzing spa-seand irregularly distributed SST data on a regular grid: The tropical Pacific Ocean. *J. Climate*, 11, 1717-1729.

Tao Shiyan, Zhang Qingyun, and Zhang Shunli, 1998: The great floods in the Changjiang River valley in 1998. *Climatical and Environmental Research*, 3, 290-298 (in Chinese).

Tao Shiyan, and Zhu Fukan, 1964: The 100 mb flow patterns in southern Asia in summer and its relation to the advance and retreat of the west-Pacific subtropical anticyclone over the far East. *Acta Meteor. Sinica*, 34, 396-407 (in Chinese).

Webster, P. T., A. M. Moore, J. P, Loschning, and R. R. Leben, 1999: Coupled ocean-atmosphere dynamics in the Indian Ocean during 1997-98. *Nature*, 401, 356-360.

WMO, ICSU, and UNESCO, 1995: CLIVAR-A Study of Climate Variability and Predictability, Science Plan, WMD / TD, No.690, WCRP-89, Geneva.

Dynamical Impact of Anomalous East-Asian Winter Monsoon on Zonal Wind over the Equatorial Western Pacific

Li Chongyin[1, 2], Pei Shunqiang[1], Pu Ye[1]

[1] LASG, Institute of Atmospheric Physics, Chinese Academy of Sciences, Beijing, China
[2] Institute of Meteorology, PLA University of Science and Technology, Nanjing, China

Abstract Zonal wind anomaly over the equatorial western Pacific plays an important role in the occurrence of ENSO. The mechanism to produce zonal wind anomaly over the equatorial western Pacific is studied in this paper. It is shown clearly that zonal wind anomaly over the equatorial western Pacific is closely related to the anomaly of East-Asian winter monsoon. Anomalous strong (weak) East-Asian winter monsoon can excite not only the westerly (easterly) anomaly over the equatorial western Pacific but also a cyclonic (an anticyclonic) circulation over the east of the Philippines. The above anomalous circulation results from dynamical impacts of anomalous pressure pattern due to the East-Asian winter monsoon. Because there is westward (eastward) pressure gradient over the equatorial western Pacific, i.e. there is $\frac{\partial p}{\partial x} < 0$ (> 0), during strong (weak) East-Asian winter monsoon.

Keywords anomalous East-Asian winter monsoon, dynamical impact, the equatorial Pacific, zonal wind.

In the late 1960s, it was advanced that the El Niño event results from air-sea interaction in the tropical Pa- cific[1]. Then a series of studies have shown that the occurrence of El Niño event is very closely related to the decrease of trade wind (appearing anomalous westerly wind) over the equatorial Pacific[2-5]. And some studies still showed that the equatorial westerly wind anomalies associated with the occurrence of El Niño event basically appear over the equatorial western Pacific at first, then they gradually expand eastwards[6-8]. Some dynamic studies also indicated that the westerly anomalies over the equatorial western Pacific play an important role in the occurrence of El Niño [9, 10]; while the El Niño event will move towards the decline and fall as the westerly stress is transformed into the easterly stress[11].

What is the physical process or reason leading the westerly wind anomaly over the equatorial western Pacific? Through the data analyses and theoretical study, it was advanced that strong East-Asian winter monsoon (activity of strong East-Asian troughs at upper troposphere) can excite anomalous westerly wind and strong convection activity (strong

Received July 28, 2004; accepted March 25, 2005.
Correspondence should be addressed to Li Chongyin (email: lcy@ lasg.iap.ac.cn)

intraseasonal oscillation) over the equatorial western Pacific, and then proceed to excite El Niño event by way of the air-sea interaction[5, 12-14]. In some studies it was also shown that there are not only the westerly anomalies over the equatorial western Pacific during the forming and developing stages of the El Niño, but also the anomalous cyclonic circulation over the east of the Philippines associated with the westerly wind anomalies over the equatorial western Pacific. While during the forming and developing stages of the La Nina, there are not only the easterly wind anomalies over the equatorial western Pacific but also the anomalous anticyclonic circulation over the east of the Philippines associated with the easterly wind anomalies over the equatorial western Pacific[15]. But the dynamical mechanism to cause the above anomalous circulation has not been understood very well.

It is the purpose of this study to discuss the dynamical impact of anomalous East-Asian winter monsoon on the formations of westerly (easterly) wind anomaly over the equatorial western Pacific and cyclonic (anticyclonic) circulation over the east of the Philippines, and further to understand the physical process of anomalous East-Asian winter monsoon exciting zonal wind anomaly over the equatorial western Pacific and the ENSO.

The data used in this study include the NCAR/NCEP reanalysis data (1958 — 1998; 2.5° × 2.5° resolution) and the OLR (outgoing long-wave radiation) data produced by the National Oceanic and Atmospheric Administration (NOAA).

1. Further Analyses on Relationship Between Zonal Wind Anomaly over the Equatorial Western Pacific and East-Asian Winter Monsoon

The composite analyses for El Niño (La Nina) cases with the ECMWF data and the NCAR/NCEP reanalysis data have shown that there are westerly (easterly) wind anomalies over the equatorial western Pacific prior to the occurrence of El Niño (La Nina), while the winter monsoon over East Asia has appeared strong (weak) anomaly before the occurrence of westerly (easterly) anomaly over the equatorial western Pacific, i.e. there were northerly (southerly) wind anomalies at first over the northwestern Pacific/East Asia (25°—35°N, 120°—135°E) region[16, 17].

In order to reveal further the relation of zonal wind anomaly over the equatorial western Pacific with anomalous East-Asian winter monsoon, temporal variations of the composite meridional wind anomaly over the northwestern Pacific/East Asia region (25° — 35°N, 120° — 135°E) and zonal wind anomaly over the equatorial western Pacific (5°S —5°N, 120°—160°E) for the 5 strongest El Niño events in the recent 50 years are shown in Fig. 1, respectively. The abscissa in Fig. 1 shows the time for 3 years, i.e. before the year of El Niño, the eruption year of El Niño and the next year of El Niño. Obviously, before the occurrence of westerly anomaly over the equatorial western Pacific, there has been northerly wind anomaly over the northwestern Pacific/East Asia region. This means that abnormal strong winter monsoon (strong northerly wind) in East Asia should play an important role in the occurrence of westerly anomalies over the equatorial western Pacific.

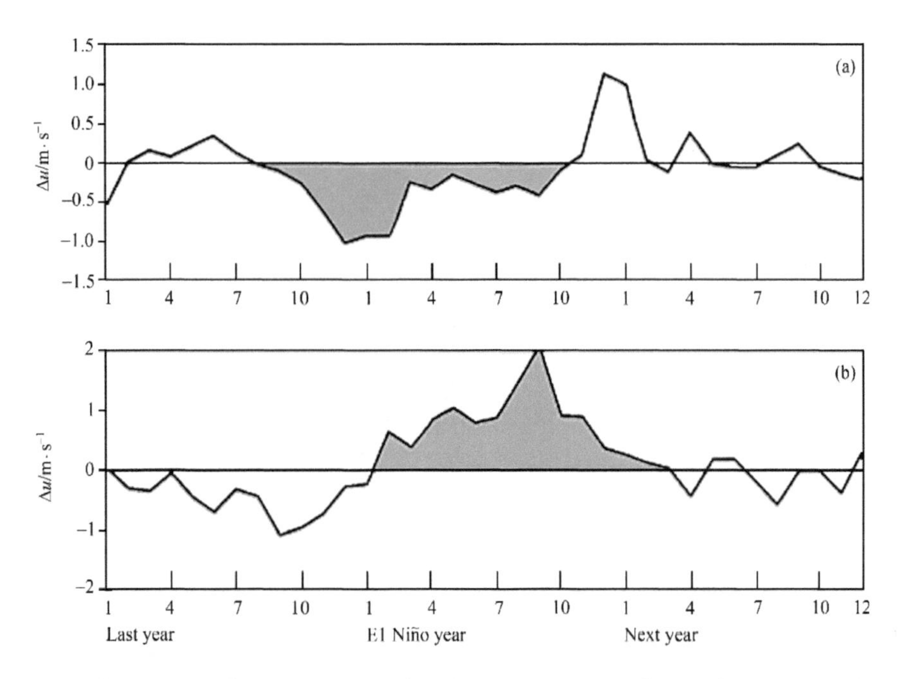

Fig. 1. Temporal variations of composite meridional wind anomaly in the northwestern Pacific/East Asia region (a) and zonal wind anomaly over the equatorial western Pacific (b) for El Niño cases.

The distribution of correlation coefficients between 850hPa zonal wind over the equatorial western Pacific and 850hPa meridional wind over the western Pacific and the tropical Indian Ocean in wintertime (Nov. — Apr.) is shown in Fig. 2. The negative correlation in the figure means anomalous northerly (southerly) wind corresponding to anomalous westerly (easterly) wind. It is clear that there is obvious negative correlation area, i.e. anomalous westerly wind over the equatorial eastern Indian Ocean and equatorial western Pacific corresponding to anomalous northerly wind (strong winter monsoon) in the northwestern Pacific/ East Asia region, while there is anomalous easterly wind over the equatorial eastern Indian Ocean and the equatorial western Pacific corresponding to anomalous southerly wind (weak winter monsoon) in the northwestern Pacific/ East Asia region. And the negative correlation area propagates eastwards as time goes on. This suggests that zonal wind anomaly over the equatorial western Pacific is strongly related to the anomalous winter monsoon in East Asia. Strong (weak) East-Asian winter monsoon corresponding to northerly (southerly) wind anomaly will lead to westerly (easterly) wind anomaly over the equatorial eastern Indian Ocean and western Pacific. In addition, anomalous zonal wind shows obviously eastward propagation feature.

2. Dynamical Impact of Anomalous East-Asian Winter Monsoon on Zonal Wind over the Equatorial Western Pacific

In some studies, it has been indicated that stronger East-Asian winter monsoon will lead the cumulus convection to enhancement over the equatorial western Pacific, and then stronger

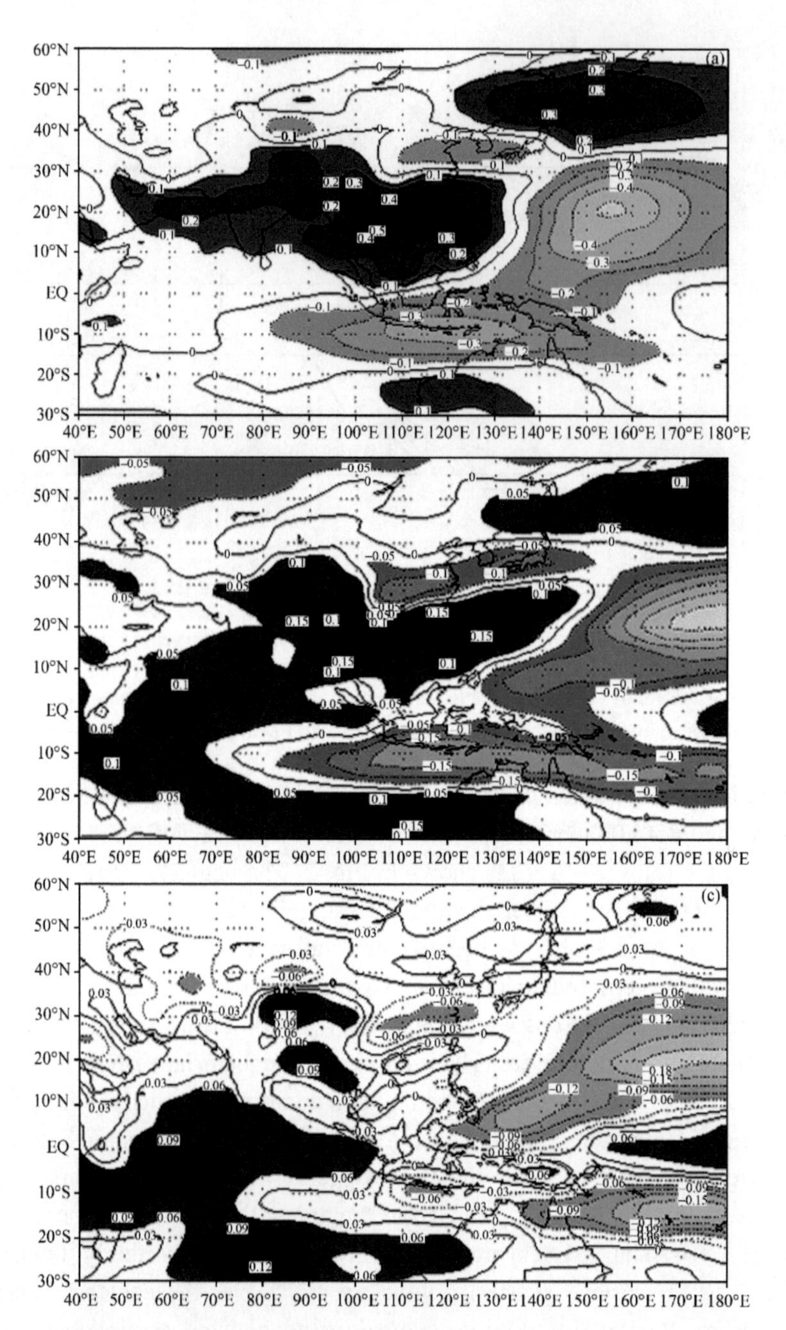

Fig. 2. Distributions of correlation coefficients between zonal wind at 850hPa over the equatorial eastern Indian/western Pacific and meridional wind over the northwestern Pacific/East Asia region in wintertime. The shading shows significance above the 95% level; (a) represents contemporary correlation; (b) and (c) represent lag correlations of zonal wind lagging meridional wind for 5 days and 10 days.

tropical intraseasonal oscillation (ISO) can be excited there[12, 18]. Through the interaction between tropical ISO and the environmental field, sustained westerly wind anomalies can be formed over the equatorial western Pacific. The distribution of correlation coefficients between 850 hPa meridional wind over the northwestern Pacific/East Asia region and the OLR in the tropics during wintertime showed an important influence of the East-Asian winter monsoon on convection activities over the equatorial western Pacific (figure is omitted). The northerly wind anomaly (strong winter monsoon) is associated with strong convection (lower OLR), while southerly wind anomaly (weak winter monsoon) is associated with weak convection (higher OLR). The data analyses also show that zonal wind anomaly over the equatorial western Pacific is closely related to the ISO activity. There is westerly (easterly) wind anomaly corresponding to strong (weak) ISO over the equatorial western Pacific. In other words, through the excited convection activity anomaly and tropical ISO anomaly, anomalous East-Asian winter monsoon can also lead to zonal wind anomaly over the equatorial western Pacific.

The above-mentioned process can be regarded as an indirect effect of anomalous East-Asian winter monsoon on zonal wind anomaly over the equatorial western Pacific. The following will discuss the direct impact of anomalous East-Asian winter monsoon on zonal wind anomaly over the equatorial western Pacific. We have indicated in the introduction that the occurrence of El Niño (La Nina) is closely related to westerly (easterly) wind anomaly over the equatorial western Pacific and anomalous cyclonic (anticyclonic) circulation over tropical western Pacific, and zonal wind anomaly over the equatorial western Pacific is closely related to anomalous East-Asian winter monsoon. In order to reveal further the important impact of anomalous East-Asian winter monsoon on exciting zonal wind anomaly over the equatorial western Pacific and atmospheric circulation anomaly over the tropical western Pacific, a simple dynamical analysis is performed as follows.

In the tropical atmosphere, the disturbance equation of zonal wind (without basic flow) can be simply written as

$$\frac{\partial u}{\partial t}-\beta yv=-\frac{\partial p}{\partial x}-D_m u, \tag{1}$$

where u and v are zonal and meridional wind speed, respectively; p is pressure, D_m is damping coefficient. Near the equator, $y \to 0$ and v is small, and as a stationary state case, eq. (1) will become

$$D_m u=-\frac{\partial p}{\partial x}. \tag{2}$$

It means that zonal wind directly depends on zonal pressure gradient near the equator. If there is $\frac{\partial p}{\partial x}<0$ then $u>0$, and anomalous westerly wind will be formed or enhanced over the equatorial area. If there is $\frac{\partial p}{\partial x}>0$ then $u<0$, and anomalous easterly wind is advantageous to forming over the equatorial area. Therefore, zonal wind anomaly over the equatorial western Pacific is directly associated with zonal pressure gradient there; the $\frac{\partial p}{\partial x}<0$ will produce or enhance anomalous westerly wind over the equatorial western Pacific, and the $\frac{\partial p}{\partial x} > 0$ will be advantageous to forming anomalous easterly wind.

Since strong (weak) winter monsoon in East Asia can lead to different zonal pressure gradients over the equatorial western Pacific, it will play an important role in the formation of anomalous zonal wind over the equatorial western Pacific. The distribution feature of anomalous sea level pressure (SLP) field over the tropical western Pacific is shown in Fig. 3, respectively for strong and weak East-Asian winter monsoon cases. Only the situation in December is shown here, since the pattern for other months in wintertime is basically similar to that in Fig. 3. It is shown that there is a low-pressure center near the 150°E, and $\frac{\partial p}{\partial x} < 0$ (the arrowhead towards the west) over the equatorial western Pacific corresponding to strong winter monsoon, but corresponding to weak winter monsoon, there is $\frac{\partial p}{\partial x} > 0$ (the arrowhead towards the east) over the equatorial western Pacific. In Fig. 4, the composite anomalous SLP fields in January for the El Niño years (1972, 1982, 1986, 1991 and 1997) and the La Nina years (1970, 1973, 1975, 1988 and 1998) are given, respectively. The anomalous characteristics of the SLP field are similar to that shown in Fig. 3 and even more outstanding. In particular there is negative (positive) pressure gradient in El Niño (La Nina) year at 0 — 10°N in the western Pacific. This is natural, because the occurrence of El Niño (La Nina) event is generally associated with strong (weak) East-Asian winter monsoon in last wintertime.

It is still shown in Fig. 3 and Fig. 4 that the zonal pressure gradient is very small in the 20°N region over the northwestern Pacific no matter whether in strong or in weak winter monsoon case. In other words, there is $\frac{\partial p}{\partial x} \approx 0$ for anomalous East-Asian winter monsoon case. Thus, as a stationary state case, eq. (1) will become

$$\beta y v \approx D_m^{\ u}. \tag{3}$$

Therefore, the southerly wind anomaly ($v > 0$) over there will lead to $u > 0$, anomalous westerly wind will be formed or enhanced; the northerly wind anomaly ($v < 0$) will lead to $u < 0$, and anomalous easterly wind will be formed or enhanced.

It can be suggested that there will be formed anomalous easterly wind in 20°N region over the northwestern Pacific/East Asia when there is northerly anomaly ($v < 0$), while anomalous westerly wind is associated with southerly anomaly ($v > 0$). It has been known that anomalous northerly (southerly) wind over the northwestern Pacific/East Asia region is always associated with strong (weak) East-Asian winter monsoon, thus strong (weak) East-Asian winter monsoon could excite anomalous easterly (westerly) wind in 20°N region over the northwestern Pacific/East Asia. This anomalous easterly (westerly) wind in 20°N region will cooperate with anomalous westerly (easterly) wind over the equatorial western Pacific and anomalous northerly (southerly) wind over the north western Pacific/East Asia caused by strong (weak)

East- Asian winter monsoon during the last wintertime prior to the occurrence of El Niño (La Nina) event. Thus, an anomalous cyclonic (anticyclonic) circulation will be formed over the east of the Philippines corresponding to strong (weak) East-Asian winter monsoon. Obviously, this kind of anomalous cyclonic (anticyclonic) circulation over the east of the Philippines will be more advantageous to maintaining anomalous westerly (easterly) wind over the equatorial western Pacific.

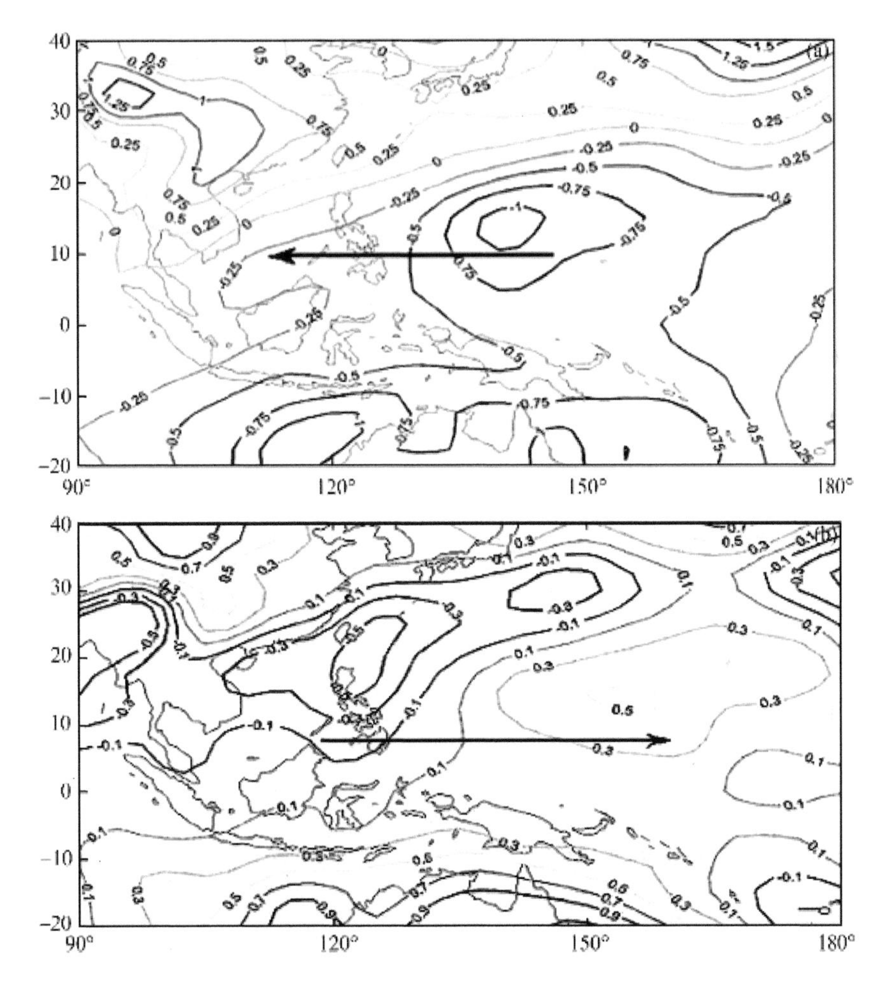

Fig. 3. Distributions of sea level pressure anomaly field (unit: hPa) over the tropical western Pacific in December for strong (a) and weak (b) East-Asian winter monsoon cases, respectively.

Therefore, strong (weak) East-Asian winter monsoon can lead to not only the occurrence of anomalous westerly (easterly) wind over the equatorial western Pacific but also anomalous cyclonic (anticyclonic) circulation over the east of the Philippines. In other words, both anomalous westerly (easterly) wind over the equatorial western Pacific and anomalous cyclonic (anticyclonic) circulation over the east of the Philippines are directly related to strong (weak) East-Asian winter monsoon.

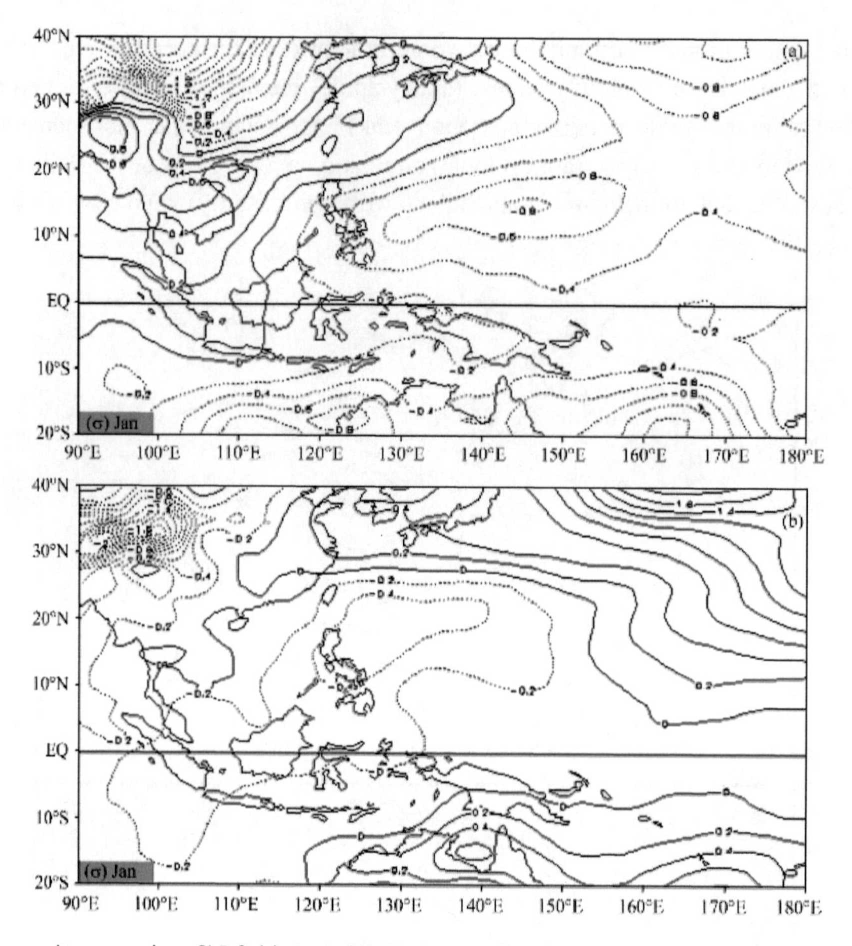

Fig. 4. Composite anomalous SLP fields (unit: hPa) in January for El Niño (a) and La Nina (b) cases.

3. Conclusion

The above analyses and discussion have revealed further the relation of westerly (easterly) wind anomalies over the equatorial western Pacific and anomalous cyclonic (anticyclonic) circulation over the east of the Philippines with strong (weak) winter monsoon in East Asia. Some major results can be summarized as follows:

1) The westerly (easterly) wind anomalies over the equatorial western Pacific and anomalous cyclonic (anti- cyclonic) circulation over the east of the Philippines, which are associated with the occurrence of El Niño (La Nina) event, are closely related to strong (weak) East-Asian winter monsoon. Generally, there are westerly (easterly) wind anomaly over the equatorial western Pacific and anomalous cyclonic (anticyclonic) circulation over the east of the Philippines prior to the occurrence of El Niño (La Nina), but before these anomalies, there had been the northerly (southerly) wind anomaly over the northwestern Pacific/East Asia region and strong (weak) winter monsoon in East Asia.

2) An anomalous pattern of the pressure field will be formed over the western Pacific and southeastern Asia for quite a long time because of the impact of anomalous East-Asian winter monsoon. The pattern of anomalous pressure field is close to reverse corresponding to strong and weak winter monsoon. The continual strong (weak) winter monsoon in East Asia is advantageous to the formation of anomalous westerly (easterly) wind over the equatorial western Pacific and anomalous cyclonic (anticyclonic) circulation over the east of the Philippines.

3) The dynamical impacts of different pressure gradients over the equatorial western Pacific and meridional wind anomaly over the northwestern Pacific/East Asia region, which result from strong or weak winter monsoon in East Asia, are important mechanism to excite anomalous westerly (easterly) wind over the equatorial western Pacific and anomalous cyclonic (anticyclonic) circulation over the east of the Philippines.

4) The result based on simple dynamical analysis in this paper is corroborated partly in some studies[19], but further numerical simulations are necessary for a thorough understanding.

Acknowledgements This study is supported by the National Natural Science Foundation of China (Grant No. 40233033) and the Blazing New Trails Program of the Chinese Academy of Sciences (Grant No. ZKCX3-SW-226).

References

1. Bjerknes, J., Atmospheric teleconnections from the equatorial Pacific, Mon. Wea. Rev., 1969, 97: 163-172.
2. Rasmusson, E. M., Carpenter, T. H., Variation in tropical sea surface temperature and surface wind field associated with the Southern Oscillation/El Niño, Mon. Wea. Rev., 1982, 110: 354-384.
3. Rasmusson, E. M., Wallace, J. M., Meteorological aspects of El Niño/Southern Oscillation, Science, 1983, 222: 1195-1202.
4. Cane, M. A., Zebiak, S. E., A theory for El Niño and Southern Oscillation, Science, 1985, 228: 1085-1087.
5. Li Chongyin, The frequent activities of stronger aerotroughs in East Asia in the wintertime and the occurrence of El Niño event, Sciention Sinica (B), 1989, 32: 976-985.
6. O'Brien, J., Busalacchi, J. J., The Pacific Ocean response to El Niño condition, WCRP Publication Series, No.1, 1983, 111-122.
7. Battisti, D. S., The dynamics and thermodynamics of a warm event in a coupled ocean-atmosphere model, J. Atmos. Sci., 1988, 45: 2889-2919.
8. Li Chongyin, Chen Yuxiang, Yuan Zhongguang, Important factor cause of El Niño event - the frequent activities of stronger cold waves in East Asia, Frontiers in Atmospheric Sciences, New York: Allenton Press INC., 1989, 156-165.
9. Huang Ronghui, Zhang Renhe, Yan Banliang, Dynamical effect of zonal wind anomaly over the tropical western Pacific on the ENSO, Science in China, Ser. D, 2001, 44: 1089-1098.
10. Cao Jiping, Cao Qinchen, Dynamics in response to the wind stress over the tropical western Pacific, Chinese J. Atmos. Sci., 2002, 26: 145-160.
11. Zhang Renhe, Huang Ronghui, Dynamical role of zonal wind stresses over the tropical Pacific on the

occurring and vanishing of El Niño, Part I: Diagnostic and theoretical analyses, Chinese J. Atmos. Sci., 1998, 22: 587-599.

12. Li Chongyin, Interaction between anomalous winter monsoon in East Asia and El Niño events, Adv. Atmos. Sci., 1990, 7: 36-46.
13. Li Chongyin, Westerly anomalies over the equatorial western Pacific and Asian winter monsoon, Proceeding of International Scientific Conference on the TOGA Programme, WCRP-91-WMO/TP, No. 717, 1995, 557-561.
14. Xu, J., Chan, J. C. L., The role of the Asian/Australian monsoon system in the onset time of El Niño events, J. Climate, 2001, 14: 418-433.
15. Zhang Renhe, Sumi, A., Kimoto, M., Impact of El Niño on the East Asian Monsoon, A diagnostic study of the 86/87 and 91/92 Events, J. Meteor. Soc. Japan, 1996, 74: 77-90.
16. Mu Mingquan, Li Chongyin, ENSO signals in interannual variability of East Asian winter monsoon, Part 1: Observed data analyses, Chinese J. Atmos. Sci., 1999, 23: 139-149.
17. Li Chongyin, Introduction of Climate Dynamics (in Chinese), Beijing: China Meteorological Press, 1995, P461.
18. Sun Boming, Li Chongyin, Relationship between the East-Asian upper trough and convection activity in the tropics, Chinese Science Bulletin (in Chinese), 1997, 42: 500-503.
19. Li Chongyin, Mu Mingquan, Numerical simulations of anomalous winter monsoon in East Asia exciting ENSO, Chinese J. Atmos. Sci., 1998, 22: 393-403.

Atmospheric Circulation Characteristics Associated with the Onset of Asian Summer Monsoon

Li Chongyin (李崇银)*, Pan Jing (潘静)

LASG, Institute of Atmospheric Physics, Chinese Academy of Sciences, Beijing

Abstract The onset of the Asian summer monsoon has been a focus in the monsoon study for many years. In this paper, we study the variability and predictability of the Asian summer monsoon onset and demonstrate that this onset is associated with specific atmospheric circulation characteristics. The outbreak of the Asian summer monsoon is found to occur first over the southwestern part of the South China Sea (SCS) and the Malay Peninsula region, and the monsoon onset is closely related to intra-seasonal oscillations in the lower atmosphere. These intra-seasonal oscillations consist of two low-frequency vortex pairs, one located to the east of the Philippines and the other over the tropical eastern Indian Ocean. Prior to the Asian summer monsoon onset, a strong low-frequency westerly emerges over the equatorial Indian Ocean and the low-frequency vortex pair develops symmetrically along the equator. The formation and evolution of these low-frequency vortices are important and serve as a good indicator for the Asian summer monsoon onset. The relationship between the northward jumps of the westerly jet over East Asia and the Asian summer monsoon onset over SCS is investigated. It is shown that the northward jump of the westerly jet occurs twice during the transition from winter to summer and these jumps are closely related to the summer monsoon development. The first northward jump (from 25°-28°N to around 30°N) occurs on 8 May on average, about 7 days ahead of the summer monsoon onset over the SCS. It is found that the reverse of meridional temperature gradient in the upper-middle troposphere (500-200 hPa) and the enhancement and northward movement of the subtropical jet in the Southern Hemispheric subtropics are responsible for the first northward jump of the westerly jet.

Keywords: the onset of Asian summer monsoon, intra-seasonal oscillation, low-frequency vortex pair, westerly jet, northward jump.

1. Introduction

The Asian monsoon is one of the most vigorous circulation systems on the planet. The

Received April 19, 2006; revised September 22, 2006.

*E-mail: lcy©lasg.jap.ac.cn

Asian summer monsoon includes the East Asian summer monsoon (EASM) system and the Indian summer monsoon (ISM) system. These two components of the Asian summer monsoon have different characteristics but are closely related. Geographically, the South China Sea (SCS, 5°-20°N, 105°-120°E) is a marginal sea located in Southeast Asia and is a junction between South and East Asia along the rim of the Asian continent. Meteorologically, the ISM and EASM systems interact over the SCS region (Tao and Chen, 1987). The Asia summer monsoon first breaks out over the SCS, next over East Asia and then over the Indian Subcontinent. To a degree, the SCS summer monsoon onset represents the Asian summer monsoon onset. The SCS summer monsoon onset not only has important circulation and climatologic features, but also salient impacts on the variability of EASM and summer precipitation in East Asia (Chen et. al., 1991; He and Luo, 1996; Liu et al., 1998; Li and Zhang, 1999; He et al., 2000). Thus, to further explore the onset characteristics has both scientific significance and practical value. The aim of this study is to investigate the possible mechanisms responsible for the Asian summer monsoon onset. Tao and Chen (1987) suggested that the onset first occurs over the SCS and this view is widely accepted by Chinese meteorologists. However, the specific region of the onset was not identified in their study due to the lack of data. Thus, a. study on the Asian summer monsoon onset from the perspective of circulation and land-sea thermal contrast is necessary. Such a study enables the examination of the physical processes, provides insight into the evolution of the monsoon system, and allows the determination of the relative contributions from the various factors.

Krishnamurti (1981) reported that the onset of ISM is associated with the vortex from the Arabian Sea. Similarly, it has been shown that a vortex pair forms over the tropical eastern Indian Ocean just prior to the SCS summer monsoon onset (Li and Qu, 1999). During the formation of the vortex pair, strong convection develops over the tropical eastern Indian Ocean (Liu et al., 2002) and the subtropical high retreats from the SCS. This is followed by the penetration of the southwesterly into the SCS region and the establishment of the SCS summer monsoon. After the onset, the vortex pair weakens and disappears (Zhang et al., 2002).

Low-frequency oscillation is a prominent feature of the Asian summer monsoon, especially the 20-day mode and the 30-60-day mode, known as the intraseasonal oscillation (ISO) (Li, 1990; Zhu and Xu, 1999; Chen et al., 2001). Song et al., (2000) showed that low-frequency oscillations play a vital role in the process of summer monsoon onset. Then, how is ISO related to the Asian summer monsoon onset? What are the mechanisms responsible for ISO to contribute to the onset? Is ISO a predictable factor for the Asian summer monsoon onset? A thorough investigation on these issues is necessary. The study of Ye et al., (1959) demonstrated that the atmospheric circulation in the Northern Hemisphere undergoes an abrupt change in June, characterized by the northward jump of the westerly jet. Tao et al. (1985) and Tao and Chen (1987) showed that mei-yu in China is closely related to this abrupt change. Is there any other northward jump of the westerly jet before June? If yes, then is this northward jump linked to the Asian summer monsoon onset? These questions are both interesting and important, which deserve a detailed study. Further, it has been suggested in some studies that the change of the land-ocean thermal contrast, arising from the heating of the Tibetan Plateau, is an important factor (He et al., 1987; Yanai et al., 1992; Li and Yanai, 1996; Wu and Zhang, 1998). The relationship between the northward jump of the westerly jet to the reversal of meridional temperature gradient in the upper troposphere also deserves investigation.

This paper is a review of our recent studies on the Asian summer monsoon onset. Our aim is to provide a better understanding of the impact of ISO and the northward jump of westerly jet on the monsoon onset and to provide a basis for future research.

2. Data

The data used in this study are mainly from the National Centers for Environmental Prediction/the National Center for Atmospheric Research (NCEP/NCAR) reanalysis. The data include the daily mean values of geopotential height, wind speed, air temperature and specific humidity from 1979 to 2003. The daily outgoing longwave radiation (OLR) data from the National Oceanic and Atmospheric Administration for 25 years (1979-2003) are also used. The horizontal resolution of these data is 2.5° latitude/longitude square. The band-filtering method (Murakami, 1984) is applied to obtain the low-frequency data.

3. Onset Dates of the Asian Summer Monsoon

Tao and Chen (1987) suggested that the Asia summer monsoon breaks out first over the SCS, next in East Asia and then over the Indian Subcontinent. However, due to the lack of data, their study did not provide detailed information on the summer monsoon onset before 10 May over the SCS. Here, we estimate the dates of the summer monsoon onset over the SCS before 10 May.

Based on the NCEP/NCAR reanalysis and the South China Sea Monsoon EXperiment (SCSMEX) data, we calculate the wind field change to determine the dates of the summer monsoon onset over the SCS before 15 May (Fig. 1). After 15 May the onset dates are similar to the results of Tao and Chen (1987): the Asian summer monsoon breaks out first over the southwestern part of the SCS and the Malay Peninsula region. The onset exhibits an abrupt change in part of SCS area during the first half of May. The Asian summer monsoon onset has been examined on basis of rainfall by Lau and Yang (1997) and of wind field by Wang et al. (2001). Their results are consistent with the results presented in Fig. 1.

4. Summer Monsoon Onset and Atmospheric Intraseasonal Oscillations

4.1 Summer monsoon onset and ISO over the SCS

Mu and Li (2000) examined the influence of ISO on the SCS summer monsoon onset and revealed that a relationship exists between the monsoon onset and ISO activities. The analyses of the SCSMEX data and the NCEP/NCAR reanalysis data show that for 1998, the onset date of the SCS summer monsoon (or the Asian summer monsoon) is about 21 May. The abrupt enhancement of the southwesterly wind and convection activities can fully describe the SCS summer monsoon onset.

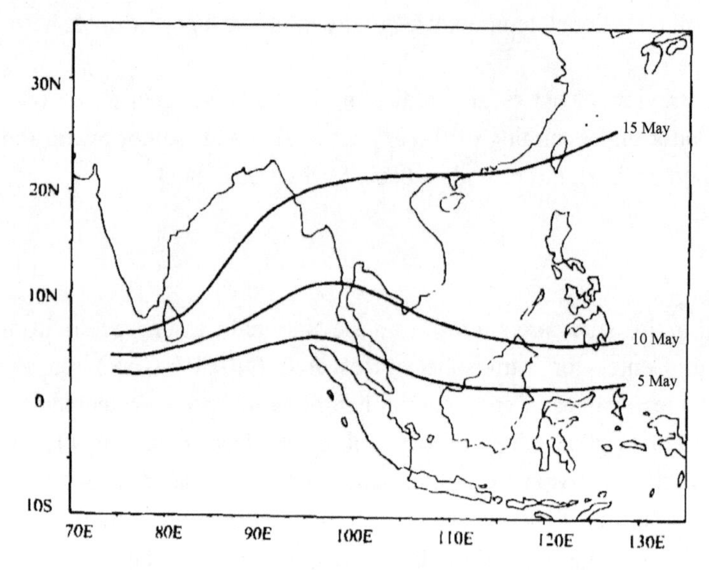

Fig. 1. The onset dates of Asian summer monsoon before May 15. (from Li and Qu, 1999)

By using the 30- to 60-day band-pass filtered wind field data, we have obtained the daily flow pattern and ISO kinetic energy distribution. The temporal variations of the mean zonal wind, ISO zonal wind and ISO kinetic energy at 850 hPa during April-June 1998 clearly show that ISO kinetic energy increased in early May and the ISO westerly wind component and the mean westerly wind all increased obviously after 20 May 1998 (not shown). This suggests that the monsoon onset in the SCS region is closely related to ISO, i.e., it follows the evident enhancement of the ISO westerly wind component (positive ISO phase) and ISO kinetic energy.

In order to unmask the relationship between the monsoon onset and ISO activities, the evolution of the ISO wind field at 850 hPa during the onset period is analyzed. As shown in Fig. 2, there was a low-frequency vortex to the east of the Philippines and another low-frequency vortex pair over the equatorial eastern Indian Ocean on 18 May. On 21-23 May 1998, the former vortex expanded into much of the SCS region and was associated with the latter vortex over the equatorial eastern Indian Ocean although the latter vortex was stationary. Therefore, the activity of low-frequency vortex to the east of the Philippines was closely related to the summer monsoon onset over the SCS. The evolution of the ISO kinetic energy at 850 hPa is shown in Fig. 3. Two ISO kinetic energy centers developed to the east of the Philippines, and these two centers expanded into the SCS region prior to the monsoon onset. On the basis of the low-frequency wind field and low-frequency kinetic energy, we suggest that the activity of the low-frequency vortex to the east of the Philippines is important to the summer monsoon onset in the SCS region.

The low-frequency vortex to the east of the Philippines prior to the summer monsoon onset

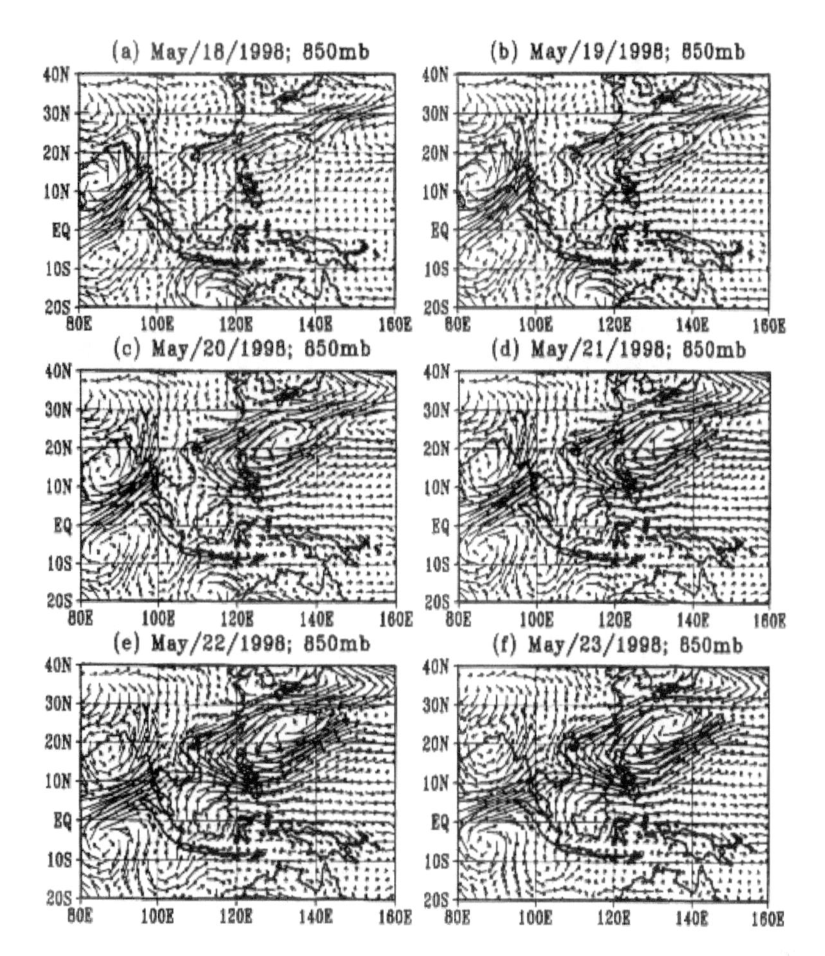

Fig. 2. 850 hPa wind fields of the ISO during summer monsoon onset period in 1998, respectively on (a) May 18, (b) May 19, (c) May 20, (d) May 21, (e) May 22 and (f) May 23. (from Mu and Li, 2000)

occurs not only in 1998, but also in other years. For example, prior to the SCS summer monsoon onset on 14 May 1980, two ISO kinetic energy centers to the east of the Philippines expanded into the SCS region (not shown) very similar the process in 1998.

The above analyses suggest that the onset of the Asian summer monsoon is closely related to the activities of atmospheric ISO at 850 hPa. In particular, the existence of a low-frequency vortex to the east of the Philippines and its expansion into the SCS region play an important role in the process of the monsoon onset.

4.2 Low frequency feature of the vortex pair over tropical eastern Indian Ocean

According to Li and Qu (1999), a vortex pair develops over the tropical eastern Indian Ocean prior to the SCS summer monsoon onset. Through the analyses of the daily stream field, we find that the vortex pair is mostly located over the region (80°-95°E, 20°S-20°N). Two cyclonic vorticity centers are found to exist symmetrically along the equator.

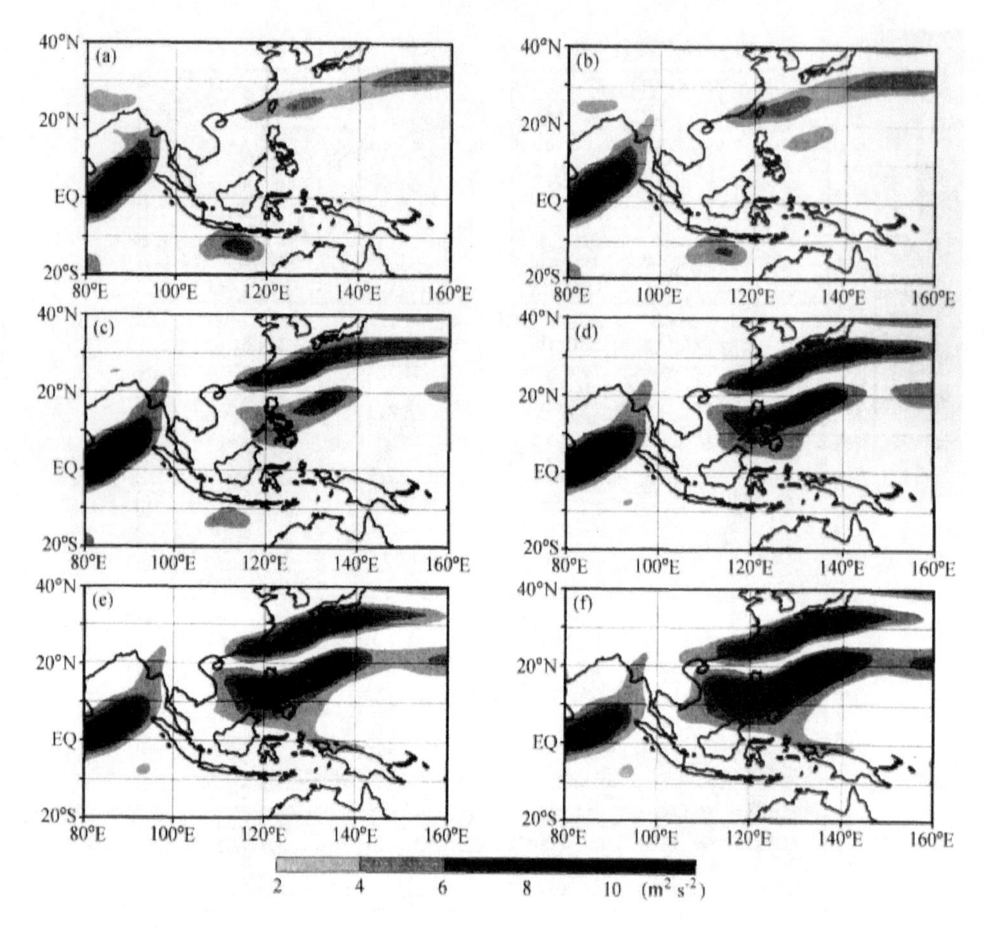

Fig. 3. Distributions of the ISO kinetic energy at 850 hPa during the onset period of the South China Sea summer monsoon in 1998, respectively on (a) May 18, (b) May 19, (c) May 20, (d) May 21, (e) May 22, and (f) May 23. (from Mu and Li, 2000)

The comparison between the occurrence date of the vortex pair over the tropical eastern Indian Ocean and the onset date of the SCS summer monsoon during 1980-2003 shows that the vortex pair indeed appears prior to the SCS summer monsoon onset. The mean occurrence date of the vortex pair is ahead of the mean monsoon onset date by about 11 days.

The vortex pair over the equatorial Indian Ocean has several prominent low-frequency features. Figure 4 presents the wavelet analysis of the area averaged zonal wind and Outgoing Longwave Radiation (OLR) over the region (80°-90°E, 5°S-5°N) from April to October for 1981, 1986 and 1994. Both the zonal wind and deep convection (as seen from OLR) display obvious low-frequency features, and the main period is 30-60 days. The patterns of the mean real wind field and the mean band-filtered (30-60 days) wind field during the pentad prior to the onset date are found to be similar (not shown). This indicates that the vortex pair over the tropical eastern Indian Ocean has a distinct feature of low-frequency oscillation. This confirms that the vortex pair is a low-frequency vortex pair.

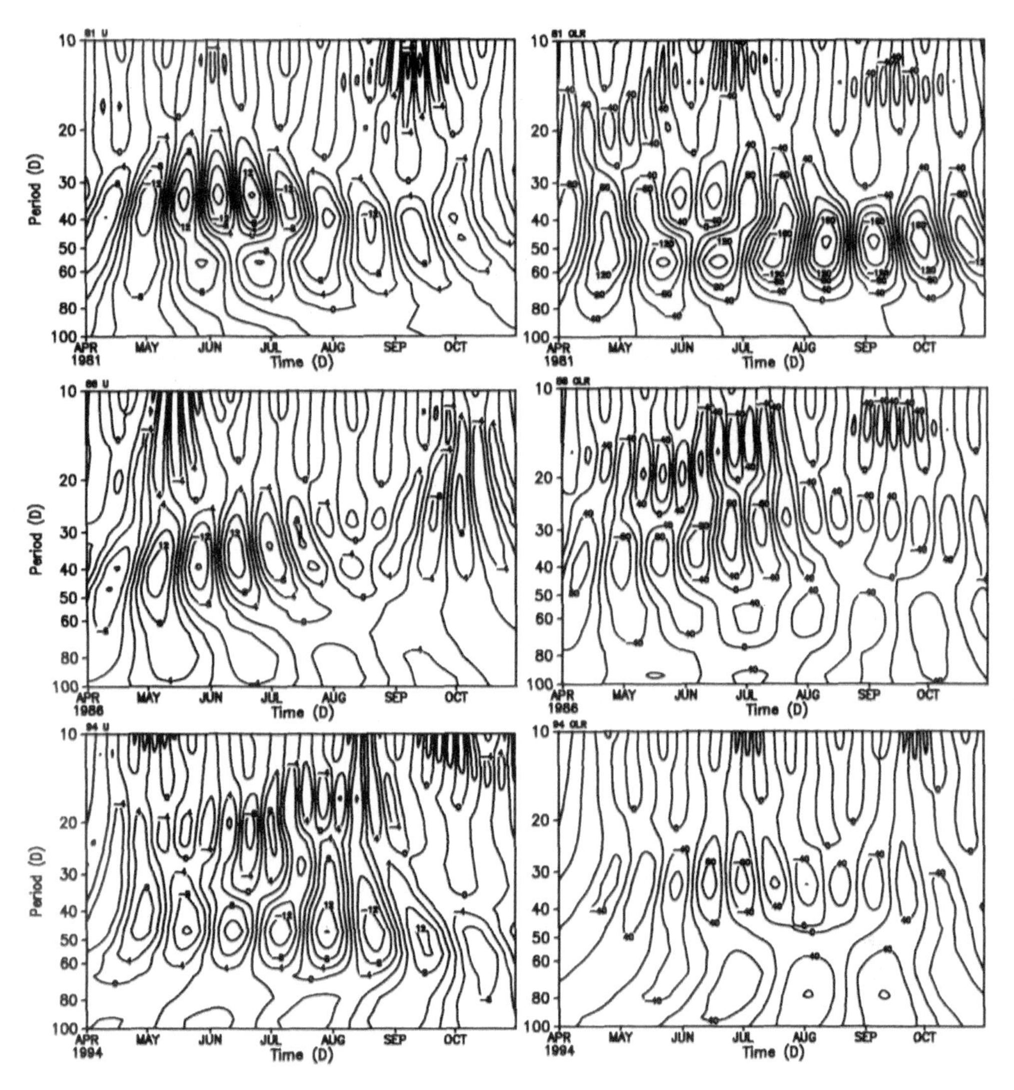

Fig. 4. The wavelet result of the area mean zonal wind of the equatorial eastern Indian Ocean (5°S-5°N, 80°-90°E) and the area mean of OLR data of the region (5°S-5°N, 80°-90°E) for 3 years (1981, 1986, 1994) from April to October, [from Pan et al. (2006), manuscript, submitted to *Adv. Atmos. Sci.*]

4.3 The summer monsoon onset and the low- frequency vortex pair

To understand its impact on the monsoon onset, the evolution of the low-frequency vortex pair is further explored. The composite method is applied here, i.e. the daily data is composed based on the onset date from 3 pentads prior to the onset to 3 pentads after the onset, roughly spanning over the life cycle of the low-frequency vortex pair.

The life cycle of the vortex pair can be divided into the pre-formation stage (about 10-15 days prior to the monsoon onset), developing stage (5-10 days prior to the monsoon onset), mature stage (0-5 days prior to the monsoon onset) and decaying stage (after the monsoon onset). The monsoon onset occurs shortly after the mature stage and after the monsoon onset, the vortex pair decays rapidly. During the life cycle of the vortex pair, the low-frequency westerly over the equatorial Indian Ocean plays a vital role.

Prior to the monsoon onset, the vortex pair appears and experiences the development and mature process. It takes about 15 days to develop into its full strength. On the -15 day (the date that is 15 days prior to the monsoon onset, Fig. 5a), there is a low-frequency anticyclone over the SCS region at 850 hPa, and the low-frequency easterlies, which comes from the anticyclone in the northern Indian Subcontinent, the SCS region and the Southern Hemisphere, prevails over the whole Indian Ocean. On the –10 day (Fig. 5b), the anticyclone in the Southern Hemisphere weakens and disappears. A low-frequency westerly band replaces the easterly over the area from eastern Africa to the equatorial eastern Indian Ocean. The westerly expands eastward, leading to the formation of the cyclonic circulation to the north of the equator with the easterlies. On the other hand, the westerly turns southward in Sumatra and forms another cyclonic to the south of the equator in the tropical eastern Indian Ocean chronologically. Thus, the vortex pair occurs and develops. However, during the formation stage, the vortex pair is not dominant and the symmetric pattern is not distinct. Just about -5 days prior to the onset of summer monsoon, the vortex pair develops and becomes mature. As shown in Fig. 5c, the low-frequency westerly expands eastward to the Malaysian Peninsula. The vortex pair develops to its full strength, the northern and southern vortex centers are located separately at (90°E, 10°N) and (90° E, 10°S). The symmetry axis of the vortex pair is located slightly to the north of the equator. Meanwhile, the northern part of the vortex pair strengthens and moves northwestward, and the cross equatorial low-frequency flow emerges. The westerly between the vortex pair then moves northward and eastward, setting a favorable environment for low-frequency southwesterly and kinetic energy to be transferred into the SCS area. The anticyclone in the SCS area then weakens, the northern part of the vortex pair moves northwestward, the westerly strengthens and turns into a southwesterly, accompanied by the cross equatorial flow. The anticyclonic flow over the SCS region is replaced by the cyclonic vortex, which represents the SCS summer monsoon onset.

To examine the movement characteristics of the vortex pair, the time-latitude section of the 850 hPa low-frequency relative vorticity along 90°E, ranging from -3 pentad prior to the onset and 3 pentad after the onset, is presented. As revealed in Fig. 6, the axis of the vortex pair is located slightly to the north of the equator. The southern part of the vortex pair forms ahead of the northern part. Consistent with the horizontal stream field, the vortex pair reaches its peak strength a pentad prior to the onset. Shortly after the monsoon onset the center of the north part of the vortex pair moves northward drastically. Ten days after the onset, the vortex pair pattern changes to the anticyclone pair pattern, which indicates the impact of the vortex pair on the onset is finished.

The relationship between the vortex pair and monsoon onset is as follows: low-frequency westerly emerges over the equatorial Indian ocean area about 15 days prior to the monsoon onset; then the vortex pair forms and develops into its mature phase; the development of the vortex pair makes the westerly stronger, transporting energy and low-frequency southwesterly into the SCS area, and hence provides a favorable condition for the monsoon onset. Shortly

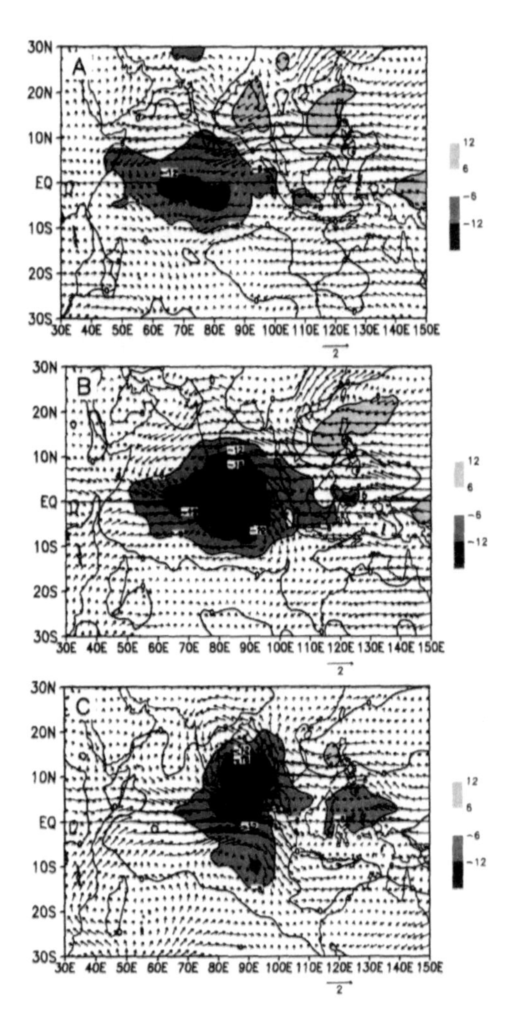

Fig. 5. Composite wind field and OLR (shaded) filtered for 30-60-day at 850 hPa prior to the monsoon onset. (a) 15 days prior to the monsoon onset; (b) 10 days prior to the monsoon onset; (c) 5 days prior to the monsoon onset.

after the monsoon onset, the vortex pair decays rapidly. In summary, the life cycle of the low-frequency vortex pair over the tropical eastern Indian Ocean leads to the SCS summer monsoon onset.

5. The Northward Jump of the East Asian Westerly Jet Related to the Monsoon Onset

5.1 First northward jump of westerly jet

Ye et al. (1959) pointed out that the circulation over the Northern Hemisphere experiences an abrupt change during June and the northward jump of westerly jet is an important feature of the abrupt change. Li and Wu (2000) suggested that there is another northward jump of the westerly jet over East Asia (referred to as the first jump hereafter) before the northward jump of the westerly jet over East Asia in June (referred to as the second jump hereafter). The first jump usually occurs prior to the SCS summer monsoon onset. It is intuitive that the first jump must be related to the monsoon onset to some extent.

The results of the data analyses are shown in Table 1. In the table, the first line is the reverse dates of meridional temperature gradient at middle-upper troposphere in the (100°-110°E) region; the second line is the dates of the first jump; the third and fourth lines are respectively the dates of sudden increases in zonal and meridional winds at 850 hPa in the SCS region, which can also be regarded as the onset dates of the SCS summer monsoon because the sudden establishment of southwesterly wind over the SCS is an important sign of the monsoon onset (Liu et al., 1998; Li and Qu, 1999); the fifth line is the onset dates of the SCS summer monsoon determined based on the index Id defined by using divergence difference between 200 hPa and 850 hPa. The first jump mentioned here means that the location of jet core jumps from south of 30°N (25°-28°N) to north of 30°N very rapidly and essentially stays to the north of 30°N. Table 1 shows that the first jump takes place before the monsoon onset. On average for the period of 1980-1999, the first jump occurs on May 8, earlier than the mean onset date (May 15) of the SCS summer monsoon by 7 days. The northward jump of the westerly jet is a manifestation of the northward contracting of the middle-high latitude system and the northward propagation of the tropical circulation system. Therefore, the first jump can be regarded as a sign of the summer monsoon onset over the SCS (Fig. 6).

Table 1. The dates (month-date) of the first northward jump of the westerly jet in East Asia and the SCS summer.

	Year									
	1980	1981	1982	1983	1984	1985	1986	1987	1988	1989
dT/dy	5-8	5-21	5-5	5-18	4-27	4-29	5-6	5-17	5-2	5-4
Jet jump	5-13	5-8	5-8	5-13	5-1	4-23	5-10	5-22	5-2	5-4
U_{850} hPa	5-15	5-9	5-22	5-16	4-27	4-28	5-10	6-9	5-19	5-16
V_{850} hPa	5-16	5-7	5-22	5-16	4-25	4-29	5-10	6-9	6-3	5-16
Monsoon onset (I_d)	5-15	5-13	5-20	6-3	4-29	4-28	5-10	6-8	5-21	5-15
Monsoon onset	5-15	5-9	5-22	5-16	4-28	4-28	5-10	6-9	5-21	5-16

In general, the timing of the first jump represents the timing of the summer monsoon onset over the SCS. For example, the first jump was early in 1985 and 1994, respectively on April 23 and April 29, the onset of the SCS summer monsoon was also early, respectively on April 28 and April 29. But in 1983, 1987, 1988 and 1999, the first jump was late, respectively on May 13, May 22, May 13 and May 19, the onset of the SCS summer monsoon was also late, respectively on May 16, June 9, May 21 and May 22.

Further, the temporal evolutions of the altitude-latitude section of mean zonal wind for two years of early monsonn onset (1985 and 1994) and two years of late monsoon onset (1983

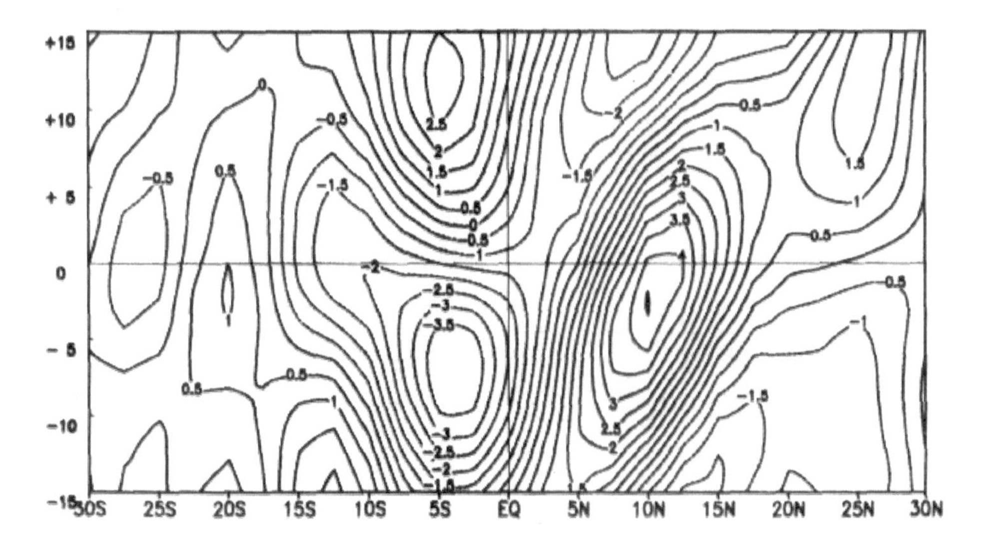

Fig. 6. Time-latitude section of composite vorticity at 850 hPa filtered for the 30-60-day period along 90°E. Unit of contour line is $0.3\times10^{-6}\ s^{-1}$.

and1997) are shown in Fig. 7 and Fig. 8, respectively. For all cases, the northward jumps of the westerly jet are evident and occur before the SCS summer monsoon onset. A jet core located in lower latitude firstly separates into the north and south cores. While the north core persists, the south core weakens and disappears. Finally, the westerly jet jumps from lower latitude to higher latitude.

The wind field at 850 hPa before the SCS summer monsoon onset is shown in Fig. 9a. The SCS region is controlled by the subtropical high ridge stretching westwards from the northwestern Pacific and the easterly/southeasterly wind are in vogue; there is a stronger northwesterly wind over the Indian Subcontinent and a westerly wind confined to the south of 5°N over the equatorial Indian Ocean region (Table 2.). The 200 hPa geopotential height indicates that a strong high pressure center is over the Indo-China peninsula, but a weak high pressure is over the southern part of the Indo-China, and a westerly wind is over the SCS and South Asia. A comparison of Figs. 9 and 10 shows that the atmospheric circulation is clearly changed with the SCS summer monsoon onset. After the onset, the subtropical high ridge has withdrawn from the SCS and the southwesterly wind, coming mainly from the equatorial Indian Ocean, dominates the SCS region. At this time, the prevailing wind over the Indian Subcontinent is still a northwesterly and the strong southwesterly wind is confined to the southern Bay of Bengal and the Indo-China peninsula. This suggests that the summer monsoon is not yet established over the Indian Subcontinent. At 200 hPa, there is a high pressure center over the northern part of the Indo-China peninsula after the summer monsoon onset, an easterly wind belt is along the 10°-15°N latitude even though the easterly jet is not quite obvious; the upper-level westerly jet over East Asia shown in Fig. 10b is evident, at about. 5 latitudes to the north of the one shown in Fig. 9b.

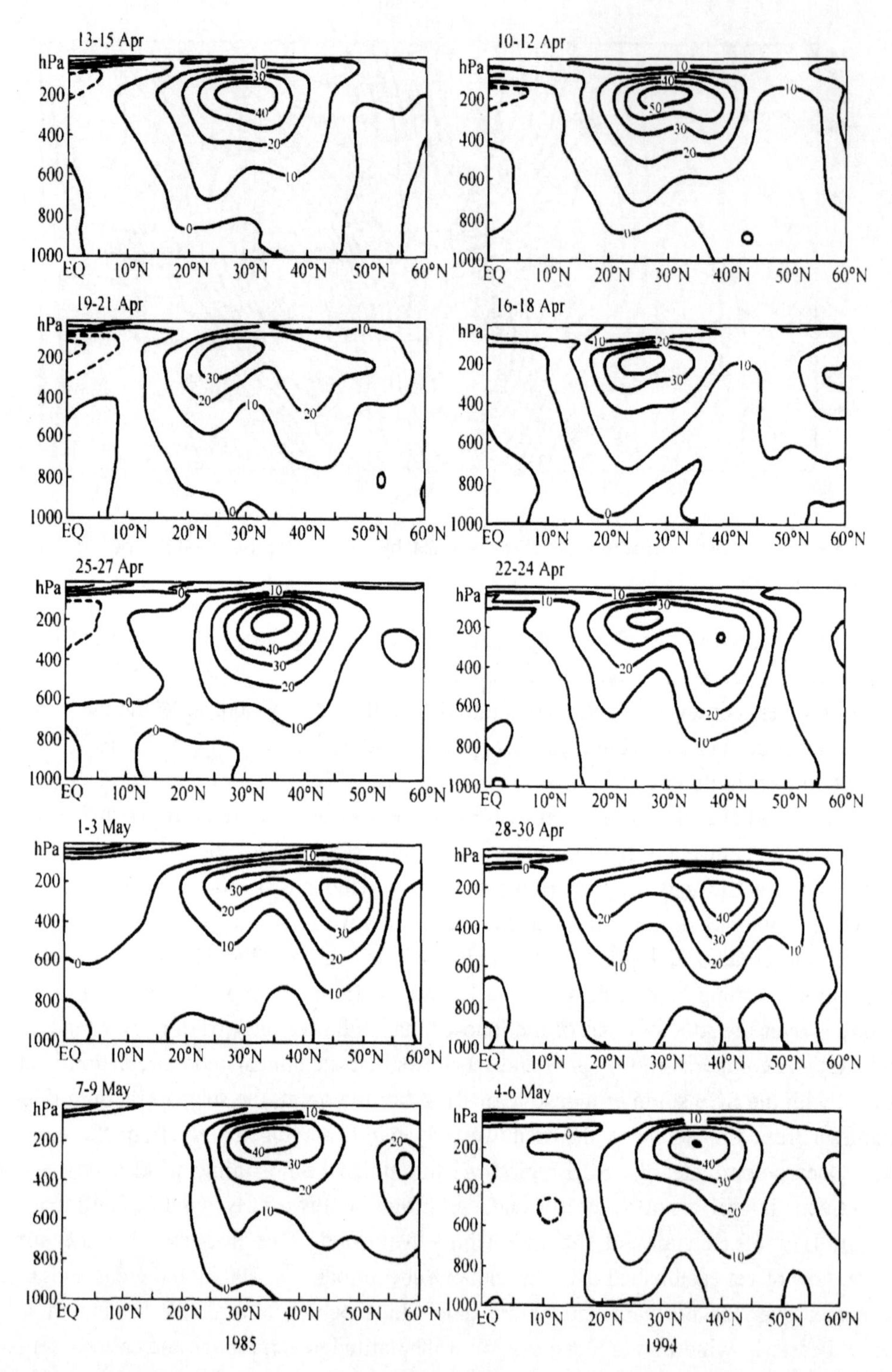

Fig. 7. Altitude-latitude sections of the averaged zonal wind (m s^{-1}) for 3 days in East Asia (105°-120°E). Left: for the case at April 13-15, April 19-21, April 25-27, May 1-3 and May 7-9 in 1985; Right: for the case at April 10-12, April 16-18, April 21-24, April 28-30 and May 4-6 in 1994. (from Li et. al., 2005)

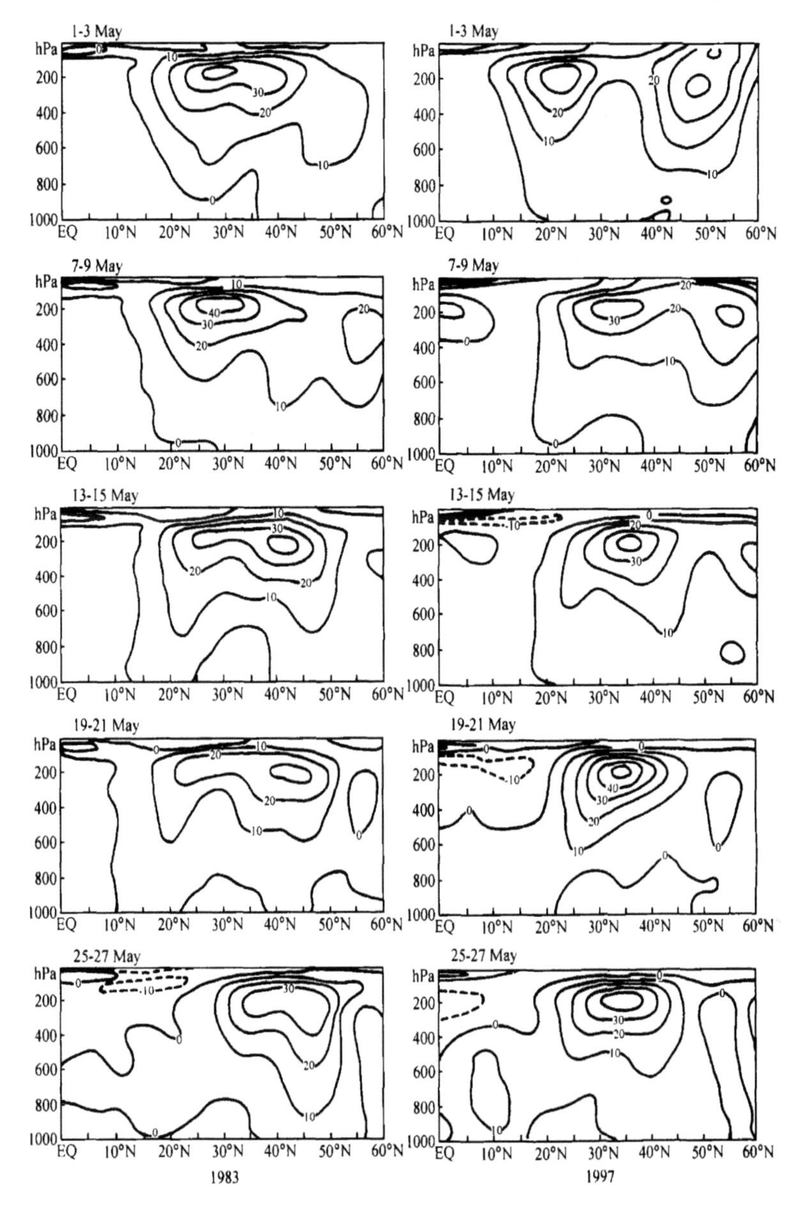

Fig. 8. Altitude-latitude sections of the averaged zonal wind (m s^{-1}) for 3 days in East Asia (105°-120°E). Left: for the case at May 1-3, May 7-9, May 13-15, May 19-21 and May 25-27 in 1983; Right: for the case at May 1-3, May 7-9, May 13-15, May 19-21 and May 25-27 in 1997. (from Li et al., 2005)

5.2 Possible mechanism responsible for the northward jump

The averaged meridional temperature gradient reversal at the upper troposphere (500 hPa-200 hPa) in South Asia is an important sign of the heating field change. The Tibetan Plateau is a heat sink in winter, the meridional temperature gradient (***dT/dy***) to the south of the Plateau is negative; but the Tibetan Plateau is a heat source in summer, there the temperature gradient is positive.

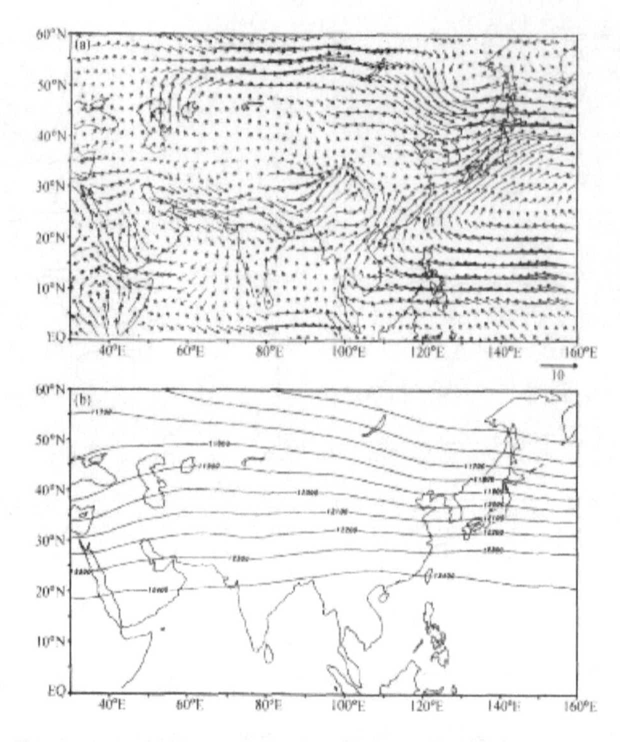

Fig. 9. (a) Composite wind fields at 850 hPa and (b) height fields at 200 hPa on 10th day before the onset of the SCS summer monsoon. (from Li et al., 2005)

Table 2. monsoon onset (from Li et al., 2005).

Year										mean
1990	1991	1992	1993	1994	1995	1996	1997	1998	1999	
5.7	4.26	5.12	5.8	4.28	5.9	4.29	5.7	5.18	4.14	5.7
5.8	4.28	5.12	4.28	4.29	5.13	5.1	5.13	5.19	5.14	5.8
5.13	6.3	5.13	5.22	4.28	5.13	5.1	5.16	5.16	5.22	5.13
5.4	5.16	5.13	5.7	4.28	5.7	4.28	5.25	5.25	5.10	5.12
5.7	5.15	5.13	5.8	4.29	5.24	5.1	5.16	5.22	5.23	5.15
5.13	5.18	5.13	5.22	4.29	5.24	5.1	5.16	5.22	5.23	5.15

It has been suggested in previous studies that the reversal of meridional temperature gradient at the upper troposphere to the south of the Tibetan Plateau is possibly related to the evolutions of the Asian summer monsoon (Yanal et al., 1992). This reversal of meridional temperature gradient can also be related to the northward jumps of the westerly jet over East Asia. To investigate why the northward jumps occur, the relationship between the meridional temperature gradient reversal in the upper troposphere and the northward jumps will be analysed using the NCEP reanalysis data for 20 years.

The first row in Table 1 shows the reversal date of the averaged meridional temperature gradient at the middle-upper troposphere (500-200 hPa) over the (100°-110°E) region, the mean date is on May 7. Comparing the reversal date of meridional temperature gradient with the date of the first jump (second line of Table 1), we find the reversal of meridional temperature gradient in the (100°-110°E) region was earlier than the first jump for most years, except 1981, 1985 and 1993. The 20-year average date of the first jump is 8 May which lags one day behind the date of the reversal of the meridional temperature gradient. The first jump is earlier than the reversal of the meridional temperature gradient in 1981, 1985 and 1993. It is found that the northward movement and enhancement of subtropical jet in the Southern Hemisphere may have played an important role in these three years. If the three years of 1981, 1985 and 1993 are excluded, the mean date of the first jump will lag 2 days behind the mean date of the reversal of the meridional temperature gradient.

According to the geostrophic adjustment theory, large-scale wind field will adjust rapidly to the pressure (temperature) field. Therefore, the change of large-scale temperature (pressure) field in Asian region, with meridional temperature gradient reversal, must lead to the adjustment of large-scale wind field. The northward jump of the westerly jet is only a form of large-scale wind field change. It can be suggested that the reversed meridional temperature gradient at the middle-upper troposphere in South Asia caused by the heating East-Asian continent is an important reason why the first northward jump of the westerly jet occurs over East Asia.

In some cases, the first jump is related to the northward propagation of subtropical jet in the Southern Hemisphere. During the seasonal transition of the atmospheric circulation from winter to summer, since strong cold air intrudes into the Southern tropics (i.e. East Asia region), the westerly jet over the Southern Hemisphere will propagate northward, and this process might be responsible for driving the first jump, as in 1989 and 1993 (not shown). The first jump on 4 May 1989 and on 1 May 1993 corresponded to the enhancement and northward jump (from 30°S to north of 25°S) of the westerly jet in the Southern Hemisphere. In other words, the first, jump can be earlier due to the forcing from the Southern Hemisphere.

In Fig. 11, we show the time-longitude sections of meridional temperature gradient (the difference between the 25°-22.5°N latitudes and the 7.5°-5°N latitude) at the middle-upper troposphere (500 hPa-200 hPa) for 1986, 1991, 1992 and 1994. It is clear that the reversal date of meridional temperature gradient at different longitude is different. The temperature gradient reversal occurs earliest in the (100°-110°E) region. From the temporal evolution of meridional temperature gradient (Fig. 11), a remarkable second reversal is found to occur in the 60°-80°E region. This second reversal of meridional temperature gradient is closely related to the second northward jump of the westerly jet over East Asia, because the mean reversal date is on 30 May, which is about 8 days earlier than the date of the second jump. This second reversal of meridional temperature gradient is probably a sign of the beginning of mei-yu in the Yangtze-Huaihe Rivers basin and the Indian summer monsoon onset.

Thus, the reversal of meridional temperature gradient at middle-upper troposphere caused by the heating in the Asian continent, particularly over the Tibetan Plateau, is a key factor for driving the northward jumps of the westerly jet over East Asia. The enhancement and

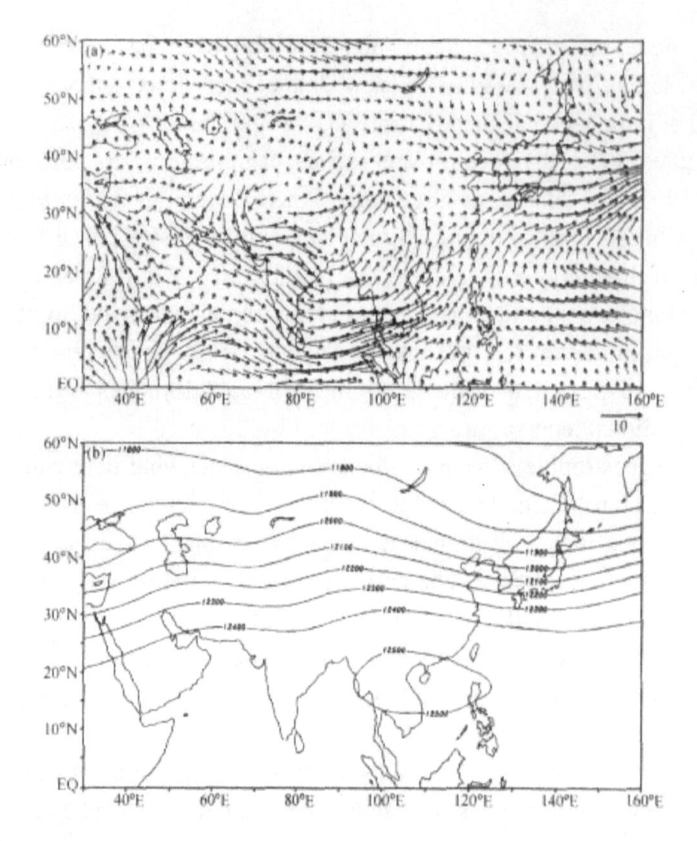

Fig. 10. (a) Composite wind fields at 850 hPa and (b) height field at 200 hPa on 10th day after the onset of the SCS summer monsoon. (from Li et al., 2005)

northward jump of the subtropical jet in the Southern Hemisphere also plays a role in the first jump, especially for early occurrence of the jump.

6. Discussion and Conclusion

In this paper, we have documented the recent progress in better understanding the atmospheric circulation evolution prior to the Asian summer monsoon onset. We emphasize that the Asian summer monsoon onset is an important event for the weather and climate in the Asian monsoon region. We have found that Asian summer nionsoon breaks out first in the southwestern part of the SCS and the Malay Peninsula region. After the summer monsoon onset over this region, the East Asian summer monsoon will step gradually from the SCS to northern China and the South Asian summer monsoon will also onset, gradually from the Bay of Bengal to northwestern India.

The onset of the Asian summer monsoon is closely related to atmospheric intraseasonal oscillations. In particular, the existence of a low-frequency vortex to the east of the Philippines and its expansion into the SCS region are of importance to the process of the SCS summer monsoon onset and hence the Asian summer monsoon onset.

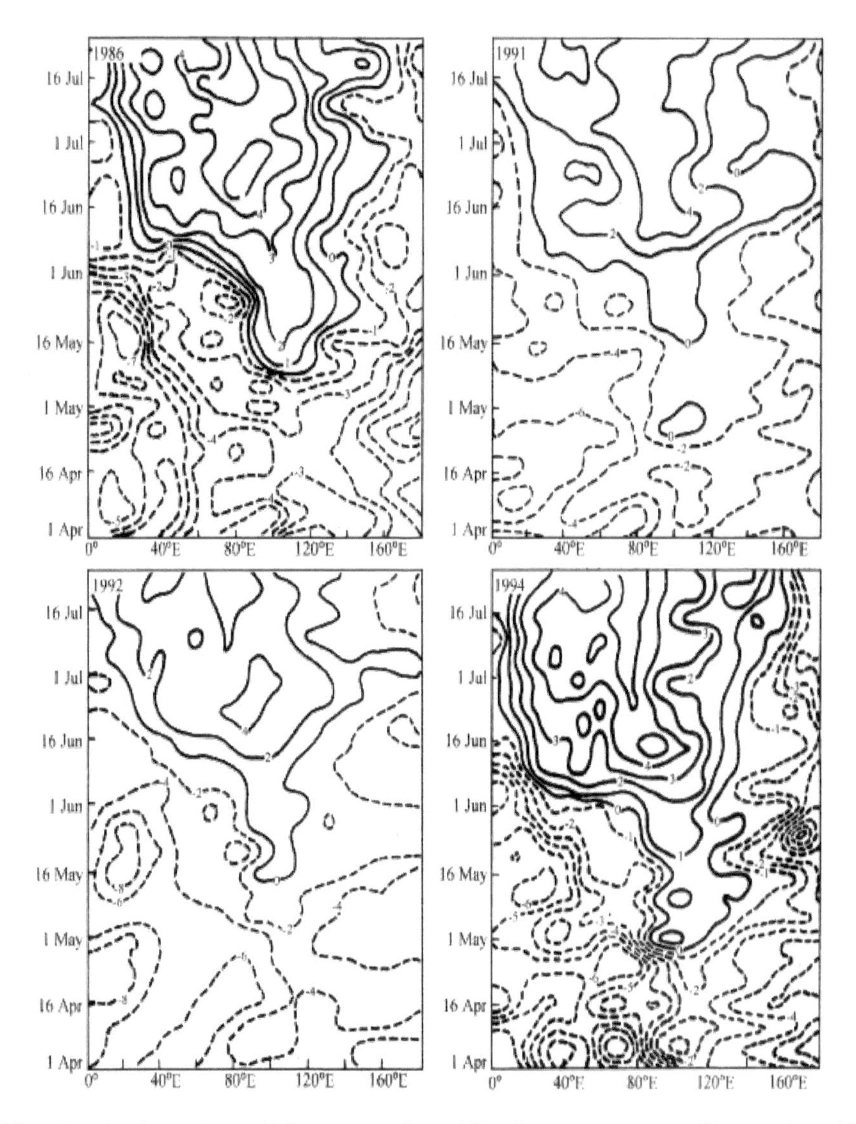

Fig. 11. Time-longitude sections of the averaged meridional temperature gradient at the middle-upper troposphere (500-200 hPa) in (a) 1986, (b) 1991, (c) 1992, and (d) 1994. The shadow is positive *dT/dy*. (from Li et al., 2005)

In addition, the SCS summer monsoon onset is found to be closely related with the vortex pair over the tropical eastern Indian Ocean, which also possesses prominent low-frequency features. This low-frequency vortex pair appears about 2 pentads prior to the monsoon onset. As the circulation develops, the stronger westerly between the vortex pair is strengthened and deep convection develops in the cyclonic circulation, which in turn promotes the westerly between the vortex pair. This is a kind of positive feedback process. With the arrival of the low-frequency westerly at the SCS area, the SCS summer monsoon onset occurs. After the monsoon onset, the vortex pair decays rapidly. The establishment of Somali jet and the enhancement of the westerly over the equatorial Indian Ocean (80°- 90°E) are important physical mechanisms responsible for the SCS summer monsoon outbreak (Li and Qu, 1999; Xu et al., 2001; Li and Wu, 2002). In terms of the low-frequency oscillation, it is suggested that the strengthening of the low-frequency westerly and the development of the low-frequency vortex pair directly lead to the SCS summer monsoon onset.

On average, the first northward jump of the westerly jet over East Asia occurs on 8 May, with the jet center shifting rapidly from 25°-28°N to the north of 30°N, while the second northward jump of the westerly jet over East Asia occurs on 7 June, with the jet center moving rapidly from 30°-33°N to the north of 35°N. Therefore, we believe the SCS summer monsoon onset is associated with the first jump, and the beginning of mei-yu in the Yangtze-Huaihe Rivers basin is related to the second jump. This phenomenon can be explained by the meridional temperature gradient reversal at middle-upper troposphere (500 hPa-200 hPa). In winter, the meridional temperature gradient is generally negative. As the atmosphere over the continent is heated rapidly in the following spring, accompanied by the convective latent release, the meridional temperature gradient becomes positive in the middle-upper troposphere over South Asia. There are two stages during the process of meridional temperature gradient reversal. The first stage occurs in the 100°- 110°E region on about 7 May ahead of the first northward jumps of the westerly jet over East Asia. The reversal of meridional temperature gradient is key to the first northward jump of the westerly jet over East Asia. Another possible mechanism for the first jump is the enhancement and northward moving of subtropical jet in the Southern Hemisphere through interaction between the two Hemispheres.

In this review, while the goal of determining the onset location of the Asian summer monsoon has been achieved, the interpretation of the predictive signals is limited due to the high-frequency variability associated with the internal dynamics of the atmosphere ocean system. Nevertheless, our capacity for the prediction of the Asian summer monsoon onset is enhanced through the improved understanding of the physical mechanisms involved. The successful prediction of the Asian summer monsoon can have profound social and economic consequences. To better understand the evolution of atmospheric circulation and the factors which give rise to the variability of the Asian summer monsoon onset remains to be a essential task for the future success in seasonal and long-term predictions.

Acknowledgments This work is supported partly by the National Natural Science Foundation of China (Grant No. 40233033) and the Chinese Academy of Sciences (KZCX3-SW-226).

References

Chen Lougxun, and Coauthors, 1991: *East Asian Summer Monsoon.* China Meteorological Press, Beijing, 362pp. (in Chinese)

Chen Longxun, Zhu Congwen, Wang Wen, and Zhang Peiqun, 2001: Analysis of the characteristics of 30-60 day low-frequency oscillation over Asia during 1998 SCSMEX. *Adv. Atmos. Sci.*, 18, 623-638.

He, H., J. W. McCinnia, Z. Song, and M, Yanai, 1987: Onset of the Asian summer monsoon in 1979 and the effect of the Tibetan Plateau. *Mon. Wea. Rev.*, 15, 1966-1995.

He Jinhai, and Luo Jingjia, 1996: Features of South China Sea Summer Monsoon Onset and Asian Summer Monsoon Establishment Sequence along with Its Individual Mechanism. *The Recent Advances in Asian Monsoon Research*, China Meteorological Press, Beijing, 74-81. (in Chinese)

He Jinhai, Xu Haiming, Zhou Bing, and Wang Lijuan, 2000: Large scale features of the SCS summer monsoon onset and its possible mechanism. *Climatic and Environmental Research*, 5, 333-344. (in Chinese)

Krishnamurti, T. N., 1981: On the onset vortex of the summer monsoon. *Mon. Wea. Rev.*, 109(2), 344-363.

Lau, K.-M., and Yang Song, 1997: Climatological and interannual variability of southeast Asian summer monsoon. *Adv. Atmos. Sci.*, 14, 141-162.

Li, C., and M. Yanai, 1996: The onset and interannual variability of the Asian summer monsoon in relation to land-sea thermal contrast. *J. Climate*, 9, 358-395.

Li Chongyin, 1990: *Low-frequency Oscillation in the Atmosphere*. China Meteorological Press, Beijing, 286pp. (in Chinese)

Li Chongyin, and Qu Xin, 1999: Atmospheric circulation evolution associated with summer monsoon onset in the South China Sea. *Chinese J. Atmos. Sci.*, 23, 311-325.

Li Chongyin, and Zhang Liping, 1999: Summer monsoon activities in the SCS and its impacts. *Chinese J. Atmos. Sci.*, 23, 378-394.

Li Chongyin and Wu Jingbo, 2000: On the onset of the South China Sea summer monsoon in 1998. *Adv. Atmos. Sci.*, 17, 193-204.

Li Chongyin, and Wu Jingbo, 2002: Important Role of the Somalian Cross-Equator flow in the onset of the South China Sea summer monsoon. *Chinese J. Atmos. Sci.*, 26(2), 185-192. (in Chinese)

Li Chongyin, Wang Jough Tai, Lin Shi Zhei, and Cho Hun Ru, 2005: The relationship between East Asian summer monsoon activity and northward jump of the upper-air westerly jet location. *Chinese J. Atmos. Sci*, 29(1), 1-20.

Liu Xia, Xie An, Ye Qian, and M. Murakami, 1998: The climate characteristics of summer monsoon onset over SCS. *Journal of Tropical Meteorology*, 14, 28-37. (in Chinese)

Liu, Y, M., J. C. L. Chen, J. Y. Mao, and G. X. Wu, 2002: The role of bay of Bengal convection in the onset of the 1998 South China Sea summer monsoon. *Mon. Wea. Rev.*, 130, 2731-2744.

Murakami, M., 1984: 30-60 day global atmospheric changes during the northern summer 1979. GARP Special Report, No. 44, 113-116.

Mu Mingquan, and Li Chongyin, 2000: On the outbreak of South China Sea summer monsoon in 1998 and activity of atmospheric intraseasonal oscillation. *Climatic and Environmental Research*, 5, 375-387. (in Chinese)

Song Yanyun, Xie An, Mao Jiangyu, and Ye Qian, 2000: Characteristics of Low-frequency Ocillations during Summer Monsoon Onset over the South China Sea. *Acta Oceanologica Sinica*, 22(2), 35-40. (in Chinese)

Tao, S. Y., and L. Chen, 1987: A review of recent research on the East Asian summer monsoon in China. *Monsoon Meteorology*, Oxford University Press, 60-92.

Tao Shiyan, Zhao Mingui, and Chen Shaoming, 1985: The relationship between the Mei-Yu period in East Asia and seasonal variation of atmospheric circulation over East Asia. *Acta Meteorologica Sinica*, 29, 119-134. (in Chinese)

Wang Anyu, and Coauthors, 2001: The climate characteristics of summer monsoon onset in Asia. *Dynamics of Atmospheric and Oceanic Circulation and Climate*, China Meteorological Press, 338-356.

Wu, G., and Y. Zhang, 1998: Tibetan Plateau forcing and the timing of the monsoon onset over South Asia and the South China Sea. *Mon. Wea. Rev.*, 126, 913-927.

Xu Haiming, He Jinhai, and Zhou Bing, 2001: Composite analysis of summer monsoon onset process over South China Sea. *Journal of Tropical Meteorology*, 17, 9-22. (in Chinese)

Yanai, M., C. Li, and Z. Song, 1992: Seasonal heating of the Tibetan Plateau and its effects on the evolutions of the Asian summer monsoon. *J. Meteor. Soc. Japan*, 70, 319-351.

Ye, D., S. Tao, and M. Li, 1959: The abrupt change of circulation over the Northern hemisphere during June and October. *The Atmosphere and the Sea in Motion*, the Rockefeller Institute Press and Oxford University Press, 249-267.

Zhang Xiuzhi, Li Jianglong, Yan Junyue, and Ding Yi hui, 2002: A study of circulation characteristics and Index of the South China Sea summer monsoon. *Climatic and Environmental Research*, 7(3), 321-331. (in Chinese)

Zhu Qiangen, and Xu Guoqiang, 1999: Features of 1998 SCS summer monsoon LFO with effect on the rainfall in Yangtze River basin in 1998. *Onset and Evolution of the SCS Monsoon and Its Interaction with the Ocean*, China Meteorological Press, 108-111.

Variation of the East Asian Monsoon and the Tropospheric Biennial Oscillation

Li ChongYin[1,2]*, Pan Jing[1], Que ZhiPing[1,3]

[1] LASG, Institute of Atmospheric Physics, Chinese Academy of Sciences, Beijing, China
[2] Meteorological College, PLA University of Science and Technology, Nanjing, China
[3] Graduate University of Chinese Academy of Sciences, Beijing, China

Abstract Study of the tropospheric biennial oscillation (TBO) has attracted significant interest since the 1980s. However, the mechanism that drives this process is still unclear. In the present study, ECMWF daily data were applied to evaluate variation of the East Asian monsoon and its relationship to the TBO. First, the general East Asian monsoon index (EAMI) was delineated on the basis of a selected area using the 850 hPa *u* and *v* components. This new index may describe not only the characteristics of summer monsoons, but also the features of winter monsoons, which is crucial to understand the transition process between summer and winter monsoons. The following analysis of EAMI shows that there is a close relationship between summer and winter monsoons. In general, strong East Asian winter monsoons are followed by strong East Asian summer monsoons, and weak winter monsoons lead to weak summer monsoons. While strong (weak) summer monsoons followed by weak (strong) winter monsoons form a kind of 2-year cycle, which may be the possible mechanism leading to the TBO over the East Asian region.
Keywords East Asian monsoons, annual variation, interaction, tropospheric biennial oscillation (TBO), dynamical mechanism.

The East Asian monsoon system is the strongest and most active monsoon system in the region. It has two important components—the winter and summer monsoon. Both of these components impose important impacts on East Asia, global atmospheric circulation and climate. Thus, research on the East Asian monsoon system is of general importance [1-3]. While most previous studies have investigated winter or summer monsoons, some have tried to evaluate their interactions in East Asia. However, the specific nature of this relationship is still unclear. Through representative case analyses, Sun et al. [4, 5] pointed out that the summer monsoon following a strong (weak) winter monsoon year tended to be weak (strong). Zhao et al. [6] found that there was an "out of phase" pattern between winter and summer in most years, and Chen et al. [7] suggested a winter monsoon index relative to a West Pacific

Received April 16, 2010; accepted July 13, 2010.
*Corresponding author (email: lcy@lasg.iap.ac.cn)

subtropical high pressure (WPSH). These authors also indicated that the WPSH shifted northward (southward) in the summer after a strong (weak) winter monsoon year in East Asia. In addition, a relationship between East Asian monsoons and the precipitation over the Yangtze River valley area in summer has been established. However, there is no unified opinion among scientists regarding the interaction between the winter and summer monsoons.

There is a biennial oscillation phenomenon in tropospheric circulation and precipitation, especially in the Asian monsoon region, in addition to the Quasi Biennial Oscillation (QBO) in the stratosphere. This phenomenon is termed the Tropospheric Biennial Oscillation (TBO) [8, 9]. Previous research has described the East Asian summer monsoon rainfall pattern as a distinct biennial oscillation feature [10, 11]. However, the dynamical mechanism of this feature is not well understood. It is possible that this process is closely related to the 2-year component of ENSO [12, 13]. Indeed, Li et al. [14] pointed out that the interaction between EAWM and ENSO may be the important mechanism to excite the TBO. In addition, the TBO has been associated with the thermodynamic conditions in the Western Pacific Ocean, especially vapor transportation [15]. Chen et al. [16] studied the mechanism of the TBO from a wider point of view, and suggested that the activity of stationary planet waves would affect the troposphere though the QBO, and then influence variation in the TBO.

In the present study, the entire East Asia monsoon system is analyzed. First, a new unified monsoon index was defined, and its variation was investigated to identify the different life cycles of the East Asian monsoon and the interaction between the summer and winter parts of the East Asian monsoon. Finally, it is suggested that the developing and shifting process of these two monsoon components in East Asia may be the important mechanism to excite the TBO over the East Asian monsoon region.

1. Data

The basic dataset employed consists of 40 years (1961—2000) of EMWCF daily averaged data archived in 23 pressure levels extending from 1000 to 1 hPa. These data had a horizontal resolution of 2.5°×2.5°. Daily anomalies in the field variables were taken as deviations from daily values of the above 40-year mean. Another dataset used was the 121-year Mei-yu dataset from 1885 to 2005 from the National Climate Center of China Meteorology Administration. This dataset was constructed by averaging the data points at five stations, Shanghai, Nanjing, Wuhu, Jiujiang and Wuhan.

2. Unified East Asian Monsoon Index

In earlier research, both the East Asian summer monsoon (EASM) and the East Asian winter monsoon (EAWM) were regarded as individual study entities. Many different indices have been applied to describe the activity and anomalies of each monsoon separately. Initially, Guo [17] defined an EASM intensity index, which could depict clearly the feature of sea level pressure (SLP) differences over the East Asian and West Pacific Oceans, and could identify the differences between land and ocean to describe the entire EASM process. Later, the South China Sea summer monsoon index was selected to represent the activity of the EASM. Different variables were applied to define this index, such as an 850-hPa wind field, outgoing longwave radiation (OLR) and the departure of the divergence between high and low tropospheric levels [18-20]. Even the zonal wind departure between tropical and subtropical areas or teleconnection wave trains were chosen as EASM indices [21, 22]. These studies concluded that it was better to apply elements of the wind field than precipitation to define a summer monsoon index in the East Asian monsoon region.

With regard to the EAWM, cold wave events often are used to denote its activity. Shi et al. [23] applied the difference of zonal belt mean SLP from 20° to 50°N, between 110° and 160°E, to define the EAWM intensity index. Other variables used as indices to describe winter monsoon activity in the East Asian region include a 500-hPa potential height, and temperature and northerly wind components [24]. However, while different variables have been used, the basic features of the EAWM have correlated with each other.

In East Asia, northeasterly winds prevail in winter, while southwesterly winds dominate in summer (Figure 1). Based on these wind conditions, we selected the southwest-northeast wind velocity as the crucial variable. The process involved averaging the domain of grid points lying within the study area 110°-122.5°E and 10°-22.5°N. Then, the normalized data were calculated as the unified EAMI:

$$\mathrm{EAMI} = V/[\sum (V)^2 / n]^{1/2},$$

where V stands for $\frac{u+v}{\sqrt{2}}$, u and v are the zonal and meridional winds, respectively, and n is the sample number (the days of the datum). Accordingly, we obtained the daily EAMI.

Figure 2 depicts the seasonal variation of daily multiyear mean EAMI data. This index shows clearly the basic circulation characteristics of both summer and winter monsoons. It is applicable not only to winter monsoons, but also to summer monsoons. This index can reflect the transition feature of winter and summer monsoons. In general, the East Asian summer monsoon onsets in late May, and retreats in October. Then, the East winter monsoon is established in mid-October and vanishes in May. Using this simple index, the basic process of the entire cycle of the East Asian monsoon system is presented. In summer, the shift from winter monsoon to summer monsoon is less abrupt than the transition from summer to winter. That is, the contour is less steep in mid-September (shift time from summer to winter), compared with that in May, which is the transition time from winter to summer.

Zeng et al. [25, 26] proposed a method to divide the global monsoon region and quantitatively estimate the atmospheric circulation transition time. The index defined in this paper is similar to their estimate.

3. Relationship Between the EAWM and EASM Anomaly

In the present study, we applied a composite analysis method to reveal the relationship

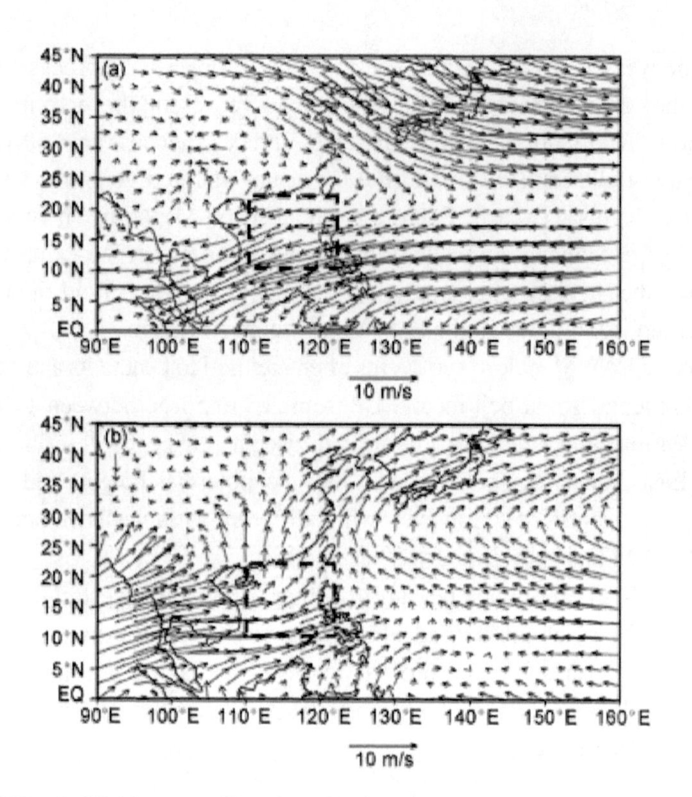

Figure 1 The 850-hPa wind field averaged between 1961-2000 in January (a) and July (b).

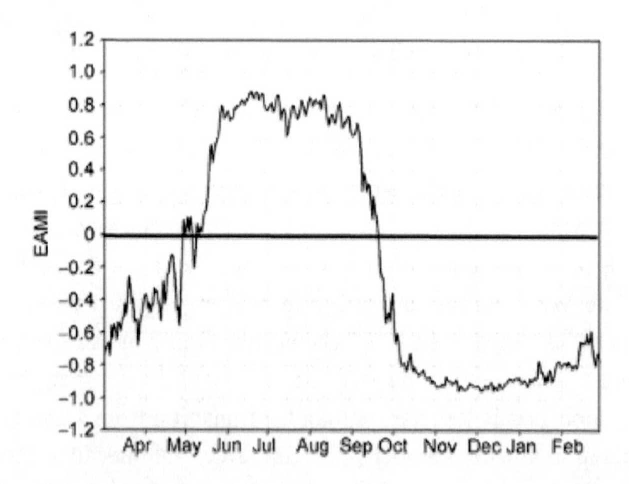

Figure 2 Seasonal variation of a multi-year (1961-2000) mean EAMI (based on a daily datum).

between the EASM and EAWM anomaly. We selected a 3-month (December, January, and February) mean EAMI to represent winter conditions, and a 3-month (June, July, and August) mean for summer conditions. Then, summer EAMI values greater than 0.8 were selected as strong summer monsoon cases. We obtained 15 strong summer monsoon cases (1967, 1968, 1972, 1974, 1975, 1981, 1982, 1984, 1985, 1990, 1991, 1993, 1994, 1997, and 1999) from 40 samples. These 15 cases were composites of strong summer monsoon cases. Then, composite results were subtracted from the multiyears mean EAMI. Finally, the strong summer monsoon anomaly of the EAMI was obtained. The variation of the summer anomaly of the EAMI is shown in Figure 3. With regard to the strong summer EAMI cases, the seasonal variation of the EAMI shows that weak winter monsoons often come after strong summer monsoon events (EAMI appears as a positive anomalous situation). Thus, this index indicates that strong summer monsoons result in weak winter monsoons.

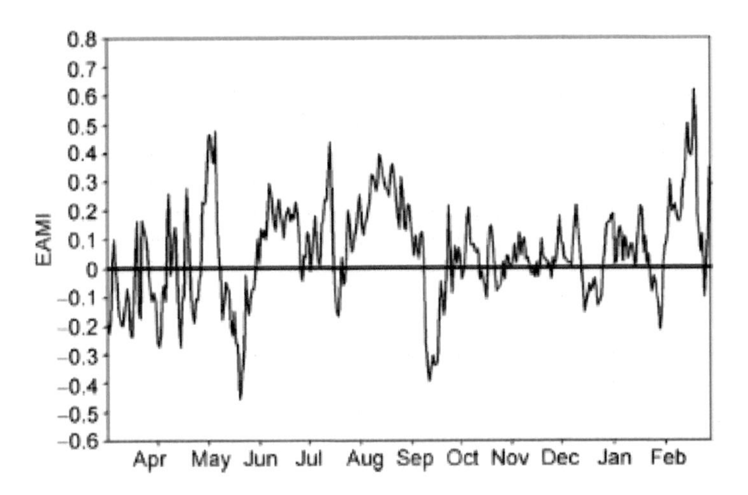

Figure 3 Variation of EAMI anomaly composite of strong summer monsoon cases.

We used the threshold value 0.66 to select weak summer monsoon years (cases where EAMI was less than 0.66). We obtained 11 weak summer cases (1964, 1966, 1969, 1971, 1980, 1983, 1988, 1989, 1995, 1996, and 1998). This method was similar to that for strong summer monsoon cases. We subtracted the multi-year mean EAMI from the composite of weak summer monsoon cases to find the weak summer monsoon anomaly. The results are shown in Figure 4. Strong winter monsoons (negative in EAMI) often follow weak summer monsoon cases, which have negative values of EAMI.

4. Variation of the East Asian Monsoon and TBO

The analysis in section 4 shows clearly that weak winter monsoon events often follow strong summer monsoon cases, while strong winter monsoon events follow weak summer monsoon cases. In other words, there is an out-ofphase pattern between summer and the following winter monsoon cases. In this section, we discuss the conditions that relate the following summer monsoon to the preceding anomalous winter monsoon. First, strong cases were selected. Then, the EAMI anomaly was achieved by subtracting the composite result from the climatological mean EAMI. The strong winter monsoon cases included 13 years (1961, 1966, 1969, 1970, 1971, 1973, 1975, 1981, 1985, 1988, 1995, 1998, and 1999). The seasonal variation of the EAMI anomaly in relation to strong winter cases can be seen in

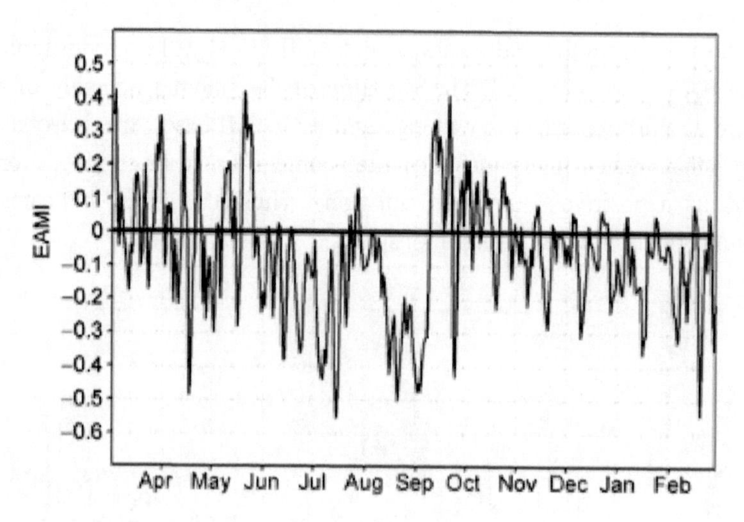

Figure 4 Variation of EAMI anomaly composite of weak summer monsoon cases.

Figure 5. After strong winter cases (negative EAMI in winter), the following summer monsoon tends to be strong. That is, the EAMI anomaly is positive in summer. With weak winter monsoon cases, the following summer monsoon is weak (negative EAMI in summer) (figure not shown). In winter monsoon composite cases, the winter monsoon and summer monsoon have an in-phase pattern. Strong (weak) winter monsoons often activate strong (weak) summer monsoons. These results are supported by previous studies [24].

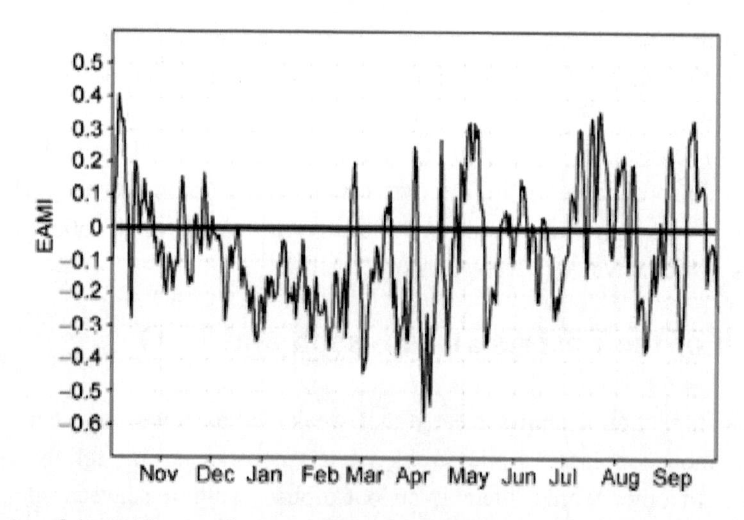

Figure 5 Variation of the EAMI anomaly composite of strong winter monsoon cases.

In the East and South Asian monsoon regions, there is a prominent periodic oscillation phenomenon extending from the troposphere to the stratosphere. This phenomenon has a period of about 2 years. In the stratosphere, this kind of oscillation is called the Quasi-Biennial Oscillation (QBO), and in the troposphere, it is called the TBO. We obtained TBO characteristics from the Mei-yu in the Yangtze River valley. Figure 6 presents the wavelet

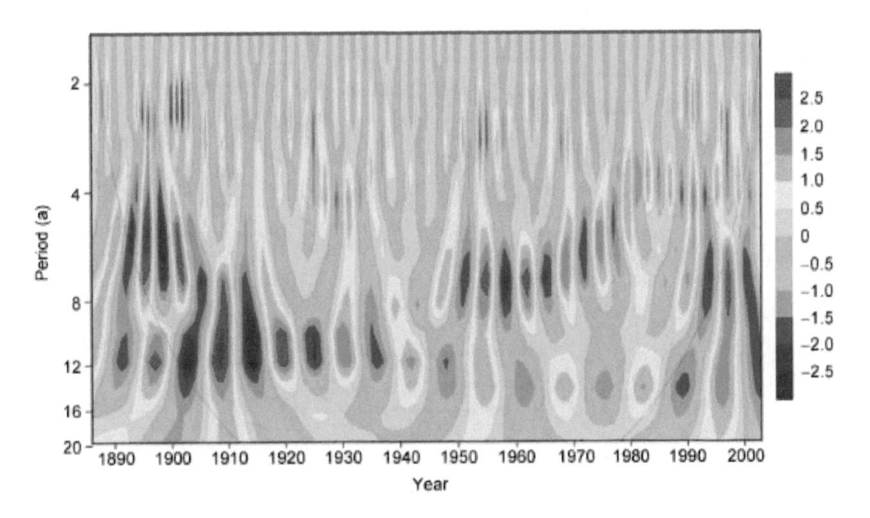

Figure 6 Wavelet results of Meiyu in the Yangtze River valley.

results of the Meiyu in the Yangtze River valley. In addition, there is high variability in Meiyu precipitation. The distinct periods were, respectively, 11, 5-8 and 2 years. Different dynamical mechanisms were used to explain the distinct activity periods. The 11-year oscillation was affected by solar activity [27] and the Pacific Decadal Oscillation (PDO) of SSTA in the North Pacific [28]. The 5-8 year oscillation was influenced by anomalous atmospheric circulation induced by the ENSO cycle [29]. The 2-year oscillation was associated clearly with some important atmospheric circulation process, especially the anomalous condition of the East Asian summer monsoon. Weak summer monsoon events bring more Mei-yu precipitation, while strong summer monsoon events bring less precipitation in the Yangtze River valley area.

The TBO phenomenon of the Meiyu precipitation in the Yangtze River valley is closely related to the East Asian summer monsoon activity in different aspects [11, 15]. The static aspects have been evaluated extensively in the literature and are not discussed in this paper. Furthermore, this paper focuses on the compact monsoon cycle that exists between a strong and weak, and a winter and summer, monsoon. The schematic diagram in Figure 7 illustrates this concept.

This monsoon cycle (Figure 7) describes the connection between winter and summer monsoons. It represents a 2-year cycle, which just fits the oscillation period of the TBO. Thus, a strong summer monsoon will induce a weak winter monsoon, and a weak winter monsoon leads to a weak summer monsoon, which is followed by a strong winter monsoon. Thus, the strong winter monsoon usually precedes a strong summer monsoon, which indicates the beginning of the monsoon cycle. The closed relationship between summer monsoons and Meiyu precipitation allows for identification of the East Asian monsoon cycle. In addition, it appears that the interaction between summer and winter monsoons may be an important dynamic mechanism that leads to the TBO.

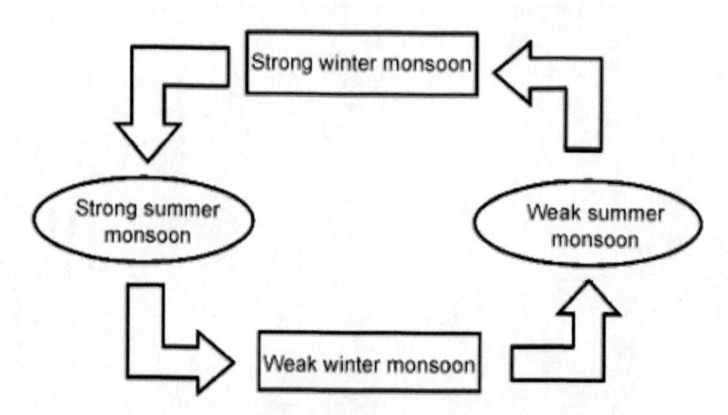

Figure 7 Schematic diagram of anomalous East Asia monsoon interaction cycle.

5. Discussion and Conclusions

To study and compare the transition features and interactions of winter and summer monsoons, a unified EAMI was defined. This index described not only the characteristics of the winter but also of the summer monsoon. This index also can be used to identify the transition features between winter and summer monsoons. The results are in agreement with the currently existing single summer or winter monsoon indices, and highlight the advantages of a simple unified EAMI.

In the previous section of this paper, the composite method was applied to different anomalous circumstances and the different interactive relationships between winter and summer monsoons. The results showed that strong winter monsoons lead to strong summer monsoons, while weak winter monsoons lead to weak summer monsoons. With summer cases, weak summer monsoons result in strong winter monsoons, while strong summer cases result in strong winter events. On the basis of these general rules, a monsoon cycle can be constructed over a 2-year period, which includes the TBO period.

The TBO is an important oscillation mode of atmospheric circulation in the East Asian region. The compact cycle of monsoons, which describes the interaction process between strong and weak, and winter and summer monsoons, may be an important mechanism to excite the TBO.

The present study focuses on East Asian monsoon activity, and discusses a possible internal mechanism for the TBO. This internal mechanism does not contradict the external possible TBO mechanism mentioned in section 1, but can be seen as a supplementary explanation to the process. From this study, we achieve further detailed knowledge about the TBO. Further studies may investigate the dynamic mechanism of the atmospheric system based on a combination of internal and external factors.

This work was supported by the National Natural Science Foundation of China (U0833602), and the National Basic Research Program of China (2007CB411805 and 2010CB950400).

References

1 Chen L X, Zhu Q G, Luo H B, et al. East Asian Monsoon (in Chinese). Beijing: China Meteorological Press, 1991. 362.

2 Institute of Atmospheric Physics, Chinese Academy of Sciences. East Asian Monsoon and Torrential Rain in China (in Chinese). Beijing: China Meteorological Press, 1998. 503.

3 Chang C P. East Asian Monsoon. Singapore: World Scientific Publisher, 2004. 416.

4 Sun S Q, Sun B M. The relationship between the anomalous winter monsoon circulation over East Asia and summer drought/flooding in the Yangtze and Huaihe River valley (in Chinese). Acta Meteorol Sin, 1995, 57: 513-522.

5 Chen J, Sun S Q. Eastern Asian winter monsoon anomaly and variation of global circulation, Part I: A comparison study on strong and weak winter monsoon (in Chinese). Chin J Atmos Sci, 1999, 23: 101-111

6 Zhao Z G. The Summer Drought/Flooding in China and Its Environment Field (in Chinese). Beijing: China Meteorological Press, 1999. 28-37.

7 Chen W, Graf H F, Huang R H. Interannual variability of East Asian winter monsoon and its relation to the summer monsoon. Adv Atmos Sci, 2000, 17: 48-60.

8 Meehl G. The annual cycle and interannual variability in the tropical Pacific and Indian Ocean region. Mon Weather Rev, 1987, 115: 27-50.

9 Ropelewski C F, Halpert M S, Wang X. Observed tropospherical biennial variability and its relationship to the Southern Oscillation. J Clim, 1992, 5: 594-614.

10 Miao J H, Lau K M. Interannual variability of East Asian monsoon rainfall. Quart J Appl Meteor, 1990, 1: 377-382.

11 Yin B Y, Wang L Y, Huang R H. The Quasi-Biennial oscillation of the East Asian summer monsoon rainfall and its possible mechanism. In: Huang R H, ed. Disastrous Climate (in Chinese). Beijing: China Meteorological Press, 1996. 196-205.

12 Lau K M, Shen P J. Annual cycle, quasi-biennial oscillation and Southern Oscillation in global precipitation. J Geophys Res, 1988, 93: 10975-10988.

13 Rasmussen E M, Wang X, Ropelewski C F. The biennial component of ENSO variability. J Mar Syst, 1990, 1: 71-90.

14 Li C Y, Sun S Q, Mu M Q. Origin of the TBO—Interaction between anomalous East-Asian winter monsoon and ENSO cycle. Adv Atmos Sci, 2001, 18: 554-566.

15 Huang R H, Chen J L, Huang G, et al. The Quasi-Biennial oscillation of summer monsoon rainfall in China and its cause (in Chinese). Chin J Atmos Sci, 2006, 30: 545-560.

16 Chen W, Li T. Modulation of northern hemisphere wintertime stationary planetary wave activity: East Asian climate relationships by the Quasi-Biennial Oscillation. J Geophys Res, 2007, 112: D20120, doi: 10.1029/2007JD008611.

17 Guo Q Y. The summer monsoon intensity index in East Asia and its variation (in Chinese). Acta Geogr Sin, 1983, 3: 207-217.

18 Liang J Y, Wu S S, You J P. The research on variations of onset of the SCS summer monsoon and its intensity (in Chinese). J Trop Meteorol, 1999, 15: 97-105.

19 He J H, Zhu Q G, Murakami M. TBB data revealed features of Asian-Australian monsoon seasonal transition and Asian summer monsoon establishment (in Chinese). J Trop Meteorol, 1996, 12: 34-42.

20 Li C Y, Zhang L P. The characteristics of South China Sea summer monsoon and its intensity index (in Chinese). Prog Nat Sci, 1999, 9: 536-541.

21 Zhang Q Y, Tao S Y, Chen L T. The inter-annual variability of East Asian summer monsoon indices and

its association with the pattern of general circulation over East Asia (in Chinese). Acta Meteorol Sin, 2003, 61: 559-568.
22 Huang G. Study of the relationship between summer monsoon circulation and anomaly index and the climatic variations in East Asia (in Chinese). Quart J Appl Meteorol, 1999, 10: 61-69.
23 Shi N, Zhu Q G, Wu B G. The east Asian summer monsoon in relation to summer large scale weather-climate anomaly in China for the last 40 years (in Chinese). Chin J Atmos Sci, 1996, 20: 575-583.
24 Yan H M, Duan W, Xiao Z N. A study on relation between Asian winter monsoon and climatic change during raining season in China (in Chinese). J Trop Meteorol, 2003, 19: 367-376.
25 Zeng Q C, Zhang B L. On the seasonal variation of atmospheric general circulation and the monsoon (in Chinese). Chin J Atmos Sci, 1998, 22: 805-813.
26 Li J P, Zeng Q C. A new monsoon index and the geographical distribution of the global monsoons. Adv Atoms Sci, 2003, 20: 299-302.
27 Pan J, Li C Y, Gu W. The possible impact of solar activity on summer rainfall anomaly in eastern China (in Chinese). Sci Meteorol Sin, 2010, 30: 6-13.
28 Li C Y, Xian P. Atmospheric anomalies related to interdecadal variability of SST in the North Pacific. Adv Atmos Sci, 2003, 20: 859-874.
29 Zong H F, Zhang Q Y, Chen L T. Temporal and spatial variation of precipitation in eastern China during the Meiyu period and their relationships with circulation and sea surface temperature (in Chinese). Chin J Atmos Sci, 2006, 30: 1189-1197.

Comparison of the Impact of two Types of El Niño on Tropical Cyclone Genesis over the South China Sea

Xin Wang[a], Wen Zhou[b*], Chongyin Li[c], Dongxiao Wang[a]

[a] State Key Laboratory of Tropical Oceanography, South China Sea Institute of Oceanology, Chinese Academy of Sciences, Guangzhou, China
[b] Guy Carpenter Asia-Pacific Climate Impact Centre, School of Energy and Environment, City University of Hong Kong, Hong Kong, China
[c] LASG, Institute of Atmospheric Physics, Chinese Academy of Sciences, Beijing, China

Abstract This study examines the impact of the cold tongue (CT) El Niño and the warm pool (WP) El Niño on tropical cyclone (TC) genesis over the South China Sea (SCS) from 1965 to 2010. During Sept-Oct-Nov (SON), the TC genesis exhibits clear interannual variability. SON TC genesis is significantly related with the WP Niño index, but not with the CT Niño index. It is found that in the past two decades the SCS TC genesis varies coherently with the WP Niño index on a timescale of approximately 4 years, which is in accordance with the recent increase in WP El Niño events. The distinctly different atmospheric teleconnection patterns related to the CT and WP El Niño over the SCS are responsible for these relationships. CT El Niño can induce anticyclone anomalies over the SCS and the western tropical Pacific WP. However, WP El Niño can result in dipolar patterns with anticyclone anomalies over the SCS and cyclone anomalies over the western tropical Pacific WP at low- and mid-level. These WP El Niño-related large-scale circulation anomalies enlarge the low-level northerlies over the SCS. This in turn enhances the vertical wind shear and thus suppresses TC genesis over the SCS.
keywords warm pool and cold tongue El Niño; tropical cyclone; South China Sea.

1. Introduction

The South China Sea (SCS), located in Southeast Asia roughly between the equator and 22°N and from 105-120°E, is a region with a high frequency of tropical cyclone (TC) genesis (Camargo *et al.*, 2007; Wang *et al.*, 2007; Yan *et al.*, 2012). Many previous studies have demonstrated that the El Niño-Southern Oscillation (ENSO) can influence the interannual variability of TC genesis over the SCS and western North Pacific to some extent (Li, 1988; Lee *et al.*, 2006; Wang *et al.*, 2007; Li and Zhou, 2012; Li *et al.*, 2012; Wang and Wang, 2013a). Wang and Chan (2002) suggested that during summer and fall between strong El

Received July 11, 2012; revised September 4, 2013; accepted October 9, 2013.
*Correspondence to: Dr. W. Zhou, Guy Carpenter Asia-Pacific Climate Impact Centre, School of Energy and Environment, City University of Hong Kong, Kowloon, Hong Kong, China. E-mail: wenzhou@cityu.edu.hk.

Niño and La Niña, the frequency of TC formation is remarkable changed in the southeast and northwest quadrants of the western North Pacific. Zuki and Lupo (2008) further elucidated that more (less) TC activity over Malaysia and regions around (in the southern SCS) during November and December is observed in La Niña (El Niño) years during 1960-2006. Besides ENSO, the TC geneses over the SCS are suggested to closely connect with the East Asian monsoon on different time scales (Wang *et al.*, 2007; Zhou and Chan, 2007; Yuan *et al.*, 2008; Feng *et al.*, 2011; Wang *et al.*, 2012; Li *et al.*, 2013).

In recent years, a number of studies have revealed a new type of El Niño, which is different from the canonical El Niño in terms of the location of the maximum sea surface temperature (SST) anomalies. Usually referred to as 'Dateline El Niño' (Larkin and Harrison, 2005), 'El Niño Modoki' (Ashok *et al.*, 2007), 'Central Pacific El Niño' (Yu and Kao, 2007) or 'warm pool El Niño' (Kug *et al.*, 2009), this new type of El Niño has significantly different El Niño-induced tropical-midlatitude teleconnections (Larkin and Harrison, 2005; Ashok *et al.*, 2007; Kao and Yu, 2009; Weng *et al.*, 2009; Yu and Kim, 2010; Wang and Wang, 2013b). In this study, the name of warm pool (WP) El Niño is used. Interest in the impact of the new type of El Niño on TC activity has been rapidly increasing (Kim *et al.*, 2009; Lee *et al.*, 2010; Chen and Tam, 2010; Kim *et al.*, 2012a; Wang *et al.*, 2013; Wang and Wang, 2013a). For example, Kim *et al.* (2009) suggested that compared to eastern Pacific warming events, central Pacific warming events are associated with an above-normal-level TC frequency and increasing landfall potential along the gulf of Mexico coast and Central America. Lee *et al.* (2010) further suggested that the Atlantic WP could readily explain the increased tropical storm frequency even in central Pacific warming years, such as 1969 and 2004. The changes in TC frequency are associated with the modulation of vertical wind shear by teleconnection pattern (Kim *et al.*, 2009) or local force (Lee *et al.*, 2010). Using a simple baroclinic model, Chen and Tam (2010) examined the significantly positive relationship between TC frequency and ENSO Modoki index, suggesting that the anomalous circulation responses to heating play roles in distinct modulations of two kinds of El Niño on TC frequency. On the basis of the results from the empirical orthogonal function of summertime TC genesis, Kim *et al.* (2010) showed the dipole oscillation between the Philippine Sea and the northern South China Sea (PS-nSCS oscillation). They suggested that the decadal modulation of ENSO Modoki could be responsible for the decadal PS-nSCS oscillation. The weaker equatorial SST gradient with central Pacific warming and western Pacific cooling associated with WP El Niño has a positive effect on the weaker trade wind in the western North Pacific. These associated large-scales responses result in the positive phase of the decadal PS-nSCS oscillation. Recently, Kim *et al.* (2011) showed the different impacts of the two kinds of El Niño on TC activity over the North Pacific by the different modulation of thermodynamic factors and large-scale circulations. Compared to the eastern Pacific warming years, the TC activity in central Pacific warming years is shifted to the west and is extended through the northwestern part of the western Pacific.

Motivated by these previous results, in this study we attempt to further compare the impacts, and investigate the physical mechanisms of the two types of El Niño on TC genesis over the

SCS. The article is organized as follows. Section 2 introduces datasets and indices used in the paper. Section 3 analyses the relationships between TC genesis over the SCS and the two types of El Niño. The large-scale environment associated with two types of El Niño is investigated in Section 4. Summary and discussion are finally given in Section 5.

2. Datasets and Methodology

The TC datasets from the website of the International Best Track Archive for Climate Stewardship (IBTrACS, v03r03) Project (http: //www.ncdc.noaa/gov/oa/ibtracs/) are used in this study. We only select TC (maximum sustained wind speed larger than 34 knots) genesis over the SCS, which is the position first recorded by the IBTrACS dataset that falls within the SCS domain (0-22°N, 105° -120°E). Although these TC data are from as early as the 19th century, the data reliability is rather low in the pre-satellite era. To avoid this data problem, only TC data with weather satellite observations during 1965—2010 is used which is similar to Wang and Chan (2002) and Kim *et al.* (2010). We also use the monthly horizontal wind and geopotential height data from the National Centers for Environmental Prediction/National Center for Atmospheric Research (NCEP/NCAR) reanalysis datasets (Kistler *et al.*, 2001), and UK Met Office/Hadley Centre's Sea Ice and SST (HadISST) (Rayner *et al.*, 2003) for the same period.

The two Niño indices proposed by Ren and Jin (2011) are used to represent the two types of El Niño, WP, and cold tongue (CT) El Niño, investigated in this study. Theses new indices can separately identify the two types of El Niño event, and capture their SST characteristics. More details about these indices can be found in Ren and Jin (2011). On the basis of the two Niño indices suggested by Ren and Jin (2011), the five strongest CT and WP El Niño events are selected for composite analysis. The strongest WP El Niño events are 1968/1969, 1990/1991, 1994/1995, 2002/2003, 2004/2005, and 2009/2010, and the strongest CT El Niño events are 1965/1966, 1972/1973, 1982/1983, 1991/1992, 1997/1998, and 2006/2007. The selected WP El Niño events are consistent with the El Niño Modoki suggested by Ashok *et al.* (2007) and the Central-Pacific El Niño suggested by Yu and Kim (2010). It is noted that the SON mean TC formation numbers over the SCS during the CT and WP El Niño developing phases are 1 and 0.6, respectively. Because of the higher correlation between the CT Niño index and the Niño3 index (Ren and Jin, 2011), the six strongest La Niña events are selected according to the CT Niño index, which are 1970/1971, 1973/1974, 1975/1976, 1988/1989, 1998/1999, 1999/2000, and 2007/2008.

3. Relationships Between TC Genesis and Warm Pool/Cold Tongue Niño Index

The June—November period is generally considered to be the active TC season over the SCS (Lee *et al.*, 2006; Wang *et al.*, 2007; Zuki and Lupo, 2008). The location of TC genesis in the SCS exhibits remarkable seasonal change, usually being north of 12°N in summer with southwest monsoon period and south of 18°N in autumn and winter with northeast monsoon period (Lee *et al.*, 2006; Wang *et al.*, 2007). Therefore, we analysed the variation in TC genesis in summer (JJA) and autumn (SON). Besides the location shift, the temporal variability in the two seasons is found to be distinctly different (Figure 1). From Figure 1(a), the interannual variability of JJA TC genesis does not appear to be remarkable, while the interdecadal variation is relatively evident, with less TC activity between the late-1970s and mid-1990s, and more afterward. The spectrum analysis results demonstrate that the interannual variation of JJA TC genesis is not significant and the interdecadal variation is clear to some extent (Figure 1(c)). The interdecadal variation of TC activity during boreal summer has been investigated in previous studies (Chen *et al.*, 2012; Kim *et al.*, 2012b). In contrast, the interannual variability of TC genesis in SON is robust, with significant oscillations of about 4-year periodicity (Figures 1(b) and (d)). This significant interannual variability may be associated with ENSO. In the followings, we will investigate and compare connections of interannual variability of SON TC genesis with the two types of El Niño.

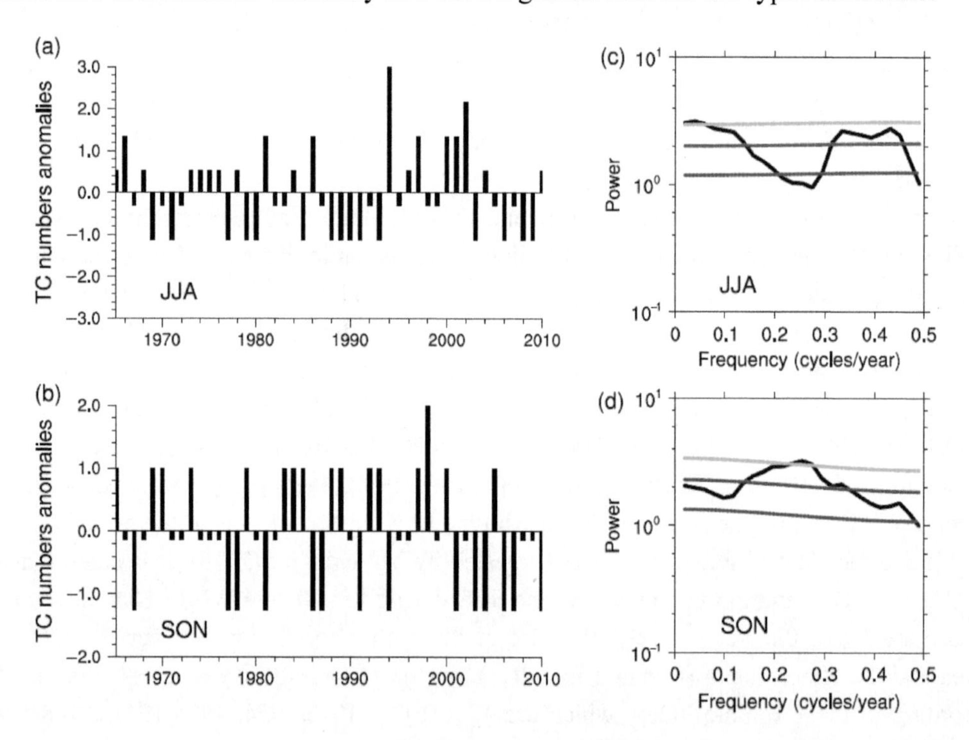

Figure 1. The time series of JJA-mean (a) and SON-mean (b) standardized anomalies of TC genesis over the SCS during 1965-2004. The spectrum analysis for JJA-mean and SON-mean TC genesis are (c) and (d), respectively. The red, blue, and green lines in (c) and (d) represent Markov red noise spectrum, 10% confidence level, and 90% confidence level.

The most evident difference between the two types of El Niño relates to the fact that the warming SST anomalies of the WP El Niño are located in the central tropical Pacific during the ENSO developing phases and peaks (Yeh *et al.*, 2009; Yu and Kim, 2010). In addition, the developing periods of the two types of El Niño are different, which the CT El Niño is usually developed in boreal spring, but the WP El Niño usually occurs in boreal autumn (Yu and Kim, 2010). Similar to CT El Niño, the WP El Niño peaks in boreal winter (Dec-Jan-Feb). Yu and Kim (2010) argued that some WP El Niño events are followed by the warm or cold phase of the CT El Niño, which is determined by the depth of the equatorial thermocline.

In this study, we focus on the relationships between the two types of El Niño and SCS TC genesis during the ENSO developing year. Table 1 gives the simultaneous relationships of the TC genesis with the CT and WP Niño indices during boreal summer (Jun-Jul-Aug) and autumn (Sept-Oct-Nov). The relationships between TC genesis and the CT and WP Niño indices in JJA are not significant. This weak relationship with the CT Niño index is consistent with the results of Wang and Chan (2002). As the WP El Niño usually occurs in autumn (SON) (Yu and Kim, 2010), the weak relationship between TC genesis and the WP Niño index in JJA is acceptable. However, in SON TC genesis is significantly related with the WP Niño index, but not with the CT Niño index, indicating that the interannual variability of TC genesis in autumn may relate to changes in the central tropical Pacific SST anomalies. It is noted that the significant relationship between the SCS TC genesis and WP Niño index was unnoticed by Kim *et al.* (2011, Figure 3). The cause for these different conclusions is that the interannual variability of TC genesis over the SCS is only remarkable during SON. However, Kim *et al.* (2011) analysed the TC during July and October.

Table 1. The simultaneous correlation coefficients between the TC genesis over the SCS and two types Niño indices.

	TC genesis	
	JJA	SON
CT Niño index	0.01	−0.1
WP Niño index	0.05	−0.42

Italic means exceeding 95% significance.

Figure 2 gives lag-lead relationships between the two Niño indices and the SON TC genesis over the SCS. Figure 2(a) shows the poor relationships between the SON TC genesis and CT Niño index. From Figure 2(b), TC genesis over the SCS in autumn significantly relates to the WP Niño index preceding 1-3 months and the correlations is increasing from preceding JJA, indicating that WP Niño index in boreal late-summer and early- autumn is a good precursory for the TC genesis over the SCS in autumn. However, for the JJA TC genesis, there are poor relationships with the WP Niño indices preceding 1-6 months (Figures not shown). Actually, Wang *et al.* (2012) suggested that the JJA TC genesis over the SCS show the significant variations, and is associated greatly with the East Asian summer monsoon.

In recent decades, the occurrence of the canonical El Niño has become less frequent (Wang *et al.*, 2009), while the WP El Niño has become more frequent (Ashok *et al.*, 2007; Kao and

Yu, 2009; Yeh *et al.*, 2009). Yeh *et al.*, (2011) suggested that the WP El Niño/CT El Niño occurrence ratio increased three times from 1950-1979 to 1980-2009. Moreover, the intensity of the WP El Niño has almost doubled in the past three decades (Lee and McPhaden, 2010). It is also noted that the increasing SST over the SCS shows significant multi-decadal variations in the 1970s (Wang *et al.*, 2010). Given the significantly negative relationship shown in Table 1, if the WP El Niño does actually influence TC genesis over the SCS, there are two hypotheses: (1) the interannual variability of TC genesis would be expected to be stronger after 1979 than before, and (2) the relationship between the WP Niño index and TC genesis would be expected to have high correlation in the past three decades because of the more frequent and stronger WP El Niño events.

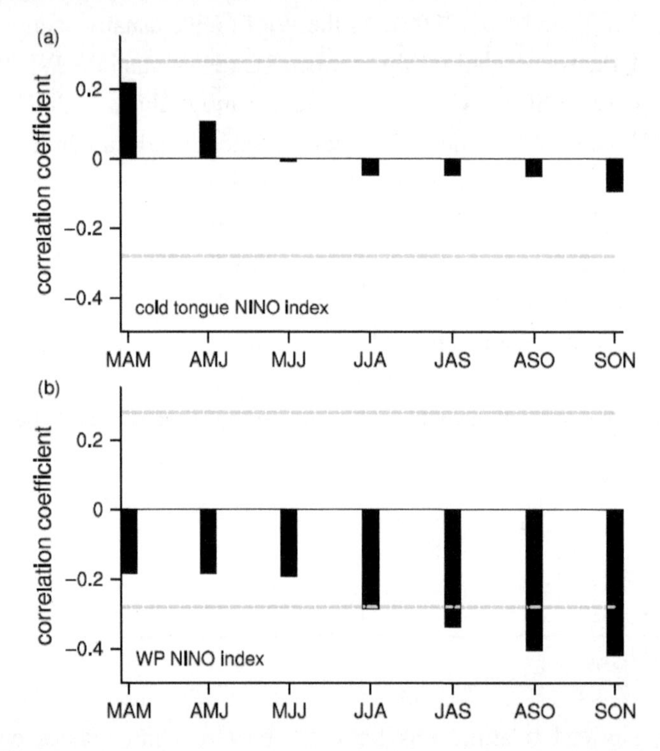

Figure 2. The relationships of SON TC genesis with the preceding 1-6 months CT (a) and WP (b) Niño index.

The first hypothesis is tested by the spectral analysis of TC genesis in two time periods (1965-1977 and 1978-2010). From Figure 3, it is very clear that the interannual variation of during 1978-2010 is remarkably significant, while it is weak during 1965-1977. The second is checked by the results of the wavelet coherence analysis between the WP Niño index and TC genesis in SON shown in Figure 4. The wavelet coherence (Grinsted *et al.*, 2004) is computed using the package supplied by http: //www.pol.ac.uk/home/research/waveletcoherence/. Figure 4 displays that the records exhibit statistically significant coherent variations of around 4-year timescales during the 1980s and 1990s (enclosed areas by thick contour). The 21-year

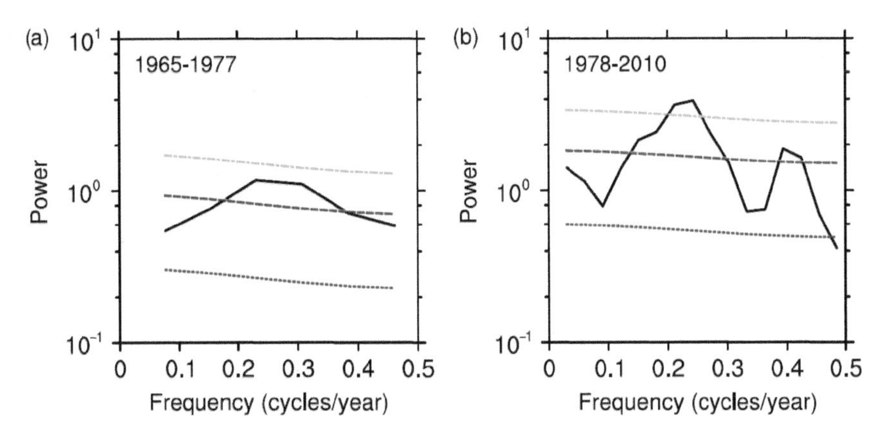

Figure 3. The spectrum analysis for SON mean TC genesis during 1965-1977 (a) and 1979-2004 (b). The red, blue, and green lines represent Markov red noise spectrum, 10% confidence level, and 90% confidence level.

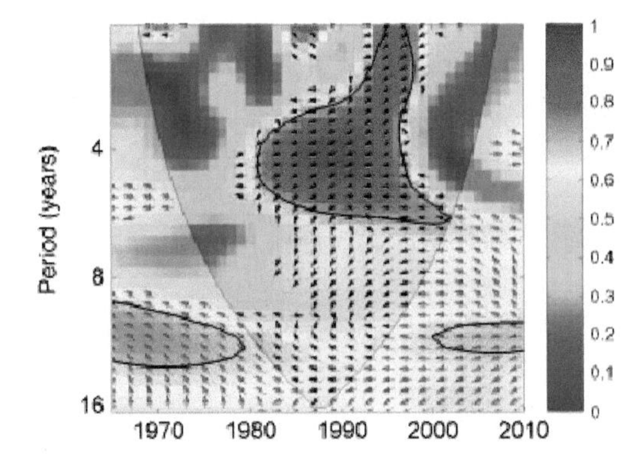

Figure 4. Squared wavelet coherence between the SON TC genesis over the SCS and the WP Niño index. Thick contours enclose the areas with correlations statistically significant at 95% confidence level against red noise. Semitransparent areas indicate the 'cone of influence' where the edge effects become important. The relative phase relationship is shown as arrows, with in-phase pointing right and anti-phase pointing left.

running correlations also show that the TC genesis over the SCS is significantly negatively related to WP Niño index during 1980s and 1990s (Figure not shown). The left orientation of the arrows in the upper part of Figure 4 suggests that on the interannual timescale the decrease in TC genesis generally corresponds to the WP El Niño events, which is consistent with the significantly negative relationship between TC genesis and WP El Niño index in Table 1.

4. Comparison of Large-Scale Environment Associated with WP and CT El Niño

The results of correlation and wavelet coherence analysis reveal that SON TC genesis is more closely linked with the WP Niño index, than the CT Niño index. The following attempts to explain why the influences of the WP and CT El Niño events on TC formation in the SCS in autumn are different.

4.1. Thermodynamic factors

It is well known that thermodynamic factors (e.g. SST and midlevel troposphere moisture) influence TC genesis. The deep convection is suggested to depend on SST of at least 28° (Graham *et al.*, 1987), and only the southeast part of SCS region reaches 28℃ in autumn. The responses of SST on the two types of El Niño are of great difference (Figure 5). The warm anomalies associated with CT Niño index are significant in the eastern part of SCS (Figure 5(a)), which results from El Niño-driven atmospheric and oceanic changes (Xie *et al.*, 2003; Qu *et al.*, 2004; Liu *et al.*, 2006; Wang *et al.*, 2006a, 2006b; Liu *et al.*, 2011). These warmer SST anomalies may supply the boundary forcing conditions of deep convection in favour of TC genesis. However, Table 1 shows the negative relationship between TC genesis and CT Niño index, indicating that other thermodynamic or dynamic factor could greatly influence TC genesis besides SST. In contrast, the impacts of WP El Niño on positive SST anomalies over the SCS are rather weak (Figure 5(b)). The significantly negative SST anomalies are observed in the east to Philippines.

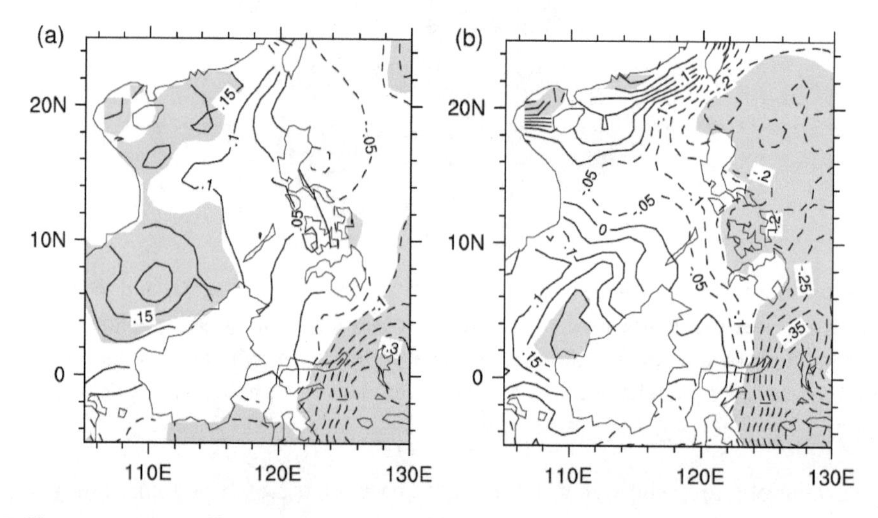

Figure 5. The regression coefficients of the SST anomalies with respect to CT Niño index (a) and WP Niño index (b). The shadings represent the SST anomalies exceeding 90% significant level.

The second thermodynamic factor related to TC genesis is the anomalous mid-tropospheric moisture (Gray, 1979), which the wet midlevels are favour of the formation and development of TC. Figure 6 shows the relationships between the relative humidity averaged between 700- and 500-hPa and Niño indices. In general, the mid level over the SCS is dry accompanying with the two types of El Niño, which is not conductive for TC occurrences. The composite pattern of the differences in the relative humidity at 500-hPa between the two types of El Niño events and the La Niña events are given in Figures 7(a) and (b). The composite difference scenarios are generally consistent with these regression patterns, which the two types of El

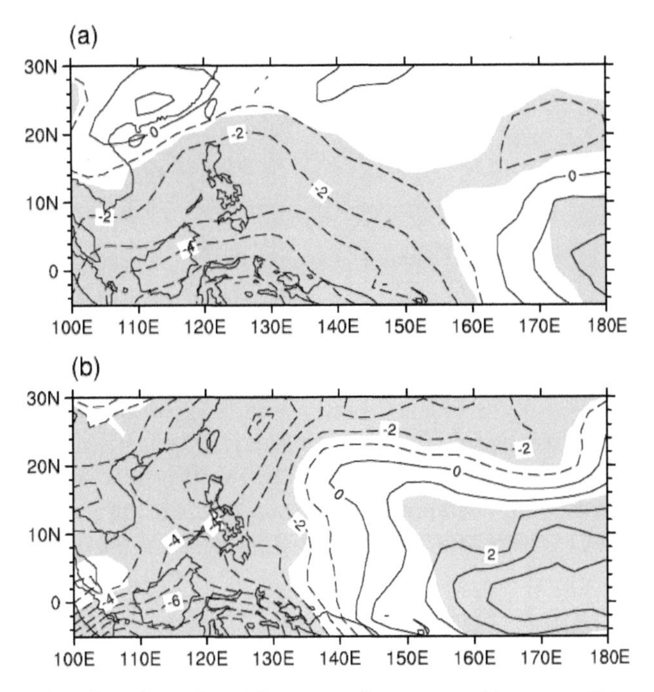

Figure 6. As in Figure 5, but for relative humidity anomalies averaged between 700 and 500-hPa.

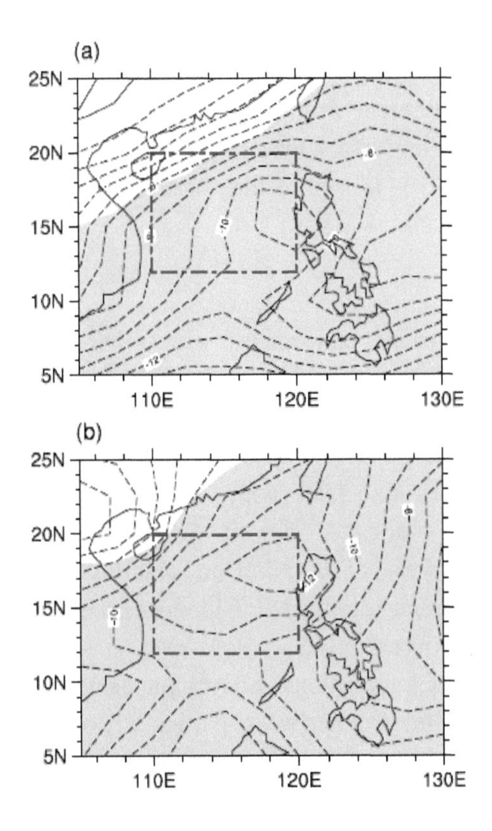

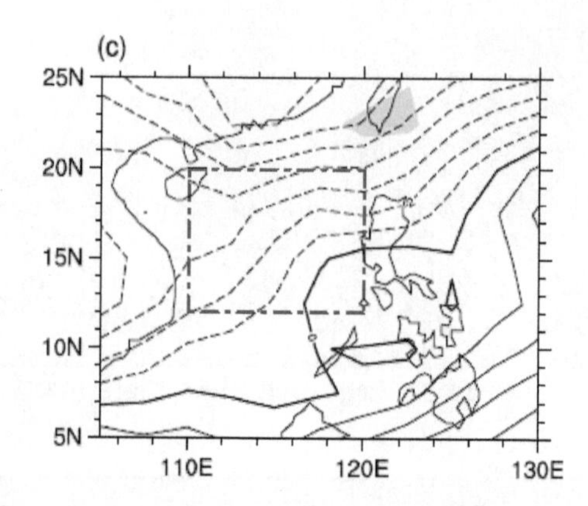

Figure 7. (a) The composite differences of relative humidity at 500-hPa between CT El Niño and La Niña events. (b) Same with (a), but for WP El Niño events. The shadings in (a) and (b) are exceeding 90% significant level. (c) The composite difference of relative humidity at 500-hPa between WP and CT El Niño events (WP El Niño minus CT El Niño). The red dash-dotted box is marked the central SCS where 73% of TC formats in autumn.

Niño can result in significantly dry midlevel. It is noted that the mid-troposphere related to WP El Niño is much drier than that related to CT El Niño over the SCS (Figure 7(c)). The impacts of SCS dry mid-troposphere related to CT El Niño in Figure 6(a) offset the positive contributions of CT El Niño-induced warm SST anomalies in the SCS on TC genesis, which thus may lead to the weakly negative relationship between CT Niño index and TC genesis over the SCS in Table 1. Different from the impacts of CT El Niño, the thermodynamic factors associated with WP El Niño suppress the TC formation over the SCS. The dry mid-troposphere over the SCS is related to the circulation anomalies, which is explained in Section 4.2.

4.2. Dynamical factors

Figure 8 gives the large-scale atmospheric circulation anomalies in the troposphere associated with the WP and CT Niño indices. At high-level, there are not significant signals over the SCS with respect to the WP and CT Niño indices. The significant anticyclone anomalies for both the WP and CT El Niños appear over the western North Pacific. It is noted that the anticyclone anomalies associated with the WP Niño index are stronger and more westward shifting than those associated with the CT Niño index. At 500-hPa, although the teleconnections of the two types of El Niño are significant in the SCS, the anomalous northerlies or northeasterlies related to the WP Niño index are remarkably stronger than those related to the CT Niño index over the SCS. At low-level, the circulations regressed by the CT Niño index exhibit a significant anomalous anticyclonic center located over the SCS, but the amplitude of the wind anomalies is rather weak, particularly over the central SCS. In contrast, the circulations with respect to the WP Niño index show significant anticyclone anomalies over the SCS and southeast China and cyclone anomalies over the western North Pacific, which in turn result in the enhancement of the anomalous northerlies over the SCS. These large-scale circulation anomalies associated with the WP Niño index are consistent with the observations and the model results (Ashok *et al.*, 2007; Chen and Tam, 2010). Figure 8 illustrates the largely distinct responses of the wind anomalies over the SCS to the two types Niño indices, suggesting that the WP and CT El Niño-induced dynamical controls of TC genesis over the SCS may be largely different. Moreover, the formation of dry mid-troposphere over the SCS related to the two types of El Niño could be explained by the responses of wind anomalies. As well known, the full moisture maintenance at mid-troposphere over the SCS depends on local evaporation or wet advection outside from the SCS. Although CT El Niño induces warm SCS SSTa in favour of local evaporation, because of the weak wind at low- and mid-levels (Figures 8(b) and (c)), the evaporation over the SCS is restricted and thus dry level is observed (Figures 6(a) and 7(a)). For WP El Niño, the stronger dry northerlies from the East Asia at low- and mid-levels (Figure 8(e) and (f)) are obviously not conductive to moisture maintenance over the SCS.

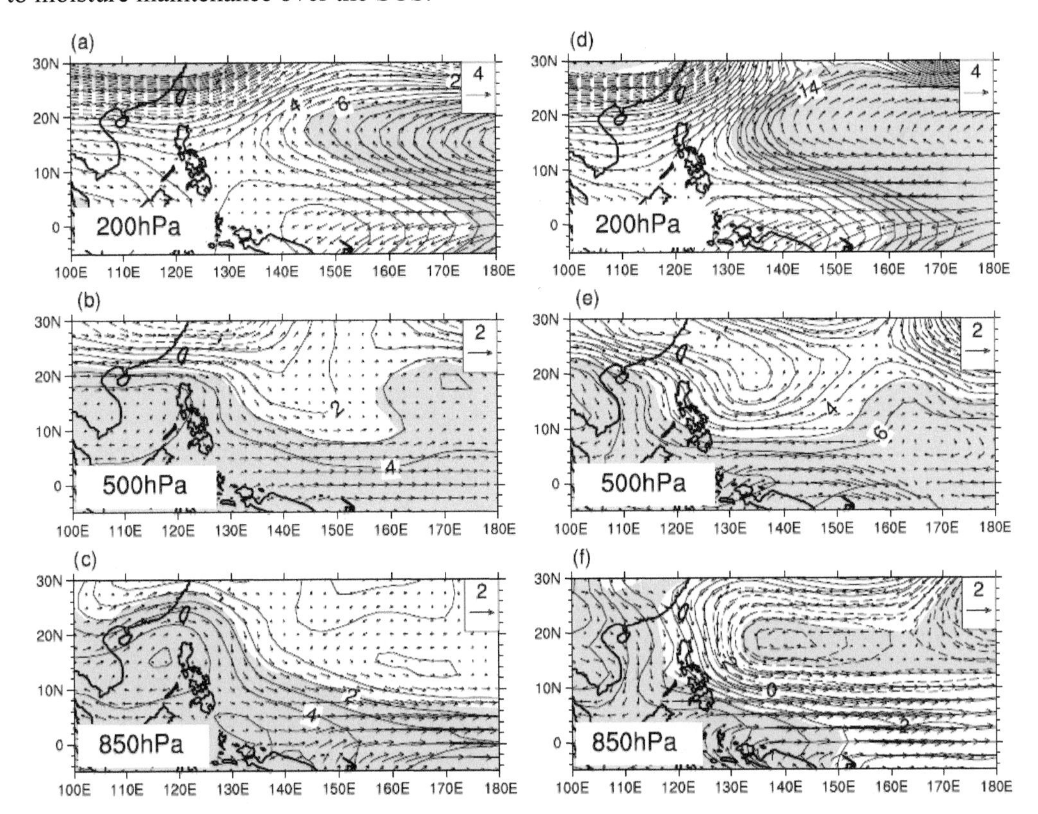

Figure 8. The regression coefficients of the wind (vector) and geopotential height (contour) with respect to CT Niño index (a, b, and c) and WP Niño index (d, e, and f). The shadings represent the geopotential height anomalies exceeding 90% significant level.

Regression analyses are employed to elucidate comparisons of the dynamic controls of vertical wind shear and 850-hPa relative vorticity in the SCS in SON related to the WP and CT Niño indices. From Figure 8(a), the significant signals of 850-hPa relative vorticity with respect to the CT Niño index are located in the southeast China coastal area and southwest SCS. However, it is noted that 73% of TC formation is in the central SCS (the red box in Figure 9) in autumn. Therefore, CT El Niño-induced low-level relative vorticity contributes little to the TC genesis over the SCS. As with the regression against the CT Niño index, the regressed 850-hPa relative vorticity against the WP Niño index is not significant over the central SCS (Figure 9(b)). These results suggest that low-level relative vorticity is not the key dynamic control of TC genesis over the SCS. However, the regressions of vertical wind shear over the SCS with respect to the CT and WP Niño indices are evidently different (Figures 9(c) and (d)). The CT Niño index shows weak connections with vertical wind shear in the central SCS (Figure 9(c)), whereas the WP Niño index is significantly related to positive vertical wind shear in the central SCS (Figure 9(d)). This indicates that the central tropical Pacific warm anomalies can result in vertical wind shear enhancement over the SCS, and thus suppress TC genesis. It is noted that the zero line of the regressions of vertical wind shear against the WP Niño index is displaced southward to that against the CT Niño index. This suggests that the area of the positive vertical wind shear anomalies related to the central tropical Pacific warming in the SCS is much larger, which evidently inhibits TC genesis over the SCS. These dynamical controls of TC genesis associated with the two Niño indices (Figure 9(a)-(d)) are actually the result of the two types of El Niño-induced anomalous circulations (Figure 8). With both the WP and CT El Niño, the horizontal shear of the wind vectors over the SCS at lower-level is rather weak (Figure 8), which explains the nonsignificant relative vorticity at 850-hPa in Figure 9(a) and (b). For vertical wind shear, there are no significant changes in horizontal wind at high-level over the SCS in the two types of El Niño (Figure 8), while the amplitudes of the wind anomalies at low-level against the WP Niño index are much larger than those against the CT Niño index over the SCS. Therefore, it is reasonable that WP El Niño-induced vertical wind shear is more significant than CT El Niño-induced over the SCS (Figure 9).

As the regression results have shown that vertical wind shear is the most important dynamical control of TC genesis over the SCS (Figures 8 and 9(a)-(d)), the composite pattern of the differences in vertical wind shear between the two types of El Niño events and the La Niña events are given in Figure 9(e) and (f). In general, the composite difference scenarios are similar to these regression patterns. Significant vertical wind shear over the central SCS is not observed in the difference pattern between the CT El Niño and La Niña events (Figure 9(e)), whereas significantly positive vertical wind shear is clearly exhibited in the difference pattern between the WP El Niño and La Niña events (Figure 9(f)). Therefore, as a result of the significant teleconnection of the enhancement of the low-level wind vector over the SCS, which enlarges the vertical wind shear, the WP El Niño could decrease the number of TC formations over the SCS significantly more than the CT El Niño.

5. Summary and Discussion

This study mainly focuses on the interannual variability of TC genesis over the SCS associated

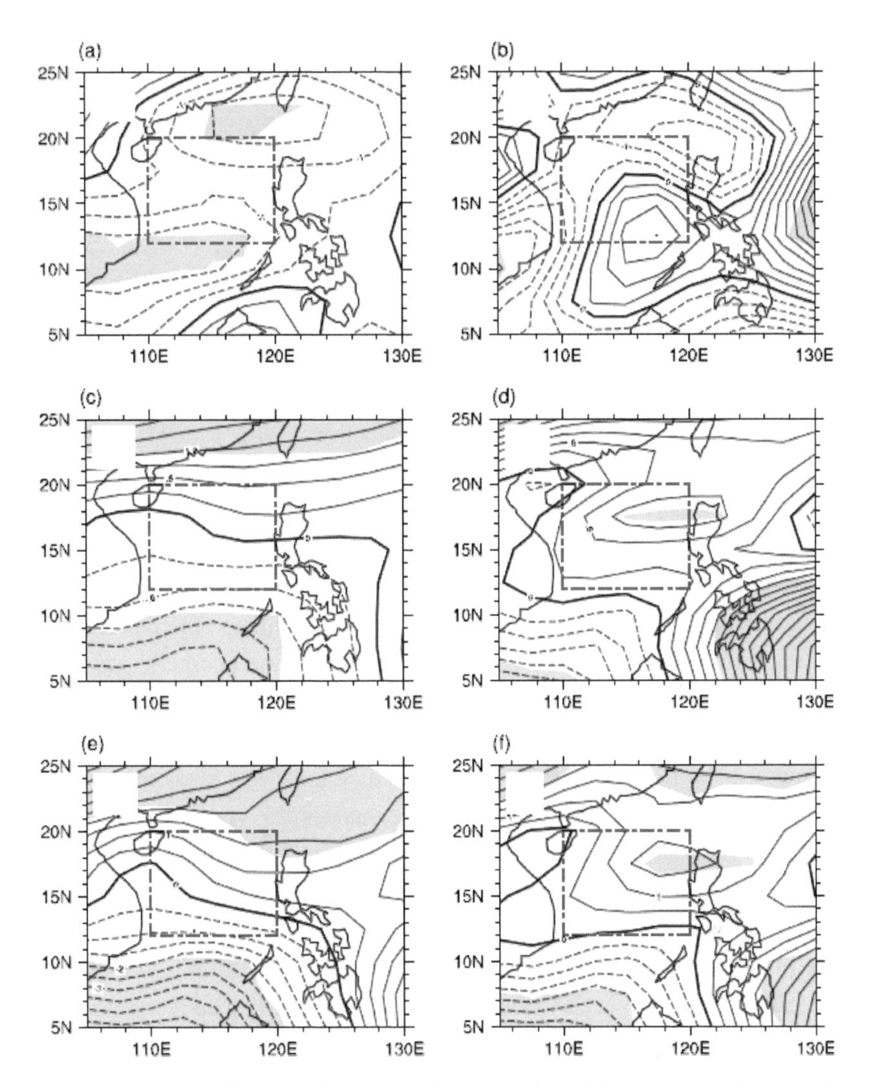

Figure 9. The regression coefficients of the 850-hPa relative vorticity (a), vertical wind shear (c) with respect to CT Niño index, and regression coefficients of the 850-hPa relative vorticity (b), vertical wind shear (d) with respect to WP Niño index. (e) The composite differences of vertical wind shear between CT El Niño and La Niña events. (f) Same with (e), but for WP El Niño events. The shadings are exceeding 90% significant level. The red dash-dotted box is marked the central SCS where 73% of TC formats in autumn. The vertical wind shear is calculated as the magnitude of the vector difference between wind at 200-hPa and 850-hPa.

with the ENSO phenomenon. Different from previous studies, which analyse the entire active TC season (Li, 1988; Wang and Chan, 2002), we separate the TC season into the two periods of summer (JJA) and autumn (SON) due to the location shift in TC formation (Lee *et al.*, 2006; Wang *et al.*, 2007). Spectral analysis illustrates that SON is the only period that exhibits significant interannual variability in TC genesis. There is increasing evidence that there are two types of El Niño: the CT and WP El Niño. The impacts of the two types of El Niño on the interannual variability of TC genesis over the SCS in SON are compared. Correlation analyses show that TC genesis over the SCS is significantly related to the WP Niño index, but not to the CT Niño index. With more frequent WP El Niño events in the last two decades, the coherent interannual variations between the WP Niño index and TC genesis are significant during this time.

To explain how different types of El Niño affect TC formation over the SCS, the variations of thermodynamic and large-scale circulations are examined. Different from the significantly warm SST anomalies in the SCS associated with CT El Niño, WP El Niño influences weakly on the SCS SST anomalies. Both the two types of El Niño can result in the dry midlevels over the SCS. Changes in atmospheric diabatic forcing over the central tropics during WP El Niño lead to the modification of the tropical teleconnections. The different responses of the large-scale circulation over the SCS and the western tropical Pacific result from the locations of warming center of the two types of El Niño. The warm SST anomalies in the equatorial central Pacific extend farther westward for WP El Niño than for CT El Niño. Based on the observation and the simple atmospheric model experiments, Wang and Wang (2013b) showed the physical progresses of how various types of El Niño influence the circulation anomalies over the western North Pacific. CT El Niño-related anticyclone anomalies over the SCS and western tropical Pacific WP are replaced by WP El Niño-related dipolar patterns with anticyclone anomalies over the SCS and cyclone anomalies over the western tropical Pacific WP at low- and mid-level. These WP El Niño-related atmospheric teleconnections enlarge the low-level northerlies over the SCS. This in turn enhances the vertical wind shear, and thus suppresses the TC genesis over the SCS. In conclusion, both thermodynamic and dynamical factors (midlevel moisture and vertical wind shear) associated with WP El Niño suppress the TC formation over the SCS. However, the CT El Niño-induced dynamical controls on TC formation are not significant over the SCS. Moreover, the CT El Niño-induced dry midlevel over the SCS offsets the positive contribution of CT El Niño-induced warm SST anomalies in the SCS on TC formation. Therefore, the TC genesis over the SCS shows weakly negative relationship with CT Niño index.

This study has found the significant relationship between WP Niño index and TC genesis over the SCS in autumn, and investigated how WP El Niño could influence TC formation over the SCS in terms of thermodynamic and large-scale circulation patterns of the large-scale environment. This study focused on the impact of the tropical Pacific on TC genesis over the SCS. Recently, it has been argued that the summer tropical Indian Ocean following El Niño could influence TC activity over the northwest Pacific, including the SCS (Du *et al.*, 2011). Therefore, the combined influence of the tropical Indian Ocean and tropical Pacific on TC genesis over the SCS needs further study. Our work helps to improve the seasonal prediction of TC activity in the SCS.

Acknowledgements This work was supported by National Basic Research Program ('973'Program) of China (Grant No. 2011CB403500), National Natural Science Foundation of China (Grant No. 41376025), and City University of Hong Kong (Strategic Research Grant

No. 7002780).

References

Ashok K, Behera SK, Rao SA, Weng H, Yamagata T. 2007. El Niño Modoki and its possible teleconnection. *J. Geophys. Res.* 112: C11007, DOI: 10.1029/2006JC003798.

Camargo SJ, Emanuel KA, Sobel AH. 2007. Use of a genesis potential index to diagnose ENSO effects on tropical cyclone genesis. *J. Climate* 20: 4819-4834, DOI: 10.1175/JCLI4282.1.

Chen G, Tam CY. 2010. Different impacts of two kinds of Pacific Ocean warming on tropical cyclone frequency over the western North Pacific. *Geophys. Res. Lett.* 37: L01803, DOI: 10.1029/2009GL041708.

Chen J, Wu R, Wen Z. 2012. Contribution of South China Sea Tropical Cyclones to an Increase in Southern China Summer Rainfall Around 1993. *Adv. Atmos. Sci.* 29(3): 585-598, DOI: 10.1007/s00376-011- 1181-6.

Du Y, Yang L, Xie SP. 2011. Tropical Indian Ocean influence on northwest Pacific tropical cyclones in summer following strong El Niño. *J. Climate* 24: 315-322.

Feng J, Chen W, Tam CY, Zhou W. 2011. Different impacts of El Niño and El Niño Modoki on China rainfall in the decaying phases. *Int. J. Climatol.* 31: 2091-2101.

Graham NE, Michaelson J, Barnett TP. 1987. An investigation of the El Niño Southern Oscillation cycle with statistical models I. Predictor field characteristics. *J. Geophys. Res.* 92: 14 251-14 270.

Gray WM. 1979. Hurricanes: their formation, structure and likely role in the tropical circulation. Meteorology over the Tropical Oceans. Shaw DB (ed). Royal Meteorological Society: Bracknell, UK; 155-218.

Grinsted A, Moore JC, Jevrejeva S. 2004. Application of the cross wavelet transform and wavelet coherence to geophysical time series. *Nonlinear Proc. Geophys.* 11: 561-566.

Kao HY, Yu JY. 2009. Contrasting eastern-Pacific and central-Pacific types of El Niño. *J. Climate* 22: 615-632, DOI: 10.1175/2008JCLI2309.1.

Kim HM, Webster PJ, Curry JA. 2009. Impact of shifting patterns of Pacific Ocean warming on North Atlantic tropical cyclones. *Science* 325: 77-80, DOI: 10.1126/science.1174062.

Kim JH, Ho CH, Chu PS. 2010. Dipolar redistribution of summertime tropical cyclone genesis between the Philippine Sea and the northern South China Sea and its possible mechanisms. *J. Geophys. Res.* 115: D06104, DOI: 10.1029/2009JD012196.

Kim HM, Webster PJ, Curry JA. 2011. Modulation of North Pacific tropical cyclone activity by three phases of ENSO. *J. Climate* 24: 1839-1849.

Kim JS, Zhou W, Wang X, Jain S. 2012a. El Niño Modoki and the summer precipitation variability over South Korea: a diagnostic study. *J. Meteor. Soc. Japan* 90: 673-684, DOI: 10.2151/jmsj.2012- 507.

Kim JH, Wu CC, Sui CH, Ho CH. 2012b. Tropical cyclone contribution to interdecadal change in summer rainfall over South China in the early 1990s. *Terr. Atmos. Ocean. Sci.* 23: 49-58, DOI: 10.3319/TAO.2011.08.26.01(A).

Kistler R et al. 2001. The NCEP-NCAR 50-Year reanalysis: monthly means CD-ROM and documentation. *Bull. Am. Meteorol. Soc.* 82: 247-267.

Kug JS, Jin FF, An SI. 2009. Two types of El Niño events: Cold tongue El Niño and warm pool El Niño. J. Climate 22: 1499-1515, DOI: 10.1175/2008JCLI2624.1.

Larkin NK, Harrison DE. 2005. Global seasonal temperature and precipitation anomalies during El Niño autumn and winter. *Geophys. Res. Lett.* 32: L16705, DOI: 10.1029/2005GL022860.

Lee T, McPhaden MJ. 2010. Increasing intensity of El Niño in the central-equatorial Pacific. *Geophys. Res. Lett.* 37: L14603, DOI: 10.1029/2010GL044007.

Lee CS, Lin YL, Cheung KKW. 2006. Tropical Cyclone Formations in the South China Sea Associated with the Mei Yu Front. *Mon. Weather Rev.* 134: 2670-2687.

Lee SK, Wang C, Enfield DB. 2010. On the impact of central Pacific warming events on Atlantic tropical storm activity. *Geophys. Res. Lett.* 37: L17702, DOI: 10.1029/2010GL044459.

Li C. 1988. Actions of typhoons over the western Pacific (including the South China Sea) and El Niño. *Adv. Atmos. Sci.* 5: 107-115.

Li CY, Zhou W. 2012. Changes in western Pacific tropical cyclones associated with the El Niño-Southern oscillation cycle. *J. Climate* 25: 5864-5878, DOI: 10.1175/JCLI-D-11-00430.1.

Li CY, Zhou W, Chan JCL, Huang P. 2012. Asymmetric modulation of the Western North Pacific cyclogenesis by the Madden-Julian Oscillation under ENSO conditions. *J. Climate* 25: 5374-5385, DOI: 10.1175/JCLI-D-11-00337.1.

Li CY, Zhou W, Li T. 2013. Influences of the Pacific-Japan teleconnection pattern on synoptic-scale variability in the Western North Pacific, *J. Clim.* in press, DOI: 10.1175/JCLI-D-13-00183.1.

Liu Q, Huang R, Wang D, Xie Q, Huang Q. 2006. Interplay between the Indonesian Throughflow and the South China Sea Throughflow. *Chin. Sci. Bull.* 51(Suppl. 2): 50-58.

Liu Q, Feng M, Wang D. 2011. ENSO-induced interannual variability in the southeastern South China Sea. *J. Oceanogr.* 67: 127-133.

Qu T, Kim Y, Yaremchuk M, Tozuka T, Ishida A, Yamagata T. 2004. Can Luzon strait transport play a role in conveying the impact of ENSO to the South China Sea? *J. Climate* 17: 3644-3657.

Rayner NA et al. 2003. Global analyses of sea surface temperature, sea ice, and night marine air temperature since the late nineteenth century. *J. Geophys. Res.* 108(D4): 4407, DOI: 10.1029/2002JD002670.

Ren HL, Jin FF. 2011. Niño indices for two types of ENSO. *Geophys. Res. Lett.* 38: L04704, DOI: 10.1029/2010GL046031.

Wang B, Chan JCL. 2002. How strong ENSO events affect tropical storm activity over the western North Pacific. *J. Climate* 15: 1643-1658.

Wang C, Wang X. 2013a. Classifying El Niño Modoki I and II by Different Impacts on Rainfall in Southern China and Typhoon Tracks. *J. Climate* 26: 1322-1338, DOI: 10.1175/JCLI-D-12-001 07.1.

Wang X, Wang C. 2013b. Different impacts of various El Niño events on the Indian Ocean Dipole. *Clim. Dyn.* in press, DOI: 10.1007/s00382-013-1711-2.

Wang C, Wang W, Wang D, Wang Q. 2006a. Interannual variability of the South China Sea associated with El Niño. J. Geophys. Res. 111: C03023, DOI: 10.1029/2005JC003333.

Wang D, Liu Q, Huang RX, DuY, Qu T. 2006b. Interannual variability of the South China Sea throughflow inferred from wind data and an ocean data assimilation product. *Geophys. Res. Lett.* 33: L14605, DOI: 10.1029/2006GL026316.

Wang G, Su J, Ding Y, Chen D. 2007. Tropical cyclone genesis over the South China Sea. *J. Mar. Syst.* 68: 318-326.

Wang X, Wang D, Zhou W. 2009. Decadal variability of twentieth century El Niño and La Niña occurrence from observations and IPCC AR4 coupled models. *Geophys. Res. Lett.* 36: L11701, DOI: 10.1029/2009GL037929.

Wang X, Wang D, Gao R, Sun D. 2010. Anthropogenic climate change revealed by coral gray values in the South China Sea. *Chin. Sci. Bull.* 55: 1304-1310, DOI: 10.1007/s11434-009-0534-3.

Wang X, Zhou W, Li CY, Wang DX. 2012. Effects of the East Asian summer monsoon on tropical cyclone genesis over the South China Sea on an interdecadal time scale. *Adv. Atmos. Sci.* 29: 249-262, DOI: 10.1007/s00376-011-1080-x.

Wang C, Li C, Mu M, Duan W. 2013. Seasonal modulations of different impacts of two types of ENSO events on tropical cyclone activity in the western North Pacific. *Clim. Dyn.*, DOI: 10.1007/s00382-012-1434-9.

Weng H, Behera SK, Yamagata T. 2009. Anomalous winter climate conditions in the Pacific Rim during recent El Niño Modoki and El Niño events. *Clim. Dyn.* 32: 663-674, DOI: 10.1007/s00382-008- 0394-6.

Xie SP, Xie Q, Wang D, Liu WT. 2003. Summer upwelling in the South China Sea and its role in regional climate variations. *J. Geophys. Res.* 108(C8): 3261, DOI: 10.1029/2003JC001867.

Yan YF, Qi YQ, Zhou W. 2012. Variability of tropical cyclone occurrence date in the South China Sea and its relationship with SST warming. *Dyn. Atmos. Oceans* 55-56: 45-59.

Yeh SW, Kug JS, Dewitte B, Kwon MH, Kirtman BP, Jin FF. 2009. El Niño in a changing climate. *Nature* 461: 511-514, DOI: 10.1038/nature08316.

Yeh SW, Kirtman BP, Kug JS, Park W, Latif M. 2011. Natural variability of the central Pacific El Niño event on multicentennial timescales. *Geophys. Res. Lett.* 38: L02704, DOI: 10.1029/2010GL045886.

Yu JY, Kao HY. 2007. Decadal changes of ENSO persistence barrier in SST and ocean heat content indices: 1958-2001. *J. Geophys. Res.* 112: D13106, DOI: 10.1029/2006JD007654.

Yu JY, Kim ST. 2010. Three evolution patterns of Central- Pacific El Niño. *Geophys. Res. Lett.* 37: L08706, DOI: 10.1029/2010GL042810.

Yuan Y, Zhou W, Yang H, Li CY. 2008. Warming in the northwestern Indian Ocean Associated with the El Niño Event. *Adv. Atmos. Sci.* 25: 246-252.

Zhou W, Chan JCL. 2007. ENSO and South China Sea summer monsoon onset. *Int. J. Climatol.* 27: 157-167.

Zuki ZM, Lupo AR. 2008. Interannual variability of tropical cyclone activity in the southern South China Sea. *J. Geophys. Res.* 113: D06106, DOI: 10.1029/2007JD009218.

Relationships Between Intensity of the Kuroshio Current in the East China Sea and the East Asian Winter Monsoon

Yin Ming[1,3], Li Xin[2,3], Xiao Ziniu[2], Li Chongyin[2,3]*

[1] Army 61936 of PLA, Haikou, China
[2] Institute of Atmospheric Physics, Chinese Academy of Sciences, Beijing, China
[3] Institute of Meteorology and Oceanography, The Army Engineering University of PLA, Nanjing, China

Abstract Based on satellite altimeter and reanalysis data, this paper studies the relationships between the intensity of the Kuroshio current in the East China Sea (ECS) and the East Asian winter monsoon (EAWM). The mechanisms of their possible interaction are also discussed. Results indicate that adjacent transects show consistent variations, and on an interannual timescale, when the EAWM is anomalously strong (weak), the downstream Kuroshio in the ECS is suppressed (enhanced) in the following year from February to April. This phenomenon can be attributed to both the dynamic effect (i.e., Ekman transport) and the thermal effect of the EAWM. When the EAWM strengthens (weakens), the midstream and downstream Kuroshio in the ECS are also suppressed (intensified) during the following year from October to December. The mechanisms vary for these effects. The EAWM exerts its influence on the Kuroshio's intensity in the following year through the tropospheric biennial oscillation (TBO), and oceanic forcing is dominant during this time. The air-sea interaction is modulated by the relative strength of the EAWM and the Kuroshio in the ECS. The non-equivalence of spatial scales between the monsoon and the Kuroshio determines that their interactions are aided by processes with a smaller spatial scale, i.e., local wind stress and heating at the sea surface.

Keywords East Asian winter monsoon, Kuroshio intensity, East China Sea, interaction, correlation analysis, composite analysis.

1. Introduction

The most important ocean current in the western North Pacific, the Kuroshio, originates from the northern branch of the North Equatorial Current (NEC). After bifurcating from the NEC off the Philippines, it passes the Luzon Strait and continues flowing northwards along

Received August 16, 2017; accepted September 26, 2017.

*Corresponding author, E-mail: lcy@lasg.iap.ac.cn

the eastern coast of Taiwan (Hsin et al., 2013). The Kuroshio enters the East China Sea (ECS) through the East Taiwan Channel, and after flowing northward along the Chinese continental shelf, the Kuroshio separates from the Japan coast and turns eastward about 35°N. Through over more than half a century of investigations, researchers worldwide have developed a preliminary understanding of the characteristics (i.e., transport, intensity, axis, depth and frequency of variability) of the Kuroshio in different regions. The many factors influencing the Kuroshio include submarine topography, mesoscale eddy, local and non-local wind stress, El Niño-Southern Oscillation (ENSO), and Pacific Decadal Oscillation (PDO) (Yang et al., 1999; Qiu, 1999; Johns et al., 2001; Chang and Oey, 2011, 2012; Qiu and Chen, 2010, 2013).

The Kuroshio is renowned for its huge transport, strong intensity, narrow width and deep depth; and its physical characteristics of high temperature, high salinity, and dark blue transparent water. Through the interchange of momentum, energy, and material, the Kuroshio exerts a remarkable influence on circulation in the China's seas, and is the main driver of offshore circulation.

In general, the formation of the monsoon system is mainly attributed to wind reversal caused by seasonal variation in the thermal contrast between land and sea (Chen et al., 2006). On the one hand, located in the monsoon region and the main area of heat loss from the world's oceans (Huang, 2012), the Kuroshio transfers huge amounts of energy to the atmosphere. This significantly affects atmospheric circulation (including the monsoon system) and climate change (Xu et al., 2008, 2011; Hu et al., 2015). On the other hand, in the mid-latitudes, especially in the boreal winter, atmospheric forcing on the ocean cannot be ignored (Wallace and Jiang, 1987; Li et al., 2011b), such that the monsoons also modulate the volume and heat transport of the Kuroshio itself. Thus, complex interactions exist between the Kuroshio and the East Asian monsoon. This work focuses on the processes operating during the boreal winter, while interactions during the summer will be reported in subsequent articles.

Many previous studies have investigated the relationships between thermal conditions in the Kuroshio, the East Asian general circulation and the climate of China (Qin and Sun, 2006; Zhang et al., 2008; Sasaki et al., 2012; Soeyanto et al., 2014). Representative of many investigations, Li and Long (1992) pointed out that positive SST anomalies cause precipitation to increase during the flood season over northern and northeastern China. Through statistical analysis, Li and Ding (2002) found that when the Kuroshio region has a positive SST anomaly, summer rainfall over the Changjiang (Yangtze) River valley and the temperature in most of Eurasia also have positive anomalies. Liang et al. (2006) reported that air-sea heat fluxes are significantly anomalous in the East Asian coastal area under the effect of the East Asian winter monsoon (EAWM). In comparison, relatively little work has been concerned with dynamic forcing of the monsoon, especially local wind stress, and most studies have focused on seasonal timescales (Chuang and Liang, 1994; Oey et al., 2010; Chang and Oey, 2012). For instance, Wang et al. (2010) suggested that different prevailing winds in winter and summer have different influences on the gap-leaping western boundary current. There are few existing studies connecting the EAWM with the Kuroshio in the ECS over interannual timescales from both thermal and dynamic perspectives.

Furthermore, because of the lack of *in situ* hydrological data, only variations in the Kuroshio's transport at specific transects can be analyzed. (Along its pathway in the ECS, the transect is usually named PN; its position is mapped in Fig. 1). However, strong eddies and internal tides in the Kuroshio may cause substantial contrast in transport for two adjacent transects. In this study, using high-resolution altimetric data, the geostrophic intensity is calculated at seven transects covering the Kuroshio in the ECS, allowing variations to be studied in detail, and revealing interannual relationships with the EAWM.

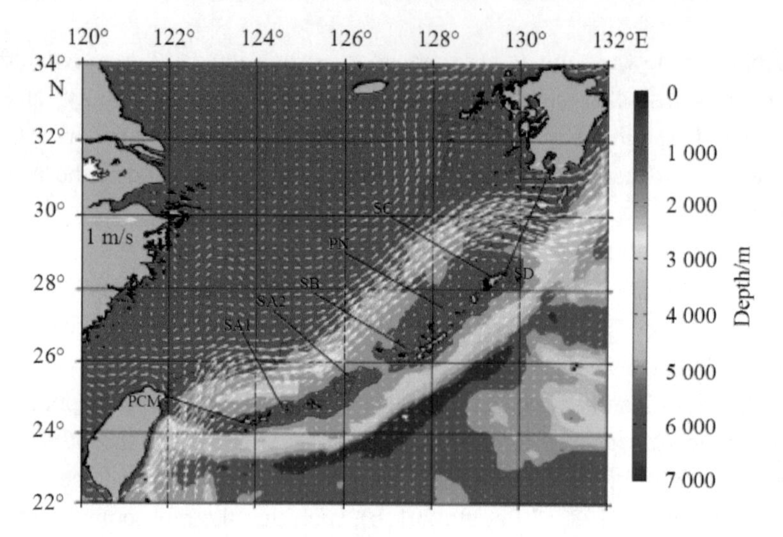

Fig. 1. The terrain of the East China Sea and adjacent waters (shading), altimeter-derived climatological mean geostrophic current field (arrows), and the selected research transects (black lines).

This paper is arranged into four sections. In Section 2, the data and methods of analysis adopted in this study are described. Results are presented in Section 3. A discussion and conclusions make up the final section.

2. Data and Methods

This paper uses the following data: state-of-the-art versions of Maps of Absolute Dynamic Topography (MADT) are provided by AVISO (Archiving Validation and Interpretation of Satellite Oceanographic data; http: //www.aviso.oceanobs.com/). The data are gridded on a 1-day time interval and a Cartesian grid of (1/4)°×(1/4)°, covering the period 1 January 1993 to 31 December 2012. The NCEP_Reanalysis 2 data of the daily wind field at 850 hPa and the monthly wind field at 10 m, provided by the NOAA/OAR/ESRL PSD, Boulder, Colorado, USA, from their web site at http://www.esrl.noaa.gov/psd/ are also utilized. Monthly ocean surface heat flux and temperature products are provided by WHOI (Woods Hole Oceanographic Institution; http://oaflux. whoi.edu/heatflux.html) (Yu and Weller, 2007). For uniformity, the timespan of all variables is consistent with the MADT data.

The intensity and transport of the Kuroshio can be expressed by both the tide gauge-derived sea surface height anomaly and the altimeter-derived geostrophic velocity (Hsin et al., 2013). The former serves as a proxy as proven by correlation analysis, while the latter is a more reliable quantitative analysis. In this paper, the second method was chosen, as proposed by Hsin et al. (2013) when they studied seasonal to interannual intensity variations in the surface Kuroshio east of Taiwan. Based on the geostrophic balance, surface geostrophic velocities in the Kuroshio in the ECS are calculated from the MADT as follows:

$$u_g(x,y,t) = -\frac{g}{f(y)}\frac{\partial h(x,y,t)}{\partial y},$$

$$v_g(x,y,t) = -\frac{g}{f(y)}\frac{\partial h(x,y,t)}{\partial x},$$

where x, y and t are longitude, latitude and time, respectively; u_g and v_g are the zonal and meridional components of the geostrophic current, respectively; h is the absolute dynamic topography; g is gravitational acceleration; $f(y) = 2\Omega \sin y$ is the Coriolis parameter, and Ω is the angular velocity of the Earth's rotation. The surface Kuroshio intensity ($\mathrm{INT_g}$) at a specific transect can be calculated by

$$\mathrm{INT_g} = \int_{Ls}^{Le} \vec{V}_g \cdot d\vec{l},$$

where $\vec{V}_g$ is the geostrophic current vector; $d\vec{l}$ dl is the infinitesimal element of the normal vector of the Kuroshio transect; L_s and L_e are the start and end points of the Kuroshio section, respectively.

To fully reflect the Kuroshio's intensity variation in the ECS, the $\mathrm{INT_g}$ at transects from PCM to SD was calculated, all of which are designed to span the Kuroshio along its path (Fig. 1). Transects PCM, SA1 and SA2 are located in the upstream ECS Kuroshio, transects SB and PN in the midstream, and transects SC and SD are downstream. These transects are chosen referring to Soeyanto et al. (2014). Transects PCM and PN are named following usual conventions (Hsin et al., 2013), and other transects are named for brevity.

Wavelet and power spectral analyses at each transect (figure omitted) indicate that the $\mathrm{INT_g}$ along the Kuroshio is dominated by intraseasonal variations at timescales below 100 d, caused by both the bottom topography and impinging westward-propagating mesoscale eddies originating in the interior Pacific (Yang et al., 1999; Qiu, 1999; Johns et al., 2001). Another significant period is annual change, due to monsoons and heat fluxes over the shelf region (Chao, 1990; Oey et al., 2010). Additionally, at SB and PN, there is a distinct oscillation of 2-3 a. However, intraseasonal and annual changes are not the main concern here. In this paper, the focus will be on interannual relationships between the Kuroshio $\mathrm{INT_g}$ in the ECS and the East Asian winter monsoon. Dynamic height noise associated with high-frequency activities can be largely filtered out by averaging (Wijffels et al., 1995), so the monthly mean of the daily $\mathrm{INT_g}$ is calculated and adopted as the parameter in this research.

The $\mathrm{INT_g}$ at transects in the same region presents a consistent variation. This is confirmed by the relatively high correlations between any two of them (Table 1), which, although not as high as expected, still pass the 95% significance test. In contrast, the INT_g shows different features in different regions, and the INT_g correlation between midstream and upstream transects is not as significant as between midstream and downstream transects.

An appropriate indicator for the East Asian monsoon needs to be applied in order to study its relationship with the Kuroshio's intensity. In this work, the unified East Asian monsoon index (EAMI) described by Li et al. (2011a) was used. The process involves averaging southwest-northeast wind velocity in the domain of grid points within the study area 10°-22.5°N and 110°-122.5°E. The normalized data are then calculated as below:

$$EAMI = \frac{V}{\sqrt{\frac{\Sigma V^2}{n}}},$$

where the regional mean V stands for $(u + v)/\sqrt{2}$, u and v are the zonal and meridional winds, respectively, and n is the sample number. Accordingly, the daily EAMI from 1 January 1993 to 31 December 2012 were obtained using the NCEP wind data at 850 hPa.

Figure 2a illustrates seasonal variation in the EAMI climatology. The basic circulation of the East Asian monsoon is consistent with that given by Li et al. (2011a). In general, the East Asian summer monsoon onsets in late May, and the East Asian winter monsoon is established in late October. The transition between summer and winter monsoons agrees with observations. Using this simple index, the relationships between the winter monsoon and the Kuroshio INT_g at the selected transects in the ECS were analyzed.

Table 1. Correlation coefficients of INT_g between pairs of transects

	PCM	SA1	SA2	SB	PN	SC	SD
PCM		0.62	0.39	0.16	0.17	0.26	0.29
SA1			0.71	0.36	0.22	0.18	0.22
SA2				0.53	0.35	0.28	0.28
SB					0.77	0.54	0.48
PN						0.74	0.63
SC							0.82
SD							

The December-to-February3-month mean EAMI were selected to represent winter conditions in a composite analysis. The time series of the normalized winter EAMI is given in Fig. 2b. Here, 1993 indicates the 1993/1994 winter, and similar labels are applied to subsequent years. It is clear that the winter monsoon possesses a pronounced inter-annual variation. Winter EAMI values greater than half the standard deviation from the average are therefore selected as weak winter monsoon cases. Four weak winter monsoon cases (1997, 2002, 2004 and 2009) were identified. Similarly, winter EAMI values less than half the standard deviation from the average are chosen as strong winter monsoon cases. The six selected strong winter monsoon cases are 1995, 1998, 1999, 2006, 2007 and 2008.

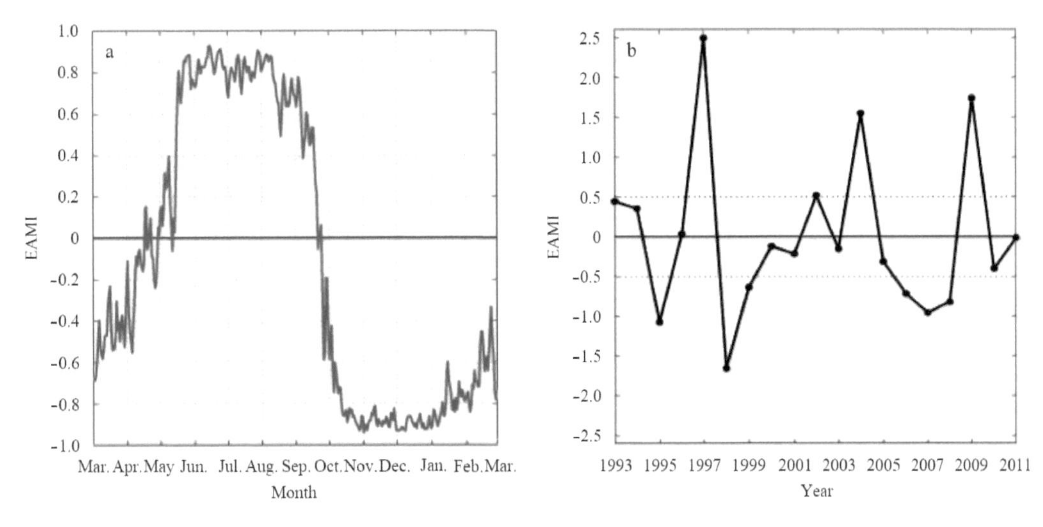

Fig. 2. Seasonal variation of the climatological EAMI (1993-2012; based on daily data) (a), and time series of the normalized winter EAMI (1993-2011) (b).

3. Results and Discussion

3.1 Lead/lag correlations between winter EAMI and INT_g

The lead/lag correlations between the winter EAMI and the Kuroshio INT_g for each transect are calculated. The procedure for calculating the correlation coefficients is as follows: (1) extract the monthly winter EAMI values from December to February and arrange them chronologically; (2) extract the monthly intensity which leads/lags particular months of the three winter months and arrange them in the same way; and (3) calculate the coefficient between the two sequences.

For equivalent periods, the performance of each transect is mostly consistent (Tables 2 and 3). For instance, leading the winter by 11 months, the INT_g from January to March at all transects is positively correlated with the winter EAMI; the INT_g leading the winter by approximately 4-5 months is negatively correlated with the winter EAMI; lagging 2 and 10 months behind the winter, the INT_g during February to April and October to December in the following year is positively correlated with the winter EAMI. At other times, although certain transects show some inconsistency, such as SA2 leading the winter by 9 months, and PCM and SA2 lagging the winter by 5 months, consistent signs are otherwise maintained for transects close to one another (e.g., SB to SD leading 10 and 6 months, PN to SD leading 3 months, and PCM to SA2 lagging 7 months). The analysis suggests that for most periods the variation in features is consistent between transects, and that transects in the same region have similar variations. However, there are periods when INT_g variations in the upstream, midstream and downstream Kuroshio are not synchronized. A detailed analysis of this will be given in the following section.

Table 2. Leading correlation coefficients of the INT_g with the winter EAMI

	−11 (Jan.)	−10 (Feb.)	−9 (Mar.)	−8 (Apr.)	−7 (May)	−6 (Jun.)	−5 (Jul.)	−4 (Aug.)	−3 (Sep.)	−2 (Oct.)	−1 (Nov.)
PCM	0.32[1)]	0.16	0.23[2)]	0.02	0.11	0.03	−0.17	−0.18	−0.01	−0.07	−0.08
SA1	0.14	0.03	0.01	−0.09	0.14	0.18	−0.09	−0.06	0.09	0.20	0.16
SA2	0.13	0.10	−0.07	−0.16	0.08	0.02	−0.07	−0.09	0.08	0.20	0.11
SB	0.23[2)]	−0.13	0.14	0.01	−0.05	−0.21	−0.23[2)]	−0.17	0.02	0.17	0.15
PN	0.20	−0.03	0.04	0.14	0.00	−0.12	−0.24[2)]	−0.25[2)]	−0.12	−0.02	0.12
SC	0.13	−0.02	0.14	−0.04	0.03	−0.09	−0.29[1)]	−0.27[1)]	−0.12	0.03	−0.01
SD	0.15	−0.10	0.21	0.18	0.13	−0.08	−0.33[1)]	−0.23[2)]	−0.04	0.14	0.17

Note: The negative signs in the column headings denote that the INT_g leads the winter EAMI, with the starting month given in brackets. 1) correlations that exceed 95%; 2) correlations that exceed 90%.

Table 3. Lag correlation coefficients of the INT_g with the winter EAMI

	0	+1 (Jan.)	+2 (Feb.)	+3 (Mar.)	+4 (Apr.)	+5 (May)	+6 (Jun.)	+7 (Jul.)	+8 (Aug.)	+9 (Sep.)	+10 (Oct.)
PCM	−0.05	0.22[2)]	0.14	−0.06	−0.32[1)]	0.03	−0.12	−0.23[2)]	−0.09	0.08	0.00
SA1	0.08	−0.12	0.09	−0.04	−0.18	−0.07	−0.05	−0.35[1)]	−0.10	−0.07	0.11
SA2	−0.14	−0.17	0.08	0.21	−0.02	0.03	0.07	−0.27[1)]	−0.14	−0.04	0.22[2)]
SB	0.05	−0.09	0.09	0.08	−0.07	−0.07	0.24[2)]	0.00	0.05	0.20	0.34[2)]
PN	0.02	0.05	0.04	−0.11	−0.20	−0.13	0.13	0.05	0.11	0.15	0.34[2)]
SC	0.04	0.12	0.32[1)]	0.04	−0.16	−0.22[2)]	0.01	−0.07	0.02	0.06	0.37[2)]
SD	0.04	0.10	0.26[1)]	0.10	−0.07	−0.10	0.12	−0.09	−0.11	0.00	0.38[2)]

Note: The positive signs in the column headings denote the INT_g lags the winter EAMI, with the starting month given in brackets. 1) correlations that exceed 95%; 2) correlations that exceed 90%.

3.2 Interaction between the Kuroshio intensity in the ECS and the EAWM

This section discusses the relationship between the Kuroshio INT_g in the ECS and the EAWM. For clarity, the main focus is on when these indices are significantly correlated—i.e., early spring (February to April) and winter (October to December).

3.2.1 Impacts of the EAWM on the Kuroshio intensity in the ECS in the following early spring

The lag-correlation results show that the winter EAMI is positively correlated with the 2-month-lag INT_g at Transects SC and SD, significant at the 90% level. This suggests that the INT_g in the downstream ECS Kuroshio is related to the EAWM to some extent. To see this clearly, the spatial distribution of correlation coefficients between the winter southwest-northeast wind speed field and the INT_g in the following February-April at seven transects is given in Fig. 3. The figure shows high positive correlations at Transects PCM, SC and SD in the monsoon region. This means that when the southwesterly winds strengthen (winter monsoon weakens), the INT_g at PCM, SC and SD intensifies, and vice versa. This result is consistent with that of the lag-correlations above. Given that the PCM transect is far from downstream, and its

2-month-lag-correlation coefficient is not high enough to be significant (Table 3), it will not be discussed here. The composites of the 3-month running mean INT_g after the strong and weak winter monsoons at SC and SD (Fig. 4) show that the INT_g at these two transects (or downstream) is indeed weaker from February to April after strong winter monsoons than after weak monsoons. To determine how the winter monsoon exerts its influence on the Kuroshio over the following months, the mechanisms from thermal and dynamic aspects were analyzed.

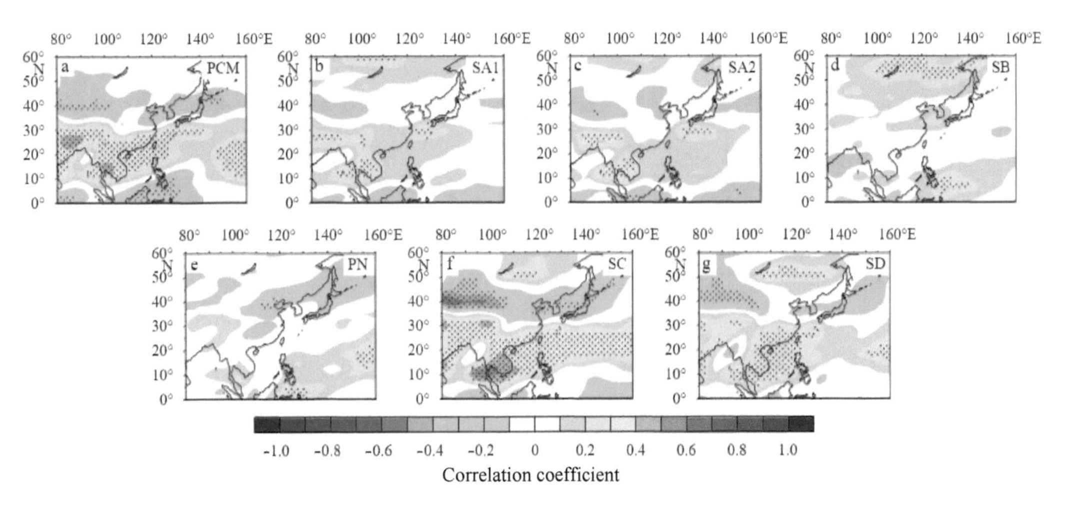

Fig. 3. Distribution of correlation coefficients between the winter southwest-northeast wind speed field and the 3-month average INT_g (February, March and April) at seven transects. The dot-shaded areas are significant at the 90% level.

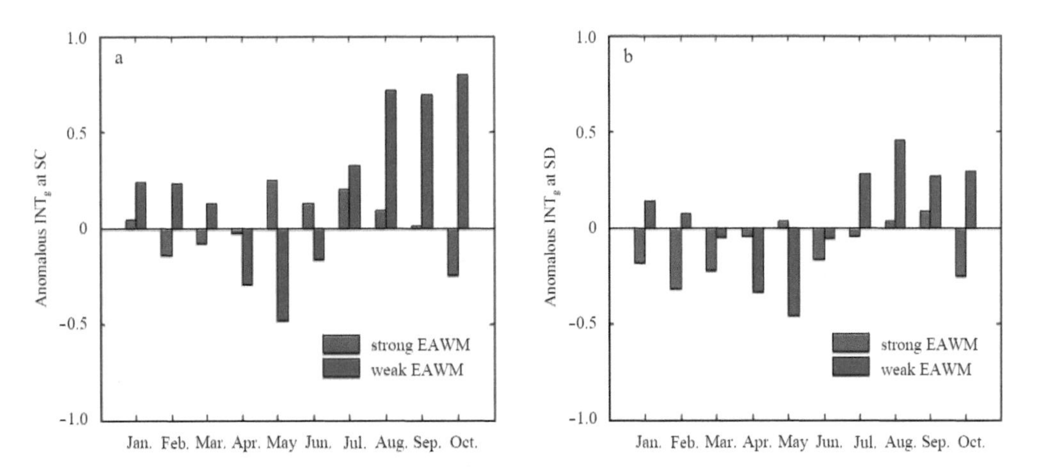

Fig. 4. Composites of the 3-month running mean INT_g anomaly (normalized) in the year following strong and weak EAWM conditions at Transects SC (a) and SD (b).

3.2.1.1 *Thermal process*

Composites of turbulent heat fluxes (sensible and latent) for different winter monsoon conditions are given in Fig. 5. There is a high-value belt where the Kuroshio releases much more heat to the atmosphere than in the neighboring vicinity (Fig. 5a). Under strong winter monsoon conditions, there is a positive anomaly covering the Kuroshio in the ECS, with extremes centered in the downstream region. Thus, the downstream Kuroshio releases more turbulent heat flux to the atmosphere in the following February-April period (Fig. 5b). In strong contrast, the anomaly is negative in weak monsoon cases and the Kuroshio in the ECS releases less heat to atmosphere (Fig. 5d).

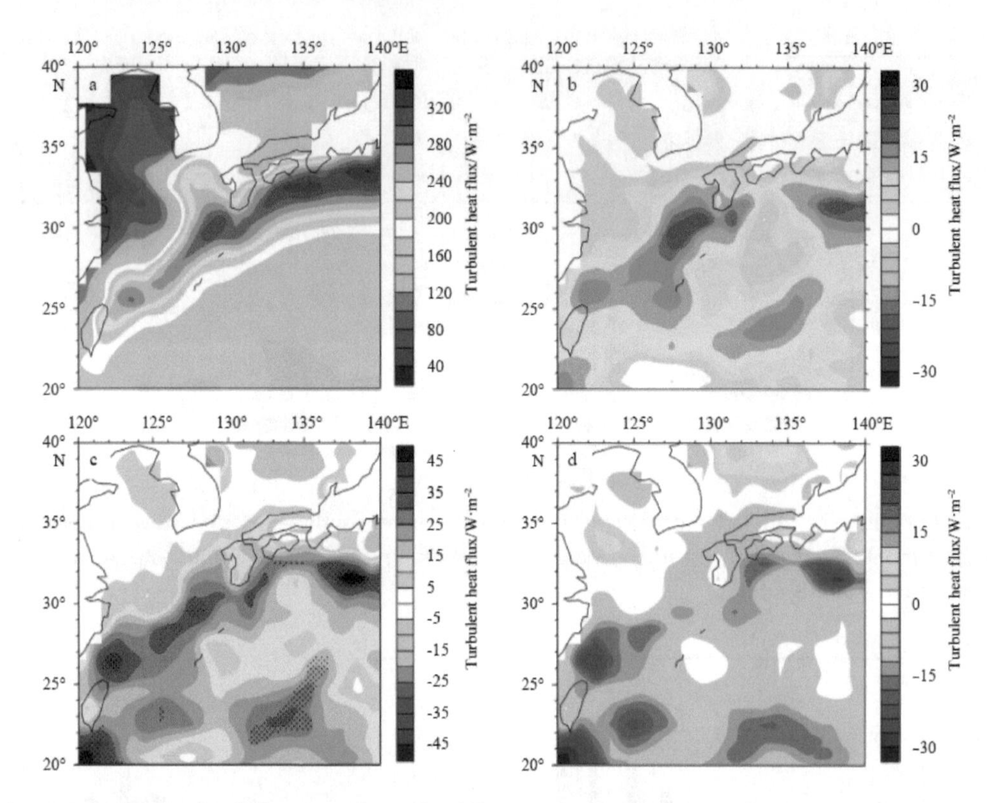

Fig. 5. Composites of turbulent heat fluxes for different winter monsoon conditions. a. Average turbulent heat fluxes (upward positive) in the Kuroshio and adjacent waters from February to April; b. composite of turbulent heat flux anomalies for strong winter monsoons; c. composite difference of turbulent heat fluxes between strong and weak winter monsoons; dotted areas are significant at 90%; and d. same as Fig. 5b, but for weak winter monsoons.

To understand the flux anomalies, the combined behavior of wind speed, the air-sea temperature difference (ΔT), and the ocean surface saturation humidity-air humidity difference (Δq) must be included. Figure 6a shows the wind anomaly under strong winter monsoon conditions. An anomalous northeasterly exists over the Kuroshio and the wind speed is greater. Under weak winter monsoon conditions (Fig. 6b), the anomaly changes to southwesterly and the wind speed is lower.

Because ΔT and Δq are correlated (cooler air is usually drier, especially outside the tropics) (Cayan, 1992), ΔT is used as the representative measure. Climatologically, the ΔT anomaly (SAT minus SST) over the Kuroshio is negative (Fig. 7a), with extremes centered in the downstream region. This distribution is reasonably similar to that of turbulent heat fluxes. When the winter monsoon is strong, the subsequent ΔT anomaly is negative (Fig. 7b), but the situation is reversed for weak cases (Fig. 7d). ΔT for strong cases is significantly smaller ($|\Delta T|$ is larger) than for weak cases, and the distribution also bears a resemblance to that in Fig. 5c, but with a reversed sign. On the other hand, the SST for strong conditions is not significantly lower than for weak conditions (figure omitted), nor is the distribution consistent with the first two elements (hear fluxes and ΔT). The dramatically cooled surface air in winter is thus seen to play the more important role in affecting turbulent fluxes. This finding implies that the primary factor in the EAWM is air-sea interaction along the Kuroshio in the ECS from February to April, generalized here as its thermal effect. A possible mechanism is as follows: higher wind speeds and $|\Delta T|$ during a strong winter monsoon increase turbulent heat fluxes vented by the ocean to the atmosphere. This corresponds to weak INT_g in the Kuroshio, especially downstream. On the contrary, wind speeds and $|\Delta T|$ are both weak during weak winter monsoons, and the resulting diminished heat fluxes are associated with a strong Kuroshio in the subsequent February-April period.

3.2.1.2 *Dynamic process*

To demonstrate further the influence of the winter monsoon on the Kuroshio, it is of interest to quantify whether the monsoon's dynamic effect is related to observed variations in the Kuroshio. Figure 8a shows the wind field climatology at 10 m and Sverdrup transport from February to April. This illustrates that, compared with the negative anomaly in adjacent areas, there is a positive band along the Kuroshio in the ECS, with extremes centered downstream. This means that the spatial distribution of wind-stress curl forces northward Sverdrup transport in the Kuroshio from February to April. The Sverdrup transport is calculated using the following formula:

$$M_y = \frac{1}{\beta} \mathrm{curl}\,\tau,$$

where $\beta \equiv (\mathrm{d} f / \mathrm{d} y)_{\Phi_0} = 2\Omega \cos \Phi_0 / a$. As generally understood, because of the friction force in western boundary currents, the Sverdrup model is not technically applicable. However, the distribution of Sverdrup transport driven by the wind field can still be used as a sketch map, to observe change in the Kuroshio.

Figure 8b presents the composite difference between strong and weak winter monsoon conditions of Sverdrup transport in the following early spring. When the winter monsoon is anomalously strong (weak), the subsequent transport is larger (smaller). This means that with respect to weak cases, the strong monsoon intensifies northward Sverdrup transport, and the Kuroshio is enhanced accordingly. Although this result fails to agree with the earlier correlation analysis, the disagreement can be understood by considering that the INT_g is derived from sea-surface data, while Sverdrup transport reflects the overall transport volume in the layer of wind-driven ocean circulation.

Now the process that weakens the surface Kuroshio INT_g downstream will be explored. If it is not the wind stress curl that reflects non-local influences, can it be the local wind field instead? To address these questions, Ekman transport along the Kuroshio was calculated,

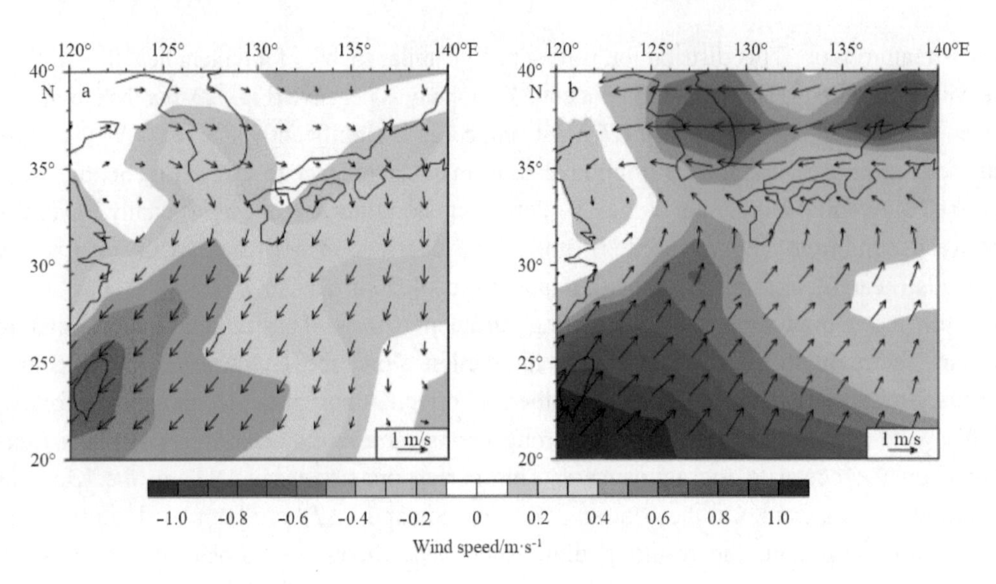

Fig. 6. Composite of wind speed (color tones) and wind field (vectors) anomalies from February to April following strong winter monsoons (a), and same as a but for weak winter monsoons (b).

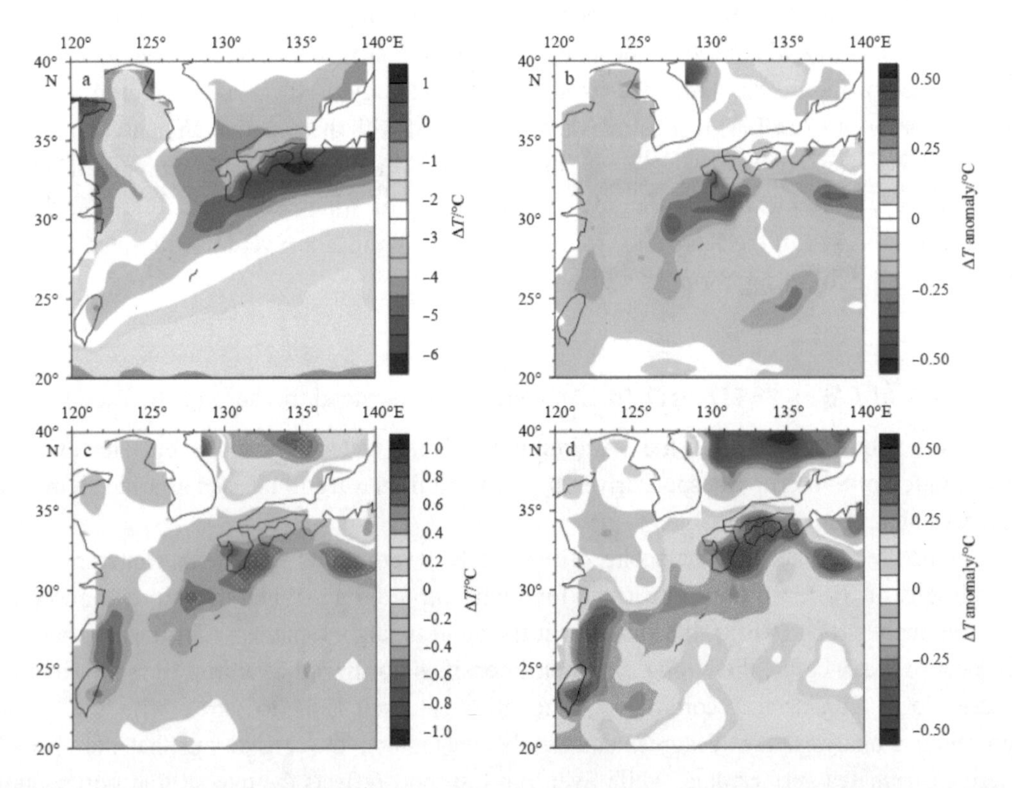

Fig. 7. Composites of ΔT for different winter monsoon conditions. a. Average ΔT over the Kuroshio and adjacent waters from February to April; b. composite of the ΔT anomaly from February to April following strong winter monsoons; c. composite difference of ΔT between strong and weak winter monsoons; dotted areas are significant at 90%; and d. same as Fig. 7b, but for weak winter monsoons.

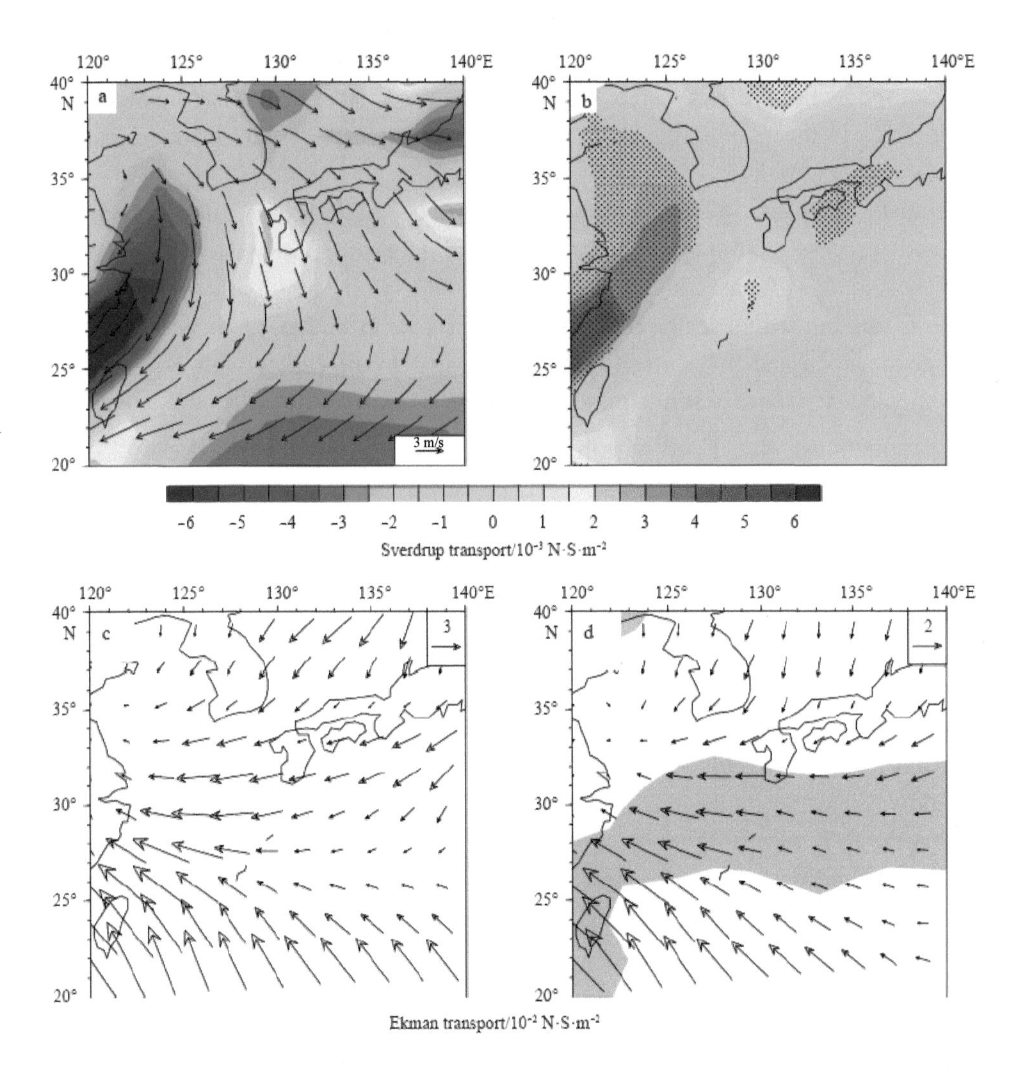

Fig. 8. Composites of Sverdrup transport for different winter monsoon conditions. a. Average Sverdrup transport (color tones) and wind field (vectors) across the Kuroshio and adjacent waters from February to April; b. composite difference between strong and weak winter monsoons of Sverdrup transport in the following February-April period; dot-shaded areas are significant at 90%; c. same as a but for Ekman transport; and d. same as Fig. 8b, but for Ekman transport; shaded areas are significant at 90%.

because it is dependent on the local wind field. In the Northern Hemisphere, the direction of Ekman transport is perpendicular and to the right of the wind stress. Figure 8c presents the climatological distribution. There is northwesterly wind in the downstream region (vectors in Fig. 8a), where southwestward-forced Ekman transport opposes the Kuroshio current direction. In the upstream region, however, the wind direction changes to the northeast and the Ekman transport becomes northwestward. The composite difference between strong and weak winter monsoons of Ekman transport from February to April is given in Fig. 8d. When the winter monsoon strengthens (weakens), there is strong (weak) southwest Ekman transport downstream, which weakens the INT_g at transects SC and SD. The result is coincident with that in Fig. 3, and explains why the INT_g at downstream transects is more notably weakened by the wind field. Thus, the dynamic effect of the winter monsoon is another of the factors that affect the Kuroshio INT_g in the ECS in the early spring, brought about by altering the local Ekman transport.

In summary, the winter monsoon subsequently affects the Kuroshio through both thermal and dynamic processes. Through midlatitude tropical interaction, the winter monsoon spreads its influence to the global scale (Chen et al., 2000). Compared with the monsoon, the Kuroshio is small in spatial scale. The differences in scale mean that the interactions between the winter monsoon and the Kuroshio INT_g are aided by processes with a smaller spatial scale-local wind stress and heating at the sea surface.

3.2.2 Impacts of the EAWM on the Kuroshio intensity in the ECS in the following early winter

The winter EAMI is significantly correlated with the 10-month-lag INT_g at all observation transects except PCM and SA1. The spatial distribution of correlation coefficients between the winter wind-speed field and the INT_g from October to December in the following year at the seven transects is mapped in Fig. 9. It shows that all transects other than PCM have high positive correlations in the monsoon region. This means that when the anomalous southwesterly winds intensify (winter monsoon weakens), the INT_g in the midstream and downstream Kuroshio is strong, and vice versa. The October-December INT_g composites for strong and weak winter monsoons (SC and SD are shown in Fig. 3; other figures omitted) reveal that the INT_g is weaker in strong winter monsoons than in weak monsoons. How then does the winter monsoon influence the Kuroshio in the subsequent early winter? Is there any difference from the mechanisms during the February - April period discussed above? This will be explored in the following section.

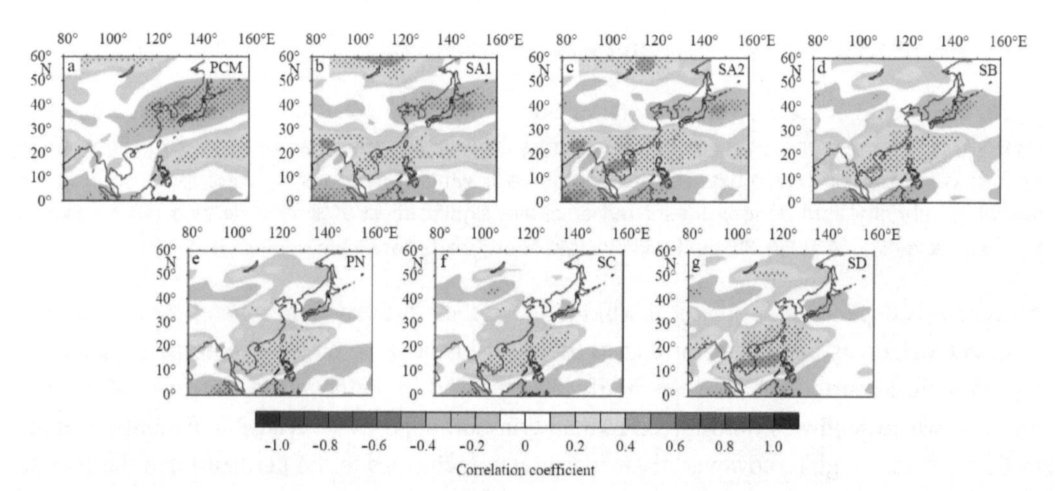

Fig. 9. Distribution of correlation coefficients between the winter southwest-northeast wind speed field and the 3-month average INT_g (October, November and December) in the following year at seven transects. The dot-shaded areas are significant at 90%.

3.2.2.1 *Thermal process*

First, the thermal process will be discussed. Composites of turbulent heat fluxes for different winter monsoon conditions are plotted in Fig. 10. The high-value belt where the Kuroshio releases much more heat to the atmosphere still exists (Fig. 10a). Under strong winter monsoon conditions, a negative anomaly covers the Kuroshio in the ECS, and thus the Kuroshio releases less turbulent heat flux to the atmosphere in the following October-December period (Fig. 10b). In addition, the extremes centered in the southeast downstream region deviate from the Kuroshio main axis. For weak monsoons, the anomaly is positive and the Kuroshio in the ECS releases more heat to the atmosphere (Fig. 10d). The extremes are also centered downstream. However, the result is opposite to the distribution from February to April (Figs 5b to d).

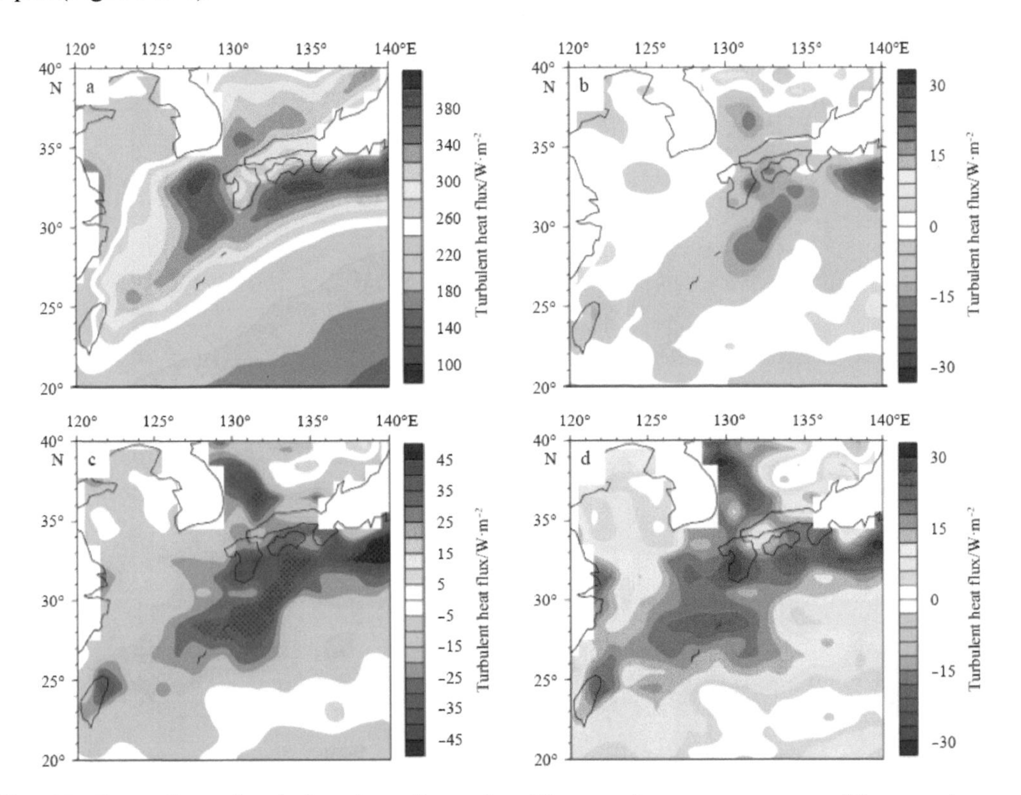

Fig. 10. Composites of turbulent heat fluxes for different winter monsoon conditions. a. Average turbulent heat fluxes (upward positive) across the Kuroshio and adjacent waters from October to December; b. composite turbulent heat flux anomalies from October to December in the year following strong winter monsoons; c. composite difference of turbulent heat fluxes between strong and weak winter monsoons; dotted areas are significant at 90%; and d. same as Fig. 10b, but for weak winter monsoons.

During the two periods examined, there are similar positive correlations with the winter EAMI. However, the composite results are opposite. Is atmospheric forcing still dominant from October to December, similar to the February-April situation? To figure out this question, composite analyses for variables concerned were performed.

Composite analysis shows that under strong winter monsoon conditions, there is a slight easterly anomaly over the Kuroshio. Under weak winter monsoon conditions, there is an anomalous northwesterly. The wind speeds are both positive anomalies with no significant difference (figure omitted). As seen from the composites of the ΔT over the Kuroshio, there is a positive anomaly under strong winter monsoons (Fig. 11b), and a negative anomaly under weak monsoons (Fig. 11d). The difference is not significant, but their distributions are similar to those of turbulent heat fluxes. Composites of SST along the Kuroshio illustrate that SST from October to December after a strong winter monsoon are lower than after a weak monsoon (Fig. 12c), and that the distribution is analogous to both ΔT and turbulent heat fluxes.

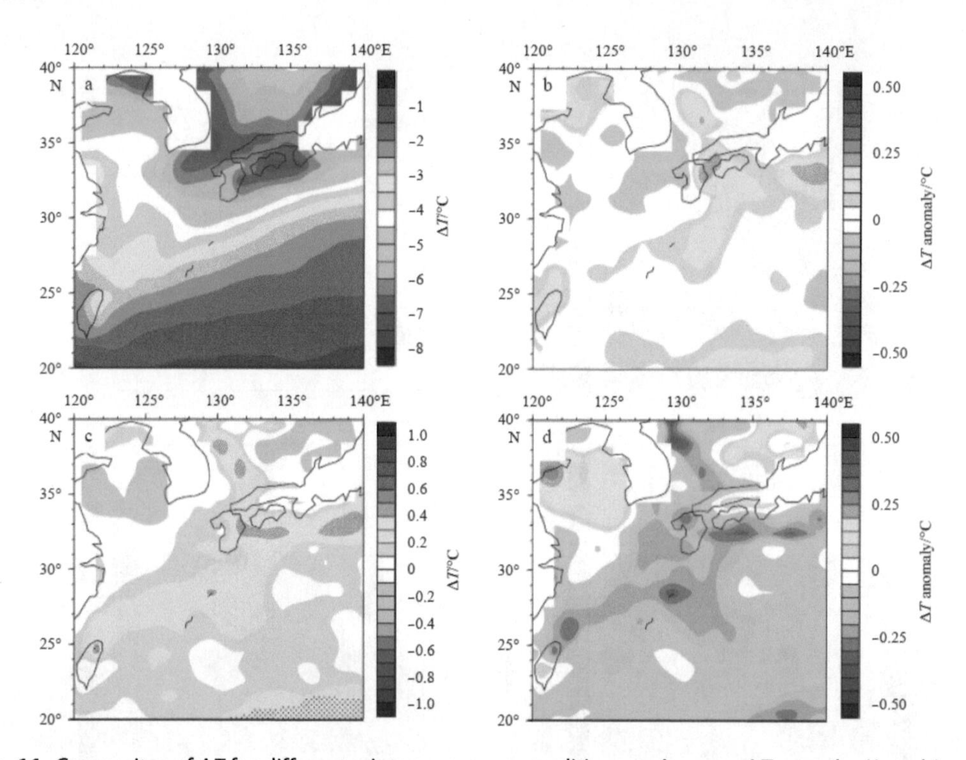

Fig. 11. Composites of ΔT for different winter monsoon conditions. a. Average ΔT over the Kuroshio and adjacent waters from October to December; b. composite of the ΔT anomaly from October to December in the year following strong winter monsoons; c. composite difference of ΔT between strong and weak winter monsoons; dotted areas are significant at 90%; and d. same as Fig. 11b, but for weak winter monsoons.

These composite results suggest a possible mechanism: from October to December in the year following the winter monsoon, oceanic forcing dominates the local air-sea interaction processes over the Kuroshio in the ECS. This means that after a strong winter monsoon, the weakened Kuroshio (corresponding to lower SST) releases less heat to the atmosphere from October to December. The decreased heating reduces the land-sea thermal contrast, which in turn weakens the subsequent winter monsoon. In contrast, after a weak winter monsoon, the strengthened Kuroshio (corresponding to higher SST) releases more heat to the atmosphere from October to December. This enhances the land-sea thermal contrast, which in turn strengthens the subsequent winter monsoon.

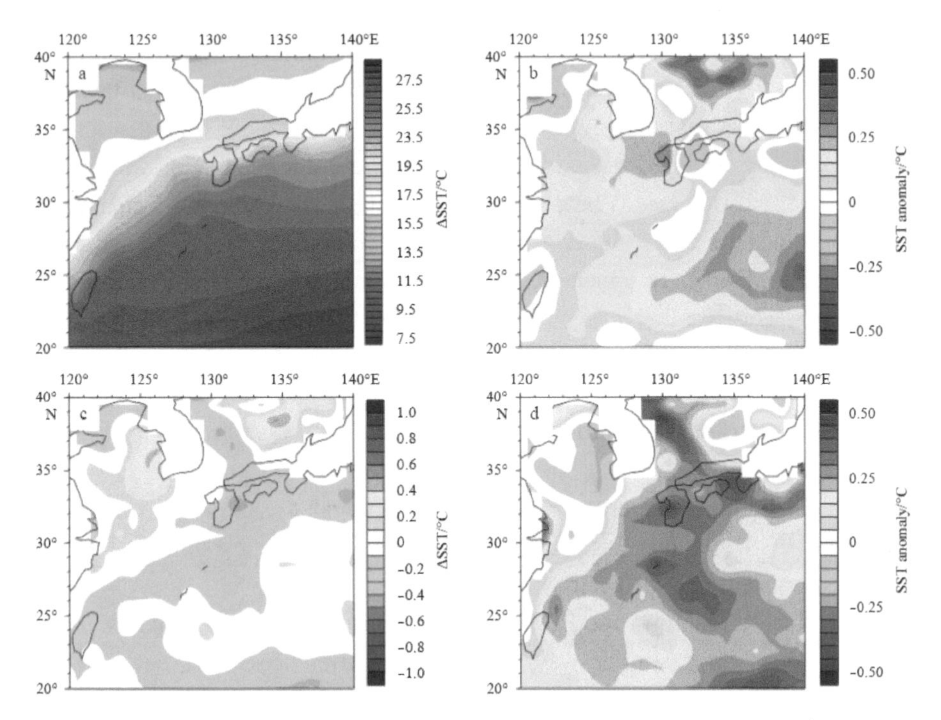

Fig. 12. Composites of SST for different winter monsoon winter conditions. a. Average SST across the Kuroshio and adjacent waters from October to December; b. composite of the SST anomaly from October to December in the year following strong winter monsoons; c. composite difference of SST (ΔSST) between strong and weak winter monsoons; and d. same as Fig. 12b, but for weak winter monsoons.

At present, there are two problems with the mechanism. One is how the winter monsoon affects the 10-month-lag INT_g; the other is why the primary factor influencing the air-sea interaction changes. Li et al. (2010) proposed that the East Asian monsoon has a tropospheric biennial oscillation (TBO), such that a strong winter monsoon precedes a weak winter monsoon. Here, through the oscillation, a strong winter monsoon may induce a weak INT_g in the midstream and downstream Kuroshio over the following October-December for a 2 a period. Due to the subsequent weakening of the winter monsoon, atmospheric forcing loses its dominance in the air-sea interaction, being replaced by oceanic forcing instead. Thus, the INT_g has an impact on the strength of the winter monsoon (as previously described), which completes a positive feedback mechanism. One point should be noted regarding this mechanism: whether atmospheric or oceanic forcing is the dominant process in the air-sea interaction of the Kuroshio in the ECS is determined by the relative strength of the EAWM and the INT_g. The INT_g in the early winter after a strong winter monsoon is weak, with its subsequent winter monsoon weaker. Under weak winter monsoon conditions, the following winter monsoon is strong, but the INT_g over the next October-December is stronger. Oceanic forcing is therefore dominant in both cases. To summarize, there may be a "critical value" in the relative strength of the EAWM and the INT_g that determines the conversion between atmospheric and oceanic forcings.

3.2.2.2 *Dynamic process*

Composite differences of both Sverdrup transport and Ekman transport along the Kuroshio in the ECS from October to December after strong and weak winter monsoons fail to pass the significance test (figure omitted). Results indicate that the dynamic process of the EAWM cannot be directly responsible for the interannual change in the Kuroshio in the ECS for the October-December period. This is because the influence of local and non-local wind stresses on the Kuroshio in the ECS is not as great on the interannual timescale as on the seasonal timescale (Yu et al., 2008). To conclude, the dynamic effect of the winter monsoon is not a dominant process. This result echoes the finding of the previous section that, from October to December, oceanic forcing is the main process in the Kuroshio in the ECS.

4. Discussion and Conclusions

Based on satellite altimeter data and reanalysis data, the relationships between the Kuroshio intensity in the ECS and the East Asian winter monsoon are examined. Their possible interaction mechanisms are also discussed.

Variability in the Kuroshio in the ECS ranges over periods of days to years. Transects in close proximity vary in a similar way, while great differences exist between upstream, midstream and downstream regions of the Kuroshio. Correlation between midstream and upstream locations is not as significant as that between midstream and downstream locations.

On the interannual timescale, there is an interaction between the Kuroshio in the ECS and the winter monsoon. When the winter monsoon strengthens, the INT_g weakens at transects SC and SD over the following February-April period. The situation is reversed for weak monsoons. The phenomenon is caused by both dynamic and thermal effects of the winter monsoon. The monsoon produces its dynamic effect by influencing Ekman transport. The thermal effect following a strong winter monsoon is that greater wind speed and $|\Delta T|$ cause the ocean to release more heat to the atmosphere, which weakens the Kuroshio INT_g, especially downstream. Under weak monsoons with lower wind speed and $|\Delta T|$, the ocean vents less heat to the atmosphere, coinciding with enhanced INT_g. During the February-April period, atmospheric forcing is therefore dominant.

When the winter monsoon strengthens (weakens), the midstream and downstream Kuroshio INT_g is also suppressed (enhanced) in the following October-December period. This process differs from that in early spring. Because of the TBO, a strong winter monsoon usually precedes a weak monsoon, and reduces the INT_g during the following October-December. Oceanic forcing takes on the leading role during this period. Consequently, a weak Kuroshio (corresponding to a lower SST) releases less heat into the atmosphere from October to December after a strong winter monsoon, and in turn reduces the land-sea thermal contrast and weakens the subsequent winter monsoon. In contrast, after a weak winter monsoon, the INT_g in the following October-December is anomalously strong (corresponding to higher SST), increasing heat transfer to the atmosphere. This enhances the land-sea thermal contrast, which in turn strengthens the subsequent winter monsoon.

Together these analyses present a complete positive feedback system, including both atmospheric and oceanic coupling processes. The schematic diagram in Fig. 13 illustrates the cycle, and describes the basic characteristics of the interaction between the EAWM and intensity of the Kuroshio in the ECS. During early winter, as determined by the relative strengths of the winter monsoon and the Kuroshio in the ECS, the dominant process over the Kuroshio is oceanic forcing, different from conditions typical in early spring when atmospheric forcing dominates. A "critical value" may determine the conversion from atmospheric to oceanic forcing processes.

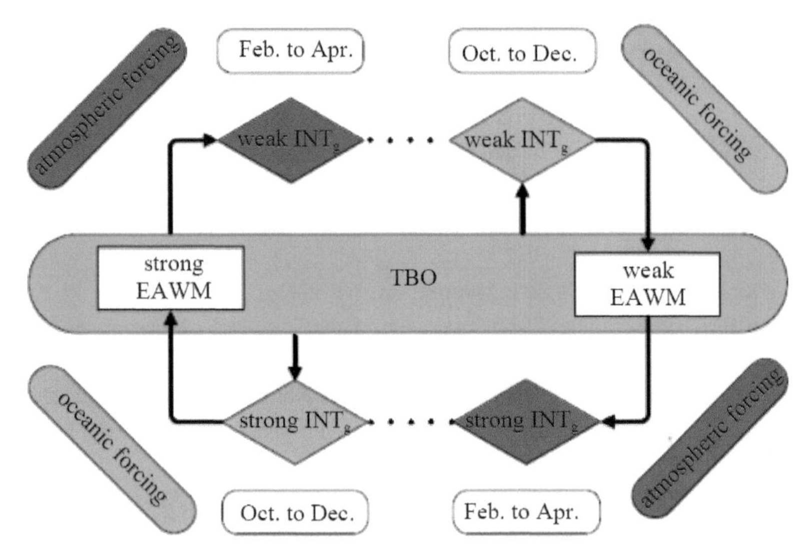

Fig. 13. A schematic diagram of the interaction between the EAWM and the Kuroshio INT_g in the ECS. The red color indicates where atmospheric forcing is dominant, blue represents oceanic forcing, and green denotes the role of TBO.

The non-equivalence in spatial scales between the EAWM and the Kuroshio means that processes with a small spatial scale (i.e., local wind stress and heating at the sea surface) must contribute to their interactions. Finally, air-sea interaction along the Kuroshio is complex and the physical mechanism of the TBO that influences the INT_g has not been well described. Numerical tests will be necessary to corroborate the results of observational analysis presented here.

References

Cayan D R. 1992. Latent and sensible heat flux anomalies over the northern oceans: driving the sea surface temperature. J Phys Oceanogr, 22(8): 859-881.

Chang Y L, Oey L Y. 2011. Interannual and seasonal variations of Kuroshio transport east of Taiwan inferred from 29 years of tide- gauge data. Geophys Res Lett, 38(8): L08603.

Chang Y L, Oey L Y. 2012. The Philippines-Taiwan oscillation: monsoonlike interannual oscillation of the subtropical-tropical western north Pacific wind system and its impact on the ocean. J Climate, 25(5): 1597-1618.

Chao S Y. 1990. Circulation of the East China Sea, a numerical study. J Oceanogr, 46(6): 273-295.

Chen Wen, Graf H F, Huang Ronghui. 2000. The interannual variability of East Asian Winter Monsoon and its relation to the summer monsoon. Adv Atmos Sci, 17(1): 48-60.

Chen Longxun, Zhang Bo, Zhang Ying. 2006. Progress in research on the East Asian monsoon. J Appl Meteor Sci (in Chinese), 17(6): 711-724.

Chuang W S, Liang W D. 1994. Seasonal variability of intrusion of the Kuroshio water across the continental shelf northeast of Taiwan. J Oceanogr, 50(5): 531-542.

Hsin Y C, Qiu Bo, Chiang T L, et al. 2013. Seasonal to interannual variations in the intensity and central position of the surface Kuroshio east of Taiwan. J Geophys Res: Oceans, 118(9): 4305-4316.

Hu Dunxin, Wu Lixin, Cai Wenju, et al. 2015. Pacific western boundary currents and their roles in climate. Nature, 522(7556): 299-308.

Huang Ruixin. 2012. Ocean Circulation: Wind-Driven and Thermohaline Processes (in Chinese). Le Kentang, Shi Jiuxin, trans. Beijing: Higher Education Press, 731.

Johns W E, Lee T N, Zhang Dongxiao, et al. 2001. The Kuroshio east of Taiwan: Moored transport observations from the WOCE PCM-1 array. J Phys Oceanogr, 31(4): 1031-1053.

Li Yuefeng, Ding Yihui. 2002. Sea surface temperature, land surface temperature and the summer rainfall anomalies over Eastern China. Climatic Environ Res (in Chinese), 7(1): 87-101.

Li Chongyin, Pan Jing, Que Zhiping. 2011a. Variation of the East Asian Monsoon and the tropospheric biennial oscillation. Chin Sci Bull, 56(1): 70-75.

Li Bo, Zhou Tianjun, Lin Pengfei, et al. 2011b. The wintertime North Pacific surface heat flux anomaly and air-sea interaction as simulated by the LASG/IAP ocean-atmosphere coupled model FGOAL_s1.0. Acta Meteor Sin (in Chinese), 69(1): 52-63.

Liang Qiaoqian, Jian Maoqiu, Peng Zhigang, et al. 2006. Impacts of East Asian winter monsoon on sea surface temperature in northwestern Pacific. Journal of Tropical Oceanography (in Chinese), 25(6): 1-7

Oey L Y, Hsin Y C, Wu C R. 2010. Why does the Kuroshio northeast of Taiwan shift shelfward in winter? Ocean Dyn, 60(2): 413-426.

Qin Zhengkun, Sun Zhaobo. 2006. Influence of abnormal East Asian winter monsoon on the northwestern Pacific sea temperature. Chin J Atmos Sci (in Chinese), 30(2): 257-267.

Qiu Bo. 1999. Seasonal eddy field modulation of the north Pacific Subtropical Countercurrent: TOPEX/Poseidon observations and theory. J Phys Oceanogr, 29(10): 2471-2486.

Qiu Bo, Chen Shuiming. 2010. Interannual variability of the North Pacific subtropical countercurrent and its associated mesoscale eddy field. J Phys Oceanogr, 40(1): 213-225.

Qiu Bo, Chen Shuiming. 2013. Concurrent decadal mesoscale eddy modulations in the western North Pacific subtropical gyre. J Phys Oceanogr, 43(2): 344-358.

Soeyanto E, Guo Xinyu, Jun O, et al. 2014. Interannual variations of Kuroshio transport in the East China Sea and its relation to the Pacific Decadal Oscillation and mesoscale eddies. J Geophys Res: Oceans, 119(6): 3595-3616.

Wallace J M, Jiang Q. 1987. On the observed structure of the interannual variability of the atmosphere-ocean climate system. In: Cattle H, ed. Atmospheric and Oceanic Variability. Bracknell: Royal Meteorological Society, 17-43.

Wang Zheng, Yuan Dongliang, Hou Yijun. 2010. Effect of meridional wind on gap-leaping western boundary current. Chin J Oceanol Limnol, 28(2): 354-358.

Wijffels S, Firing E, Toole J. 1995. The mean structure and variability of the Mindanao Current at 8°N. J Geophys Res, 100(C9): 18421-18435.

Xu Haiming, Wang Linwei, He Jinhai. 2008. Observed oceanic feedback to the atmosphere over the Kuroshio Extension during spring time and its possible mechanism. Chin Sci Bull, 53(12): 1905-1912.

Xu Haiming, Xu Mimi, Xie Shangping, et al. 2011. Deep atmospheric response to the spring Kuroshio over the East China Sea. J Climate, 24(18): 4959-4972.

Yang Y, Liu C T, Hu J H, et al. 1999. Taiwan current (Kuroshio) and impinging eddies. J Oceanogr, 55(5): 609-617.

Sasaki Y N, Minobe S, Asai T, et al. 2012. Influence of the Kuroshio in the East China Sea on the early summer (Baiu) rain. J Climate, 25(19): 6627-6645.

Yu Lisan, Weller R A. 2007. Objectively analyzed air-sea heat Fluxes for the global ice-free oceans (1981-2005). Bull Am Meteor Soc, 88(4): 527-539.

Yu Fan, Wang Qi, Liu Yulong. 2008. The seasonal and interannual variations of the upper Kuroshio circulation in the East China Sea and their relationship with local wind stress. Periodical of Ocean University of China (in Chinese), 38(4): 533-538.

Zhang Qilong, Hou Yijun, Qi Qinghua. 2008. Variations in the Kuroshio heat transport in the East China Sea and meridional wind anomaly. Adv Marine Sci (in Chinese), 26(2): 126-134.

4

Climate Variability

The Quasi-Decadal Oscillation of Air-Sea System in the Northwestern Pacific Region①

Li Chongyin(李崇银)

LASG, Institute of Atmospheric Physics, Chinese Academy of Sciences, Beijing

Abstract The data analyses found at first that the air-sea system in the northwestern Pacific region has clear systematical quasi-decadal oscillation, such as the surface air temperature, the subtropical high activities over the northwestern Pacific and the SSTA which has different time-scale features from the temporal variation with 3-4 years period of SSTA in the equatorial Pacific.

In East Asia, the climate variations, such as the surface air temperature, the precipitation and the beginning date of Mei-yu in the Yangtze River basin, also have clear quasi-decadal oscillation. They can be regarded as the influences of quasi- decadal oscillation of air-sea system in the northwestern Pacific region.

Keywords Quasi-decadal oscillation, Air-sea system.

Ⅰ. Introduction

In recent years, the decadal variability of climate has been paid much attention and it becomes an important part of the science plan for CLIVAR (CLIVAR-Dec Cen). But the existed studies are mostly variability in the ocean. For example, the change of SST in the North Atlantic (Deser and Blackmon, 1993; Kushnir, 1994) and in the North Pacific (Trenberth and Hurrell, 1994; Graham, 1994); the variability in the ocean's interior (Read and Gould, 1992; Gammelsroed et al., 1992); and the variability of the ocean's ice cover (Mysak et al., 1990; Hill and Jones, 1990). In fact, the decadal variation is also obvious in the atmosphere (Li Chongyin, 1992; Hurrell and Van Loon, 1994). It is quite necessary to study decadal variabilities in the climate and atmospheric circulation.

It is known that ENSO - a typical interannual climate variation has been regarded as a result of tropical air-sea interaction. Is it still important that the air-sea interaction excites decadal climate variability? Particularly in the middle latitudes, the study is more necessary.

In present paper, we will analyze the quasi-decadal variability of the air-sea system in the northwestern Pacific region. Then, the influence of this quasi-decadal oscillation on the climate variation in East Asia will be discussed. And a simple dynamic analysis in relation to

---

Received October 15, 1996; revised September 5, 1997.

① This research was supported by the Chinese Academy of Sciences and '95' National Project "Short-Range Climate Production System".

this oscillation is also shown in this paper.

II. Quasi-Decadal Oscillation of Air-Sea System in the North Western Pacific Region

1. Temporal Variation of SST

The SST and its variation (anomaly) are very important to the climate change. Therefore, we should study interannual variation of SST based on the COADS; and in order to make comparison, the analyses are not limited in the northwestern Pacific.

In the equatorial Pacific, as we know, the ENSO cycle (it is also called ENSO mode) with time-scale 3-4 years is a fundamental feature of SST interannual variation. The temporal variation of monthly SSTA in the Niño 3 region (5°S-5°N, 150°W-90°W) and its power spectrum are shown in Fig.l, respectively. A spectral peak with 3-4 years period is very prominent in Fig.l, so that the ENSO cycle is again proved to be a fundamental feature in interannual variations of SST.

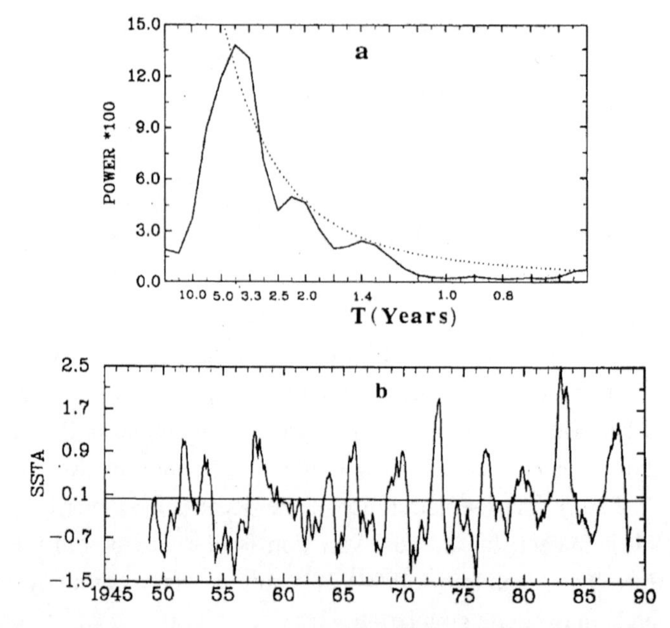

Fig. 1. Temporal variation of monthly SSTA in Niño 3 (5°S-5°N, 150°W-90°W) region (b) and its power spectrum (a). The dotted line represents the significance level 95%.

In the Kuroshio region (17.5°-32.5°N, 122.5°-152.5°E), the temporal variation of monthly SSTA is quite different from that in the Niño 3 region (Fig.2). A spectral peak with about 17 years period is very prominent and implies decadal variability of SST being fundamental feature in there. In order to further prove the existence of decadal variability of SST in the region above-mentioned, the temporal variation of monthly SSTA and its power spec trum in the western Pacific (7.5°-17.5°N, 132.5°-167.5°E) located between the Niño 3 region and Kuroshio region, is shown in Fig.3. It is very clear that there are two spectral peaks respectively with 7-10 years and 3-4 years period. In other words, temporal variations of SST in the western Pacific have two major quasi-periodic oscillations, one is the ENSO mode like, another belongs to decadal mode.

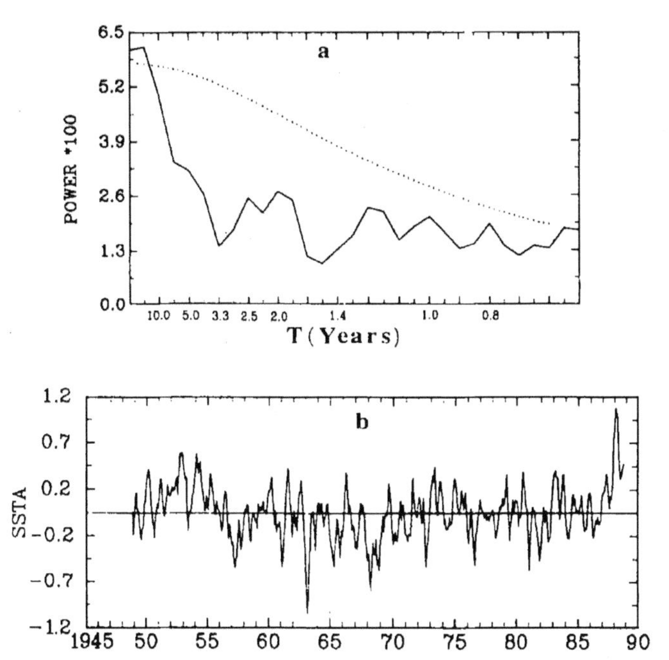

Fig. 2. Temporal variation of monthly SSTA in the Kuroshio (17.5°-32.5°N, 122.5°-152.5°E) region (b) and its power spectrum (a). The dotted line represents the significance level 95%.

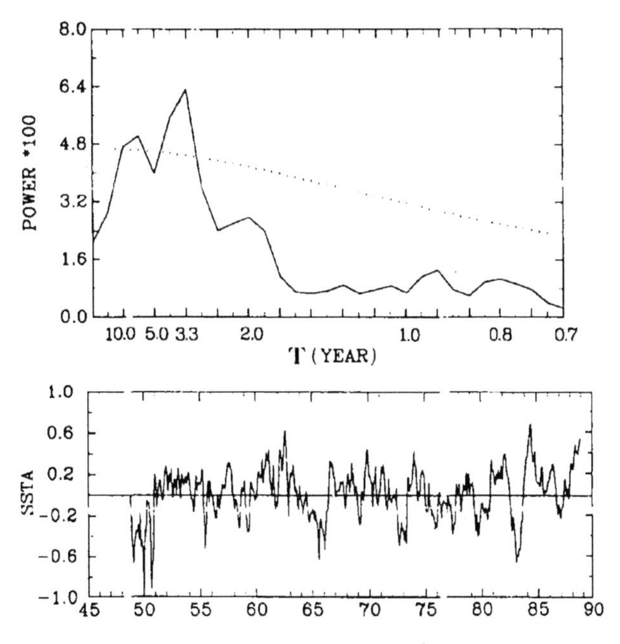

Fig. 3. Temporal variation of monthly SSTA in the western Pacific (7.5°-17.5°N, 132.5°- 167.5°E) region (b) and its power spectrum (a). The dotted line represents the significance level 95%.

The temporal variation of monthly SSTA in the northern Pacific (37.5°-47.5°N, 152.5°E-142.5°W) and its power spectrum are given in Fig.4. A spectral peak with about 7-8 years period is very prominent and implies the temporal variation of SST in the northern Pacific having a major quasi-periodic oscillation, which belongs to decadal mode and differs from the ENSO mode.

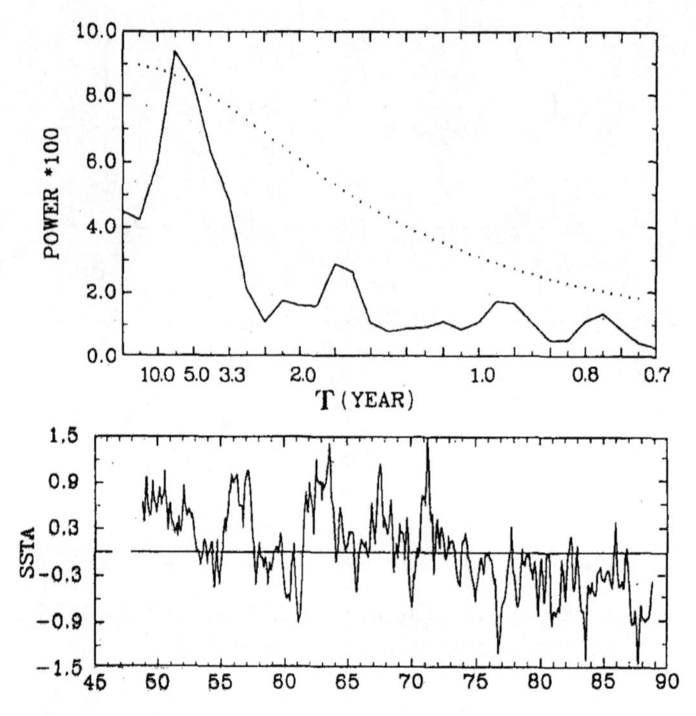

Fig. 4. Temporal variation of monthly SST in the northern Pacific (37.5°-47.5°N, 152.5°E-142.5°W) region (b) and its power spectrum (a). The dotted line represents the significance level 95%.

According to above-mentioned analyses in relation to temporal variations of SSTA, it is clearly shown that the SST in the northwestern Pacific has a quasi-decade oscillation. This is a fundamental oceanic feature in the northwestern Pacific and different from that in the equatorial Pacific in which the ENSO cycle is fundamental.

2. Decadal Variabilities in the Atmosphere over the Northwestern Pacific

The data analyses also showed that the atmospheric circulation in the northwestern Pacific region has clear decadal oscillation as well as the SST. In Fig.5, the interannual variation of the surface air temperature anomaly at Naze and Ishigaki in winter (Dec.-Feb.) and its power spectrum are shown. It is very obvious that a spectral peak with 10-20 years period is prominent and implies the surface air temperature over the northwestern Pacific having a quasi-decadal oscillation.

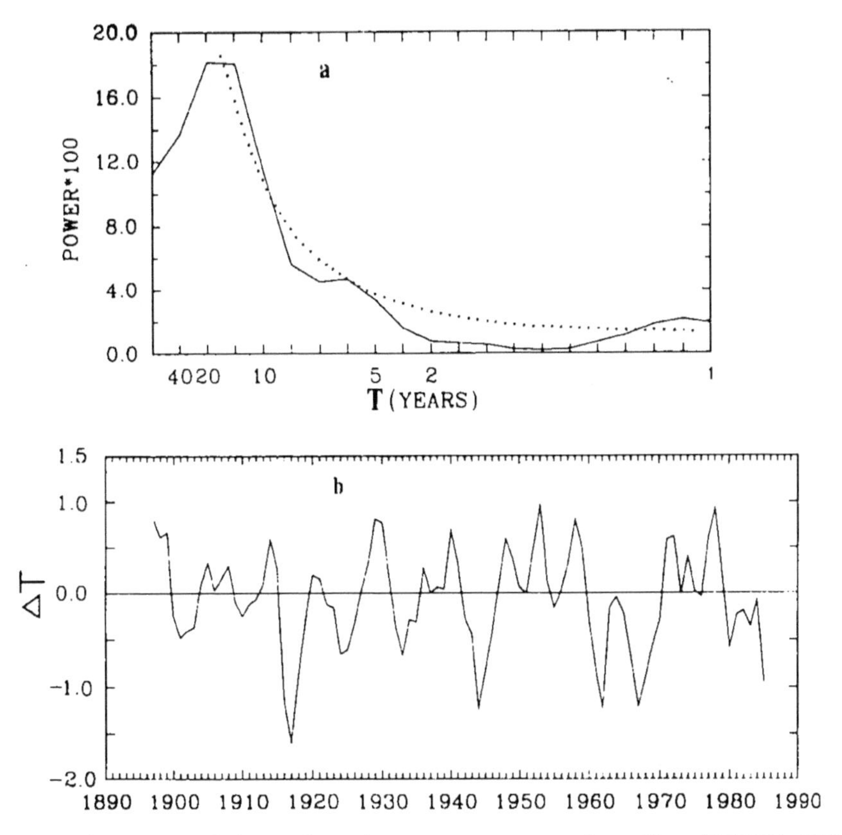

Fig. 5. Interannual variation of the surface air temperature anomaly at Naze (28°23'N, 129°30'E) and Ishigaki (24°20'N, 124°10'E) Islands averaged in December-February (b) and its power spec trum (a). The dotted line represents the significance level 95%.

The subtropical high is an important atmospheric circulation system in the northwestern Pacific. The intensity and latitude location play an significant effect in the climate change in the northwestern Pacific and East Asia regions. Therefore, studying the temporal variation feature of the subtropical high should be necessary and important. The difference between the averaged height at 500 hPa in (20°-30°N, 120°-150°E) region and the zonal mean in 20°-30°N latitudes is able to represent relative intensity of subtropical high over the northwestern Pacific. In Fig.6, temporal evolution (1948-1989) of relative intensity anomaly of subtropical high over the northwestern Pacific and its power spectrum are shown. Two major oscillation periods are also clear, one is about 3.5 years which shows the influence of ENSO on the subtropical high as well as that indicated in some studies (Chen, 1977; Fu and Zeng, 1986), another is 10-15 years which shows a quasi-decadal oscillation feature of the subtropical high over the northwestern Pacific.

Interannual variation of the latitude-location of the ridge-line of subtropical high at 500 hPa over the northwestern Pacific in summer (June-August) is given in Fig.7. And the dashed line shows that it is determined by using OLR data in order to prove reliability of the analyses. The good consistency of the solid line and dashed line means that the Fig.7 is able to describe temporal variation of the ridge-line of subtropical high over the northwestern Pacific. It is clear, particularly based on the dotted line which is obtained by using 3-year moving average, the quasi-decadal oscillation is also existent in the ridge-line location variability of subtropical high over the northwestern Pacific.

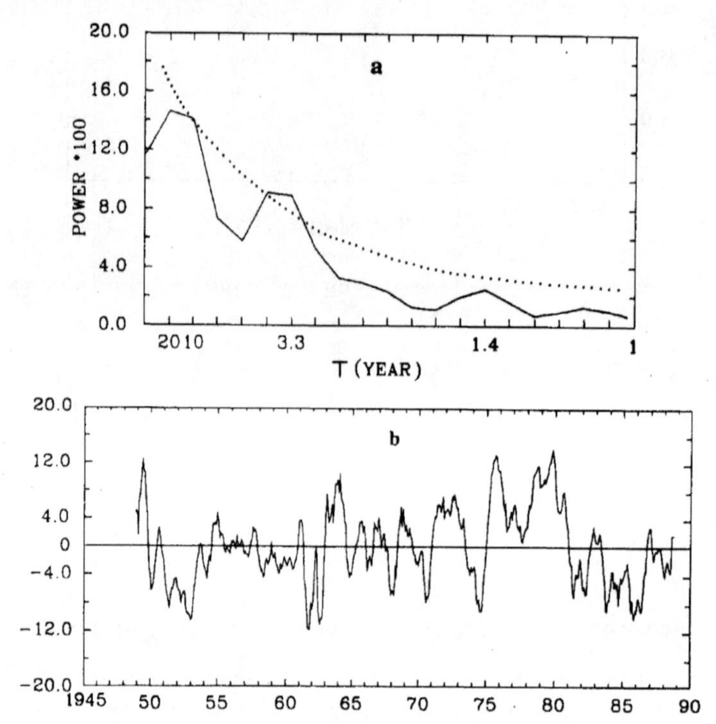

Fig. 6. Temporal variation of relative intensity anomaly of subtropical high over the northwestern Pacific (b) and its power spectrum (a). The dotted line represents the significance level 90%.

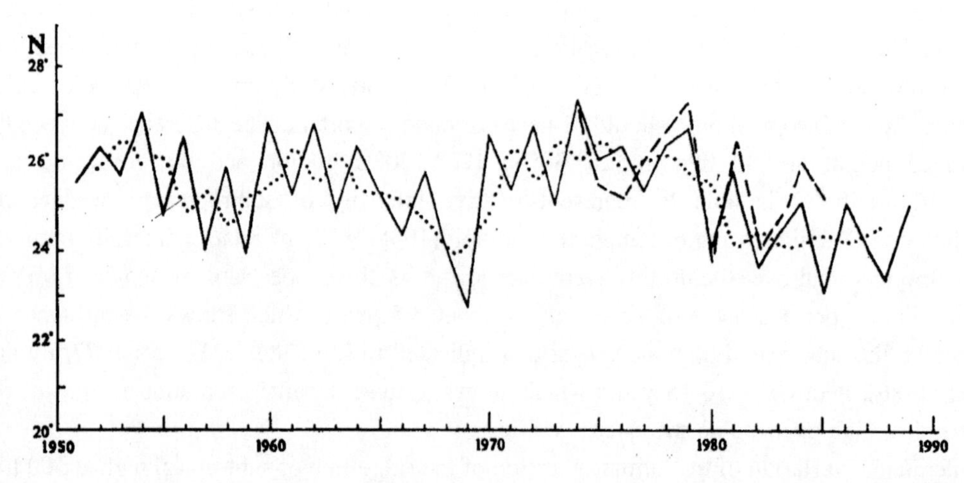

Fig. 7. Interannual variation of the latitude-location of the ridge-line of subtropical high over the northwestern Pacific in summer (June-August). The solid line and dashed line are respectively determined by using the geopotential height at 500 hPa and the OLR data; The dotted line is obtained by using 3-year moving average.

Comparing Fig.7 with Fig.6, it is shown that stronger subtropical high corresponds to its ridge-line to the north. This result is consistent with the weather analyses and they represent an active feature of subtropical high over the northwestern Pacific.

Comparing Fig.7 with Fig.2, it can be found that the subtropical high to the north (south) is basically associated with positive (negative) SSTA in the northwestern Pacific region.

Above-mentioned analyses show that the SST and the atmospheric circulation in the northwestern Pacific region all exhibited a quasi-decadal oscillation feature and the relationship is also shown. Therefore, we can suggest that the quasi-decadal oscillation is possibly a fundamental characteristic of the air-sea system in the northwestern Pacific region.

Ⅲ. Its Influences on Climate in East Asia

In the last part, analyzing the quasi-decadal oscillation of the air-sea system in the northwestern Pacific region has partly included the climate variation feature, such as the surface air temperature. The following discussion will focus to study the climate variability in China as the possible effect of the air-sea system in the northwestern Pacific.

In Fig.8, interannual variation of the precipitation anomalies (%) in summer in Huabei region (including 14 stations: Beijing, Tianjin, Hohhot, Zhangjiakou, Taiyuan, Xixian, Shijiazhuang, Baoding, Xingtai, Zhenzhou, Anyang, Dezhou, Jinan and Hezhe). It is clear that two relative dry periods (1965-1971 and 1980-1987) are obvious and the quasi-decadal varition is also existent. In order to expose the quasi-decadal oscillation of the precipitation in Huabei area, the longer-time data at Beijing, Tianjin, Baoding and Jinan are analyzed. The interannual variation of the summer rainfall at those stations and its power spectrum are shown in Fig.9. A major spectral peak locates at about 12 years and means the quasi-decadal oscillation being a fundamental feature of the summer rainfall in Huabei area.

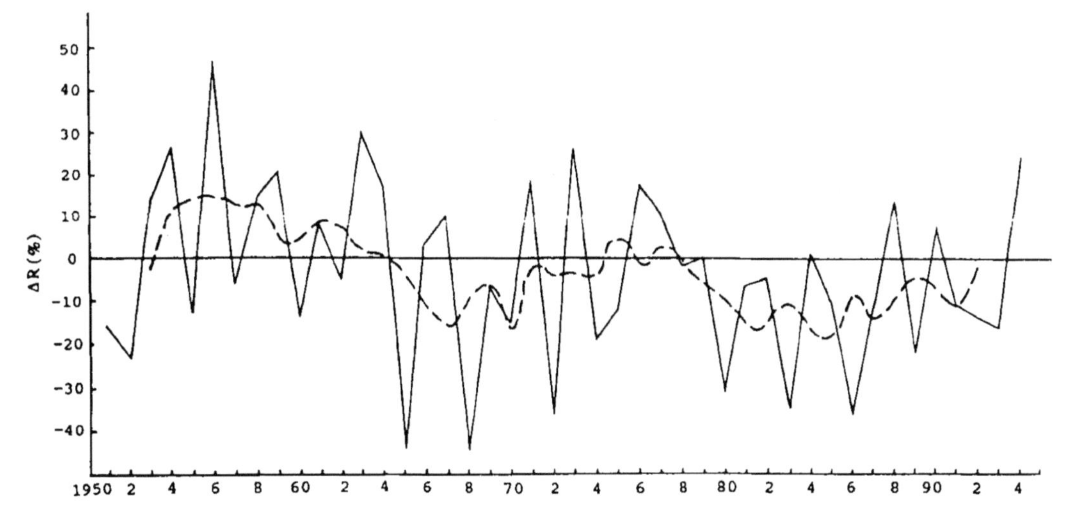

Fig. 8. Interannual variation of the precipitation anomalies (%) in Huabei region in summer (June-August). The dashed line is obtained by using 5-year moving average.

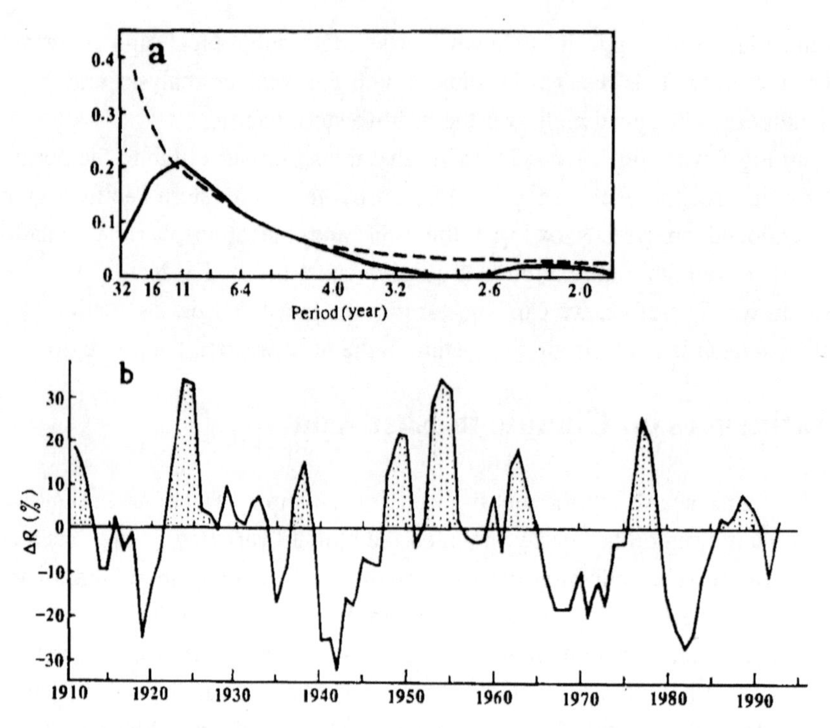

Fig. 9. Interannual variation of the precipitation anomalies (%) in summer (June-August) averaged in Beijing, Tianjin, Baoding and Jinan (b) and its power spectrum (a). The dashed line represents the significance level 95%.

As well as the summer rainfall in Huabei area, the surface air temperature in East China has the quasi-decadal oscillation feature. For example, interannual variation feature of the surface air temperature anomaly in winter (December-February) in Beijing, Qingdao and Nanjing is given in Fig. 10; A spectral peak with about 16 years period is very obvious.

It is well known that the Mei-yu (bai-u in Japan) is very important climate event in East Asia, because its anomaly can lead to the drought and flooding disasters in wide scope. The beginning date of the Mei-yu in the Yangtze River basin is usually closed to the drought or flooding in this area. Analyzing the interannual variation of beginning date of the Mei-yu, it is clearly shown that the early and later beginning date has obvious fluctuation feature and a major period is about 10-15 years (figure omitted).

In Fig.11, the interannual variation of the circulation index at 500 hPa averaged during June-August in (40°-60°N, 90°-170°E) area is shown. It is very clear that the circulation index in middle latitudes of the Asia-northwestern Pacific area has an obvious oscillation feature with about 10 years period.

Comparing above some figures, we can still find that the quasi-decadal oscillations shown in this paper are associated with each other, particularly, consistent very well with that of SSTA in the western Pacific (7.5°-17.5°N, 132.5°-167.5°E) region (Fig.3). These mean that temporal variations with decadal time scale shown in this paper are systematic and they represent a quasi-decadal oscillation feature of air-sea system in the northwestern Pacific region

and the climate variation in East Asia will be affected.

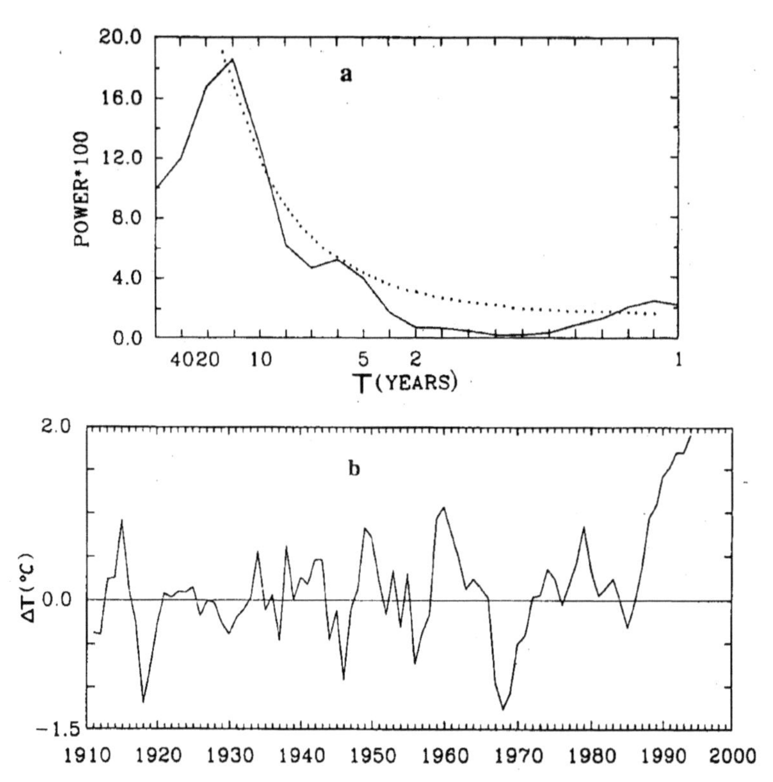

Fig. 10. Interannual variation of mean temperature anomaly in Beijing, Qingdao and Nanjing in winter (b) and its power spectrum (a). The dashed line represents the significance level 95%.

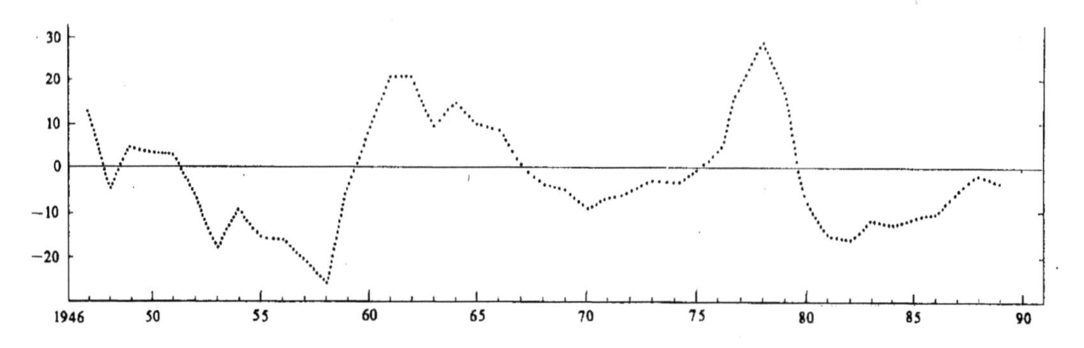

Fig. 11. Temporal variation of the circulation index at 500 hPa averaged during June-August in (40°-60°N, 90°-170°E) area.

Ⅳ. Summary and Discussion

In precent paper, the quasi-decadal oscillations of air-sea system in the northwestern Pacific region and of the climate in East Asia are studied through diagnostic and theoretical analyses. Some important results are obtained, which include the variation feature and possile mechanism.

1) The air-sea system in the northwestern Pacific region has clear systematical quasi-decadal oscillation, such as the surface air temperature, the subtropic high activities and the SSTA which has different time-scale feature from the temporal variation with 3-4 years mean period of SSTA in the equatorial Pacific.

2) The climate variations in East Asia, such as the air temperature, the precipitation, the beginning date of Mei-yu in the Yangtze River basin and so on, also have clear quasi-decadal oscillation. They are related to each other and associated with the decadal variation of air-sea system in the northwestern Pacific region. They can be regarded as the influences of quasi-decadal oscillation of the northwestern Pacific air-sea system.

3) The air-sea coupled interaction is a possible mechanism to excite decadal variations of air-sea system in the northwestern Pacific region and the climate in East Asia. This not only depends on the results of data analyses, which show that there are all the decadal variations in the atmosphere and in the SST and they are closely related with each other; And a simple dynamical study has preliminary shown that the decadal mode can be produced in a simple air-sea coupled model and the effect of the deep ocean process still reduces the frequency of the coupled mode.

The author would like to thank Mr. Liao Qinghai for his work to complete partly computing of this paper and also thank Miss Wang Xuan for typing this manuscript.

References

Chen Lieting (1977), The effect of the anomalous sea-surface temperature of the equatorial eastern Pacific ocean on the tropical circulation and rainfall during the rainy period in China, *Scientia Atmospherica Sinica*, 1: 1-12.

Deser, C. and M.L. Blackmon (1993), Surface climate variations over the North Atlantic ocean during winter: 1900- 1989, *J. Climate*, 6: 1743-1753.

Fu Congbin and Zeng Zaimei (1986), Ten years experiment on long range prediction of the Northwest Pacific high according to sea surface temperature anomalies, *W MO/TD*, 147: 77-86.

Gammelsroed, T., S. Osterhus and O. Godoy (1992), Decadal variations of ocean climate in the Norwegian Sea observed at Ocean Station "Mike" (66°N, 2°E), *ICES Mar. Sci. Symp.*, 195: 68-75.

Graham, N.E, (1994), Decadal-scale climate variability in the tropical and North Pacific during the 1970s and 1980s: Observations and model results, *Climate Dynamics*, 9: 135-162.

Hill, B.T. and S. Jones (1990), The new found land ice extent and the solar cycle from 1860 to 1988, *J. Geophys. Res.*, 95: 5385-5394.

Hurrell, J.W., and H. Van Loon (1994), A modulation of the atmospheric annual cycle in the Southern Hemisphere, *Tellus*, 46A: 325-338.

Kushnir, Y. (1994), Interdecadal variations in the North Atlantic sea-surface temperature and associated atmosheric conditions, *J. Climate*, 7: 141-157.

Li Chongyin (1992), An analytical study on the precipitation in the flood period over Huabei area, *Acta Meteorologica Sinica*, 50: 41-49.

Mysak, L.A., D.K. Manak and R.F. Marsden (1990), Sea-ice anomalies observed in the Greenland and Labrador Seas during 1901-1984 and their relation to an interdecadal Arctic climate cycle, *Climate*

Dynamics, 5: 111-113.

Read, J.F. and W.J. Gould (1992), Cooling and freshening of the subpolar North Atlantic Ocean since the 1960s, *Nature*, 360: 55-57.

Trenberth, K. E. and J. W. Hurrelle (1994), Decadal atmosphere-ocean variations in the Pacific, *Climate Dynamcis*, 9: 303-319.

A Review of Decadal/Interdecadal Climate Variation Studies in China

Li Chongyin*[1] (李崇银), He Jinhai[2] (何金海), Zhu Jinhong[3] (朱锦红)

[1] LASG, Institute of Atmospheric Physics, Chinese Academy of Sciences, Beijing
[2] Nanjing Institute of Meteorology, Nanjing
[3] Department of Atmospheric Sciences, School of Physics, Peking University, Beijing

Abstract Decadal/interdecadal climate variability is an important element in the CLIVAR (Climate Variability and Predictability) and has received much attention in the world. Many studies in relation to interdecadal variation have also been completed by Chinese scientists in recent years. In this paper, an introduction in outline for interdecadal climate variation research in China is presented. The content includes the features of interdecadal climate variability in China, global warming and interdecadal temperature variability, the NAO (the North Atlantic Oscillation)/NPO (the North Pacific Oscillation) and interdecadal climate variation in China, the interdecadal variation of the East Asian monsoon, the interdecadal mode of SSTA (Sea Surface Temperature Anomaly) in the North Pacific and its climate impact, and abrupt change feature of the climate.
Keywords decadal/interdecadal climate variation, abrupt change, east-Asian monsoon, sea surface temperature anomaly.

1. Introduction

In the 1990s, the research on interdecadal variation originally focused on the oceanic state, because the oceanic variability was thought to be a slower process and its interdecadal features more evident. Some studies have shown that the sea surface temperature (SST) variation in the North Atlantic Ocean has a clear interdecadal character and it is related with the NAO (the North Atlantic Oscillation) (Bacon and Carter, 1993; Hurrell, 1995). In the Pacific Ocean, interdecadal variation of the ENSO has been studied (Wang, 1995; Qian et al., 1998) and the EOF analysis of SST in the North Pacific still shows an interdecadal variation feature. In the EOF analyses, the primary part of the main EOF components, which is similar with the ENSO variation, was regarded as the representative of interannual variation (Tanimoto et al., 1993); the remaining part of the main EOF components, which is similar to the ENSO mode, was regarded as the interdecadal variation and named the "ENSO-like mode" (Zhang et al., 1997) or the Pacific Decadal Oscillation (PDO) (Mantua et al., 1997).

Received April 2, 2003; revised November 19, 2003.
*E-mail: lcy@lasg.iap.ac.cn

The studies also indicated the existence of interdecadal variation in the North Pacific with other data analyses (Trenberth and Hurrelle, 1994; Li and Liao, 1996; Li, 1998) and it was still clear in the thermocline variation (Zhang and Levitus, 1997). Naturally, the interdecadal variability of the North Pacific SST and its impact on the climate became important parts of the international CLIVAR (A study of Climate Variability and Predictability) program (WMO et al., 1995). The studies also indicated that the interdecadal climate variation is still shown in the atmospheric circulation variability (Li and Li, 1999a; Li, 2000).

In fact, interdecadal climate variation in China and its jump feature have been studied early. The drought and flood variations in China, and the long-term variations of the summer rainfall and surface temperature in China have been studied and some interesting results were found (Wang and Zhao, 1979; Wang et al., 1981; Wang, 1990; Jiang et al., 1999; Huang et al., 1999; Chen, 1999; Lu, 1999). Some studies and major results in recent years will be shown in this paper in broad outline, and we regret that it is hard to avoid the omission of some studies.

2. Features of Interdecadal Climate Variation in China

Some studies have shown that there is a clear interdecadal variability of summer rainfall over eastern China during the second half of the 20th century (Zhao, 1999; Wang, 2001). Power spectrum analyses for summer rainfall showed a significant peak at 26.7 years. A decreasing trend in precipitation variations has been found based on the observations since 1951, and it seems to end in the 1980s. A weak increasing trend was observed in the 1990s, but their characteristics were different for different areas in China. It is shown that summer rainfall over eastern China, especially over North China, is above normal during the 1950s. Rainfall was slightly above normal north of the Huaihe River and drought occurred along the lower- middle Yangtze River basin and South China during the 1970s. The floods occurred along the Yangtze River basin and droughts were predominant in South and North China during the 1980s. In the 1990s, summer rainfall was above normal along the Yangtze River basin and South China. At the same time, North China was still facing a prolonged drought period.

The characteristics of interdecadal variability of annual precipitation are similar to those of summer rainfall over eastern China. There have been five drought spells since 1880. The first one was from the end of the 19th century to the beginning of the 20th century, the second from the second half of the 1920s to the beginning of the 1930s, the third and the fourth for the whole period of the 1940s and 1960s, and the fifth from the end of the 1970s to the beginning of the 1980s. No linear trend occurred during the period from 1880 to 1999. The power spectrum analysis shows a significant peak around 30 years. A 20-40-yr periodicity is predominant for the whole series. The anomalies relative to the period of 1880-1999 for 6 flood and 5 drought spells are shown in Table 1.

The studies on summer rainfall in North China showed that the interdecadal variation is also evident (Chen, 1999; Huang et al., 1999; Li et al., 2002). The major variation periods are about 20 years and about 40 years. It was indicated that these inter-decadal variations of summer precipitation are related to the anomalies of the intensity of the East Asian summer monsoon and the latitude of the subtropical high ridge over the west Pacific in summer (Fig. 1). For the strong (weak) East Asian summer monsoon period, there is abundant (sparse) rainfall in North China and for the north (south) side of the subtropical high ridge over the west Pacific in summer, there is abundant (sparse) rainfall in North China.

Table 1. The precipitation anomalies averaged for 35 stations relative to the period of 1880-1999 during 6 flood and 5 drought periods (Wang, 2001).

NO	Flood		Drought	
	Period	Anomaly (mm)	Period	Anomaly (mm)
1	1881-1885 1888-1892	42.2 72.0	1899-1902	−109.6
2	1911-1915 1918-1922	94.5 65.2	1925-1929	−75.5
3	1931-1935	45.7	1942-1946	−19.9
4	1950-1954	90.0	1963-1968	65.1
5	1972-1976	62.2	1978-1982	−30.4
6	1990-1994	43.2		

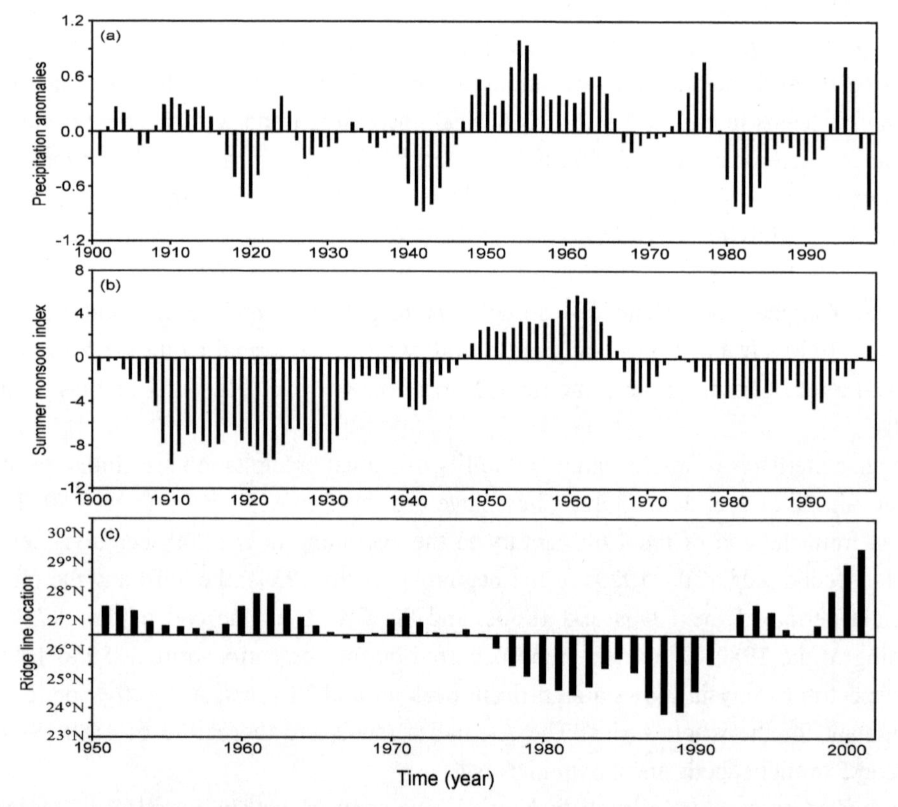

Fig. 1. (a) 9-year running mean histograms for the summer precipitation in North China, (b) the intensity of the East Asian summer monsoon, and (c) the latitude of the subtropical high ridge over the west Pacific in summer. (Li et al., 2002)

Precipitation variation is quite different for eastern China and western China. There was no linear trend during 1880-1999 and a 20-40-yr period variation was predominant for the whole series of eastern China precipitation. On the contrary, the increasing trend of precipitation in west China was very noticeable in the second half century, especially during the last 30 years. It also showed tremendous drought in the 1920s-1930s in west China.

A decadal-centennial variability is studied based on rainfall coded-level data in 1470-1999 at 25 stations over the eastern part of China. The power spectrum analyses of 530 years of rainfall data demonstrate that an 80-yr oscillation exists in some areas of the eastern part of China and reaches the 95% significance level (Fig. 2). The 80-yr oscillation component of summer rainfall in North China even explains 27% of the variance in the low frequency band. This component of summer rainfall over North China, the lower-middle Yangtze River valley, and South China shows that the phase of this component over North China is precisely consistent with that over South China and out of phase with that along the lower-middle Yangtze River valley.

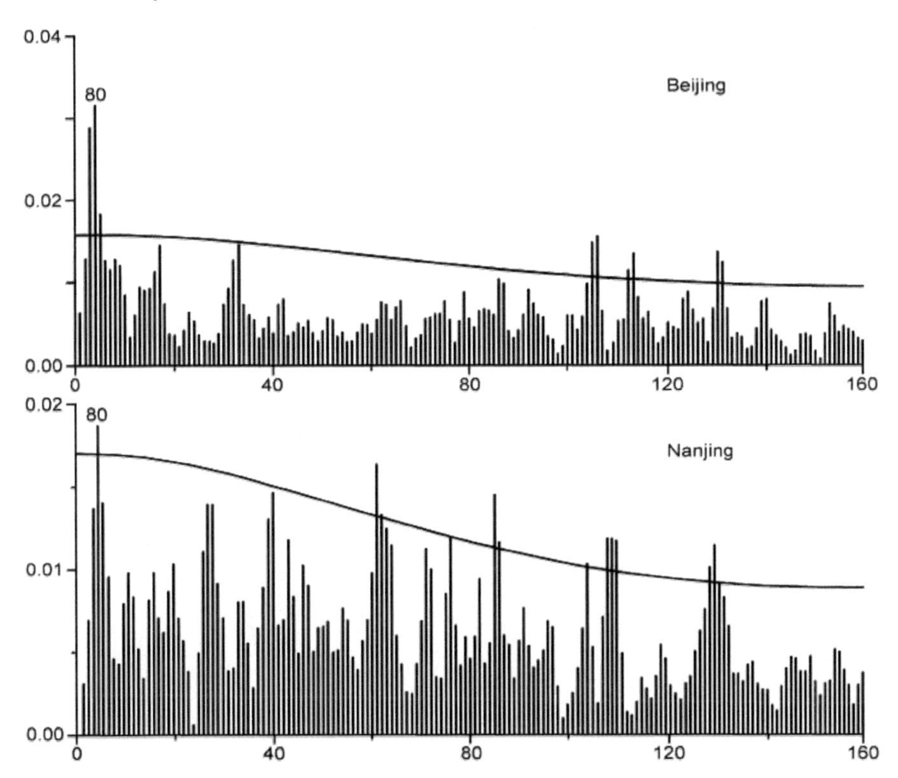

Fig. 2. Power spectra for the rainfall graded data (1470-1999) at Beijing and Nanjing. The solid line is the 95% significance level. (Zhu and Wang, 2001)

3. Interdecadal Variations of the East Asian Monsoon

The East Asian Summer Monsoon (EASM) is one of the important factors, which control summer rainfall over the east part of China. The rainfall in North China is also sensitive to the intensity of EASM. The subtropical high and ITCZ usually move to the north if the summer monsoon is strong and active, then the precipitation will be above normal over North and South China. In the opposite case, the subtropical High and ITCZ are displaced to the south when the summer monsoon is weaker, droughts will be found over North and South China, and floods will occur along the middle and lower reaches of the Yangtze River. The studies also showed that the EASM and summer rainfall over eastern China have variability features with multiple timescales, including interannual variation, decadal/interdecadal variation, and an 80-yr quasi-period oscillation (Zhang, 1999; Lu, 1999; Yang and Song, 1999; Zhu and Wang, 2001).

Li and He (2000) examined the correlation between the EASM and the SSTA over the China off-sea area, the western Pacific subtropical high ridge, and the equatorial eastern-central Pacific SST in the preceding winter and spring. The result showed that the relation between the EASM and equatorial eastern-central Pacific SST displays a strong interdecadal change, with a higher correlation after 1976 than before 1976. From the time evolution of the East-Asian summer meridional cell index, the zonal Walker cell index, and the equatorial eastern-central Pacific SSTA (Sea Surface Temperature Anomaly) in summer, it is clear that the coupling between the East-Asian summer meridional cell and zonal Walker cell exhibits an interdecadal change, although there is a stable, significant relation between the Walker cell and the equatorial eastern-central Pacific SSTA (Fig. 3). It is interesting to note that the negative correlation coefficients between the index of the East Asian summer meridional cell and zonal Walker cell is greater in magnitude than –0.77 in 1976-1993 at the 99.9% significance level, much higher in magnitude than that of –0.41 in 1958-1975. Therefore, after 1976 the warmer equatorial eastern- central Pacific SSTA leads to a weaker zonal Walker cell (weaker westerly flows in the upper troposphere and weaker easterlies in the lower one). These changes weaken the East Asian summer meridional cell (weaker northerly flows in the

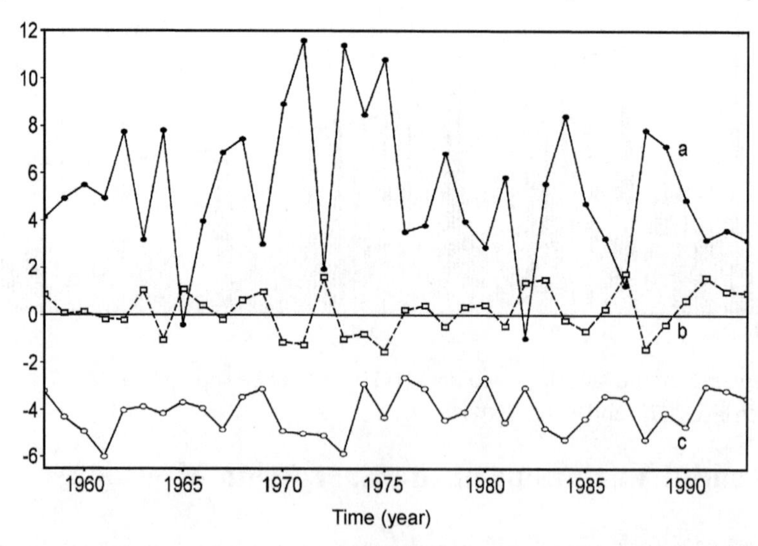

Fig. 3. Time evolution of the East-Asian summer meridional cell index (solid line c), zonal Walker cell index (solid line a), and the equatorial eastern-central Pacific SSTA in summer (dashed line b). (Li and He, 2000)

upper troposphere and weaker southerlies in the lower one) due to its close coupling with the zonal Walker cell, accompanied by the weaker summer monsoon and the southward shifting of the subtropical high, resulting in enhanced rainfall in the low-mid reaches of the Yangtze River.

4. Global Warming and Interdecadal Temperature Variability

The annual mean temperature anomaly series of ten regions in China are studied and obtained for the period of 1880 to 1999, which are determined relative to the normals of 1961-1990 (Wang et al., 1998). The temperature series of China are averaged over ten regions considering the regional weights (figure omitted). It is indicated that the warming in the 20th century started in 1920 and was interrupted in the 1950s and 1960s. Positive anomalies over China during the period of the 1920s-1940s are noticeable. The temperature increased persistently since the end of the 1960s. The linear trend for the period of 1880-1999 is 0.62℃ $(100\ \text{yr})^{-1}$, a little greater than that of the globe 0.60℃ $(100\ \text{yr})^{-1}$. 1998 was the warmest year in China since 1880. Studies of the relationship between temperature and precipitation indicated no consistent correlation.

On the basis of multi-taper spectral analysis (Jiang et al., 2001), the statistical analysis of the monthly mean temperature time series in the Northern and Southern hemispheres from 1856 to 1998 showed that the warming trend played a dominant role in mean temperature variability in the Northern and Southern hemispheres during the last 150 years. However, there is significant interdecadal variation with periods of about 40 and 60-70 years, which are superimposed on a linear warming trend for the Northern Hemisphere mean temperature (Fig. 4). This situation leads to the diminishing of the linear warming rate with its significance and stability, as opposed to that in the Southern Hemisphere, especially in summer. Moreover, in comparing surface temperature on the land to the sea, interdacadal variations detected in the latter are more remarkable than those in the former, in contrast to the linear warming rate. Furthermore, in terms of the GCM results from the HadCM2 model, a preliminary analysis implied that the interdacadal variation may be the inherent oscillation of the ocean and atmosphere system, but that warming trends are not related to natural variability.

5. NAO/NPO and Interdecadal Climate Variation in China

In recent years, some studies have indicated that the interdecadal variation of the NAO exhibited a rising trend (Hurrell, 1995; Jones et al., 1997). Through data analyses, it is very evident that both the NAO index and NPO index experience variation suddenly in the 1960s, such that their common characteristics were represented by the abnormal rising of the amplitude, where the amplitude after the 1960s was about 2-3 times greater than before the 1960s (Li and Li, 1999a). The wavelet analyses for temporal variations of the NAO and NPO respectively show that interdecadal variations of the NAO and NPO are very clearly represented (figure omitted). First, the amplitudes of the NAO and NPO increased abnormally since the 1960s. Second, it is very evident that the interannual variations with a 3-4-year period were fundamental both the NAO and the NPO before the 1960s, but the decadal variations with an 8-15-year period have been fundamental since the 1960s. Therefore, both the NAO and the NPO experienced anomalous variations, which were not only represented in the amplitude increasing but also in the changing of the period for the dominant mode from 3-4 years to 8-15 years. In other words, the increasing of amplitude since the 1960s is not only represented in the NAO but also in the NPO. Therefore, this kind of interdecadal variation not only exists in the North Atlantic region, but it seems to also occur in other regions.

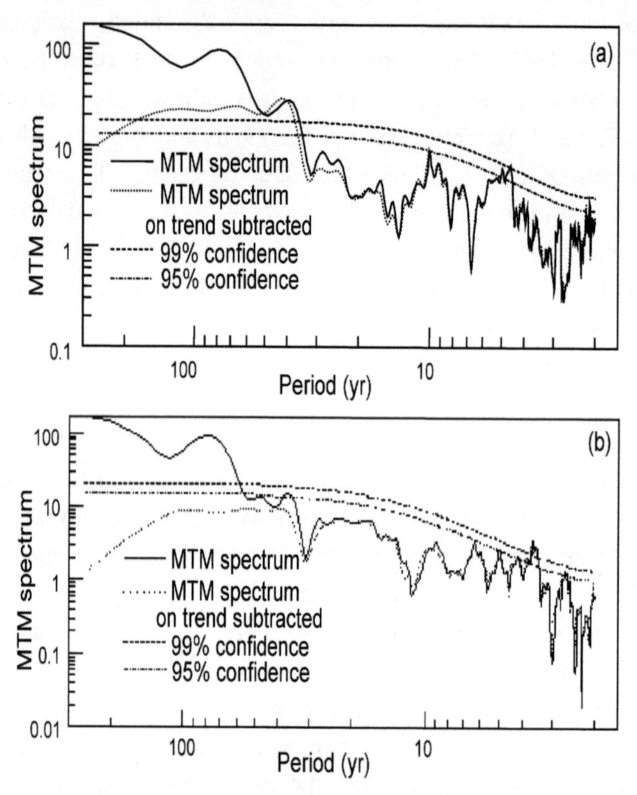

Fig. 4. MTM spectral estimation of the (a) Southern and (b) Northern hemispheres mean surface temperature time series (MTM spectrum based on original series (solid line), MTM spectrum based on time series with trend subtracted off (dotted lines) and the 95% (dot-dashed lines) and 99% (dashed lines) confidence limits based on a robust, red-noise fit to the spectrum). (Jiang et al., 2001).

The climate jump in the China in 1960s was indicated in some studies (Yamamoto et al., 1986; Yan et al., 1990). It still appeared clearly in the summer (June-August) precipitation anomaly (%) in North China, where there were mainly positive precipitation anomalies before 1964 but mainly negative precipitation anomalies from and after 1964, and where the averaged summer precipitation changed suddenly from being above normal to being below normal (Li, 1992). The surface air temperature anomaly in winter (December-February) in Sichuan also changed into a cold period (negative temperature anomalies) since 1962, even though positive anomalies during shorter time periods were in existence (Li and Li, 2000). Obviously, it is very evident that a climate jump occurred in the 1960s and the interdecadal climate variation in China was demarcated in the 1960s. These results can suggest that the interdecadal variations of the NAO and NPO are closely related to the climate jump in the 1960s. Although it is difficult to say that the climate jump in the 1960s (or interdecadal climate variation) in China resulted from atmospheric circulation variation, particularly from the interdecadal variation of the NAO and NPO; but at least, the above analysis results can suggest that the climate jump in the 1960s, or the interdecadal climate variation in China, is closely related to the interdecadal variation of the NAO and the NPO.

The influence of the NAO variation on the East-Asian monsoon and climate was studied preliminarily by Wu and Huang (1999) who indicated that the Siberian cold high would be affected by the NAO variation at first in winter, then the anomalous Siberian cold high could affect the cold waves (winter monsoon) and the climate (including summer rainfall) in East Asia. This occurs because a strong (weak) East-Asian winter monsoon is closely related to a strong (weak) Siberian cold high and a strong (weak) NAO index.

Based on the climate jump in the 1960s and its relationship to the anomalies of the NAO and NPO, it can be suggested that the atmospheric circulation anomaly, which is represented by the NAO and NPO, is also a possible important factor causing interdecadal climate variation in China.

6. Interdecadal Mode of SSTA in the North Pacific and Its Impact

In order to understand the interdecadal variation of the SSTA in the North Pacific Ocean (meaning the Pacific Ocean north of 10°S latitude in this study), the interdecadal mode of the North Pacific SST and its evolution features are investigated further by using the Hadley Center monthly data (1900-1997) but in a different way from EOF analysis (Xian and Li, 2003). The spectrum analyses of the SSTA in the North Pacific showed that two common, main spectrum peaks can be found, one with a period of about 7-10 years and the other with a period of about 25-35 years. The wavelet analysis results of the SSTA in the North Pacific also showed that the 7-10-yr and 25-35-yr periods are two fundamental periods. Therefore, the variations with 7-10-yr and 25-35-yr periods can be regarded as two major interdecadal modes, although there are still other periods.

In order to show the pattern of the two inter-decadal modes of the SSTA variation in the North Pacific, band-pass filterings of the SST in the North Pacific with a 7-10-year filter and a 25-35-year filter are respectively performed and the patterns of the two modes are obtained. In Fig. 5, the basic situations of the positive phase and negative phase of the 25-35-yr mode and 7-10-yr mode are shown. For the positive phase, there is positive SSTA in the area of 30°-50°N and west of 140°W; but negative SSTA in the area south of 30° N and along the coast of North America. For the negative phase, there is positive SSTA in the area south of 30° N and along the coast of North America; but negative SSTA in the area of 30°-50°N and west of 140°W. Although the pattern of the 7-10-yr mode is similar to that of the 25-35-yr mode, we do not want to compose them into one. They exist independently in the spectrum analysis results, so it is unsuitable to compose them into one artificially.

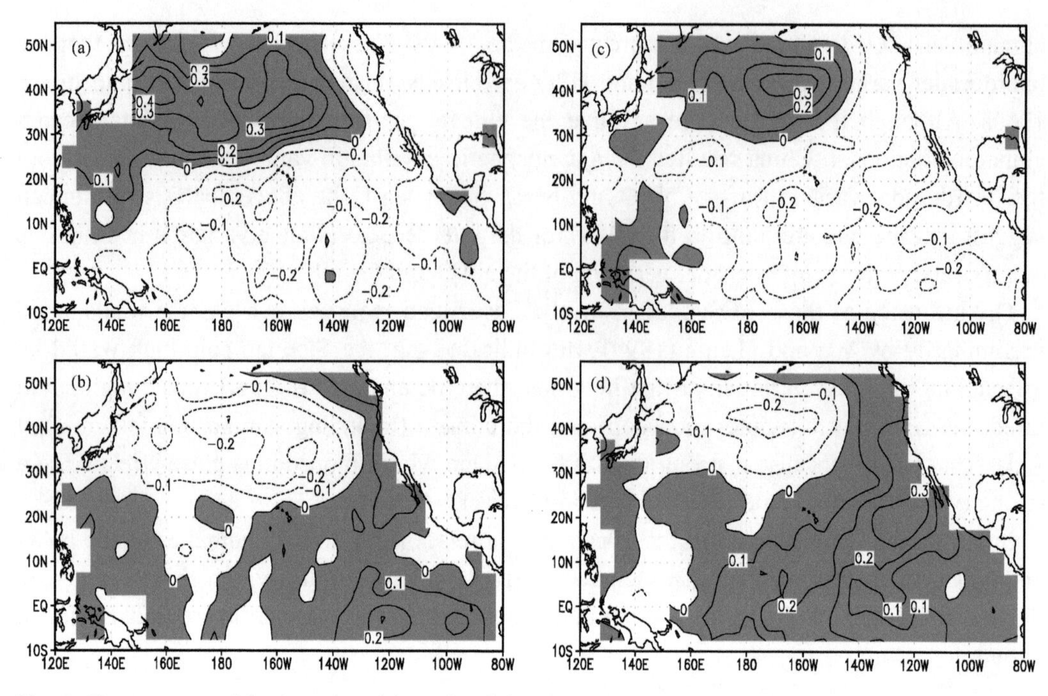

Fig. 5. The patterns of the interdecadal mode of the North Pacific SST in positive phase (a) and negative phase (b) for 25-35-yr mode; in positive (c) and negative (d) phases for 7-10-yr mode. (Xian and Li, 2003)

The fundamental patterns of the interdecadal mode of the North Pacific SSTA are different from the "ENSO-like mode" although the above modes also showed that the signal is stronger in the mid-latitudes than in the Tropics. The basic character of the ENSO mode should be as follows: positive (negative) SSTA within the limited scope of the equatorial eastern Pacific and maximum SSTA nearby the equator; and a band-type negative (positive) SSTA in a southwest-northeast direction from the equatorial western Pacific to the northeastern Pacific but positive (negative) SSTA in the northwestern Pacific. However, the pattern of the interdecadal mode of the North Pacific SSTA has a consistent symbol SSTA in the equatorial Pacific, the western Pacific, and the northeastern Pacific, so the features of the ENSO mode are not clear there.

The impacts of the interdecadal mode of the North Pacific SSTA on the climate were studied by using data analysis (Li and Xian, 2003). The composite analyses of sea level pressure (SLP) corresponding to the 25-35-yr mode of the North Pacific SSTA showed that during winter when the positive phase of the North Pacific SSTA 25-35-yr mode appears, positive SLP anomalies are found north of 30°N in the Pacific region, with a maximum of 4hPa located near the Aleutian Islands, indicating a weak Aleutian low in that period (Fig. 6a). Positive anomalies are also found over Siberia and the North Atlantic Ocean indicating a strengthened Siberian high and a weak Icelandic low; furthermore, negative anomalies with a smaller amplitude appear on the North American continent, suggesting a weak North American high in that period. During winters when the negative phase of the 25-35-yr mode is present, an opposite SLP anomaly pattern over the North Pacific emerges. There are negative anomalies over the North Pacific, centered at the Aleutian Islands with a maximum of 4 hPa, smaller negative anomalies over the North Atlantic, positive anomalies over most parts of the North American continent, and weak positive anomalies over the North Eurasian continent (Fig. 6b). The SLP anomalies over the North Pacific are most directly affected by SSTA. Therefore when SSTA changes its polarity from positive to negative, the sign of the SLP anomalies is also changed correspondingly.

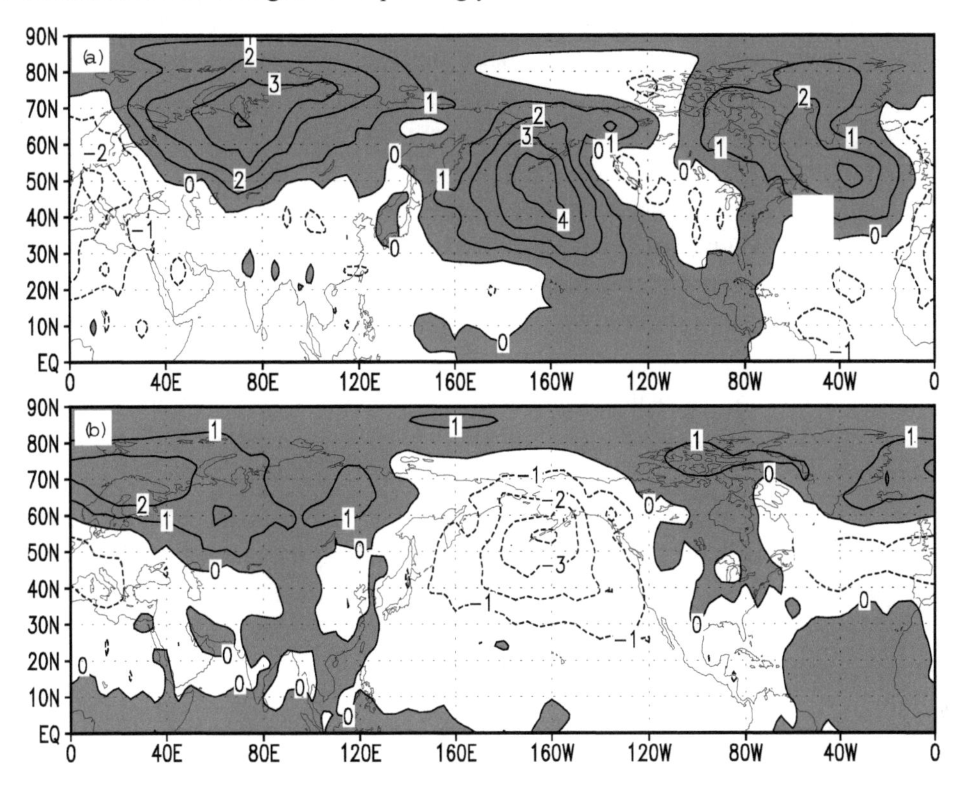

Fig. 6. The SLP anomalies (hPa) in winter corresponding to the (a) positive and (b) negative phases of the SSTA 25-35-yr mode in the North Pacific. (Li and Xian, 2003)

During winter when the positive (negative) phase of the North Pacific SSTA 7-10-yr mode appears, the anomalous patterns of the SLP field are similar to those of 25-35-yr mode positive (negative) phase. This means that corresponding to the positive or negative phase of the 7-10-yr mode or 25-35-yr mode of the North Pacific SSTA, the global sea level pressure field generally has identical responses. In other words, similar anomalous SLP patterns correspond to similar SSTA distributions in the North Pacific, which fully reveals the significant impacts of interdecadal SSTA modes (variations) in the North Pacific on the atmospheric circulation and climate.

The anomalous patterns of the 500-hPa height and 1000-hPa wind field in wintertime corresponding to the positive/negative phases for the 25-35-yrmode and 7-10-yr mode are analyzed. It can be shown that anomalous patterns of the SLP and 500-hPa fields are very similar to each other, and the anomalous wind field is systematically coordinated to the anomalous SLP field. Furthermore, comparing the spatial signatures of the response fields of the SLP and 500-hPa height in the extratropical North Pacific and tropical Pacific, we find a positive correlation (response) in the extra-tropical region and a negative correlation (response) in the tropical region, i.e., a positive (negative) SLP anomaly in the extratropical region results from positive (negative) SSTA, but a negative (positive) SLP anomaly in the tropical region results from positive (negative) SSTA.

The anomalous field of annual global land precipitation is analyzed corresponding to the two phases of each interdecadal mode (Li and Xian, 2003). The results clearly show that the global precipitation pattern is closely related to the interdecadal modes of the North Pacific SSTA. During years when the positive phase of the SSTA 25-35-yr mode in the North Pacific occurs, there is more precipitation in eastern and southeastern Asia, but less precipitation in southern North America. In Australia, there is more precipitation in the east, and less in the west. During years when the negative phase of the SSTA 25-35-yr mode in the North Pacific appears, there is less precipitation in eastern and southeastern Asia, but more precipitation in southern North America. In Australia there is less precipitation in the east, and more in the west (Fig. 7).

A similar precipitation anomalous field is also found corresponding to the North Pacific SSTA 7-10-yr mode. During years when the positive phase is present, there is more precipitation in eastern China; for North America, there is less precipitation in the east and south, and more in the west. There is less precipitation in central and southern South America, and more precipitation in eastern Australia and central Africa (figure omitted). During year when the negative phase appears, there is less precipitation in eastern China, more precipitation in eastern and southern North America, less in northwestern North America, and more in central and southern South America and less in central Africa.

7. Abrupt Climate Change

The climate variation, particularly the long-term climate variation (change), usually shows an abrupt change feature. Some data analyses have indicated that there are evident climate (temperature and precipitation) abrupt changes in China during the 1920s and during the 1960s (Yan et al., 1990; Yan, 1992; Ye and Yan, 1993). A statistical test method named the Mann-Kendall Rank examination is available to determine climate abrupt change. As an example, by using the Mann-Kendall test result of the drought index in eastern China during 1887-1986, it is shown that the abrupt changes occurred at 1922 and 1965, which represented a variation from the relatively moist period to the relatively dry period in eastern China (figure omitted).

Some studies have also shown that the jump variations of summer precipitation in eastern China are clear. The data analyses using the recent 50 years of data showed that summer precipitation over the Huabei region displayed two jump variations, in the middle of the 1960s and at the end of the 1970s, respectively (Chen, 1999). Further analysis showed that the jump variation around 1965 was mainly the reduction of precipitation over the Huabei region, and the character of jump variation around 1976 was the increase of precipitation in the Yangtze River basin but continued decrease of precipitation over the Huabei region (Huang et al., 1999).

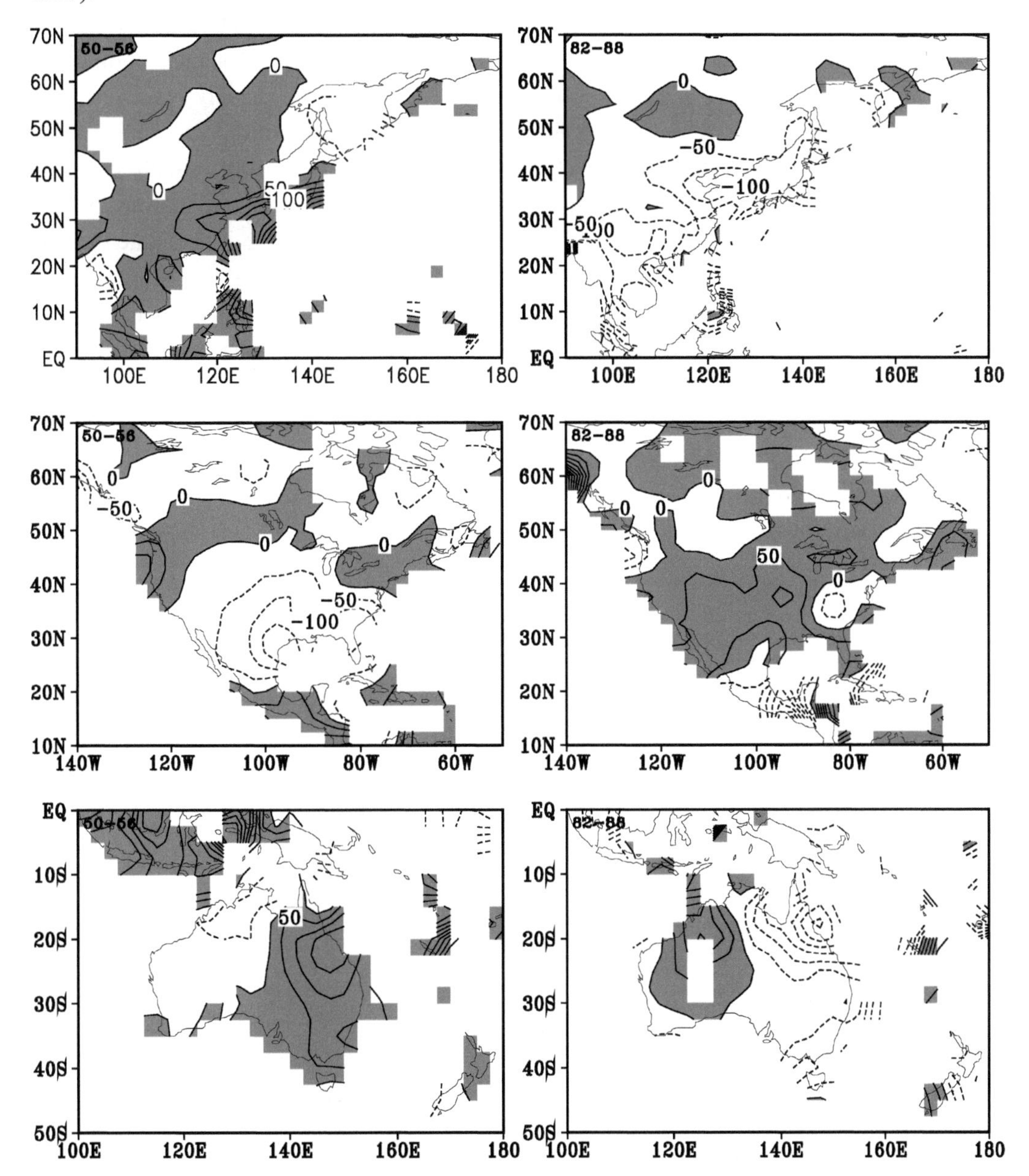

Fig. 7. Annual precipitation anomalies corresponding to the positive (left) and negative (right) phases of the North Pacific SSTA 25-35-yr mode. Top: in East Asia; middle: in North America; bottom: in Australia. (shading represents positive anomaly). (from Li and Xian, 2003)

Using the Mann-Kendall Rank examination, the abrupt change of summer rainfall over North China and the low-mid Yangtze River basin was also found around the mid-1970s (Li and He, 2000). This abrupt change was related with the anomaly of the East Asian summer monsoon (EASM). The stronger EASM lead to more summer rainfall amount in North China, as contrasted to the low-mid Yangtze River basin before 1976. And the opposite situation was observed after 1976. The analysis still showed that the EASM anomaly is closely correlated to SSTA in the North Pacific, which affected the interannual variation of summer rainfall in North China before the mid-1970s. After the mid-1970s, the EASM anomaly was closely related to SSTA in the equatorial eastern-central Pacific instead of the North Pacific SSTA, and the impact of the equatorial eastern-central Pacific SSTA on summer rainfall over the low-mid Yangtze River basin was enhanced. Furthermore, the analyses still showed that the air-sea temperature difference over the North Pacific displayed a significant interdecadal change (Fig. 8). Before the mid-1970s, there was a greater air-sea temperature difference over the North Pacific, which means the SSTA have a stronger impact upon the atmosphere, so that the relation of the North Pacific SSTA with the EASM circulation was enhanced. But after the mid-1970s, the air-sea temperature differences were lower and the effect of SSTA on the atmosphere was insignificant, and there was a weaker relation between the North Pacific SSTA and the ESAM.

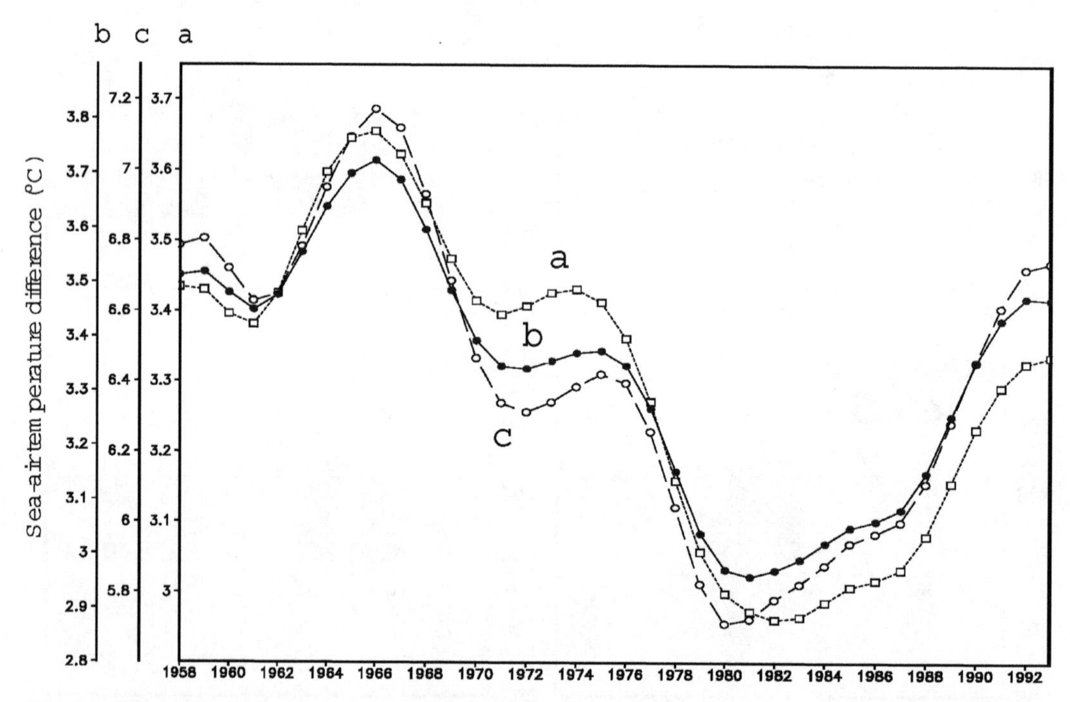

Fig. 8. Time series of a 9-point running mean air-sea temperature difference over the North Pacific: (a) SST-$T_{925\ hPa}$; (b) SST-$T_{1000\ hPa\text{-}850\ hPa}$; (c) SST-$T_{850\ hPa}$. (from Li and He, 2000)

8. Conclusions

(1) The long-term climate (precipitation and temperature) variability in China has multiple

timescale features. Except for interannual variation, the decadal/interdecadal variation, the 20-40-yr, around 10-yr and 60-80-yr, are major periods. These decadal/interdecadal climate variations are related to the SSTA in the Pacific.

(2) The East Asian summer monsoon (circulation and precipitation) also has quite evident decadal/interdecadal variation.

(3) The warming trend plays a dominant role in the mean temperature variability in the Northern and Southern hemispheres during the last 150 years. However, the significant interdecadal variation was superimposed on a linear warming trend of mean temperature, particularly, in the Northern Hemisphere.

(4) The temporal variations of the NAO and NPO very clearly show that the amplitudes of these two oscillations increased suddenly in the 1960s and their main period of interannual variations changed from 3-4 years to 8-15 years. These evident variations of the two oscillations in the 1960s represented a fundamental anomaly in the atmospheric circulation in the 1960s. The climate jump that occurred in the 1960s was related to the anomalies of the NAO and NPO, and it can be suggested that the atmospheric circulation anomaly, which is represented by the NAO and NPO, is also a possible important factor causing the interdecadal climate variation in China. The atmospheric circulation anomalies can also be regarded as an important way to understand interdecadal climate variation and the climate jump.

(5) The interdecadal variation of the North Pacific SST has two fundamental modes, the 7-10-year mode and the 25-35-year mode. These two interdecadal modes have similar patterns, particularly their fundamental patterns for positive phase and for negative phase. Whether in the positive phase or negative phase, the pattern of the interdecadal mode is different from the "ENSO-like mode". From the analyses, it can be suggested that the evident oscillation in the northwest-southeast direction and the clockwise rotation along the Pacific Ocean basin are the common evolution features of the interdecadal mode of the North Pacific SST.

(6) Corresponding to the positive and negative phases of the interdecadal mode of the SSTA in the North Pacific, the anomalous patterns of the atmospheric circulation/climate are very different. This means the impact of the interdecadal mode of the SSTA in the North Pacific on the atmospheric circulation/climate is very clear. The global SLP field has similar responses to the same phases of the 25-35-yr mode and the 7-10-yr mode of SSTA in the North Pacific. The 500-hPa height anomaly patterns are similar to the SLP anomaly patterns, which implies the response of the extratropical atmosphere to the interdecadal modes of SSTA in the North Pacific exhibits a barotropical structure, but the response of the tropical atmosphere shows a baroclinic structure. The global 1000-hPa wind field has an analogous response corresponding to the two interdecadal modes of SSTA in the North Pacific. The anomalous wind field has a systematic structure and is very coherent with the anomalous SLP field.

The impact of the interdecadal mode of SSTA in the North Pacific on regional annual precipitation is not negligible. During years when the positive (negative) phase of the interdecadal modes appears, some regions have less or more precipitation. For example, during the positive (negative) phase period, there is more (less) precipitation over eastern China, less (more) over southern North America, and a band of increased precipitation in eastern (western) Australia.

(7) The climate variation, particularly the longterm (decadal/interdecadal scale) climate variation (change) usually shows an abrupt change feature. The Mann-Kendall Rank analyses showed that in the last century three abrupt changes occurred respectively in 1920-1925, 1960-1965, and the mid-1970s.

Acknowledgments This study was partly supported by the National Natural Science Foundation of China (Grant No. 40233033) and the Chinese Academy of Sciences (ZKCX2-SW-210 and KZCX2-203).

References

Bacon, S., and D. J. T. Carter, 1993: A connection between mean wave height and atmospheric pressure gradient in the North Atlantic. *Int. J. Climatology*, 13, 423-436.

Chen Lieting, 1999: Regional features of interannual and interdecadal variations in summer precipitation anomalies over North China. *Plateau Meteorology*, 18, 477-485. (in Chinese)

Huang Ronghui, Xu Yuhong, and Zhou Liantong, 1999: The interdecadal variation of summer precipitations in China and the drought trend in North China. *Plateau Meteorology*, 18, 465-476. (in Chinese)

Hurrell, J. W., 1995: Decadal trends in the North Atlantic Oscillation: Regional temperatures and precipitation. *Science*, 269, 676-679.

Jiang Zhihong, Ding Yuguo, and Tu Qipu, 1999: Interdecadal spatial structure and evolution of extreme temperatures in winter and summer over China during the past 50 years. *Quarterly Journal of Applied Meteorology*, 10(Suppl.), 97-103. (in Chinese)

Jiang Zhihong, Tu Qipu, and Shi Neng, 2001: The multitaper spectral analysis method and its application in the global warming research. *Acta Meteorologica Sinica*, 54, 480-490.

Jones, P. M., T. Jonsson, and D. Wheeler, 1997: Extension to the North Atlantic Oscillation using early instrumental pressure observations from Gibralter and SW Iceland. *Int. J. Climatology*, 17, 1433-1450.

Li Chongyin, 1992: An analysis study on the flood season precipitation in the Huabei region. *Acta Meteorologica Sinica*, 50, 41-49. (in Chinese)

Li Chongyin, 1998: The quasi-decadal oscillation of air-sea system in the northwestern Pacific region. *Adv. Atmos. Sci.*, 15, 31-40.

Li Chongyin, 2000: Decadal and interdecadal climate variation. *Introduction to Climate Dynamics (second edition)*, China Meteorological Press, Beijing, 421-448. (in Chinese)

Li Chongyin, and Liao Qinghai, 1996: Quasi-decadal oscillation of climate in East Asia and Northwestern Pacific region. *Climatic and Environmental Research*, 1, 133-140. (in Chinese)

Li Chongyin, and Li Guilong, 1999a: Variation of the NAO and NPO associated with climate jump in the 1960s. *Chinese Science Bulletin*, 44, 1983-1986.

Li Yaoqing, and Li Chongyin, 1999b: Relationship between temperature decrease in Sichuan region and the SSTA in the tropical western Pacific in recent 40 years. *Climatic and Environmental Research*, 4, 388-395. (in Chinese)

Li Chongyin, and Li Guilong, 2000: The NPO/NAO and interdecadal climate variation in China. *Adv. Atmos.*

Sci., 17, 555-561.

Li Chongyin, and Xian Peng, 2003: Atmospheric anomalies related to interdecadal variability of SST in the North Pacific. *Adv. Atmos. Sci.*, 20, 859-874.

Li Chun, Sun Zhaobo, and Chen Haishan, 2002: Interdecadal variation of North China summer precipitation and its relation with East Asian general circulation. *Journal of Nanjing Institute of Meteorology*, 25, 455-462. (in Chinese)

Li Feng, and He Jinhai, 2000: Interdecadal variation of the interaction between SSTA in the North Pacific and the summer monsoon in East Asia. *Journal of Tropical Meteorology*, 23, 378-387. (in Chinese)

Lu Riyu, 1999: Interdecadal variations of precipitations in various months of summer in North China. *Plateau Meteorology*, 18, 509-519. (in Chinese)

Mantua, N. J., S. R. Hare, Y. Zhang, J. M. Wallalce, and R. C. Francis, 1997: A Pacific interdecadal climate oscillation with impacts on salmon production. *Bull. Amer. Meteor. Soc.*, 78, 1069-1079.

Qian Weihong, Ye Qian, and Zhu Yafen, 1998: Monsoonal oscillation revealed by the upper-troposphere water vapor band brightness temperature. *Chinese Science Bulletin*, 43, 1489-1494.

Tanimoto, Y., N. Iwasaka, K. Hanawa, and Y. Toba, 1993: Characteristic variations of sea surface temperature with multiple time scales in the North Pacific. *J. Climate*, 6, 1153-1160.

Trenberth, K. E., and J. W. Hurrelle, 1994: Decadal atmosphere-ocean variations in the Pacific. *Climate Dyn.*, 9, 303-319.

Wang, B., 1995: Interdecadal changes in El Niño onset in the last four decades. *J. Climate*, 8, 267-285.

Wang Shaowu, 1990: The variation tendency of temperature in China and the globe during the last one hundred years. *Meteorology*, 16, 11-15. (in Chinese)

Wang Shaowu, 2001: *Advances of Modern Climatology Research in China*. China Meteorological Press, Beijing, 458pp. (in Chinese)

Wang Shaowu, and Zhao Zongci, 1979: 36yr period of the drought/flood in China and its mechanism. *Acta Meteorologica Sinica*, 37, 64-73. (in Chinese)

Wang Shaowu, Zhao Zongci, and Chen Zhehua, 1981: Reconstruction of the summer rainfall regime for the last 500 years in China. *J. Geography*, 5, 117-122. (in Chinese)

Wang Shaowu, Ye Jinling, Gong Daoyi, and Zhu Jinhong, 1998: Construction of mean annual temperature series for the tast last one hundred years in China. *Quarterly Journal of Applied Meteorology*, 9, 392-401. (in Chinese)

WMO, ICSU, and UNESCO, 1995: *CLIVAR—A Study of Climate Variability and Predictability*. WMO/TD No.690, Geneva, 172pp.

Wu Bingyi, and Huang Ronghui, 1999: Effects of the extremes in the North Atlantic Oscillation on East Asia winter monsoon. *Chinese J. Atmos. Sci.*, 23, 641-651.

Xian Peng, and Li Chongyin, 2003: Interdecadal modes of sea surface temperature in the North Pacific Ocean and its evolution. *Chinese J. Atmos. Sci.*, 27, 118-126.

Yamamoto, R., T. Iwashima, and N. K. Sanga, 1986: An analysis of climatic jump. *J. Meteor. Soc. Japan*, Ser.II, 64, 273-281.

Yan Zhongwei, 1992: A primary analysis of the process of the 1960s northern hemispheric summer climatic jump. *Chinese J. Atmos. Sci.*, 16, 111-119. (in Chinese)

Yan Zhongwei, Ji Jinghun, and Ye Duzheng, 1990: Northern hemispheric summer climatic jump in the 1960's, I: Precipitation and temperature. *Science in China* (*B*), 1, 97-103.

Yang Hui, and Song Zhengshan, 1999: Multiple time scales analysis of water resources in North China. *Plateau Meteorology*, 18, 496-508. (in Chinese)

Ye Duzheng, and Yan Zhongwei, 1993: Climate Jumps in the history. *Climate Variability*, China

Meteorological Press, Beijing, 3-14. (in Chinese)

Zhang Qingyun, 1999: The variations of the precipitation and water resources in North China since 1880, *Plateau Meteorology*, 18, 486-495. (in Chinese)

Zhang, R. H., and S. Levitus, 1997: Structure and cycle of decadal variability of upper ocean temperature in the North Pacific. *J. Climate*, 10, 710-727.

Zhang, Y., J. M. Wallace, and D. S. Battisti, 1997: ENSO- like interdecadal variability: 1900-93. *J. Climate*, 10, 1004-1020.

Zhao Zhenguo, 1999: *Summer Drought and Flood in China and the Circulation Fields*, China Meteorological Press, Beijing, 214pp. (in Chinese)

Zhu Jinhong, and Wang Shaowu, 2001: 80a-oscillation of summer rainfall over the east part of China and East- Asian summer monsoon. *Adv. Atmos. Sci.*, 18, 1043-1051.

Interdecadal Variation of the Relationship Between Indian Rainfall and SSTA Modes in the Indian Ocean

Xin Wang[a, b, *], Chongyin Li[a, c], Wen Zhou[a]

[a] LASG, Institute of Atmospheric Physics, Chinese Academy of Sciences, Beijing, China
[b] Graduate School of the Chinese Academy of Sciences, Beijing, China
[c] Institute of Meteorology, PLA University of Science and Technology, Nanjing, China

Abstract This paper examines the relationships between Indian rainfall and the sea-surface temperature anomalies (SSTA) in different areas, including the Arabian Sea, the equatorial Indian Ocean, the southern Indian Ocean and the equatorial eastern Pacific. Their relationships have clear temporal and spatial variabilities. Before the 1980s, the correlation between Indian summer rainfall and ENSO was much stronger than the correlation between Indian summer rainfall and other SSTA. Thus, Indian summer rainfall was mainly affected by ENSO during that period. But in recent decades, ENSO has become less decisive and the Indian rain is influenced by combinations of SSTA in all the regions. The influences of the Arabian Sea on the Indian summer rainfall are affected by ENSO and are much weaker than those of the equatorial Indian Ocean and the southern Indian Ocean. SSTA in the equatorial Indian Ocean could affect rainfall over India independently. When the amplitude of SSTA in the southern Indian Ocean is large enough, SSTA in the southern Indian Ocean can play an important role in controlling rainfall, which is evident in the late 1980s.

Keywords SSTA in the Indian Ocean, ENSO, Indian rainfall.

1. Introduction

Indian rainfall has been impacted by the sea-surface temperature (SST) in the Indian Ocean and the Arabian Sea (e.g. Shukla, 1975; Shukla and Mrsra, 1977; Shukla and Mooley, 1987). Shukla (1975) suggested that colder sea-surface temperature anomalies (SSTA) over the western Arabian Sea may cause reduction of summer monsoon rainfall over India and adjoining areas. However, Weare (1979) illustrated that the correlation coefficient between SST in the Arabian Sea and Indian rainfall is weakly negative. But Rao and Goswami (1988) pointed that Weare ignored the large seasonal variations of SST in his study, and furthermore found a homogeneous region in the southeastern Arabian Sea over which the March-April

Received April 14, 2005; revised September 13, 2005; accepted October 4, 2005.
* Correspondence to: e-mail: wangxin@mail.iap.ac.cn

SSTA are significantly positively correlated with the seasonal (June-September) rainfall over India. Clark *et al.* (2000) found that in the boreal fall and winter preceding the summer Indian monsoon, SST throughout the tropical Indian Ocean correlates positively with subsequent monsoon rainfall over India, and the high correlation of SST in the Arabian Sea with the Indian summer rain is largely unaffected by the removal of the ENSO signal, whereas the correlation of the tropical central Indian Ocean with the Indian rain is reduced. The different types of zonal distributions of SSTA in the Arabian Sea and the South China Sea may result in different variations of the Indian and east Asian monsoon systems (Chen, 1991).

In general, the deficit/surplus of Indian summer rainfall relates to El Niño/La Nina. But in 1997, when a strong ENSO occurred, the Indian rain was not significantly reduced. This inverse relationship is noteworthy by many experts. It was pointed out in 1965 that the connection between Asian monsoon and ENSO actually appeared to be statistically nonstationary (Troup, 1965). Troup's conclusion was further validated by Torrence and Webster (1999). Kumar *et al.* (1999) suggested that the inverse relationship between ENSO and the Indian summer monsoon has broken down in recent decades, and this could be due to the southeastward shifting of the Walker circulation anomalies associated with ENSO and the increased surface temperatures over Eurasia in winter and spring. Recently, Chang *et al.* (2001) showed that the weakening relationship between Indian monsoon rainfall and ENSO results from the strengthening and poleward shift of the jet stream over the North Atlantic.

Clark *et al.* (2000) showed that the SST indices in the Arabian Sea (ASI) and in the northwest of Australia (NWAI) significantly correlate (positively) with the rainfall index over the Indian subcontinent (AIRI). But according to the definition and method of Clark *et al.* (2000), although the 18-year sliding means of ASI and NWAI are positive, the rainfall over India was negative during 1964-1975 (Figure 1). Apparently, the relationship between SSTA in the Indian Ocean and summer rainfall over India has different features in different periods, which shows distinct interdecadal variation.

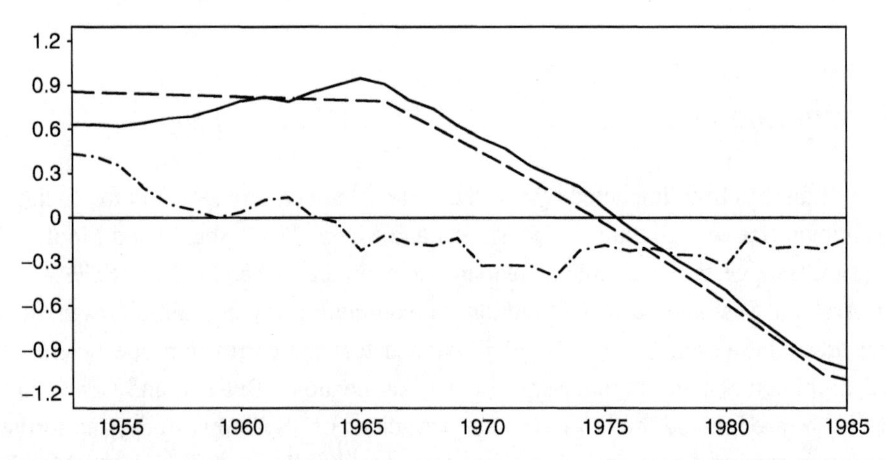

Figure 1. The variations of 18-year sliding means of SSTA in the Arabian Sea index (solid line), SST in the northwest of Australia index (dashed line) and the index of rainfall over the Indian subcontinent (dotted-dashed line)

The previous studies have shown that the correlation between SST and rainfall over India is complicated. The purpose of this paper is to reexamine the impacts of SST in different ocean areas on monsoon rainfall over India, including the Arabian Sea, the equatorial Indian Ocean and the southern Indian Ocean. Data and methodology are first introduced in Section 2. Results of the correlation analysis are then presented in Section 3. The possible linear regressions between Indian rainfall and the SST index in different regions are examined in Section 4. Concluding remarks are given in Section 5.

2. Data and Methodology

The SST data set used in this study is the global sea-ice and sea-surface temperature (GISST) data set compiled from ship records. These data are in the form of 1° latitude-longitude monthly mean temperature for the years 1900-1999. The SST in the GISST data set is processed using the empirical orthogonal function (EOF) interpolation. The rain data is from the mean rain of the area over the Indian region for the period 1900-1999.

The Niño3 SST index is used to represent the strength of ENSO; Niño3 is the area-averaged SSTA over the eastern equatorial Pacific (5°S-5°N, 150°-90°W) shown in Figure 2 as Niño3 Index. The Arabian Sea index (ASI) is defined as the area-averaged SSTA in the ASI area shown in Figure 2, normalized by its annual average. The residual of the normalization by the annual average of area-averaged SSTA in the equatorial Indian Ocean index (EIDI)_1 area and EIDI_2 area in Figure 2 is the dipole mode of the EIDI. The strength of the dipole mode in the southern Indian Ocean index (SIDI) is depicted by the residual of the normalization by annual average of area-averaged SSTA in SIDL_1 area to SIDI_2 area in Figure 2. It illustrates that the dipole mode appears in the southern Indian Ocean (Behera and Yamagata, 2001).

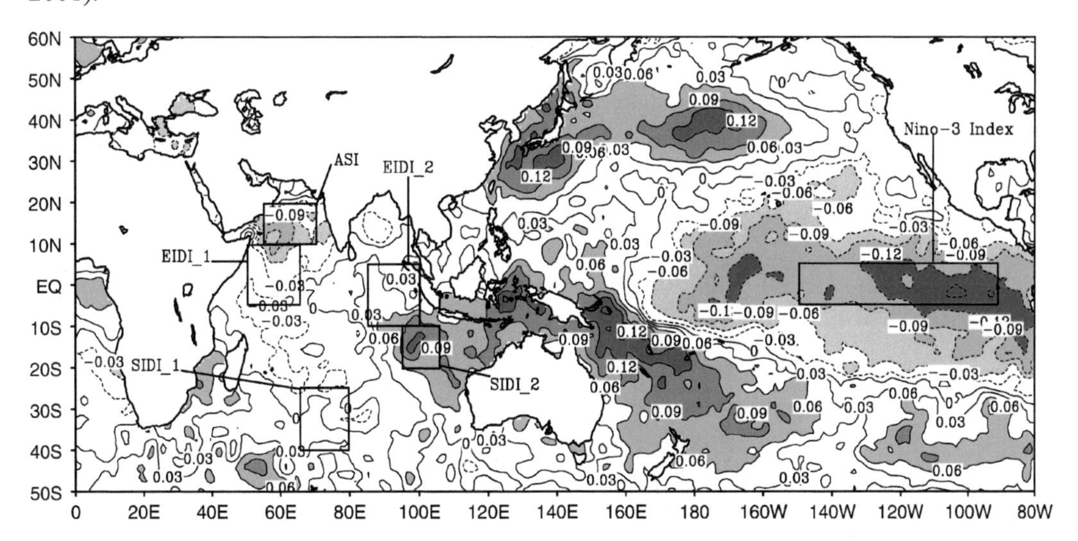

Figure 2. Correlation between monthly Indian rain and SSTA in the Indian Ocean and Pacific. Correlations significant at 5% and above are shaded.

We compute the correlation coefficients between SST and Indian rainfall and use the *t*-test to determine their significance. Since Chang *et al.* (2001) and Kumar *et al.* (1999) used 21-year sliding correlations between Indian summer monsoon rain and Niño3 SST, we will also use the 21-year sliding correlations for a fair comparison with their results. Besides, we have utilized different data to develop line regression equations using the method of least square:

$$Y = a_0 + a_1X_1 + a_2X_2 + \ldots + a_m X_m \quad (1)$$

where Y denotes the Indian summer rainfall, a_0, a_1, a_2, …a_m are the constants and X_1, X_2, …X_m are the different predictor parameters. The multiple correlation coefficients are computed and the *F*-test is used to check whether the rains derived from the line regression equations are significant or not.

3. Results of the Correlation Analysis

Ashok *et al.* (2001) elucidated that the Indian Ocean dipole (IOD) (Saji *et al.*, 1999) plays an important role as a modulator of the Indian monsoon rainfall, influencing the correlation between the Indian summer monsoon rain and ENSO. They discovered that the Indian summer monsoon rain anomalies depend on the relative intensities of the IOD and the El Niño/La Nina events. A relatively stronger Niño3 SSTA, with a normalized value of 1.7, as compared to the concurrent positive intensities of the Indian Ocean dipole (IODMI) with a normalized value of 1.4, may have caused a deficit in rain in India. On the other hand, a stronger positive IOD in 1983 (as compared to the concurrent El Niño) may have caused a surplus in the monsoon rainfall. In our study, we analyze the spatial variability of correlation between Indian rainfall and the SST in different areas. As shown in Figure 2, the rainfall over the Indian region is significantly negatively (at the 95% level) correlated with the SST in Niño3 area and in the Arabian Sea and is significantly positively correlated with the SST in the North Pacific, the tropical western Pacific, the eastern Indian Ocean and the southwest Indian Ocean. The correlation of Indian rainfall and ENSO and its temporal variability have many beneficial results. In comparison with these results, we mainly study the impacts on the Indian rainfall with regard to the SST in the Arabian Sea, the dipole mode of the equatorial Indian Ocean and the dipole mode of the southern Indian Ocean.

3.1 Temporal variability of correlations of SST in different areas and Indian rain

The long-recognized negative correlation between Indian monsoon rain and ENSO has obviously weakened since the late 1970s. Likewise, the correlations of the Indian rainfall and the ASI, the EIDI and the SIDI have temporal variations (Figure 3). The relationship between the Niño3 index and Indian rainfall (Figure 3), which has weakened since the 1980s and has had a small positive correlation in the late 1980s, is consistent with the results of Kumar *et al.* (1999) and Chang *et al.* (2001). During 1930-1960, the Indian rainfall had a significant positive correlation with the SIDI and a negative correlation with the ASI, but these two correlations seem to have weakened from the late 1960s to the 1980s. While in 1970-1980, the correlation of Indian rainfall and the EIDI was very small and fluctuated between positive and negative, the correlation of ASI and Indian rainfall changed from negative to positive in the early 1980s and increased until the 1990s. Obviously, the correlation coefficient between the Indian rainfall and the EIDI was mostly negative, except for the years 1950-1960, 1970 and after 1985. Note that before 1980 the correlation of the Indian rain with the Niño3 index was much larger than those coefficients of the Indian rain with the ASI, the SIDI and the EIDI, respectively, but magnitudes of the correlations of the Indian rainfall with the SST index are similar after 1980, suggesting that the Indian rainfall was mainly impacted by ENSO before 1980s and was jointly affected by the ENSO, SST in the Arabian Sea, the dipole mode of the equatorial Indian Ocean and the dipole mode of the southern Indian Ocean. It seems that the SST in the Arabian Sea, the dipole mode of the equatorial Indian Ocean or the dipole mode of the southern Indian Ocean do not simply positively (or negatively) correlate with the Indian rainfall as mentioned in previous studies, but they have different effects on the Indian rain during different periods. The changing correlation between Indian rain and ENSO might be influenced by the Indian Ocean dipole (Ashok *et al.*, 2001), by Atlantic circulation (Chang *et al.*, 2001) or by Eurasian warming (Kumar *et al.*, 1999). These explanations about the changing relationships, however, have no definite conclusions.

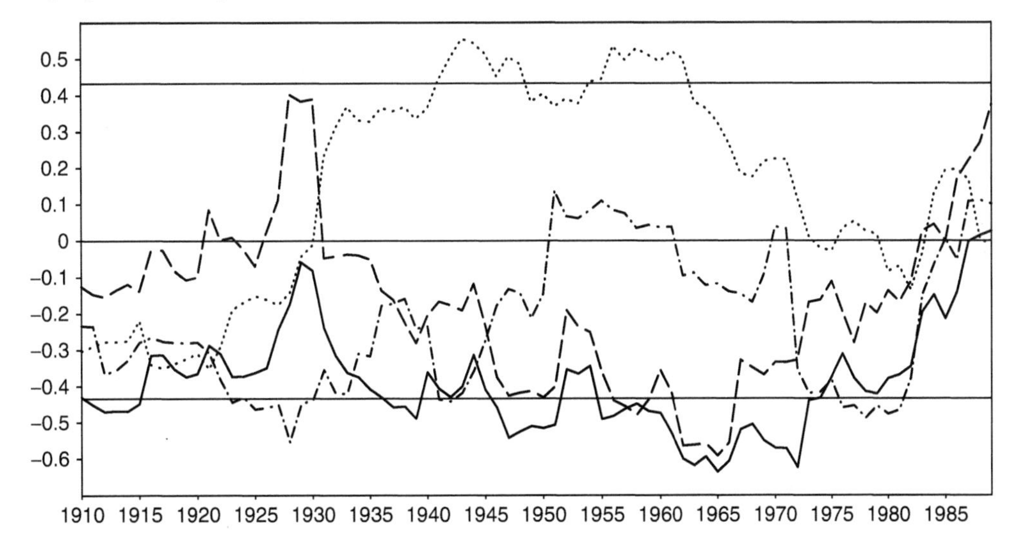

Figure 3. Correlation (based on a 21-year sliding window) of Indian rainfall and Niño3 index (solid line), ASI (dashed line), SIDI (dotted line) and EIDI (dotted-dashed line). The 5% significant levels are indicated as two horizontal lines

In order to synthetically analyze the amplitudes of the Niño3 index, ASI, EIDI and SIDI and the influences of these indices on the Indian rain, we use the 21-year sliding means of the Niño3 index, ASI, EIDI and SIDI to multiply the corresponding 21-year sliding correlation coefficients of these indices and the Indian rain. Figure 4 shows the different effects of these indices on rainfall. The variation of the ENSO-related Indian rainfall is similar to the original

Indian rainfall. The amplitude of the ENSO-related Indian rainfall varies a lot during 1950-1970 but not very much in the late 1980s, which may be due to the weakening of its correlation coefficient in the late 1980s (Figure 3) and the decrease of the amplitude of the Niño3 index after 1980 (Figure 5). In other words, there is a consistent negative impact on Indian rainfall from the ENSO signal; the Indian rainfall varies when the amplitude of the ENSO signal changes.

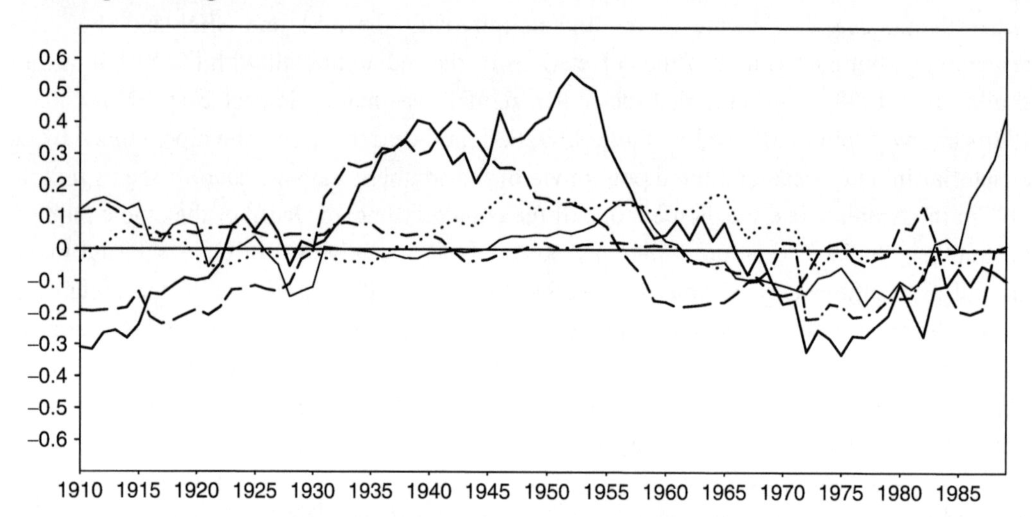

Figure 4. Arithmetic products using the 21-year sliding means of Niño3 index (dotted), ASI (thin solid), EIDI (dotted-dashed) and SIDI (dashed) multiplied with the corresponding 21-year sliding correlation coefficients of these indices and the Indian rain. The thick solid line is the result of the 21-year sliding mean of normalized Indian rain by annual average

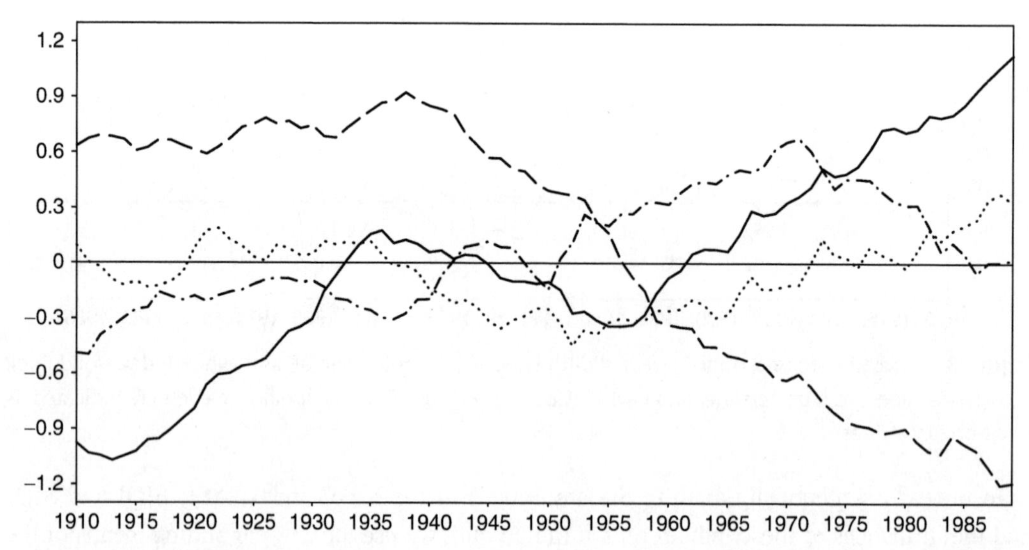

Figure 5. The variations of the 21-year sliding means of ASI (solid line), SIDI (dashed line), Niño3 index (dotted) and EIDI (dotted-dashed line)

Unlike the impact of the Niño3 index, the ASI-related Indian rainfall does not correlate positively with the original Indian rainfall (Figure 4). The ASI-related Indian rainfall and the original Indian rainfall were in phase during 1945-1963 and 1967-1983, but were out of phase during other periods, particularly after 1983. This antiphase indicates that the impact of ASI on Indian rain is very weak. Figure 4 also shows that the influence of EIDI on Indian rainfall has distinct interdecadal variability. Before 1930, the EIDI-related Indian rainfall and the original Indian rainfall were antiphase, but were strongly in phase after 1930. Thus, the effect of the EIDI on Indian rain changed and appeared to be stronger after 1930, and the influence of EIDI appeared to be weak again in the late 1970s. It is also interesting to note that the SIDI-related Indian rainfall had a similar variation as the original Indian rainfall most of the time except for two short periods (1956-1967 and 1980-1983). The fact that the SIDI significantly correlates with Indian rain has not been noticed in previous studies. To some extent, the SIDI might play an important role in the Indian rainfall.

3.2 Correlations of SST (removal of the ENSO signal) in different areas and Indian rain

It is well known that the ENSO signal could influence global sea temperatures. To highlight the effects of the sea temperatures in the equatorial Indian Ocean, south Indian Ocean and the Arabian Sea on Indian rainfall, the method of Clark *et al.* (2000) is employed to remove the ENSO signal, and the correlation of the 21-year moving average of EIDI, SIDI and ASI with the 21-year moving average of Indian rainfall is computed and shown in Figures 6, 7 and 8 respectively. Clark *et al.* (2000) performed a regression analysis at each point in a related ocean area to remove the ENSO signal from SST. The regression equation is derived from the definition of the correlation coefficient. Except for a smaller value, there is no significant change of the correlation between EIDI and Indian rainfall (Figure 6), which means that the equatorial Indian dipole affects Indian rainfall in a similar way as ENSO. For the SIDI, the correlation is very different after the removal of the ENSO signal (Figure 7), especially during the 1930-1950s when the correlation coefficients were of a different sign. The correlation coefficient also changed from negative to positive in 1950 and back to negative in 1973. During 1950-1970, the original SIDI and that after the removal of the ENSO signal correlated positively with Indian rainfall, but the correlation was weaker in the original; after the 1980s, the correlation became opposite in sign, with a negative correlation for the SIDI after the removal of the ENSO signal, and the correlation was stronger (larger absolute value of correlation coefficient) than that of the original SIDI. In brief, the significant difference of correlations before and after the removal of the ENSO signal implies that the SST effects of the tropical eastern Pacific could not be overlooked when we investigated the relationship between the south Indian Ocean SST and Indian rainfall. Figure 8 shows the significant difference in the correlation of ASI and Indian rainfall before and after the removal of the ENSO signal. For the ENSO-removed ASI, the correlation coefficient was positive most of the time during 1920-1989, consistent with Shukla (1975). But the correlation coefficient was negative for the original ASI during 1930-1982, and became positive after 1983. The amplitude of the oscillation of the correlation coefficient is also different from that of the ENSO-removed ASI. These results indicate that the effect of the Arabian Sea SST on Indian rainfall is heavily modified by ENSO. When the ENSO influence on the Arabian Sea is not considered, the Arabian Sea SST correlates positively with Indian rainfall. When the ENSO influence is considered, the correlation becomes negative.

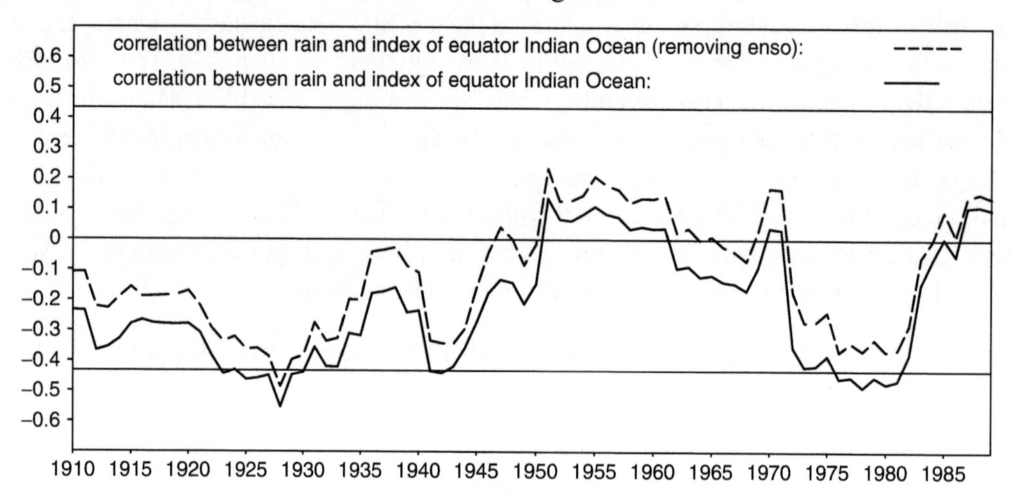

Figure 6. Correlations (based on a 21-year sliding window) of Indian rain and EIDI (solid), EIDI with the removal of the ENSO signal (dashed). The 5% significant levels are indicated as two horizontal lines

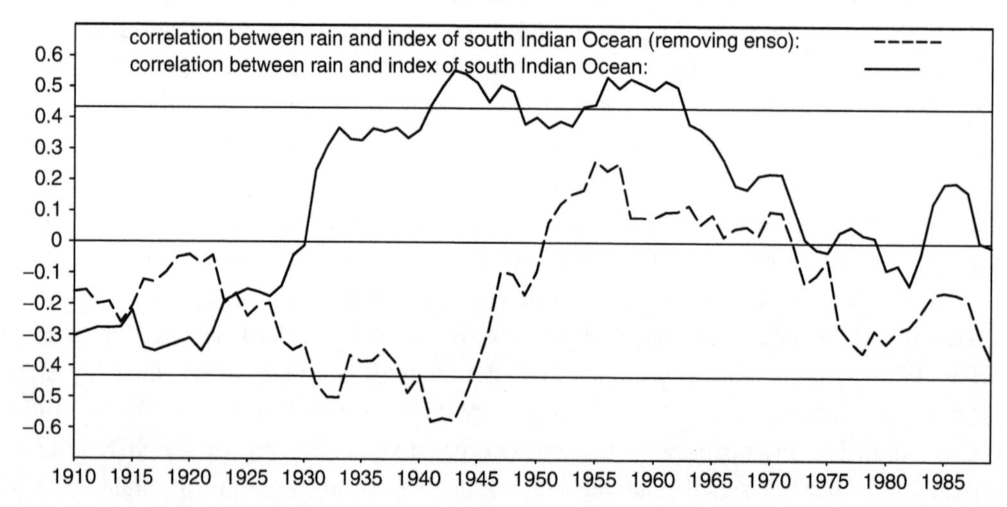

Figure 7. Correlations (based on a 21-year sliding window) of Indian rain and SIDI (solid), SIDI with the removal of the ENSO signal (dashed). The 5% significant levels are indicated as two horizontal lines

In order to have an even better understanding of the ENSO-removed ASI, EIDI and SIDI on Indian rainfall, computation as in Figure 4 is performed (Figure 9). The ENSO-removed ASI correlates negatively with observed Indian rainfall, which means that changes in the Arabian Sea have little influence on Indian rainfall. The impact of EIDI on Indian summer rainfall has little change before and after the removal of ENSO (comparing Figures 4 and 9).

This means that ENSO has little influence on the correlation between the equatorial Indian Ocean dipole and Indian rainfall. The ENSO-removed SIDI and observed Indian rainfall were antiphase during 1953-1973 but in phase after the 1980s. In comparison with Figure 4, it could be deduced that the south Indian Ocean can not only influence the Indian Ocean through the control of ENSO but that it can also influence Indian rainfall by itself. The amplitude of SIDI oscillation determines the magnitude of the effect of the south Indian Ocean on Indian rainfall. The amplitude of SIDI is larger after 1980 as shown in Figure 10.

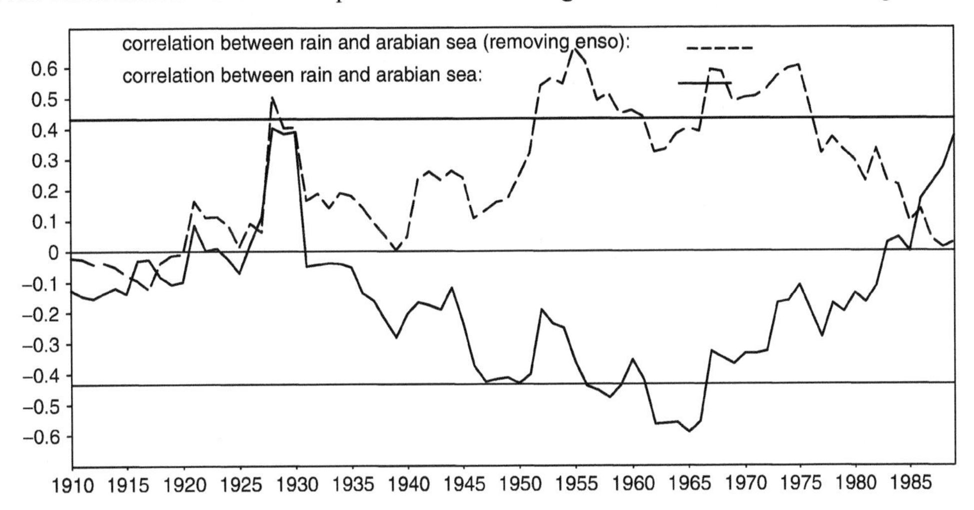

Figure 8. Correlations (based on a 21-year sliding window) of Indian rain and ASI (solid), ASI with the removal of the ENSO signal (dashed). The 5% significant levels are indicated as two horizontal lines

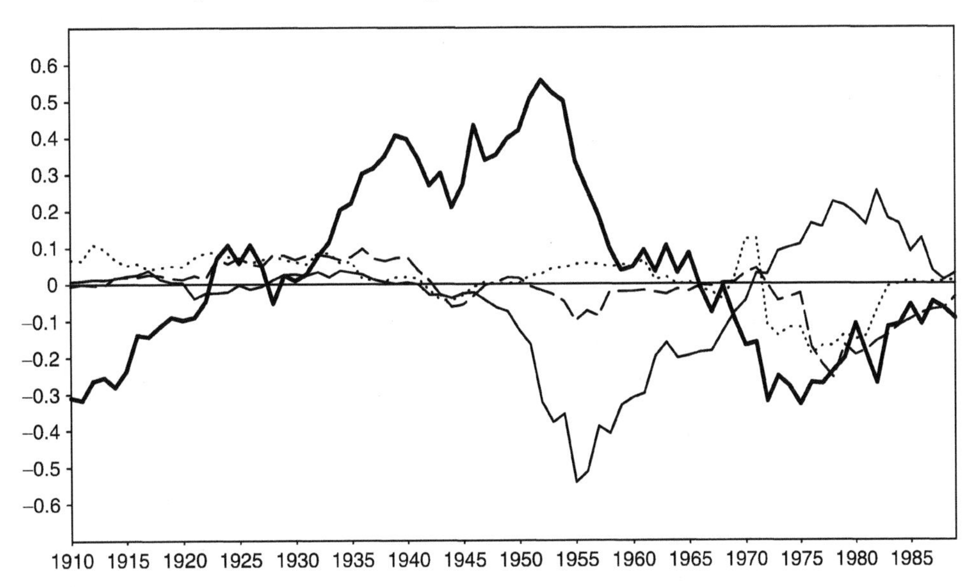

Figure 9. Arithmetic product using the 21-year sliding means of ASI (thin solid), EIDI (dotted and SIDI (dashed) with the removal of the ENSO signal multiplied with the corresponding 21-year sliding correlation coefficients of these indices and the Indian rain. The thick solid line is the result of the 21-year sliding mean of normalized Indian rain

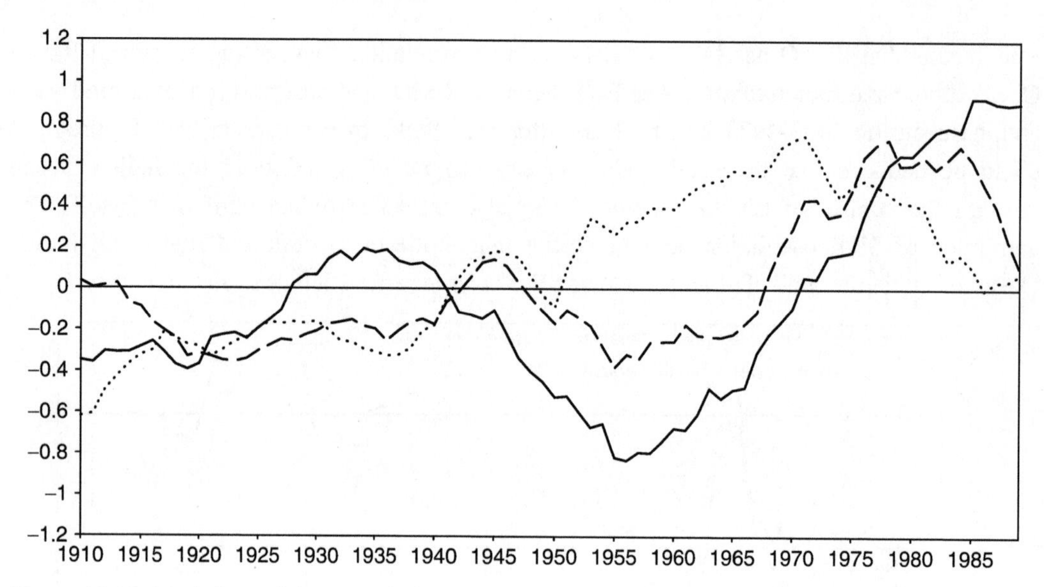

Figure 10. The variations of the 21-year sliding means of ASI (solid), SIDI (dashed) and EIDI (dotted) with the removal of the ENSO signal

4. A Linear Regression of Indian Rain and Sst Indices in Different Areas

Shukla and Mooley (1987) developed a regression equation to predict the summer rainfall over India by using the pressure in Darwin and the latitudinal position of a 500-hPa ridge along 75 °E as two quasi-independent predictor parameters. In order to compare the impacts of SST in the Arabian Sea, SIDI, EIDI and ENSO on Indian rainfall in a specific way, we try to establish the line regression equation of Indian rainfall by using the above SST indices as predictor parameters.

Figure 11 shows the comparison of the observed rainfall and the predicted rainfall by the linear regression equation. The predictor parameters are the 21-year sliding means of the Niño3 index, ASI, EIDI and SIDI multiplied by the corresponding 21-year sliding correlation coefficients of these indices and the Indian rainfall, referred to as the Niño3, the AS, the EID and the SID. The observed rain and the rainfall regressed by Niño3, AS, EID and SID are significantly correlated and their multiple correlation coefficient is as high as 0.963 (Figure 11(a)). The line regression equation of Figure 11(a) is

$$rain = -0.0252 - 0.181 \times AS + 0.768 \times SID + 0.871 \times EID + 1.656 \times Niño3 \quad (2)$$

The weight coefficient of the AS is the least in Equation (2), indicating that the impact of the ASI on rain is smallest compared to those of the other SST indices. If the AS is removed, the multiple correlation coefficient of about 0.951 is very close to 0.963 (Figure 11(b)); thus, the ASI impact and non-ASI impact have little differences. Figure 11(c) and (d) shows the observed rainfall and regression result using the SID (EID) and the Niño3. It is obvious that the regression result of the SID and the Niño3 is much better than that of the EID and the

Niño3. Note that the regression results shown in Figure 11(a), (b) and (c) are better before 1980 but worse after 1980 when compared to those in Figure 11(d); thus the EIDI impact on Indian rainfall appeared to be stronger after 1980.

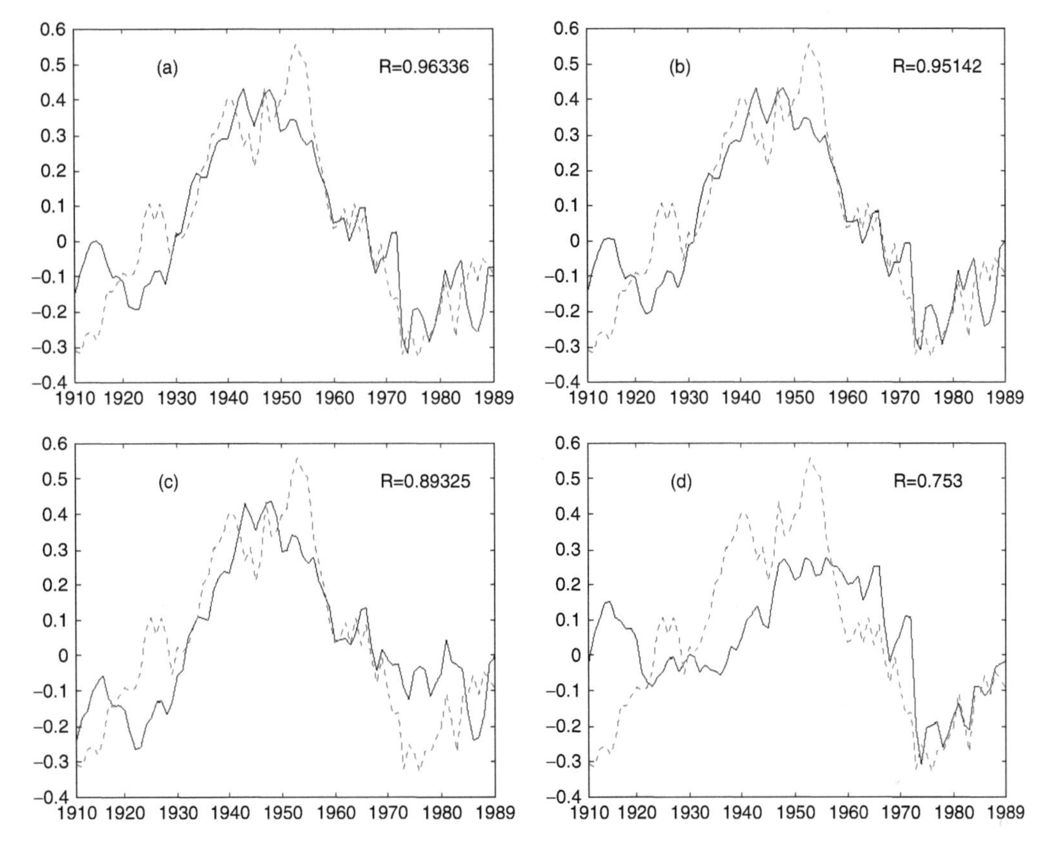

Figure 11. The different line regression results using different predictor parameters. The dashed line denotes the observed rain and the solid line denotes the result of regression. (a) Line regression using Niño3, SID, AS and EID; (b) line regression using SID, EID and Niño3; (c) line regression using SID and Niño3 and (d) line regression using EID and Niño3

Furthermore, we remove the ENSO signal from the SST in different areas according to the method (Clark *et al.*, 2000) and then also use the 21-year sliding means of the Niño3 index, ASI, EIDI and SIDI multiplied the corresponding 21-year sliding correlation coefficients of these indices and the Indian rain as the predictor parameters. The multiple correlation coefficients in Figure 12 are smaller than the most corresponding results in Figure 11 except in Figure 12(d). Therefore, the impacts of SSTA in the Arabian Sea and the south Indian Ocean on Indian rainfall are modulated by the ENSO signal to some extent, and ENSO can amplify the influence of EIDI on Indian rain. When studying the correlation of SSTA in the Arabian Sea, the south Indian Ocean and the equatorial Indian Ocean with Indian rain, the effect of ENSO should be taken into account in different ways.

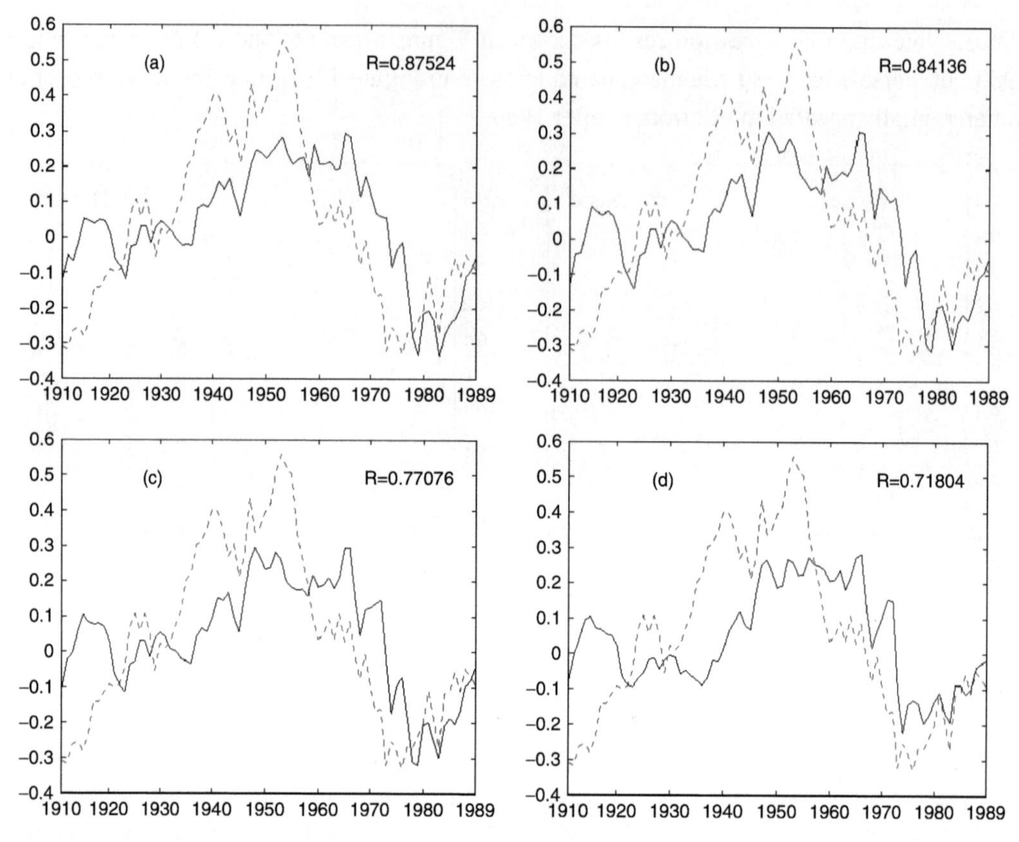

Figure 12. The different line regression results using different predictor parameters. The ENSO signal is removed from all SST indices. The dashed line denotes the observed rain and the solid line denotes the result of regression. (a) Line regression using Niño3, SID, AS and EID; (b) line regression using SID, EID and Niño3; (c) line regression using SID and Niño3 and (d) line regression using EID and Niño3

5. Discussion and Conclusions

It has been pointed out that the impact of SST in different areas on the Indian summer rainfall is very significant. Using the data of the century, this paper reaffirms that Indian rain significantly correlates with the SST in the Niño3 area, the Arabian Sea, the equatorial Indian Ocean and the southern Indian Ocean and discovers that the dipole mode in the southern Indian Ocean also affects the Indian rainfall.

Previous studies only analyze the effect of SST in different areas on rainfall, and the differences of the influence of SST in different regions on rain are omitted. The authors show the different impacts of SST in different areas on Indian rainfall by analyzing the variations of correlation coefficients and amplitudes of SSTA. ENSO always negatively correlates with the Indian rainfall, but its effect on rain has reduced clearly after 1980. The amplitude of ENSO determines the impact of ENSO on rain. SSTA in the Arabian Sea correlates more weakly with Indian rain than SSTA in southern Indian Ocean and ENSO. The Indian rainfall is affected by ENSO, SIDI and EIDI. EIDI may directly affect Indian rain, and the ENSO can influence the rainfall not only directly but also indirectly through the SSTA in the southern Indian Ocean. However, if the amplitude of SIDI is large enough, the SSTA in the southern Indian Ocean can play an important role in controlling rainfall, which was evident in the late 1980s.

SIDI includes two area-averaged SSTs, one (SIDI_1, Figure 2) located in the area near Mascarene's high and the other (SIDI_2, Figure 2) in the northwest of Australia. The SST in these two areas affects the cross-equatorial flow and influences summer rain over India. The higher temperature in the area of SIDI_1 and lower temperature in the area of SIDI_2 lead to decreased cross-equatorial flow because Mascarene's high is weakened by warmer SST, and the colder surface temperature would decrease the fresh water in flow. Therefore, it seems that SIDI correlates negatively with summer rainfall (dash line in Figure 7). But ENSO is related to the Indian Ocean by the atmospheric zonal (Walker) circulation on the equator (Li and Mu, 2001) and could decrease the impact of SIDI on rain. The amplitude of ENSO is larger during 1930-1970 than in other periods (Figure 5); thus, SIDI positively correlates with rainfall from 1930 to 1970. Since the amplitude of ENSO was small and the amplitude of SIDI was large before 1930 and after 1970, the correlation between SIDI and rain is small and even negative. Soo-Hyun Yoo *et al.* 2005 have also shown that the southern Indian Ocean (south of 10 °S) SST is related to the Asian summer monsoon (Indian monsoon) more closely than the northern Indian Ocean SST.

In our study, the SST in the Arabian Sea is negatively correlated with Indian rainfall, which is the same with the conclusions of Weare (1979) and in contradiction to others' results (Shukla, 1975; Rao and Goswami, 1988; Clark *et al.*, 2000). The following are the explanations. When he analyzed the correlation of SST in the Arabian and Indian rain using the noncoupled ocean-atmosphere model, Shukla (1975) only used the SST in the Arabian Sea and merely took it as the boundary condition. Therefore, he concluded that SST in the Arabian Sea is significantly positively correlated with the Indian rain, which is the same as the results obtained with the removal of the ENSO signal from SST in this paper. It is obvious that Shukla's numerical experiment focused on the variation of SST in the Arabian Sea and did not take into account the impact of ENSO on it. When analyzing the relation of SSTA in the Arabian Sea and Indian rainfall, Clark *et al.* (2000) and Rao and Goswami (1988) removed the long-term trend from SST; the former removed the strong warming trend of SST in different periods, while the latter subtracted the ten-year running mean from the raw time series. Since the interdecadal variability of SST in the Arabian Sea has been subtracted, the impact of ENSO on the SST in the Arabian Sea is removed partly and the correlation of SST with Indian rain is significantly positive, which is the same with our results obtained with the removal of the ENSO signal. But, once ENSO is considered, the correlation between Arabian Sea and rain changes. The conventional description of the ENSO-induced teleconnection response in the monsoon is through the large-scale east-west shifts in the tropical Walker circulation (Kumar *et al.*, 1999). During an El Niño event, the tropical convection and the associated rising limb of the Walker circulation normally located in the western Pacific shifts toward the anomalously warm waters in the central and eastern Pacific. Consequently, there is an anomalous subsidence extending from the western Pacific region to the Indian subcontinent. This subsidence suppresses convection and precipitation over the western Pacific and the Indian subcontinent. It is associated with anomalously high pressure over the western Pacific-eastern Indian Ocean sector and with anomalously low pressure over the eastern and central Pacific. Therefore, SST in the Arabian Sea weakly affects the Indian rainfall, and the impact of SST in the Arabian Sea on rainfall is controlled strongly by ENSO.

As is well known, ENSO has quasi-periodic variability of 3-7 years. Torrence and Webster (1999) found that there is interdecadal variation of ENSO. Therefore, the impact of ENSO on climate also shows interdecadal variation, including the relation of it and Indian rain. The results of this paper also show that the Niño3 index is not the only mode of ENSO and that it cannot reflect the impact of ENSO on climate variation completely.

Acknowledgements This work was supported partly by the National Nature Science Foundation of China (Grant No. 40233033) and the Chinese Academy of Sciences (KZCX3-SW-226).

References

Ashok K, Guan ZY, Yamagata T. 2001. Impact of the Indian Ocean dipole on the relationship between the Indian monsoon rainfall and ENSO. *Geophysical Research Letters* 28: 4499-4502.

Behera SK, Yamagata T. 2001. Subtropical SST dipole events in the southern Indian Ocean. *Geophysical Research Letters* 28(2): 327-330.

Chang CP, Harr P, Ju JH. 2001. Possible roles of Atlantic circulations on the weakening Indian monsoon rainfall- ENSO relationship. *Journal of Climate* 14: 2376-2380.

Chen LT. 1991. Effect of zonal difference of sea surface temperature anomalies in the Arabian Sea and the South China Sea on summer rainfall over the Yangtze River. *Chinese Journal of Atmospheric Science* 15(1): 33-42.

Clark CO, Cole JE, Webster PJ. 2000. Indian Ocean SST and Indian summer rainfall: predictive relationships and their decadal variability. *Journal of Climate* 13: 2503-2519.

Kumar KK, Rajagopalan B, Cane MA. 1999. On the weakening relationship between the Indian monsoon and ENSO. *Science* 284: 2156-2159.

Li CY, Mu MQ. 2001. The dipole in the equatorial Indian Ocean and its impact on climate. *Chinese Journal of Atmospheric Science* 25: 433-443.

Rao KG, Goswami BN. 1988. Interannual variations of sea surface temperature over the Arabian Sea and the Indian monsoon: A new perspective. *Monthly Weather Review* 116: 558-568.

Saji NH, Goswami BN, Vinayachandran PN, Yamagata T. 1999. A dipole mode in the tropical Indian Ocean. *Nature* 401: 360-363.

Shukla J. 1975. Effect of Arabian Sea-surface temperature anomaly on Indian summer monsoon: a numerical experiment with the GFDL Model. *Journal of the Atmospheric Sciences* 32: 503 -511.

Shukla J, Mrsra BM. 1977. Relationships between sea surface temperature and wind speed over the central Arabian Sea and monsoon rainfall over Indian. *Monthly Weather Review* 105: 998-1002.

Shukla J, Mooley DA. 1987. Empirical prediction of the summer monsoon rainfall over India. *Monthly*

Weather Review 115: 695-703.

Soo-Hyun Yoo, Song Yang, Chang-Hoi Ho. 2005. *Variability of the Indian Ocean SST and Its Impacts on Asian-Australian Monsoon Climate, under review in JGR.*

Torrence C, Webster PJ. 1999. Interdecadal changes in the ENSO-monsoon system. *Journal of Climate* 12: 2679-2690.

Troup AJ. 1965. The Southern Oscillation. *Quarterly Journal of the Royal Meteorological Society* 91: 490-506.

Weare BC. 1979. A statistical study of the relationships between ocean surface temperatures and the Indian monsoon. *Journal of the Atmospheric Sciences* 36(12): 2279-2290.

Possible Connection Between Pacific Oceanic Inter-decadal Pathway and East Asian Winter Monsoon

Wen Zhou,[1,2], Chongyin Li[1], Xin Wang[1,3]

[1] LASG, Institute of Atmospheric Physics, Chinese Academy of Sciences, Beijing, China
[2] Laboratory for Atmospheric Research and CityU-IAP Laboratory for Atmospheric Sciences, Department of Physics and Materials Science, City University of Hong Kong, Hong Kong, China
[3] Graduate School of the Chinese Academy of Sciences, Beijing, China

Abstract This paper highlights the SST interdecadal variability over the Eastern North Pacific and its connection with East Asian winter monsoon (EAWM) on decadal timescales. Aside from PDO pattern, the SST interdecadal variation is the most significant over IP region, where there is the intense air-sea interaction involved with EAWM, including both wind stress and latent heat flux. A possible explanation for the phase change of PDO is related to the localized atmospheric forcing and the "upstream" winter monsoon. SSTA interdecadal variation over the IP regions is the 'bridge' linking the coupled ocean-atmosphere interaction between midlatitude Pacific and the tropical Pacific. Citation: Zhou, W., C. Li, and X. Wang (2007), Possible connection between Pacific Oceanic interdecadal pathway and east Asian winter monsoon, *Geophys. Res. Lett.*, *34*, L01701, doi: 10.1029/2006GL027809.

1. Introduction

The far-reaching effects of the PDO (Pacific Decadal Oscillation) phenomenon on the interdecadal variability of climate in the Pacific have been explained in detail by *Zhang et al.* [1997] and *Mantua et al.* [1997]. *Mestas-Nuñez and Enfield* [1999] found that the second rotated EOF mode of non-ENSO SST filed (referred as Eastern North Pacific interdecadal mode) has interdecadal variability with positive phase in the eastern North Pacific, the eastern tropical south Pacific and western tropical North Pacific, negative phase in the central North Pacific. They speculated that the origin of PDO might be in the midlatitude eastern North Pacific.

Some observational evidences found that the flow of thermocline water move southwesterly from midlatitudes to tropics [e.g., *McCreary and Lu*, 1994; *Liu et al*., 1994; *Lu and McCreary*, 1995]. However, a consensus has not yet been reached concerning whether the North Pacific interdecadal signal could propagate to the tropical. The air- sea coupled model result of *Pierce et al.* [2000] suggested that the midlatitude decadal SST anomalies force a change in the

Received August 8, 2006; revised October 19, 2006; accepted November 1, 2006; published January 3, 2007.

tropical trade wind system, and the changes in surface wind stress might lead to decadal modulation of tropical SST variability. But the North Pacific equatorward ventilation in the interior pathway is largely blocked at about 10°N due to the wind stress associated with the intertropical Convergence Zone [*Lu and McCreary*, 1995; *Liu and Huang*, 1998]. *Schneider et al.* [1999] pointed out that thermal anomalies south of 18°N arise from local forcing rather than the equatorward propagation of midlatitude anomalies. *Latif and Barnett* [1994] proposed that the low frequency climate variability might be attributed to a cycle involving unstable air-sea interactions between the subtropical gyre circulation in the North Pacific and the Aleutian low-pressure system. They thought the existence of this cycle provides a basis for long-range climate forecasting over the western United States at decadal time scales.

It is generally recognized that the midlatitude Pacific Ocean influences the climate over the "downstream" regimes, such as North America [*Latif and Barnett*, 1994; *Gershunov and Barnett*, 1998; *McCabe and Dettinger*, 1999; *Higgins and Shi*, 2000], but seldom focus on the connection with the climate over the "upstream" regimes, such as East Asia. This raises the obvious question of whether there is an air-sea interaction between the midlatitude decadal Sea Surface Temperature (SST) anomalies and "upstream" winter monsoon, and how the atmosphere mediates the response between the tropics and midlatitudes on decadal timescale. Present work tries to establish a connection between East Asian winter monsoon and Eastern North Pacific interdecadal mode.

The data analyzed are described in section 2. The most significant interdecadal variations in the Pacific are then examined in section 3. Possible connections between East Asian Winter Monsoon (EAWM) and Eastern North Pacific interdecadal mode and their possible roles in changing PDO phase are discussed in section 4. A summary of the results and possible future work are presented in section 5.

2. Datasets

The observed datasets used in current study include the monthly mean sea surface temperature (HadISST1) (1870-2003) and the monthly mean sea level pressure (SLP) (1899-2004) from the British Meteorological Office with the horizontal resolution of 1° latitude/longitude square and 5° latitude/longitude square, respectively [*Rayner et al.*, 2003], the monthly latent flux (1948-2002) T62 Gaussian grid with 192 × 94 points and monthly SLP (1948-2003) with 2.5° × 2.5° resolution from NCEP/NCAR reanalysis dataset. The monthly surface wind stress wind with 2.5° × 2.5° resolution from the European Centre for Medium-Range Weather Forecasts (ECMWF) ERA-40 reanalysis (1958-2001).

3. Results

3.1. Interdecadal Pathway

Much of the previous works have proposed the most significant interdecadal mode in the North Pacific [e.g., *Zhang et al.*, 1997; *Mantua et al.*, 1997], but the variance difference over the eastern North Pacific shown in Figure 1 is obviously positive from the difference between the variance of the 8-35 yr band filtered SST anomalies and that of 2-7 yr band filtered SST anomalies, suggesting that the contribution of the eastern North Pacific at about 20°N region on decadal timescales is most obvious than other regions. We speculate this region with the most significant interdecadal variation might be the interdecadal pathway (IP) linking the midlatitudes and tropics. Note that the IP region is very similar with that large positive amplitudes regions of the eastern North pacific interdecadal mode mentioned by *Mestas-Nuñez and Enfield* [1999]. Thus, we use the area-averaged SST anomalies over 8 points [(12°N, 160°E), (14°N, 170°E), (16°N, 180°), (18°N, 170°W), (20°N, 160°W), (22°N, 150°W), (24°N, 140°W), (30°N, 130°W)] in the extratropical region to represent IP index. Furthermore, the wavelet analysis is applied to identify the main oscillation in the monthly IP index. Only about 20-30 yr peak remain significant from the global wavelet power (Figure 2b). The variance time series averaged at 2-7 yr timescale and 8-35 yr timescale show consistent interdecadal changes, but only the variance at period of 8-35 yr is significant at 95% confidence level between 1940 and 1980 (Figure 2a). Thus the fluctuation in power of IP index over a range of 8-35 yr band is more evident than that of 2-7 yr band.

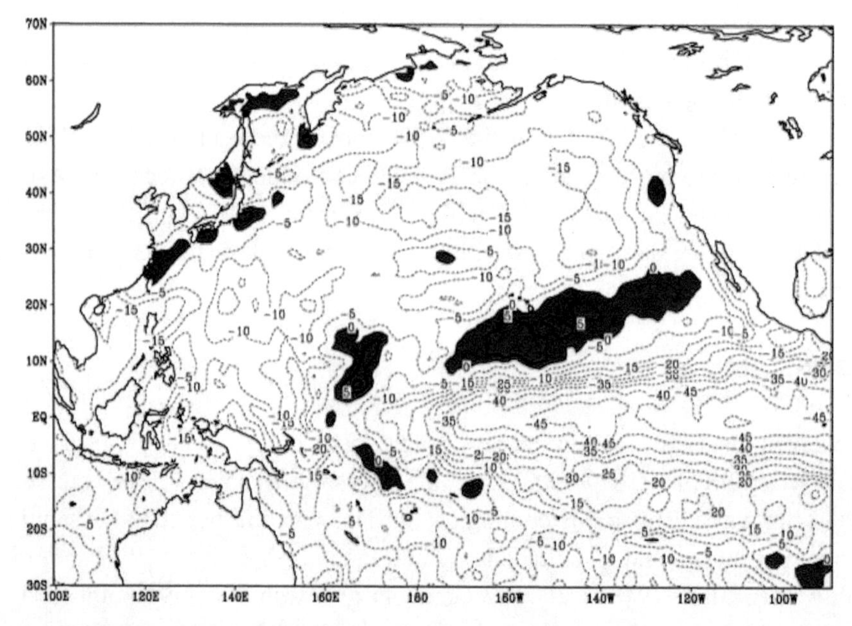

Figure 1. The difference between the variance of 8-35 yr band filtered SST anomalies and that of 2-7 yr band filtered SST anomalies. Shaded area indicates the positive values.

Moreover, the first SVD mode of SLP (sea level pressure) and latent heat flux during winter season with explained variance about 31% represents the interannual signal (not shown), while the explained variance of the second SVD mode is about 17%. As is expected, two spectral peaks appear on the MTM (multitaper method) spectrum. Both two peaks, centered at the period of about 2-3 yr and ~20 yr, stand out clearly from the red noise at the 95% significant level (Figure 3b). Besides the interannual part, the interdecadal signal is very obvious in the second mode. A significant maximum band of latent heat flux in the IP regions further indicates that the air-sea exchange is most evident and such latent heat flux distribution is accompanied by the land-sea SLP difference and also an anomalous east-west SLP gradient in the tropical Pacific (Figure 3 a). Here, the land-sea SLP difference refers to the developed (diminished) Siberian high at the surface, indicating the phenomenon of EAWM [*Jhun and Lee*, 2004]. While the east-west SLP gradient with relatively higher (lower) SLP over the eastern Pacific area and lower (higher) SLP over the western warming pool is present, it helps maintain (weaken) the strength of trade winds and surface drag, thus leads to the east-west SST gradient changed in the tropical region. The regression of SST and surface stress against the time coefficient of the second SVD mode (Figure 3c) shows a PDO (or ENSO-like) pattern with the wind stress only significant over East Asian and the western ocean of Hawaii. Thus, the variations of EAWM, PDO and SST over IP region are interrelated shown in Figure 3, further illuminating the coupled nature of SST, surface heat flux and winds over the Pacific and the land. EAWM could affect the heat flux exchange between the midlatitude and tropics over the IP region. The consequent water mass along the IP region propagates to the western Pacific and further changes the east-west gradient of SST. In turn, the resultant east-west gradient of SLP and trade winds might lead to the interdecadal variability of ENSO [*Tourre and Kushnir*, 1999].

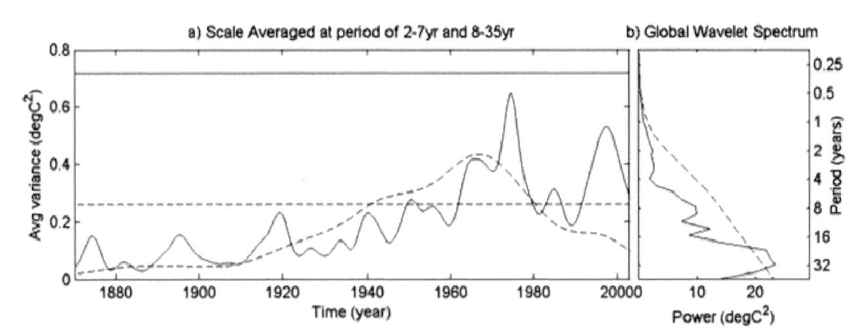

Figure 2. (a) Scale-averaged wavelet power of the IP index over the 2-7 yr band (solid) and 8-35 yr band (dashed). The thin solid line is the 95% confidence level for the 2-7 yr band, while the thin dashed line is the 95% level for the 8-35 yr band. (b) Global wavelet power spectrum (solid), the thin dashed line is the 95% confidence level.

3.2 Possible Impact of EAWM on IP

On the interannual timescales, EAWM causes the variation of trade wind and cumulus convection in the equatorial middle-western Pacific, thus influences the occurrence of El Niño [*Li*, 1990]. However, on interdecadal timescales, the surface temperature over IP regions might be interacted with the "upstream" EAWM surface winds. As mentioned before, the IP regions is the most large amplitude regions in the eastern North Pacific interdecadal mode [*Mestas-Nuñez and Enfield*, 1999; *Enfield and Mestas-Nuñez*, 1999], Figure 3 suggests a robust relationship between EAWM and IP on interdecadal timescales. It seems the interannual interacting region mainly locates in the northwest to equatorial Pacific, while the interdecadal interacting region mainly locates in the IP regions. The question here is how the EAWM links to the interdecadal variation of this IP regions.

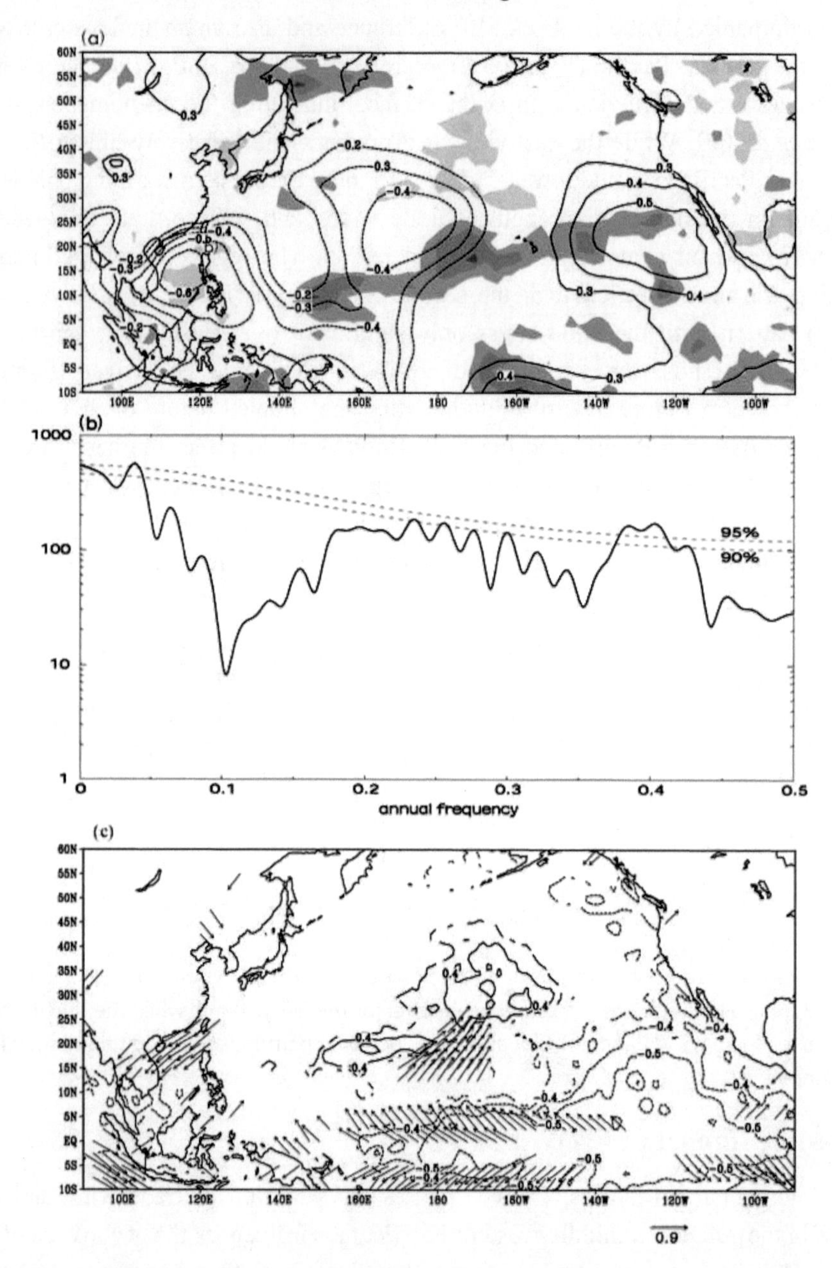

Figure 3. (a) The second SVD mode of SLP (contour) and latent heat flux (shaded) during winter season. (b) MTM Spectrum of the second SVD time coefficient of latent heat flux. The two smooth dashed curves indicate the 90% and 95% significant levels. (c) The regression of SST (contour) and surface wind stresses (vector) against the time coefficient of the second SVD mode in Figure 3b. Only significant values exceeding the 95% level have been plotted.

From the two leading EOF models (empirical orthogonal functions) of winter mean (DJF) SLP, a similar spatial structure within 110-160°E, an obvious difference between Eurasian continent and the Pacific Ocean, indicates that Asian Winter monsoon is dominated by the Siberian High and the Aleutian Low over the Eurasian continents and the northern Pacific at mid and high latitudes (figures not shown). The EAWM index is therefore defined as the two area-averaged SLP anomaly difference between (20-45°N, 110-120°E) and (20-45°N, 150-160°E).

The lagged regression of the 10-yr low-pass filtered SSTA, wind stress anomaly and SLP anomaly against the same filtered EAWM index respectively shown in Figure 4 illustrates the EAWM-SST-wind connection on decadal timescales. When there is a strong EAWM leading SST anomaly 10 years, the cooling phase is found to be significant from the Kuroshio extension to PDO region, and the significant warming phase is along the equator, particularly over the equatorial Eastern Pacific. Note that the SLP anomaly is a multi-pole pattern with positive anomaly over Asian mainland and equatorial eastern Pacific, negative anomaly in between. The northerly wind of strong winter monsoon tends to be westerly wind when reaching the shore, the offshore westerly might enhance the cooling phase over the Kuroshio Current and its extension, thus at least maintain the PDO cooling phase at lag –10 and –8. The negative SLP over north Pacific tends to be weakened and move northward at lag –6, the cooling phase over PDO region is thus weakened. However, the cooling phase over IP region is strengthened partly because of subduction by the westerly at the southern part of this low pressure [*Gu and Philander*, 1997], or partly because of the PDO cooling phase itself leaking through IP regions. At the same time, the maximum latent heat flux shown in Figure 3a might help strengthen the cooling phase over IP region. Though the regression appears to be more significant in the Kuroshio extension and tropical region at lag –10 and lag –8, the Eastern subtropical Pacific or IP region seems to play some roles in the phase switch at lag –6. Note that the higher surface pressure over the equatorial eastern Pacific will help the warm phase propagate westward. Obviously, this high pressure over Asian continent moves to the sea at lag –4, and the SLP anomaly over the equatorial eastern Pacific decreases, thus the zonal SLP gradient seems to reverse, the warming phase originally equatorial eastern Pacific is now over the western Pacific, while the incessant warming signal from the southern Pacific will form a supplement to the warming phase over western Pacific [*Wang and Liu*, 2000]. The warming phase over western Pacific will be brought to the PDO regions by the Kuroshio Current at lag –2. Seeing that the easterly anomalies near the IP regions due to the high pressure there, the cooling phase over IP regions is supposed to be reinforced at lag –4 and reached the maximum at lag –2 and lag 0. Particularly, this maximum cooling phase over IP regions infiltrates into Niño 3.4 regions (170°W-120°W, 5°S-5°N) from lag –2 and strengthens later. The SST anomalies in the subtropics have been initiated at lag –6 and then persist, and finally lead to the change of the tropics. In this way, the phase turnabout over PDO regions and ENSO regions may take at least 10 years. And more likely, the reversed zonal SLP difference indicates a weaker EAWM about 10 years later. Therefore, the delaying SST feedback leads to a weaker EAWM when the SST anomalies lag a stronger EAWM, showing a coupled ocean-atmosphere interaction.

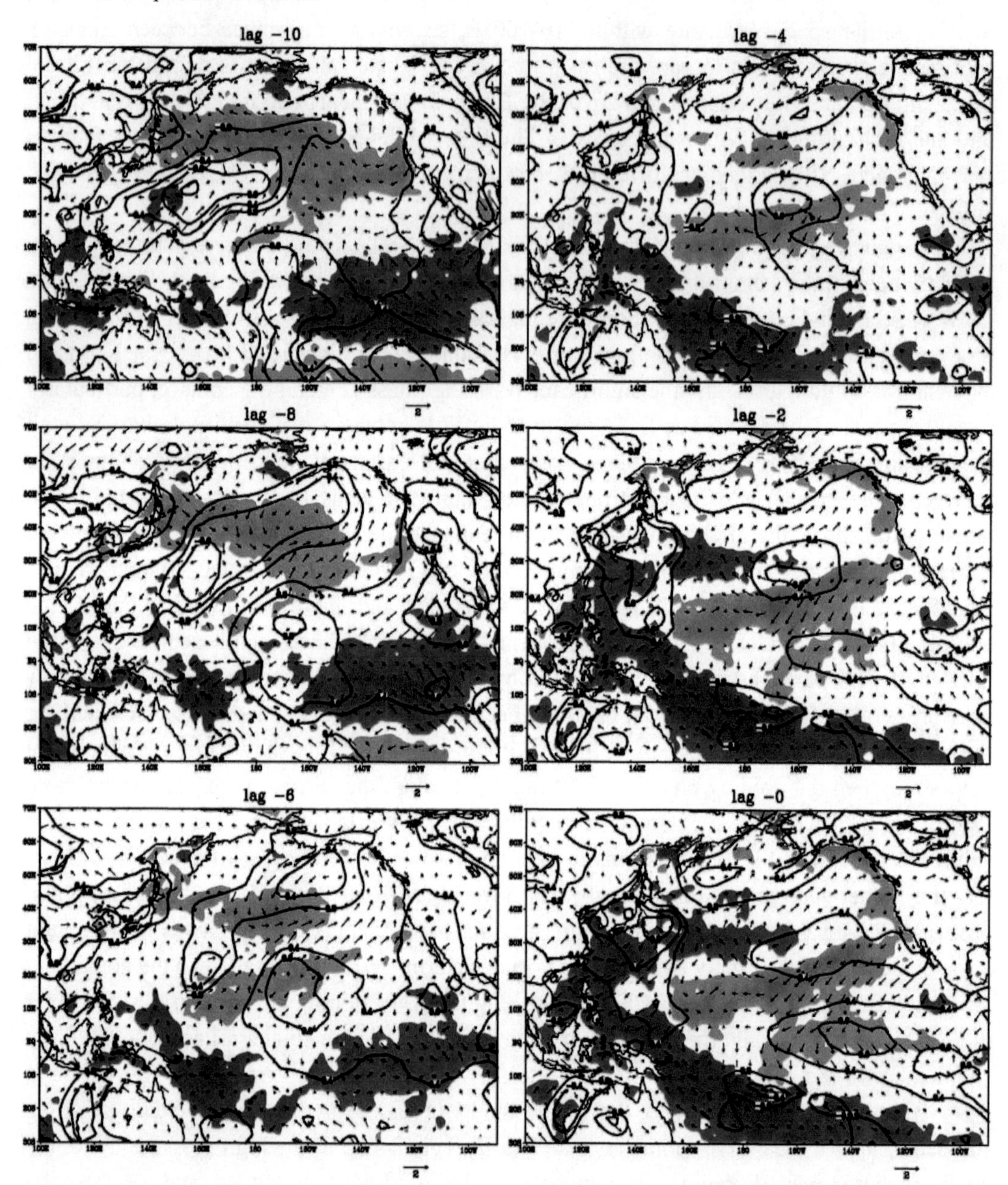

Figure 4. Lagged regression of the 10-yr low-pass filtered SSTA/SLP/wind stress (shading/contour/vector) against the same filtered EAWM, *lag*(−Δ*t*) means the EAWM index leads SST by Δ*t* year (s). Dark (light) shadings are for positive (negative) values. Only values exceeding the 95% significance level have been plotted.

4. Conclusions and Discussions

This paper proposes that the most obvious interde-cadal variation of Sea Surface

Temperature (SST) anomalies is over the Eastern North Pacific at about 20°N, and this interdecadal pathway (IP) has maximum latent heat flux, which might play an important role in the connection of midlatitude and tropical SST anomalies. As shown in Figure 3a, the response of the land-sea SLP difference indicates the role of the atmospheric forcing can not be ignorable. Furthermore, depending on the lagged regression of SST anomalies against EAWM, the winter monsoon impacts on the SST anomalies on the decadal timescales. To some extent, the "atmospheric forcing" in the midlatitude acting as the "upstream effect" is much obvious when there is strong EAWM. Our present work highlights the "bridge" role of the IP regions linking the coupled ocean-atmosphere interaction between midlatitude Pacific and the tropical Pacific.

The cause of PDO change has remained controversial. Recent observational and modeling studies indicate that the change in the PDO may depend on the fresh water flux into the outcrops [*Huang et al.*, 2005] suggested that the slow response of ocean temperature due to anomalous net atmospheric freshwater may play an important role in the PDO or ENSO-induced atmospheric anomalies (*Bjerknes* [1966] proposed that the intensified local Hadley circulation thereby influenced the wind over the North Pacific Ocean during winter). In our study, we speculate that through the air-sea interaction over IP region, the change of PDO phase depends on the variation of EAWM amplitude on decadal timescale. That is, when a strong EAWM leading, the cooling phase over PDO regions might be intensified at first few years and then reach a peak. Later the maximum of cooling phase would shift eastward and weaken with the EAWM-induced westward wind stress. Concurrently, the cooling phase over IP regions develops increasingly due to subduction over midlatitude Pacific and leaking from midlatitude Pacific. Further, the warming phase over western Pacific, which is originally from equatorial eastern Pacific, will be brought to the PDO regions by the Kuroshio Current. Thus the PDO phase is marching in a period of at least 10 years. But why the interdecadal variation over IP regions is most significant and what is the physical mechanism responsible for the leaking from PDO region and IP region, remains the subject of ongoing research.

Acknowledgments This research is supported by the National Nature Science Foundation of China (grant 40675051) and the research grant KZCX3-SW-226 of Chinese Academy of Sciences. The work of Wen Zhou is partly supported by City University of Hong Kong Research Scholarship Enhancement Scheme and City University of Hong Kong Grant 7001825. Two anonymous reviewers are highly appreciated for their useful suggestions and comments for improving this paper.

References

Bjerknes, J. (1966), A possible response of the atmospheric Hadley circulation to equatorial anomalies of ocean temperatures, *Tellus*, 18, 820-829.

Enfield, D. B., and A. M. Mestas-Nuñez (1999), Multiscale variabilities in global sea surface temperatures and their relationships with tropospheric climate patterns, *J. Clim.*, 12, 2719-2733.

Gershunov, A., and T. P. Barnett (1998), Interdecadal modulation of ENSO teleconnections, *Bull. Am. Meteorol. Soc.*, 79, 1715-2725.

Gu, D., and S. G. H. Philander (1997), Interdecadal climate fluctuations that depend on exchanges between the tropics and extratropics, *Science*, 275, 805-807.

Higgins, R. W., and W. Shi (2000), Dominant factors responsible for interannual variability of the summer monsoon in the south-western United States, *J. Clim.*, 13, 759-775.

Huang, B. Y., V. M. Mehta, and N. Schneider (2005), Oceanic response to idealized net atmospheric freshwater in the Pacific at the decadal time scale, *J. Phys. Oceanogr.*, 35, 2467-2486.

Jhun, J.-G., and E.-J. Lee (2004), A new east Asian winter monsoon index and associated characteristics of the winter monsoon, *J. Clim.*, 17, 711-726.

Latif, M., and T. P. Barnett (1994), Causes of decadal climate variability over the North Pacific and North America, *Science*, 266, 634-637.

Li, C. (1990), Interaction between anomalous winter monsoon in east Asia and El Niño events, *Adv. Atmos. Sci.*, 7, 36-46.

Liu, Z., and B. Huang (1998), Why is there a tritium maximum in the central equatorial Pacific thermocline?, *J. Phys. Oceanogr*, 28, 1527-1533.

Liu, Z., S. G. H. Philander, and R. C. Pacanowski (1994), A GCM study of the tropical-subtropical upper-ocean circulation, *J. Phys. Oceanogr*, 24, 2606-2623.

Lu, P., and J. P. McCreary (1995), Influence of the ITCZ on the flow of the thermocline water from the subtropical to the equatorial Pacific Ocean, *J. Phys. Oceanogr.*, 25, 3076-3088.

Mantua, N. J., et al. (1997), A Pacific interdecadal climate oscillation with impacts on salmon production, *Bull. Am. Meteorol. Soc.*, 78, 1069-1079.

McCabe, G. J., and M. D. Dettinger (1999), Decadal variations in the strength of ENSO teleconnections with precipitation in the western United States, *Int. J. Climatol.*, 19, 1399-1410.

McCreary, J. P., and P. Lu (1994), Interaction between the subtropical and equatorial ocean circulations: The subtropical cell, *J. Phys. Oceanogr.*, 24, 466-497.

Mestas-Nuñez, A. M., and D. B. Enfield (1999), Rotated global modes of non-ENSO sea surface temperature variability, *J. Clim.*, 12, 2734-2746.

Pierce, D. W., T. P. Barnett, and M. Latif (2000), Connections between the Pacific Ocean tropics and midlatitudes on decadal timescales, *J. Clim.*, 13, 1173-1194.

Rayner, N. A., D. E. Parker, E. B. Horton, C. K. Folland, L. V. Alexander, D. P. Rowell, E. C. Kent, and A. Kaplan (2003), Global analyses of sea surface temperature, sea ice, and night marine air temperature since the late nineteenth century, *J. Geophys. Res.*, 108(D14), 4407, doi: 10.1029/2002JD002670.

Schneider, N., A. J. Miller, and M. A. Alexander (1999), Subduction of decadal North Pacific temperature anomalies: Observations and dynamics, *J. Phys. Oceanogr*, 29, 1056-1070.

Tourre, Y. M., and Y. Kushnir (1999), Evolution of interdecadal variability in sea level pressure, sea surface temperature and upper ocean temperature over the Pacific Ocean, *J. Phys. Oceanogr.*, 29, 1528-1541.

Wang, D. X., and Z. Liu (2000), The pathway of interdecadal variability in the Pacific Ocean, *Chin. Sci. Bull.*, 45, 1555-1561.

Zhang, Y., J. M. Wallace, and D. S. Battisti (1997), ENSO-like interdecadal variability: 1900-1993, *J. Clim.*, 10, 1004-1020.

Interdecadal Unstationary Relationship Between NAO and East China's Summer Precipitation Patterns

Wei Gu[1], Chongyin Li[2,3], Weijing Li[1], Wen Zhou[4], Johnny C. L. Chan[4]

[1] Laboratory for Climate Studies, National Climate Center, China Meteorological Administration, Beijing, China
[2] Meteorological College, PLA University of Science and Technology, Nanjing, China
[3] LASG, Institute of Atmospheric Physics, Chinese Academy of Sciences, Beijing, China
[4] Department of Physics and Materials Science, City University of Hong Kong, Hong Kong, China

---

Abstract Based on a 120-year Chinese station rainfall dataset and historical NAO indices, the interdecadal unstationary relationship between the North Atlantic Oscillation (NAO) and east China's summer precipitation patterns is revealed for the period 1880-1999. It is shown that on the interannual timescale, the March NAO is closely related to the first leading precipitation mode which exhibits an out- of-phase variation of the precipitation between the Yangtze River Valley and Southeast China, and the January NAO is closely related to the third leading mode which depicts the anomalous rainfall in North China and its out-of-phase variation in the Yangtze River Delta. Both of the two relationships are characterized by clear and consistent interdecadal variations, with almost the same transition points occurring in about 1905, 1925 and 1950, respectively. Further analyses imply that these interdecadal changes of the relationship are possibly associated with the SSTA in North Pacific and North Atlantic.
Citation: Gu, W., C. Li, W. Li, W. Zhou, and J. C. L. Chan (2009), Interdecadal unstationary relationship between NAO and east China's summer precipitation patterns, *Geophys. Res. Lett.*, *36*, L13702, doi: 10.1029/ 2009GL038843.

---

1. Introduction

The summer precipitation is of great social and economic importance for China and many other East Asian countries [e.g., *Huang et al.*, 2003, 2007; *Ding and Chan*, 2005]. Its variability and associated influencing factors therefore are hot issues among the East Asian climate studies. As a major mode of the global atmospheric circulation, the North Atlantic Oscillation (NAO)/Arctic Oscillation (AO) is found to be able to influence the East Asian summer precipitation in some particular regions on the interannual timescale. For example, *Sung et al.* [2006] found a delayed impact of December NAO on summer precipitation in

Received April 23, 2009; revised May 22, 2009; accepted June 4, 2009; published July 2, 2009.

Korea and part of east China for the period 1951 -2000. With a high-pass-filtered dataset, *Gong and Ho* [2003] identified a close interannual relationship between May AO and the following total summer rainfall in the Yangtze River Valley (YRV) and southern Japan for a longer period 1900-1998. These results provide potentials to predict the summer precipitation of particular East Asian regions through the NAO/AO signal which leads several months.

On the other hand, the validity of certain potential predictors may experience significant interdecadal variations, that is, these potential predictors are quite significant during some decades while not in the others [e.g., *Gershunov and Barnett*, 1998; *Kumar et al.*, 1999; *Wang et al.*, 2007, 2008]. Although the reason for such interdecadal variations is still not clear, it does urge that such unstationary features should be taken into account in practical short-term climate predictions [*Gershunov and Barnett*, 1998; *Wang et al.*, 2008]. Therefore, the purpose of this study is to evaluate the unstationarity of the NAO-east China's summer precipitation relationship on the interdecadal timescale.

2. Data and Methods

The datasets used in this study are listed in Table 1. All the data are monthly means except for the seasonal mean 35-station rainfall dataset, which was collected and established using observations and some proxy data for the period 1880-1999 [*Wang et al.*, 2000]. The 35 stations cover most of east China with good spatial distributions (Figure 1a), and can well describe the variability of precipitation in east China [*Wang et al.*, 2000]. The observed rainfall data from 160 China stations since 1951 are employed to compare with the results derived from the 35-station dataset. The summer means are considered throughout this paper, and they are constructed by averaging June, July and August.

When considering the NAO-east China's summer precipitation relationship, we focus on the precipitation patterns instead of the average rainfall in a particular region. This pays more attention to the whole east China and is slightly different from those of *Gong and Ho* [2003] and *Sung et al.* [2006]. The empirical orthogonal function (EOF) analysis is employed to extract the leading precipitation patterns.

3. Results

Figures 1b-1d presents the first three EOF leading modes of the summer precipitation derived from the 35- station dataset for the period 1880-1999. They explain 16.3%, 13.3% and 8.3% of the total variance, respectively, and are well separated from the remaining modes according to the criteria of *North et al.* [1982]. The EOF1 mode reflects the out-of-phase variations of precipitation in between the YRV and the coastal region in Southeast China (Figure 1b). The EOF2 mode clearly illustrates the intensity of precipitation over South China (Figure 1c). The EOF3 mode mainly describes the rain band in North China and its out-of-phase relationship with the precipitation in the Yangtze River Delta (Figure 1d). Since the distribution of the 35 stations in east China is relatively sparse, we also performed the same analysis on the more densely distributed 160-station rainfall data for the period 1951-2006 to compare with the current results. It reveals that the first three leading modes are almost identical between the two datasets (not shown). In addition, the corresponding principal components (PCs) between the two datasets are highly correlated, with the correlation coefficients for the first three PCs being 0.93, 0.88 and 0.64 for the period 1951-1999, respectively, all exceeding the 99% confidence levels. This result suggests that the leading summer precipitation patterns in east China and their time variations can be well captured by the 35-station dataset. Moreover, these patterns are stationary and do not change with time.

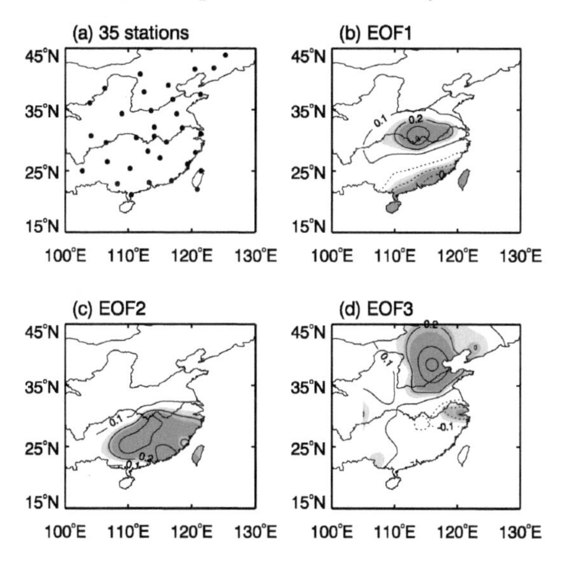

Figure 1. (a) The distribution of 35 stations in east China. The (b) first, (c) second and (d) third EOF mode of the 35- station-based summer precipitation. Light, middle and dark shading indicates 90%, 95% and 99% confidence level, respectively.

In order to investigate the relationship between summer precipitation patterns and preceding NAO signals, we calculated the correlation coefficients between the first three summer precipitation PCs and monthly NAO indices from the preceding winter to the simultaneous summer. When calculating the correlation, all the indices are 8-year-high- pass filtered through the Fourier transformation method to exclude the possible influence of the interdecadal component. The first and last four years are omitted after filtering to avoid the edge effect. This process is similar to that of *Gong and Ho* [2003]. The result reveals that significant relationship exists between PC3 and preceding December/ May NAO indices, with the correlation coefficients being 0.22/–0.23 for 1884-1995, exceeding the 95%/95% confidence level. It suggests that NAO of certain month may be used as a potential predictor for the EOF3 pattern of east China's summer precipitation, consistent with previous results of *Sung et al.* [2006] and *Gong and Ho* [2003], respectively.

As mentioned in the introduction, the validity of certain potential predictors may experience significant interdecadal variations [*Kumar et al.*, 1999; *Wang et al.*, 2008]. This leads to a point that although no significant relation exists between NAO and summer precipitation patterns over east China for 1880-1999 except December/May NAO and EOF3, it is possible that the NAO-precipitation relationship is quite significant during some period, and this relationship may be weakened or even reversed during other periods, which results in an insignificant correlation for the total data period. Therefore, a sliding correlation analysis with a 21-year moving window [*Kumar et al.*, 1999] is performed between the leading east China's summer precipitation PCs and monthly NAO indices to check this possibility. It reveals that in addition to the December/May NAO-PC3 relationship, two other interesting correlations emerge between March NAO and PC1, and January NAO and PC3.

Figure 2 shows the 21-year sliding correlation between March NAO and PC1, and January NAO and PC3. It reveals that both the March NAO-PC1 and the January NAO-PC3 relationship experienced obvious and similar interdecadal variations, with positive correlations during some decades and negative correlations during other decades (Figure 2). The interannual correlations are significant at 95% or higher confidence levels during the peaks of the interdecadal phases. Therefore, during these periods with high confidence level, NAO are closely related with certain summer precipitation patterns over east China, and can be used as a potential predictor. However, this prediction potential can not be seen for the longer period, since the positive correlations may counteract with the negative ones. During recent decades, both correlations are in their negative interdecadal phases, and seem to experience another transition. It is worth noting that the transition points of the two curves are roughly the same, which are around 1905, 1925 and 1950. This similarity implies that both the January NAO-PC3 relationship and the March NAO-PC1 relationship seem to be associated with some common interdecadal signals.

Table 1. Data Sets Used in This Study

	Resolution	Period of Record	Source
Precipitation	35 stations	1880-1999	*Wang et al.* [2000]
NAO index	-	1880-1999	http: //www.cgd.ucar.edu/cas/jhurrell/indices.html
HadISST1	1°× 1°	1880-1999	*Rayner et al.* [2003]
Precipitation	160 stations	1951-1999	China Meteorological Administration

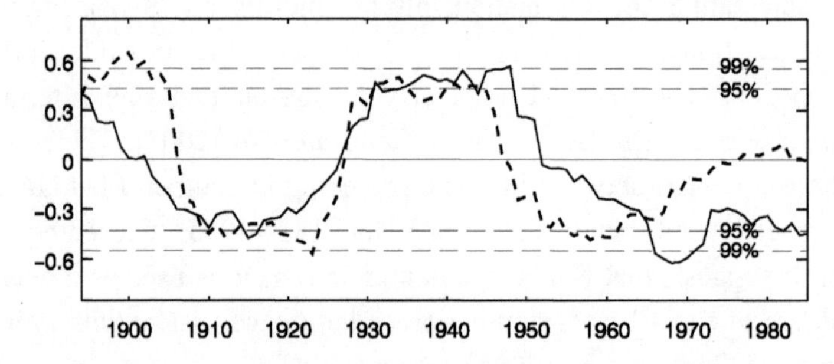

Figure 2. The sliding correlation between March NAO and PC1 (dashed line), and between January NAO and PC3 (solid line) with 21-year moving window. The x-label indicates the central year of the moving window.

In order to illustrate further the interannual relationship between March (January) NAO and PC1 (PC3) and the interdecadal variation of this relationship, based on the sliding correlations in Figure 2, composite difference of east China's summer precipitation according to the NAO indices are calculated for four epochs: 1884-1904, 1905-1925, 1932-1952, 1956-1976, each of which consists of 21 years. The period 1884-1904 and 1932-1952 are in the positive interdecadal phases (Figure 2). During these periods, positive NAO in January generally corresponds to significant more precipitation over North China (Figure 3 a). Meanwhile, positive NAO in March is usually accompanied with significant more precipitation over the YRV and less precipitation over Southeast China (Figure 3c). For the negative interdecadal phase of 1905-1925 and 1956-1976, the aforementioned relationships are generally reversed (Figure 3b for January NAO and Figure 3d for March NAO). The composite results further confirm that the January and the March NAO are closely related to east China's summer precipitation pattern, and the NAO-precipitation relationships vary on the interdecadal timescale. Moreover, these results also indicate that if the interdecadal phases in which the significant NAO-precipitation relationship is located can be confirmed, the NAO signals can then be used as a good potential predictor during these periods.

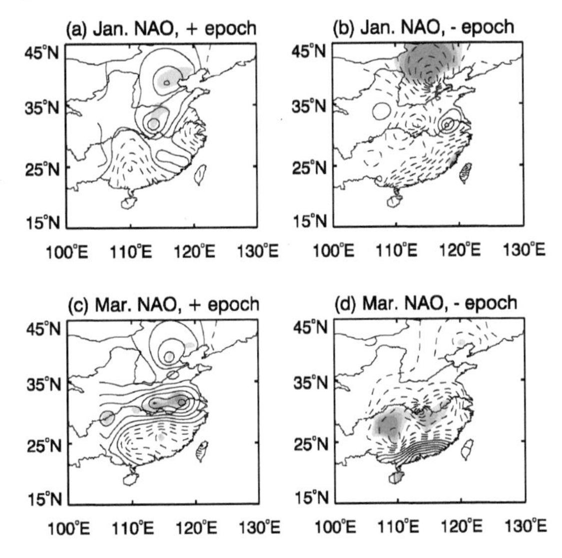

Figure 3. Composite difference of summer precipitation between typical positive and negative January NAO years during (a) 1884-1904 and 1932-1952, and(b) 1905-1925 and 1956-1976. (c and d) Same as Figures 3a and 3b, but for March NAO. Contour intervals are 30mm. Zero contour lines are omitted. Light, middle and dark shading indicates 90%, 95% and 99% confidence level, respectively.

As mentioned previously, there exists significant relationship between December/May NAO and PC3 for the total period of 1880-1999, it is therefore also meaningful to evaluate whether these two relationships are stationary. The 21-yr sliding correlation indicates that the correlation between December (May) NAO and PC3 is either not stable. The December NAO is positively correlated with PC3 for most of the period, but the correlation reaches the 95% confidence level only around 1910 and 1950 for very short periods (not shown). It suggests that although the December NAO-PC3 is significantly correlated for the total period 1880-1999, the validity of this relationship is relatively weak for shorter (e.g., about 21 year) periods, which reduces its usefulness as a potential predictor for east China's summer precipitation. As for the May NAO, it is negatively correlated with PC3 for about 3/4 of the total period, with high (over 99%) confidence level during about 1930-1947 and 1969-1976 (not shown), implying that the May NAO can be used as a good predictor for east China's summer precipitation around these periods. However, the variation of May NAO-PC3 correlation is not similar to those of January NAO-PC3 and March NAO-PC1 relationships, so we will not discuss it here.

4. Summary and Discussions

In this paper, the interdecadal unstationary relationship between preceding NAO and east China's summer precipitation patterns is investigated based on a 120-year (1880-1999) observed precipitation dataset from 35 Chinese stations and historical NAO indices. The precipitation patterns are extracted through EOF methods and their relationship with NAO is evaluated by correlation and sliding correlation analyses. Two interesting relationships are presented for the first time between March/January NAO and the out-of-phase pattern in precipitation between the YRV and Southeast China (EOF1)/the out-of-phase pattern in precipitation between North China and Yangtze River Delta (EOF3). It is found that although the relationships between March (January) NAO and PC1 (PC3) are quite weak for the total data period (1880-1999), their relationships are significantly positive during some decades and significantly negative during other decades, which exhibit clear and consistent interdecadal variations as revealed by the 21-yr sliding correlation analyses. The most interesting finding is that the interdecadal transition points of the March NAO-PC1 and the January NAO-PC3 relationships are similar, which occur around 1905, 1925 and 1950, respectively. This result suggests that if the interdecadal phase in which the NAO-precipitation relationship is located can be confirmed, the NAO signals can be used as a good potential predictor during these certain periods.

An important issue that leaves open is what causes the interdecadal change in the relationships between NAO and east China's summer precipitation patterns. This is a very complex problem and no clear answer has been proposed. As suggested by *Power et al.* [1999] and *Wang et al.* [2008], this kind of interdecadal variations may be related with the changes in the background climatic conditions such as the sea surface temperature (SST). Based on the transitions shown in Figure 2, the composite difference of SST from preceding winter to simultaneous summer between positive epochs (1884-1904 and 1932-1952) and negative epochs (1905-1925 and 1956-1976) is presented (Figure 4). Such composite is reasonable because the composite SST anomalies resemble each other among these three seasons (not shown). It is illustrated that significant SST anomalies emerge over North Pacific and North Atlantic (Figure 4). The SST pattern over North Pacific resembles the Pacific Decadal Oscillation [*Mantua et al.*, 1997] with positive SST anomalies around the Kuroshio-Oyashio extension region and negative SST anomalies to its north and along the west coast of North America. The SST anomalies over North Atlantic are mainly negative with most significant signals over mid-high latitudes. It suggests that the background SST is quite different over Northern Hemisphere between negative and positive epochs. This changed background condition may alter the process through which the NAO influences the east China's summer precipitation, and therefore cause the unstationary interannual relationship between NAO and certain precipitation patterns. However, the exact mechanism is still not clear and it should be investigated by using a coupled atmospheric and oceanic general circulation model in the future.

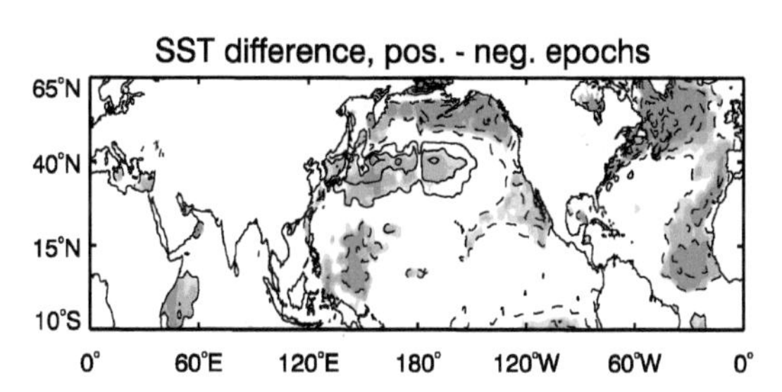

Figure 4. Composite difference of SST from preceding winter to simultaneous summer (DJFMAMJJA) between positive epochs (1884-1904 and 1932-1952) and negative epochs (1905-1925 and 1956-1976). Contour intervals are 0.1℃. Zero contour lines are omitted. The light, middle and dark shading indicates 90%, 95% and 99% significance level, respectively.

Acknowledgments We thank the two anonymous reviewers for their valuable suggestions in improving the quality of this paper. This research is supported by the National Key Technology R&D Program (2006BAC02B04, 2008BAK50B02), the 2009 CMA operation project (the operational system setup of seasonal climate prediction) and the 973 Program of China (2006CB403600).

References

Ding, Y., and J. C. L. Chan (2005), The East Asian summer monsoon: An *overview*, *Meteorol. Atmos. Phys.*, 89, 117-142.

Gershunov, A., and T. P. Barnett (1998), Interdecadal modulation of ENSO teleconnections, *Bull. Am. Meteorol. Soc.*, 79, 2715-2725.

Gong, D.-Y., and C.-H. Ho (2003), Arctic oscillation signals in the East Asian summer monsoon, *J. Geophys. Res.*, 108(D2), 4066, doi: 10.1029/2002JD002193.

Huang, R. H., L. T. Zhou, and W. Chen (2003), The progresses of recent studies on the variabilities of the East Asian monsoon and their causes, *Adv. Atmos. Sci.*, 20, 55-69.

Huang, R. H., J. L. Chen, and G. Huang (2007), Characteristics and variations of the East Asian monsoon system and its impacts on climate disasters in China, *Adv. Atmos. Sci.*, 24, 993-1023.

Kumar, K., B. Rajagopalan, and M. A. Cane (1999), On the weakening relationship between the Indian monsoon and ENSO, *Science*, 284, 2156-2159.

Mantua, N. J., S. R. Hare, Y. Zhang, J. M. Wallace, and R. C. Francis (1997), A Pacific decadal climate oscillation with impacts on salmon, *Bull. Am. Meteorol. Soc.*, 78, 1069-1079.

North, G. R., T. L. Bell, and R. F. Cahalan (1982), Sampling errors in the estimation of empirical orthogonal functions, Mon. *Weather Rev.*, 10, 699-706.

Power, S., T. Casey, C. Folland, A. Colman, and V. Mehta (1999), Interdecadal modulation of the impact of ENSO on Australia, *Clim. Dyn.*, 15, 319-324.

Rayner, N. A., D. E. Parker, E. B. Horton, C. K. Folland, L. V. Alexander, D. P. Rowell, E. C. Kent, and A. Kaplan (2003), Global analyses of sea surface temperature, sea ice, and night marine air temperature since the late nineteenth century, *J. Geophys. Res.*, 108(D14), 4407, doi: 10.1029/ 2002JD002670.

Sung, M.-K., W.-T. Kwon, H.-J. Baek, K.-O. Boo, G.-H. Lim, and J.-S. Kug (2006), A possible impact of the North Atlantic Oscillation on the East Asian summer monsoon precipitation, *Geophys. Res. Lett.*, 33, L21713, doi: 10.1029/2006GL027253.

Wang, L., W. Chen, and R. Huang (2007), Changes in the variability of North Pacific Oscillation around 1975/1976 and its relationship with East Asian winter climate, *J. Geophys. Res.*, 112, D11110, doi: 10.1029/ 2006JD008054.

Wang, L., W. Chen, and R. Huang (2008), Interdecadal modulation of PDO on the impact ofENSO on the East Asian winter monsoon, *Geophys. Res. Lett.*, 35, L20702, doi: 10.1029/2008GL035287.

Wang, S. W., D. Gong, J. Ye, and Z. Chen (2000), Seasonal precipitation series of eastern China since 1880 and the variability, *Acta Geogr. Sin.*, 55, 281-293.

Arctic Oscillation Anomaly in Winter 2009/2010 and Its Impacts on Weather and Climate

Li Lin[1]*, Li Chongyin[1,2], Song Jie[2]

[1] Institute of Meteorology, PLA University of Science and Technology, Nanjing, China
[2] Institute of Atmospheric Physics, Chinese Academy of Sciences, Beijing, China

Abstract Using the Arctic Oscillation (AO) index, the exceptional winter (DJF) of 2009 has been analyzed. The middle-to-high latitudes of the Northern Hemisphere suffered from a nearly zonally symmetric anomaly of temperature and pressure. This situation revealed that two negative AO events occurred in the winter of 2009/2010, with unprecedented low values in January 2009 and February 2010. The negative AO event in January 2009 can be further divided into two stages: the first stage was mainly driven by enhanced upward-propagating planetary waves, which led to a weak stratospheric polar vortex associated with a downward-propagating negative AO signal; the second stage was caused by a lower tropospheric positive temperature anomaly in the high latitudes, which maintained the positive geopotential height anomaly of the first stage. The two successively occurring stages interacted and caused the lower troposphere to experience a strong and lengthy persistence of the negative AO event. We consider that the second event of negative AO in February 2010 is related to the downward-propagating negative AO after sudden stratospheric warming. Eleven long- persistence negative AO events were analyzed using reanalysis data. The results suggest that the negative AO in the troposphere might have been caused by stratospheric sudden warming, a downward-propagating weak stratospheric circulation anomaly or dynamic processes in the troposphere. Further study shows that the negative phase of the AO in the winter of 2009/2010 corresponded to a wide range of temperature and precipitation anomalies in the Northern Hemisphere. Therefore, to improve the accuracy of weather forecasting and climate prediction, more attention should be paid to the AO anomaly and its impact.

Keywords Arctic Oscillation, stratospheric circulation anomaly, planetary waves, weather and climate extremes, SSW.

The Arctic Oscillation (AO) is the leading atmospheric mode of the non-tropical Northern Hemisphere and is characterized by anti-phasic fluctuations of sea-level atmospheric pressure (SLP) between the polar and mid latitudes [1]. When there are positive (or negative) pressure

Received March 1, 2011; accepted June 1, 2011; published online January 2, 2012.
*Corresponding author (email: llen1983@163.com)

anomalies in the polar region, there are corresponding zonally systematic negative (or positive) pressure anomalies in the mid latitudes. Accordingly, the AO is also called the Northern Hemisphere annular mode (NAM), while the North Atlantic Oscillation (NAO) is considered a manifestation of the AO in the North Atlantic region [2]. In a fundamental study of the AO, without using empirical orthogonal function (EOF) analysis, Li et al. [3, 4] investigated the basic essence of the AO and represented a clearer physical mechanism of this phenomenon. The AO not only exists in the lower troposphere, but it can also extend from the high stratosphere to the lower troposphere in winter with a quasi-barotropicstructure [5]. The AO typically exhibits its anomaly in the stratosphere and can then propagate downward, reaching the ground in approximately three weeks [6]. This clearer physical description of the mechanism of this phenomenon means that the AO can be used to improve the quality of tropospheric extended-range weather forecasting [7]. Subsequently, a series of studies have focused on the coupling between the stratosphere and troposphere in the winter and its significant influence on weather and climate [8-12]. Some studies have found that October Eurasian snow cover may be a precursor to stratospheric AO anomalies [13, 14].

The close relationship between the East Asian winter climate and the AO has been demonstrated on an inter-annual scale. Strong AO was found to be connected with higher temperatures and precipitation during winter throughout most of China [15, 16]. Moreover, when the AO index exceeds one standard deviation, the summer precipitation from the entire Yangtze River valley to Southern Japan will decrease by 3%-9%, although it can increase by approximately 3%-6% from Northern China to the far east of Russia. Through changing quasi-stationary planetary wave activity, the AO can influence the Siberian High, affecting the strength of the East Asian winter monsoon [17]. The affect of the NAM on temperature has also been studied [18]. Yang et al. [19] suggested that in winter, the westerly and subtropical circulations are distinct in different AO phases. In a positive AO phase, the East Asian monsoon is weak and the temperature is high in Eastern China, Northeast China, Xinjiang and Inner Mongolia. The AO in March may, in turn, exert an influence on summertime circulation in East Asia, which is closely related to the summertime thermal conditions of the troposphere over East Asia, leading to an anomalous convergence/divergence in the Yangtze River valley and the anomaly of Mei-Yu precipitation [20].

In the winter of 2009/2010, the Northern Hemisphere experienced record-breaking low temperatures and snowstorms, resulting in enormous human and economic losses. Figure 1 shows the precipitation anomaly percentage of China in the winter of 2009/2010. Northern China suffered a significant precipitation anomaly, which, in many areas, was more than twice that of normal years. Meanwhile, in the Eurasian high latitudes and southern and central North America, the temperature was lower by approximately 1-3℃ compared with normal years. Most significantly, the temperature in northern and central Russia and northeastern Europe decreased by 3-6℃.

Large-scale weather and climate anomalies in the Northern Hemisphere during the winter of 2009/2010 were considered to be closely related to significant negative AO anomalies during the same period. Therefore, we concentrated on an analysis of the AO anomaly in the winter of 2009/2010, and explored the possible reasons for its formation. We also investigated the influence it had on weather and climate. Further understanding of the evolution of such negative AO events and their impacts may be helpful for weather forecasting and climate prediction.

Figure 1 Precipitation anomalies (as percentages) in China in the winter of 2009/2010.(See the original text for details.)

1. Data and Methods

NCEP/NCAR daily reanalysis data from 1 January 1948 to 28 February 2010 were used. The data have a 2.5°×2.5° horizontal resolution and extend from 1000 to 10 hPa with 17 vertical pressure levels.

The monthly and daily AO index was obtained from the National Climate Prediction Center of America (CPC) (http: //www.cpc.noaa.Gov), and the rainfall data from 160 stations of the National Climate Center of China (http: // ncc.cma.gov.cn/cn/).

The EP flux was used to investigate planetary wave activity following Chen et al. [21]. The components of EP flux (F) and its divergence (D_F) are

$$F = \left(-\rho a \cos\varphi \overline{u'v'}, \rho a \cos\varphi \frac{Rf}{HN^2} \overline{v'T'} \right),$$

$$D_F = \frac{\nabla \cdot \vec{F}}{\rho a \cos\varphi},$$

where ρ is air density, a is the radius of the Earth, φ is the latitude, R is the Universal Gas Constant, f is the Coriolis parameter, H is a constant-scale height, u and v are zonal and meridional wind-speeds and T is air temperature. The zonal wave numbers 1 to 3 are used to represent stationary planetary waves.

The AO evolution following Baldwin et al. [22] was used to illustrate downward signals from the stratosphere during Negative AO events.

2. Negative AO Events in the Winter of 2009/2010

Figure 2(a) shows the monthly AO index in December from 1950 to 2009 and February from 1951 to 2010. It is evident that the monthly AO index varies significantly on an inter-annual timescale. Basically, the values are typically between –3 to +3 over the record. However, in December 2009 and February 2010, the AO indices are –3.41 and –4.27 respectively. These are the lowest values in the last 60 years. Such low AO indices indicate that in winter of 2009/2010 there was an extremely negative AO event.

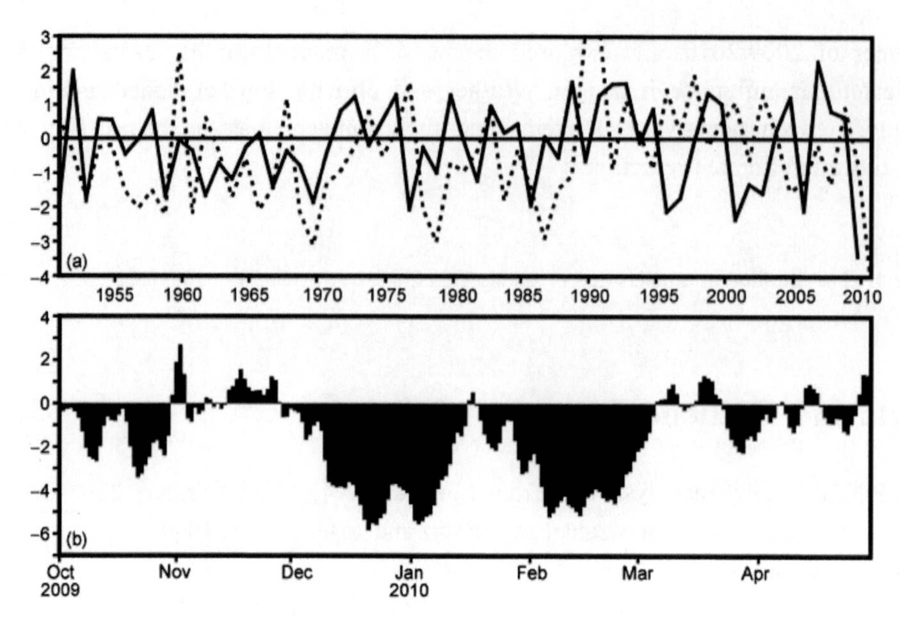

Figure 2 (a) Monthly AO index in December from 1950 to 2009 and in February from 1951 to 2010; (b) daily AO index from October 2009 to April 2010.

Figure 2(b) shows the daily AO variations in the winter of 2009/2010. The Figure demonstrates there were two obvious negative AO events which occurred during this period. The first negative AO event started in early December 2009, when the AO index gradually turned negative, and on December 23, it reached a minimum (–5.557). Subsequently, this minimum was slightly improved. In early January 2010, the index became stronger, and then on January 4, the index fell to a second minimum (–5.403) followed by a gradual decay of the negative anomalies. By the middle of January, the first negative AO event had ended, having lasted 48 days. Experiencing an unusual event similar to the first, the second negative AO event began in the middle of January 2010 and reached its minimum at –5.205. This event ended in early March, having persisted for 47 days. The two successively occurring events caused the lower troposphere of the Northern Hemisphere to experience a strong and lengthy negative AO over the whole of winter 2009/2010.

The AO exhibited antiphase fluctuations between the high and mid-latitudes. Figure 3 shows the geopotential height anomalies at 1000 hPa in December 2009 and February 2010. The distribution of anomalies was similar, with strong positives at high latitudes and strong negatives in the North Atlantic and North Pacific at mid-latitudes, corresponding to a classic negative AO pattern. Comparing the size and intensity of the positive and negative anomalies between the two months, consistent with the monthly AO index the AO in February 2010 seems to be stronger than that in December 2009.

As the leading mode of the Northern Hemisphere, in midwinter the AO is a deep system which exists from the lower troposphere to the high stratosphere. The AO anomaly can propagate downward from the stratosphere and influence the weather and climate in the lower troposphere [7]. The AO evolution following Baldwin [22] is used to illustrate the downward

AO process. It is obvious that from December 2009 to March 2010, there were two negative AO events coinciding with the daily AO index variation. Although there were distinctions between the two events, the two negative AO processes in the lower troposphere both had some connection with the stratospheric AO anomaly (Figure 4).

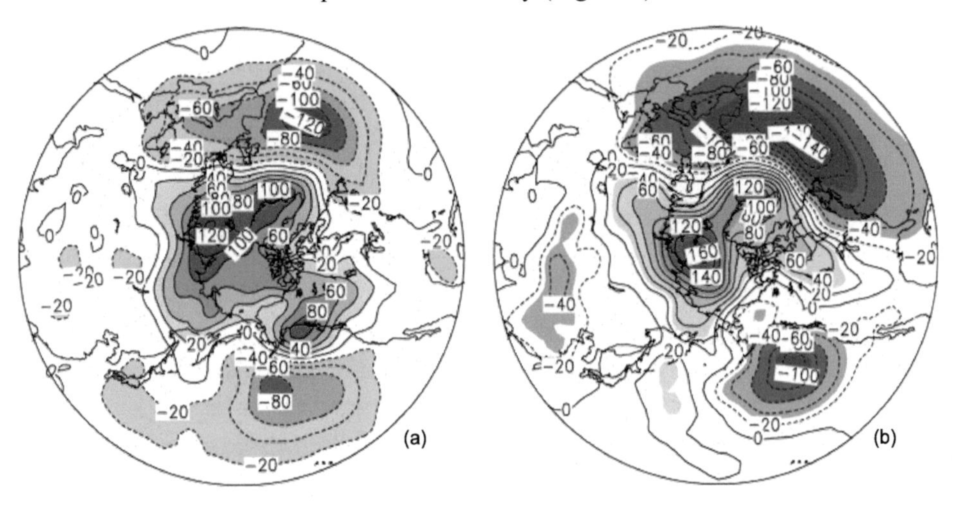

Figure 3 1000 hPa geopotential height anomalies (contours in gpm) in December 2009 (a) and February 2010 (b). Shading is below −20 gpm and over 20 gpm.

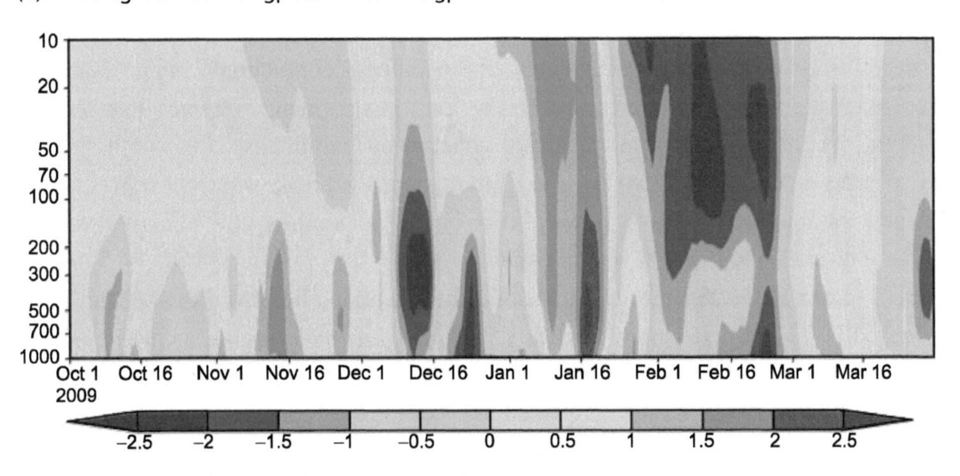

Figure 4 AO evolution from October 2009 to April 2010.

The first negative AO event began in early December 2009 and ended in early January 2010. That event can be divided into two stages according to its evolution. The first stage was in early to mid-December. A negative AO extended from the stratosphere to the troposphere with its minimum at 500 to 200 hPa. The second stage was from late December 2009 to early January 2010. The minimum negative AO was in the lower troposphere, starkly contrasting the positive AO in the stratosphere. These two stages exactly corresponded to the two peaks existing in the daily AO index in December. The second negative AO event started in mid-January, with the strongest negative AO in the stratosphere. A negative AO was first observed in the stratosphere as the precursor of the negative AO in the troposphere. It should be noted that the daily AO index at 1000 hPa in Figure 2(b) was generated by projecting the daily data onto the leading EOFs of geopotential height anomalies, while the AO evolution in Figure 4 was obtained, following Baldwin et al. [22], by calculating the EOFs of the zonal-mean geopotential height at each level. The results show that the main features are duplicated across both methods, although there are minor differences. This analysis shows that the strong negative AO in the winter of 2009/2010 had two significant independent processes, each of which has different characteristics.

3. Causes of the AO Anomaly in the Winter of 2009/2010

Two negative AO events were identified in the winter of 2009/2010. According to the AO index evolution, we suggest that the first negative AO event had two stages from December 2009 to early January 2010. In the first stage, in early December, the AO had a consistent variation from the troposphere to the stratosphere. However, an opposite AO anomaly developed in the troposphere and stratosphere during the second stage, from middle December to early January. The evolution of the negative AO in the first process is reflected by the zonal average geopotential height and zonal wind (Figure 5). It can be seen that consistently positive geopotential height anomalies and negative zonal wind anomalies extended from the troposphere into the stratosphere in early December, representing a weakened polar vortex and a negative AO anomaly. Subsequently, the negative geopotential height anomalies and positive zonal wind anomalies emerged in the high stratosphere. It seems that the positive geopotential height anomalies and negative zonal wind anomalies in the high stratosphere tended to propagate downward. However, simultaneously the lower troposphere and lower stratosphere maintained positive geopotential height anomalies and negative zonal wind anomalies anti-phasic with the high stratosphere.

Figure 6 shows the zonal mean geopotential height anomalies from the second pentad of December 2009 to the first pentad of January 2010. During the second pentad of December 2009, consistent positive geopotential height anomalies extended from the stratosphere to the troposphere at high latitudes accompanied by negative geopotential height anomalies at mid-latitudes. The negative and positive geopotential height anomalies between the high and middle latitudes formed a negative AO anomaly, corresponding to the negative AO index from the stratosphere to troposphere in early December. In the third pentad of December, positive geopotential height anomalies emerged in the mid latitudinal high stratosphere (Figure 6(b)). Then, in the fourth pentad of December, negative geopotential height anomalies occurred at high latitudes in the high stratosphere (Figure 6(c)). Therefore, the positive and negative geopotential height anomalies between the high and middle latitudes in the high stratosphere formed an opposite AO anomaly against that in the troposphere. The maximum negative geopotential height anomaly in the mid latitudes occurred in the upper troposphere and lower stratosphere. Thus, at the same level, the geopotential height anomalies could cause a strong negative AO anomaly, which is consistent with the maximum negative AO index in the high troposphere. The negative geopotential height anomalies at the high latitudes in the high stratosphere in the fourth pentad of December propagated downward slowly and reached the lower troposphere by early January 2010 (figure omitted). It seems that the positive geopotential height anomalies occurring in the lower troposphere and lower stratosphere blocked the downward negative geopotential signal, which led to a stable negative AO anomaly controlling the entire Northern Hemispheric troposphere, forming the second stage of the first negative AO event.

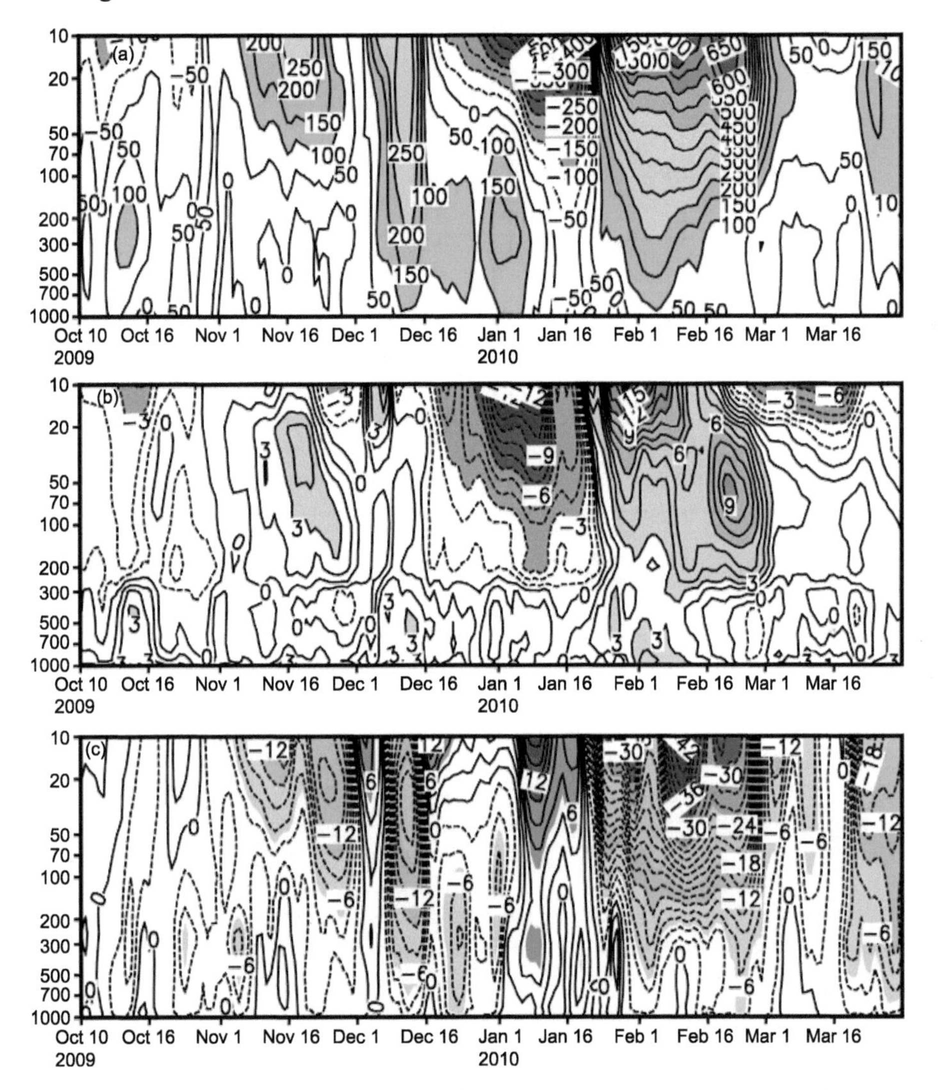

Figure 5 Time-height development of geopotential height (a), temperature (b) and zonal wind (c) averages north of 60°N from October 2009 to March 2010. Contour intervals are 50 gpm in (a), 1 K in (b) and 2 m s^{-2} in (c).

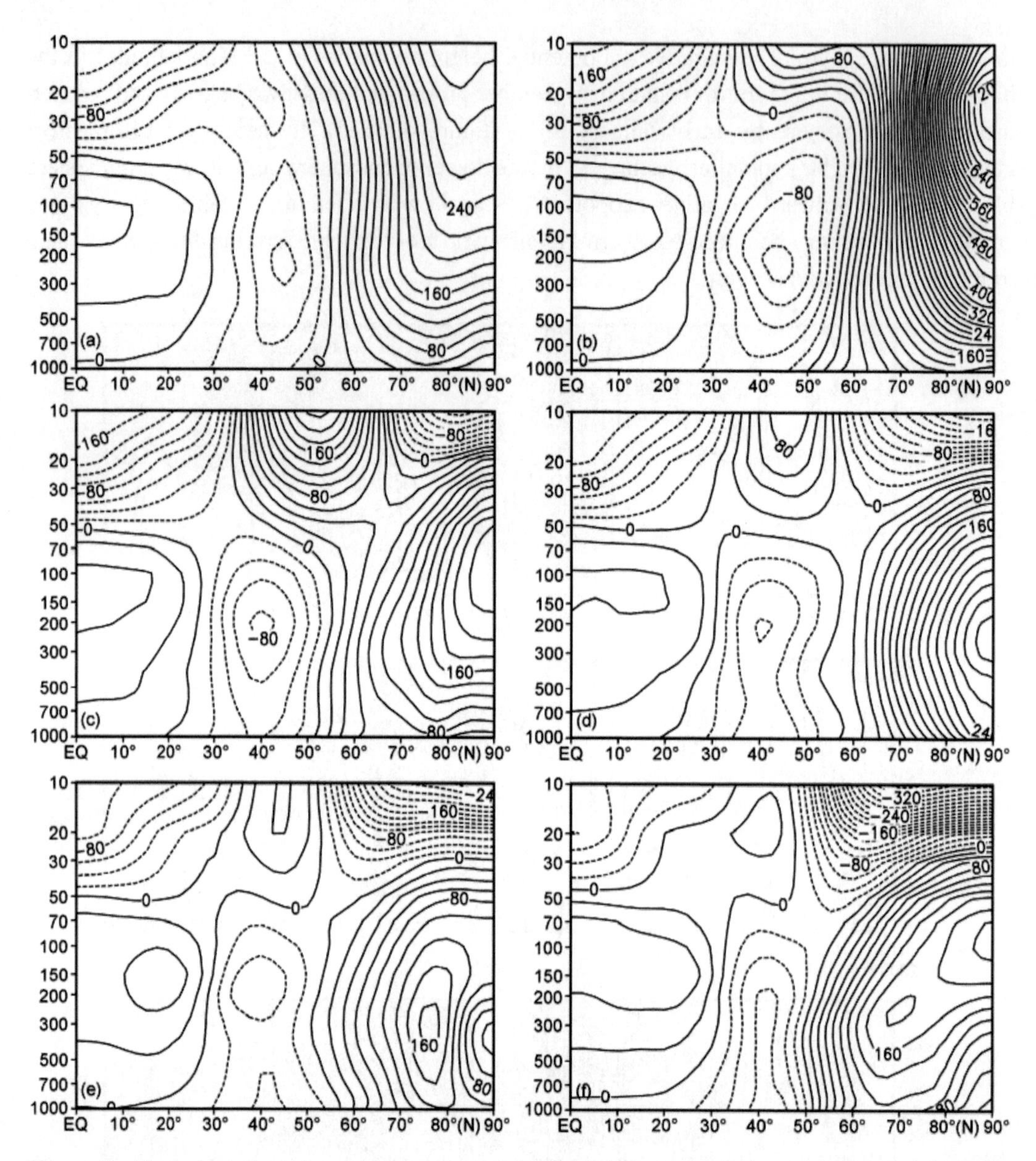

Figure 6 Vertical section of zonal average geopotential height anomaly (contours in gpm) from the second pentad of December 2009 to the first pentad of January 2010.

The zonal mean temperature anomalies are shown in Figure 7 from the second pentad of December 2009 to the first pentad of January 2010. For all of December and early January, positive temperature anomalies controlled the high latitudes in the troposphere with a maximum at 1000 hPa. Negative temperature anomalies at the high latitudes in the stratosphere began in the second pentad of December 2009. Although the downward propagation of these negative temperature anomalies was noticeable, it did not reach the lower troposphere. Because negative anomalies cannot propagate downward, the existence of positive temperature anomalies at high latitudes in the lower troposphere may be the reason for the persistence of the positive geopotential height at high latitudes in the lower troposphere. Consequently, the positive geopotential height anomalies formed the stable negative AO in the second stage of the first negative AO event.

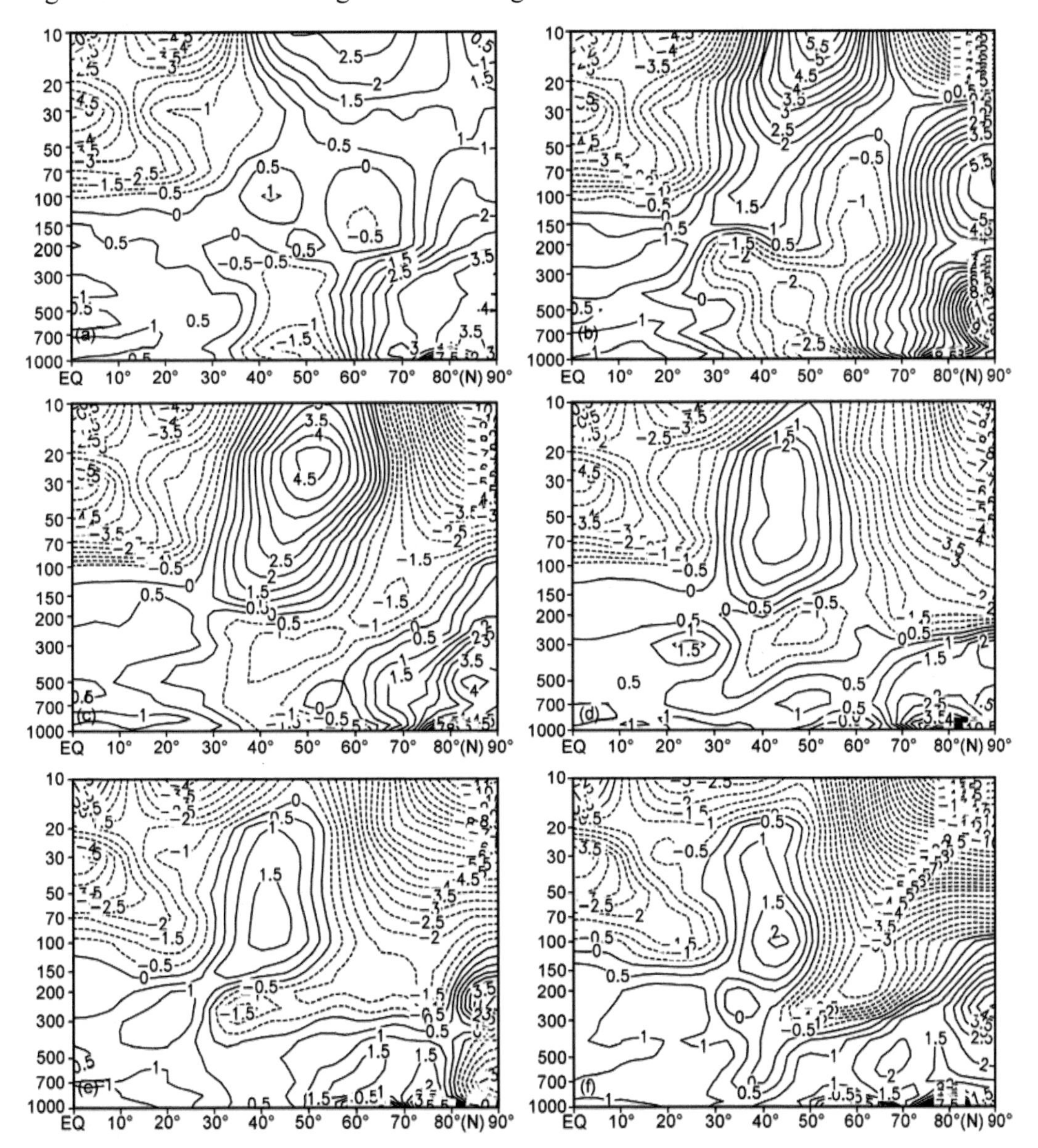

Figure 7 Vertical section of the zonal average temperature (contours in K) anomaly from the second pentad of December 2009 to the first pentad of January 2010.

To understand the lasting positive temperature anomalies at high latitudes during the second stage of the first negative AO event, we analyzed the geopotential height at 500 hPa from the second pentad of December 2009 to the first pentad of January 2010. A block event was noticeable in the North Atlantic, lasting for more than 20 days during this period (figure omitted). The block structure extended to the upper troposphere to approximately 200 hPa. Because blocks can cause the strong meridional exchange of large-scale air masses and heat, we believe that the duration of the Atlantic block may be the main reason for the positive temperature anomalies at the high latitudes in the lower troposphere during the second pentad of December 2009 to the first pentad of January 2010.

A quasi-stationary planetary wave was shown to be the main cause for the circulation variation in the winter stratosphere, as it propagated to the stratosphere and affected the stratosphere circulation by wave-circulation interaction [23]. Figure 8 shows the normalized vertical component of the EP flux averaged north of 60°N from 200 hPa to 10 hPa. Waves 1-3, 1 and 2 were calculated to represent the change of the quasi-stationary planetary wave. Evidently, wave 1-3 had a monthly oscillation. The peaks in late November and late January mean that the upward quasi-stationary planetary waves increased in these two periods. As wave 1 had a consistent variation with wave 1-3, we suggest it played the main role. The quasi-stationary planetary waves traveled upward to the stratosphere and dissipated there,

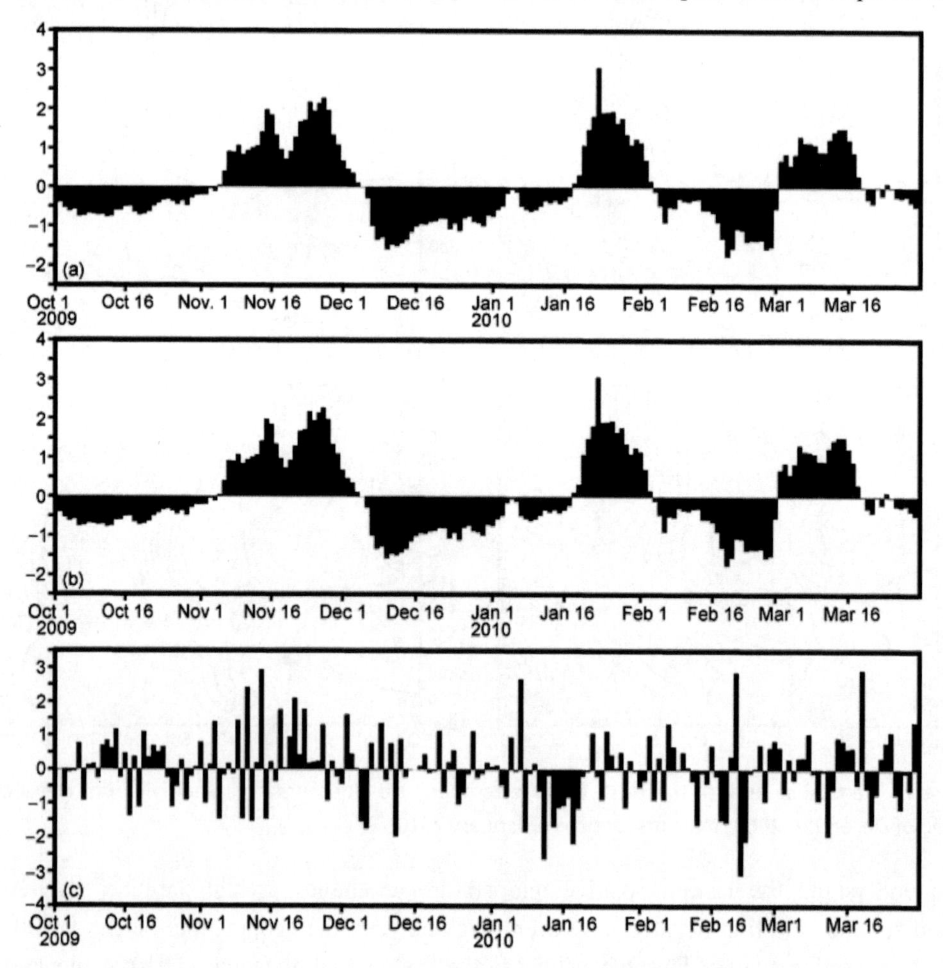

Figure 8 Normalized vertical component of EP flux average north of 60°N from 200 to 10 hPa of wave 1 3 (a), wave 1 (b), and wave 2 (c). The unit of EP flux is m3 s^{-2}.

producing warming by decelerating the mean flow [24]. Consequently, the enhanced upward planetary wave identified in November could have led to a weakened polar vortex in the stratosphere. The weakened polar vortex, as with the noticeable positive geopotential height anomalies at the high latitudes in early December, caused the first stage of the first negative AO event. After the strong upward propagation, the quasi-stationary planetary wave decreased in mid-December, mainly because of the decreasing westerly due to the weakened polar vortex. Subsequently, the polar vortex recovered from the input of diabatic heat. Therefore, the negative geopotential anomalies can be identified at high latitudes in the high stratosphere in the fourth pentad of December (Figure 5(d)).

Compared with the first negative AO event, the second was relatively simple. Figure 9 shows the time series of the zonal average temperature and the geopotential height gradient between the inside and outside of the polar vortex from October 2009 to March 2010 at 20 hPa. It is evident that the zonal average temperature and geopotential height gradient reversed in late January, which is consistent with the main feature of Stratospheric Sudden Warming (SSW). Also, there were strong positive geopotential height and temperature anomalies beginning in the late pentad of January (Figure 4). Meanwhile, the mean zonal wind field showed an easterly flow in the stratosphere lasting over one month. All signs illustrate that a SSW event happened in late January 2010. The quasi-stationary planetary wave variation shows there were enhanced upward planetary waves before the early pentad of January. Matsuno [25] pointed out that strong upward planetary waves could cause an SSW in mid-winter. After the strong SSW, the stratospheric circulation always undergoes a significant change, leading to a negative AO anomaly [11]. From the AO evolution (Figure 4), the negative AO propagated downward from the stratosphere to the troposphere in early February, causing the second negative AO event in the lower troposphere.

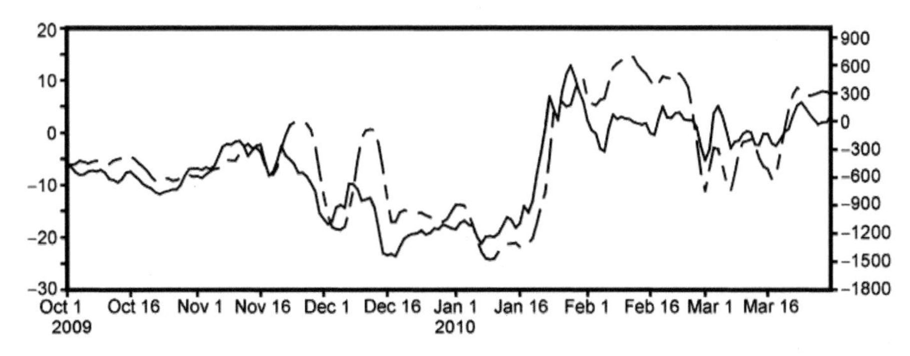

Figure 9 Time scries of zonal average temperature (dashed line, unit: K) and geopotential height (solid line, unit: gpm) gradient between 60°N and 90°N at 20 hPa.

In above analysis, we suggest that the strong upward planetary waves were the reason for the negative AO anomalies in the stratosphere, and the negative AO anomalies propagated downward, leading to the two negative AO events in the lower troposphere. The two negative AO events during December 2009 and February 2010 demonstrate that the negative AO in the troposphere might be caused by SSW, a downward-propagating weak stratospheric circulation anomaly or dynamic processes in the troposphere and their combined effects.

We define a negative AO event as a negative AO index lasting more than 30 days and the minimum of the index as lower than -4. According to this definition, 11 negative AO events were selected from 1948 to 2010: January 1951, February 1955, January 1959, February 1966, January 1969, February 1969, February 1977, February 1979, January 1996, February 1998 and February 2001. Figure 10 gives the AO evolution and the daily AO index of the 11 negative AO events.

Five cases with downward AO from the stratosphere to the troposphere were identified in the 11 negative AO events. Two cases were confirmed to be strong SSW events in February 1977 and February 1998 (Figure 10). Although the circulation change in the stratosphere was not as strong as with a SSW, the other three cases with downward AO also had downward positive geopotential anomalies at high latitudes from the stratosphere to the troposphere. The downward AO anomaly in the three cases was enhanced when it reached the lower troposphere. In addition to these five events, the opposite AO signal between the troposphere and the stratosphere emerged in the other six events, which means that the dynamic tropospheric process played an important role in the maintenance of the negative AO process.

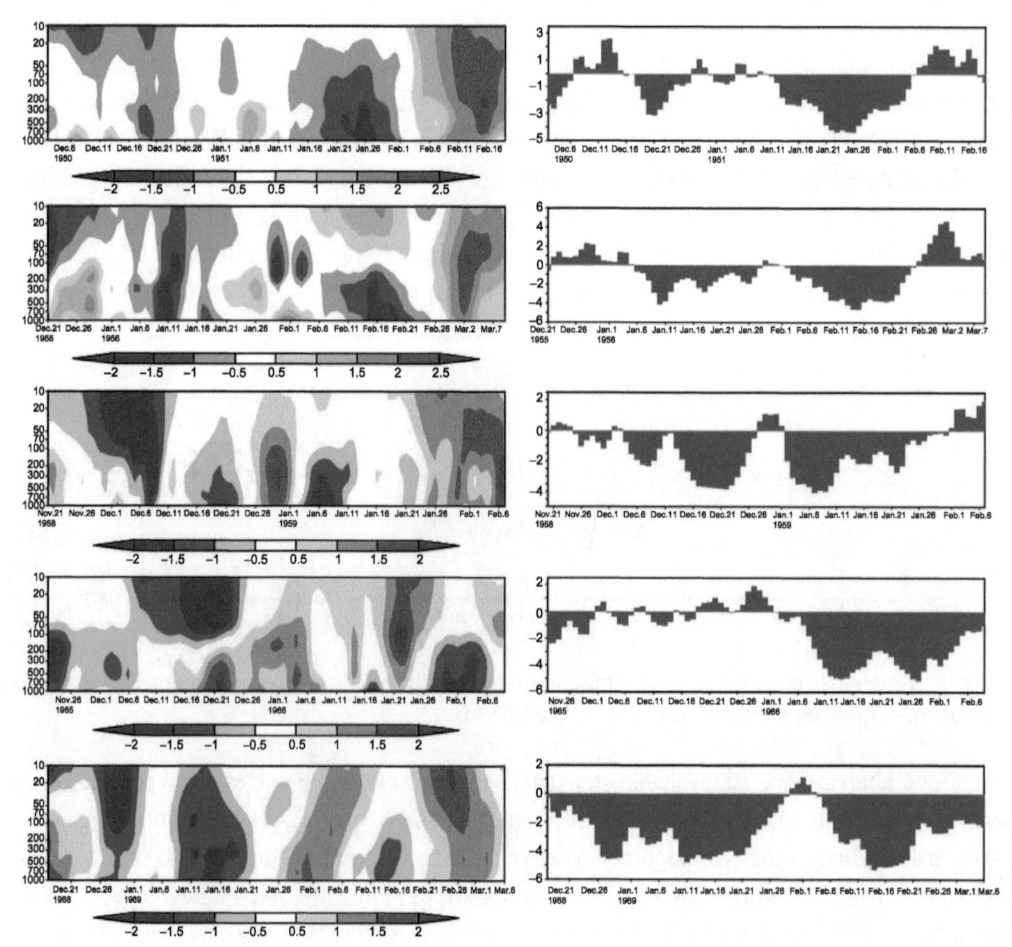

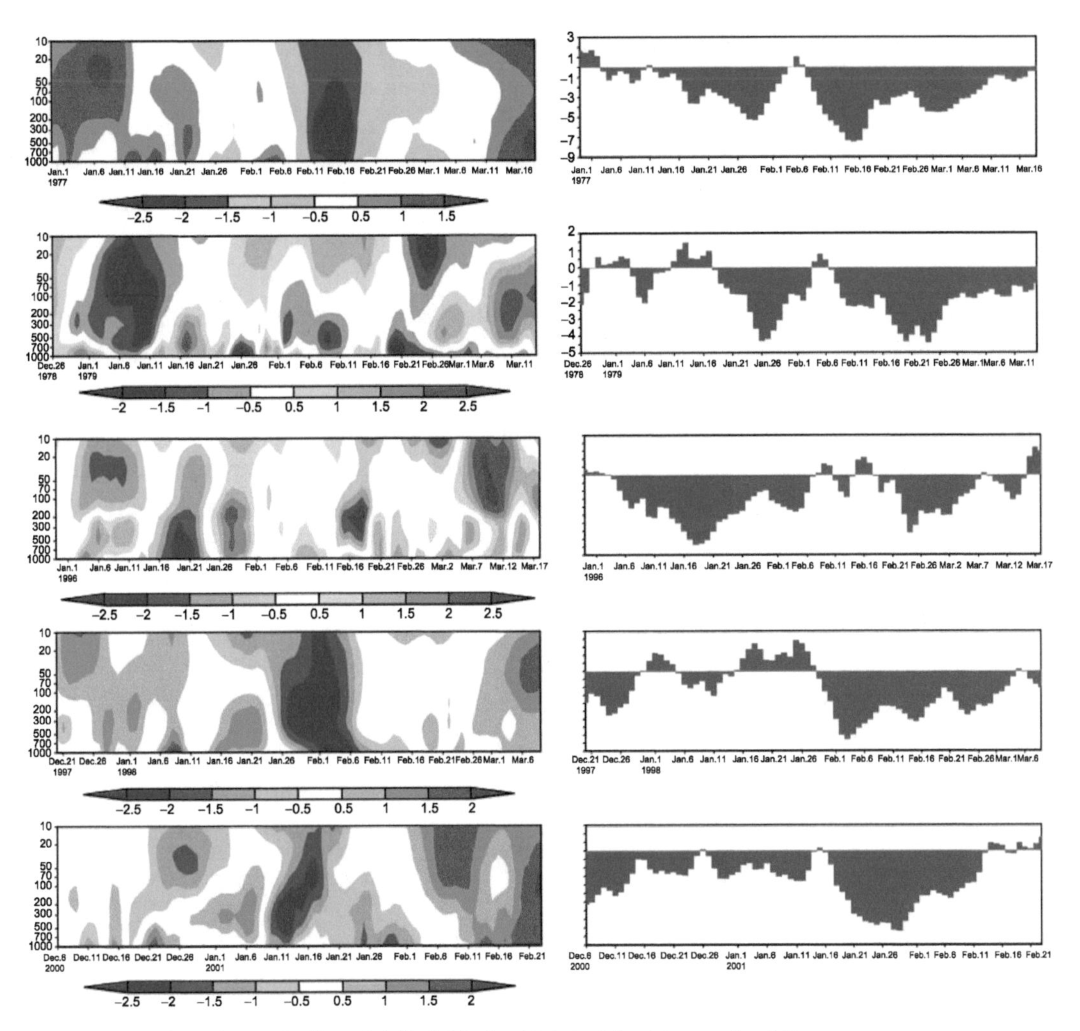

Figure 10 AO evolution (left) and daily AO index (right) of the 11 negative AO events.

4. Influence of AO in the Winter of 2009/2010

China suffered serious climate anomalies in the winter of 2009/2010, indeed, most regions of the Northern Hemisphere experienced significant climate anomalies. Large- scale Eurasian high-latitude regions and central and southern North America had temperatures 1 to 3℃ lower than normal. The temperatures were approximately 3℃ to 8℃ colder in North-central Russian and northeastern Europe, while they were higher by approximately 1 to 2℃ than usual in northern North America and South Asia, where northeast and northwest North America had temperatures 4℃ higher. Previous work has shown that, as the strongest mode in the Northern Hemisphere, AO anomalies should have an important influence on the weather and climate. This was certainly the case with the strong and persistent negative AO anomalies in the winter of 2009/2010.

On the global scale, a negative (positive) AO phase corresponds to the mid-latitude jet stream slanting equatorward (polarward) as the polar vortex expands (shrinks) and weakens (strengthens). A negative AO is usually accompanied with an apparent trough and ridge in the middle latitudes, which can enhance the exchange of cold and warm air. A meridional wind anomaly at 300 hPa in winter 2009/2010 (Figure 11(b)) shows the impact of the negative AO anomaly on the global atmospheric circulation. The evident meridional wind anomalies at the middle latitudes (approximately 40°N-70°N) indicate a trough and ridge system was active in the winter of 2009/2010. With such an enhanced trough and ridge system, the cold and warm air exchange between north and south was intensified, making the polar latitudes warmer and the middle and high latitudes of Eurasia colder (Figure 11(a)). In addition to the circulation anomaly caused by the AO anomaly, the AO anomaly may also have led to the low-frequency oscillation anomaly, thus affecting weather and climate. Figure 12 shows the correlation coefficient between the daily AO index and the low-frequency (20-70 days) oscillation kinetic energy distribution from December 2009 to February 2010. The correlation coefficient exceeded 0.8 in some regions. As the background of the synoptic scale system, the low

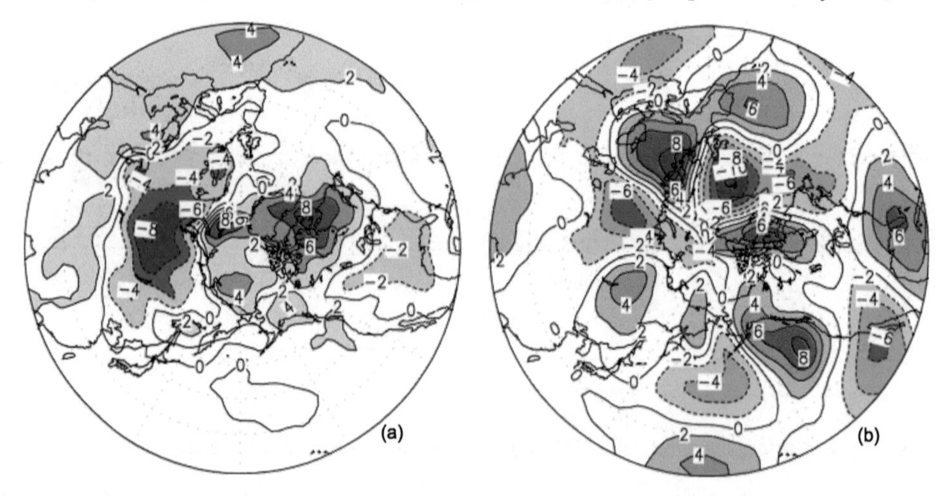

Figure 11 Average temperature (contours in K) at 1000 hPa (a) and meridional wind (contours in m s^{-1}) at 300 hPa from December 2009 to February 2010.

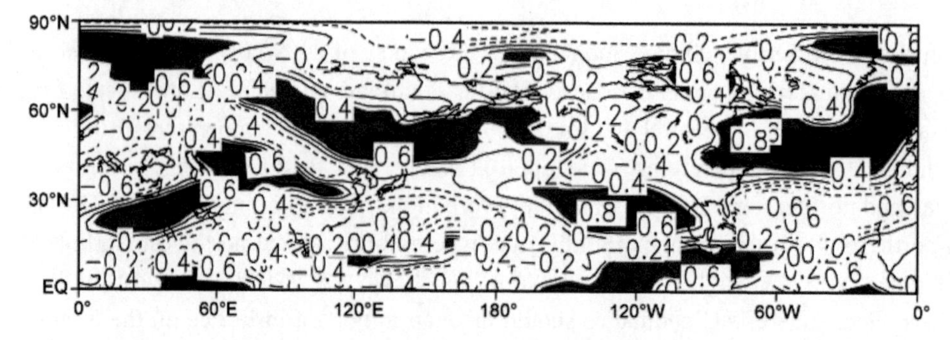

Figure 12 Correlation coefficient between the daily AO index and the low-frequency (20-70 days) oscillation kinetic energy distribution from December 2009 to February 2010.

low-frequency oscillation can impact the weather and climate through the interaction of different scale circulation systems. Consequently, the influence of the AO on the low-frequency oscillation could ultimately affect the weather and climate. It is worth mentioning that the correlation of the AO and the synoptic scale (3-8 days) oscillation is lower than that of the AO and the low-frequency oscillation, indicating that the impact of the AO anomaly is mainly on the low-frequency system because of the spatial and time scales of the AO.

Corresponding to the circulation and temperature anomalies, there were also large-scale precipitation anomalies in the Northern Hemisphere. The precipitation anomaly percentage in the winter of 2009/2010 is shown in Figure 13. The most obvious anomalies were in the Northern Hemisphere. The mid-latitudes of Asia (including Northeast and Northwest China), central and southern North America and equatorial Africa had strong positive anomalies (2-4 times normal). However, Southern Asia (including southwest China), North Africa, Russia and other parts of central and northwestern North America had serious negative precipitation anomalies that were approximately 30%-80% less than normal years. Figure 13 also clearly indicates that China's Yunnan Province and nearby areas suffered severe drought in the winter of 2009/2010. The specific causes of the drought (including the impact of the abnormal AO) will be the subject of further work.

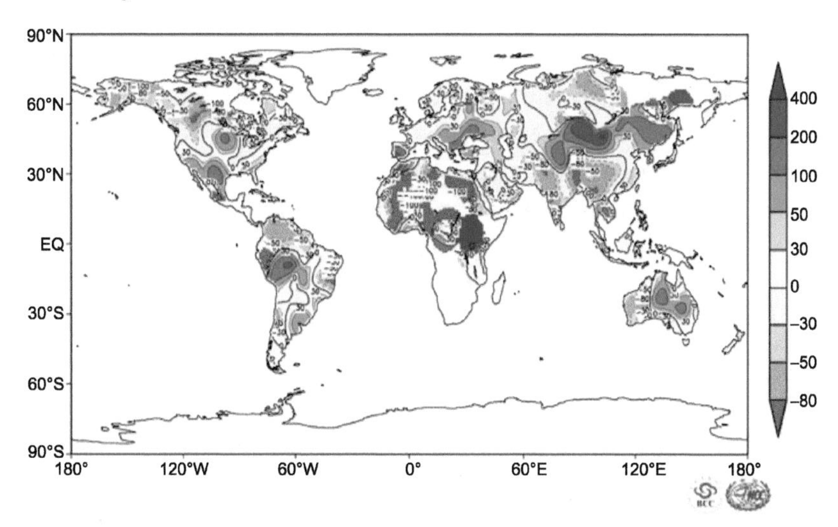

Figure 13 Global precipitation anomaly percentage in the winter of 2009/2010.

Figures 11 and 13 clearly show serious weather and climate anomalies in the Northern Hemisphere in the winter of 2009/2010 in terms of both strength and geographic range. Although these weather and climate anomalies were not all caused by the abnormal AO, the significant impact of the Arctic Oscillation (AO) anomaly on the weather and climate in the Northern Hemisphere is worthy of attention.

5 Discussion and Conclusion

The AO experienced two negative events in the winter of 2009/2010. The upward quasi-stationary planetary wave played an important role in the two events. The first negative AO event from December 2009 to early January 2010 can be divided into two stages. The first stage was the result of downward stratospheric anomalies after the enhanced upward quasi-stationary planetary wave. Positive, persistent temperature anomalies at the high latitudes in the troposphere were found to be caused by a block in the North Atlantic, leading to the second stage. The two successive stages interacted and hence the lower troposphere experienced a strong and persistent negative AO event. Meanwhile, the downward circulating anomalies after an SSW event produced the second negative AO event in February 2010.

Through the analysis of the two negative AO events, we suggest that a weak stratospheric polar vortex anomaly may also have had an impact on the troposphere, although the mechanism is not well understood. In this study, a negative AO event was defined, and 11 cases were identified from 1948 to 2010. Statistically, two of the 11 negative AO events were caused by a strong SSW, 3 cases were formed by weak downward stratospheric anomalies, and the other 6 cases were produced by dynamic processes in the troposphere itself. The negative process from December 2009 was formed by the combination of the weak downward stratospheric anomalies and the dynamic processes in troposphere. This is an example of how complex negative AO events can be. Prediction of such negative AO events should be a priority.

The enduring negative AO anomalies in the winter of 2009/2010 led to an active trough and ridge system at the middle latitudes (approximately 40°-70°N), causing polar warming and colder Eurasian middle and high latitudes and intensifying air exchange between north and south. In addition, the mid-latitudes of Asia, central and southern North America and equatorial Africa had strong positive precipitation anomalies. With its serious impact on the weather and climate in the Northern Hemisphere, the strong AO anomaly in winter deserves more attention in weather forecasting and climate prediction.

We would like to thank two anonymous referees for their constructive suggestions and the editor for polishing English.

References

1 Thompson D W J, Wallace J M. The Arctic Oscillation signature in the wintertime geopotential height and temperature fields. Geophys Res Lett, 1998, 25: 1297-1300.
2 Wallace J M. North Atlantic Oscillation/Annular Mode: Two para-digms-one phenomenon. Q J R Meteorol Soc, 2000, 126: 791-805.
3 Li J P, Wang J X L. A modified zonal index and its physical sense. Geophys Res Lett, 2003, 30: 1632.
4 Li J P. Physic senses of Arctic Oscillation and its relationship with East Asian circulation (in Chinese). In: Yu Y Q, Chen W, et al, eds. Impact of Sea-Air Interaction on Climate Change in China. Beijing: Meteorological Press, 2005. 169-176.
5 Thompson D W J, Wallace J M. Annular modes in the extratropical circulation. Part I: Month-to-month variability. J Clim, 2000, 13: 1000-1016.
6 Baldwin M P. Dunkerton T J. Propagation of the Arctic Oscillation from the stratosphere to the troposphere. J Geophys Res, 1999, 104: 937-946.

7 Baldwin M P, Stephenson D B, Thompson D W J, et al. Stratospheric memory and skill of extended-range weather forecasts. Science, 2003, 301: 636-639.
8 Thompson D W J, Baldwin M P, Wallace J M. Stratospheric connection to Northern Hemisphere wintertime weather: Implications for prediction. J Clim, 2002, 15: 1421-1428.
9 Black R X. Stratospheric forcing of surface climate in the Arctic Oscillation. J Clim, 2002, 15: 268-277.
10 Baldwin M P, Thompson D W J. Shuckburgh E F, et al. Weather from the stratosphere? Science, 2003, 301: 317-318.
11 Li L, Li C Y, Tan Y K, et al. Stratospheric sudden warming impacts on the weather/climate in China and its role in the influences of ENSO (in Chinese). Chin J Geophys, 2010, 53: 1529-1542.
12 Wang L, Chen W. Downward Arctic Oscillation signal associated with moderate weak stratospheric polar vortex and the cold December 2009. Geophys Res Lett, 2010, 37: L09707.
13 Satio K, Cohen J. The potential role of snow cover in forcing interannual variability of the major Northern Hemisphere mode. Geophys Res Lett, 2003, 30: 302.
14 Cohen J, Barlow M, Kushner P J, et al. Stratosphere-troposphere coupling and links with Eurasian land-surface variability. J Clim, 2007, 20: 5335-5343.
15 Gong D Y, Wang S W, Zhu J H. East Asian winter monsoon and Arctic Oscillation. Geophys Res Lett, 2001, 28: 2073-2076.
16 Gong D Y, Zhu J H, Wang S W. Significant relationship between spring AO and the summer rainfall along the Yangtze River. Chin Sci Bul. 2002, 47: 948-951.
17 Chen W, Kang L H. Linkage between the Arctic Oscillation and winter climate over East Asia on the interannual timescale: Roles of quasi-stationary planetary wave (in Chinese). Chin J Atmos Sci, 2006, 30: 863-870.
18 Ren R C, Cai M. Meridional and vertical out-of-phase relationships of temperature anomalies associated with the Northern Annular Mode variability. Geophys Res Lett, 2007, 34: L07704.
19 Yang H, Li C Y. Influence of Arctic Oscillation on temperature and precipitation in winter (in Chinese). Clim Envi Res, 2008, 13: 395-404.
20 Li C Y, Gu W, Pan J. Mei-yu, Arctic Oscillation and stratospheric circulation anomalies (in Chinese). Chin J Geophys, 2008, 51: 1632-1641.
21 Chen W, Takahashi M, Graf H F. Interannual variations of stationary 25 planetary wave activity in the northern winter troposphere and stratosphere and their relations to NAM and SST. J Geophys Res, 2003, 108: 4797.
22 Baldwin M P, Thompson D W J. A critical comparison of stratosphere-troposphere coupling indices. Q J R Meteorol Soc, 2009, doi: 10.1002/qj.479.
23 Charney J G, Drazin P G. Propagation of planetary-scale disturbances from the lower into the upper atmosphere. J Geophys Res, 1961, 66: 83-109.
24 Andrews D G, Holton J R, Leovy C B. Middle Atmosphere Dynamics. New York: Academic Press, 1987. 1-489.
25 Matsuno T. A dynamical model of the stratospheric sudden warming. J Atmos Sci, 1971, 28: 1479-1494.

Observed Relationship of Boreal Winter South Pacific Tripole SSTA with Eastern China Rainfall During the Following Boreal Spring

Gang Li[1], Chongyin Li[2], Yanke Tan[1], Xin Wang[3]*

[1] College of Meteorology and Oceanography, PLA University of Science and Technology, Nanjing, China

[2] College of Meteorology and Oceanography, PLA University of Science and Technology, Nanjing, and LASG, Institute of Atmospheric Physics, Chinese Academy of Sciences, Beijing, China

[3] State Key Laboratory of Tropical Oceanography, South China Sea Institute of Oceanology, Chinese Academy of Sciences, Guangzhou, China

Abstract The present study investigates the relationships between the December-February (DJF) South Pacific tripole (SPT) sea surface temperature anomaly (SSTA) pattern and the following March-May (MAM) rainfall over eastern China based on multiple datasets. It is found that the relationships between the DJF SPT and the following MAM rainfall over eastern China are modulated by the El Niño-Southern Oscillation (ENSO). When the ENSO signal is removed, the positive DJF SPT is significantly associated with more rainfall over eastern China during the following boreal spring. However, such significant relationships disappear if ENSO is considered. After removing ENSO impacts, the possible mechanisms through which the DJF SPT impacts the following MAM rainfall over eastern China are investigated. The positive DJF SPT is associated with the significantly positive SSTA in the tropical western Pacific, which can persist to the following MAM. In response to the positive SSTA in the tropical western Pacific, a wave-like train in the low-level troposphere extends from the tropical western Pacific (an anomalous cyclone) to the western North Pacific (an anomalous anticyclone) during the following MAM. The anomalous anticyclone over the western North Pacific enhances the anomalous southwesterly over eastern China, which can bring more moisture and favor anomalous increased rainfall. It should be pointed out that La Nina (El Niño) could induce an anomalous cyclone (anticyclone) over the western North Pacific, which offsets the MAM anomalous anticyclone (cyclone) caused by the positive (negative) SPT in the preceding DJF and thus weakens the relationship between the SPT and the rainfall over eastern China.

Manuscript received January 16, 2014, in final form June 28, 2014.
*Corresponding author address: E-mail: wangxin@scsio.ac.cn

1. Introduction

Many previous studies have revealed that the ocean- atmosphere coupling systems in the Southern Hemisphere play an important role in climate variability over East Asia. Many studies have suggested that the circulation changes in the Southern Hemisphere are one of the origins of the East Asian monsoon (Li 1955; Tao and Chen 1987; Li and Wu 2002; Xue 2005). However, the scarcity of high-quality oceanic and atmospheric observations in the Southern Hemisphere greatly inhibited the study of the relationships between climate variability over East Asia and ocean-atmosphere coupling systems in the Southern Hemisphere. Now, with the increase of satellite-derived atmospheric and oceanic datasets available starting in the late 1970s, this situation has been changed significantly.

The southern annular mode (SAM) is the dominant pattern of atmospheric circulation in the Southern Hemisphere on interannual time scales, and it is characterized by a nearly zonally symmetric atmospheric mass seesaw between the subtropical latitudes, centered at 45°S, and the Antarctic continent (Gong and Wang 1999; Thompson and Wallace 2000). It is well known that the SAM has a significant impact on climate variability in China. Previous studies found that when the SAM is in its positive (negative) phase during boreal spring, rainfall over the Yangtze River valley in China tends to increase (decrease) during the following boreal summer (Nan and Li 2003; Nan et al. 2008; Sun et al. 2009). The influence of the SAM may be transmitted to the Yangtze River valley by the "ocean bridges" of the Indian Ocean (Nan and Li 2005; Nan et al. 2008) or by a convection bridge over the Maritime Continent (Sun et al. 2009). Wu et al. (2009) further noted that the SAM in boreal autumn could exert a pronounced influence on the China winter monsoon in the following boreal winter. Zheng and Li (2012) suggested that the SAM in boreal winter can influence rainfall variability in southern China during the following boreal spring through sea surface temperature (SST) anomalies (SSTA) in the middle and high latitudes of the Southern Hemisphere. In addition, the SAM may be a possible factor for the forecast of tropical cyclone activity in the western North Pacific (Ho et al. 2005) and the dust weather frequency in northern China (Fan and Wang 2004).

The Southern Ocean is thought to be one of the most energetic regions in the world's oceans (Wu et al. 2011). Many studies have indicated that the Southern Ocean plays an important role in climate variability in China. The subtropical dipole pattern (Behera and Yamagata 2001; Jia and Li 2005; Yan et al. 2009), which is a remarkable climate pattern in the southern Indian Ocean, has a significant influence on climate variations not only over its subtropical continents (Reason 2001; Suzuki et al. 2004; De Almeida et al. 2007), but also in China (Jia and Li 2005; Yang and Ding 2007; Yang 2009). Jia and Li (2005) found that the anomalous increased rainfall in China during boreal summer is closely linked with a positive subtropical dipole during the preceding boreal summer and autumn. They found that the sub-tropical dipole pattern could result in a weak South China Sea summer monsoon in the following boreal summer. They suggested that the western Pacific warm pool plays an important role in the relationship between the subtropical dipole pattern and the following boreal summer rainfall in China. Yang and Ding (2007) and Yang (2009) investigated the impacts of the subtropical dipole pattern in boreal spring on rainfall variability in China during the following boreal summer via a low-frequency wave train of the circum-Pacific pattern (Wang 2005).

As opposed to the southern Indian Ocean, the South Pacific is a vast region covering almost half of the Southern Ocean. Previous studies focused on the relationship between SST variability in the South Pacific and the El Niño-Southern Oscillation (ENSO) (Kidson and Renwick 2002; Giese et al. 2002; Luo et al. 2003; Huang and Shukla 2006; Wang et al. 2007; Shakun and Shaman 2009; Terray 2011; Li et al. 2012). It has been shown that ENSO has a significant impact on the South Pacific, while the South Pacific can also affect ENSO to some extent. However, compared with these studies that have been devoted to the influence of the southern Indian Ocean on climate variability in China, the South Pacific has received less attention. Exceptions are a few recent studies by Liu et al. (2008), Zhou and Cui (2011), Hsu and Chen (2011), and Zhou (2011). It is found that the SSTA to the eastern coast of Australia may significantly influence the tropical cyclone frequency over the western North Pacific and rainfall over the Yangtze River valley in China by an interhemispheric teleconnection (Liu et al. 2008; Zhou and Cui 2011; Zhou 2011). Hsu and Chen (2011) conjectured that during July-October in the second half of the twentieth century, the 10-20-yr South Pacific (inter) decadal oscillation can lead to an anomalous Hadley-like circulation in the western Pacific and indirectly affect the convection activity in the Philippine Sea, which in turn impacts the rainfall in the Philippines, Taiwan, and Korea.

Given the significant interannual variability of the South Pacific SSTA during boreal winter (Terray 2011), as well as its possible influence on climate variability in China, as mentioned above, the question arises of what is the possible influence of the South Pacific in boreal winter on rainfall variability in China during the following boreal spring. Although the variability of boreal summer rainfall in China and its causes have been extensively studied (i.e., Huang and Wu 1989; Chang et al. 2000; Huang et al. 2003; Wu et al. 2003; Yang and Lau 2004; Chan and Zhou 2005; Li et al. 2006; Zhou and Chan 2007), boreal spring rainfall has a significant contribution to the annual rainfall in China (Feng and Li 2011). In addition, boreal spring rainfall variability in China has an important impact on agricultural activities. Therefore, it is necessary to study the contributing factors to and physical processes of rainfall variability in China during the boreal spring.

The organization of this paper is as follows. The dataset and methodology are described in section 2. A South Pacific tripole pattern index is defined in section 3. In section 4, the relationship between the winter South Pacific tripole pattern and rainfall variability in China during the following spring is investigated. Possible physical mechanisms are investigated in section 5. A summary and discussion are given in section 6.

2. Dataset and Method

Several datasets are used in this study. The monthly SST data are from the Met Office

Hadley Centre Sea Ice and Sea Surface Temperature (HadISST) dataset (Rayner et al. 2003) with a horizontal resolution of 1° × 1°. The monthly atmospheric circulation data with a horizontal resolution of 2.5°× 2.5° are derived from the National Centers for Environmental Prediction (NCEP) and National Center for Atmospheric Research (NCAR) (Kalnay et al. 1996).

The high-resolution rainfall datasets with a spatial resolution of 0.5°× 0.5° are derived from two different sources in this study. The first dataset is the Full Data Reanalysis version 6.0 (V6) precipitation dataset from the Global Precipitation Climatology Centre (GPCC) (Schneider et al. 2011). The second dataset is the Climatic Research Unit (CRU) Time-Series (TS) version 3.21 precipitation dataset from the University of East Anglia Climatic Research Unit (Jones and Harris 2013). In addition, we use the accumulated monthly station rainfall data at 160 stations in China, which are obtained from the Chinese Meteorological Data Center.

Considering the quality and quantity of South Pacific SST data, this study focuses on the period during January 1979 through December 2010, in which the satellite- derived SST is available (Ciasto and Thompson 2008; Terray 2011). Anomalies of all variables are obtained by removing the climatological means during 1979 through 2010. The boreal winter and spring in the study are referred to as the seasonal means during December- February (DJF), and March-May (MAM), respectively. Here, the boreal winter of 1980 refers to the average between December 1979 and February 1980. All of the anomalies are detrended by the linear regression method before analyses to remove the possible influence of global warming.

ENSO is described by the normalized Niño3.4 index, which is defined by the SSTA averaged over the region (5°S-5°N, 170°-120°W).

Empirical Orthogonal Function (EOF) analysis is used to detect the domain pattern of variables. Correlation and partial correlation is also used in this study. For partial correlation,

$$r_{12,3}=\frac{r_{12}-r_{13}r_{23}}{\sqrt{(1-r_{13}^2)(1-r_{23}^2)}} \tag{1}$$

where $r_{12,3}$ indicates the partial correlation coefficient between x_1 and x_2 after removing the influence of x_3; and r_{12}, r_{13}, and r_{23} indicate the correlation coefficient between x_1 and x_2, x_1 and x_3, and x_2 and x_3, respectively. All statistical significance tests for correlations are performed using the two-tailed Student's t test. When examining the statistical significance of the correlation and partial correlation, the effective number of degrees of freedom is considered according to Davis (1976).

3. Tripole Pattern of SSTA in the South Pacific

An EOF analysis is applied to boreal winter SSTA in the South Pacific (60°-20°S, 144°-294°E) during the period 1980-2010. The domain region of EOF analysis is selected because of the following reasons: The selected spatial domain could reduce the impact of ENSO on the South Pacific SSTA; on the other hand, this region covers the large standard deviations of the South Pacific, as suggested by Terray (2011). It should be pointed out that the choice of the northern boundary for EOF analysis is somewhat arbitrary. However, the EOF results don't depend on the selection of the northern boundary. The EOF results have few changes, even if the north boundary is extended to 10°S. Figure 1 displays the leading EOF (EOF1) of the South Pacific SSTA (Fig. 1a) and its standardized principal component (PC1) time series for the period 1980-2010. The percent variance explained by EOF1 is 40.4%. There is a clear separation between EOF1 and the higher EOF modes according to the criterion of North et al. (1982), indicating that EOF1 is robust and stable.

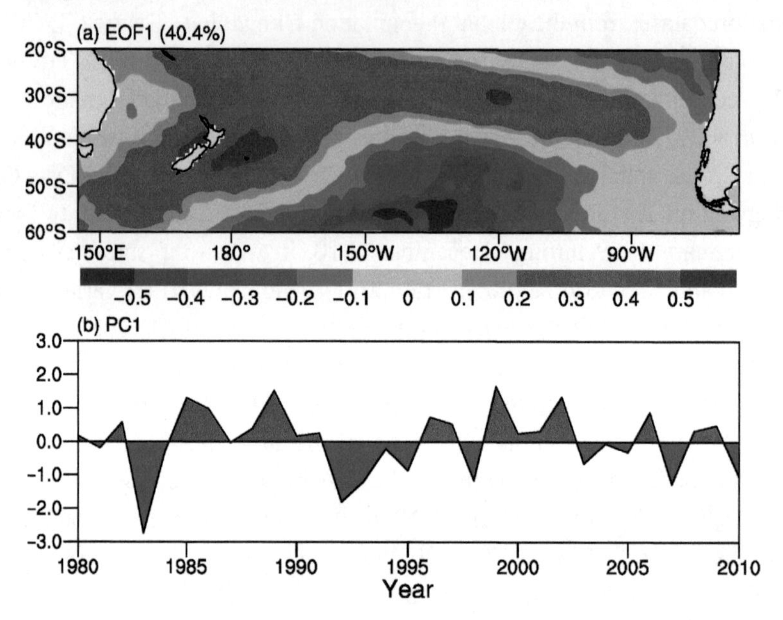

FIG. 1. (a) EOF1 mode of the South Pacific (60°-20°S, 144°- 294°E) SSTA for the period 1980-2010 during the boreal winter. (b) PC1 time series of the EOF1 for the period 1980-2010.

From Fig. 1a, it is clear that the spatial pattern of EOF1 is characterized by a tripole-like structure. Significantly positive eigenvector loadings are observed in the subtropical and southwestern South Pacific, with two centers around 30°S, 120°W and New Zealand, which are referred to as the subtropical South Pacific pole (SSP) and the New Zealand pole (NZP) in this study, respectively. The pronouncedly negative eigenvector loadings are seen in the northeastern and high latitudes of the South Pacific centered near 580S, 1350W, which is referred to as the high-latitude pole (HLP). Although there are some differences in the choice of region and periods of data, the EoF1 bears a strong resemblance to the leading EOF pattern revealed by previous studies (i.e., Kidson and Renwick 2002; Wang et al. 2007; Shakun and Shaman 2009; Terray 2011; Li et al. 2012). The PC1 shows clear variations on interannual-to-decadal time scales (Fig. 1b). Note that the simultaneous correlation coefficient between PC1 and the Niño –3.4 index is –0.8 (exceeding the 95% confidence level), indicating that the tripole SSTA mode has a close relationship with ENSO.

Based on the EOF1 pattern presented in Fig. 1a, we define the areas of the SSP, NZP, and HLP. The SSP and NZP are delineated as (35°-25°S, 235°-245°E) and (50°-35°S, 170°-190°E), and the HLP is delineated as (60°-50°S, 205°-240°E). The area-averaged SSTA in these three regions are referred to as SSPI, NZPI, and HLPI, shown in Fig. 2. Correlation analysis shows that SSPI has a correlation of 0.58 with NZPI, while it has a correlation of –0.77 with HLPI (Table 1). Moreover, NZPI has a correlation of –0.84 with HLPI (Table 1). This indicates that the variations of HLPI are out of phase with those of SSPI and NZPI, and the variations of SSPI and NZPI are in phase. Therefore, the correlation analysis further demonstrates the existence of a tripole SSTA structure in the South Pacific suggested by the EOF1 pattern shown in Fig. 1a.

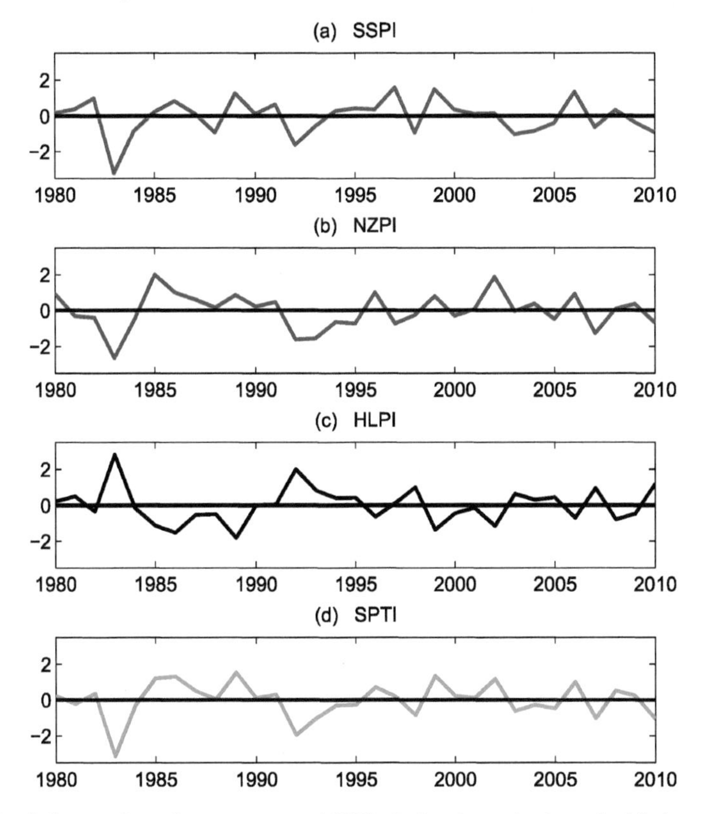

FIG. 2. Standardized time series of area-averaged SSTA during boreal winter in (a) the subtropical South Pacific pole (350-250S, 235°-245°E), (b) the New Zealand pole (50°-35°S, 170°-190°E), and (c) the high latitudes of the South Pacific pole (60°-50°S, 205°- 240°E). (d) Standardized time series of the SPT during boreal winter.

Table 1. Cross correlations between the standardized time series of area-averaged SSTA over the subtropical South Pacific (35°- 25°S, 235°-245°E), New Zealand (50°-35°S, 170°-190°E), and the high latitudes of the South Pacific (600-500S, 2050-2400E). All of the correlation coefficients are significant at the 90% confidence level based on the Student's t test.

	SSP	nzp	HLP
SSP	1	-	-
NZP	0.58	1	-
HLP	–0.77	–0.84	1

To investigate the relationships of the tripole SSTA mode with climate variability, we derive a South Pacific tripole (SPT) index (SPTI), which is defined as follows:

$$\mathrm{SPTI} = 0.5[\mathrm{SSTA}]_{\mathrm{ssp}} + 0.5[\mathrm{SSTA}]_{\mathrm{nzp}} - [\mathrm{SSTA}]_{\mathrm{hlp}}. \quad (2)$$

The square brackets in Eq. (2) represent area-averaged SSTA in SSP, NZP, and HLP, which are defined above. Figure 2d shows the standardized time series of SPTI. The correlation coefficient of the SPTI and PC1 is as high as 0.97, indicating that the SPTI is appropriate to represent the variability of SPT. In addition, the SPTI is significantly related to the Niño3.4 index (correlation coefficient is –0.77, exceeding the 95% confidence level), suggesting that a positive (negative) SPT is associated with the concurrent La Nina (El Niño) event over the tropical central-eastern Pacific.

4. Influences of the DJF SPT on the Following MAM Rainfall Variability over Eastern China

The relationships between the DJF SPT and the following MAM rainfall over eastern China based on the GPCC V6 dataset are shown in Fig. 3a, which shows negative relationships over most of the East Asia landmass. The significant negative correlation appears in south China, indicating that it tends to be wet (dry) during the boreal spring in south China, when the preceding DJF SPTI is in a positive (negative) phase. It is noted that the DJF SPT has no significant relationship with the following MAM rainfall over eastern China (Fig. 3a).

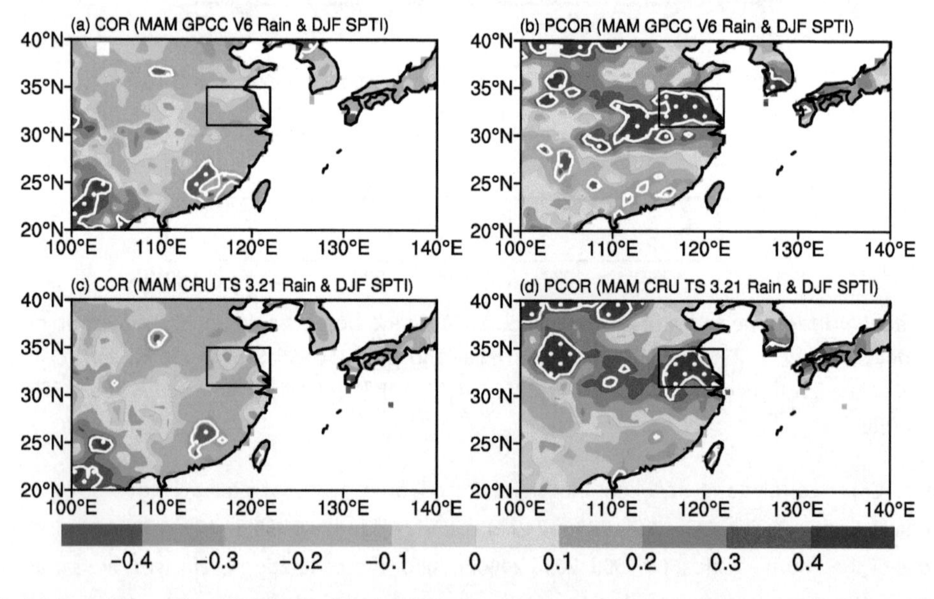

Fig. 3. (a) Correlation of the DJF SPTI with the following MAM rainfall anomalies over East Asia from 1980 to 2010 based on the GPCCV6 rainfall dataset. (b) Partial correlation of the DJF SPTI with the following MAM rainfall anomalies over East Asia with the ENSO signal removed based on the GPCCV6 rainfall dataset. (c), (d) As in (a), (b), but based on the CRU TS 3.21 rainfall dataset. The white contours filled with dots indicate a correlation exceeding the 90% confidence level based on the Student's t test. The black box is the domain for defining the YHRI.

In section 3, we have shown that SPT is closely associated with ENSO. As is well known, ENSO has significant impacts on rainfall over eastern China (Wu et al. 2003; Lin and Lu 2009; Wang et al. 2012; Wang and Wang 2013). Therefore, it is inferred that ENSO has the potential to contaminate the relationships between the DJF SPT and the following MAM rainfall over eastern China. To examine the hypothesis of ENSO influencing the relationship between the DJF SPT and the following MAM rainfall over eastern China, partial correlation is used to investigate the relationship of the DJF SPT with the eastern China rainfall in the following boreal spring (Fig. 3b). After removing the ENSO signal, the positive correlations appear north of 27°N. In particular, the significant positive correlations appear in the middle latitudes of East Asia between 30° and 35°N, extending from central China to the southern Korean Peninsula and western Japan, indicating that more rainfall anomalies occur in these regions during the positive SPT years without ENSO impacts. On the other hand, it is noted that the negative correlation over south China decreases remarkably, compared with that before removing the ENSO signal (Fig. 3a). Therefore, it is suggested that the relationship between the DJF SPT and the following MAM rainfall over eastern China becomes significant after removing the ENSO signal.

To further investigate the stability of the relationship between the DJF SPT and the following MAM rainfall over eastern China, we repeated our correlation analysis using the CRU TS 3.21 rainfall dataset for the same period (1980-2010). The results based on the CRU TS 3.21 rainfall dataset (Figs. 3c, d) are similar to those based on the GPCC V.6 rainfall dataset. The consistency in both rainfall datasets implies a stable and robust relationship between the DJF SPT and the following MAM rainfall over eastern China without ENSO impacts.

From Fig. 3, it is found that the significantly positive correlation appears over eastern China, especially over the Yangtze-Huaihe valley (31°-35°N, 115°-122°E; the boxed area in Fig. 3), after removing the ENSO signal. To show the rainfall variability over the Yangtze-Huaihe valley, the Yangtze-Huaihe valley rainfall index (YHRI) is defined as the station-averaged rainfall anomalies from 9 stations (Xinpu: 34.36°N, 119.10°E; Qingjiang: 33.60°N, 119.03°E; Xuzhou: 34.19°N, 117.22°E; Bangbu: 32.95°N, 117.37°E; Fuyang: 32.93°N, 115.83°E; Dongtai: 32.87°N, 120.32°E; Nanjing: 32.00°N, 118.80°E; Hefei: 31.87°N, 117.23°E; Shanghai: 31.41°N, 121.46°E) obtained from the China Meteorological Administration. Figure 4a presents the original standardized time series of the DJF SPTI and the following MAM YHRI. The correlation coefficient between SPTI and YHRI is only –0.02, indicating a weak relationship between the DJF SPTI and the following MAM YHRI. Because of the impacts of ENSO, we then use a linear regression method [Eq. (3)] to remove the influence of ENSO on SPTI and YHRI. This method has been widely used (i.e., Clark et al. 2000; Wang et al. 2006).

$$y_r = y - cx\left[\frac{\sigma(y)}{\sigma(x)}\right] \tag{3}$$

Here, x and y are two time series, $\sigma(x)$ and $\sigma(y)$ are the standard deviation of x and y, c is the correlation coefficient between x and y, and y_r is the residual value, from which the influence of x has been removed. After removing the ENSO signal, the correlation coefficient between the DJF SPTI and the following MAM YHRI has increased to 0.56 (significant at the 95% confidence level), indicating the significant in-phase relationship between the DJF SPTI and the following MAM YHRI after removing the ENSO signal (Fig. 4b).

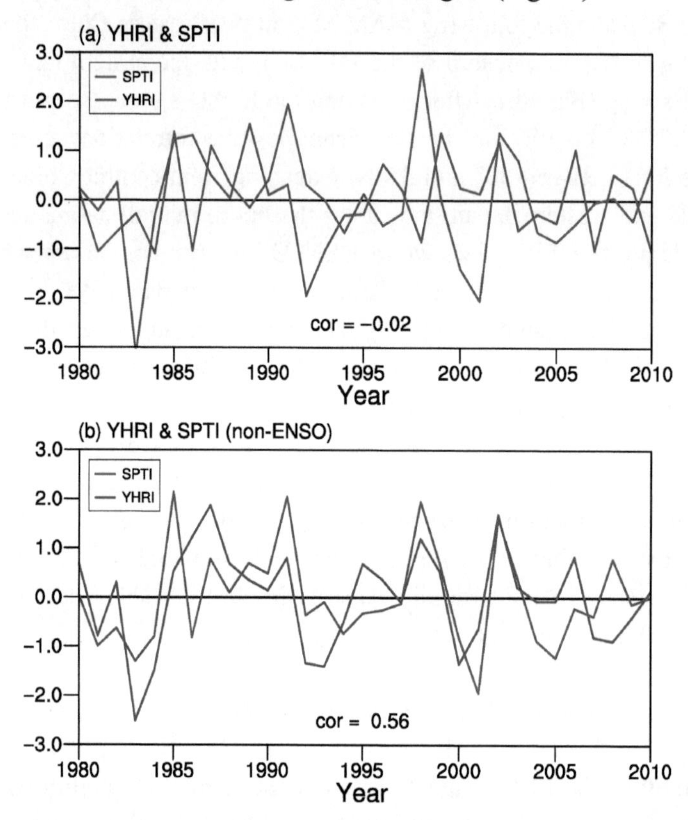

FIG. 4. (a) Standardized time series of the DJF SPTI (blue line) and the following MAM YHRI (red line). (b) As in (a), but removing the ENSO signal.

To further demonstrate the relationship between the DJF SPT and the following MAM rainfall over eastern China, we composite the rainfall anomalies for the positive and negative SPTI years. Considering the influences of ENSO, we remove the ENSO signal in SPTI and rainfall using Eq. (3) and then composite the rainfall anomalies for positive and negative SPTI years. Here, the positive and negative SPTI years are identified by the criterion that the values of standard deviation of SPTI exceed ± 0.5. The positive and negative SPTI years before and after removing ENSO are listed in Table 2. Before removing the ENSO signal (Fig. 5a), the positive rainfall anomalies mainly appear over the Yangtze River valley and the southeast coast of China during the positive SPTI years, and the negative rainfall anomalies largely occur over north and south China. During negative SPTI years (Fig. 5b), positive rainfall anomalies are observed over most of the East Asia landmass. Figure 5c shows that there are no significant differences between the positive and negative SPTI years over eastern China, implying the weak relationship between the DJF SPTI and the following MAM rainfall over eastern China. In contrast, after removing the ENSO signal, during the positive SPTI years (Fig. 5d), the significantly positive rainfall anomalies are seen over eastern China. Nevertheless, rainfall anomalies nearly show an opposite distribution during the negative SPTI years (Fig. 5e). From Fig. 5f, the differences of the MAM rainfall anomalies over eastern China between the positive and negative SPTI years are statistically significant (exceeding the 90% confidence level).

Table 2. Positive and negative SPTI years before and after removing the ENSO signal. The positive and negative SPTI years are identified by the criterion that the values of standard deviation of SPTI exceed ±0.5.

	SPTI before removing ENSO	SPTI after removing ENSO
Positive years	1985, 1986, 1987, 1989, 1996, 1999, 2002, 2006, and 2008	1980, 1985, 1986, 1987, 1988, 1991, 1995, 1998, 1999, 2002, and 2006
Negative years	1983, 1992, 1993, 1998, 2003, 2007, and 2010	1981, 1983, 1984, 1992, 1993, 1994, 2000, 2001, 2007, and 2008

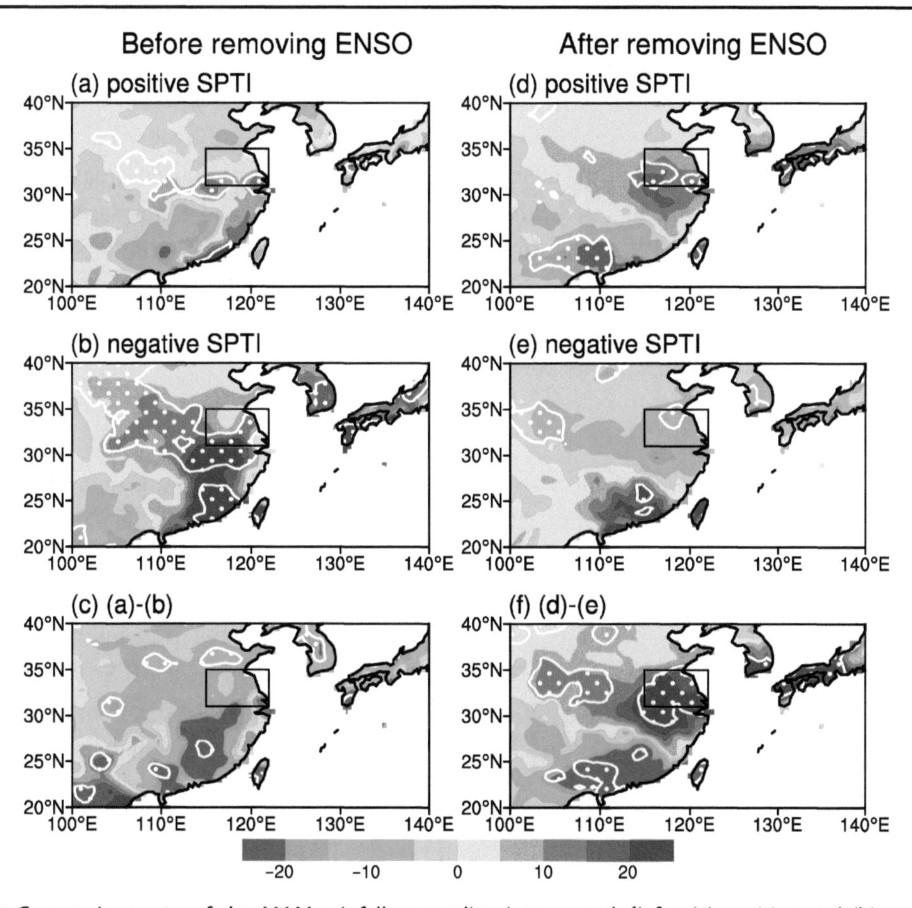

Fig. 5. Composite maps of the MAM rainfall anomalies (mm month^{-1}) for (a) positive and (b) negative preceding DJF SPTI years, and (c) the composite difference between the positive and negative preceding DJF SPTI years. (d)-(f) As in (a)-(c), but removing the ENSO signal. The white contours filled with the dots indicate the correlation exceeding the 90% confidence level based on the Student's *t* test. The black box is the domain for defining the YHRI.

Therefore, the relationships between the DJF SPTI and the following MAM rainfall over eastern China are significant if the ENSO impacts are removed.

Based on the above analyses, the relationship between the DJF SPT and the following MAM rainfall over eastern China can be influenced by ENSO. If there is an ENSO event in boreal winter, SPT has no significant relation to rainfall over eastern China during the following boreal spring. In contrast, in boreal winter without ENSO, a close relationship is established between the DJF SPT and the following MAM rainfall over eastern China (i.e., corresponding to the positive SPT in boreal winter, anomalous increased rainfall can be found over eastern China during the following boreal spring, especially in the Yangtze-Huaihe valley, and vice versa). Therefore, we may conclude that ENSO can weaken the relationship between the DJF SPT and the following MAM rainfall over eastern China.

5. Possible Physical Mechanisms

The previous section has shown that the DJF SPT has a significant positive relationship with the following MAM rainfall over eastern China without ENSO impacts, especially over the Yangtze-Huaihe valley. One question is raised: what are the possible physical mechanisms through which the DJF SPT can influence the following MAM rainfall over eastern China? To answer this question, we examine below the variations of SSTA and atmospheric circulation anomalies associated with SPT. It should be pointed out that the focus of the present study is placed on the relationship between SPT and rainfall over eastern China without ENSO impacts. Therefore, we mainly use partial correlation to illustrate the variations of SSTA and atmospheric circulation anomalies associated with the DJF SPT.

5.1 Evolution of SSTA

Figure 6 shows a partial correlation of the DJF SPTI with SSTA during DJF and following MAM, which removes the ENSO impacts. During the positive phase of the DJF SPTI, the significantly positive SSTA is mainly in the southern Pacific (south of the equator), extending eastward from the western to the eastern South Pacific, forming a “Y” shape (Fig. 6a). In addition, the significantly positive SSTA is also observed over the tropical northwestern Pacific, the northern North Pacific, and the southern Indian Ocean near 45°S. On the other hand, the significantly negative SSTA dominates over the east of the tropical and subtropical southern Pacific, the high latitudes of the South Pacific, and the region to the south of Australia (Fig. 6a).

For the following boreal spring (Fig. 6b), both positive and negative correlations south of the equator weaken significantly. The significantly positive correlations observed over the tropical northwestern Pacific during the boreal winter shift southward and expand eastward to south of Hawaii, although it becomes weak to some extent. Over the North Pacific, a significantly positive correlation band appears, extending and tilting from the central to western North Pacific.

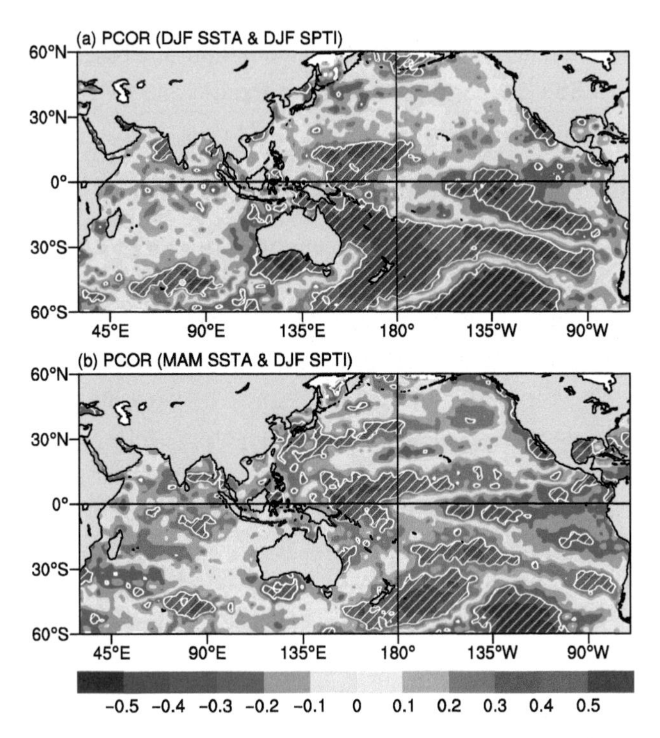

FIG. 6. Partial correlations of the DJF SPTI with seasonal SSTA during (a) boreal winter and (b) the following boreal spring with the ENSO signal removed. The hatched areas represent correlations exceeding the 90% confidence level based on the Student's *t* test.

Based on the above analyses, we list the following characteristics. When the DJF SPT is in a positive phase, pronounced positive SSTA appears over the tropical western Pacific. Moreover, this positive SSTA is characterized by significant persistence from the boreal winter to the following boreal spring. In fact, some previous studies have suggested that the air-sea interaction in the Southern Hemisphere can indirectly influence climate variability over East Asia by the SST over the tropical western Pacific (Sun et al. 2009; Hsu and Chen 2011; Zhou 2011). Moreover, previous studies have also suggested the possible linkages between SST in the tropical western Pacific and SST in the South Pacific. Giese et al. (2002) showed that the abrupt climate shift in 1976 originates with subsurface temperature anomalies in the south tropical Pacific Ocean that can propagate along 10°S to the tropical western Pacific. Both the observation (Luo and Yamagata 2001) and model experiments (Luo et al. 2003, 2005; Yu and Boer 2004) suggested that the subsurface anomalies in the South Pacific could propagate to the tropical western Pacific. Transportation of the subsurface anomalies signal by advection (or isopycnal advection) is the possible mechanism. In addition, some studies showed that the anomalous signal in the sea surface of the South Pacific can also propagate to the tropical western Pacific (Gu and Philander 1997; Wang and Liu 2000; Wang et al. 2007). Wang et al. (2007) suggested that the westward propagating Rossby waves play important roles, which is different from the possible mechanism of transportation of subsurface anomalies. However, this study focuses on the impacts of SST in the South Pacific on rainfall in eastern China and investigates the modulation of ENSO. Therefore, discussions of the linkage between SST in the tropical western Pacific and SST in the South Pacific are beyond the scope of this study.

5.2 Atmospheric circulation anomalies

Figure 7 shows the partial correlation between the DJF SPTI and the following MAM geopotential height anomalies at 500 hPa after removing the ENSO impacts. During the positive SPTI, the pattern is characterized by a significant negative-positive-negative wave-like train from low to high latitudes in Fig. 7. In the low latitudes of the North Pacific, the significantly negative correlation band extends and tilts from the subtropical northeastern Pacific to the tropical western North Pacific, with a negative center to the west of Hawaii. To the northwest of the negative correlation band, there is a pronounced positive correlation band extending southwestward from the midlatitude North Pacific to eastern China, with two positive centers: one is over the midlatitude North Pacific, and the other is to the east of China. A significantly negative correlation center is seen in eastern Russia. The positive correlation center to the east of China means that the SPTI is closely related to the low-level atmospheric circulation anomalies over there, which can influence rainfall variations over eastern China.

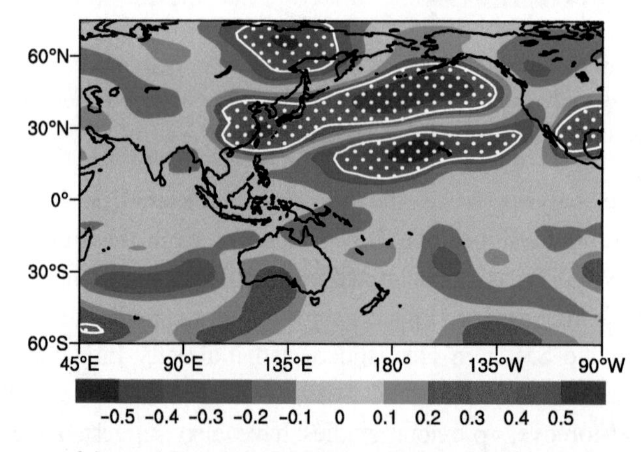

Fig. 7. Partial correlations of the DJF SPTI with the following MAM 500-hPa geopotential height anomalies with the ENSO signal removed. The white contours filled with dots indicate correlations exceeding the 90% confidence level based on the Student's *t* test.

The variations of low-level (850 hPa) wind associated with the DJF SPTI without ENSO impacts are shown in Fig. 8. During the positive phases of the DJF SPTI, an anomalous cyclone is found to the west of Hawaii in the tropical northern Pacific in the following boreal spring. This elongated anomalous cyclone extends southwest ward, with a smaller scale and weak anomalous cyclone to the east of the Philippines, which is associated with the underlying positive SSTA (Fig. 6b). To the north, a strong anomalous anticyclone is in the northern North Pacific (north of 30°N). The significant northeasterly anomalies prevail in the North Pacific around 30°N, which are superposed on the climatological westerly wind and thus can lead to a decrease in wind speed, resulting in the positive SSTA band extending from the central to the western North Pacific (Fig. 6b). Particularly, a weak anomalous anticyclone is seen to the east of China. The weak anomalous anticyclone to the east of China can cause a significant anomalous southwesterly over eastern China, which brings anomalous increased moisture into eastern China and favors rainfall over this region.

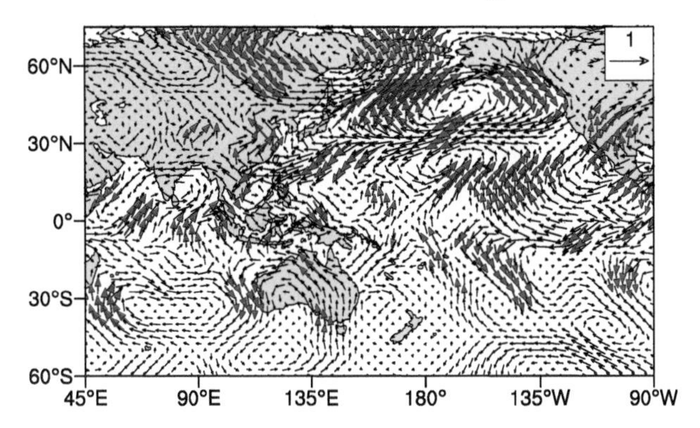

Fig. 8. As in Fig. 7, but for 850-hPa wind anomalies. The red vectors indicate meridional wind exceeding the 90% confidence level based on the Student's *t* test.

Because the moisture transport is closely related to rainfall variations, we examine the vertically integrated moisture flux and its divergence during the following spring associated with winter SPT. The vertically integrated moisture flux is calculated according to the equation:

$$Q = -\frac{1}{g}\int_{\text{psfc}}^{300\text{hPa}} (q\mathbf{V})dp, \tag{4}$$

where g is gravity, psfc is the surface pressure, q is the specific humidity, $\mathbf{V}$ is the wind field, and p is the pressure. Figure 9 shows a partial correlation of the winter SPTI with vertically integrated moisture flux and its divergence anomalies during the following spring, after

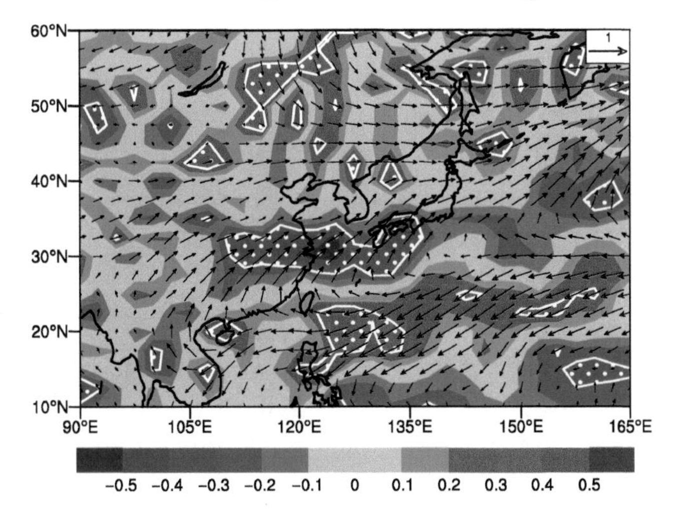

Fig. 9. As in Fig. 7, but for vertically integrated moisture flux anomalies (vector) and moisture divergence anomalies (shading) in the troposphere. The white contours filled with dots indicate the correlation exceeding the 90% confidence level based on the Student's *t* test.

removing the influence of ENSO. Corresponding to the positive SPTI, the significantly anomalous moisture convergence is seen in eastern China, which is consistent with more rainfall over this region.

To explore the linkage of how the SPTI influences rainfall over eastern China, we calculate the partial correlation of the DJF SPTI with the latitude-pressure cross section of vertical circulation anomalies averaged along 110°-120°E (Fig. 10) during the following MAM without ENSO impacts. From Fig. 10, significantly anomalous ascent motions are observed over eastern China, especially the Yangtze-Huaihe valley (30°- 35°N), which could provide a pronounced dynamic condition for more rainfall. However, no significant meridional cell patterns are found over the tropical (0°-20°N) and midlatitude (40°-60°N) regions. Therefore, it can be inferred that the anomalous ascent motions may not result from the regionally anomalous meridional circulation. To explore the possible mechanism of the formation of anomalous ascent motions over the Yangtze-Huaihe valley, we show in Fig. 11 the partial correlation of the DJF SPTI with the following MAM anomalous pressure vertical velocity at different levels (300, 500, and 850hPa) without ENSO impacts. The correlation patterns at different levels exhibit a significantly negative-positive-negative wave-like train pattern extending from the tropical northwestern Pacific (negative) through the western North Pacific (positive) to eastern China (negative), indicating significant anomalous ascent motions extending from Hawaii to east of the Philippines and over eastern China, as well as significant anomalous descent motions over the western North Pacific. It is noted that the significant anomalous ascent motions extending from Hawaii to east of the Philippines are related to the underlying positive SSTA in the tropical northwestern Pacific (Fig. 6b). Based on the above analyses, it can be inferred that this pattern plays an important role in propagating the impacts

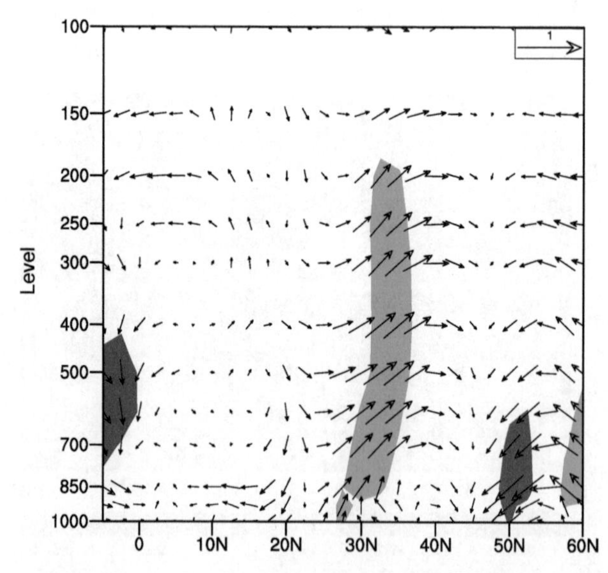

FIG. 10. Partial correlations of the DJF SPTI with the following MAM meridional circulations along 110°-120°E with the ENSO impacts removed. The red and green shaded areas indicate the descent and ascent motions exceeding the 90% confidence level based on the Student's *t* test.

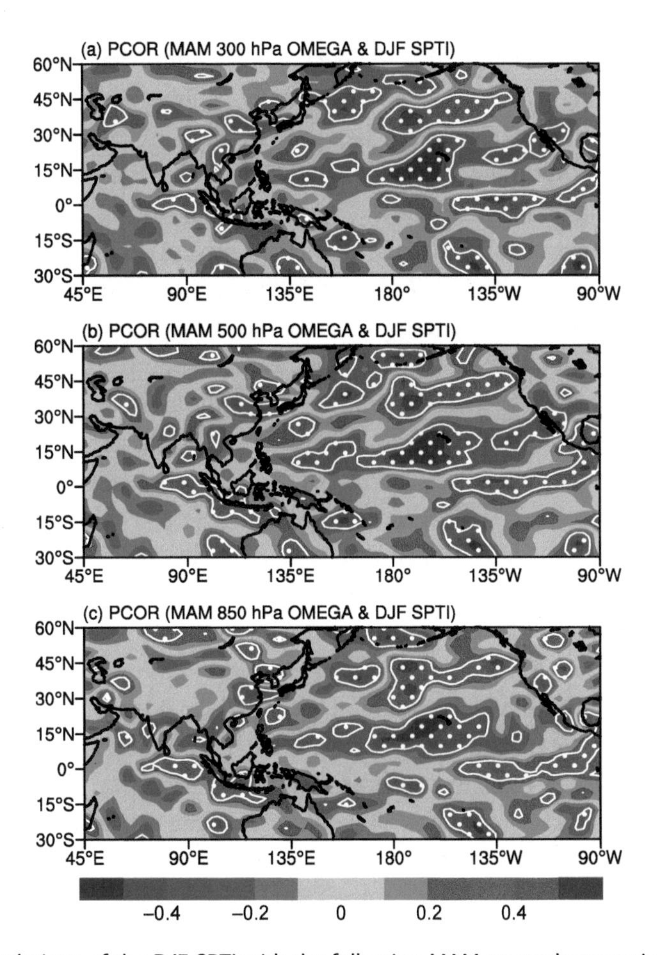

FIG. 11. Partial correlations of the DJF SPTI with the following MAM anomalous vertical pressure velocity at different levels for (a) 300, (b) 500, and (c) 850hPa with the ENSO signal removed. The white contours filled with dots indicate the correlation exceeding the 90% confidence level based on the Student's *t* test.

of heat sources (positive SSTA) in the tropical northwestern Pacific on rainfall over eastern China. Therefore, the anomalous ascent motions over eastern China should be caused by heat sources (positive SSTA) in the tropical northwestern Pacific (Fig. 6b) through the wave-like train pattern.

6. Conclusions and Discussion

Based on multiple datasets, we define an index in this study in order to represent the variations of the South Pacific tripole (SPT) mode, which is the first leading EOF (EOF1) pattern of SSTA in the South Pacific (60°- 20°S, 144°-294°E).

Using correlation and partial correlation methods, this study investigates the relationships of the DJF SPT with the following MAM rainfall variations over eastern China. It is found that the relationships of the DJF SPT with the following MAM rainfall variations over eastern China are modulated by ENSO, which could offset the influences of the SPT on rainfall.

Before removing the ENSO signal, the DJF SPT has no significant relation to the following MAM rainfall over eastern China. However, a closely positive relationship appears between them after removing the ENSO signal. When the DJF SPT is in its positive phase, significantly more rainfall is observed over eastern China during the following boreal spring, especially over the Yangtze-Huaihe valley.

The possible physical mechanisms of how the positive DJF SPT results in anomalous increased rainfall over eastern China during the following boreal spring are examined. When the DJF SPT is in a positive phase, a significantly positive SSTA appears over the tropical western Pacific. Such positive SSTA can persist to the following boreal spring and extend eastward to south of Hawaii. In response to this underlying positive SSTA, a wave-like train at 850hPa extends from the tropical western Pacific to the western North Pacific, with an anomalous cyclone over the tropical western Pacific and an anomalous anticyclone to the east of China. To the west flank of the anomalous anticyclone is the significant anomalous southwesterly over eastern China, which can bring more moisture and favors more rainfall. Moreover, the positive SSTA over the tropical northwestern Pacific can excite significantly anomalous ascent motions, which can indirectly result in pronounced anomalous ascent motions over the Yangtze-Huaihe valley through a wave-like train, providing an important dynamic condition, which is favorable for more rainfall over eastern China.

The possible mechanism of how ENSO weakens the linkage between the boreal winter SPT and the following spring rainfall over eastern China is discussed. We have found that the positive/negative SPT in the preceding DJF could induce the anomalous anticyclone/cyclone over the western North Pacific in MAM and thus result in positive/negative rainfall anomalies in eastern China. However, a positive (negative) SPT is closely associated with the concurrent La Niña (El Niño) event, which leads to cyclonic (anticyclonic) circulation anomalies over the western North Pacific (Wang et al. 2000). That is to say, ENSO could offset the influences of DJF SPT over the western North Pacific. An anomalous anticyclone over the western North Pacific induced by the positive SPT in boreal winter is weakened by the cyclonic circulation anomalies resulting from the teleconnections of the concurrent La Nina. Therefore, ENSO has a significant impact on the relationship between South Pacific SST and East Asia rainfall variability.

It is noted that the method of partial correlation is used to remove the influence of ENSO. Although this is a very common method in many studies, this method cannot fully remove the influence of ENSO because of the nonlinearity of ENSO. To clarify the relationship between the South Pacific and precipitation over East Asia without ENSO influence, we calculate the composite rainfall anomalies for positive and negative SPTI years with a neutral ENSO phase (figures not shown). Cases are selected in which the standard deviation of Niño –3.4 is neutral (between –1 and 1) and the standard deviation of SPTI exceeds ±1.0. Based on the definition, the positive SPTI years are 1986, 2002, and 2006, and the negative SPTI years are 1993 and 2007. For positive SPTI years, strongly positive rainfall anomalies appear over eastern China, which are similar to those in Fig. 5c. During the negative SPTI years, negative rainfall anomalies appear over eastern China, which resembles those in Fig. 5d. Such results reconfirm the relationship between the SPTI and rainfall over eastern China.

In this study, we do not consider the variations and influences of SST in the central tropical Pacific, which have been discussed widely in recent years (i.e., Ashok et al. 2007; Feng and Li 2011; Wang and Wang 2013). This issue will be discussed in future study to compare the influences of the central Pacific El Niño and eastern Pacific El Niño on the relationships between the South Pacific and the East Asian climate.

Acknowledgments The authors thank the editor and three anonymous reviewers, who provided valuable and constructive comments on our manuscript which were helpful for improving the quality of the manuscript. This research is sponsored by the Chinese Key Developing Program for Basic Sciences (2010CB950400 and 2013CB956200), the Strategic Priority Research Program of the Chinese Academy of Sciences (XDA11010403), and the National Natural Science Foundation of China (41422601 and 41376025).

References

Ashok, K., S. K. Behera, S. A. Rao, H. Weng, and T. Yamagata, 2007: El Niño Modoki and its possible teleconnection. *J. Geophys. Res.*, **112**, C11007, doi: 10.1029/2006JC003798.

Behera, S. K., and T. Yamagata, 2001: Subtropical SST dipole events in the southern Indian Ocean. *Geophys. Res. Lett.*, **28**, 327-330, doi: 10.1029/2000GL011451.

Chan, J. C. L., and W. Zhou, 2005: PDO, ENSO and the early summer monsoon rainfall over south China. *Geophys. Res. Lett.*, **32**, L08810, doi: 10.1029/2004GL022015.

Chang, C.-P., Y. Zhang, and T. Li, 2000: Interannual and interdecadal variations of the East Asian summer monsoon and tropical Pacific SSTs. Part II: Meridional structure of the monsoon. *J. Climate*, **13**, 4326-4340, doi: 10.1175/1520-0442(2000)013<4326: IAIVOT>2.0.CO; 2.

Ciasto, L. M., and D. W. J. Thompson, 2008: Observations of large scale ocean-atmosphere interaction in the Southern Hemisphere. *J. Climate*, **21**, 1244-1259, doi: 10.1175/2007JCLI1809.1.

Clark, C. O., J. E. Cole, and P. J. Webster, 2000: Indian Ocean SST and Indian summer rainfall: Predictive relationships and their decadal variability. *J. Climate*, **13**, 2503-2519, doi: 10.1175/ 1520-0442(2000)013< 2503: IOSAIS>2.0.CO; 2.

Davis, R.E., 1976: Predictability of sea surface temperature and sea level pressure anomalies over the North Pacific Ocean. *J. Phys. Oceanogr.*, **6**, 249-266, doi: 10.1175/1520-0485(1976)006<0249: POSSTA> 2.0.CO; 2.

De Almeida, R. A. F., P. Nobre, R. J. Haarsma, and E. J. D. Campos, 2007: Negative ocean-atmosphere feedback in the South Atlantic convergence zone. *Geophys. Res. Lett.*, **34**, L18809, doi: 10.1029/ 2007GL030401.

Fan, K., and H. J. Wang, 2004: Antarctic Oscillation and the dust weather frequency in north China. *Geophys. Res. Lett.*, **31**, L10201, doi: 10.1029/2004GL019465.

Feng, J., and J. Li, 2011: Influence of El Niño Modoki on spring rainfall over south China. *J. Geophys. Res.*, **116**, D13102, doi: 10.1029/2010JD015160.

Giese, B. S., S. C. Urizar, and N. S. Fučkar, 2002: Southern Hemisphere origins of the 1976 climate shift. *Geophys. Res. Lett.*, **29**, doi: 10.1029/2001GL013268.

Gong, D., and S. Wang, 1999: Definition of Antarctic oscillation index. *Geophys. Res. Lett.*, **26**, 459-462, doi:

10.1029/1999GL900003.

Gu, D. F., and S. G. H. Philander, 1997: Interdecadal climate fluctuations that depend on exchanges between the tropics and extratropics. *Science*, **275**, 805-807, doi: 10.1126/ science.275.5301.805.

Ho, C.-H., J.-H. Kim, H.-S. Kim, C.-H. Sui, and D.-Y. Gong, 2005: Possible influence of the Antarctic Oscillation on tropical cyclone activity in the western North Pacific. *J. Geophys. Res.*, **110**, D19104, doi: 10.1029/2005JD005766.

Hsu, H.-H., and Y.-L. Chen, 2011: Decadal to bi-decadal rainfall variation in the western Pacific: A footprint of South Pacific decadal variability? *Geophys. Res. Lett.*, **38**, L03703, doi: 10.1029/2010GL046278.

Huang, B., and J. Shukla, 2006: Interannual SST variability in the southern subtropical and extra-tropical ocean. Center for Ocean-Land-Atmosphere Studies Tech. Rep. 223, 20 pp.

Huang, R., and Y. Wu, 1989: The influence of ENSO on the summer climate change in China and its mechanism. *Adv. Atmos. Sci.*, **6**, 21-32, doi: 10.1007/BF02656915.

——, L. Zhou, and W. Chen, 2003: The progresses of recent studies on the variabilities of the East Asian monsoon and their causes. *Adv. Atmos. Sci.*, **20**, 55-69, doi: 10.1007/BF03342050.

Jia, X., and C. Li, 2005: Dipole oscillation in the southern Indian Ocean and its impacts on climate (in Chinese). *Chin. J. Geo- phys.*, **48**, 1323-1335, doi: 10.1002/cjg2.780.

Jones, P., and I. Harris, 2013: CRU TS3.21: Climatic Research Unit (CRU) Time-Series (TS) version 3.21 of high resolution gridded data of month-by-month variation in climate (Jan. 1901-Dec. 2012). Climate Research unit data, doi: 10.5285/ D0E1585D-3417-485F-87AE-4FCECF10A992.

Kalnay, E., and Coauthors, 1996: The NCEP/NCAR 40-Year Reanalysis Project. *Bull. Amer. Meteor. Soc.*, **77**, 437-471, doi: 10.1175/1520-0477(1996)077<0437: TNYRP>2.0.CO; 2.

Kidson, J. W., and J. A. Renwick, 2002: The Southern Hemisphere evolution of ENSO during 1981-1999. *J. Climate*, **15**, 847-863, doi: 10.1175/1520-0442(2002)015<0847: TSHEOE>2.0.CO; 2.

Li, C., and J. Wu, 2002: Important role of the Somalian crossequator flow in the onset of the South China Sea summer monsoon (in Chinese). *Chin. J. Atmos. Sci.*, **26**, 185-192.

——, W. Zhou, X. Jia, and X. Wang, 2006: Decadal/interdecadal variations of the ocean temperature and its impacts on climate. *Adv. Atmos. Sci.*, **23**, 964-981, doi: 10.1007/s00376-006-0964-7.

Li, G., C. Y. Li, Y. K. Tan, and T. Bai, 2012: Principal modes of the boreal wintertime SSTA in the South Pacific and their relationships with the ENSO (in Chinese). *Acta Oceanol. Sin.*, **34**, 48-56.

Li, X., 1955: Study of typhoon (in Chinese). *Monograph of Modern Science in China: Meteorology (1919-1949)*, Science Press, 119-146.

Lin, Z. D., and R. Y. Lu, 2009: The ENSO's effect on eastern China rainfall in the following early summer. *Adv. Atmos. Sci.*, **26**, 333-342, doi: 10.1007/s00376-009-0333-4.

Liu, G., Q. Zhang, and S. Sun, 2008: The relationship between circulation and SST anomaly east of Australia and the summer rainfall in the middle and lower reaches of the Yangtze River (in Chinese). *Chin. J. Atmos. Sci.*, **32**, 231-241, doi: 10.3878/ j.issn.1006-9895.2008.02.04.

Luo, J.-J., and T. Yamagata, 2001: Long-term El Niño-Southern Oscillation (ENSO)-like variation with special emphasis on the South Pacific. *J. Geophys. Res.*, **106**, 22 211-22 227, doi: 10.1029/2000JC000471.

——, S. Masson, S. Behera, P. Delecluse, S. Gualdi, A. Navarra, and T. Yamagata, 2003: South Pacific origin of the decadal ENSO-like variation as simulated by a coupled GCM. *Geophys. Res. Lett.*, **30**, 2250, doi: 10.1029/2003GL018649.

Luo, Y., L. M. Rothstein, R.-H. Zhang, and A. J. Busalacchi, 2005: On the connection between South Pacific subtropical spiciness anomalies and decadal equatorial variability in an ocean general circulation model. *J. Geophys. Res.*, **110**, C10002, doi: 10.1029/2004JC002655.

Nan, S., and J. Li, 2003: The relationship between the summer precipitation in the Yangtze River valley and the boreal spring Southern Hemisphere annular mode. *Geophys. Res. Lett.*, **30**, 2266, doi: 10.1029/2003GL018381.

——, and ——, 2005: The relationship between the summer precipitation in the Yangtze River valley and the boreal spring Southern Hemisphere annular mode: The role of the Indian Ocean and South China Sea as an "ocean bridge"(in Chinese). *Acta Meteor. Sin.*, **63**, 847-856.

——, ——, X. Yuan, and P. Zhao, 2008: Boreal spring Southern Hemisphere annual mode, Indian Ocean sea surface temperature, and East Asian summer monsoon. *J. Geophys. Res.*, **114**, D02103, doi: 10.1029/2008JD010045.

North, G. R., T. L. Bell, R. F. Cahalan, and F. J. Moeng, 1982: Sampling errors in the estimation of empirical orthogonal functions. *Mon. Wea. Rev.*, **110**, 699-706, doi: 10.1175/ 1520-0493(1982)110<0699: SEITEO>2.0.CO; 2.

Rayner, N. A., D. E. Parker, E. B. Horton, C. K. Folland, L. V. Alexander, D. P. Rowell, E. C. Kent, and A. Kaplan, 2003: Global analyses of sea surface temperature, sea ice, and night marine air temperature since the late nineteenth century. *J. Geophys. Res.*, **108**, 4407, doi: 10.1029/2002JD002670.

Reason, C. J. C., 2001: Subtropical Indian Ocean SST dipole events and southern African rainfall. *Geophys. Res. Lett.*, **28**, 2225-2227, doi: 10.1029/2000GL012735.

Schneider, U., A. Becker, P. Finger, A. Meyer-Christoffer, B. Rudolf, and M. Zeise, 2011: GPCC Full Data Reanalysis version 6.0 (at 0.5°, 1.0°, 2.5°): Monthly land-surface precipitation from rain gauges built on GTS-based and historic data. GPCC Data Rep., doi: 10.5676/DWD_GPCC/FD_M_V6_050.

Shakun, J. D., and J. Shaman, 2009: Tropical origins of North and South Pacific decadal variability. *Geophys. Res. Lett.*, **36**, L19711, doi: 10.1029/2009GL040313.

Sun, J., H. Wang, andW. Yuan, 2009: A possible mechanism for the co-variability of the boreal spring Antarctic Oscillation and theYangtze River valley summer rainfall. *Int. J. Climatol.*, **29**, 1276-1284, doi: 10.1002/joc.1773.

Suzuki, R., S. K. Behera, S. Iizuka, and T. Yamagata, 2004: Indian Ocean subtropical dipole simulated using a coupled general circulation model. *J. Geophys. Res.*, **109**, C09001, doi: 10.1029/ 2003JC001974.

Tao, S. Y., and L. X. Chen, 1987: Areviewofrecent research on the East Asian summer monsoon in China. *Review of Monsoon Meteorology*, C. P. Chang and T. N. Krishnamurti, Eds., Oxford University Press, 60-92.

Terray, P., 2011: Southern Hemisphere extra-tropical forcing: A new paradigm for El Niño-Southern Oscillation. *Climate Dyn.*, **36**, 2171-2199, doi: 10.1007/s00382-010-0825-z.

Thompson, D.W.J., and J. M. Wallace, 2000: Annular modes in the extratropical circulation. Part I: Month-to-month variability. *J. Climate*, **13**, 1000-1016, doi: 10.1175/1520-0442(2000)013<1000: AMITEC>2.0.CO; 2.

Wang, B., R. Wu, and X.-H. Fu, 2000: Pacific-East Asian teleconnection: How does ENSO affect East Asian climate? *J. Climate*, **13**, 1517-1536, doi: 10.1175/1520-0442(2000)013<1517: PEATHD>2.0.CO; 2.

Wang, C., and X. Wang, 2013: Classifying El Niño Modoki I and II by different impacts on rainfall in southern China and typhoon tracks. J. *Climate*, **26**, 1322-1338, doi: 10.1175/JCLI-D-12-00107.1.

Wang, D.X., and Z. Y. Liu, 2000: The pathway of the interdecadal variability in the Pacific Ocean. *Chin. Sci. Bull.*, **45**, 1555-1561, doi: 10.1007/BF02886211.

Wang, H., 2005: The Circum-Pacific Teleconnection Pattern in meridional wind in the high troposphere. *Adv. Atmos. Sci.*, **22**, 463-466, doi: 10.1007/BF02918759.

Wang, X., C. Li, and W. Zhou, 2006: Interdecadal variation of the relationship between Indian rainfall and SSTA modes in the Indian Ocean. *Int. J. Climatol.*, **26**, 595-606, doi: 10.1002/ joc.1283.

——, ——, and ——, 2007: Interdecadal mode and its propagating characteristics of SSTA in the South

Pacific. *Meteor. Atmos. Phys.*, **98**, 115-124, doi: 10.1007/s00703-006-0235-2.

——, D. X. Wang, W. Zhou, and C. Y. Li, 2012: Interdecadal modulation of the influence of La Niña events on mei-yu rainfall over the Yangtze River valley. *Adv. Atmos. Sci.*, **29**, 157-168, doi: 10.1007/ s00376-011-1021-8.

Wu, R., Z.-Z. Hu, and B. P. Kirtman, 2003: Evolution of ENSO- related rainfall anomalies in East Asia. *J. Climate*, **16**, 3742-3758, doi: 10.1175/1520-0442(2003)016<3742: EOERAI>2.0.CO; 2.

Wu, L., Z. Jing, S. Riser, and M. Visbeck, 2011: Seasonal and spatial variations of Southern Ocean diapycnal mixing from Argo profiling floats. *Nat. Geosci.*, **4**, 363-366, doi: 10.1038/ NGEO1156.

Wu, Z., J. Li, B. Wang, and X. Liu, 2009: Can the Southern Hemisphere annular mode affect China winter monsoon? *J. Geophys. Res.*, **114**, D11107, doi: 10.1029/2008JD011501.

Xue, F., 2005: Influence of the Southern circulation on East Asia summer monsoon (in Chinese). *Climatic Environ. Res.*, **10**, 401-408.

Yan, H., C. Li, and W. Zhou, 2009: Influence of subtropical dipole pattern in southern Indian Ocean on ENSO event (in Chinese). *Chin. J. Geophys.*, **52**, 2436-2449.

Yang, F., and K.-M. Lau, 2004: Trend and variability of China precipitation in spring and summer: Linkage to sea-surface temperatures. *Int. J. Climatol.*, **24**, 1625-1644, doi: 10.1002/joc.1094.

Yang, M., and Y. Ding, 2007: A study of the impact of south Indian Ocean dipole on the summer rainfall in China (in Chinese). *Chin. J. Atmos. Sci.*, **31**, 685-694.

Yang, Q., 2009: Impact of the Indian Ocean subtropical dipole on the precipitation of east China during winter monsoons. *J. Geophys. Res.*, **114**, D14110, doi: 10.1029/2008JD011173.

Yu, B., and G. J. Boer, 2004: The role of the western Pacific in decadal variability. *Geophys. Res. Lett.*, **31**, L02204, doi: 10.1029/ 2003GL018471.

Zheng, F., and J. Li, 2012: Impact of preceding boreal winter Southern Hemisphere annular mode on spring precipitation over south China and related mechanism (in Chinese). *Chin. J. Geo phys.*, **55**, 3542-3557, doi: 10.6038/j.issn.0001-5733.2012.11.004.

Zhou, B., 2011: Linkage between winter sea surface temperature east of Australia and summer precipitation in the Yangtze River valley and a possible physical mechanism. *Chin. Sci. Bull.*, **56**, 1821-1827, doi: 10.1007/s11434-011-4497-9.

——, and X. Cui, 2011: Sea surface temperature east of Australia: A predictor of tropical cyclone frequency over the western North Pacific? *Chin. Sci. Bull.*, **56**, 196-201, doi: 10.1007/ s11434-010-4157-5.

Zhou, W., and J. C. L. Chan, 2007: ENSO and South China Sea summer monsoon onset. *Int. J. Climatol.*, **27**, 157-167, doi: 10.1002/joc.1380.

5

The Oceanic Wave Activity and Renewable Energy Resource over the Sea

Variation of the Wave Energy and Significant Wave Height in the China Sea and Adjacent Waters

Chongwei Zheng[a, b, c], Chongyin Li[a, b, *]

[a] College of Meteorology and Oceanography, PLA University of Science and Technology, Nanjing, China
[b] LASG, Institute of Atmospheric Physics, Chinese Academy of Sciences, Beijing, China
[c] PLA Dalian Naval Academy, Dalian, China

Abstract Given the current background of ongoing environmental and resource issues, the increased exploitation of clean and renewable energy could help to alleviate the energy crisis, as well as contributing to emissions reduction and environmental protection, and so promote future sustainable development. This study explores to reveal the climatic long term trends of the China Sea wave power and significant wave height (SWH) for the period 1988-2011, using a WAVEWATCH-III (WW3) hindcast wave data. The regional difference and seasonal difference of the variation are also presented firstly. Results show that, (1) The China Sea exhibits a significant overall increasing trend in the wave power density (0.2012 (kW/m)/yr) and the SWH (1.52 cm/yr) for the period 1988 to 2011. (2) There is a noticeable increasing trend in most parts of the China Sea, of 0.1-0.7 (kW/m)/yr in wave power density and 0.5-4.5 cm/yr in SWH. Areas with strong increasing trend distribute in the Ryukyu Islands waters, Taiwan Strait and north of the South China Sea (SCS), especially in the Dongsha Islands waters. (3) There is a noticeable seasonal difference in the variation of both SWH and wave power density. The variation in different waters is dominated by the different seasons. The increasing trend of SWH in DJF and MAM is obviously stronger than that in JJA and SON. The increasing trend of wave power density in DJF is stronger than that in other seasons.

Keywords China Sea, wave power, significant wave height, long term trend, regional difference, seasonal difference, dominant season.

1. Introduction

The resource crisis causes a serious impact on the sustainable development of human beings, and even causes serious environmental crisis, regional conflicts. The development and utilization of new energy to alleviate the energy crisis, environmental protection has important significance [1]. Wave energy as one of the most important new energy, has advantage in

Received May 3, 2014; revised August 28, 2014; accepted November 1, 2014.
*Correspondence address: Tel.: + 86 18640814027; fax: + 86 041185883108; E-mail address: cy@lasg.iap.ac.cn (C.Y. Li).

large reserves, wide distribution, pollution-free and renewable. Wave energy has the highest marine energy density in the coastal areas [2]. In the power supply difficult remote island highlight the advantages of wave power. Resource evaluation in advance can make contribution to orderly development of wave energy, and avoid blind construction.

Nomenclature

CC	correlation coefficient
CCMP	Cross-Calibrated, Multi-Platform
DJF	December, January, February
ECMWF	European Centre for Medium-Range Weather Forecasts
ERA-40	40-year ECMWF re-analysis
JJA	June, July, August
MAM	March, April, May
NRMSE	normalized root mean square error
RMSE	root mean square error
SCS	South China Sea
SON	September, October, November
SWH	significant wave height
SWAN	Simulated WAves Nearshore
WW3	WAVEWATCH-III

Previous researchers have made great contributions to the wave energy resource assessment. Early in the 70 s of last century, scholars have analyzed the global coastal wave energy resources using scarce observation wave data. Results show that the global ocean wave energy mainly rich in the northeastern of the North Atlantic Ocean, the northeastern of the Pacific Ocean, western coast of the North America, the southern coast of Australia, Chile in the South America and the southwest coast of the South Africa [3-5]. With the rapid development of numerical simulation method, we can realize large-scale, high resolution research on the wave energy. Kamranzad et al. [6] have analyzed the wave energy characteristic in the Persian Gulf for the period 1984-2008 using SWAN (Simulated WAves Nearshore) wave model driven by ECMWF (European Centre for Medium-Range Weather Forecasts) wind data. Three points in the western, central and eastern of the Persian Gulf were selected, and the time series of energy were evaluated at these points. The results show that there are both seasonal and decadal variations in the wave energy trends in all considered points due to the climate variability. They also pointed out that a small variation in the wind speed can cause a large variation in the wave power. In 2009, Roger [7] successfully forecast the wave energy in the east coast of the Pacific Ocean using WW3 wave model. Akpamar and Komurcu [8] have analyzed the Black Sea wave energy resource based on 15-year SWAN hindcast data. Results show that the areas with relatively abundant wave energy resources were distributed in the southwestern areas of the Black Sea. They pointed out that the south-west coasts of the Black Sea are suggested as the best site. Zheng et al. [9, 10] pointed out that although the China Sea does not locate in the rich area of global ocean wave energy, the wave energy can also be used to utilization. China is a big energy consuming country; the full development of new energy will contribute to alleviate the energy crisis and environmental protection [11, 12]. Using 22-year WW3 hindcast data, Zheng et al. [9] have analyzed the China Sea wave energy. Synthetically considering the value of wave power density, probability of exceedance of wave power density level, exploitable SWH, the stability of wave power density, total storage and exploitable storage of energy resources, they found that the relative wave energy rich area locates in the north waters of the SCS and surrounding waters of the Taiwan Island.

But until now, there is little research about the long term trend of the wave energy, which is one of the important key points in the wave power plant selection. Increasing in the wave power density is helpful for the development of resource; the opposite is not conducive to long term development and utilization of resource. This study presents the long term trend of the China Sea wave energy using a 24-year WW3 hindcast data, including the regional and seasonal differences. Hope to provide reference for the long-term plan of wave energy development.

2. Data and Methodology

Drive WW3 wave model with CCMP (Cross-Calibrated, Multi-Platform) ocean surface wind to simulate the 3-hourly China Sea wave field from 0000 UTC on January 1st, 1988 to 1800 UTC on December 31st 2011. Using this WW3 hindcast wave data, we analyze the overall long term trends of the SWH and wave power density, and their regional differences and seasonal differences.

The special range of the data we set is 3.875°S-41.125°N, 97.125°-135.125°E (Fig. 1). The spatial resolution is 0.25° × 0.25°. The time resolution is three hours. The time range is 0000 UTC on January 1st, 1988 to 1800 UTC on December 31st 2011. Contrasting this data with buoy data from Japan "SATA Cape", "Fukue Island" and Korea "Cheju Island", we have found a good consistency between simulated wave data and observed data [9]. In this study, we verify the precision of the simulated wave data by using Japan Fukue Island buoy data and Korea station 22001 buoy data, as shown in Figs. 2 and 3. The simulated SWH and observed SWH show a good consistency on the curve trend regardless of the site difference. The observed SWH curve shows jumpy phenomenon while the simulated SWH curve appears slightly smooth. And the simulated SWH is slightly smaller than the observed SWH.

Fig. 1. Topography of the China Sea and adjacent waters, and the buoy stations. A: Cheju Island station; B: Fukue Island station; C: SATA Cape station; D: station 22001. (See the original text for details.)

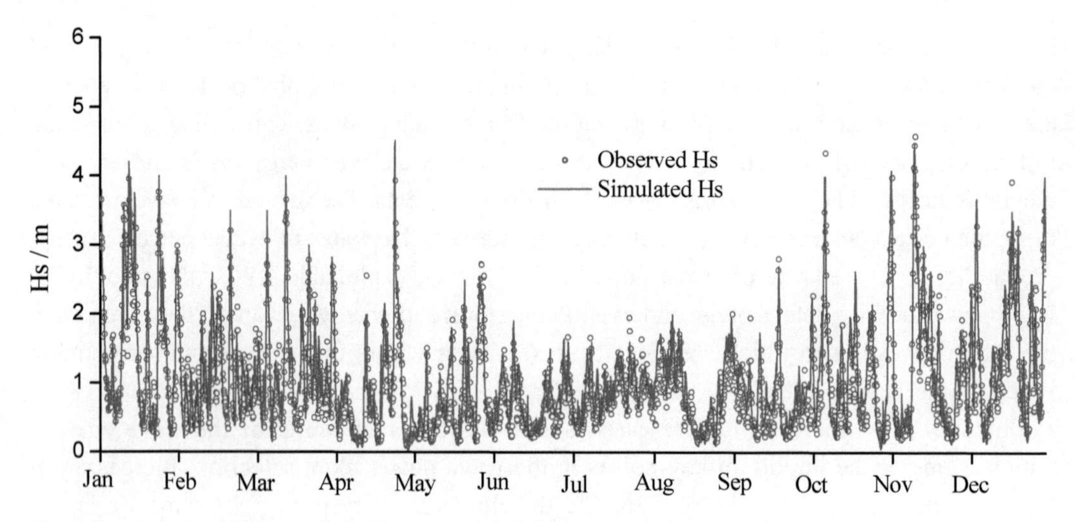

Fig. 2. Simulated and observed significant wave height in Japan Fukue Island in 2009.

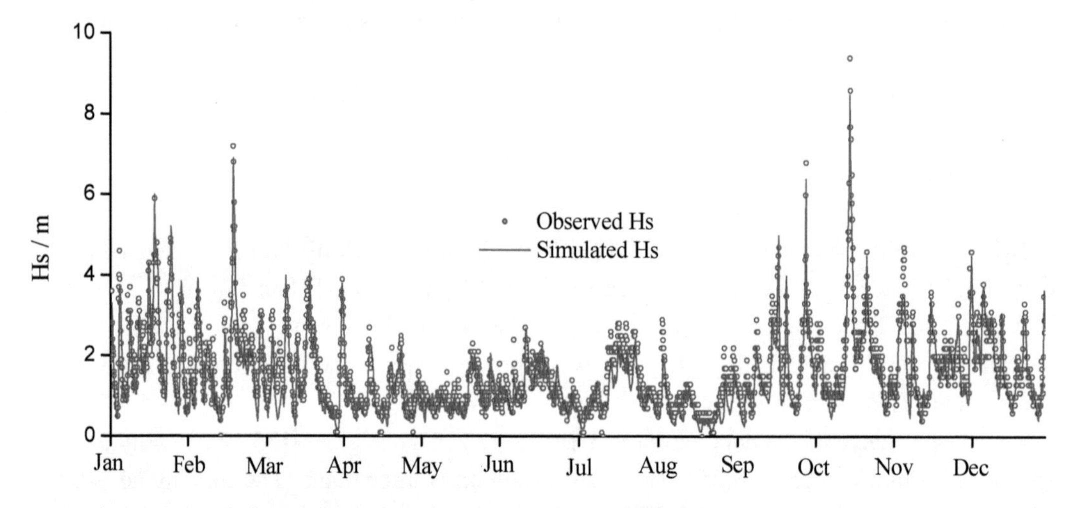

Fig. 3. Simulated and observed significant wave height in Korea station 22001 in 1998.

In order to analyze the precision of simulated SWH quantitatively, the correlation coefficient (CC), the Bias, the root mean square error (RMSE) and the normalized root mean square error (NRMSE) are calculated and presented in Table 1.

Table 1 Precision of the simulated significant wave height.

Station	Location		Time	CC	Bias	RMSE	NRMSE
	Longitude/°E	Latitude/°N					
Cheju Island	126.03	33.08	October, 2009	0.60	0.01	0.36	0.32
Fukue Island	128.63	32.76	2009	0.86	−0.11	0.36	0.15
SATA Cape	130.75	31.05	October, 2009	0.83	−0.09	0.40	0.30
22001	126.33	28.17	1998	0.92	−0.09	0.40	0.11

$$CC = \frac{\sum_{i=1}^{n}(x_i - \overline{x})(y_i - \overline{y})}{\sqrt{\sum_{i=1}^{n}(x_i - \overline{x})^2 \sum_{i=1}^{n}(y_i - \overline{y})^2}} \tag{1}$$

$$Bias = \overline{y} - \overline{x} \tag{2}$$

$$RMSE = \sqrt{\frac{1}{N}\sum_{i=1}^{n}(y_i - x_i)^2} \tag{3}$$

$$NRMSE = \sqrt{\frac{\sum_{i=1}^{n}(y_i - x_i)^2}{\sum_{i=1}^{n} x_i^2}} \tag{4}$$

where, x_i represents the observed data, y_i represents the simulated data, $\overline{x}$ and $\overline{y}$ are average value of observed data and simulated data, N for the total sample.

From Bias, we find that the simulated SWH is slightly smaller than observed SWH. Judging from CC, there is a close relationship between simulated and observed data. The error of simulated data remains low when analyzed by RMSE and MAE. Previous studies also show a good ability of WW3 on wave field simulation in the China Sea [13]. From Figs. 2 and 3 and Table 1, we find that the simulated data is reliable.

3. Long Term Variations of SWH and Wave Power Density

3.1 Calculation method of wave power density

Refer to the calculation and evaluation method of Iglesias and Carballo [14], Cornett [15] and Vosough [16], we obtained the 3- hourly China Sea wave power density for the period January, 1988 to December, 2011, using the 24-yr hindcast wave data.

In deep water, calculation method is as follows,

$$P_w = \frac{\rho g^2}{64_{\pi e}} H_{m0}^2 T_e = 0.49 H_{m0}^2 T_e \tag{5}$$

In shallow water, calculation method is as follows,

$$P_w = \frac{\rho g}{16} H_{m0}^2 \sqrt{gd} \tag{6}$$

where, P_w is wave power (unit: kW/m), T_e is the energy period (unit: s), d is the water depth (with resolution of 1′× 1′, available at http: //www.ngdc.noaa.gov/mgg/global/global.html), ρ is the sea water mass density (~1 028 kg/m^3) [15], H_{m0} is the significant wave height (unit: m), H_s is a characteristic wave height commonly used to describe a given sea state. H_s is defined as the average height of the highest 1/3 of zero crossing waves for a given sample and is determined by analysis of a surface elevation record [14, 17]. In real seas H_{m0} over estimates H_s by 1.5-8% [18].

The significant wave height can be estimated from the frequency domain as

$$H_{m0} = 4\sqrt{m_0} \tag{7}$$

where, H_{m0} is the spectral estimate of significant wave height, m_0 is the zeroeth moment of variance spectrum. The nth order moments of the variance spectrum are calculated as

$$T_e \equiv T_{-10} = \frac{m_{-1}}{m_0} \quad (8)$$

$$m_n = \sum_i f_i^n S_i \Delta f_i \quad (9)$$

where, m_n is the spectral moment of nth order, S_i is the directional spectrum.

3.2 Distribution characteristics of SWH and wave power density

Do average of SWH from 0000 UTC on January 1st, 1988 to 1800 UTC on December 31st 2011 at every 0.25° × 0.25° grid point, we obtain the annual mean SWH in the China Sea, as shown in Fig. 4a. Similarly, the annual mean wave power density is obtained, as shown in Fig. 4b. The spatial distribution characteristic of wave power density has a good consistency with that of SWH. Since locates in the edge of the ocean, the annual average SWH in the China Sea is obviously lower than that in the ocean, basically within 2.0 m. Large value regions of SWH are found in the Ryukyu Islands - the Luzon Strait - southeastern area of the Indochina Peninsula, showing a northeast - southwest zonal distribution, of about 1.4-2.0 m in SWH and 12-22 kW/m in wave power density. The large center locates in the Luzon Strait and its west adjacent waters, of about 1.8-2.0 m in SWH and 18-22 kW/m in wave power density.

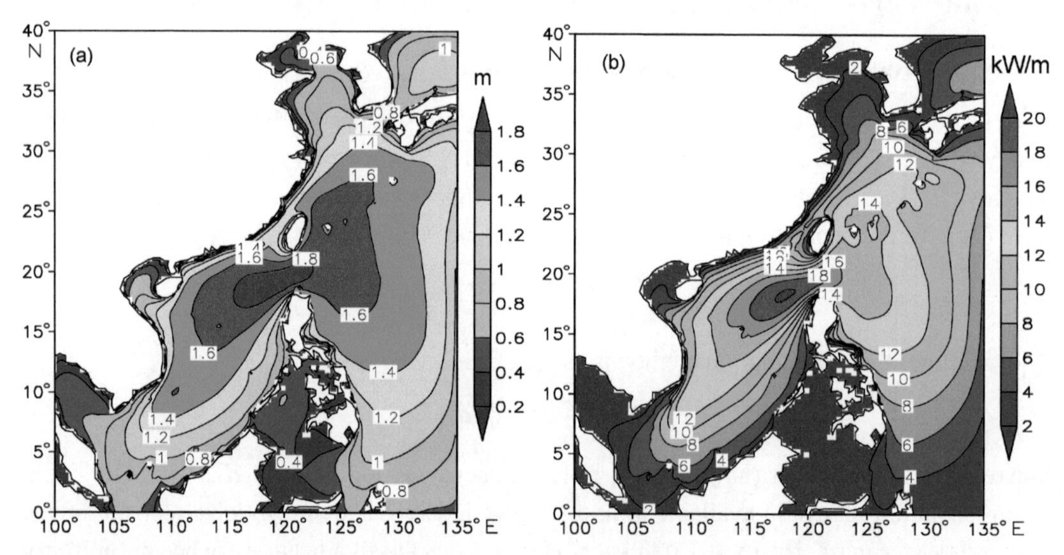

Fig. 4. Annual average significant wave height (a) and wave power density (b) in the China Sea.

The Bohai Sea and the north area of the Yellow Sea is the low value center, of below 0.6 m in SWH and below 2 kW/m in wave power density. To the central and southern Yellow Sea, as the sea becomes broader, the ocean wave is fully grown, the SWH and wave power density increase significantly. The SWH and wave power density in the East China Sea is much larger than that in the Bohai Sea and Yellow Sea, of above 1.0 m in SWH, and above 8 kW/m in wave power density. A more optimistic China Sea wave power density than the traditional valuation 2-7kW/m [19] is found.

3.3 Overall variation of SWH and wave power density

Do average of SWH from 0000 UTC on January 1st, 1988 to 1800 UTC on December 31st 1988, a yearly average value of SWH at every 0.25° × 0.25° grid point in the China Sea is obtained. Then we obtained a zonal average value of SWH through Thiessen polygon method. Using the same method, we obtain 24 zonal and yearly average values of SWH. Then the overall variation of the China Sea SWH is analyzed using linear regression method, as shown in Fig. 5a. Similarly, the overall variation of the China Sea wave power density is also analyzed, as shown in Fig. 5b.

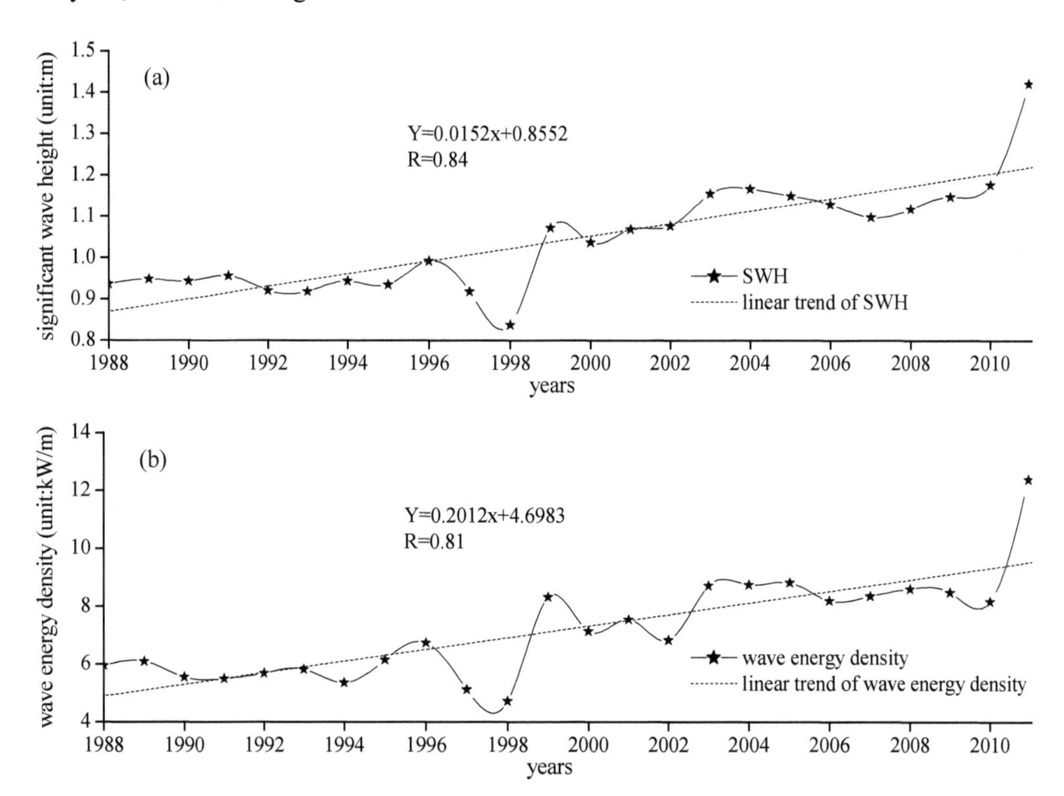

Fig. 5. Long term trends of the China Sea SWH (a) and wave power density (b) with regional average.

As shown in Fig. 5a, correlation coefficient (R) of the SWH is 0.84, significant at the 99.9% level t-test ($|R| = 0.84 > r_{0.001} = 0.51$). The regression coefficient is 0.0152. It means that the SWH exhibits a significant increasing trend of 0.0152m/yr (1.52 cm/yr) in the China Sea as a whole for the past 24 years. Similarly, the China Sea wave power density has a significant increasing trend, of about 0.2012 (kW/m)/yr.

For the period 1988-1997, there is a small variation in both SWH and wave power density, with annual average value about 0.95 m in SWH, and 6.0 kW/m in wave power density. The lowest point appears in 1998 for the past 24 years. Since 1998, the SWH and wave power density increase strongly. From April 1997 to May 1998, a strong El Niño phenomenon happened, accompanying the weaken of East-Asian Monsoon [20, 21]. This caused the China Sea SWH and wave power density reaching the trough in 1998. For the period 1998.09-2000.07, a strong La Nina phenomenon happened, accompanying the strengthen of East-Asian Monsoon [20, 21]. This should be due to the abrupt increase of the China Sea SWH and wave power density from 1998 to 1999.

3.4 Regional difference of the long term trends

In order to exhibit the regional differences of the long term trend in SWH and wave power density, we also calculate the variation at every 0.25° × 0.25° grid point, as shown in Fig. 6.

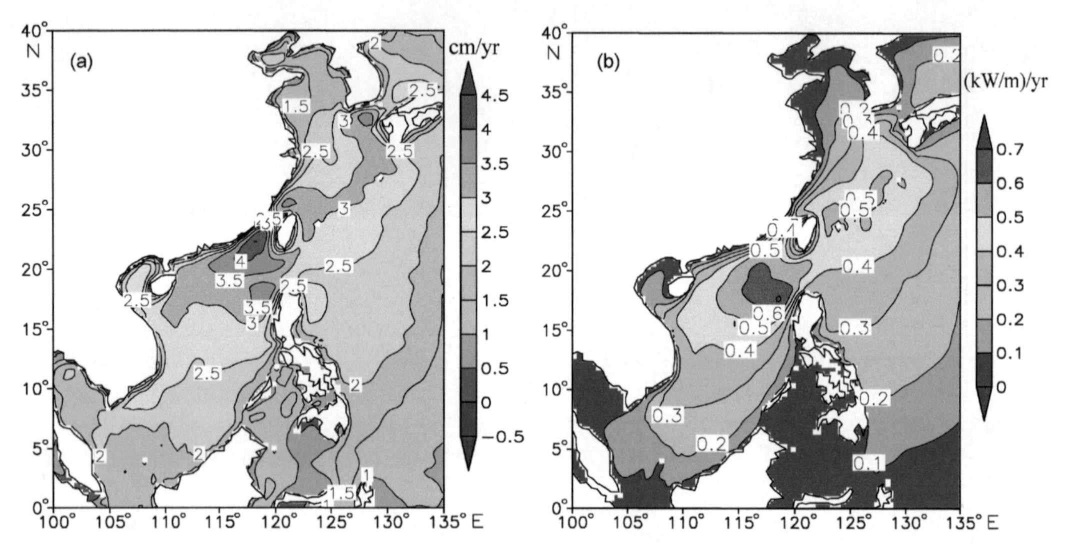

Fig. 6. Long term trends of significant wave height (a) and wave power density (b) in the China Sea. Only area significant at 95% level is presented.

From Fig. 6a, it is clearly that the SWH in most parts of the China Sea exhibit a significant increasing trend, of about 0.5-4.5 cm/yr. Large areas distribute in the Ryukyu Islands, the Luzon Strait and the north area of the SCS, especially in the Dongsha Islands, even up to 4.5 cm/yr. Young et al. [22] found a significant increasing trend of SWH in most of the global ocean, of about 2.64-4.50 cm/yr in the China Sea and adjacent waters, using a 23- year database of calibrated and validated satellite altimeter measurements. The increasing trend of SWH in the China Sea in this study is slightly smaller than the result reported by Young et al. [22] and slightly larger than the result reported by Semedo et al. (0.5-1.5 cm/yr) [23]. The data used by Young et al. [22] is GEOSAT satellite altimeter measurements. With the rapid development of marine remote sensing, the precision of satellite altimeter measurements is high. Though contrasting to the buoy data, we find that the hindcast wave data in this study is reliable and slightly smaller than the observed data. This should be due to the difference between our result and Young et al. The data used by Semedo et al. [23] is the ECMWF ERA-40 (40-year ECMWF reanalysis) wave re-analysis data, which is the first global reanalysis production using a WAM wave model coupled to a general circulation model and assimilated observation data. In ERA-40 low wave heights tend to be overestimated and high wave heights tend to be underestimated. This feature is a global characteristic, and not a peculiarity of a particular location [24]. This should be due to the difference between our result and Semedo et al. Only a few waters with a decreasing trend or without significant variation are distributed sporadically.

From Fig. 6b, we also find a noticeable increasing trend in wave power density in most parts of the China Sea, of about 0.1-0.7 (kW/m)/yr. And this phenomenon is benefit for the development of wave power. Areas with stronger increasing trend mainly locate in the Ryukyu Islands, Taiwan Strait, Luzon Strait and the north waters of the SCS, above 0.4 (kW/m)/yr. The increasing trend in the Luzon Strait and its west waters can be up to 0.6 (kW/m)/yr. Contrasting Fig. 4a and b, it is easy to find that the increasing trend in the wave power density is mainly caused by that in SWH.

Zheng and Pan [25] have pointed out that most of global ocean sea surface wind speed has a obvious increasing trend for the period 1988-2011, about 3-11 cm/s/yr. The wind sea accounts for a large proportion in the mixed wave for the China Sea locates in the edge of the Northwest Pacific Ocean [10]. Kamranzad et al. [6] have also pointed out that a small variation in the wind speed can cause a large variation in the wave power.

Using the 6-hourly CCMP wind data for the period 1988-2011, we statistic the gale occurrence of wind speed greater than class 6 (wind speed above 10.8 m/s), as shown in Fig. 7. It is obvious that the variation of the China Sea wind speed should be due to the increasing of gale occurrence.

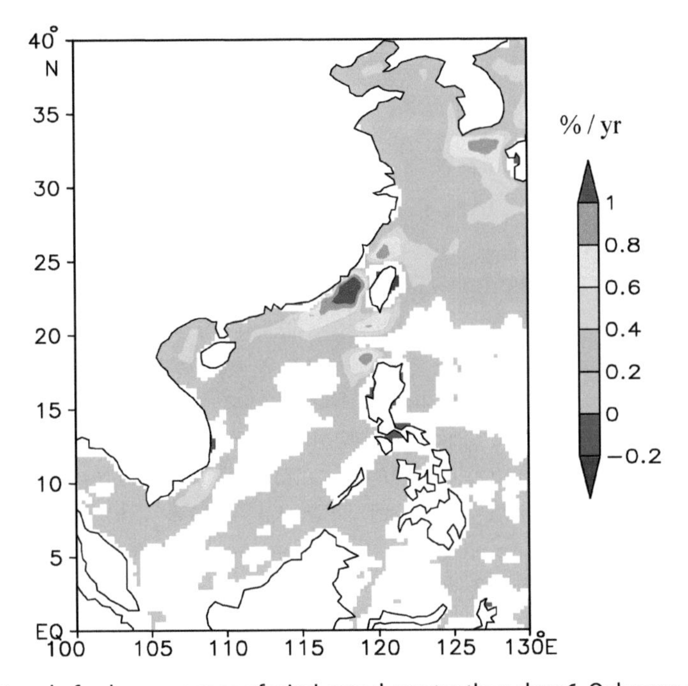

Fig. 7. Long term trend of gale occurrence of wind speed greater than class 6. Only area significant at 95% level is presented.

Previous researches show that El Niño phenomenon has a significant influence on the gale occurrence in the China Sea and surrounding waters [26-28]. In the China Sea and surrounding waters, the gale occurrence with the contemporaneous Niño3 index and two months delayed gale occurrence with Niño3 index both show a significant negative correlation, as shown in Fig. 8. The negative correlation is strongest and the area significant at the 95% level is largest when the gale occurrence delays for two months. The strongest negative correlation (the coefficient takes value of –0.6) is located in surrounding areas of Taiwan, the northern SCS and waters to the east of Philippines. The negative correlation in central areas of the Yellow Sea is significant as well when the coefficient takes value of –0.5.

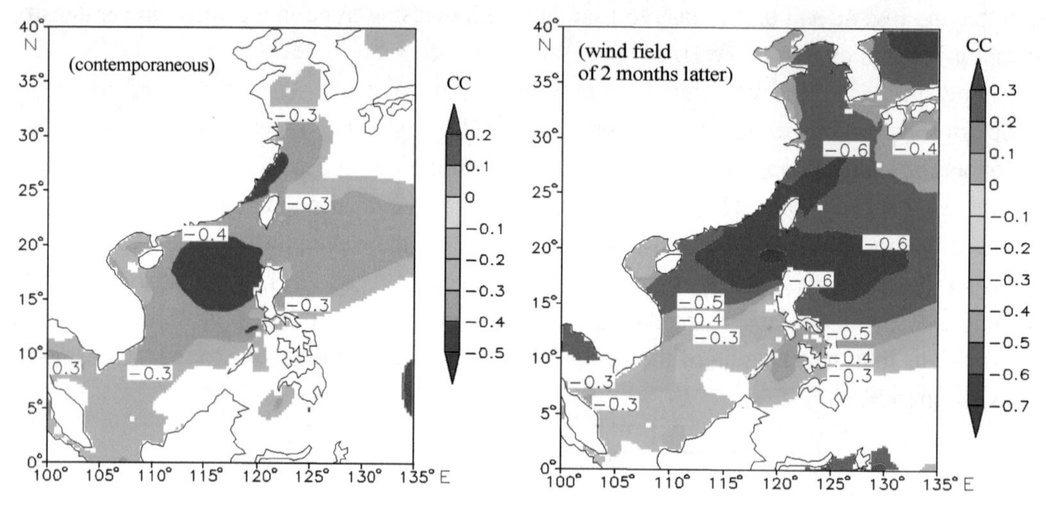

Fig. 8. Correlation coefficient (CC) between wind field and Niño3 index. Only area significant at 95% level is presented.

The China Sea often suffers from the cold air with the prevailing wind direction north and northeast. The wind speed decreases in the transition season of monsoon. Li [26] and Chen [27] pointed out that the inter-annual variability of the East Asian monsoon is well related to ENSO. The East Asian monsoon is always relatively weak in the year of El Niño and relatively strong in the year of La Nina. It may be the reason for the significant negative correlation between the gale occurrence in the China Sea and the Niño3 index at the corresponding period. Mirzaei et al. [28] have found that the SCS SWH correlated negatively with Niño3.4 index during winter, spring and autumn, but became positive in the summer monsoon. Such correlations correspond well with the surface wind anomalies over the SCS during El Niño events. It is concluded that the increasing trend of the sea surface wind speed in the China Sea may be mainly affected by the atmospheric circulation, the monsoon variation and ENSO et al.

3.5 Seasonal difference of the long term trend

Calculating the variations of SWH and wave power density in MAM (March, April, May),

JJA (June, July, August), SON (September, October, November) and DJF (December, January, February), we analyze the seasonal differences of the changing trend, as shown in Fig. 9.

From Figs.9a-d, there is an obvious seasonal difference of the long term trend of the China Sea SWH. Stronger increasing trend in DJF and MAM is obviously higher than that in JJA and SON. And the area scale with increasing trend is also wider in DJF and MAM. Similarly, the long term trend of the China Sea wave power density in DJF is stronger than that in other seasons, as shown in Figs.9e-h.

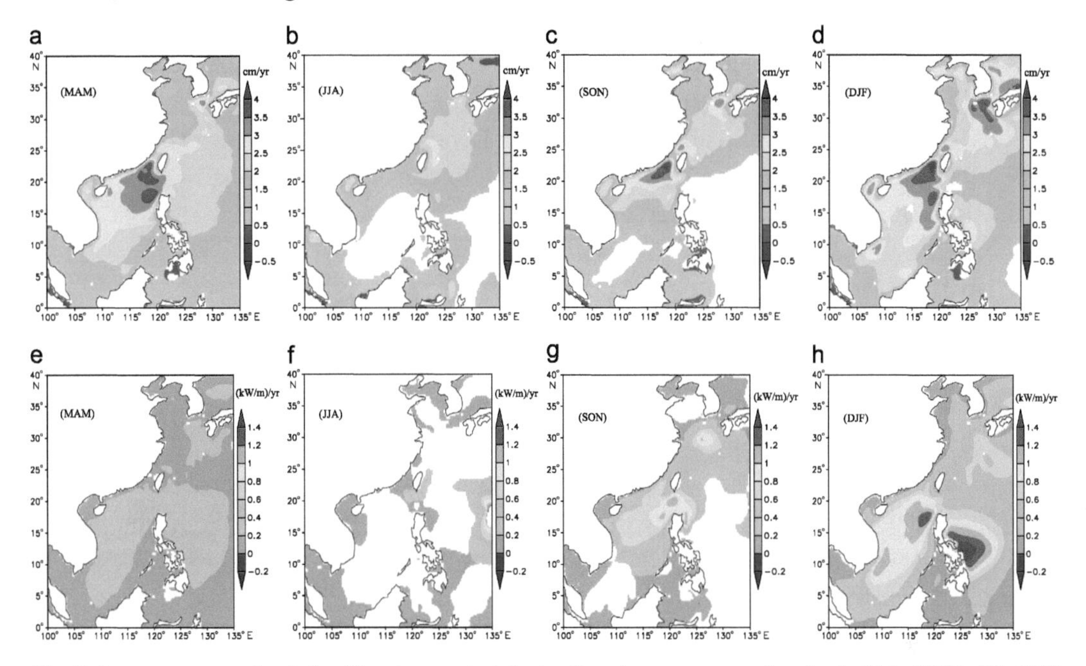

Fig. 9. Long term trends of significant wave height (a-d) and wave power density (e-h) in MAM, JJA, SON and DJF in the China Sea. Only area significant at 95% level is presented.

In MAM, almost all of the China Sea has a significant increasing trend, of about 1.0-4.0 cm/yr in SWH (Fig. 9a), and 0.0-0.6 (kW/m)/yr in wave power density (Fig. 9e). The regional difference of long term trend in SWH is much more obvious than that in wave power density. Especially in the Dongsha Islands, Luzon Strait and its west waters, the increasing trend of SWH is very noticeable, even up to 3.0 cm/yr.

In JJA, the scope of area with increasing trend in SWH narrowed down. Areas with increasing trend in SWH mainly locate in the Bohai Sea, Yellow Sea, East China Sea, north of the SCS, Beibu Gulf and the Gulf of Thailand, of 0.5-3.0 cm/yr, as shown in Fig. 9b. The scope of area with increasing trend in wave power density is smallest all year round. Large scale waters do not have significant variation. Only some sporadic waters has significant increasing trend in wave power density, as shown in Fig. 9f.

In SON, the increasing trend within the first island chain is noticeable, of 0.5-4.0cm/yr in SWH and 0.2-1.2 (kW/m)/yr in wave power density. The increasing trend of SWH in the Dongsha Islands is very strong, can even up to 4.0 cm/yr. Waters in the east of Philippines do not have significant variation in SWH. Areas without increasing trend in wave power density mainly distribute in the middle of the Yellow Sea, low-latitude waters of the South China Sea, and east waters of Philippines.

In DJF, the increasing trend in SWH and wave power density is strongest, and the scope of area with increasing trend is the widest all year round. The variation in most parts of the China Sea is 1.5-4.5 cm/yr in SWH and 0.2-1.6 (kW/m)/yr in wave power density. Large centers of increasing trend in SWH locate in the Tsushima Strait, Dongsha Islands waters, Luzon Strait and its west waters, above 3.5 cm/yr. The increasing trend of wave power density in most of the South China Sea (above 0.8 (kW/m)/yr) and the east waters of Philippines (above 1.0 (kW/m)/yr) is very noticeable.

Contrasting the variations in different seasons, it is very clearly that the increasing trend of SWH is mainly dominated by DJF and MAM, while the increasing trend of wave power density is mainly dominated by DJF.

4. Conclusions

Based on a 24-year (1988-2011) WW3 hindcast wave data in the China Sea, we analyzed the overall long term trends of the SWH and wave power density, and their regional differences and seasonal differences. The results show that,

(1) A more optimistic China Sea wave power density than the traditional valuation (2-7 kW/m) is found. Large areas locate in the belt of the Ryukyu Islands - the Luzon Strait - southeastern area of the Indochina Peninsula, of about 1.4-2.0 m in SWH and 12-22 kW/m in wave power density.

(2) For the past 24 years, the China Sea exhibits a overall increasing trend in SWH (1.52 cm/yr) and wave power density (0.2012 (kW/m)/yr). Especially during the period 1998-2011, the increasing trends are very noticeable.

(3) Both the SWH and wave power density have a significant increasing trend in most parts of the China Sea for the past 24 years, of 0.5-4.5 cm/yr in SWH and 0.1-0.7 (kW/m)/yr in wave power density. Areas with strong increasing trend in SWH and wave power density distribute in the Ryukyu Islands waters, Taiwan Strait and north of the South China Sea, especially in the Dongsha Islands waters can be up to 4.0 cm/yr in SWH and 0.6 (kW/m)/yr in wave power density.

(4) There is a noticeable seasonal difference in the variation of both SWH and wave power density. The increasing trend of SWH in DJF and MAM is obviously stronger than that in JJA and SON. The increasing trend of wave power density in DJF is stronger than that in other seasons. The scope of area with increasing trend of both SWH and wave power density in DJF is the widest all year round.

Acknowledgments This work was supported by the National Key Basic Research Development Program Astronomy and Earth Factor on the Impact of Climate Change (Grant No. 2012CB957803), the Special Fund for Public Welfare Industry (Meteorology) (Grant No.

GYHY201306026), and the program titled Asian Regional Sea-Air Interaction Mechanism and its Role in Global Change (Grant No. 2010CB950400).

References

[1] Rashid A, Hasanzadeh S. Status and potentials of offshore wave energy resources in Chahbahar area (NW Omman Sea). Renew Sust Energ Rev 2011; 15(9): 4876-83.

[2] Saidur R, Islam MR, Rahim NA, Solangi KH. A review on global wind energy policy. Renew Sust Energ Rev 2010; 14(7): 1744-62.

[3] Tornkvist R. Ocean wave power station, report 28, Swedish Technical Scientific Academy: Helsinki Finland; 1975.

[4] Hulls K. Wave power. New Zeal Energ J 1977; 50: 44-8(April).

[5] Glendenning I. Ocean wave power. Appl Energ 1997; 3(3): 197-222.

[6] Kamranzad B, Etemad-shahidi A, Chegini V. Assessment of wave energy variation in the Persian Gulf. Ocean Eng 2013; 70: 72-80.

[7] Roger B. Wave energy forecasting accuracy as a function of forecast time horizon. EPRI-WP-013, <http: // oceanenergy.epri.com/attachments/wave/reports/013_Wave_Energy_Forecasting_Report.pdf>; 2009.

[8] Akpamar A, Komurcu MI. Assessment of wave energy resource of the Black Sea based on 15-year numerical hindcast data. Appl Energ 2013; 101: 502-12.

[9] Zheng CW, Pan J, Li JX. Assessing the China Sea wind energy and wave energy resources from 1988 to 2009. Ocean Eng 2013; 65: 39-48.

[10] Zheng CW, Shao LT, Shi WL, et al. An assessment of global ocean wave energy resources over the last 45 a. Acta Oceanologica Sinica 2014; 33(1): 92-101.

[11] Liu W, Lund H, Mathiesen BV, Zhang XL. Potential of renewable energy systems in China. Appl Energ 2011; 88(2): 518-25.

[12] Zhang B, Chen GQ, Li JS, Tao L. Methane emissions of energy activities in China 1980-2007. Renew Sust Energ Rev 2014; 29: 11-21.

[13] Chu PC, Qi YQ, Chen YC, Shi P, Mao QW, South China Sea. wind-wave characteristics. Part I: validation of Wavewatch-III using TOPEX/Poseidon data. J Atmos Ocean Tech 2004; 21: 1718-33.

[14] Iglesias G, Carballo R. Choosing the site for the first wave farm in a region: a case study in the Galician Southwest (Spain). Energy 2011; 36(9): 5525-31.

[15] Cornett AM. A global wave energy resource assessment. Proceedings of the eighteenth international offshore and polar engineering conference held in Canada. 2008. p. 318-326.

[16] Vosough A. Wave power. Int J Multidiscipl Sci Eng 2011; 2(7): 60-3.

[17] Lenee-Bluhm P, Paasch R, Ozkan-Haller HT. Characterizing the wave energy resource of the US Pacific Northwest. Renew Energ 2011; 36: 2106-19.

[18] Ochi MK. Ocean waves. Cambridge: Cambridge University Press; 1998.

[19] Wang CK, Lu W. Analysis methods and reserves evaluation of ocean energy resources. Beijing: Ocean Press; 2009.

[20] Tao SY, Zhang QY. Response of the Asian winter and summer monsoons to ENSO events. Chinese J Atmos Sci 1998; 22: 399-407.

[21] Li CY, Sun SQ, Mu MQ. Origin of the TBO-interaction between anomalous East-Asian winter monsoon and ENSO cycle. Adv Atmos Sci 2001; 18(4): 554-66.

[22] Young IR, Zieger S, Babanin AV. Global trends in wind speed and wave height. Science 2011; 332(6028): 451-5.

[23] Semedo A, Suselj K, Rutgersson A, Sterl A. A global view on the wind sea and swell climate and variability from ERA-40. J Climate 2011; 24: 1461-79.

[24] Caires S, Sterl A. 100-year return value estimates for ocean wind speed and significant wave height from the ERA-40 data. J Climate 2005; 18: 1032-48.

[25] Zheng CW, Pan J. Assessment of the global ocean wind energy resource. Renew Sust Energ Rev 2014; 33: 382-91.

[26] Li CY. Interaction between anomalous winter monsoon in East Asia and El Niño events. Adv Atmos Sci 1990; 7(1): 36-46.

[27] Chen W, Hans FG, Huang RH. The interannual variability of East Asian winter monsoon and its relation to the summer monsoon. Adv Atmos Sci 2000; 17: 46-60.

[28] Mirzaei A, Tangang F, Juneng L, Mustapha MA, Husain ML, Akhir MF. Wave climate simulation for southern region of the South China Sea. Ocean Dyn 2013; 63(8): 961-77.

An Overview of Global Ocean Wind Energy Resource Evaluations

Chongwei Zheng[a, b, c*], Chongyin Li[a, b], Jing Pan[b], Mingyang Liu[a], Linlin Xia[a]

[a] College of Meteorology and Oceanography, PLA University of Science and Technology, Nanjing, China
[b] LASG, Institute of Atmospheric Physics, Chinese Academy of Sciences, Beijing, China
[c] PLA Dalian Naval Academy, Dalian, China

Abstract With the rapid development of human society, the demand for energy has accordingly increased, and along with this increasingly serious energy and environmental crises have developed. Many countries have been focusing on new energy resources to combat these crises, and offshore wind energy resources are especially attractive; they are safe, non-polluting, renewable, and widely distributed with large reserves, which has made them become the focus of developed countries. However, the distribution of wind energy has strong regional and seasonal differences, which determines the success and efficiency of wind energy developments. Therefore, there is a clear need for "resource evaluation and planning in advance" in the wind energy development. Previous research has made a great contribution to the evaluation of offshore wind energy resources, mostly through analysis of the climatic characteristics of wind energy. In the actual development process of wind energy resources, these analyses of the climatic characteristics of wind energy provide a reference for site selection. However, after constructing wind farms, to aid their operation, there needs to be a more comprehensive understanding of other factors, such as the short-term forecasting and medium- to long-term predictions of wind energy. This paper reviews the research progress of the wind energy resource evaluations, and then considers where future research needs to focus, for the evaluation of wind energy resources. This mainly includes further analyses of the climatic characteristics of wind energy, short-term forecasting, medium-to long-term predictions, early disaster warning systems, the establishment of a wind energy development index (WEDI) and an integrated application system, in hope of providing a reference for offshore wind power generation, seawater desalination and other wind energy resource developments, and accelerating the industrialization and utilization of offshore wind energy. Doing this will alleviate the energy and environmental crises, and promote the sustainable development of human society.

Keywords offshore wind energy; resource evaluation; short-term forecasting; medium- to long-term prediction; wind energy development index; integrated application system

Received January3, 2015; revised May 29, 2015; accepted September 17, 2015; available online November 10, 2015.
*Correspondence to: Tel.: + 86 18640814027. E-mail address: chinaoceanzcw@sina.cn (C.W. Zheng).

1. Introduction

The environmental crises have been attracting much attention in recent years. With shortages of conventional energy, such as coal and oil, predicted in the near future, humans have been focusing on new energy resources, researching which resource is the best for coping with climatic change and the shortages of conventional energy; this has been a common strategy, adopted by many countries. There has been a gradual move towards the industrialization and large-scale utilization of solar and onshore wind energy resources, but their use is restricted severely by their uneven spatial distributions. Nuclear energy can provide significant amounts of energy, but also poses a potentially big threat to human life.

Many developed countries have encouraged the development of new energy resources by legislating, as well as reducing or exempting tax, amongst other measures [1]. For example, in order to support the production and use of renewable energy, Turkey implemented a new law in 2010 [2], within which the principles of "Wind Power Plant Supporting Mechanisms", such as price, time and payments were determined. Only 1% of the licensing costs during the application process are paid by corporate entities in Turkey, and these entities do not pay annual licensing costs for the first 8 years [3]. Past renewable energy policies in Japan have had a weak market focus on wind resources, so there has been no increase in the wind energy share in Japan. As a result, only 1% of the global total installation of wind energy has occurred in Japan, which is far behind the Unites States, China, Germany, Spain, and many other countries [4]. Within the European Union (EU), each member state is supposed to reach a mandatory 20% share of renewable resources in their total energy consumption by 2020 [5]. In the United Kingdom, wind power is recognized to be the main renewable energy resource that will enable the country to achieve the European Union 2020 renewable energy targets. Currently, over 50% of renewable power in the United Kingdom is generated from onshore wind, with a large number of offshore wind projects in development [6].

The advantages of offshore wind energy have made it become extremely attractive as a potential resource. These advantages include that it is a safe, non-polluting, renewable resource, with large reserves, across a wide distribution, and that its development will not take up land resources. The biggest challenge for utilizing offshore wind resources is the relatively high cost of development. However, according to a study commissioned by the European Wind Energy Association (EWEA) in 2015, with the rapid development of science and technology, offshore wind costs could be reduced to EUR 100/MW h by 2020 and EUR 90/MW h by 2030 [7]. Junginger and Faaij [8] and Blanco [9] have shown that despite the recent increase in the capital costs of wind power generation, the long-term trends have indicated a substantial reduction, due to the rapid development of technology. Considering the costs of power generation, environmental pollution [10], wind energy conversion, and other external costs synthetically [11, 12], wind power will become a significant competitor in the energy market.

Wind power generation is the main way to develop and utilize wind energy resources, which are also used widely in desalination, navigation, irrigation, wind-heating and other projects [13, 14]. Coastal areas around the world tend to have advanced economies and, as such, are often also centers of high demand for electric power; therefore, dependent on the local conditions, exploiting and making full use of the offshore wind energy in these areas, could effectively alleviate the energy crisis and promote sustainable development. Remote islands and deep sea areas have a particularly urgent demand for electricity. Taking advantage of the marine wind resources and developing offshore wind energy in these areas could not only solve the power dilemma, but will also protect the environment of these ecologically sensitive islands, and avoid destruction brought about by diesel power generation (e.g. pollution).

Onshore wind power generation technologies have been established worldwide, but the technology for offshore wind power generation has only really been developed in a few European countries [15, 16]. In 2014, the cumulative capacity of global offshore wind power rose to 8759 MW; the added capacity in that year was 1713 MW. More than 91% of all the offshore wind installations worldwide are in European waters, particularly in the North Sea (5094 MW: 63.3%) [7]. Thus, the offshore wind power generation potential in many countries is still basically undeveloped.

The distribution of wind energy resources has strong regional and seasonal differences. Therefore, for the large-scale development of wind energy resources, there needs to be clear resource evaluation and planning in advance. To realize the ordered and efficient exploitation of wind energy resources, there is a need for comprehensive evaluations of wind energy resources, along with the creation of strategic plans for wind power development and the construction of associated power networks, through detailed investigations of the wind energy resources. This paper reviews the research progress and presents some prospects for offshore wind energy resource evaluations, in the hope of making a contribution to alleviate the energy and environmental crises.

2. Comparison Between Offshore and Onshore Wind Energy

2.1 Common advantages

There are some common advantages between offshore and onshore wind energy resources. Firstly compared with conventional energy resources (coal, oil, natural gas, etc.), both onshore and offshore wind energy resources are clean, non-polluting, renewable and widely distributed. The utilization of wind energy does not produce toxic or greenhouse gases, which benefits the environment. Secondly, compared with nuclear energy, wind energy is safer. Although it has the advantages of high efficiency, long functional time-scales and high energy outputs, nuclear energy comes with a high risk and can result in environmental and humanitarian disasters, such as the 1986 Soviet Chernobyl nuclear disaster, caused by operator error, and the 2011 Japanese Fukushima nuclear leakage, caused by a tsunami. Finally, solar energy can only be generated during the daytime, but wind energy can be generated round-the-clock.

Nomenclature

ANN	Artificial Neural Network
ARIMA	autoregressive integrated moving average
ARPS	Advanced Regional Prediction System
CCMP	Cross-Calibrated, Multi-Platform
CLLJ	Caribbean low-level jet
CMIP5	fifth phase of Coupled Model Intercomparison Project
DJF	December, January, February
DOE	Department of Energy of the United State
EEMD	Ensemble Empirical Mode Decomposition
EWEA	European Wind Energy Association
EU	European Union
GIS	Geographic Information System
JJA	June, July, August
MCP	Measure-Correlate-Predict
NASA	National Aeronautics and Space Administration
NREL	National Renewable Energy Laboratory
PO.DAAC	Physical Oceanography Data Active Archive Center
RAN	Resource Allocating Network
SAR	Synthetic Aperture Radar
SVM	support vector machine
WAsP	Wind Atlas Analysis and Application Program
WEDI	Wind Energy Development Index
WEST	Wind Energy Simulating Toolkit
WPD	Wind power density
WPPT	Wind Power Prediction Tool
WRF	Weather Research and Forecasting Model

2.2 Differences

Compared with offshore wind energy, onshore wind energy has several advantages. It is easier to install wind energy generators onshore, which inherently means that the construction costs are much lower; the cost of constructing a wind farm offshore is 1.5-2 times greater than that of constructing one onshore. In addition, the maintenance and repair of offshore wind farms are more challenging, due to their location [17]. The application of onshore wind energy technology is well established, and it is easy to incorporate the electricity generated into local networks. It is also easy to obtain observational data onshore, which is beneficial for evaluations of the wind energy resource, and can improve the precision of simulated data, to optimize the results of resource evaluations.

There are also several advantages to the development of offshore, compared with onshore, wind energy. As a resource, offshore wind energy is attractive, because offshore winds are higher in velocity, more reliable and more consistent [18]; the sea surface wind speeds 10 km off the coast are usually 25% greater than those on land, and offshore wind energy resources can generally be utilized to generate electricity for 2-3 times longer, during the same time period, than those onshore [19, 20]. The fact that offshore wind energy developments are offshore means less use of land resources, no human migration as a result of development, and a lower public visual impact. The offshore wind turbines are inherently far away from the public, so that the issue of noise, which generally results in public complaints with onshore wind developments, can be ignored [17, 21]. Due to the low roughness and small friction on the sea surface, the variation of wind speed with height is small, which reduces the cost of wind turbines because they do not need to be as high as those on land. In addition, with the low turbulence intensity and friction on the sea surface, the fatigue loads of the wind turbines, caused by changing winds is reduced, which extends the life of the wind turbines; the foundations are also reusable, with a design life that can be up to 50 years [22]. Finally, with the large area of sea adjacent to land masses around the world, offshore wind energy resources have an effectively limitless development space.

Although it has not been developed as much as onshore wind energy, offshore wind power generation has been established in certain European countries, so the technology is available. Observational data at sea has also become increasingly abundant, and simulations have improved, which is beneficial for the evaluations of wind energy resources across large-scale marine areas.

3. Research Progress

Benefiting from abundant observational data, there have been numerous research projects into onshore wind energy resources. In 1980, the United States generated a distribution diagram of its wind energy resources using surface wind data from 975 meteorology stations, and in 1986 a further 270 meteorology stations were added to the data set, to generate a more comprehensive distribution diagram of the onshore wind energy resources. The Risoe National Laboratory in Denmark collected weather observational data from 220 meteorology stations across 12 European countries, and, by incorporating the impacts of the buildings, the terrain surface conditions, and surface roughness at all the weather stations, a distribution diagram of European onshore wind energy densities was generated. In India, observational data from 570 anemometer towers, installed since 1987, were used by Indian Energy Consultants Ltd. to generate a distribution diagram of Indian local onshore wind energy resources. The China Meteorological Administration conducted wind energy resource censuses in the 1980s and 1990s, respectively, which analyzed the historical wind data using statistical methods, and from these they generated a distribution diagram of Chinese onshore wind energy resources.

Due to the difficulties of collecting ocean observational data, there is a great shortage of data, so there are relatively fewer evaluations of offshore wind energy resources and the research scope is relatively small. Offshore wind energy resource evaluations have gone through the following stages: (1) offshore wind energy resource evaluations based on limited surface observational data; (2) application of satellite observational data into the offshore wind energy resource evaluations; (3) application of numerical simulation methods into the offshore wind energy resource evaluations; and (4) application of reanalysis data into offshore wind energy resource evaluations. These stages are discussed in detail below.

3.1 Offshore wind energy resource evaluations based on surface observations

Towards the end of the last century, scientific researchers began to focus on offshore wind energy; despite the great shortage of oceanic data, their research made a great contribution to current knowledge of offshore wind energy, through exploring the methods of evaluating offshore wind energy resources. Researchers utilized the extremely limited site observational data and ship reported data to evaluate the offshore wind energy resources at single stations or small-scale coastal sea areas. From these studies, near shore wind energy resource development started in European and American developed countries.

Using surface observational data, Youm et al. [23] analyzed the distribution characteristics of wind energy resources in the northern coastal sea area of Senegal, and found that the annual average wind speed and wind power density (WPD) in the area was, 3.8 m/s and 158 W/m^2, respectively. In 2004, Musial and Butterfield [24] evaluated the development state of the United States offshore wind energy resources, and, based on their study, Michael [25] examined the wind energy resources in the seas surrounding the United States; Michael [25] presented a detailed analysis of the wind power around the coastline, which provides a scientific basis for the development of offshore wind energy resources in the United States (Fig. 1). Kucukali and Dinckal [26] analyzed the wind energy resources at Izmit, located in the Western Black Sea (a coastal region of Turkey), based on data from measurement masts 50 m high, covering the period 06/2008-06/ 2009. González-Longatt et al. [27] created a wind resource atlas of Venezuela, based on 32 weather stations. Their results showed that the best wind energy resources were located in the northern coastal area of Venezuela. In 2005, Archer and Jacobson [28] were the first to perform an evaluation of global onshore and coastal wind power, at heights of 10 m and 80 m above sea level.

China and other Asian countries started researching into their offshore wind energy resources after the European and American countries, but are making rapid progress on their evaluations [29, 30]. In China, Chen et al. [31] analyzed the distribution characteristics of wind resources in Lianyungang and its coastal sea area, using 30 years of meteorological data. Their study showed that the area had abundant and relatively stable wind energy resources, with broad prospects for wind energy resource development.

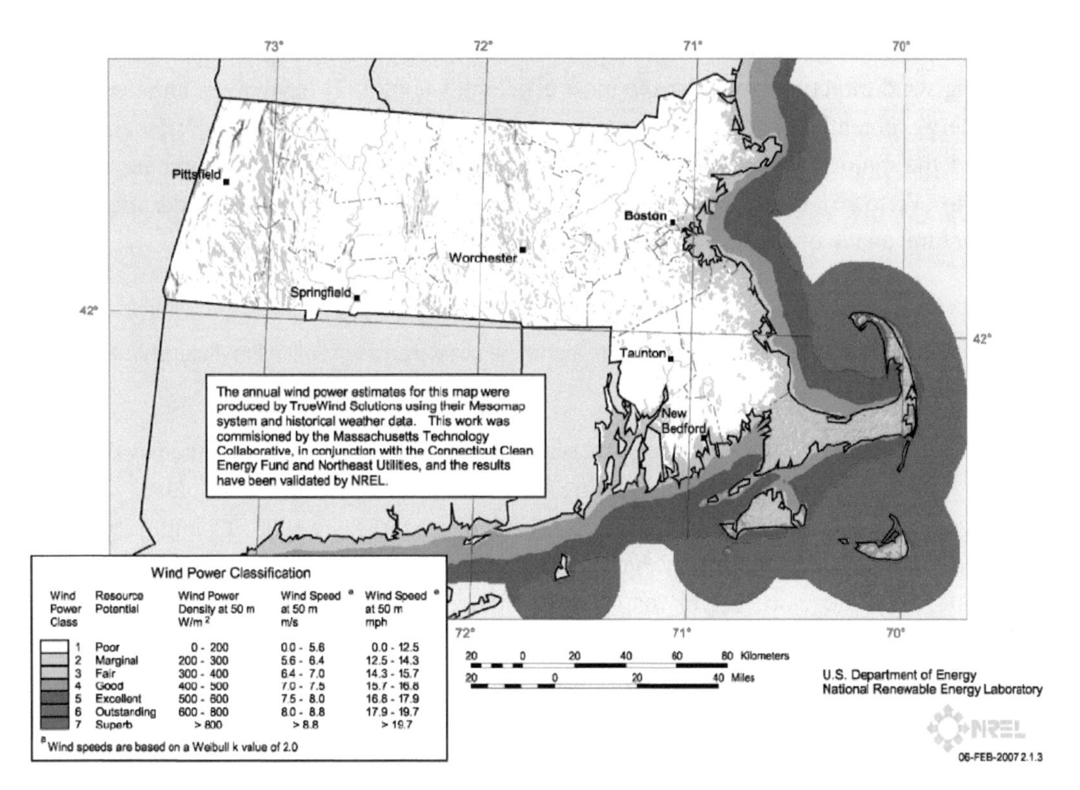

Fig. 1. Wind resource assessment map of the state of Massachusetts, including offshore areas. Wind speed and wind power density are shown at a 50 m height. This map was created by the National Renewable Energy Laboratory for the Department of Energy. Source: Windpoweramerica.gov.

A comprehensive study was performed by Wen et al. [32], who analyzed the wind energy resources in the coastal areas of Fujian Province, using 70 m height observational data from 18 anemometer towers (Fig. 2). Their results showed that wind energy in the coastal area of Fujian Province was abundant, and the maximum energy resources were located in the coastal area, from central and southern Fujian to the south of Quanzhou. The annual effective WPD was calculated to be 517-930 W/m^2, and more than 930 W/m^2 in Pingtan Island. The wind energy resources were also found to be very rich in the town of Chichu, in Zhangpu county, in southern Fujian, where the annual effective WPD was more than 510 W/m^2. In the coastal areas of Fujian, there were 7837.3 annual average effective wind energy hours, that is, 89.4% of the time wind energy power generation would be effective. The development of wind energy resources for power generation in the coastal area from central and southern Fujian to the south of Quanzhou, was classified as "best", with the remaining areas classified as between "better" and "good".

Surface observational data remains important for wind energy evaluations in current research. Based on observational wind data, Oh et al. [21] analyzed the wind energy potential at a demonstration offshore wind farm in the sea to the southwest of the Korean Peninsula. They found that the wind energy potential of the demonstration wind farm was Wind Class 3, in terms of the energy density. Since the major wind direction was distinct (the prevailing wind was northwest), they showed that a wind farm layout that takes into account the prevailing wind direction would be the most efficient. Luong [33] have given an overview of wind energy potential, its current application and future development in Vietnam. They identified the major barriers that need to be addressed for the future development of wind energy in Vietnam; these include technical and technological issues, under developed infrastructure, a lack of financial resources and services, and institutional issues.

Fig. 2. Wind power densities at a height of 70 m along the coastal areas of Fujian Province, China. (See the original text for details.)

There is a large deficiency in surface observational data for wind energy resource evaluations. Due to the relatively backward measuring methods and the shortage of data, scientific researchers can only evaluate the offshore wind energy resources at a single station or small-scale coastal sea areas. However, extensive wind energy resource evaluations over large-scale areas are more useful for selecting the regions with the best energy resources available.

For the construction of wind energy developments, the height of the wind turbines is directly related to the input costs. Consequently, comparing the results of wind energy resource evaluations at different height levels is also useful, to reduce the cost of the development of wind farms. Wind energy resources can generally only be studied at a fixed level, when based on the limited surface observational data; for instance, ship reported data is generally located 10 m above the ocean surface, and buoy data is also generally at a fixed altitude.

A further issue is that the installment time of anemometer towers is often different, which leads to the disunity of the period of wind data, and this in turn affects the accuracy of the wind energy resource evaluations. Installing anemometer towers also requires manpower and material resources; this makes it impossible to establish an intensive observation network on a large-scale and conduct regular observations, like with onshore weather stations. In accordance with all the issues highlighted, it is infeasible to evaluate regional wind energy resources solely on the basis of surface observational data.

3.2 Application of satellite observational data into offshore wind energy resource evaluations

With the progress of ocean measuring methods, more and more satellite data have been used to study marine resources. During 2004-2006, the Risoe National Laboratory in Denmark and several other research institutions conducted the SAT-WIND research program and verified the possibility of applying satellite-derived data, including surface wind distribution data derived from passive microwave remote sensors, altimeters, Scatterometers and Synthetic Aperture Radars (SAR), into offshore wind energy resource evaluations. The final results showed that it was feasible to evaluate the offshore wind energy resources using the satellite-derived wind speed distributions. To provide further verification, using the ocean surface wind speed data derived from SAR, Charlotte et al. [34] studied wind energy resources over the Baltic Sea. They compared the wind speed data derived from SAR and surface observational wind speed data, and found that the former had high accuracy in the Baltic Sea, in which the wind energy density was 300-800 W/m^2.

The National Aeronautics and Space Administration (NASA) [35] used wind speed data derived from QuikSCAT satellite data to generate the distribution of WPDs in the oceans globally, during JJA (June, July, August) and DJF (December, January, February) (Fig. 3). During JJA, the WPDs in the southern ocean were significantly greater than those in the northern oceans. In the oceans of the Northern Hemisphere, the sea around Somalia had significantly high WPDs, which were probably caused by the strong southwest monsoon. During DJF, due to the powerful cold air in the Northern Hemisphere, the WPDs in the oceans of the Northern Hemisphere were significantly greater than in the Southern Hemisphere oceans. Thus, they found that, in general, the WPDs in the winter hemisphere were greater than those in the summer hemisphere.

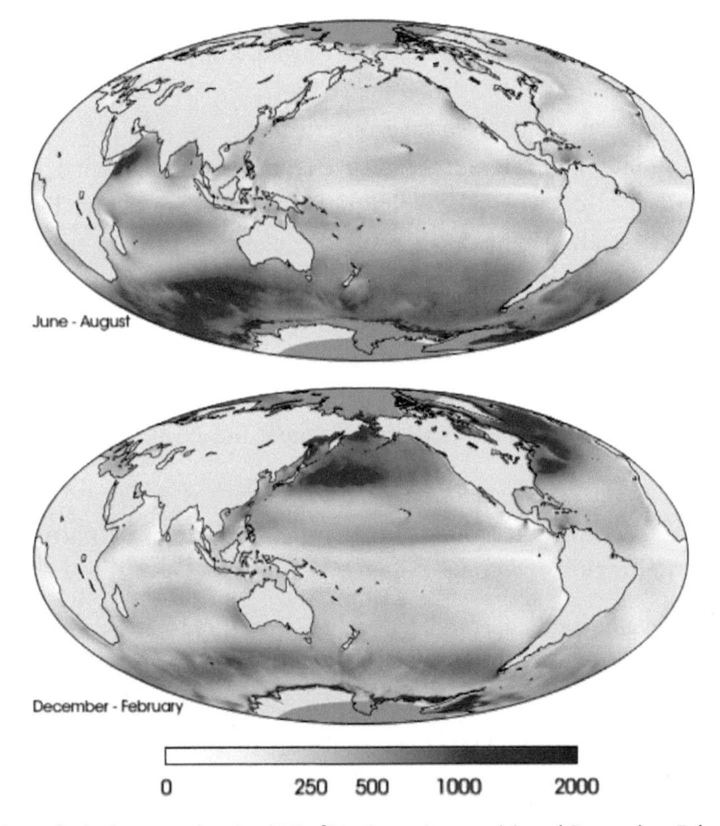

Fig. 3. Distribution of wind power density (W/m^2) in June-August (a) and December-February (b), derived by NASA.

Liu et al. [36] also analyzed the global ocean WPDs in winter and summer, based on the QuikSCAT Level-2 data at a 12.5 km resolution, which were obtained from the Physical

Oceanography Data Active Archive Center (PO.DAAC). A good agreement between the results of Liu et al. [36] and NASA [35] was found. Capps and Zender [37, 38] evaluated the global ocean wind power potential and global ocean 80 m wind energy, accounting for surface layer stability. They demonstrated that during 2000-2006, available 80 m high WPDs, between 100 and 500 W/m^2, existed over approximately 50% of the ice-free ocean surface area.

Satellite observational data has also been used widely to assess local wind energy. Jiang et al., [39] presented a distribution study of offshore wind power in China using QuikSCAT Level-2 satellite measurements, at a 0.5° horizontal resolution, over the past 9 years. They found that the coastal region along Fujian Province had the best wind resources, compared with other offshore regions of China; this agreed well with the results presented by Wen et al. [32]. Based on more than 1000 satellite SAR images, Hasager et al. [40] found a WPD range from 300 to 800W/m^2 for the 14 existing and 42 planned wind farms in the Baltic Sea. Based on the synergetic use of Envisat ASAR, ASCAT and QuikSCAT data, Hasager et al. [41] identified that significant coastal wind speed gradients were identified with SAR data. In regard of eight new offshore wind farm areas in Denmark, they showed that the spatial variability of the mean energy density, based on the SAR data, ranged from 347W/m^2 in Sejerøbugten to 514 W/m^2 at Horns Rev 3.

Compared with the conventional surface observations, satellite data can cover a wider spatial range and allows more comprehensive evaluations of the wind energy resources offshore, but the data also come with some deficiencies. Due to the limited number of satellites and orbits, satellite data have some deficiencies in time synchronization and spatial resolution; namely they cannot cover large-scale areas at the same observation time point. In addition, while there is a clear need to compare evaluations of wind energy resources at different levels (to reduce the development costs of wind energy resources, as already discussed), satellite-derived data cannot reflect variations in the wind field at different levels. Therefore, the data cannot fully meet the needs of assessments for wind power development.

3.3 Application of numerical simulation methods into offshore wind energy resource evaluations

With the rapid development of computer technology, more and more numerical models have been used for wind energy resource evaluations; such models have enabled the gradual realization of fine-scale evaluations of wind energy resources in local sea areas, as well as the study of regions for which observational data is lacking. In general, the numerical simulations have been found to produce reliable data [42, 43]. A number of advanced wind energy simulation software programs have been developed in American and in European countries. For example, in the 1980s and 1990s, the Risoe National Laboratory in Denmark developed resource analysis software for the micro-site selection of wind farms, called the 'Wind Atlas Analysis and Application Program'(WAsP). The United States TrueWind Solutions Company [44] developed the MesoMap and SiteWind wind resource evaluation systems, which have been applied in wind energy resource evaluations across more than 20 countries and regions. The SiteWind system can correct wind maps using real wind field data, which can greatly reduce model errors. The Canadian Meteorological Bureau established the Wind Energy Simulating Toolkit (WEST) numerical model [45], by combining the mesoscale model MC2 and the small-scale model Ms-micro.

Currently, researchers mostly apply integrated model systems to evaluate wind energy resources; such systems are composed of a mesoscale meteorological numerical model, usually the Weather Research and Forecasting Model (WRF) or Mesoscale Model 5, and a complex terrain dynamical diagnosis model, usually the California Meteorological Model or The Advanced Regional Prediction System (ARPS). Carvalho et al. [46] forced the WRF model with different initial and boundary conditions (NCEP-R2, ERA-Interim, NCEP-CFSR, NASA-MERRA, NCEP-FNL and NCEP-GFS) and conducted ocean surface wind simulations to assess which one of the data sets provided the most accurate ocean surface wind simulation and offshore wind energy estimates. The simulation results of the models driven by the NCEP-FNL and NCEP-GFS analyses were better than those produced using the NCEP-CFSR and NASA-MERRA reanalyses.

It is feasible to apply numerical models to simulate wind in a region, which has a number of advantages. Firstly, simulation results can be used as auxiliary information to wind energy resource surveys, and so can contribute to evaluations of wind energy resources in a particular region for which there are no wind observational data. Simulated results can especially make up for the deficiency of oceanic surface observational data, which can guide the site selection of wind farms. The numerical simulation methods can also generate wind energy data of a higher spatial resolution, which enables the accurate determination of the area of available wind energy resources and the available wind energy reserves at the height of wind turbines. A key drawback of observational data (both surface and satellite) is that the altitude is usually fixed, which limits the wind energy resource evaluations to some extent (as discussed above); by applying numerical simulation methods in wind energy resource evaluations, the wind energy parameters for all grid points on the three-dimensional space within the research area can be obtained, and so a more comprehensive evaluation of the wind energy resources can take place [47]. However, there are some deficiencies of numerical modeling. Namely, the influences of topography, among other factors, mean that the simulation data are not very good in some complex regions, which needs further improvement.

3.4 Application of reanalysis data into offshore wind energy resource evaluations

Thanks to the rapid development of technology to produce observational data and model simulations, more and more reanalysis data have been applied widely for wind energy resource evaluations. For example, through statistical analysis of 10 m NCEP/DOE reanalysis wind data, during 1979-2010, Chadee and Clarke [48] derived a regional annual wind resource map, which showed that the Caribbean low-level jet (CLLJ) region was an area with superb WPDs, of 400-600 W/m^2. In addition, they identified the eastern Caribbean and the Netherland Antilles as locations where there were excellent wind energy resources (300-400W/m^2), and the Greater Antilles and the Bahamas as areas with very good wind resources (200-300W/m^2).

Based on multiple satellite data (SAR and Scatterometer ASCAT images) and WRF modeling data, Chang et al. [49] analyzed the offshore wind resources over the South China Sea (Fig. 4). Their results showed that the South China Sea was rich in wind energy, at both 10 m and 100 m heights. The areas with the best WPDs were located in the Beibu Gulf and waters to the east of Hainan Island. Over a 30-year period (1980-2010), daily averaged high altitude wind data was extracted using the NCEP/DOE AMIP-II Reanalysis (Reanalysis-2) by Ban et al. [50], who analyzed the high altitude wind energy in the southeast of Europe. Their results showed the WPDs at an altitude of 2.5 km, to identify the best places to position ground stations in the southeast of Europe. However, the highest WPDs were found offshore, with increasing tendency towards the south, between Italy and Greece.

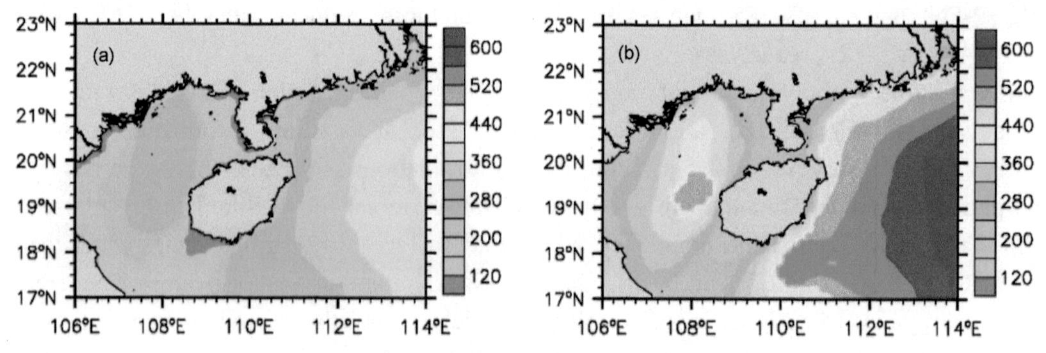

Fig. 4. Wind power density at 10 m (a) and 100 m (b) height over the South China Sea, unit: W/m^2.

Previous research has made a great contribution to the scientific understanding of WPDs and the stability of wind energy, but there has been little research on the class division of ocean wind energy resources globally. In October 2005, the United States Department of Energy, National Renewable Energy Laboratory (NREL) [51] presented a wind power class map of the global oceans, based on QuilSCAT wind data (Fig. 5). The study showed that satellite-derived estimates of wind resources in near shore, coastal, and island areas do not always agree with observational data.

In 2014, using Cross-Calibrated, Multi-Platform (CCMP) wind data, Zheng and Pan [53] analyzed the global ocean wind energy resources. The CCMP data possess the characteristics of a long-term series, high spatial resolution and high precision [52]. Zheng and Pan [53] comprehensively considered the magnitude of the WPD, energy level occurrence (wind power can generally be utilized when the WPD is above 100 W/m^2, and energy-rich regions have WPDs above 200W/m^2; these levels are classified according to their 'energy level occurrence'), gale occurrence (strong wind occurrence), occurrence of the effective wind speed (wind speeds between 3 and 25 m/s are regarded as 'available'), stability of the WPD (including the coefficient of variation, monthly variability index and seasonal variability index), long-term trends of the WPD, and resource storage (including total storage per unit area, effective storage, and exploitable storage), amongst other aspects. According to the technical standards for the development of wind energy resources, created by the Department of Energy of the United States [54] the and National Development and Reform Commission of China [55], Zheng and Pan [53] produced a class division of global ocean wind energy resources (Table 1; Fig. 6).

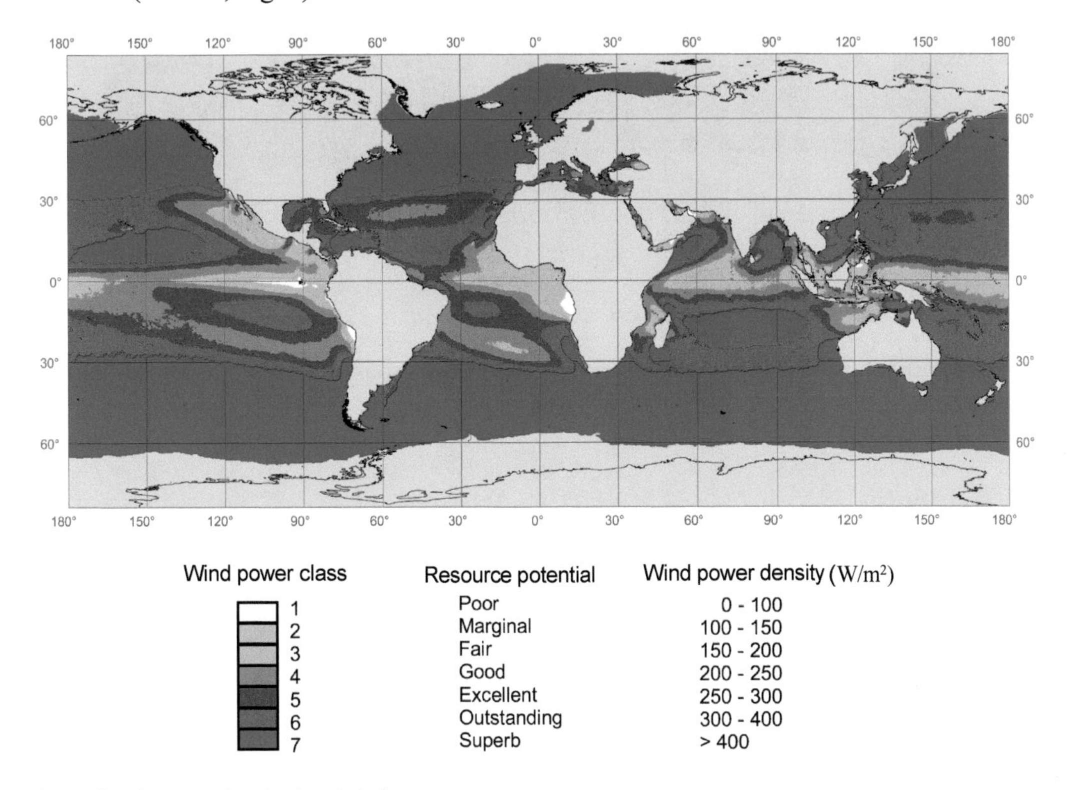

Fig. 5. Wind power class in the global ocean.

Table 1 Standards for wind power classification [54, 55].

Wind power class	Annual average wind speed (m/s)	Annual average wind power density (W/m^2)		Significant interval (h)	Wind energy division
		Method 1	Method 2		
1	0.0-4.4	<100	< 50	< 2000	Indigent area
2	4.4-5.1	100-150	50-150	2000-3000	Available area
3	5.1-5.6	150-200	150-200	3000-5000	Subrich area
4	5.6-6.0	200-250	200-250	> 5000	Rich area
5	6.0-6.4	250-300	250-300		
6	6.4-7.0	300-400	300-400		
7	7.0-9.4	400-1000	400-1000		

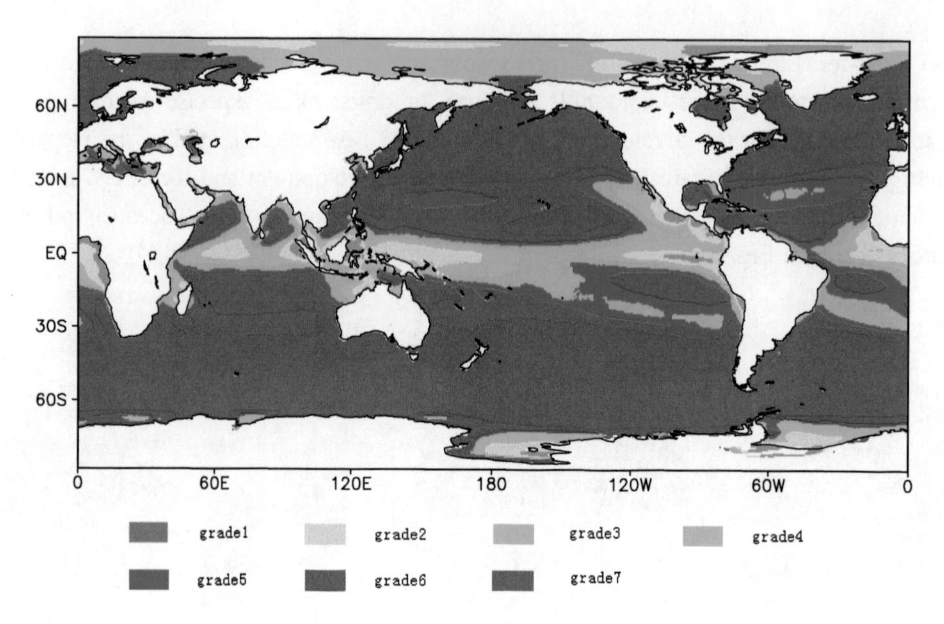

Fig. 6. Grade classification map of the global ocean wind energy resources. Note: a higher grade means a richer resource.

4. Prospects

Previous research has made a great contribution to scientific understanding of the offshore wind energy resources, but the research has mostly concentrated on analyzing the climatic characteristics of wind energy. In terms of the actual development of the wind energy resources, these advanced analyses of the climatic characteristics of wind energy can provide a reference for site selection. However, for the operational working of wind farms, different types of information are needed and the focus of wind energy resource evaluations in the future should not only include further research into the climatic characteristics of wind energy, but also consider: (1) short-term forecasting of wind energy; (2) medium-to long-term predictions of wind energy; (3) early warning systems for the prevention and reduction of natural disasters to wind farms; and (4) development of a Wind Energy Development Index (WEDI) and integrated application system.

4.1 Further research into climatic characteristics of wind energy

Systematic analysis of the climatic characteristics of wind energy includes: the magnitude of the WPD, energy level occurrence, gale occurrence, the occurrence of effective wind speeds, the stability and long-term trends of WPD, resource storage, and other aspects, hereafter collectively referred to as the "wind energy climatic factors". In this study, a "wind energy rose" (as opposed to a wind rose) was created, which summarizes the wind energy climatic factors at a specific site (rather than just wind speed and direction; Fig. 7). From the wind energy rose, it can clearly be seen that the gross energy at station 42001 (25.89°N,

89.66°W) depended on the wind direction and magnitude of the WPD. In February, the wind energy source mainly came from the SE-SSE direction, followed by a northerly direction, but the large levels of wind energy came from the NNW direction. In May, the wind energy mostly came from the E-SE direction, but was largest when the wind came from the SE or NW direction. In August, the wind energy mainly came from the NE-SE direction, especially the ESE direction. In November, the wind energy mainly came from the NE-E direction, and when coming from the NE and NNE directions, it was strongest. And this "wind energy rose" can also be used in the short term forecasting of wind energy resource.

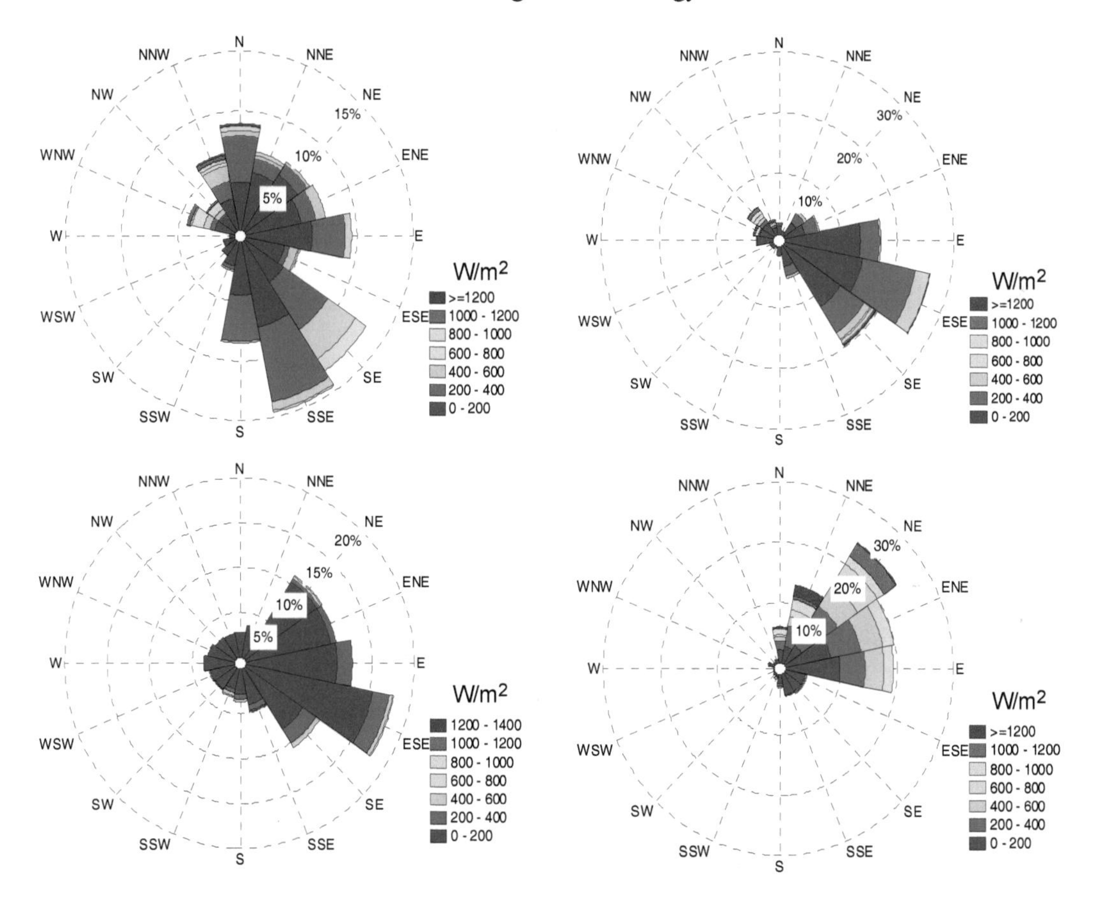

Fig. 7. Wind energy rose at station 42001 (NDBC buoy) in February, May, August and November 2013.

In terms of the wind energy grade division, on the basis of the traditional macroscopic annual resource grade division, in the above analysis, this study emphasizes the grade division in different months and the microscopic grade division. Although Zheng and Pan [53] assessed the wind energy grade division in oceans globally, by comprehensively considering the wind energy climatic factors, their results mostly concentrated on the annual resource grade division. However, in the actual site selection process, the resource grade division during different months is important, and this needs to be especially considered in the future. Assessments of the resource grade division at the large-scale ocean can provide reference for large-scale site selection, but a high-resolution grade division for small-scale ocean areas is also required, to provide reference for microscopic site selection. This work could be completed by the integration of observation, satellite and simulation data. These ideas need to be realized during future research endeavors.

4.2 Short-term forecasting of wind energy

Short-term forecasting of wind energy can provide reference for wind farm operators, which can then improve the collection and conversion efficiency of wind energy and provide an accurate basis for regulating short-term electricity generation. The typical methods for predicting wind energy include numerical weather prediction and wind forecasting, ensemble forecasting, physical methods, statistical and learning approach methods, benchmarking and uncertainty analysis, and hybrid methods, amongst others [56, 57]. Shi et al. [58] compared hybrid forecasting methods using three major single prediction models, that is, the autoregressive integrated moving average (ARIMA) model, the artificial neural network (ANN) models, and the support vector machine (SVM), to predict hourly wind speeds and power generation. Their results showed that the hybrid methodology was always superior to the single models.

Han et al. [59] proposed a forecasting model for wind power based on the Resource Allocating Network (RAN). Compared with observational data, their results showed that the selfreconstruction structure of the RAN made it possible to forecast with a look-ahead horizon of up to more than one day, for different types of wind turbines at different locations. Tascikaraoglu and Uzunoglu [60] outlined the combined forecasting approaches and presented an up-to date annotated bibliography of the wind forecasting literature; they also proposed combined approaches composed of several spatial correlation models, which have not been included in this study. The evaluation and comparison of these models should be considered in a future study, because of their successful implementation.

Wind energy forecasting software has also developed rapidly in recent decades [61]. In the early 1990s, a number of European countries started to develop wind energy forecasting systems, which were used for their forecasting services [62]. Forecasting techniques mostly use medium-term prediction models, which nest high-resolution effective area models (or nest higher resolution regional models), and generation capacity models, to forecast the generation capacity of wind farms. For instance, the Denmark Predictor forecasting system has been applied for short-term wind energy business operations in Denmark, Spain, Ireland and Germany, while the Wind Power Prediction Tool (WPPT) is also used for wind energy forecasting business operations in some European areas.

In the mid-1990s, the United States True Wind Solutions Company began their commercial wind energy forecasting service, through which they developed the wind energy forecast software 'eWind', which consists of high-resolution mesoscale meteorological numerical models and statistical models that have been used to forecast the wind field and generation capacities. eWind and Predictor are currently both used for the forecasting service of two large wind farms in California [63, 64]. In October 2002, the European Commission started the program ANEMOS, which aims to develop advanced forecasting models that are better than the existing ones, with emphasis on the condition of complex terrains and extreme weather in the forecasts, as well as the development of nearshore wind energy forecasting [65, 66]. By combining MC2 and WAsP, the Canadian wind energy resource numerical evaluation and forecasting software, WEST, generates an atlas of wind energy, with a resolution of 100-200 m [67, 68]. In addition, Previento (developed in Germany), LocalPred and RegioPred (from Spain), and HIRPOM (developed in Ireland and Denmark), are proposed combination models that could also be applied for wind energy forecasting [69].

4.3 Medium- and long-term predictions of wind energy

The medium- to long-term trends of wind energy resources are important for the long-term planning of wind energy development and the power market [70, 71]. Three methods are usually applied for medium and long-term predictions of wind energy. The first method employed is the use of the long-term forecasting wind field, such as the fifth phase of the Coupled Model Intercomparison Project (CMIP5), to produce long-term wind energy resource predictions. Although not widely used currently, this method will be popular in the future. Wang et al. [72] analyzed the changes in global ocean wave heights, projected with multi-model CMIP5 simulations. The second method is to use some regular and strong astronomical and earth factors, to assist the medium and long-term prediction of wind energy. The third method uses the ANN, SVM, ARIMA, RAN [73], Measure-Correlate-Predict (MCP), or Ensemble Empirical Mode Decomposition (EEMD) [74] methods to produce medium and long-term predictions of wind energy.

4.4 Early warning systems for the prevention and reduction of natural disasters

For the development of offshore wind power, apart from wind energy evaluations, it is necessary to accurately analyze the extreme wind speeds [75]. Disaster reduction and provision is especially important in remote islands, because they are far away from the continent. In the engineering design of wind farms, it is necessary to calculate the gale occurrence (occurrence of strong wind speeds) and extreme wind speeds, to enable the provision of equipment that will cope with such high wind speeds [76, 77]. With the temporary provision of equipment, it is highly possible to cause dangers which could have been avoided.

Calculating the annual gale occurrence and extreme wind speeds provides a reference for engineering design and can extend the lifespan of wind turbines. Accurate calculations of gale occurrence and extreme wind speeds in different months must be made, because these factors will affect the actual construction and operational working processes. For example, during medium-to long-term planning, it is necessary to use the extreme values as a reference to determine whether to reinforce the equipment of wind farms, in case of adverse sea conditions, for certain months. The annual largest gale occurrence and largest extreme values can be used, however, their use will cause a great amount of avoidable waste, due to the high cost of offshore construction, if construction could occur in a month when the extreme events are very unlikely to occur.

4.5 WEDI and integrated application systems

The construction of offshore wind farms have many constraints, such as the water depth, the distance from the coast, sea-route, fishing areas, the distance from the nearest port, the seabed conditions, the impacts of waves, sea level, the occurrence of typhoons or earthquakes, and military facilities, amongst others. Lee [78] described eight key factors, which were the wind power, earthquake occurrence, the land use, shore type, typhoon occurrence, the distance to shoreline, flight safety and the water depth, that should be considered during the process of offshore wind energy resource evaluations. On the basis of the distribution diagram of offshore wind energy resources, and by combining these restraining factors, the offshore areas suitable for wind farm developments in the future can be identified [79].

This study proposes the establishment of a WEDI, which includes the factors of WPD, occurrence of effective wind speed, occurrence of WPD > 100 W/m^2, occurrence of WPD > 200 W/m^2, gale occurrence, and water depth, amongst other factors. Each factors should be weighted differently, according to its importance. For example, the magnitude of the WPD is closely related to the richness of wind energy, which should occupy a key positive weight. In addition, a WPD > 100 W/m^2 and a WPD > 200 W/m^2, are needed for the resource to be considered 'available' and 'rich', respectively, so the occurrences of these two factors should be positively weighted. In contrast, increases in the gale occurrence and water depth will cause difficulties in the development and utilization of the wind resource, so these two factors should be negatively weighted.

Building an integrated application system for the development of wind energy is also very important, much more important than a platform for wind farm site selection. For example, in order to support the decision making process effectively and efficiently, an application system should be established, which includes the WEDI, wind energy climatic characteristics, location queries, short-term forecasting, medium-to long-term predictions, and automated risk early warning systems, amongst other factors, based on a Geographic Information System (GIS).

The combined exploitation of wave and offshore wind energies has also become an attractive research topic recently [80]. The ocean is the cradle of human life, and a huge treasure of resources. Completing the resource evaluations, along with the rational development of marine energy resources, will effectively alleviate the energy and environmental crises that we face today.

5. Conclusion

In modern society, humans are being significantly affected by energy and environmental crises. It is undoubtedly an ideal choice to develop renewable resources, to cope with the energy and environment crises; this needs to be done according to local conditions, thus promoting the sustainable development of human society. Offshore wind energy is attractive as a renewable nonpolluting, resource with large reserves, and a wide distribution. However, its instability makes it a complex resource to exploit and utilize. In order to develop offshore wind energy resources efficiently and properly, there need to be comprehensive evaluations of the resources, such as analyses of the climatic characteristics and grade divisions for different months of the year, short-term forecasting and medium-to long-term predictions of wind energy, early warning systems to prevent and reduce damage from natural disasters, WEDI and integrated application systems.

Just as a coin has two sides, offshore wind energy has both advantages and disadvantages. Leung and Yang [17] found that offshore wind turbines impact on marine animals, like dab, salmon, and migratory birds, and can cause changes to the climate. Tabassum-Abbasi et al. [81] showed that, with the rapidly increasing utilization of wind energy, environmental concerns also rise, such as the visual, noise and wildlife impacts. Saidur et al. [82] have also shown that wind turbines have significant impacts on wildlife, such as birds, bats, and raptors. In the development of offshore wind energy, the impacts of utilizing the wind energy on the environment, in addition to the wind energy evaluations themselves, should be carefully considered, before any wind farm is constructed, or a decision is made. Only the rational and scientific exploitation of offshore wind energy resources will promote the sustainable and harmonious development between humans and the environment.

Acknowledgments This work was supported by the National Key Basic Research Development Program Astronomy and Earth Factor on the Impact of Climate Change (Grant nos. 2015CB453200, 2013CB956200, Nos. 2012CB957803, 2010CB950400) and National Nature Science Foundation of China (Grant nos. 41490642, 41275086, 41475070).

References

[1] Saidur R, Islam MR, Rahim NA, Solangi KH. A review on global wind energy policy. Renew Sustain Energy Rev 2010; 14: 1744-62.

[2] Dursun B, Gokcol C. Impacts of the renewable energy law on the developments of wind energy in Turkey. Renew Sustain Energy Rev 2014; 40: 318-25.

[3] Kaplan YA. Overview of wind energy in the world and assessment of current wind energy policies in Turkey. Renew Sustain Energy Rev 2015; 43: 562-8.

[4] Mizuno E. Overview of wind energy policy and development in Japan. Renew Sustain Energy Rev 2014; 40: 999-1018.

[5] Hadžić N, Kozmar H, Tomić M. Offshore renewable energy in the Adriatic Sea with respect to the Croatian 2020 energy strategy. Renew Sustain Energy Rev 2014; 40: 597-607.

[6] Higgins P, Foley A. The evolution of offshore wind power in the United Kingdom. Renew Sustain Energy Rev 2014; 37: 599-612.

[7] GWEC (Global Wind Energy Council). Global wind report: annual market update 2014; March 2015.

[8] Junginger M, Faaij A, Turkenburg WC. Cost reduction prospects for offshore wind farms. Wind Eng 2004; 28: 97-118.

[9] Blanco MI. The economics of wind energy. Renew Sustain Energy Rev 2009; 13: 1372-82.

[10] Varun G, Prakash R, Bhat IK. Energy, economics and environmental impacts of renewable energy systems. Renew Sustain Energy Rev 2009; 13(9): 2716-21.

[11] Diaf S, Notton G. Evaluation of electricity generation and energy cost of wind energy conversion systems in southern Algeria. Renew Sustain Energy Rev 2013; 23: 379-90.

[12] McKenna R, Hollnaicher S, Fichtner W. Cost-potential curves for onshore wind energy: a high-resolution analysis for Germany. Appl Energy 2014; 115: 103-15.

[13] Markus F, Fredrik M, Fernando D. Feasibility study on wind-powered desalination. Desalination 2007; 203: 463-70.

[14] Zheng CW, Pan J, Li JX. Assessing the China Sea wind energy and wave energy resources from 1988 to 2009. Ocean Eng 2013; 65: 39-48.

[15] Ackermann T, Soder L. An overview of wind energy-status 2002. Renew Sus-tain Energy Rev 2002; 6(1-2): 67-127.

[16] Thompson S, Duggirala B. The feasibility of renewable energies at an off-grid community in Canada. Renew Sustain Energy Rev 2009; 13(9): 2740-5.

[17] Leung DYC, Yang Y. Wind energy development and its environmental impact: a review. Renew Sustain Energy Rev 2012; 16: 1031-9.

[18] Kaplan YA. Overview of wind energy in the world and assessment of current wind energy policies in Turkey. Renew Sustain Energy Rev 2015; 43: 562-8.

[19] Tambke J, Lange M, Focken U. Forecasting offshore wind speeds above the North Sea. Wind Energy 2005; 8: 3-16.

[20] Wang JZ, Qin SS, Jin SQ, Wu J. Estimation methods review and analysis of offshore extreme wind speeds and wind energy resources. Renew Sustain Energy Rev 2015; 42: 26-42.

[21] Oh KY, Kim JY, Lee JK, Ryu MS, Lee JS. An assessment of wind energy potential at the demonstration offshore wind farm in Korea. Energy 2012; 46: 555-63.

[22] Li XY, Yu Z. Developments of offshore wind power. Acta Energiae Sol Sin 2004; 25(1): 78-84.

[23] Youm I, Sarr J, Sall M. Analysis of wind data and wind energy potential along the northern coast of Senegal. Renew Energy 2005; 8: 95-108.

[24] Musial W, Butterfield S. Future for offshore wind energy in the United States. In: Proceedings of the energyocean proceedings NREL/CP-500-36313. Palm Beach, Florida; 28-29 June 2004.

[25] Michael CR. Renewable energy technologies for use on the outer continental shelf. Colorado: National Renewable Energy Lab; 2006.

[26] Kucukalí S, Dinçkal Ç. Wind energy resource assessment of Izmit in the West Black Sea Coastal Region of Turkey. Renew Sustain Energy Rev 2014; 30: 790-5.

[27] González-Longatt F, González JS, Payán MB, Santos JMR. Wind-resource atlas of Venezuela based on on-site anemometry observation. Renew Sustain Energy Rev 2014; 39: 898-911.

[28] Archer CL, Jacobson MZ. Evaluation of global wind power. J Geophys Res 2005; 110: 5462-81. http: // dx.doi.org/10.1029/2004JD005462 D12110.

[29] Chang LX, Zhan FS. Wind energy in China: current scenario and future perspectives. Renew Sustain Energy Rev 2009; 13(8): 1966-74.

[30] Zhang D, Zhang XL, He JK, Chai QM. Offshore wind energy development in China: current status and future perspective. Renew Sustain Energy Rev 2011; 15(9): 4673-84.

[31] Chen F, Ban X, Qi X. Evaluation of wind energy resources on the coastland and adjacent sea of Lianyungang. Sci Meteorol Sin 2008; 28: 101-6.

[32] Wen MZ, Wu B, Lin XF. Distribution characteristics and assessment of wind energy resources at 70 m height over Fujian coastal areas. Resour Sci 2011; 33 (7): 1346-52.

[33] Luong ND. A critical review on potential and current status of wind energy in Vietnam. Renew Sustain Energy Rev 2015; 43: 440-8.

[34] Charlotte BH, Metrete B, Alfredo P. SAR-Based wind resource statistics in the Baltic Sea. Remote Sens 2011; 3: 117-44.

[35] NASA. Global Ocean Wind energy potential. From http: //earthobservatory. nasa.gov/IOTD/view.php? id =8916; 2008.

[36] Liu WT, Tang WQ, Xie XS. Wind power distribution over the ocean. Geophys Res Lett 2008; 35: L13808. http: //dx.doi.org/10.1029/2008GL034172.

[37] Capps SB, Zender CS. Global ocean wind power sensitivity to surface layer stability. Geophys Res Lett 2009. http: //dx.doi.org/10.1029/2008GL037063 L09801.

[38] Capps SB, Zender CS. Estimated global ocean wind power potential from QuikSCAT observations, accounting for turbine characteristics and siting. J Geophys Res 2010; 115: 12679-91. http: //dx.doi. org/10.1029/2009JD012679 D09101.

[39] Jiang D, Zhuang DF, Huang YH, Wang JH, Fu JY. Evaluating the spatio-temporal variation of China's offshore wind resources based on remotely sensed wind field data. Renew Sustain Energy Rev 2013; 24: 142-8.

[40] Hasager CB, Badger M, Peña A, Larsén XG, Bingöl F. SAR-based wind resource statistics in the Baltic Sea. Remote Sens 2011; 3: 117-44.

[41] Hasager CB, Mouche A, Badger M, Bingöl F, Karagali I, Driesenaar T, Stoffelen A, Peña A, Longépé N. Offshore wind climatology based on synergetic use of Envisat ASAR, ASCAT and QuikSCAT. Remote Sens Environ 2015; 156: 247-63.

[42] Dvorak MJ, Archer CL, Jacobson MZ. California offshore wind energy potential. Renew Energy 2009; 1(1-11). http: //dx.doi.org/10.1016/j.renene.2009.11.022.

[43] Al-Yahyai S, Charabi Y, Gastli A. Review of the use of numerical weather prediction (NWP) Models for wind energy assessment. Renew Sustain Energy Rev 2010; 14(9): 3192-8.

[44] Ayotte KW, Davy Robert J, Coppin PA. A simple temporal and spatial analysis of flow in complex terrain in the context of wind energy modeling. Bound-Layer Meteorol 2001; 98: 275-95.

[45] Yu W, Benoit R, Girard C. Wind Energy Simulation Toolkit (WEST): a wind mapping system for use by the wind-energy industry. Wind Eng 2006; 30 (1): 15-33.

[46] Carvalho D, Rocha A, Gómez-Gesteira M, Santos CS. Offshore wind energy resource simulation forced by different reanalyses: comparison with observed data in the Iberian Peninsula. Appl Energy 2014; 134: 57-64.

[47] Lee ME, Kim G, Jeong ST, Ko DH, Kang KS. Assessment of offshore wind energy at Younggwang in Korea. Renew Sustain Energy Rev 2013; 21: 131-41.

[48] Chadee XT, Clarke RM. Large-scale wind energy potential of the Caribbean region using near-surface reanalysis data. Renew Sustain Energy Rev 2014; 30: 45-58.

[49] Chang R, Zhu R, Badger M, Hasager CB, Xing XH, Jiang YR. Offshore wind resources assessment from multiple satellite data and WRF modeling over South China Sea. Remote Sens 2014; 6: 1-21.

[50] Ban M, Perković L, Duić N, Penedo R. Estimating the spatial distribution of high altitude wind energy potential in Southeast Europe. Energy 2013; 57: 24-9.

[51] NREL. QuikSCAT Annual wind power density at 10m. From<http: //en. openei.org/w/index.php?title= File: QuikSCAT-_Annual_Wind_Power_Den sity_at_10m.pdf&page =1>; 2005.

[52] Atlas R, Hoffman RN, Ardizzone J, Leidner SM, Jusem JC, Smith DK, Gombos D. A cross-calibrated, multiplatform ocean surface wind velocity product for meteorological and oceanographic applications. Bull Am Meteorol Soc 2011; 92: 157-74.

[53] Zheng CW, Pan J. Assessment of the global ocean wind energy resource. Renew Sustain Energy Rev 2014; 33: 382-91.

[54] Wind Energy Resource Atlas of the United States (WERA). the National Renewable Energy Laboratory; 1986. Available from.http: //rredc.nrel.gov/ wind/pubs/atlas.

[55] National Development and Reform Commission. Technical requirements for wind energy resource assessment. Beijing: Chinese Standard Press; 2004 GB/T 18710-2002.

[56] Foley AM, Leahy PG, Marvuglia A, McKeogh EJ. Current methods and advances in forecasting of wind power generation. Renew Energy 2012; 37: 1-8.

[57] Xiao L, Wang JZ, Dong Y, Wu J. Combined forecasting models for wind energy forecasting A case study in China. Renew Sustain Energy Rev 2015; 44: 271-88.

[58] Shi J, Guo JM, Zheng ST. Evaluation of hybrid forecasting approaches for wind speed and power generation time series. Renew Sustain Energy Rev 2012; 16: 3471-80.

[59] Han L, Romero CE, Yao Z. Wind power forecasting based on principle com-ponent phase space reconstruction. Renew Energy 2015; 81: 737-44.

[60] Tascikaraoglu NA, Uzunoglu M. A review of combined approaches for pre-diction of short-term wind speed and power. Renew Sustain Energy Rev 2014; 34: 243-54.

[61] Mederos ACM, Padrón JFM, Lorenzo AEF. An offshore wind atlas for the Canary Islands. Renew Sustain Energy Rev 2011; 15: 612-20.

[62] Lars L. Short-term prediction of the power production from wind farms. J Wind Eng Ind Aerodyn 1999; 90: 207-20.

[63] Bailey B, Brower MC, Zack J. Short-term wind forecasting. In: Proceedings of the European wind energy conference. (ISBN 1902916X) [c]//. Nice, Frace; March 1–5, 1999. p. 1062-65.

[64] Milligan M, Schwartz M, Wan Y. Statistical wind power forecasting for U.S. wind farms. In: Proceedings of the conference WINDPOWER 2003. Austin; May 18-21, 2003.

[65] Focken U, Lange M, Waldl HP. Previento–a wind power prediction system with an innovative upscaling alogrithm. In: Proceedings of the European wind energy conference. Copenhagen, Denmark; June 2–6, 2001. p. 826-29.

[66] Kariniotakis G, Moassafir J, Usaola J. ANEMOS: development of a next gen-eration wind resource forecasting system for the large-scale integration of onshore and offshore wind farms. In: Proceedings of the European wind energy conference and exhibition, EWEC 2003. Madrid. Spain; 2003.

[67] Pinard JP, Benoit R, Yu WA. West wind climate simulation of the mountainous Yukon. Atmos-Ocean 2005; 43: 259-82.

[68] Nielsen TS, Madsen H, Nielsen HA. Short-tem wind power forecasting using advanced statistical models. In: Proceedings of the European wind energy conference; 2006.

[69] Xiao L, Wang JZ, Dong Y, Wu J. Combined forecasting models for wind energy forecasting: a case study in China. Renew Sustain Energy Rev 2015; 44: 271-88.

[70] Tascikaraoglu A, Uzunoglu M. A review of combined approaches for prediction of short-term wind speed and power. Renew Sustain Energy Rev 2014; 34: 243-54.

[71] Zheng CW, Li CY. Variation of the wave energy and significant wave height in the China Sea and adjacent waters. Renew Sustain Energy Rev 2015; 43: 381-7.

[72] Wang XL, Feng Y, Swail VR. Changes in global ocean wave heights as projected using multimodel CMIP5 simulations. Geophys Res Lett 2014; 41: 1026-34. http: //dx.doi.org/10.1002/2013GL058650.

[73] Beccali M, Cellura M, Brano VL, Marvuglia A. Short-term prediction of household electricity consum-ption: assessing weather sensitivity in a Medi-terranean area. Renew Sustain Energy Rev 2008; 12(8): 2040-65.

[74] Carta JA, Velázquez S, Cabrera P. A review of measure-correlate-predict (MCP) methods used to estimate long-term wind characteristics at a target site. Renew Sustain Energy Rev 2013; 27: 362-400.

[75] Wang JZ, Qin SS, Jin SQ, Wu J. Estimation methods review and analysis of offshore extreme wind speeds and wind energy resources. Renew Sustain Energy Rev 2015; 42: 26-42.

[76] Wu J, Wang ZX, Wang GQ. The key technologies and development of offshore wind farm in China. Renew Sustain Energy Rev 2014; 34: 453-62.

[77] Perveen R, Kishor N, Mohanty SR. Off-shore wind farm development: present status and challenges. Renew Sustain Energy Rev 2014; 29: 780-92.

[78] Lee TL. Assessment of the potential of offshore wind energy in Taiwan using fuzzy analytic hierarchy process. Open Civil Eng J 2010; 4: 96-104.

[79] Dong J, Feng TT, Yang YS, Ma Y. Macro-site selection of wind/solar hybrid power station based on ELECTRE-II. Renew Sustain Energy Rev 2014; 35: 194-204.

[80] Pérez-Collazo C, Greaves D, Iglesias G. A review of combined wave and off-shore wind energy. Renew Sustain Energy Rev 2015; 42: 141-53.

[81] Tabassum-Abbasi Premalatha M, Abbasi T, Abbasi SA. Wind energy: increasing deployment, rising environmental concerns. Renew Sustain Energy Rev 2014; 31: 270-88.

[82] Saidur R, Rahim NA, Islam MR, Solangi KH. Environmental impact of wind energy. Renew Sustain Energy Rev 2011; 15: 2423-30.

Numerical Forecasting Experiment of the Wave Energy Resource in the China Sea

Chongwei Zheng[1,2,3], Chongyin Li[1,2], Xuan Chen[2], Jing Pan[1]

[1] LASG, Institute of Atmospheric Physics, Chinese Academy of Sciences, Beijing, China
[2] College of Meteorology and Oceanography, PLA University of Science and Technology, Nanjing, China
[3] PLA Dalian Naval Academy, Dalian, China

Abstract The short-term forecasting of wave energy is important to provide guidance for the electric power operation and power transmission system and to enhance the efficiency of energy capture and conversion. This study produced a numerical forecasting experiment of the China Sea wave energy using WAVEWATCH-III (WW3, the latest version 4.18) wave model driven by T213 (WW3-T213) and T639 (WW3-T639) wind data separately. Then the WW3-T213 and WW3-T639 were verified and compared to build a short-term wave energy forecasting structure suited for the China Sea. Considering the value of wave power density (WPD), "wave energy rose, " daily and weekly total storage and effective storage of wave energy, this study also designed a series of short-term wave energy forecasting productions. Results show that both the WW3-T213 and WW3-T639 exhibit a good skill on the numerical forecasting of the China Sea WPD, while the result of WW3-T639 is much better. Judging from WPD and daily and weekly total storage and effective storage of wave energy, great wave energy caused by cold airs was found. As there are relatively frequent cold airs in winter, early spring, and later autumn in the China Sea and the surrounding waters, abundant wave energy ensues.

1. Introduction

In the current world where human beings are severely plagued by the problems of environment and resources, full exploitation and utilization of new energy resources will effectively alleviate the energy crisis and contribute to global energy-saving, emission reduction, and environmental protection, thus promoting sustainable development. Abundant wave energy has become a particular area of interest for all developed countries [1-4]. Previous studies have made great contributions to the development of wave energy resources [5-9]. The use of observational data and numerical simulation to assess and divide the class of the wave energy

Received January 29, 2016; revised June 4, 2016; accepted June 29, 2016.
Correspondence should be addressed to ChongYin Li; lcy@lasg.iap.ac.cn

resource has provided better references for site selection of wave energy development, for example, wave power plant location. Akpinar and Kömürcü [10] have analyzed the Black Sea wave energy resource based on 15-year hindcast data by using the SWAN (Simulated WAves Nearshore) wave model. Applying the SWAN wave model, Neill and Hashemi [11] made an assessment of the offshore wave energy in Northwestern Europe. The results suggest that wave energy resources in the area have a relatively close relationship with the North Atlantic Oscillation (NAO). In 2009, Roger [12] carried out offshore wave energy forecasting in the East Pacific using the WW3 wave mode. Comprehensively considering the value and stability of the WPD, energy level frequency, and resource reserve, Zheng et al. [13, 14] have analyzed the overall characteristics of the China Sea wave energy resource. Results show that while the China Sea is not located in areas with the most abundant wave energy, they do contain suitable wave energy resources for exploitation, especially in the East China Sea and the northern South China Sea. Zheng and Li [15] have revealed the climatic long-term trends of the China Sea wave power and significant wave height (SWH) for the period 1988-2011 based on 24-year hindcast wave data, which can provide reference for the long-term plan of wave energy resource development.

Most of the previous researches focus on the evaluation of the wave energy climatic characteristics rather than numerical forecasting. Assessment of wave energy resource can provide a reference for site selection of wave energy development (such as wave power plant location). But until now, there is little research on the forecasting of the China Sea wave energy resource. After a power generating device is installed, the wave energy forecasting is especially important to provide routine support to enhance the efficiency of energy capture and conversion. As there are relatively frequent cold airs in winter, early spring, and late autumn in the China Sea and the surrounding waters, abundant wave energy ensues. Accurate forecasting of wave energy will make contribution to the development of wave energy resource.

Previous researchers have made great contribution to the numerical wave model. In the previous researches, the reanalysis wind data is usually used to force the numerical wave model to simulate the past wave field, which overall results in a good simulation outcome attribute to the relative high precision of the reanalysis wind data. However, in the process of numerical forecasting of the future wave energy, the driven field is forecasting wind data. The quality of forecasting wind data is usually not as good as the reanalysis wind data, which may greatly increase the difficulty of wave energy forecasting. In this study, we produced a numerical forecasting experiment of the China Sea wave energy during two cold air processes on 12-17 March 2013, using WW3 wave model driven by T213 (WW3-T213) and T639 (WW3- T639) wind data separately. Then the WW3-T213 and WW3- T639 were verified and compared to build a short-term wave energy forecasting structure suited for the China Sea. We also designed a series of short-term wave energy forecasting productions, including the size of WPD, "wave energy rose, " daily total storage of wave energy, daily effective storage of wave energy, and total storage and effective storage of wave energy. We hope this work can provide reference for the development of wave energy and alleviate the energy crisis, as well as contributing to emissions reduction and environmental protection, and so promote future sustainable development.

2. Methodology and Data

2.1 Numerical Simulation Method of Wave Energy

This study aims to build a numerical forecasting structure of wave energy suited for the China Sea. Two strong cold air processes for the period 8 to 16 March 2013 were selected to produce a numerical forecasting experiment. Firstly, the WW3 (the latest version 4.18) wave model was driven by T213 (WW3- T213) and T639 (WW3-T639) wind data separately. Then the WW3-T213 and WW3-T639 were verified and compared. Considering the value of wave power density (WPD), "wave energy rose, " and daily and weekly total storage and effective storage of wave energy, this study also designed a series of short-term wave energy forecasting productions.

The related calculation data are as follows: in order to eliminate the boundary effect to improve the precision of simulation wave data, we set the simulating area as10°S~55°N, 95°S~160°E and then extract concerned area as 0°~41°N, 100°~ 135°E which contains the main part of the China Sea and surrounding waters; the topography is shown in Figure 1. The value of 0.1°× 0.1°was taken as the spatial resolution. The duration of 300 s was taken as the time step of calculation. The results were outputted each hour. The computing time was from 00: 00 8 March 2013 to 2100 18 March 2013.

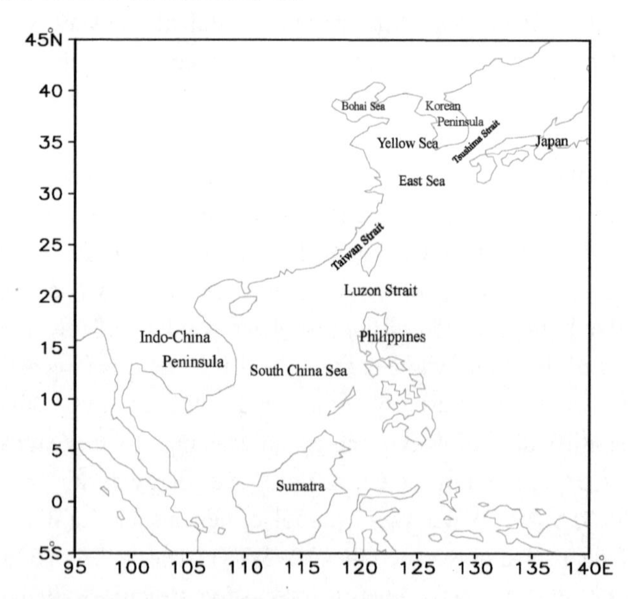

Figure 1. Topography of the China Sea and surrounding waters.

2.2 Wave Power Density Calculation Method

Wave power density calculation method is as follows [16-18].

In shallow water, calculation method is as follows:

$$P_w = \frac{\rho g}{16} H_s^2 \sqrt{gd}. \tag{1}$$

In deep water, calculation method is as follows:

$$P_w = \frac{\rho g^2}{64\pi} H_s^2 T_e = 0.49 H_s^2 T_e, \tag{2}$$

where P_w is wave power density (unit: kW/m), H_s is the significant wave height (unit: m), T_e is the energy period (unit: s), and d is the water depth.

2.3 Wind Field Data and Topographic Data

Topographic data input of the wave model is obtained from the ETOPO1 high resolution data set of global ocean of NOAA (1′ * 1′), and the coastline data was obtained from the GSHHS global high resolution coastline database.

The T213 and T639 forecasting wind data were used as the driving field of the WW3 wave model separately. The T213 model was introduced from ECMWF in 1997. The T213 model has 31 vertical resolution layers, with top of 10 hpa. The spatial resolution is 0.5625°×0.5625°. T639 is the abbreviation of the global medium-term numerical weather prediction model improved from T213. Compared to the T213 product, the T639 data has the advantages of factor abundance and high temporal-spatial resolution [19]. The T639 mode has 60 vertical resolution layers. The temporal resolution is 3 hours. The spatial resolution is 0.3°×0.3°. The spatial scope covers the entire Northern Hemisphere. The mode was put into operation in 2008 by the National Meteorological Center of China. In addition, Ma et al. [20] have also found that the T639 model can depict the tropical cyclone much better than the T213 model.

2.4 Observed Wave Data

Currently, the worldwide observed wave data is relatively scarce, especially in China. There are no NDBC buoys in the China Sea. Although the accuracy of the significant wave height from the satellite is widely recognized, the number of satellite orbits over the China Sea is small and the repetition period is long (e.g., the *T/P* altimeter cycle is 10 d). This results in many drawbacks that the wave data from satellite is short in spatial resolution and temporal resolution [13, 14]. In light of the shortage of observed wave data, we extensively collected buoy data from the South Korea, Japan, and Chinese Taiwan Region to verify the effectiveness of the forecasting data.

3. Validation of the Forecasting Wave Power Density

Comparing the curves of the forecasting WPD and the observed WPD, it is possible to visually make out the accuracy of the forecasting data, as shown in Figures 2-4. During the two processes of cold air invasion, in the waters surrounding the Korean Peninsula, Japan, and Taiwan Island, the forecasting and observed WPD exhibit a good consistency. And the observed value is slightly larger than the forecasting value. The observed WPD curve shows noticeable jumpy phenomenon while the forecasting WPD curve appears slightly smooth.

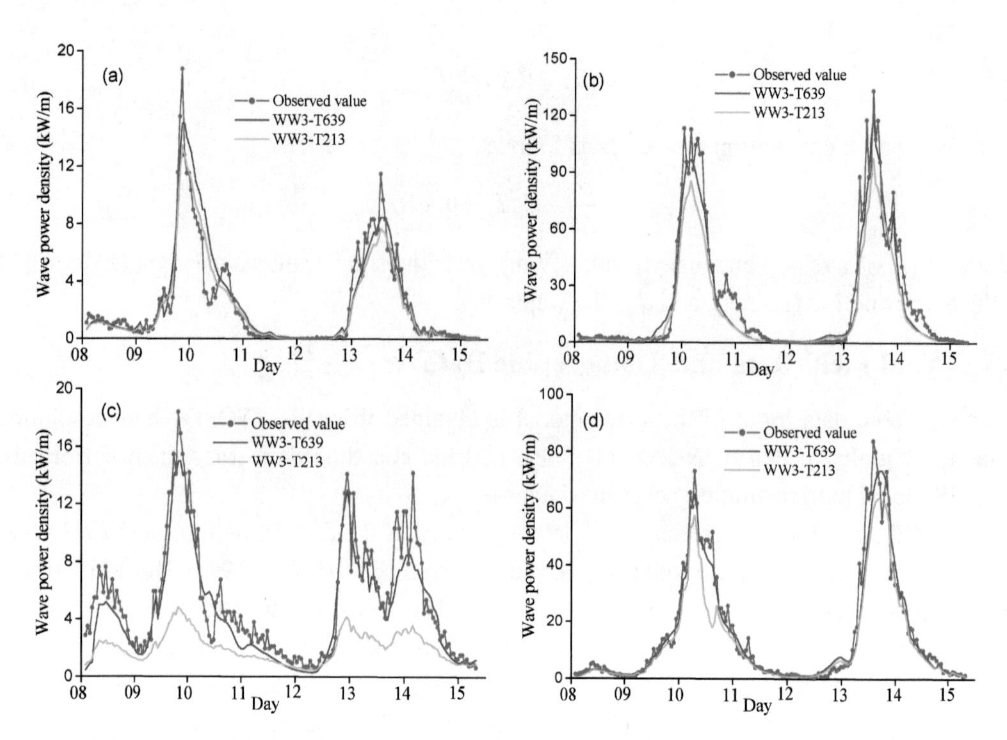

Figure 2. Observed and forecasting wave power density in March 2013 around the Korean Peninsula. (a)-(d): denoting stations 22101, 22102, 22103, and 22017.

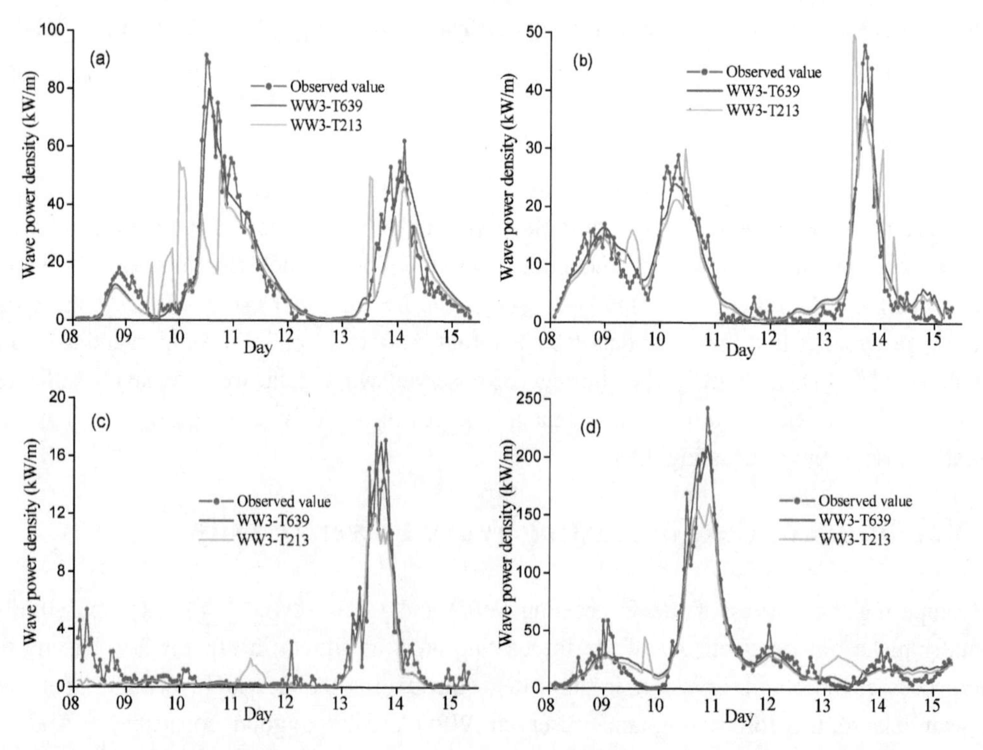

Figure 3. Observed and forecasting wave power density in March 2013 around Japan. (a)-(d): denoting stations Kyogamisaki, Irozaki, Karakuwa, and Kaminokuni.

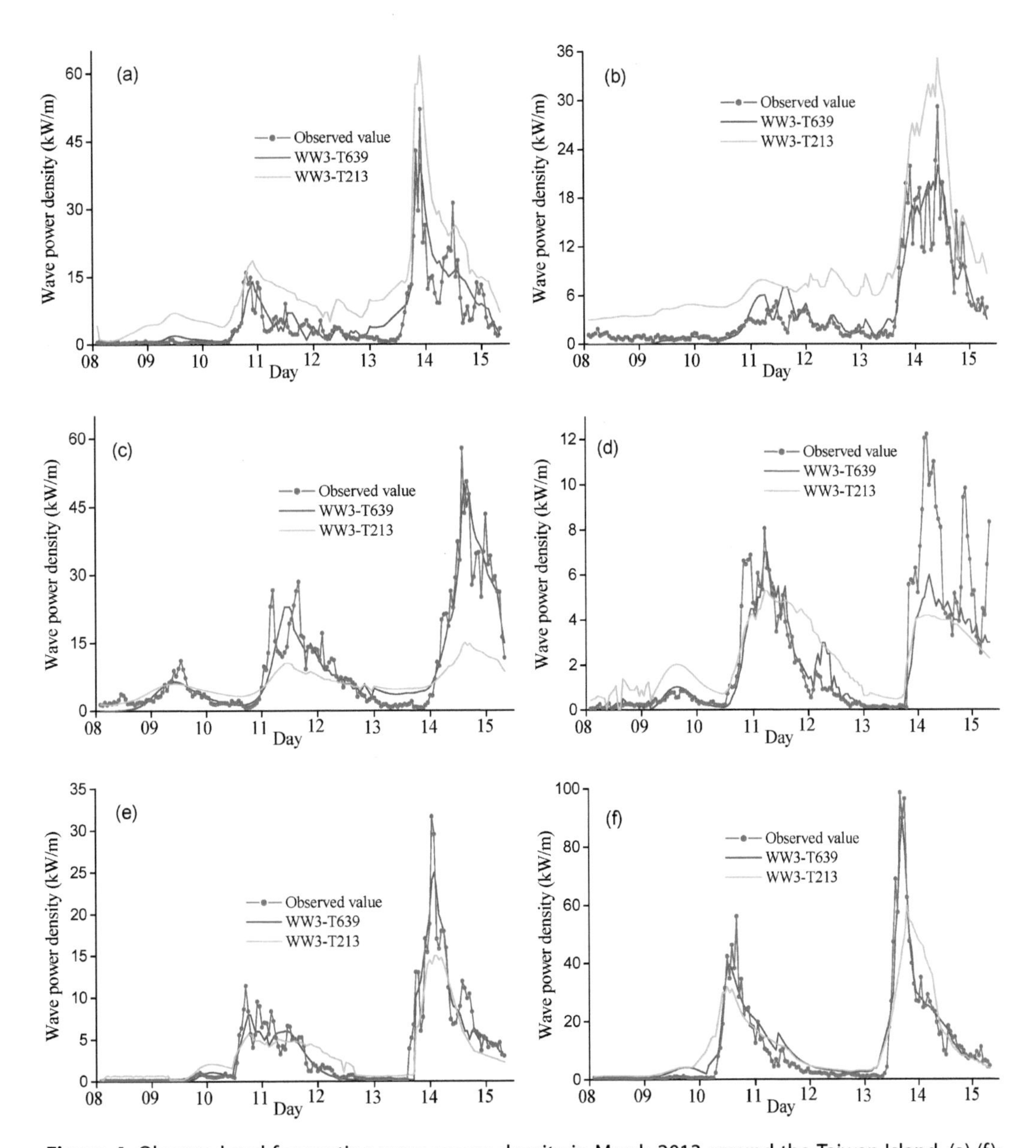

Figure 4. Observed and forecasting wave power density in March 2013 around the Taiwan Island. (a)-(f): denoting stations Longdong, Hualian, Dongsha, Qigu, Jinmen, and Mazu.

In the waters surrounding the Korean Peninsula (Figure 2), judging from the trend of the curves of the forecasting and observed values, they are in better agreement with each other. The WPD of WW3-T639 is closer to the observed WPD at each station, while the WPD of WW3-T213 is smaller than the WPD of WW3-T639, especially at station 22103 during the second cold air process. The forecasting values are able to effectively demonstrate the course of the increase in the WPD caused by two processes of cold air invasion, especially that for the first process of cold air invasion (10 March 2013). During the cold air invasion period, the forecasting WPD is slightly smaller than the observed data, suggesting that the measured

WPD is richer than the forecasting data. Obviously, the cold air process can bring very rich wave energy to the waters of stations 22102 and 22107, of up to 120 kW/m, while the wave energy caused by the cold air at stations 22101 and 22103 is smaller, which can be attributed to the effect of geography characteristic. In the development of wave energy, WPD greater than 2 kW/m is regarded as available and greater than 20 kW/m is regarded as rich. It is clear that the growth of wave energy caused by the cold air process is promising.

In the waters surrounding Japan (Figure 3), the trend of the curves of the forecasting values and observed values also retained relatively good consistency. There is a good agreement between the observed WPD and WPD of WW3-T639. The WPD of WW3-T213 has obvious jumping phenomenon, which results in a larger error. The WPD increased to 80kW/m at station Kyogamisaki, to 40kW/m at stations Irozaki, and even up to about 250kW/m at station Kaminokuni. The cold airs do not affect station Karakuwa significantly, especially the period of the first cold air invasion. This should be due to the effect of topography.

There is an excellent agreement between the forecasting and observed WPD at all stations in the waters surrounding Taiwan Island, except Qigu station, as shown in Figure 4. Obviously, the relationship between observed WPD and WPD of WW3-T639 is much better than that between observed WPD and WPD of WW3-T213. The noticeable error of forecasting value may be due to the block on the wave field of Taiwan Island. In the waters surrounding the islands like stations Longdong, Dongsha, and Mazu, the WPD could increase to 60-80kW/m under the influence of cold air; particularlyin station Mazu, the WPD can be up to 100 kW/m during the period of the second cold air invasion. In the regions surrounding stations Hualian and Jinmen, the WPD could increase to 30-40kW/m during the cold air invasion period.

In order to analyze the accuracy of the forecasting WPD quantitatively, we also calculated the correlation coefficient (CC), Bias error, root mean square error (RMSE), and mean absolute error (MAE), as shown in Table 1.

From the correlation coefficient (CC), the forecasting and observed WPD have good correlation (significant at the 99% level) regardless of the sea areas surrounding the Korean Peninsula, Japan, or Chinese Taiwan Region. The CC between observed WPD and WPD of WW3-T639 is above 0.90 at each station, while the CC between observed WPD and WPD of WW3-T213 is obviously smaller. It is also worth noting that the forecasting results of the WW3 mode in the sea areas surrounding the Korean Peninsula and Japan are relatively better than that in the sea area surrounding Taiwan Island. About the Bias, RMSE, and MAE, the errors are relatively small at most stations. Only the errors in the sea area at stations 22102, 22107, Kyogamisaki, Kaminokuni, and Mazu are relatively large but within the acceptable range, because the wave power densities at these 5 stations are remarkably larger than that at other stations (as shown in Figures 2-4). From the Bias, the forecasting WPD is overall smaller than the observed values, meaning that the actual wave energy is more optimistic than the forecasting condition. As a whole, the forecasting WPD is more accurate, suggesting that the WW3 model can appropriately simulate the China Sea WPD under the influence of two processes of cold air invasion. Previous researches also proved a good skill of WW3 wave model on the numerical simulation of the China Sea wave field [13, 14, 21-23]. The results of this paper provide confirmatory evidence that the WW3 wave model can accurately forecast the China Sea WPD very well.

Table 1: Precision of the forecasting wave power density.

Site number		Longitude /°E	Latitude /°N	CC		Bias		RMSE		MAE	
				WW3-T639	WW3-T213	WW3-T639	WW3-T213	WW3-T639	WW3-T213	WW3-T639	WW3-T213
Waters surrounding the Korean Peninsula	22101	126.0	37.3	0.97	0.94	−0.14	−0.26	1.91	1.27	0.98	0.74
	22102	125.8	34.7	0.97	0.91	−3.13	−6.48	8.21	12.49	5.15	7.33
	22103	127.0	34.2	0.92	0.88	−0.73	−3.23	1.65	4.40	1.17	3.23
	22107	126.0	33.1	0.98	0.96	−0.87	−3.16	4.44	6.94	2.53	3.74
Waters surrounding Japan	Kyogamisaki	135.3	25.7	0.94	0.65	−0.39	−2.10	6.76	15.59	4.60	8.29
	Irozaki	138.9	34.6	0.96	0.87	−0.20	0.36	2.74	4.87	2.16	2.89
	Karakuwa	141.6	38.4	0.94	0.93	−0.45	−0.85	1.29	1.45	0.76	0.83
	Kaminokuni	140.1	41.8	0.97	0.95	−1.47	−2.24	11.00	17.45	6.64	11.37
Waters surrounding Taiwan Island	Longdong	121.9	25.1	0.92	0.89	−0.10	6.80	2.83	8.68	1.88	6.87
	Hualian	121.6	24.0	0.90	0.92	−0.08	4.91	2.13	5.84	1.31	4.96
	Dongsha	118.8	21.0	0.93	0.91	−0.19	−3.44	3.93	9.40	2.57	5.64
	Qigu	120.0	23.1	0.91	0.74	−0.15	−0.29	2.11	2.11	1.20	1.47
	Jinmen	118.4	24.4	0.93	0.89	−0.25	−0.42	1.80	2.71	0.97	1.67
	Mazu	120.5	26.4	0.94	0.83	−1.03	1.07	4.72	9.94	2.96	5.51

4. Forecasting of Wave Energy

4.1 Spatial-Temporal Distribution of the Forecasting Wave Power Density

Based on the simulation wave data and calculation method of WPD, the hourly China Sea WPD from 00: 00 on 8 March 2013 to 21: 00 on 18 March 2013 is obtained. The spatial-temporal distribution characteristics of the forecasting WPD field in the period of cold air invasion are presented in Figure 5. At 12: 00 on 12 March, the cold air moved into the Bohai Sea, bringing about a WPD of over 40 kW/m in most of the Bohai Sea. In the center with high values, the density can even be up to 80 kW/m. At 12: 00 on 13 March, the cold air moved into the mid-south area of the Yellow Sea, and the WPD induced is from 50 to 120 kW/m. At 12: 00 on 14 March, when the cold air moved into the Ryukyu Islands waters, the waters became opener; a wide range of significant waves grew fully, resulting in the enlargement of the range of WPD greater than 40 kW/m. At 12: 00 on 15 March, cold air moved into the Taiwan Island waters. The scale of WPD greater than 40 continues to broaden. At 12: 00 on 16 March, cold air moved into the north area of the South China Sea, bringing about a cold surge. However, because the latitude of the South China Sea is lower, the strength of the cold air was greatly weakened after the air entered therein so that the scope with the WPD above 40 kW/m was apparently shrunken. At 12: 00 on 17 March, the cold air moving into the mid-south area of the South China Sea generated apparent cold surge and increasing process of the WPD, but the intensity is not as strong as that in the East China Sea and the northern South China Sea.

It is usually assumed that areas with WPD greater than 20 kW/m can be classified as energy-rich regions, such as the North Sea in Europe. From Figure 5, it is very clear that the cold air can result in a WPD of >20 kW/m in large scale area. Even more, as there are relatively frequent cold airs in winter, early spring, and late autumn in the China Sea and the surrounding waters, abundant wave energy ensues.

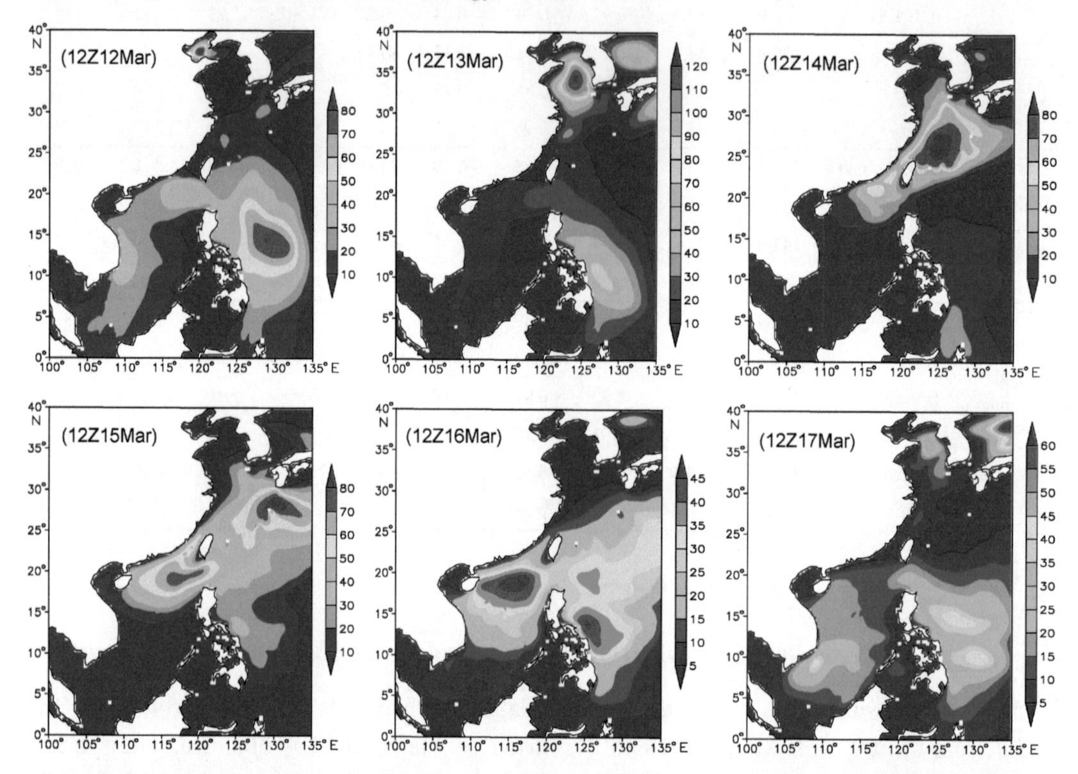

Figure 5. Wave power density in the China Sea during the cold air process in March 2013, unit: kW/m.

4.2 Forecasting of Wave Energy Storage

Calculating the storage of energy can provide decision-making for the electric power operation and power transmission system. In this present study, the per-unit-area total storage, effective storage, and technological storage of the China Sea wave energy at each 0.1°×0.1° bin were obtained. We mainly calculated the daily total storage of wave energy (Figure 6), daily effective storage of wave energy (Figure 7), and daily exploitable storage of wave energy (figure omitted) from 12 to 17 March 2013. The total storage and effective storage of wave energy in the next week start from 12 March 2013 were also exhibited, as shown in Figure 8.

The energy storage calculation method is as follows:

$$E_{PT} = \overline{P} * H, \tag{3}$$

$$E_{PE} = \overline{P} * H_E, \tag{4}$$

$$E_{PD} = E_{PE} * C_e, \tag{5}$$

where E_{PT} is the daily total storage of wave energy, $\overline{P}$ is the daily average WPD, and H = 24h in one day. In (4), E_{PE} is the daily effective storage of wave energy and H_E is the hours of effective significant wave height in one day. In (5), E_{PE} is the daily technological development volume of wind energy resources, showing some discrepancy along with the difference of installation. C_e is the absorption ability of installation.

As shown in Figure 6, the cold air brings wave energy storage of 400-800 kWh/m on 12 March in the Bohai Sea. On 13 March, the cold air moved into the Yellow Sea, resulting in a rapid increase in the wave energy storage, of >800 kWh/m in most areas of the Yellow Sea and even greater than 1800kWh/m in the large center. On 14 March, when the cold air moves into the East Sea, with the broader of waters, a wide range of significant waves grew fully. The scales of wave energy storage >800 kWh/m and >1800 kWh/m enlarge significantly. On 15 March, with the cold air moving into the Taiwan Island waters, the scale of wave energy storage >800 kWh/m continues to broaden. On 16 March, the cold air moved into the north area of the South China Sea, where the wave energy storage is 600-1200kWh/m. On 17 March, the cold air moving into the mid-south area of the South China Sea generated wave energy storage of 400-600 kWh/m, for the intensity is not as strong as that in the East China Sea and the northern South China Sea.

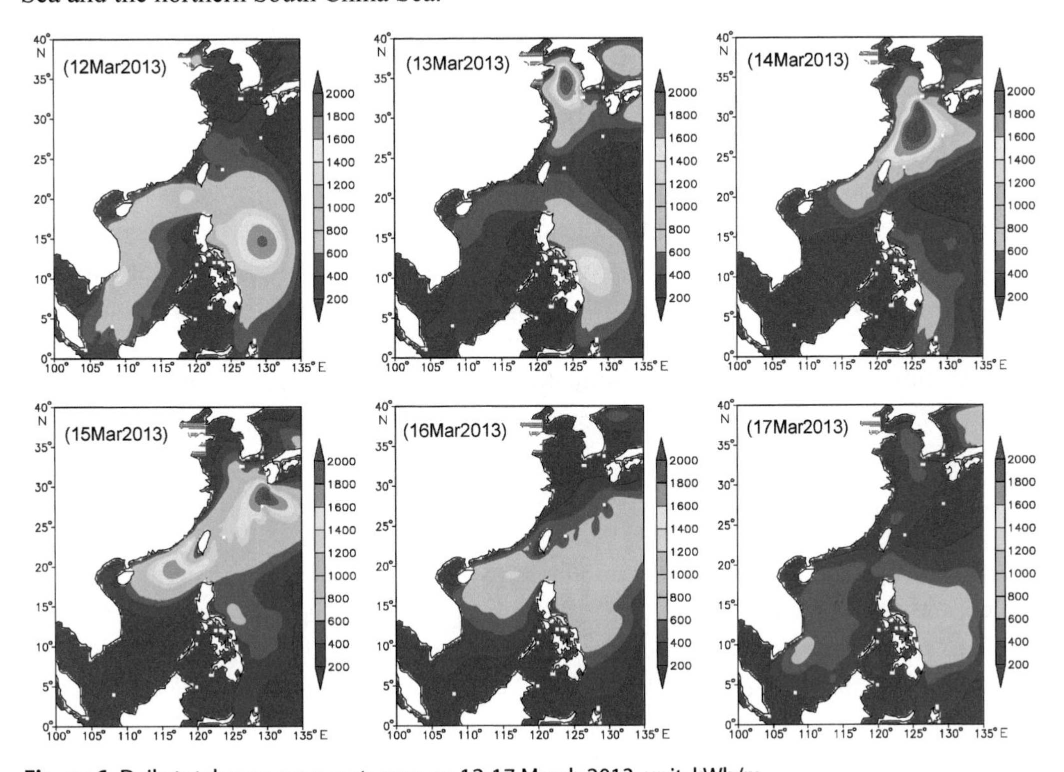

Figure 6. Daily total wave energy storage on 12-17 March 2013, unit: kWh/m.

From Figure 7, we find that there is a good agreement between the daily effective storage of wave energy and daily total storage of wave energy daily (Figure 6). This should be due to the high occurrence of the significant wave height >1.0 m.

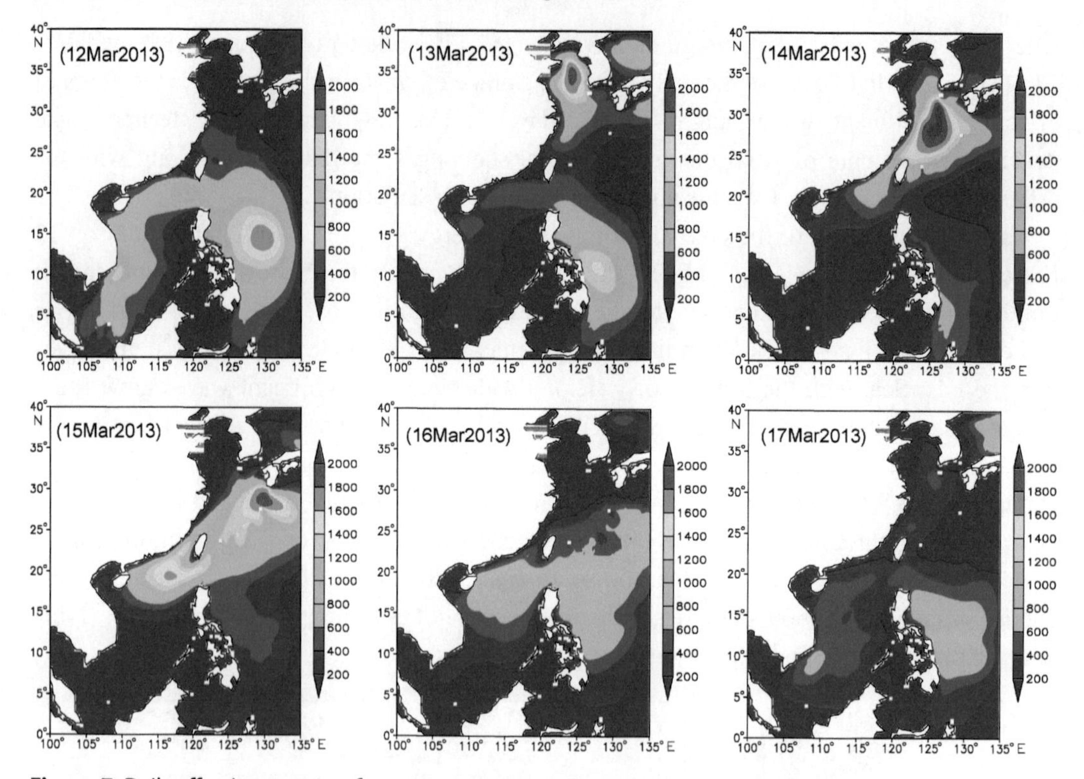

Figure 7. Daily effective storage of wave energy on 12-17 March 2013, unit: kWh/m.

The total storage and effective storage of wave energy in the next week start from 12 March 2013 were presented in Figure 8. The total storage of wave energy is promising during this cold air process, of >2000 in large scale of the China Sea. The large areas are mainly distributed in the middle of the Yellow Sea, Ryukyu Islands waters, north of the South China Sea, and east nearshore of Philippine, of above 4000 kWh/m and even greater than 6000 kWh/m in the large center. The distribution characteristic of effective storage is similar to that of total storage, due to the high occurrence of the significant wave height >1.0 m during the cold air process.

4.3 Station Forecasting

In the development of wave energy resource, we usually pay more attention to the energy condition at some certain sites. For the sake of providing reference for the electric power operation, we produced station forecasting of wave energy. Here, we select two stations at will as the study object: station A (30°N, 130°E, at the Tsushima Strait) and station B (18°N, 115°E, near the Dongsha Islands).

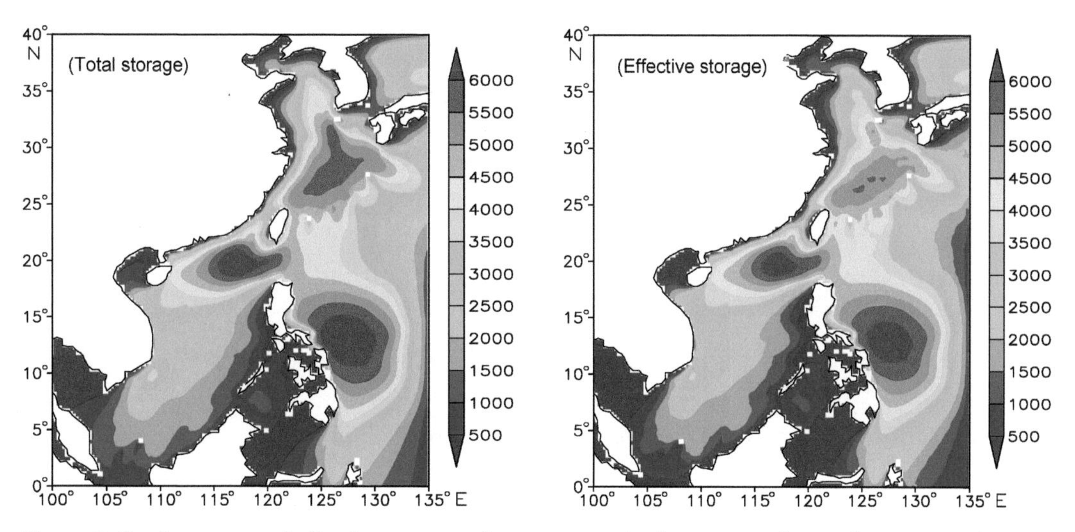

Figure 8. Total storage and effective storage of wave energy in the next week start from 12 March 2013, unit: kWh/m.

We present the 3-hourly WPD at the two stations, as shown in Figure 9. During this cold air process, the peak value of WPD at station A is higher than that at station B. But the stability at station B is better.

The wave direction and grade of wave energy are the key factors in the exploitation of wave energy. We perform the first attempt to exhibit the "wave energy rose" during this cold air process on 12-17 March 2013, from which we can easily find the main contribution of wave direction and WPD to the wave energy, as shown in Figure 10. It is very clear that the wave energy at station A is mainly contributed by the NW direction wave; particularly the occurrences of WPD 40-50kW/m and 50-60kW/m are high. At station B, the wave energy is mainly contributed by NE and ENE direction wave. It should be noticed that, during the cold air process, the wave direction exhibits an arc shape along the coastal line.

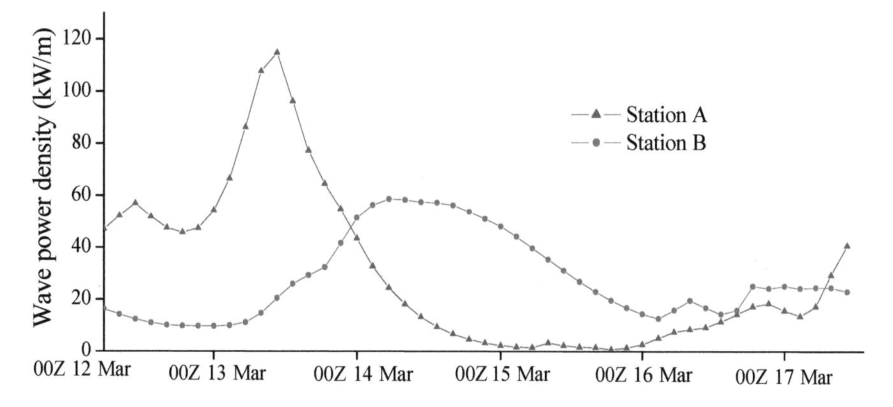

Figure 9: Wave power density at station A and station B on 12-17 March 2013.

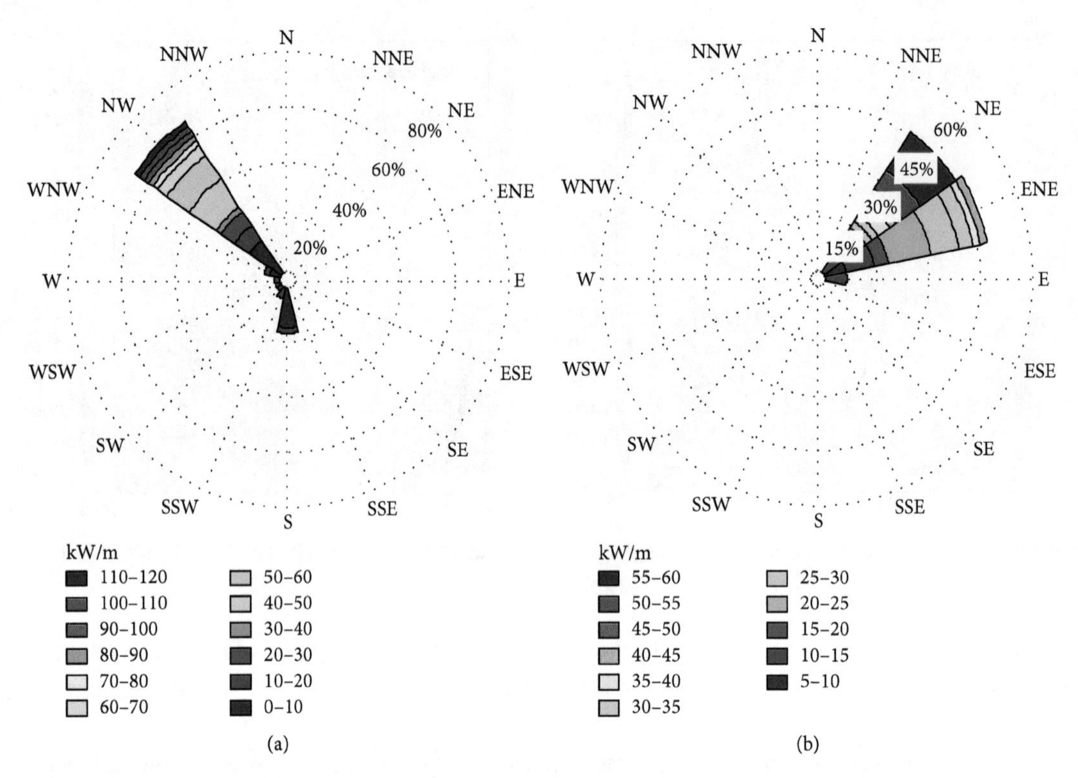

Figure 10: Wave energy rose at station A (a) and station B (b) on 12-17 March 2013.

5. Conclusions

This study produced a numerical forecasting experiment of the China Sea wave energy during two cold air processes on 12-17 March 2013, using WW3 wave model driven by T213 (WW3-T213) and T639 (WW3-T639) wind data separately. Then the WW3-T213 and WW3-T639 were verified and compared to build a short-term wave energy forecasting structure suited for the China Sea. Comprehensively considering the WPD, "wave energy rose, " daily total storage of wave energy, daily effective storage of wave energy, and total storage and effective storage of wave energy in the next week, we also designed a series of short-term wave energy forecasting productions. The results show the following:

(1) Forced by the T213 and T639 forecasting wind fields, the WW3 wave model exhibits a good skill on the numerical forecasting of the China Sea WPD, while the result of WW3-T639 is much better. During the cold air process, the forecasting WPD has a good agreement with the observed WPD, except stations affected by the topography significantly, such as Qigu station in the southwest of Taiwan Island. The simulation ability of WW3 mode in the sea areas surrounding the Korean Peninsula and Japan is relatively better than that in the sea area surrounding Taiwan Island. The forecasting WPD is overall slightly smaller than the observed values, meaning that the actual wave energy is more promising than the forecasting condition.

(2) The cold air can bring a WPD of >20 kW/m (which is regarded as rich energy) in large scale area of the China Sea. Even more, as there are relatively frequent cold airs in winter, early spring, and late autumn in the China Sea and the surrounding waters, abundant wave energy ensues, meaning abundant wave energy resources in the China Sea.

(3) Judging from the daily total storage of wave energy, daily effective storage of wave energy, and the total storage and effective storage of wave energy in the next week start from one cold air process, abundant wave energy was found in the China Sea during the cold air process.

(4) From the "wave energy rose, " it is clear that the wave energy at station A (at the Tsushima Strait) is mainly contributed by the NW direction wave; particularly the occurrences of WPD 40-50kW/m and 50-60 kW/m are high. At station B (near the Dongsha Islands), the wave energy is mainly contributed by NE and ENE direction wave.

Nomenclature

CC: Correlation coefficient
CCMP: Cross-Calibrated Multiplatform
MAE: Mean absolute error
NAO: North Atlantic Oscillation
RMSE: Root mean square error
SWAN: Simulated WAves Nearshore
T/P: TOPEX/Poseidon
WW3: WAVEWATCH-III.

Competing Interests

The authors declare that they have no competing interests.

Acknowledgments

This work was supported by the National Science Foundation for Young Scientists of China (Evaluation and Prediction of Wave Energy Resource in the South China Sea and North Indian Ocean), the National Key Basic Research Development Program, astronomy and earth factor on the impact of climate change (Grant no. 2012CB957803), and Asian regional sea air interaction mechanism and its role in global change (Grant no. 2010CB950400).

References

[1] A. Rashid and S. Hasanzadeh, "Status and potentials of offshore wave energy resources in Chahbahar area (NW Omman Sea), " *Renewable and Sustainable Energy Reviews*, vol. 15, no. 9, pp. 4876-4883, 2011.

[2] F. Chen, S.-M. Lu, K.-T. Tseng, S.-C. Lee, and E. Wang, "Assessment of renewable energy reserves in Taiwan, " *Renewable and Sustainable Energy Reviews*, vol. 14, no. 9, pp. 2511-2528, 2010.

[3] L. Hammar, J. Ehnberg, A. Mavume, B. C. Cuamba, and S. Molander, "Renewable ocean energy in the Western Indian Ocean, " *Renewable and Sustainable Energy Reviews*, vol. 16, no. 7, pp. 4938-4950, 2012.

[4] K. Gunn and C. Stock-Williams, "Quantifying the global wave power resource, " *Renewable Energy*, vol. 44, pp. 296-304, 2012.

[5] R. Tornkvist, "Ocean wave power station, " Report 28, Swedish Technical Scientific Academy, Helsinki, Finland, 1975.

[6] K. Hulls, "Wave power, " *The New Zealand Energy Journal*, vol. 50, pp. 44-48, 1977.

[7] P. K. Robert and R. Mitchell, "Environmental implications of wave energy proposals for the outer hebrides and Moray Firth, " *Ocean Engineering*, vol. 10, no. 6, pp. 459-469, 1983.

[8] M. T. Pontes, R. Aguiar, and H. O. Pires, "A nearshore wave energy atlas for Portugal, " *Journal of Offshore Mechanics and Arctic Engineering*, vol. 127, no. 3, pp. 249-255, 2005.

[9] M. Folley, T. J. T. Whittaker, and A. Henry, "The effect of water depth on the performance of a small surging wave energy converter, " *Ocean Engineering*, vol. 34, no. 8-9, pp. 1265-1274, 2007.

[10] A. Akpinar and M. I. H. Kömürcü, "Assessment of wave energy resource of the Black Sea based on 15-year numerical hindcast data, " *Applied Energy*, vol. 101, pp. 502-512, 2013.

[11] S. P. Neill and M. R. Hashemi, "Wave power variability over the northwest European shelf seas, " *Applied Energy*, vol. 106, pp. 3146, 2013.

[12] B. Roger, "Wave energy forecasting accuracy as a function of forecast time horizon, " Tech. Rep. EPRI-WP-013, 2009, http: //www.doc88.com/p-5334126229545.html.

[13] C.-W. Zheng, J. Pan, and J.-X. Li, "Assessing the China Sea wind energy and wave energy resources from 1988 to 2009, " *Ocean Engineering*, vol. 65, pp. 39-48, 2013.

[14] C. W Zheng, G. Lin, and L. T. Shao, "Frequency of rough sea and its long-term trend analysis in the China Sea from 1988 to 2010, " *Journal of Xiamen University*, vol. 52, no. 3, pp. 395-399, 2013 (Chinese).

[15] C. W. Zheng and C. Y. Li, "Variation of the wave energy and significant wave height in the China Sea and adjacent waters, " *Renewable and Sustainable Energy Reviews*, vol. 43, pp. 381-387, 2015.

[16] G. Iglesias and R. Carballo, "Choosing the site for the first wave farm in a region: a case study in the Galician Southwest (Spain), " *Energy*, vol. 36, no. 9, pp. 5525-5531, 2011.

[17] A. M. Cornett, "A global wave energy resource assessment, " in *Proceedings of the 18th International Offshore and Polar Engineering Conference (ISOPE '08)*, pp. 318-326, Vancouver, Canada, July 2008.

[18] A. Vosough, "Wave power" *International Journal of Multidisciplinary Sciences and Engineering*, vol. 2, no. 7, pp. 60-63, 2011.

[19] D. Xiao, L. T. Deng, J. Chen, and J. K. Hu, "Tentative verification and comparison of WRF forecasts driven bydata from T213 and T639 models, " *Torrential Rain and Disasters*, vol. 29, no. 1, pp. 20-29, 2010.

[20] S. H. Ma, Y. Wu, A. X. Qu, T. G. Xiao, and X. Li, "Comparative analysis on tropical cyclone numerical forecast errors of T213 and T639 models, " *Journal of Applied Meteorological Science*, vol. 23, no. 2, pp. 167-173, 2012.

[21] P. C. Chu, Y. Qi, Y. Chen, P. Shi, and Q. Mao, "South China Sea wind-wave characteristics. Part 1: validation of wavematch-III using TOPEX/Poseidon data, " *Journal of Atmospheric and Oceanic Technology*, vol. 21, no. 11, pp. 1718-1733, 2004.

[22] M. K. Li, Y. J. Hou, B. S. Yin, J. B. Song, and W. Zhao, "Numerical simulation of scatterometer assimilated wind and ocean wave in eastern China seas and adjacent waters, " *Chinese Journal of*

Oceanology and Limnology, vol. 24, no. 1, pp. 42-47, 2006.

[23] C. W. Zheng, J. Pan, and G. Huang, "Forecasting of the China Sea ditching probability using WW3 wave model, " *Journal of Beijing University of Aeronautics and Astronautics*, vol. 40, no. 3, pp. 314-320, 2014 (Chinese).

An Overview of Medium- to Long-Term Predictions of Global Wave Energy Resources

Chongwei Zheng[a, b, c, *], Qing Wang[a], Chongyin Li[b]

[a] Ludong University, Yantai, China
[b] College of Meteorology and Oceanography, PLA University of Science and Technology, Nanjing, China
[c] PLA Dalian Naval Academy, Dalian, China

Abstract Against a backdrop of increasing energy demand, the development of wave energy technology is a logical means of both meeting this demand and mitigating the environmental degradation associated with conventional power generation. Previous research has made considerable progress in the climatic characterization and short-term forecasting of wave energy. However, medium- to long-term predictions of wave energy resources, which are central to the development of future operating and trading strategies, remain scarce. This study provides an overview of long-term climatic trends and medium- to long-term predictions of wave energy, before discussing the focus of future predictions. Finally, a new method is proposed for predicting wave energy resources on a medium- to long-term basis that incorporates the swell index and propagation characteristics of swell energy. This model was developed with the aim of improving the precision of wave energy predictions, thereby providing a reference for the effective utilization of wave resources. The results of this study demonstrate that long-term climatic trend analysis should include not only variations in wave power density (WPD), but also long-term variability in wave energy stability, energy level occurrence, and variability in the occurrence of effective significant wave height (SWH). The medium- to long-term prediction of wave energy should also synthetically consider the above factors. We conclude that monitoring the propagation of swell energy and calculating the swell index constitutes a robust theoretical basis for predicting the WPD of mixed wave.
Keywords Wave energy; Climatic long-term trend, Medium- to long-term prediction, Wave power density, Wave energy stability, Energy level occurrence.

---

1. Introduction

In light of the growing demand for energy and the environmental impact of conventional power generation, the development of renewable energy sources will play a vital role in the

Received March 15, 2016; revised February 14, 2017; accepted May 18, 2017.
* Corresponding author at: E-mail address: chinaoceanzcw@sina.cn (C.W. Zheng).

future sustainability of human society. The advantages of wave energy are numerous and include its broad geographic viability, large storage capacity, and lack of pollution, as well as the conservation of terrestrial resources and needing no immigration, all of which make this resource an attractive option for developed countries [1-4]. Wave power generation is the principal stimulus for the development and utilization of wave-energy resources, which are also used in desalination, hydrogen production, pumping, and heating processes, as well as to provide energy for marine aquaculture, offshore weather buoys, oil platforms, offshore lighthouses, and islets. However, before wave energy can be harnessed for power production, the characteristics of this energy source must be fully evaluated, as does its economic and technical feasibility [5-10]. For example, instability, which is defined as clear seasonal and regional differences in wave energy, is a complicating factor in energy development and must be accounted for prior to extraction.

To date, numerous studies have assessed the characteristics and feasibility of wave power generation, with most adhering to the following procedure: (1) evaluation of wave energy resources based on limited observational data; (2) incorporation of satellite-based observational data into those evaluations; (3) numerical simulation of wave energy resources; and (4) evaluations based on reanalysis data. Each stage is discussed in detail below.

In 1986, Denis [11] presented a global distribution of coastal wave power density (WPD) based on limited observational data. Over time, however, the rapid development of ocean measuring methods means that satellite data are increasingly being used to evaluate wave energy potential. For instance, Barstow et al. [12] used significant wave height (SWH) data, derived from two-year TOPEX/Poseidon (T/P) satellite altimetry, to calculate WPD values for hundreds of sites and, ultimately, establish the global distribution of offshore WPD. Similarly, Wan et al. [13] presented an analysis of wave energy in the China seas based on merged multi-satellite radar altimeter data. With the rapid growth in computing power, hindcast wave datasets such as Wave Model (WAM), Simulating WAves Nearshore (SWAN), and WAVEWATCH-III (WW3) are increasingly being used to evaluate wave energy resources [14-19]. Using the NWW3 wave model, for example, Cornett [20] simulated global ocean WPD for the period 1997-2006, whereas Folley and Whittaker [21] and Iglesias and Carballo [22] used third-generation wave model data to analyze changes in nearshore wave energy off Scotland and Spain, respectively. Similarly, Gunn and Stock-Williams [23] used the WW3 wave model to simulate global WPD and estimated global wave power resources to be 2.11±0.05 TW. Numerical simulation is an effective means of interpolating missing values, particularly when dealing with extended gaps in the wave buoy record, and data reanalysis is a valuable tool for assessing the distribution of global wave energy. Using the 45-year (September 1957 to August 2002) European Centre for Medium-Range Weather Forecasts (ECMWF) Reanalysis (ERA-40) wave reanalysis, Zheng et al. [24] modeled WPD, wave energy level occurrence, effective SWH occurrence, stability of and long-term trends in WPD, swell index, and wave energy storage to analyze and regionalize global wave energy resources. In addition, Arinaga and Cheung [25] presented an atlas of global wave energy based on a decade of reanalysis and WW3 hindcast wave data. The atlas reveals that, above 30°N, monthly median WPD from wind-driven waves is 17-130 kW/m, whereas below this latitude WPD is more consistent (50-100 kW/m).

Nomenclature

ANFIS	an enhanced type of Takagi-Sugeno-based fuzzy inference system
ANN	artificial neural network
AO	Arctic Oscillation
ARIMA	Autoregressive Integrated Moving Average
CC	correlation coefficient
CMIP5	Coupled Model Intercomparison Project Phase 5
Cv	coefficient of variation
DJF	December-January-February
ECMWF	European Centre for Medium-Range Weather Forecasts
EEMD	Ensemble Empirical Mode Decomposition
EP-NP	East Pacific-North Pacific
ERA-40	ECMWF Reanalysis
GGA-ELM	grouping genetic algorithm-Extreme Learning Machine
GOW	Global Ocean Waves
H^s	swell wave height
ICOADS	International Comprehensive Ocean-Atmosphere Data Set
JJA	June-July-August
MAE	mean absolute error
MCDM	multi-criteria decision making
MCP	Measure-Correlate-Predict
Mv	monthly variability index
NAO	North Atlantic Oscillation
PDO	Pacific Decadal Oscillation
PNA	Pacific-North America
RMSE	root mean square error
SCS	South China Sea
Sv	seasonal variability index
SVM	Support Vector Machine
SWAN	Simulating WAves Nearshore
SWH	significant wave height
T/P	TOPEX/Poseidon
WAM	Wave Model
WPD	wave power density
WW3	WAVEWATCH-III

Previous research has made a considerable contribution to our understanding of wave energy resources and provides a valuable reference point for the siting of power plants and seawater desalination plants [26]. Nonetheless, robust data for the medium- to long-term prediction of wave energy resources remain scarce, thereby impeding the implementation of power generation and trading strategies. Here, this study provides an overview of the long-term climatic trends and medium- to long-term predictions of wave energy, before discussing the focus of future predictions. Finally, a new method is proposed for predicting wave energy resources on a medium- to long-term basis that incorporates the swell index and propagation characteristics of swell energy. The ultimate goal of this model is to improve the precision of wave energy predictions, thereby providing a reference for the effective utilization of wave resources.

2. Progress of Research Into Medium- to Long-Term Wave Energy Prediction

Previous investigations have greatly improved our ability to forecast marine and meteorological parameters, and the availability of accurate medium- to long-term prediction methods has increased. These methods typically involve two steps: qualitative prediction and quantitative prediction. The former includes forecasts made by experienced meteorologists and subjective probability forecasting methods. In contrast, quantitative prediction employs such approaches as regression analysis modeling, time-series prediction, grey system prediction modeling, artificial neural networks (ANN), and coupling prediction modeling to forecast future trends. Although these methods are widely used today for medium- to long-term prediction of marine and meteorological parameters, relatively little is known about predicting wave energy on a medium- to long-term basis.

The medium- to long-term analysis of wave energy trends involves long-term trend analysis of climatic variables and medium- to long-term prediction, which are discussed in Sections 2.1 and 2.2, respectively. Analytical data for long-term trends in wave energy are relatively abundant and include general trends, regional and seasonal differences in long-term trends, and relationships between wave energy and various key indices. This analytical approach also provides a theoretical basis for refining the precision of medium- to long-term prediction capabilities, which is requisite if we are to meet the needs of wave energy utilization strategies.

2.1 Long-term climatic trends in wave energy

To date, considerable attention has been paid to long-term climatic trends in WPD and SWH. For example, using the 61-year Global Ocean Waves (GOW) reanalysis dataset, which provides an hourly time series of wave parameters between 1948 and 2008, Reguero et al. [27] characterized mean global WPD in addition to monthly and seasonal variability. They presented the differences in mean WPD for the decades 1990-2000 and 2001-2008 relative to mean wave-energy levels in 1981-1990, and their results indicated a sustained increase in the Southern Ocean and concomitant decrease in the North Atlantic. This disparity was most pronounced during the last decade, as shown in Fig. 1.

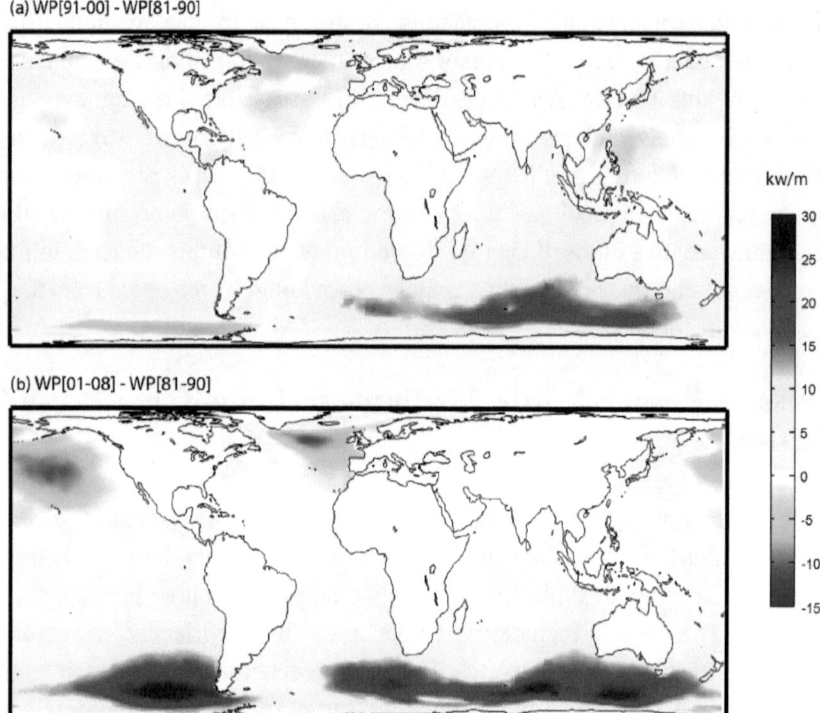

Fig. 1. Differences in the mean wave power density during the periods 1990–2000 (a) and 2001–2008 (b) with respect to the mean resources in the period 1981–1990.

On the basis of numerical wind-wave hindcasting, Bertin et al. [28, 29] observed an increase in SWH of 0.01 m/yr over the entire North Atlantic Ocean north of 50°N, which equates to an increase of between 20% and 40% over the course of the 20th Century. Moreover, the magnitude of SWH variability increases gradually from low to high latitudes (Fig. 2), a pattern that is attributed to enhanced wind speed. For example, 20th Century

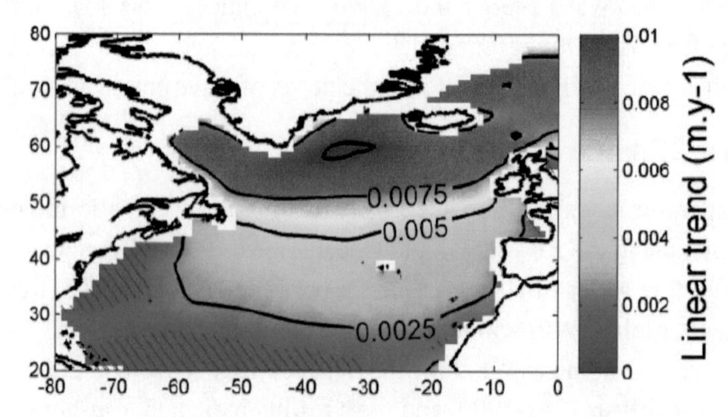

Fig. 2. Linear trend of significant wave height in the North Atlantic Ocean.

reanalysis data reveal an 8 ± 2% increase in wind speed over the northeast North Atlantic during the period 1900-2008, which is equivalent to an annual increase of approximately 0.07%. As we discuss in subsequent sections, WPD is also related to other North Atlantic climate indices.

On the basis of a 57-year hindcast of SWH data, Dodet et al. [30] reported a significant increasing trend of up to 0.02 m/yr in the northeastern North Atlantic over the period 1953-2009. Similarly, Zheng and Li [31] used a 24-year (1988-2011) hindcast to calculate long-term climatic trends in WPD for the China seas. Their results document a clear positive trend of 0.1-0.7 (kW/m)/yr over much of the region, with the highest values occurring around the Ryukyu Islands, the Taiwan Strait, and north of the South China Sea (SCS; Fig. 3). That study also revealed an obvious seasonal difference in long-term WPD, with the most pronounced increase occurring in December-January- February (DJF).

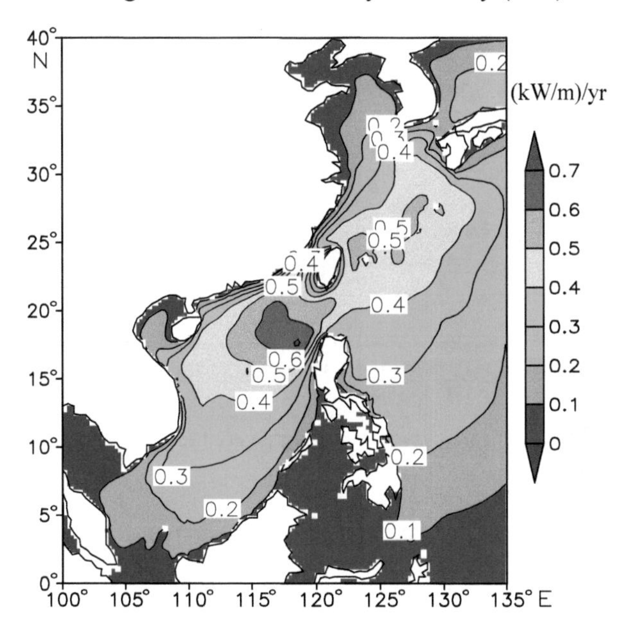

Fig. 3. Long term trend of wave power density in the China Sea. Only area significant at 95% level is presented.

According to the analysis of the International Comprehensive Ocean-Atmosphere Data Set (ICOADS) data, Gulev and Grigorieva [32] reported that wintertime SWH in the mid-latitude North Pacific and North Atlantic oceans increased by 10-40 cm/decade between 1958 and 2002. This pattern was also reported by Semedo et al. [33], who used ERA-40 wave reanalysis data for the period 1957-2002. In their large-scale assessment of the ERA-40 reanalysis data, Zheng et al. [34] concluded that global SWH increased by up to 12 cm/decade between 1958 and 2001 (Fig. 4), with the most significant changes occurring in the southern and northern westerlies.

Based on six decades (1948-2008) of WW3 hindcast data, Bromirski et al. [35] observed a uniformly positive trend in both SWH and WPD over the North Pacific, particularly in the middle latitudes. Furthermore, they reported that the trend is more pronounced in winter months (November-March) than during summer (May-September). In contrast, the hindcast also revealed a recent decrease in WPD across much of the North Pacific. In their global assessment, Zheng et al. [36] described a significant linear increase (28 cm/decade) in ocean wave swell height for the period 1958-2001, a pattern that is consistent with the long-term trend in SWH. They also pointed out that swell plays a dominant role in the mixed wave composition of most global oceans.

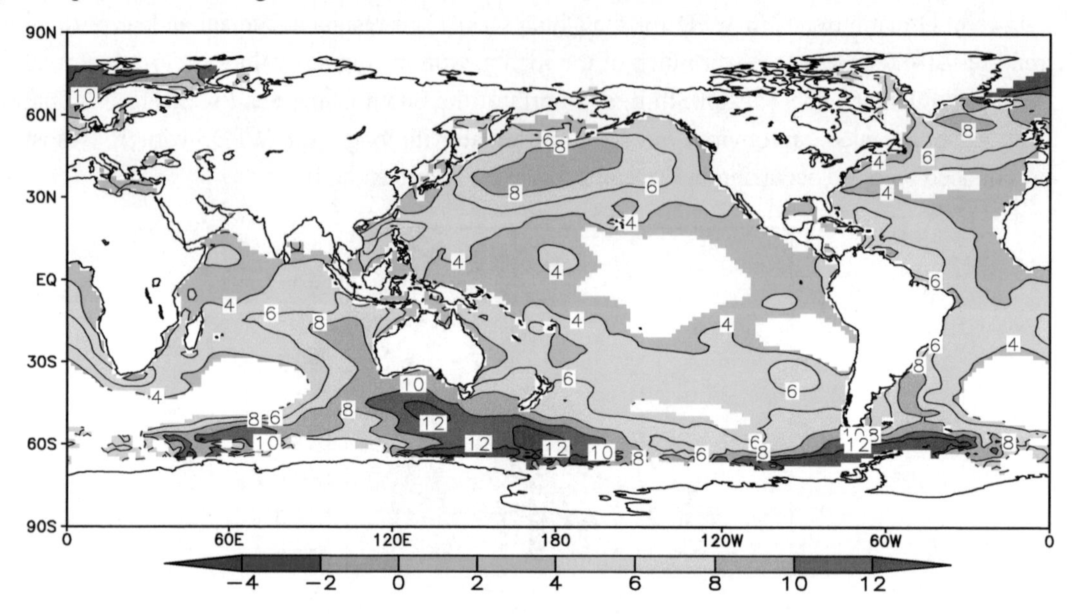

Fig. 4. Long-term trends in significant wave height between 1958 and 2001 (cm/decade). Only trends significant at the 95% level are shown.

In summary, analysis of long-term WPD and SWH trends has improved our understanding of how these parameters vary and provides a foundation for improving medium- to long-term predictions.

2.2 Medium- to long-term prediction of wave energy resources

Whereas short-term forecasting protocols are relatively well developed for wind and wave energy resources [37-40], longer-term predictions for periods of weeks to months remain impractical, and this has implications for the effective planning of wave energy utilization. For example, when considering whether it is possible to manage the wave power requirements of a remote reef population up to a year in advance, the medium- to long-term prediction of wave energy must play a key role in the decision-making process. In their analysis of wave energy in Cornwall, United Kingdom, Reeve et al. [41] presented a quantitative comparison of future (2061-2100) conditions relative to a modern-day (1961-2000) control. Their results show available WPD increasing by between 2% and 3% in the A1B scenario and decreasing by 1-3% in the B1 scenario, a pattern that, if true, suggests efforts to lower greenhouse gas emissions may also reduce wave energy resources. In a similar study, Pinson et al. [42] used a log-normal assumption to project wave energy flux for 13 locations in North America over a 15-month period, and their results showed that this is a viable approach (Fig. 5).

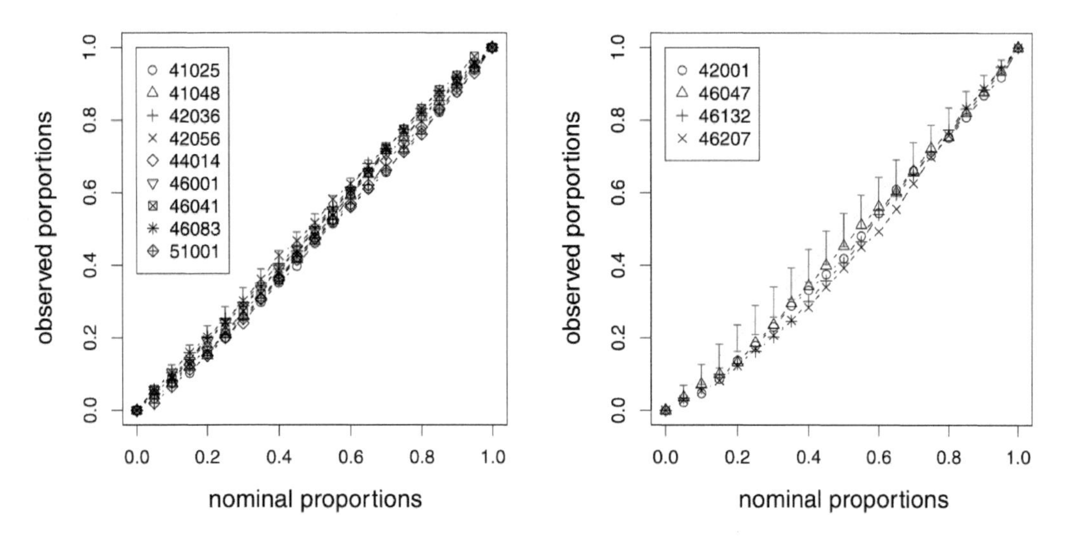

Fig. 5. Probability integral transform diagrams for the evaluation of the reliability of probabilistic forecasts (for lead times of 48 h ahead).

As highlighted by Ghosh et al. [43], there is a distinct lack of objective, relative, and cognitive tools available with which to estimate the suitability of sites for wave energy production. To address this shortcoming, the authors proposed a new approach involving multicriteria decision making (MCDM) cascaded to ANN techniques to predict an index for site suitability. The index was then applied to two separate locations, which had different wave energy characteristics, with promising results. In a different study, Hashim et al. [44] used an enhanced version of the Takagi-Sugeno-based fuzzy inference system (ANFIS) to predict SWH in the western North Atlantic Ocean. Noting that most previous studies used only the sea surface wind speed (U) and wind direction (θ), Hashim et al. [44] concluded that the addition of air temperature (T_a) produced the most effective set of input parameters for their SWH predictions at three stations. In his predictive modeling study, Özger [45] argued that fuzzy logic modeling provides the possible non-linear relationship between ocean wave energy and meteorological variables, such as wind speed, air temperature, and sea temperature. This constraint enables wave energy predictions to include all possible uncertainties in system behavior. Finally, Cornejo-Bueno et al. [46] recently proposed a Grouping Genetic Algorithm-Extreme Learning Machine (GGA-ELM) approach to predicting significant wave height and wave energy flux. Experimentation using this methodology produced favorable results.

Of the studies described above, the majority analyzed historical wave energy data. To realize the full global potential of wave energy, however, it will be necessary to predict the availability of resources several weeks to months in advance. Therefore, as pointed out by Widén et al. [47], a priority of future research must be the development of protocols that are able to evaluate renewable energy forecasts effectively and easily.

3. Future Prediction of Medium- to Long-Term Wave Energy Resources

3.1 Climate prediction methods

Existing methods of climate prediction include regression analysis modeling, time series and grey system prediction, ANN, support vector machine (SVM) classification, autoregressive integrated moving average (ARIMA) modeling [48-52], measure-correlate-prediction (MCP), and ensemble empirical mode decomposition (EEMD) [53]. Ghosh et al. [43] have attempted to propose a new method which is both objective and cognitive to identify suitable locations where the optimal amount of wave energy can be produced. They used the MCDM cascaded to ANN techniques to predict an index that directly represents the suitability of locations for wave energy development. And then the index was applied to two different locations with varied levels of wave energy potential. As a result, the model proposed by Ghosh et al. [43] was found to be accurate and due to its platform independency can be embedded and used for various purpose of coastal and ocean related surveys. They also pointed out that the same model can also be included as a real time monitoring system to monitor the wave energy potential changes within various coastal regions through online monitoring frameworks.

Yet, although these methods generally produce viable results, most are used to predict only meteorological and oceanic conditions, as well as wind energy resources, with relatively few being used to predict wave energy resources. In this regard, the recent development of a site assessment protocol [43] represents an important step towards predicting and harnessing wave energy resources. Going forward, we suggest that medium- and long-term prediction of wave energy could be greatly improved by implementing the mature climate prediction methodologies outlined above.

3.2 Medium- to long-term prediction based on correlation with climate indices

Prior research has established correlations between WPD (or SWH) and various climate indices that can be incorporated into medium- to long-term predictions of wave-energy resources. For instance, Reguero et al. [27] reported close relationships between WPD and the Arctic Oscillation (AO), the Pacific-North America (PNA) index, the North Atlantic Oscillation (NAO), and the East Pacific-North Pacific (EP-NP) pattern (Fig. 6). Ultimately, these geographically specific relationships can facilitate the accurate prediction of wave energy in their respective ocean basins.

Similarly, Bertin et al. [28] reported a correlation between the NAO and North Atlantic SWH. By calculating the Pearson correlation coefficient between mean-annual SWH and the NAO index (Fig. 7), they demonstrated that the NAO partially controls the interannual variability of SWH, resulting in a positive (negative) correlation in the northeastern (southwestern) sector of their study area. A similar relationship between SWH and the NAO

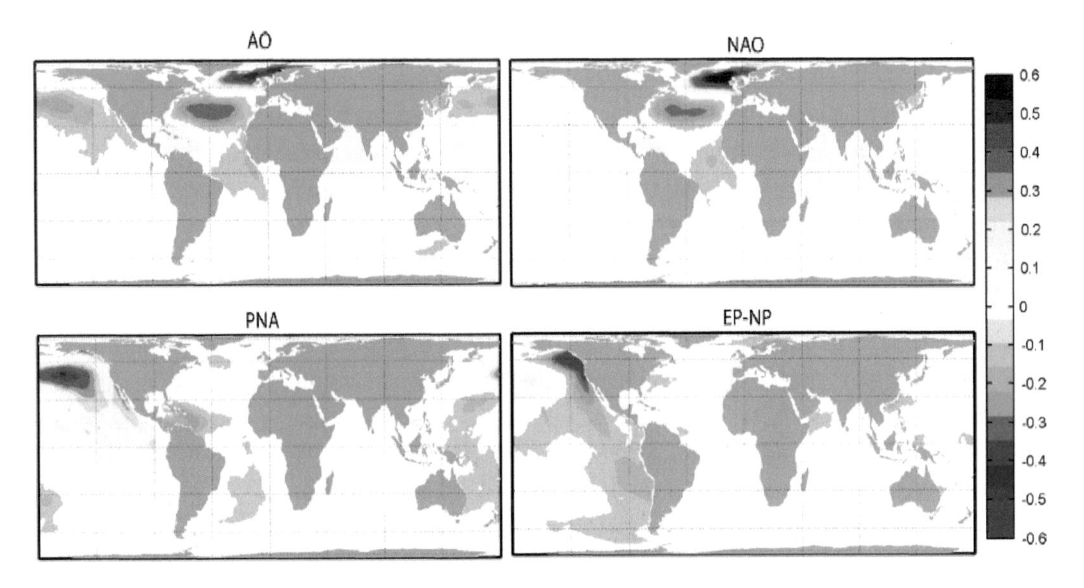

Fig. 6. Monthly correlation of wave power density with climate indices. Values correspond to the linear correlation coefficient.

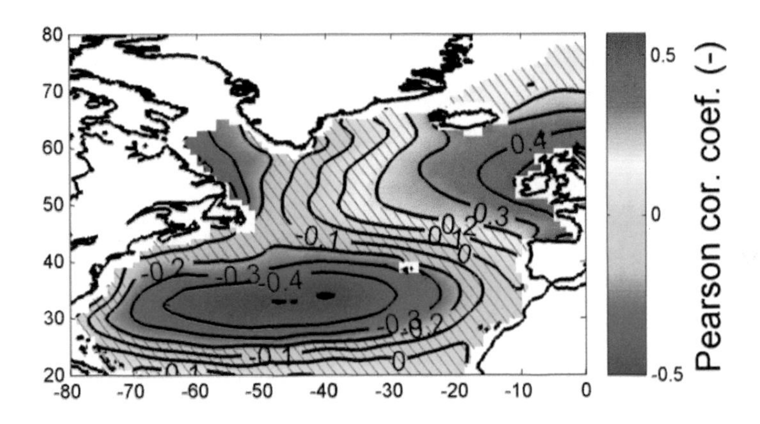

Fig. 7. Pearson correlation coefficient between yearly SWH and NAO index in the North Atlantic Ocean.

was observed by Dodet et al. [30], who reported a maximum correlation coefficient of R = 0.91 in the Northern Hemisphere and a negative correlation (up to R = –0.6) in the Southern Hemisphere. These researchers also reported a well- defined relationship between the NAO index and wave peak period (Tp), as shown in Fig. 8. In the Pacific Ocean basin, Bromirski et al. [35] suggested that the recent decline in WPD and SWH is associated with the cool phase of the Pacific Decadal Oscillation (PDO), which developed after the late 1990s. In summary, we suggest the demonstrated correlations between wave energy and key climate indices will be instrumental in developing medium- to long-term predictions of wave energy resources.

3.3 Numerical simulation methods

Thanks to the rapid development of computer-based numerical modeling, wave models such as WAM, WW3, and SWAN are now widely used to forecast short-term wave energy resources [54] and simulate the climatic characteristics of wave energy [31, 55]. However, to date, few of these models have been incorporated into medium- to long-term wave energy predictions. In their statistical projections of global SWH change, Wang et al. [56] employed 21st Century sea level pressure data from 20 Coupled Model Intercomparison Project Phase 5 (CMIP5) global climate models. By comparing their results to SWH calculated using ERA40 and ERA-Interim wave reanalysis, Wang et al. [56] reported good agreement for the period 1980-1999 (Fig. 9). Similarly, Zheng et al. [57] concluded that the CMIP5-based projections can also be used to make reliable predictions of future offshore wind energy resources. Zheng et al. [58] have proposed a prediction system of wave energy resource by using the CMIP5 projection wind data to drive the WW3 wave model. The China seas are selected as a case study. Firstly, the China seas' wave energy in 2015 is predicted, using the CMIP5 projection wind data to drive the WW3 wave model. Then the multi-year average status of the China seas' wave energy is also presented based on a 24-year hindcast wave data. Comparing the prediction values with the multi-year average values of wave energy parameters could make a positive contribution to the planning of wave energy utilization, as shown in Fig. 10. In the future work, it is possible to enlarge the prediction spatial range and time series of wave energy by using this method. For example, it is possible to use the CMIP5 projection wind data to drive the WW3 wave model to predict the wave energy for the period 2020-2059 globally. Together, these findings support the use of CMIP5 data to help project medium- to long-term wave energy resources on a global scale.

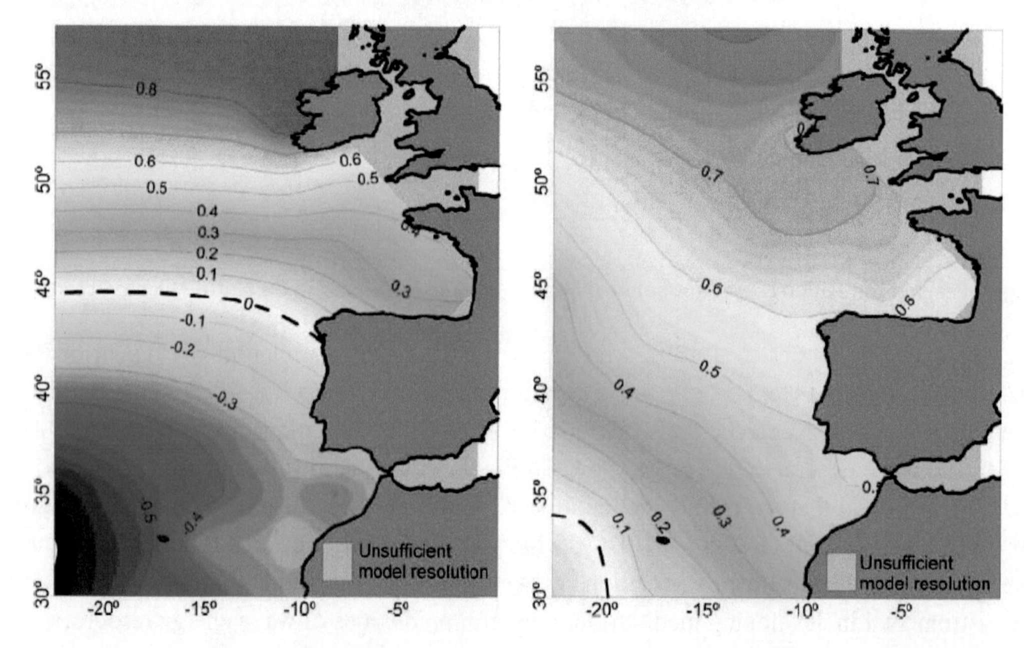

Fig. 8. Pearson correlation coefficient between NAO index and (left) Hs90 and (right) winter-mean Tp.

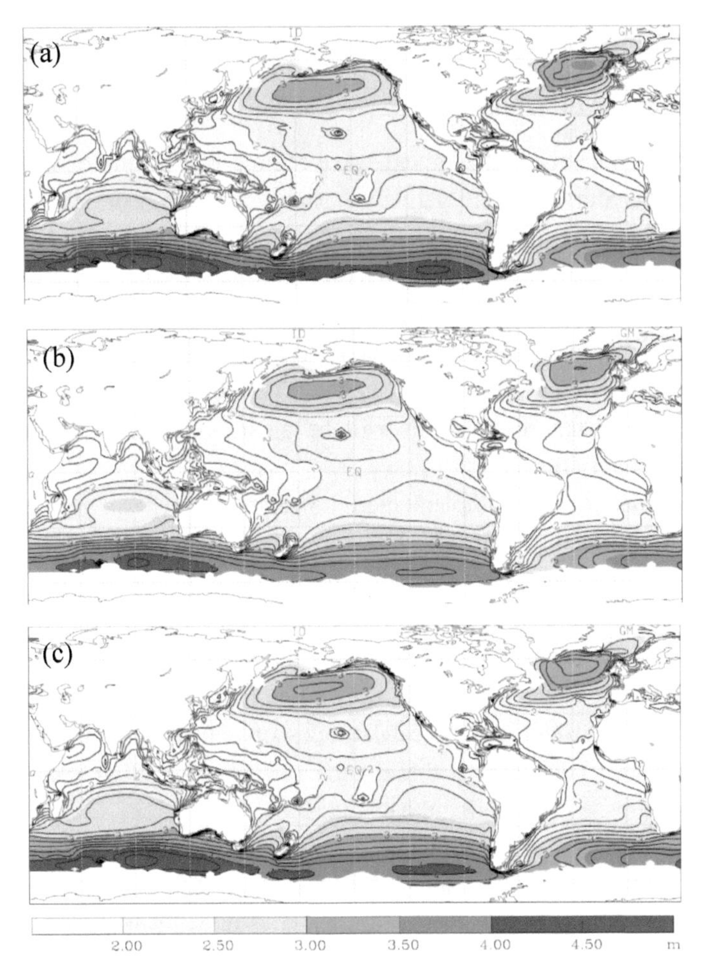

Fig. 9. CMIP5 ensemble mean significant wave height (a), ERA40 significant wave height (b) and ERA-Interim significant wave height (c) (adapted from Wang et al. [56]).

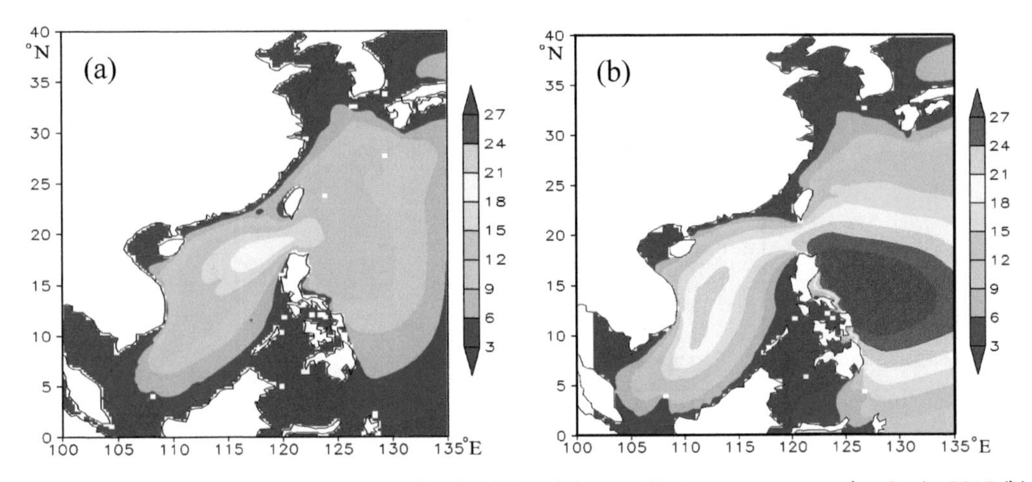

Fig. 10. Multi-year average wave power density (a) and the prediction wave power density in 2015 (b), unit: kW/m.

3.4 Prospects for wave energy prediction based on swell propagation characteristics and swell index

It has long been established that a mixed-wave regime is made up of wind-sea and swell, and that the wind-sea component is relatively unstable. However, prior research [59-62] has also demonstrated the persistence of swell waves, which can propagate thousands of kilometers along circular trajectories with little attenuation. In their calculation of the global swell index (defined as the proportion of swell energy in the mixed-wave energy budget), Chen et al. [63] observed widespread high values, which reflect the dominant role played by swell energy in the global mixed-wave energy budget. Therefore, the combination of the swell index and the persistent nature of swells will be an important component in longer-term predictions of wave energy resources.

Before these parameters can be implemented, however, two criteria must be satisfied. First, the swell index must be calculated to a high temporal and spatial resolution. If referring to the index of Chen et al. [63], for example, it will be necessary to calculate values for specific months and, if possible, increase the temporal resolution from months to weeks. Second, the stability of swell energy can be calculated via its propagation characteristics. For instance, Alves [59] reported that westerly swells in the southern Indian Ocean commonly propagate eastwards into the tropical and subtropical latitudes of both the Indian and Pacific oceans. Similarly, Zheng et al. [64] observed these same swells propagating great distances to the north and, in some cases, even crossing the equator.

Despite the significant advances in our understanding of swell propagation, quantitative data concerning propagation routes and velocity remain scarce. In the present study, we selected two regions of the Indian Ocean to investigate swell-propagation characteristics. Region A covers 40-60°S, 80-100°E in the central Indian Ocean and Region B covers 10°S-10°N, 70-90°E, between Java and Australia. Fig. 11 shows the lead and lag correlation

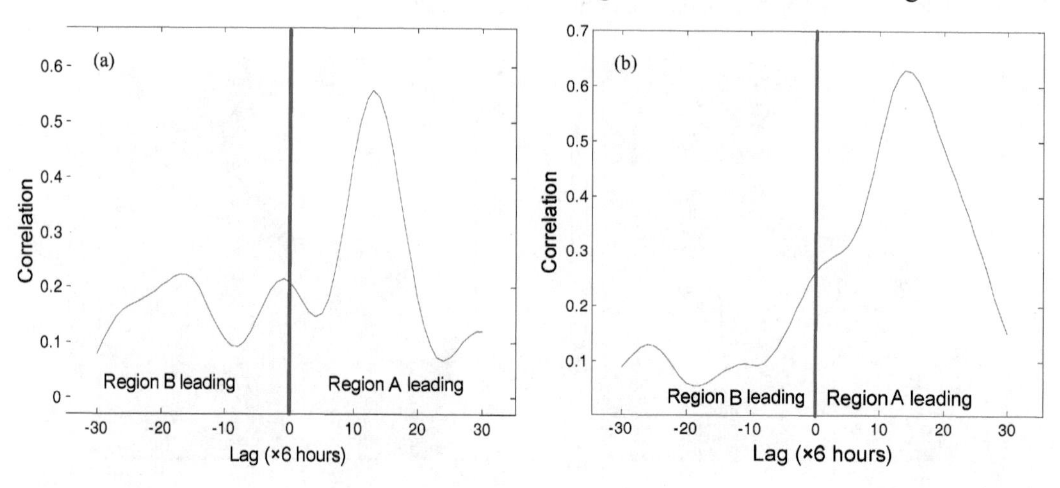

Fig. 11. Simultaneous, leading and lagging correlations between swell in regions A and B in JJA 2001 (a) and DJF 2001 (b).

coefficients (CC) calculated between the 6-hourly swell wave heights (H^s) for both regions during JJA (June-July-August), 2001, and DJF, 2001. As shown in Fig. 11a, the simultaneous CC is weak, whereas the CC reaches its peak value when H^s of region B lagging 13 intervals, of 0.56 (at 99.9% significance). A similar pattern occurs in DJF.

To illustrate the relationship between swells in Regions A and B, we have contoured the H^s values in Region B of lagging 13 intervals and H^s of Region A for the JJA period (Fig. 12a). Similarly, we have contoured the values of H^s in Region B of lagging 14 intervals and H^s of Region A for the DJF period (Fig. 12b). Our comparison demonstrates the close agreement in H^s values between both regions and shows that the swell requires approximately 13-14 intervals (ca. 78-84h) to propagate west from the central southern Indian Ocean to the vicinity of Java and Australia.

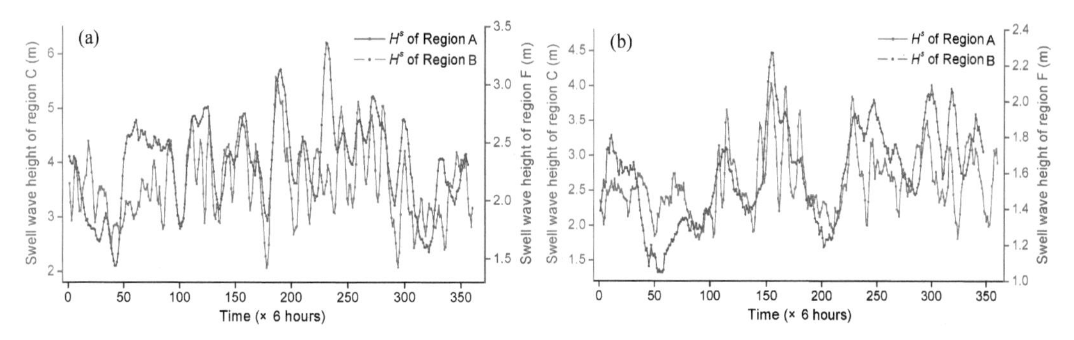

Fig. 12. Swell in region B of lagging 13 intervals and swell in region A in JJA 2001 (a), swell in region B of lagging 14 intervals and swell in region A in DJF 2001 (b)

We observed a similar pattern in the Pacific Ocean, where we used the ERA-40 wave reanalysis data and a time-backward inference method to investigate the source of swell energy. Our experiment assumed that there is a wave power plant located at Clipperton Island (black rectangle in Fig. 13a) off the west coast of Mexico in what we term Region E. This location serves as the terminus for the propagating swell. We then used the time-backward inference method to detect the swell's propagation source. First, we calculated and contoured the simultaneous CC between 6-hourly swell energy (E) of region E (E_E) and E of the Pacific Ocean at each 1.5°×1.5° bin for JJA, 2001 (Fig. 13a). We then calculated the CC between the 6-hourly E_E of lagging 24, 48, 72, 96, 120, 144, 168, and 192h separately and E at each 1.5°×1.5° bin for the same period (Fig. 13b-i). Our results indicate that the area of significance (at the 0.001 confidence level) exhibits a southeast-northwest propagation pattern, particularly in the centre of our study area. Furthermore, this assessment suggests the region north of Hawaii (black rectangle in Fig. 13i) is the principal source of swell energy for Clipperton Island.

Based on these findings, we are able to calculate the swell index, propagation route, and propagation speed of the swell energy. Additionally, we are able to monitor the swell quantitatively, thereby providing a robust foundation for making medium- to long-term predictions of wave-energy distributions.

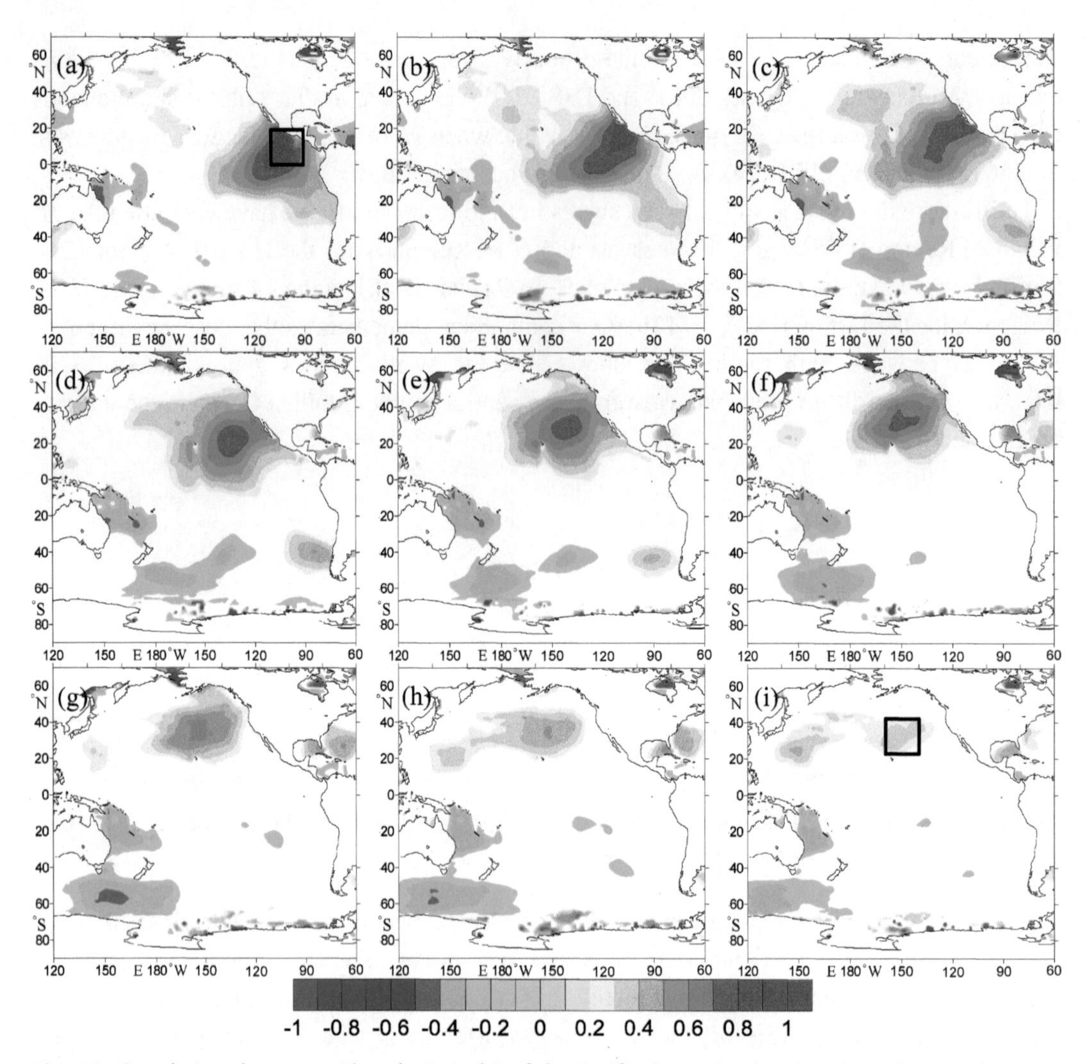

Fig. 13. Correlations between 6-hourly E_E and E of the Pacific Ocean in JJA 2001. In a, the simultaneous correlation coefficients between 6-hourly E_E and E of the Pacific Ocean at each 1.5°×1.5° bin. In b, c, d, e, f, g, h and i the correlations between the 6-hourly E_E of lagging 24, 48, 72, 96, 120, 144, 168, and 192 h separately and E at each 1.5°×1.5° bin. With the areas significant at 0.001 confidence level colored. (For interpretation of the references to color in this figure legend, the reader is referred to the web version of this article.)

4. Conclusions

Detailed information concerning long-term climate trends and the medium- to long-term predictability of wave energy remains scarce. In this study, we have discussed the principal challenges to making longer-term predictions and presented several areas for improvement. For instance, whereas the majority of previous investigations have considered only the size of WPD, the effective development of wave energy requires additional parameters. These include the stability of wave energy, the energy level occurrence, and the available SWH of wave-energy resources (termed "effective SWH"). Stability is closely related to the capture efficiency of wave energy and the lifespan of a specific wave energy device, and to calculate the stability of wave and wind energy requires the coefficient of variation (Cv), monthly variability index (Mv), and seasonal variability index (Sv) [65-67].

The occurrence of effective SWH is directly related to the utilization rate of wave energy [65, 68], whereas the energy level occurrence (e.g., the occurrence of WPD $\geqslant$ 20 kW/m) determines the energy richness [69]. We conclude that the climatic trend analysis of wave energy should include the long-term variability of WPD, wave energy stability (Cv, Mv, and Sv), energy-level occurrence, and the occurrence of effective SWH.

Typically, medium- to long-term predictions of wave energy resources constitute a single projection of WPD. For future investigations, we suggest that predictions should include additional parameters, including Cv, Mv, Sv, the energy level occurrence, and the occurrence of effective SWH.

In predicting ocean waves, the majority of studies to date have focused on the mixed wave and ignored the influence of swell. Yet previous research demonstrates that swell waves are persistent [70-73] and that swell energy plays a dominant role in the global mixed- wave energy budget [36, 63]. Therefore, the combination of the swell index and the persistence characteristic can provide a quantitative reference point for medium- to long-term predictions of wave energy resource. Here, we have proposed a prediction protocol that incorporates both swell parameters. By monitoring the propagation of swell energy and calculating the swell index, we show that it is possible to predict the WPD of mixed waves to a high degree of precision.

In summary, we conclude that the prediction capacity of wave energy can be improved by incorporating climate prediction methods; assessing correlations between wave energy and climate indices; and the careful application of numerical simulation methods, swell propagation characteristics, and the swell index.

Acknowledgments

This work was supported by the Junior Fellowships for CAST Advanced Innovation Think-tank Program (entitled "Evaluation of the Oceanic Dynamic resources of the 21st century Maritime Silk Road and its strategic points", No. DXB-ZKQN-2016-019), the Major Research Topic of the Propaganda Department of CAST (entitled "Problems and Countermeasures and Suggestions of the Strategic Points Construction of the 21st century Maritime Silk Road), the National Key Basic Research Development Program (Grant No. 2012CB957803), and the National Nature Science Foundation of China (No. 41490642). All the authors would like to thank an anonymous referee and the editor for providing their excellent comments and valuable advice in relation to improving this paper.

References

[1] Pérez-Collazo C, Greaves D, Iglesias G. A review of combined wave and offshore wind energy. Renew

Sustain Energy Rev 2015; 42: 141-53.
[2] Bonar PAJ, Bryden IG, Borthwick AGL. Social and ecological impacts of marine energy development. Renew Sustain Energy Rev 2015; 47: 486-95.
[3] Lin YG, Bao JW, Liu HW, Li W, Tu L, Zhang DH. Review of hydraulic transmission technologies for wave power generation. Renew Sustain Energy Rev 2015; 50: 194-203.
[4] Heras-Saizarbitoria I, Zamanillo I, Laskurain I. Social acceptance of ocean wave energy: a case study of an OWC shoreline plant. Renew Sustain Energy Rev 2013; 27: 515-24.
[5] Vasilis F, James EM, Ken Z. The technical, geographical, and economic feasibility for solar energy to supply the energy needs of the US. Energy Policy 2009; 37(2): 387-99.
[6] US Department of the Interior Bureau of Ocean Energy Management. Offshore wind and wave energy feasibility mapping for the outer continental shelf off the State of Oregon. From 〈〈https: //www. boem.gov/2014-658/〉〉; 2014.
[7] Astariz S, Iglesias G. The economics of wave energy: a review. Renew Sustain Energy Rev 2015; 45: 397-408.
[8] Deane JP, Dalton G, Ó Gallachóir BP. Modelling the economic impacts of 500 MW of wave power in Ireland. Energy Policy 2012; 45: 614-27.
[9] Ozkop E, Altas IH. Control, power and electrical components in wave energy conversion systems: a review of the technologies. Renew Sustain Energy Rev 2017; 67: 106-15.
[10] Falcão AFDO. Wave energy utilization: a review of the technologies. Renew Sustain Energy Rev 2010; 14(3): 899-918.
[11] Denis M. Wave climate and the wave power resource. Hydrodynamics of Ocean Wave-Energy Utilization 1986; p. 133-56.
[12] Barstow S, Haug O, Krogstad H. Satellite Altimeter Data in Wave Energy Studies. Proc. Waves'97, ASCE 1998; 2: p. 339-54.
[13] Wan Y, Zhang J, Meng JM, Wang J. A wave energy resource assessment in the China's seas based on multi-satellite merged radar altimeter data. Acta Oceanol Sin 2015; 34(3): 115-24.
[14] Li Y, YuYH. A synthesis of numerical methods for modeling wave energy converter- point absorbers. Renew Sustain Energy Rev 2012; 16(6): 4352-64.
[15] Rusu E, Onea F. Evaluation of the wind and wave energy along the Caspian Sea. Energy 2013; 50: 1-14.
[16] Akpinar A, van Vledder GP, Kömürcü MI, Özger M. Evaluation of the numerical wave model (SWAN) for wave simulation in the Black Sea. Cont ShelfRes 2012; 50- 51: 80-99.
[17] Zheng CW, Zhou L, Jia BK, Pan J, Li X. Wave characteristic analysis and wave energy resource evaluation in the China Sea. J Renew Sustain Energy 2014; 6: 043101. http: //dx.doi.org/10.1063/ 1.4885842.
[18] Zheng CW, Li CY. Development of the islands and reefs in the South China Sea: wind power and wave power generation. Period Ocean Univ China 2015; 45(9): 7-14.
[19] Morim J, Cartwright N, Etemad-Shahidi A, Strauss D, Hemer M. Wave energy resource assessment along the Southeast coast ofAustralia on the basis of a 31-year hindcast. Appl Energy 2016; 184: 276-97.
[20] Cornett AM. A global wave energy resource assessment. In: Proceedings of the Eighteenth International Offshore and Polar Engineering Conference held in Canada, 2008; p. 318-26.
[21] Folley M, Whittaker TJT. Analysis of the nearshore wave energy resource. Renew Energy 2009; 34(7): 1709-15.
[22] Iglesias G, Carballo R. Wave energy resource in the Estaca de Bares area (Spain). Renew Energy 2010;

35: 1574-84.

[23] Gunn K, Stock-Williams CS. Quantifying the global wave power resource. Renew Energy 2012; 44: 296-304.

[24] Zheng CW, Shao LT, Shi WL, Su Q, Lin G, Li XQ, Chen XB. An assessment of global ocean wave energy resources over the last 45 a. Acta Oceanol Sin 2014; 33(1): 92-101.

[25] Arinaga RA, Cheung KF. Atlas of global wave energy from 10 years of reanalysis and hindcast data. Renew Energy 2012; 39: 49-64.

[26] Rashid A, Hasanzadeh S. Status and potentials of offshore wave energy resources in Chahbahar area (NW Omman Sea). Renew Sustain Energy Rev 2011; 15(9): 4876-83.

[27] Reguero BG, Losada IJ, Méndez FJ. A global wave power resource and its seasonal, interannual and long-term variability. Appl Energy 2015; 148: 366-80.

[28] Bertin X, Prouteau E, Letetrel C. A significant increase in wave height in the North Atlantic Ocean over the 20th century. Glob Planet Change 2013; 106: 77-83.

[29] Bertin X, Fortunato AB, Oliveira A. Simulating morphodynamics with unstructured grids: description and validation of an operationalmodel for coastal applications. Ocean Model 2009; 28: 75-83.

[30] Dodet G, Bertin X, Taborda R. Wave climate variability in the North-East Atlantic Ocean over the last six decades. Ocean Model 2010; 31(3-4): 120-31.

[31] Zheng CW, Li CY. Variation of the wave energy and significant wave height in the China Sea and adjacent waters. Renew Sustain Energy Rev 2015; 43: 381-7.

[32] Gulev SK, Hasse L. Variability of the winter wind waves and swell in the North Atlantic and North Pacific as revealed by the voluntary observing ship data. J Clim 2006; 19: 5667-85.

[33] Semedo A, Suselj K, Rutgersson A. A global view on the wind sea and swell climate and variability from ERA-40. J Clim 2011; 24: 1461-79.

[34] Zheng CW, Zhou L, Shi WL, Li X, Huang CF. Decadal variability of global ocean significant wave height. J Ocean Univ China 2015; 14(5): 778-82.

[35] Bromirski PD, Cayan DR, Helly J, Wittmann P. Wave power variability and trends across the North Pacific. J Geophys Res: Oceans 2013; 118: 6329-48.

[36] Zheng CW, Zhou L, Huang CF, Shi YL, Li JX, Li J. The long-term trend of a sea surface wind speed and a (wind wave, swell, mixed wave) wave height in global ocean during the last 44 a. Acta Oceanol Sin 2013; 32(10): 1-4.

[37] Al-Yahyai S, Charabi Y, Gastli A. Review of the use of Numerical Weather Prediction (NWP) Models for wind energy assessment. Renew Sustain Energy Rev 2010; 14(9): 3192-8.

[38] Tascikaraoglu NA, Uzunoglu M. A review of combined approaches for prediction of short-term wind speed and power. Renew Sustain Energy Rev 2014; 34: 243-54.

[39] Roger B. Wave energy forecasting accuracy as a function of forecast time horizon. From 〈〈www.epri.com/oceanenergy/〉〉; 2009.

[40] Reikard G. Integrating wave energy into the power grid: simulation and Forecasting. Ocean Eng 2013; 73: 168-78.

[41] Reeve DE, Chen Y, Pan S, Magar V, Simmonds DJ, Zacharioudaki A. An investigation of the impacts of climate change on wave energy generation: the Wave Hub, Cornwall, UK. Renew Energy 2011; 36(9): 2404-13.

[42] Pinson P, Reikard G, Bidlotc JR. Probabilistic forecasting of the wave energy flux. Appl Energy 2012; 93: 364-70.

[43] Ghosh S, Chakraborty T, Saha S, Majumder M, Pal M. Development of the location suitability index for wave energy production by ANN and MCDM techniques. Renew Sustain Energy Rev 2016; 59:

1017-28.

[44] Hashim R, Roy C, Motamedi S, Shamshirband S, Petković D. Selection of climatic parameters affecting wave height prediction using an enhanced Takagi-Sugeno- based fuzzy methodology. Renew Sustain Energy Rev 2016; 60: 246-57.

[45] Ödzger M. Prediction of ocean wave energy from meteorological variables by fuzzy logic modeling. Expert Syst Appl 2011; 38: 6269-74.

[46] Cornejo-Bueno L, Nieto-Borge JC, Garcia-Diaz P, Rodriguez G, Salcedo-Sanz S. Significant wave height and energy flux prediction for marine energy applications A grouping genetic algorithm-Extreme Learning. Renew Energy 2016; 97: 380-9.

[47] Widén J, Carpman N, Castellucci V, Lingfors D, Olauson J, Remouit F, Bergkvist M, Grabbe M, Waters R. Variability assessment and forecasting of renewables: a review for solar, wind, wave and tidal resources. Renew Sustain Energy Rev 2015; 44: 356-75.

[48] Beccali M, Cellura M, Brano VL, Marvuglia A. Short-term prediction of household electricity consumption: assessing weather sensitivity in a Mediterranean area. Renew Sustain Energy Rev 2008; 12(8): 2040-65.

[49] Ahmad AS, Hassan MY, Abdullah MP, Rahman HA, Hussin F, Abdullah H, Saidur R. A review on applications of ANN and SVM for building electrical energy consumption forecasting. Renew Sustain Energy Rev 2014; 33: 102-9.

[50] Gairaa K, Khellaf A, Messlem Y, Chellali F. Estimation of the daily global solar radiation based on Box-Jenkins and ANN models: a combined approach. Renew Sustain Energy Rev 2016; 57: 238-49.

[51] Xiao L, Wang JZ, Dong Y, Wu J. Combined forecasting models for wind energy forecasting: a case study in China. Renew Sustain Energy Rev 2015; 44: 271-88.

[52] Cuadra L, Salcedo-Sanz S, Nieto-Borge JC, Alexandre E, Rodriguez G. Computational intelligence in wave energy: comprehensive review and case study. Renew Sustain Energy Rev 2016; 58: 1223-46.

[53] Carta JA, Velazquez S, Cabrera P. A review of measure-correlate-predict (MCP) methods used to estimate long-term wind characteristics at a target site. Renew Sustain Energy Rev 2013; 27: 362-400.

[54] Zheng CW, Li CY, Chen X, Pan J. Numerical forecasting experiment of the wave energy resource in the China Sea. Adv Meteorol 2016. http: //dx.doi.org/10.1155/ 2016/5692431.

[55] Agarwal A, Venugopal V, Harrison GP. The assessment of extreme wave analysis methods applied to potential marine energy sites using numerical model data. Renew Sustain Energy Rev 2013; 27: 244-57.

[56] Wang XL, Feng Y, Swail VR. Changes in global ocean wave heights as projected using multimodel CMIP5 simulations. Geophys Res Lett 2014; 41: 1026-34.

[57] Zheng CW, Li CY, Liu MY, Xia LL. An overview of global ocean wind energy resources evaluation. Renew Sustain Energy Rev 2016; 53: 1240-51.

[58] Zheng CW, You ZJ, Gao CZ, Chen X. Prediction of wave energy resource: a case study of the China seas. J Mar Sci Technol-Taiwan 2017, [accepted].

[59] Alves JHGM. Numerical modeling of ocean swell contributions to the global wind- wave climate. Ocean Model 2006; 11: 98-122.

[60] Ardhuin F, Chapron B, Collard F. Observation of swell dissipation across oceans. Geophys Res Lett 2009; 36: L06607. http: //dx.doi.org/10.1029/2008GL037030.

[61] White A, McTigue J, Markides C. Wave propagation and thermodynamic losses in packed-bed thermal reservoirs for energy storage. Appl Energy 2014; 130: 648-57.

[62] Bhowmick SA, Kumar R, Chaudhuri S, Sarkar A. Swell propagation over Indian Ocean Region. Int J Ocean Clim Syst 2011; 2(2): 87-99.

[63] Chen G, Chapron B, Ezraty R, Vandemark D. A global view of swell and wind sea climate in the ocean by satellite altimeter and scatterometer. J Atmos Ocean Technol 2002; 19: 1849-59.
[64] Zheng CW, Li CY. Temporal and spatial distribution of the windsea, swell and mixed wave in the Indian Ocean (in Chinese). J PLA Univ Sci Technol (Nat Sci Ed) 2016; 36(2): 1-7.
[65] Zheng CW, Gao ZS, Liao QF, Pan J. Status and prospect of the evaluation of the Global wave energy resource. Recent Pat Eng 2016; 10(2): 98-110.
[66] Zheng CW, Pan J. Assessment of the global ocean wind energy resource. Renew Sustain Energy Rev 2014; 33: 382-91.
[67] Justin TT, Barve KH, Ranganath LR, Dwarakish GS. Assessment of wave energy potential along South Maharashtra coast. Int J Earth Sci Eng 2016; 9(3): 26-31.
[68] Zheng CW, Pan J, Li JX. Assessing the China Sea wind energy and wave energy resources from 1988 to 2009. Ocean Eng 2013; 65: 39-48.
[69] Zheng CW, Zhuang H, Li X, Li XQ. Wind energy and wave energy resources assessment in the East China Sea and South China Sea. Sci China Technol Sci 2012; 55(1): 163-73.
[70] Ardhuin F, Chapron B, Collard F. Observation of swell dissipation across oceans. Geophys Res Lett 2009; 36: L06607. http: //dx.doi.org/10.1029/2008GL037030.
[71] Zheng CW, Li CY. Analysis of temporal and spatial characteristics of waves in the Indian Ocean based on ERA-40 wave reanalysis. Appl Ocean Res 2017; 63: 217-28.
[72] Aboobacker VM, Vethamony P, Rashmi R. "Shamal" swells in the Arabian Sea and their influence along the west coast of India. Geophys Res Lett 2011; 38: L03608. http: //dx.doi.org/10.1029/2010GL045736.
[73] Li XM. A new insight from space into swell propagation and crossing in the global oceans. Geophys Res Lett 2016; 43: 5202-9.

Propagation Route and Speed of Swell in the Indian Ocean

Chongwei Zheng[1,2,3], Chongyin Li[1,4*], Jing Pan[4]

[1] College of Meteorology and Oceanography, National University of Defense Technology, Nanjing, China
[2] State Key Laboratory of Estuarine and Coastal Research, Shanghai, China
[3] Navigation department, PLA Dalian Naval Academy, Dalian, China
[4] LASG, Institute of Atmospheric Physics, Chinese Academy of Sciences, Beijing, China

Abstract The characteristics of swell propagation play an important role in the forecasting of ocean waves as well as on research on global climate change, wave energy development, and disaster prevention and reduction. To reveal the propagation routes, terminal targets and speeds of swells that originate from the southern Indian Ocean westerly (SIOW), an intraseasonal swell index (SI) was defined based on the 45 year (September 1957 to August 2002) ERA-40 wave reanalysis data product from the European Center for Medium-Range Weather Forecasts (ECMWF). The results show that the main body of the SIOW-related swells typically spread to the waters off Sri Lanka and Christmas Island, while the branches spread to the Arabian Sea and other waters. The propagation speeds of swells originated in the SIOW were fastest in May and August, followed by November, and were slowest in February. Swells usually required 4-6 days to propagate from the western part of the SIOW to the waters off Sri Lanka and Christmas Island, whereas swells usually required 2-4 days to propagate from the eastern part of the SIOW to the waters off Christmas Island.
Keywords A new method is designed to detect the propagation route and speed of swells; Propagation route and speed of the southern Indian Ocean westerly swells are revealed; Propagation destination of the southern Indian Ocean westerly swells is presented.

1. Introduction

Swells in the ocean can often be surprisingly destructive and lead to phenomena such as hogging and sagging, which can cause serious damage to ships. After being generated by a storm, waves can propagate very long distances with little attenuation until they break and dissipate upon reaching a coast (Alves, 2006; Ardhuin et al., 2009; Munk et al., 1963; Semedo, 2010; Snodgrass et al., 1966). These characteristics make swells an indicator of various atmospheric phenomena such as tropical cyclones, distant storms, or even large-scale sea breezes such as those related to monsoons. Swells also have significant impacts on the transport and dispersion of oil plumes within the ocean mixed layer, ocean surface roughness, wind stress, and other things (Chen et al., 2016; Hwang, 2008; Wu et al., 2016). Because of their substantial energy and good stability, energy production from swell waves has received increasing attention. Studies have shown that swells have a dominant status in a mixed wave (Semedo et al., 2011; Zheng et al., 2017), which means that swells also have a significant impact on air-sea interactions and global climate change. As a result, in-depth study of the characteristics of swells has practical value for swell wave power generation, ocean wave numerical simulation and forecasting, and studies of global climate change (Remya & Kumar, 2013; Remya et al., 2012).

Although data on swells are extremely scarce, previous researchers have provided important insights into swell generation and propagation. Arinaga and Cheung (2012) performed a global analysis of wind-sea and swells based on a 10 year simulation of wave data; their Figure 5 shows that the swell wave height (H^s) gradually decreases from south to north in each month and that contours have a northward salient. Zheng et al. (2014) analyzed the seasonal characteristics of the wave power densities of wind-sea, swells, and mixed waves; their Figure 2 shows that the Indian Ocean swell wave power density increases from west to east and tends to pile up to the west of Australia and then spreads to the north. Semedo (2010) noted that mixing of the global ocean by waves is dominated by swells; they also found that the winter swell in the North Atlantic Ocean displays north-south propagation. Semedo et al. (2011) presented the global seasonal characteristics of wind-sea, swells, and mixed waves, including wave direction and wave height. Zheng and Li (2017) contoured the wave direction and wave height of Indian Ocean swells based on the ERA-40 wave reanalysis and noted that the H^s contour has an obvious northward salient and that the dominant swell wave direction is south throughout the year. However, wave direction cannot indicate the long-distance route of swell movement. For example, in the low-latitude waters of the southern Indian Ocean, the swell wave direction is southeast, but this does not mean that the swell's propagation route in this region is from southeast to northwest. Alves (2006) noted that many southern Indian Ocean westerly (SIOW) swells freely propagate eastward into the tropical and subtropical latitudes of the Indian and Pacific Oceans. By monitoring the wave height and peak period, Remya et al. (2016) determined that swells propagated from the SIOW to northern Indian Ocean during 14-21 May 2005. By tracking the swell observations using SAR wave mode data, Jiang et al. (2016) found that swells generated by a storm in 58°S, 132°W could propagate to the coast of Mexico.

To date, studies of swell propagation routes, terminal targets, and propagation speeds have been relatively rare. In this study, the SIOW was divided into four regions. The swell index (SI) of each region was then defined to analyze the swell propagation characteristics, primarily their propagation routes, terminal targets, and propagation speeds, to provide a reference for ocean wave forecasting, energy development from swells, studies of air-sea interaction, and other applications.

2. Materials and Methods

2.1 Materials

The data set used in this study were the ERA-40 wave reanalysis from ECMWF, which is the first reanalysis product produced by coupling simulation results from a wave-atmosphere model (WAM) with assimilated observational data. Its range in space is 90°S-90°N and 180°W-180°E, its spatial resolution is 1.5° × 1.5°, and it covers the period from September 1957 to August 2002 at a time step of 6 h. The biggest advantage of this data set is that it separates wind-sea from swell waves. Because the ERA-40 wave reanalysis covers a long period and the entire global ocean, these data are widely used to analyze the character of global ocean waves, especially in the North Atlantic, North Pacific, and Southern Oceans (Caires et al., 2005; Hemer et al., 2007; Semedo et al., 2011). The ERA-40 wave reanalysis is divided into four different periods. However, problems in the observational data from December 1991 to May 1993 lead to the data assimilation time errors being relatively larger.

The criteria used for sea-swell separation is as follows (Semedo et al., 2009). The WAM model output is the two-dimensional wave energy spectrum $F(f, \theta)$ (Komen et al., 1994). Here, f is frequency and θ is direction. From these spectra, several derived integrated wave parameters can be obtained. The mean variance of the sea-surface elevation (the *zeroth* moment) is statistically related to the significant wave height: $SWH \cong H_s = 4.04\sqrt{m_0}$, where $m_0 = \int\int f^0 F(f, \theta) df d\theta$ is the variance or the zeroth moment. By weighting $F(f, \theta)$, θ_m is defined in the WAM mode as $\theta_m = a\tan(SF/CF)$, where the weights are defined as $SF = \int\int sin(\theta) F(f, \theta) df d\theta$ and $CF = \int\int cos(\theta) F(f, \theta) df d\theta$. The significant wave height, mean periods, and mean wave directions of wind-sea and swell waves are computed by separating the one-dimensional (1D) spectrum into wind-sea and swell components. The separation frequency is defined as the frequency corresponding to wave phase speed $\hat{c}$ where $33.6 \times (u_*/\hat{c}) \cos(\theta - \varphi) = 1$. The wind-sea and swell integrated parameters (in the present case m_0 and m_1) are computed by integrating over the respective 1-D spectral part.

Caires and Sterl (2005) validated the significant wave height (H_s) from ERA-40 against GEOSAT altimeter measurements from 1988, ERS-1 off-line (OPR) altimeter observations for June to December 1993, TOPEX altimeter measurements from 1993 onward, and ERS-2 OPR for June 1995 to May 1996. Using the method of Caires and Sterl (2005), the ERA-40 wave reanalysis data were validated against TOPEX/Poseidon altimeter swath measurements (available at http: //www.aviso.altimetry.fr/en/data.html), as shown in Figure 1. The ERA-40 H_s and observed H_s were consistent during both JJA and DJF 2001. The correlation coefficient (CC), mean error (bias), root mean square error (RMSE), and scatter index (SI) were calculated to quantitatively analyze the accuracy of the ERA-40 H_s. There was an evident close relationship between ERA-40 H_s and observed H_s based on the results of CC, bias, RMSE, and SI. Additionally, Caires et al. (2004) determined that ERA-40 wave data are of better quality than similar wave reanalysis products. As a result, it can be concluded that ERA-40 wave reanalysis data are reliable in the Indian Ocean.

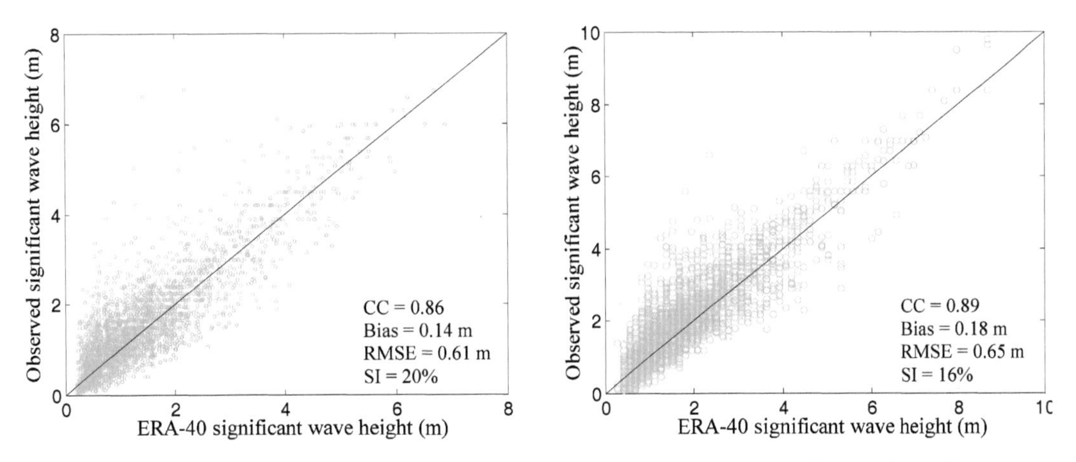

Figure 1. Correlation coefficients between ERA-40 significant wave height and observed significant wave height during (left) June-July-August (JJA) and (right) December-January-February (DJF) 2001.

2.2 Methods

The SIOW was first divided into four regions as shown in Figure 2: A (40°-60°S, 40°-60°E), B (40°-60°S, 60°-80°E), C (40°-60°S, 80°-100°E), and D (40°-60°S, 100°-120°E). The propagation characteristics of swells that originated in these four regions were then examined in the following analysis. Regions T1 (18°-28°S, 60°- 80°E), T2 (10°-25°N, 80°-100°E), and T3 (10°-25°N, 55°-70°E) were randomly selected experimental areas chosen to determine if the SIOW-related swell could propagate to regions T1, T2, and T3. Regions E (10°S-10°N, 70°-90°E) and F (5°-25°S, 100°-120°E) were the key regions of swell propagation terminal

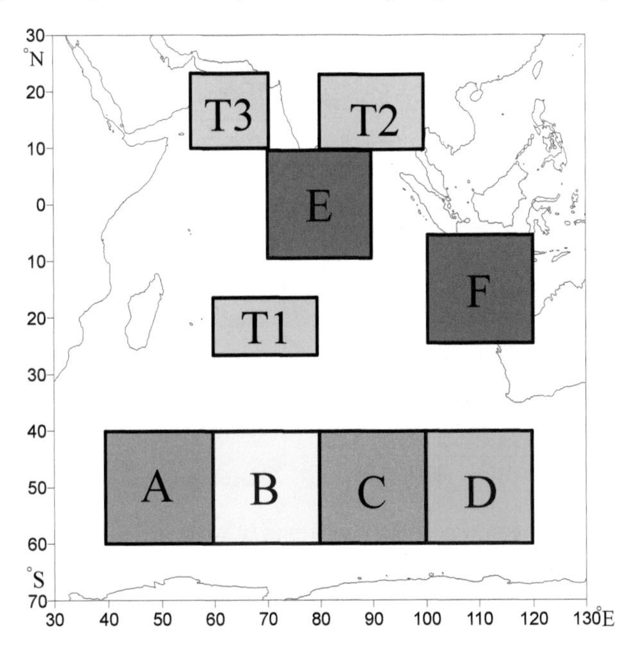

Figure 2. Geographical features of the Indian Ocean and the important regions in this study.

targets. Then, the *SIs* of regions A, B, C, and D were calculated to analyze the characteristics of swell propagation. Calculation of the simultaneous, leading, and lagging correlations between *SI* and H^s in each 1.5° × 1.5° bin then revealed the propagation route, terminal target, and propagation speed.

The *SIs* were defined as follows. H^s in region A at 00:00 1 January 1958 was first regionally averaged to acquire the regional mean H^s, which represented the current swell index of region A (represented as SI_A). The 6 hourly SI_A values for the period between September 1957 and August 2002 were similarly obtained. Using the same method, values of SI_B, SI_C and SI_D for the period between September 1957 and August 2002 were obtained at 6 h intervals.

3. Results

3.1 Zonal Mean Characteristics

The 6 hourly values of sea surface wind speed (WS), wind sea wave height (H^w), and mixed wave height (H_s) for July 2001 and January 2002 were selected. A 6 hourly zonal average for each element was then calculated to exhibit the northward propagation phenomenon of each element, as shown in Figure 3.

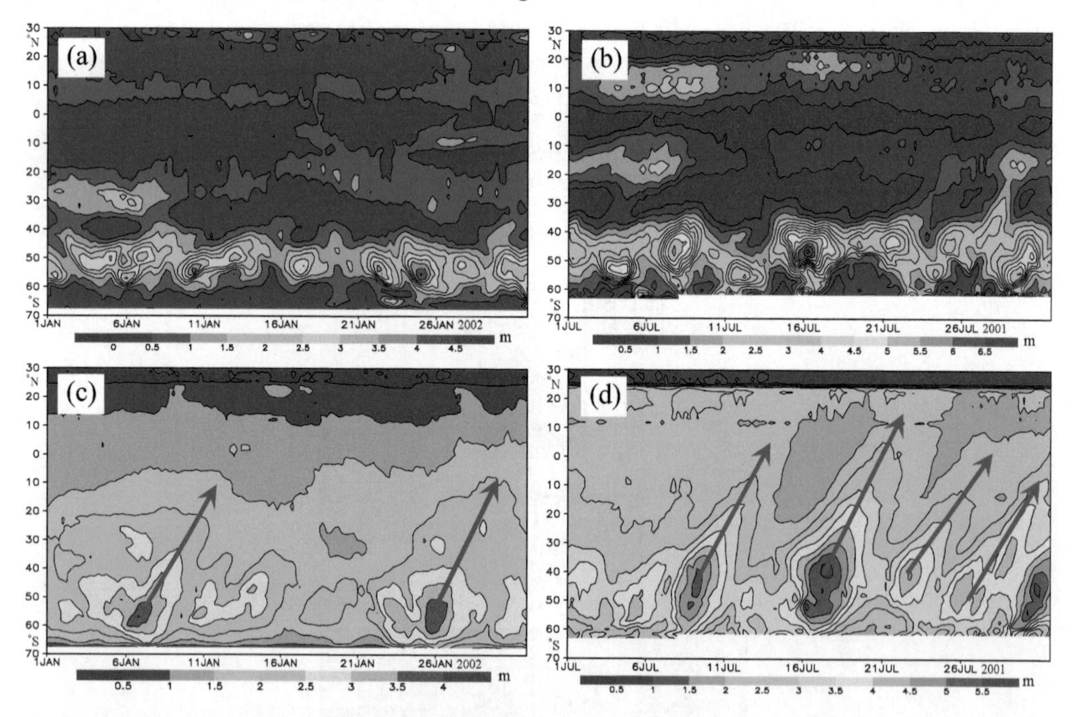

Figure 3. The 6 hourly zonal mean wave height of (a and b) wind-sea and (c and d) swell waves in (left) January 2002 and (right) July 2001. (a) Do zonal average of wind-sea wave height of the Indian Ocean at 00:00 1 January 2002 to obtain the current zonal mean value; 6 hourly zonal mean values of wind-sea wave height of the Indian Ocean in January 2002 were similarly obtained. The *x* axis is time, and the *y* axis is latitude. Using the same method of Figure 3a, the 6 hourly zonal mean values of (a and b) wind-sea wave height and (c and d) swell wave height of the Indian Ocean in January 2002 and July 2001 were obtained.

The zonal mean characteristics of WS (not shown) were similar to those of H^w (Figures 3a and 3b). As shown in Figure 3a, the 0.5 m contour of H^w is clearly truncated in the equatorial region and near 40°S in January. As shown in Figure 3b, the 0.5 m contour of H^w is clearly truncated in the equatorial region and near 30°S in July. This result indicates that H^w could not propagate northward through the two obstacles. The zonal mean characteristics of H^s are shown in Figures 3c and 3d. Obviously, the 2.0 m contour of H^s in January exhibits an obvious northward propagation and southward shrinkage characteristic. Similar phenomenon can also be found in July through the 2.5 m contour of H^s. The zonal mean characteristics of H_s were similar to that of H^s, not shown here. Bhowmick et al. (2011) presented an overview of the swell in the Indian Ocean propagating from the Southern Ocean during the year 2005, as seen from WAM model simulations. Their results showed that the swells of the southern Indian Ocean can propagate across the equator. Our results agreed well with those of Bhowmick et al. (2011). Moreover, swells even propagate beyond 10°N, as shown in Figure 2d. Just as the previous researchers pointed out, once generated, the swell wave can propagate very long distances with little attenuation until they break and dissipate upon reaching a coast (Alves, 2006; Munk et al., 1963; Semedo, 2010; Snodgrass et al., 1966). However, the wind-sea is quite different. According to the distribution of sea surface wind field (Figures omitted), it is not hard to find that the equator waters of the Indian Ocean is an obvious transition belt of wind direction from northeast monsoon to southeast trade wind in January, while it an obvious transition belt of wind direction from southwest monsoon to southeast trade wind in July. Similarly, near 40°S in January and near 30°S in July of the Indian Ocean are the transformation belts of wind direction from southeast trade wind to westerly. The significant variation of sea surface wind direction could disturb the wind-sea obviously.

3.2 Northward-Propagation Test of Swells

Figure 3 clearly shows the northward propagation of SIOW-related swells using 6 hourly data. Two experimental areas were randomly selected (regions T1, T2, and T3 in Figure 2) in this study for analysis to determine if SIOW-related swells could propagate to regions T1, T2, and T3. First, we selected a single JJA period (00: 00 on 1 June 2001 through 18: 00 on 31 August 2001); then, we drew 6 hourly curves of SI_A and H^s in region T1 (represented as $H^s{}_{T1}$), H^s in region T2 (represented as $H^s{}_{T2}$), and H^s in region T3 (represented as $H^s{}_{T3}$) to allow assessment of SI_A propagation to regions T1, T2, and T3. The results are shown in Figures 4-6.

As can be seen in Figure 4a, 6 hourly SI_A and $H^s{}_{T1}$ did not show good simultaneous correlation, although there was an apparent lagging or leading correlation. Therefore, we quantitatively calculated the simultaneous, leading, and lagging correlations between the 6 hourly SI_A and $H^s{}_{T1}$, shown in Figure 4b. It is clear that the simultaneous correlation was very poor; the correlation coefficient (R) was close to 0. When a lag of 10 intervals (60 h) relative to $H^s{}_{T1}$ was used, the R reached its peak value of 0.7, which is significant at the 0.001 level. To display the strong lagged correlation more clearly, curves of $H^s{}_{T1}$ using a 10 interval lag and SI_A are shown in Figure 5. The curves in Figure 5 show a high degree of similarity, which means that swells that originated in region A could propagate to region T1 (Madagascar waters) in JJA. It is obvious from Figures 6a and 6b that the simultaneous, leading, and lagging correlations between SI_A and $H^s{}_{T2}$ were very poor, within ±0.2, which means that swells that originated in region A did not propagate to region T2, which is in the central-north part of the Bay of Bengal. Similarly, as shown in Figures 6c and 6d, swells that originated in region A did not propagate to region T3 in this experiment, which is in the top of the Arabian Sea.

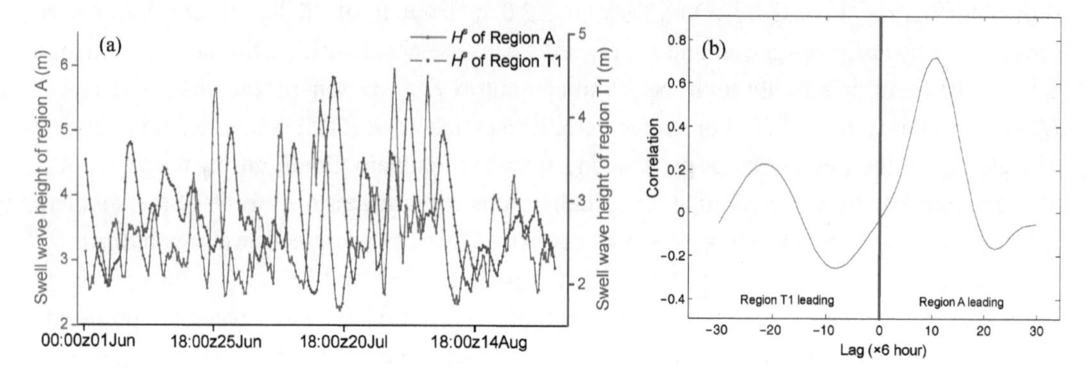

Figure 4. (a) 6 hourly H^s of regions A and T1 in JJA 2001 and (b) their simultaneous, leading, and lagging correlation coefficients.

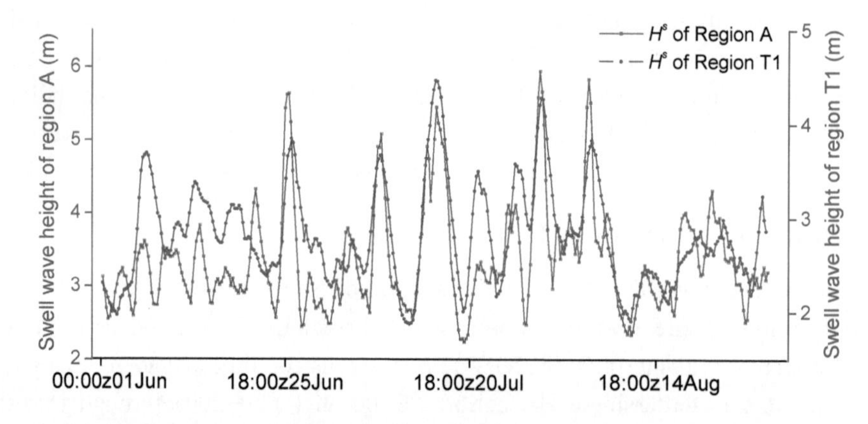

Figure 5. 6 hourly H^s of region A and H^s of region T1 (lagged by 10 intervals) in JJA 2001.

3.3 Main Propagation Routes of Swells

We found through the preceding analyses (Figures 4-6) that swells that originated in region A clearly propagated northward. We calculated the simultaneous, leading, and lagging correlations between 6 hourly SI_A and H^s in each 1.5° × 1.5° grid cell. Similarly, the correlations between 6 hourly SI_B and H^s, SI_C and H^s, and SI_D and H^s were calculated in each 1.5° × 1.5° grid cell.

3.3.1 Exploration of Main Propagation Routes

We selected a single JJA period (00:00 on 1 June 2001 through 18:00 on 31 August 2001) and then calculated the simultaneous, leading, and lagging correlations between 6 hourly SI_A and H^s in each 1.5° × 1.5° grid cell to find the main propagation route of swells that originated in region A. The leading correlation of H^s was not significant (not shown). The correlation coefficients between 6 hourly SI_A and simultaneous H^s in each 1.5° ×1.5° grid cell in JJA 2001 were calculated and are shown in Figure 7a. The correlation coefficients between the 6 hourly values of H^s (lagged by 48, 96, 120, 144, and 168 h) and SI_A in JJA 2001 were calculated for each 1.5° × 1.5° grid cell and are shown in Figures 7b-7f.

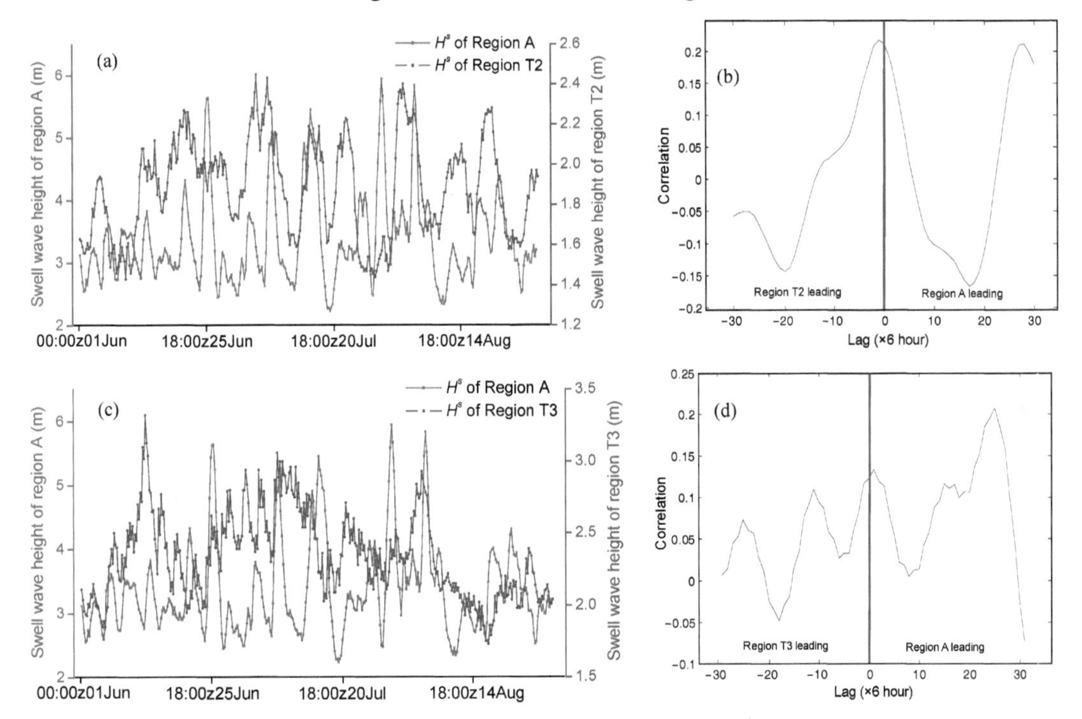

Figure 6. 6 hourly H^s of (a) regions A and T2 and (b) region A and T3 in JJA 2001 and (b and d) their simultaneous, leading, and lagging correlation coefficients separately.

The values of H^s in an approximately circular area near region A show good simultaneous correlation with SI_A (the areas that passed the 0.001 significant reliability level are colored) and are above 0.8 in the large center. After 21 h (i.e., with H^s lagged by 24 h), the shape of the region that was significant at the 0.001 reliability level changed from approximately circular to the northwest-southeast direction. During the process of propagation, the area that passed the significance threshold expanded, and the value of the CC diminished, which means that the swells that originated in region A gradually became diffuse and decreased during the propagation process. From 0 to 24 h, the large area of high CC moved northeast to Kerguelen Island. From 24 to 48 h, the large area continued to move northeast to the east of Madagascar. By 48 h, the swell had propagated northeast to the Bay of Bengal. The area that passed the significance threshold gradually shrank and the CC value decreased during the propagation process. By 168 h, the swell had propagated to the middle of the Bay of Bengal. Note that 120 h later, one of the branches had spread to the waters between the island of Java and Australia. As a result, it is not hard to find that the main body of the SIOW-related swells spread to the Sri Lanka waters, while partial of the SIOW-related swells can also spread to the south of the

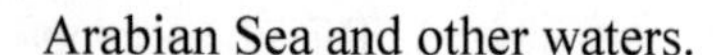

Arabian Sea and other waters.

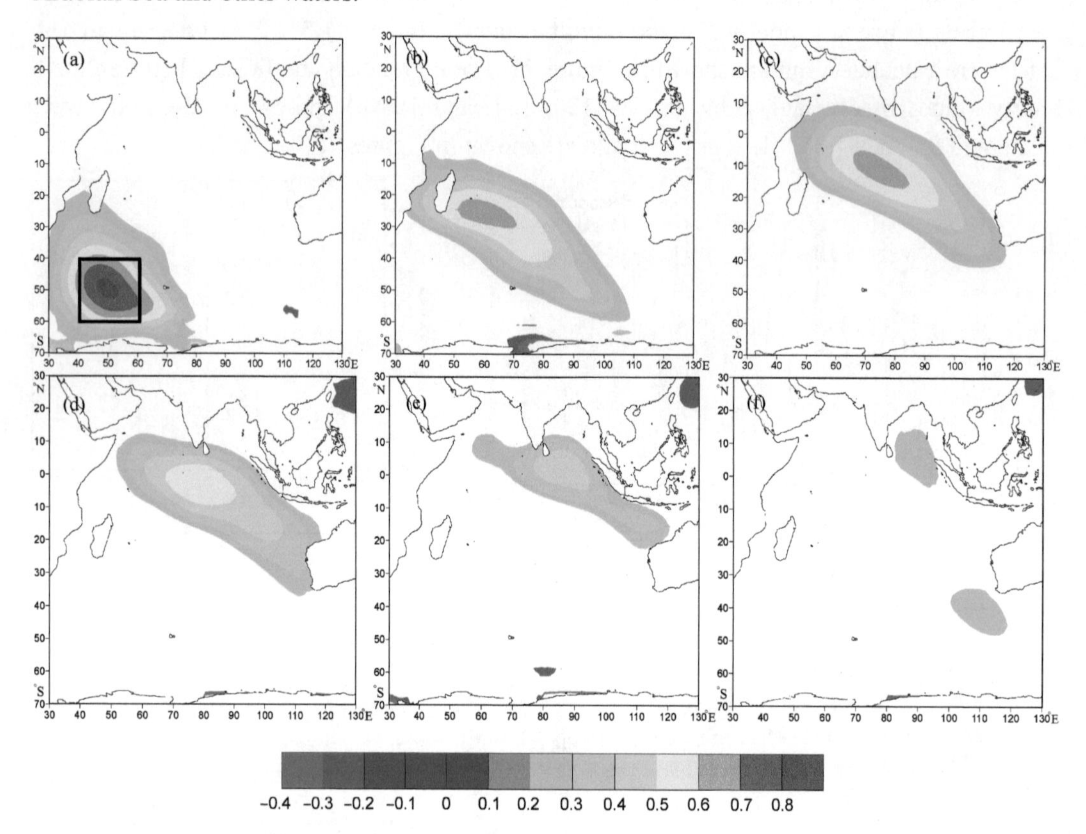

Figure 7. The calculated and contoured correlations between 6 hourly SI_A and H^s for each 1.5° × 1.5° bin in JJA 2001. (a) The correlation coefficients (CC) between 6 hourly SI_A and simultaneous H^s at each 1.5° × 1.5° bin in JJA 2001. (b) The CC between the 6 hourly values of H^s at each 1.5° × 1.5° bin for a lag of 48 h and SI_A in JJA 2001. (c, d, e, and f) The CC between the 6 hourly values of H^s at each 1.5° × 1.5° bin for separate lags of 96, 120, 144, and 168 h and SI_A in JJA 2001. In this figure, only the areas significant at a 0.001 reliability level are colored.

By using the observations, Remya et al. (2016) found 10 high swell events in North Indian Ocean (NIO) (also named Kallakkadal events) during 2005, which are caused by the swells propagating from south of 30°S. In all cases, 3-5 days prior to the high swell events in NIO, they observed a severe low pressure system, called the Cut-Off Low (COL) in the Southern Ocean. These COLs provides strong (about 25 m/s) and long duration (about 3 days) surface winds over a large fetch; essential conditions for the generation of long-period swells. The intense equator ward winds associated with COLs in the Southern Indian Ocean (SIO) trigger the generation of high waves, which propagate to NIO as swells. Furthermore, these swells cause high wave activity and sometimes Kallakkadal events along the NIO coastal regions. The overall propagation characteristic of SIOW-related swells in this studyis agreed with Remya et al. (2016).

3.3.2 Propagation Route in JJA and DJF

Figure 7 roughly shows the propagation of swells that originated in region A in JJA but does not show a clear route. To display the propagation route more clearly, the relatively large areas in each plot of Figure 7 were first highlighted and then connected with a single line, as shown in Figure 8a, where the red arrow represents the main propagation route. Similarly, the propagation routes of swells that originated in regions A, B, C, and D in JJA and DJF are presented in Figure 8.

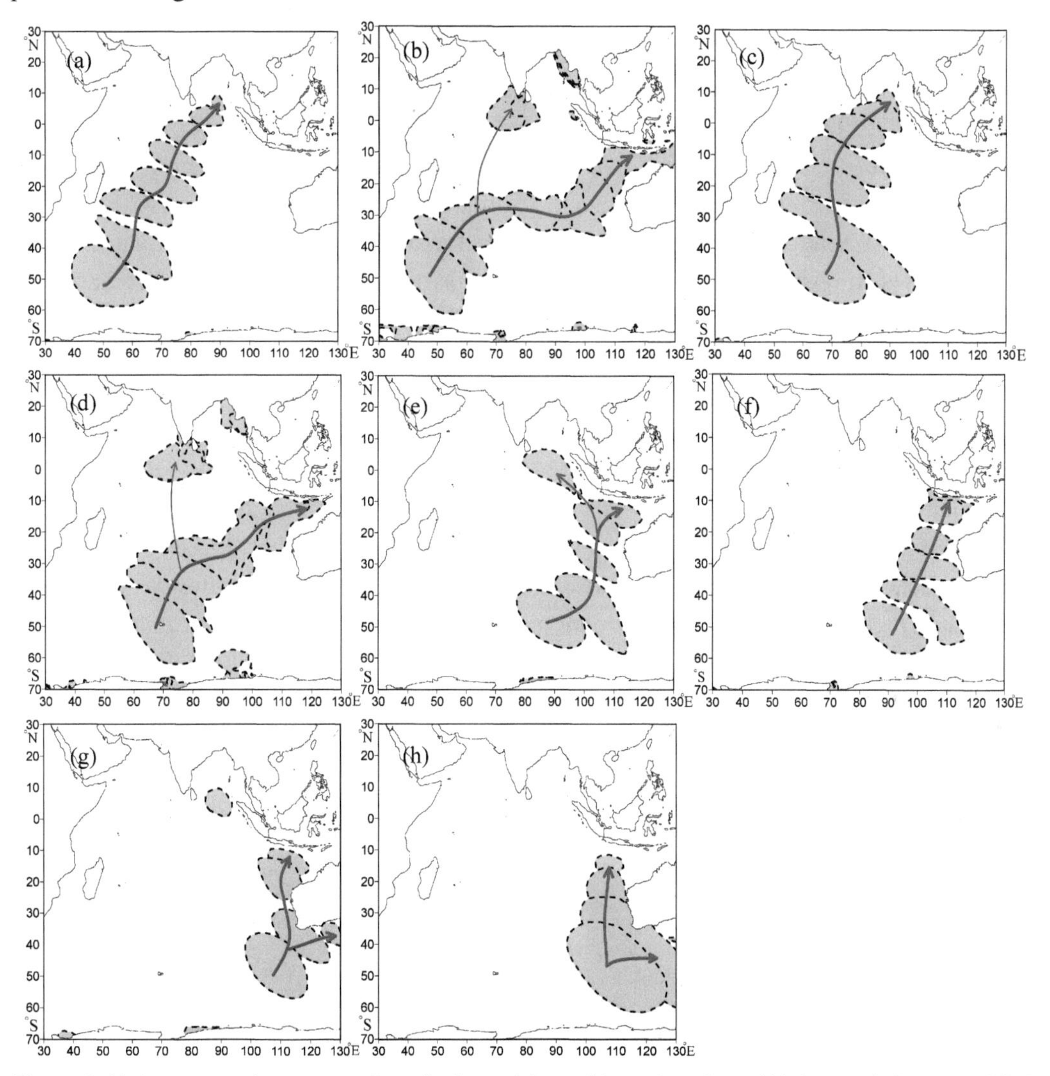

Figure 8. Main propagation routes of swells that originated in regions (a and b) A, (c and d) B, (e and f) C, and (g and h) D in (left) JJA 2001 and (right) DJF 2001. (a) The correlation coefficients (CC) between 6 hourly SI_A and simultaneous H^s at each 1.5° × 1.5° bin in JJA 2001 were calculated and contoured. Then, the relative large value of CC $\geqslant$ 0.5 was contoured with the dotted line as the edge and filled with blue color. The CC between H^s of separate 24, 48, 72, 96, 120, and 144 h lags with SI_A were similarly calculated. Then, the relative large values of CC $\geqslant$ 0.5 were also separately contoured with the dotted line as the edge and filled with blue color. The relatively large values of CC were connected by the red solid line. The swell propagation routes in region A were thus obtained. The arrow represents the propagation direction. (b-h) The same method used in (a) was used to plot the swell propagation routes in regions B, C, and D in JJA and DJF.

3.3.3 JJA

Swells that originated in region A mainly spread to the east of Sri Lanka (Sri Lanka waters) in a north- northeasterly direction and gradually became diffuse and diminished during the propagation process. Swells that originated in region B first propagated north-northeast for a short distance. At 40°S, the propagation direction turned to the north, and at 10°S, the propagation direction turned to the northeast; the propagation target was near Sri Lanka. Swells that originated in region C mainly spread to 105°E in an east- northeasterly direction for the first 24 h. Then, the main body of the swells moved northward along the western coast of Australia because of terrain effects. At 20°S, one branch spread in a northeasterly direction to the waters between the island of Java and Australia (the waters off Christmas Island), and another branch spread along the south coast of Sumatra to the waters east of Sri Lanka. Swells that originated in region D mainly spread in a northeasterly direction for the first 24 h. The swells were cut into two parts by land when they reached the southwest corner of Australia; one branch moved to the east along the south coast of Australia, whereas the other branch moved northward to the waters between the island of Java and Australia.

Comparison of the four routes in JJA (Figure 8, left) clearly shows that the propagation routes of swells that originated in regions C and D in JJA were significantly affected by the Australian landmass. The corresponding sea surface wind field in JJA 2001 is also presented in Figure 9a. Combined with Figures 8a and 8c and 9a, it is clear that the sea surface wind field plays an important leading role in determining the propagation routes of swells that originate in regions A and B. One important region is the southeast trade wind zone in the low latitudes of the southern Indian Ocean (shown by the red solid line in Figure 9a). Relatively strong southeast trade winds, which have an average wind speed of 7-10 m/s, caused the first change in swell propagation direction (from northeast to north) to pass through the equator. When spreading to Madagascar's northeastern coast, the strong southwest monsoon in the

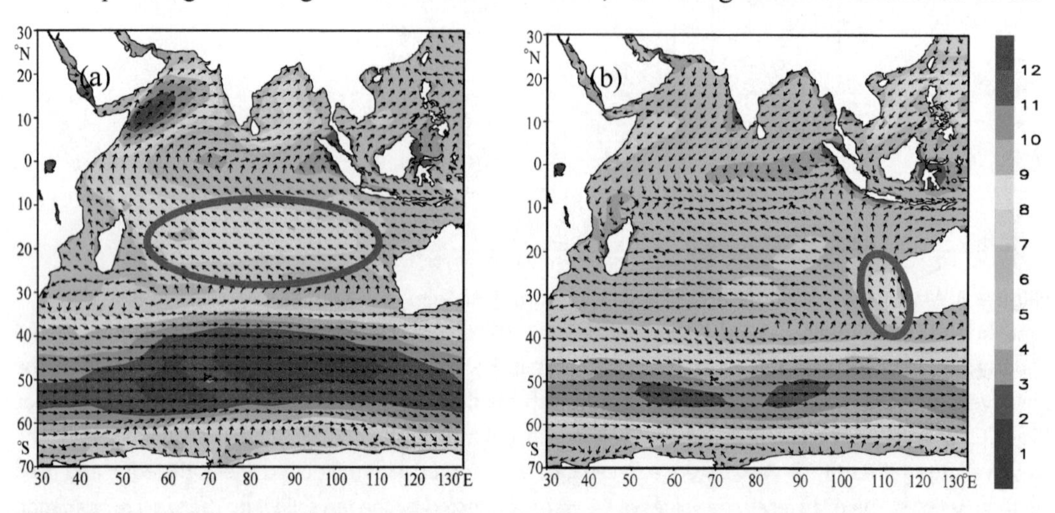

Figure 9. Sea surface wind field in (a) JJA 2001 and (b) DJF 2001. Colors represent wind speed (m/s) and unit arrows represent wind direction.

northern Indian Ocean had a significant effect on the propagation direction of swells and caused the direction of propagation to change from north to northeast.

3.3.4 DJF

Swells that originated in region A mainly spread in a north-northeasterly direction to 30°S, where they abruptly turned to the east-southeast. The propagation direction strongly changed again, to the northeast, when they spread to 100°E. The main body of the swells spread to the waters off Christmas Island 192 h later. The propagation route of swells that originated in region B was relatively simple; they propagated northeast and spread to the waters off Christmas Island approximately 144 h later. The propagation route of swells that originated in region C was also simple; they propagated north-northeast and spread approximately 120 h later to the waters off Christmas Island. Swells from region D were cut into two parts by land when they spread to the southwest corner of Australia; one branch moved to the east along the south coast of Australia, and the other branch moved northward to the waters off Christmas Island.

Comparing the four routes seen in DJF (Figure 8 right), it is clear that the propagation routes of swells that originated in regions A and B were different from those in JJA. Combined with the sea surface wind field (Figure 9b) and propagation route, it is clear that the southeast trade winds of the low-latitude southern Indian Ocean in DJF are not as strong as those in JJA, with the result that the southeast trade wind cannot lead the main body of swells from the west region of the SIOW to cross the equator. Additionally, the frequent inputs of cold air from the northern Indian Ocean may suppress the northward swell propagation. Note that the relatively large area of high wind speeds west of Australia (shown by the red solid line in Figure 9b) may play a positive role in the northward propagation of swells in the eastern region of the SIOW. The propagation route of swells from region D is significantly affected by the terrain of Australia.

Samiksha et al. (2012) found that a series of very high swells that originated at 40°S, off the southern tip of South Africa, propagated to the northeast and broke over the island of Reunion in the subtropical waters of the southern Indian Ocean. A similar phenomenon was also reported by Alves (2006). Our results agree with those of Samiksha et al. (2012) and Alves (2006), although the northward propagation is more obvious in our analysis. Aboobacker et al. (2011) and Glejin et al. (2013) determined that the predominant northward swell propagation (coming from the south) in the midwest of the Arabian Sea is disrupted during DJF and that the Shamal swells play an important role in this disruption. Similarly, the swells and wind-seas generated by the cold airs in DJF in the northern Indian Ocean may have disrupted the northward propagation of the swells. The southern hemisphere is winter in JJA, and the strength of SIOW wind speed in summer is weaker than that in winter. The above two phenomena determined that the swell in the west of the SIOW in DJF could not propagate northward as far as that in JJA.

3.4 Swell Propagation Speed and Terminal Target

It can clearly be seen in Figure 8 that swells that originated in regions A, B, C, and D in

JJA and DJF often propagated to two areas: the waters off Sri Lanka (region E in Figure 2) and Christmas Island (region F in Figure 2; the waters between the island of Java and Australia). The multiyear average CCs from January to December between swells of the above regions were calculated; only the CCs between regions C and F are presented in Figure 10.

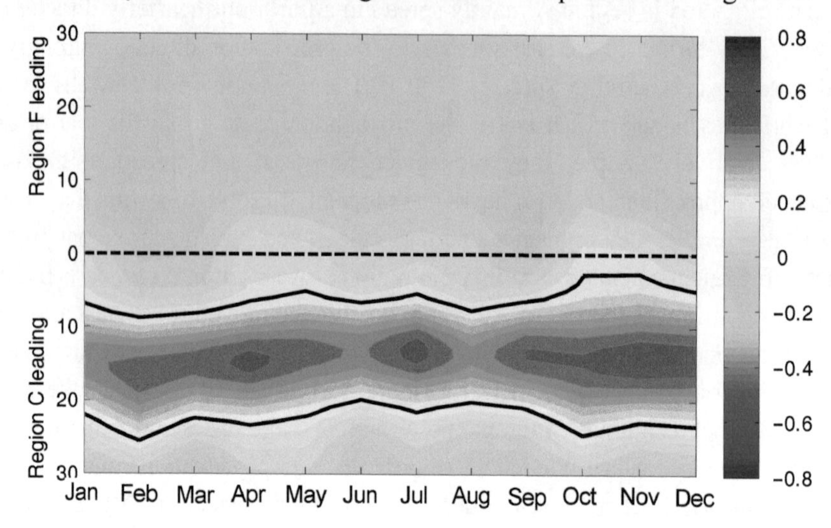

Figure 10. Monthly characteristics of leading and lagging correlations between swells in regions C and F. First, the CCs between 6 hourly SI_C and $H^S{}_F$ in September 1957 were calculated, including leads and lags of 30 intervals. Using the same method, the monthly correlations between 6 hourly SI_C and $H^S{}_F$ for the 540 months between September 1957 and August 2002 were obtained. Then, the multiyear average CCs from January to December were obtained. The area in the black solid line is significant at the 0.05 reliability level.

SI_C clearly plays a leading role in each month. The correlation coefficient (CC) usually reaches the peak value and SI_C leads by 10-20 intervals (60-120 h). July was selected as an example to analyze the speed of swell propagation. In July, the CC reached its peak value of 0.53 (significant at the 0.001 level) when SI_C led by 14 intervals (84 h). This result means that the propagation of swells from region C to region E required approximately 84 h. The swell propagation speed from region C to region E in each month could be obtained in the same way.

Using the method shown in Figure 10, the multiyear average leading and lagging correlations from January to December between 6 hourly SI_A and $H^S{}_E$, SI_B and $H^S{}_E$, SI_C and $H^S{}_E$, SI_D and $H^S{}_E$, SI_A and $H^S{}_F$, SI_B and $H^S{}_F$, SIC and $H^S{}_F$, and SI_D and $H^S{}_F$ were calculated. Then, the seasonal characteristics of the above correlations were analyzed as shown in Table 1, with February, May, August, and November selected as the representative months for DJF, March-April-May (MAM), JJA, and September-October-November (SON), respectively.

3.4.1 Propagation Speed of Swells That Originated in Region A

The CCs between SI_A and lagged $H^S{}_E$ were best for February (0.55) and November (0.50), followed by May (0.42), and were smallest in August (0.38). The CCs between SI_A and $H^S{}_E$ usually reached their peak value when $H^S{}_E$ was lagged by 22-24 intervals (132-144 h). This result means that the propagation of swells from region A to region E required 132-144 h. The CCs between H^S_F and SI_A were much smaller than those between H^S_E and SI_A for each month. The propagation of swells from region A to region F required 27 intervals (162 h) in February and 23-24 intervals (138-144 h) in May, August, and November, which means that the speed of swell propagation from region A to region F in February was much slower than in other months.

Table 1 Correlation Coefficients (CCs) Between Swells in Regions A/B/C/D and Regions E/F (Arrows (→) indicate Swell Propagation from Regions A/B/C/D to E/F; Figures Without Underlining Indicate Significance at the 0.001 Level; Figures With Underlining Are not Statistically Significant)

Swells' propagation direction	February		May		August		November	
	CC	Lagging intervals of E/F (×6h)	CC	Lagging intervals of E/F(×6h)	CC	Lagging intervals of E/F(×6 h)	CC	Lagging intervals of E/F(×6h)
A→E	0.55	24	0.42	22	0.38	22	0.50	23
A→F	0.30	27	0.31	23	0.26	23	0.36	24
B→E	0.59	20	0.43	18	0.40	18	0.54	19
B→F	0.39	22	0.42	18	0.35	18	0.46	19
C→E	0.47	15	0.32	14	0.33	14	0.47	14
C→F	0.48	17	0.47	13	0.41	14	0.54	15
D→E	0.34	10	0.17	7	0.24	8	0.34	7
D→F	0.47	12	0.43	9	0.38	9	0.52	9

3.4.2 Propagation Speed of Swells That Originated in Region B

The propagation speed and terminal target of swells that originated in region B were similar to those of swells that originated in region A. Additionally, they usually required 4-5 more intervals (24-30 h) to propagate from region A to regions E and F.

3.4.3 Propagation Speed of Swells That Originated in Region C

The CCs between H^S_F and SI_C were much greater than the CCs between H^S_E and SI_C for each month, which means that the swells that originated in region C mainly propagated to the waters off Christmas Island, although a small branch could spread to the waters off Sri Lanka. The propagation of swells from region C to region E or F usually required 13-15 intervals (78-90 h).

3.4.4 Propagation Speed of Swells That Originated in Region D

The CCs between H^S_F and SI_D were much greater than those between H_E^S and SI_D for each month, which means that swells that originated in region D mainly propagated to the waters off Christmas Island. It is worth noting that CCs between H^S_E and SI_D in February, August, and November were also significant at the 0.05 level. Combining this result with Figures 8g and 8h, we found that swells that originated in region D did not generally spread to Sri Lanka waters, although the correlation between SI_D and H^S in a small region near Sri Lanka passed the significance threshold. This result occurred because region D and regions A, B, and C are all located in the SIOW, which produced similar H^s values in these four regions. Moreover, the H^s values of regions A, B, and C could spread to region E, which would cause swells from region E to have the features of regions A, B, and C. Finally, although swells that originated in regions D and E appeared to have some correlation with one another, no significant propagation from region D to region E occurred.

Overall, the propagation of swells from SIOW to region E and F usually required 3-6 days. Remya et al. (2016) found that 10 high swell events in North Indian Ocean (NIO) (also named Kallakkadal events) during 2005 are caused by the swells propagating from south of 30°S. In all cases, 3-5 days prior to the high swell events in NIO. Nayak et al. (2013) found that the low-frequency swells from the Southern Ocean reach the southern tip of Indian mainland in about 4 days without much energy dissipation. Our results agreed well with those of Remya et al. (2016). In addition, the propagation speed of swells in the Indian Ocean was fastest (required the shortest time) in May and August, followed by November. It was slowest (required the longest time) in February. This result occurred because DJF in the southern Indian Ocean corresponds to summer, whereas JJA corresponds to winter. The intensity of westerly winds in winter was obviously stronger than that in summer, which resulted in a faster swell propagation speed in JJA. Remya et al. (2016) have provided important insights into the SIO meteorological conditions prior to the NIO high wave events. They pointed out that the strong equator ward surface winds associated with COL development. These strong and persistent winds generated high waves at the COL active region. The strong south south-westerly surface wind fields associated with the COL, generated waves that propagate toward north north-east into the NIO in a band of frequencies with a peak wave period of around 20 s. And in all cases, 3-5 days prior to the high swell events in NIO, they observed a severe low pressure system, called the COL in the Southern Ocean. Zheng and Li (2017) analyzed the interannual and interdecadal variabilities and intraseasonal oscillations of WS, H^w, H^s, and H_s in the Roaring Forties and tropical waters of the Indian Ocean. Their results show that the WS and H^s in the Roaring Forties and H^s in the tropical waters of the Indian Ocean share a common period of approximately 8 days (weekly oscillation) on an intraseasonal scale. And approximately 132-138 h are required for H^s to propagate from the Roaring Forties to the tropical waters of the NIO. Based on the results from Remya et al. (2016) and Zheng and Li (2017), it means that the natural hazards along the NIO coasts can be forecasted at least 2 days in advance if the meteorological conditions of the SIO are properly monitored.

Note that the CCs in Table 1 were usually largest in February and November and smallest in August, although the Southern Hemisphere westerlies are stronger in August than in February and November. The southwest monsoon in the northern Indian Ocean is also very strong in this season. Swells generated by the strong southwest monsoon can also affect the waters off Sri Lanka. As a result, JJA swells in the waters off Sri Lanka had the signals of swells from both the SIOW and the southwest monsoon, which resulted in a weaker relationship between swells of different regions in August. The Southern Hemisphere westerlies in February and November are weak because it is summer, which does not promote the long-distance northward propagation of swells as in August. The wind-seas and swells generated by cold air from the northern Indian Ocean may have disrupted the northward propagation of the swell. However, the intensity of cold air was much weaker than the southwest monsoon in JJA, which meant less impact on the northward swell propagation. As a result, the CCs between swells of different regions in February and November were greater than in August.

3.4.5 Propagation Terminal Target

Examination of Figure 8 and Table 1 clearly shows that most SIOW-related swells in DJF mainly propagated to the waters off Christmas Island. Moreover, a small portion of swells that originated in the west of the SIOW (regions A and B) could spread to Sri Lanka waters. In JJA, swells that originated in the west of the SIOW mainly propagated to Sri Lanka waters, whereas swells that originated in the east of the SIOW (regions C and D) mainly propagated to the waters off Christmas Island.

Combining Figure 8 and Table 1, it is clear that there was a close relationship among swells in the SIOW and the waters off Sri Lanka and Christmas Island. To directly display this close relationship, we contoured the values of SI_A, SI_B, SI_C, SI_D, and lagged H^S_F in DJF 2001 (this period was selected at random; similar phenomena can also be found in other time periods), as shown in Figure 11. Obviously, the curves in Figure 11 are quite consistent. This result again shows that swells from the SIOW can propagate to the waters off Christmas Island.

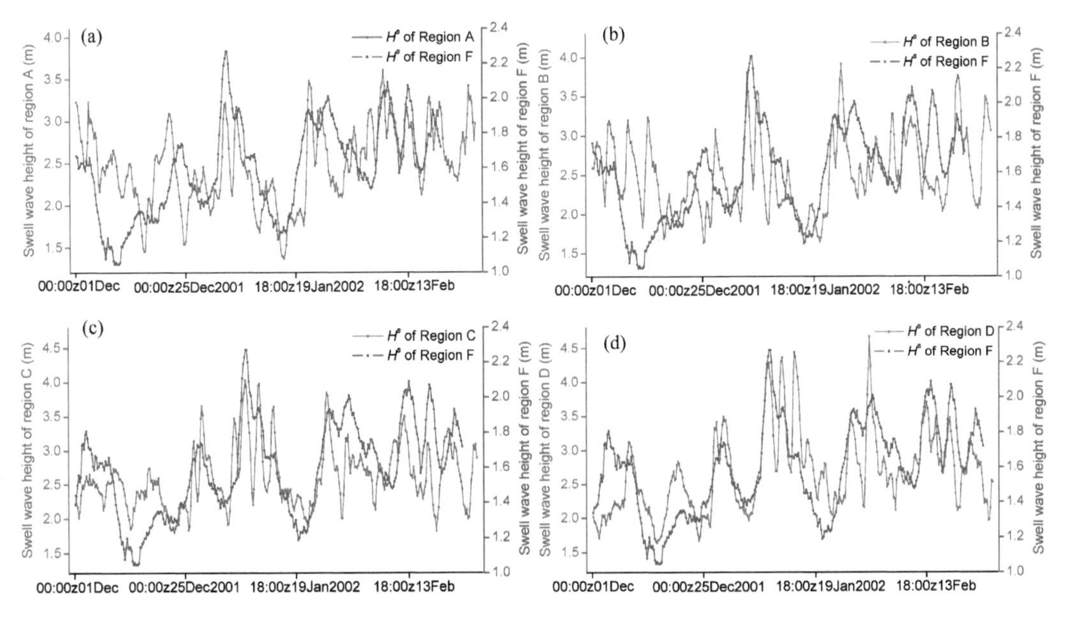

Figure 11. 6 hourly H^s of regions A, B, C, and D and lagged H^s of regions F in DJF 2001.

4. Conclusion and Prospects

The propagation routes, terminal targets, and propagation speeds of SIOW-related swells were analyzed based on the ERA-40 wave reanalysis from ECMWF. The results were as follows:

The main body of the SIOW-related swells mainly spread to two regions, the waters off Sri Lanka and Christmas Island, while the branches spread to the Arabian Sea and other waters. In JJA, swells from the western part of the SIOW (40°-80°E) usually propagated to the waters off Sri Lanka, whereas swells from the east region (80°-120°E) usually propagated to the waters off Christmas Island. In DJF, swells from the entire region of the SIOW mainly propagated to the waters off Christmas Island.

The Australian landmass had a significant impact on swell propagation routes in the eastern region of the SIOW. When spreading to the southwest corner of Australia, swells were cut into two parts by the land; one branch moved to the east along the southern coast of Australia, whereas the other branch moved northward to the waters off Christmas Island.

The sea surface wind field had an obvious leading or inhibitory effect on swell propagation. The southeast trade winds in the low-latitude southern Indian Ocean had an important leading role in the propagation route of swells in the middle and western regions of the SIOW, especially in JJA. Relatively strong southerly winds in the west of Australia played a positive role in the northward propagation of swells, especially in DJF. The wind-sea and swell generated by the cold air of the northern Indian Ocean disrupted the northward propagation of the swell in DJF.

The swell propagation speed in the Indian Ocean was fastest in May and August, followed by November, and was slowest in February. Swell usually required 132-144 h to propagate from region A to regions E or F, 108-132 h to propagate from region B to regions E or F, 78-90 h to propagate from region C to E/F, and 54 h to propagate from region D to region F.

The relationships between swells in the SIOW and the waters off Sri Lanka, as well as those between swells in the SIOW and the waters off Christmas Island, were usually best in February and November, followed by May, and were smallest in August.

This study presented a method to exhibit the propagation route of swells, and JJA 2001 and DJF 2001 were taken as a case study. In the future actual application, it is necessary to present the propagation features of swells in each month, thus to establish a practical reference for swell wave power generation, ocean wave forecasting, research on global climate change, and other applications.

Acknowledgments

This work was supported by the Open Research Fund of State Key Laboratory of Estuarine and Coastal Research (grant SKLEC-KF201707), the Nature Science Foundation of China (41490642 and 41775165), the National Basic Research Program of China (2013CB956203 and 2015CB453200), the Key Laboratory of Renewable Energy, Chinese Academy of Sciences (Y707k31001), and the Junior Fellowships for CAST Advanced Innovation Think-tank Program (DXB-ZKQN-2016-019). All the authors would like to thank anonymous referees and the editor for providing their excellent comments and valuable advice for improving this paper. All the authors would like to thank ECMWF for providing the ERA-40 wave reanalysis (Available at http: //apps.ecmwf.int/ datasets/data/era40-daily/levtype=sfc/).

References

Aboobacker, V. M., Vethamony, P., & Rashmi, R. (2011). Shamal" swells in the Arabian Sea and their influence along the west coast of India.

Ardhuin, F., Chapron, B., & Collard, F. (2009). Observation of swell dissipation across oceans. *Geophysical Research Letters*, *36*, L06607. https: //doi.org/10.1029/2008GL037030.

Arinaga, R. A., & Cheung, K. F. (2012). Atlas of global wave energy from 10 years of reanalysis and hindcast data. *Renewable Energy*, *39*, 49-64.

Bhowmick, S. A., Kumar, R., Chaudhuri, S., & Sarkar, A. (2011). Swell propagation over Indian Ocean Region. *International Journal of Ocean and Climate Systems*, *2*(2), 87-99.

Caires, S., &Sterl, A. (2005). Validation and non-parametric correction of significant wave height data from the ERA-40 reanalysis. *Journal of Atmospheric and Oceanic Technology*, *22*, 443-459.

Caires, S., Sterl, A., Bidlot, J. R., Graham, N., & Swail, V. (2004). Intercomparison of different wind wave re-analyses. *Journal of Climate*, *17*(10), 1893-1913.

Caires, S., Sterl, A., & Gommenginger, C. P. (2005). Global ocean mean wave period data: Validation and description. *Journal of Geophysical Research*, *110*, C02003. https: //doi.org/10.1029/2004JC002631.

Chen, B., Yang, D., Meneveau, C., &Chamecki, M. (2016). Effects of swell on transport and dispersion of oil plumes within the ocean mixed layer. *Journal of Geophysical Research: Oceans*, *121*, 3564-3578. https: //doi.org/10.1002/2015JC011380.

Geophysical Research Letters, 38, L03608. https: //doi.org/10.1029/2010GL045736 Alves, J. H. (2006). Numerical modeling of ocean swell contributions to the global wind-wave climate. *Ocean Modelling*, *11*, 98-122.

Glejin, J., Kumar, V. S., Nair, T. M. B., Singh, J., & Mehra, P. (2013). Observational evidence of summer Shamal swells along the west coast of India. *Journal of Atmospheric and Oceanic Technology*, *30*, 379-388.

Hemer, M. A., Church, J. A., & Hunter, J. R. (2007). Waves and climate change on the Australian coast. *Journal of Coastal Research*, *50*, 432-437.

Hwang, P. A. (2008). Observations of swell influence on ocean surface roughness. *Journal of Geophysical Research*, *113*, C12024. https: //doi. org/10.1029/2008JC005075.

Jiang, H., Stopa, J. E., Wang, H., Husson, R., Mouche, A., Chapron, B., & Chen, G. (2016). Tracking the attenuation and nonbreaking dissipation of swells using altimeters. *Journal of Geophysical Research: Oceans*, *121*, 1446-1458. https: //doi.org/10.1002/2015JC011536.

Komen, G. J., Cavaleri, L., Doneland, M., Hasselmann, K., Hasselmann, S., & Janssen, P. A. E. M. (Eds.). (1994). *Dynamics and modelling of ocean waves*. Cambridge, UK: Cambridge University Press.

Munk, W. H., Miller, G. R., Snodgrass, F. E., & Barber, N. F. (1963). Directional recording of swell from distant storms. *Philosophical Transactions of the Royal Society London*, *A255*, 505-584.

Nayak, S., Bhaskaran, P. K., Venkatesan, R., & Dasgupta, S. (2013). Modulation of local wind-waves at Kalpakkam from remote forcing effects of Southern Ocean swells. *Ocean Engineering*, *64*, 23-35.

Remya, P. G., & Kumar, R. (2013). Impact of diurnal variation of winds on coastal waves off South East Coast of India. *International Journal of Ocean and Climate Systems*, *4*(3), 171-179.

Remya, P. G., Kumar, R., Basu, S., & Sarkar, A. (2012). Wave hindcast experiments in the Indian Ocean using

MIKE 21 SW model. *Journal of Earth System Science, 121*(2), 385-392.

Remya, P. G., Vishnu, S., Praveen Kumar, B., Balakrishnan Nair, T. M., & Rohith, B. (2016). Teleconnection between the North Indian Ocean high swell events and meteorological conditions over the Southern Indian Ocean. *Journal of Geophysical Research: Oceans, 121*, 74767494. https: //doi.org/10.1002/ 2016JC011723.

Samiksha, S. V., Vethamony, P., Aboobacker, V. M., & Rashmi, R. (2012). Propagation of Atlantic Ocean swells in the north Indian Ocean: A case study. *Natural Hazards and Earth System Sciences, 12*, 3605-3615.

Semedo, A. (2010). *Atmosphere-ocean interactions in swell dominated wave fields* (pp. 53-54). Uppsala, Stockholm: Uppsala University Press.

Semedo, A., Suseelj, K., & Rutgersson, A. (2009). *Variability of wind sea and swell waves in the North Atlantic based on ERA-40 re-analysis*, Paper presented at Proceedings of the 8th European Wave and Tidal Energy Conference, Uppsala, Sweden.

Semedo, A., Suseelj, K., Rutgersson, A., & Sterl, A. (2011). A global view on the wind sea and swell climate and variability from ERA-40. *Journal of Climate, 24*, 1464-1479.

Snodgrass, F. E., Groves, G. W., Hasselmann, K. F., Miller, G. R., Munk, W. H., & Powers, W. H. (1966). Propagation of swell across the Pacific. *Philosophical Transactions of the Royal Society London, A259*, 431-497.

Wu, L., Rutgersson, A., Sahlee, E., & Guo Larsen, X. (2016). Swell impact on wind stress and atmospheric mixing in a regional coupled atmosphere-wave model. *Journal of Geophysical Research: Oceans, 121*, 4633-4648. https: //doi.org/10.1002/2015JC011576.

Zheng, C. W., & Li, C. Y. (2017). Analysis of temporal and spatial characteristics of waves in the Indian Ocean based on ERA-40 wave reanalysis. *Applied Ocean Research, 63*, 217-228.

Zheng, C. W., Shao, L. T., Shi, W. L., Su, Q., Lin, G., Li, X. Q., & Chen, X. B. (2014). An assessment of global ocean wave energy resources over the last 45 a. *Acta Oceanologica Sinica, 33*(1), 92-101.

Zheng, C. W., Wang, Q., & Li, C. Y. (2017). An overview of medium- to long-term predictions of global wave energy resources. *Renewable and Sustainable Energy Reviews, 79*, 1492-1502.

6

Stratosphere and Near Space Environment

Evolution of QBO and the Influence of Enso

Li Chongyin (李崇银), Long Zhenxia (龙振夏)

LASG, Institute of Atmospheric Physics, Academia Sinica, Beijing

Keywords: quasi-biennial oscillation (QBO), El Niño /Southern oscillation (ENSO).

Ⅰ. Introduction

QBO (quasi-biennial oscillation), originally referred to the quasi-periodic oscillation of the wind field with about 27 months in the tropical lower stratosphere. Afterwards, a lot of investigations indicated that QBO is an important phenomenon for the atmospheric circulation in the tropical lower stratosphere. It is not only related to the circulation and environment (such as the mount of O_3) in the stratosphere, but also has some influence on the variations of the global weather and climate. The studies have shown that this kind of oscillation between the easterly and westerly always occurs at 30 km altitude, and propagates downwards with a constant amplitude above 23 km. The mean period of the transition between the easterly and the westerly is about 27 months. The oscillation phenomenon is symmetrical to the equator[1-5]. The generation mechanism and propagation property of the QBO are regarded as an upward-propagating theory of the planetary wave[6, 7]. In other words, the upward propagation of the mixed Rossby-gravity wave and Kelvin wave in the tropical troposphere and the interaction between them and the wave with half year period in the stratosphere can lead to QBO. The evolution of the QBO and its relation to ENSO are not clear yet. In this article, they will be investigated by using data analyses.

Ⅱ. Evolution of QBO

Where is the origin in the global tropics? Through analyzing the wind data at 50 hPa, Belmont suggested that QBO originates in the tropical America and then it spreads both eastwards and westwards[8].

According to the generation mechanism of QBO, Kelvin wave and mixed Rossby-gravity wave in the tropical troposphere are the basic factors. Holton[9] and Li Chongyin[10] have still indicated in their theoretic studies that Kelvin wave and mixed Rossby-gravity wave are excited by the cumulus convection heating in the tropics. The strongest convection heating within the global tropics is in South Asia and the equatorial western Pacific area and not in the tropical America. Therefore, Belmont's result does not correspond to the generation mech-

Received March 20, 1990.

anism of QBO. Recently, Li Chongyin also indicated that the vertical propagation of the planetary wave is directly related to vertical wind shear. The stronger the vertical wind shear, the more favourable to the upward propagation of the planetary wave[11]. Since there is the strongest vertical wind shear in South Asia, it is favourable for the upward propagation of the planetary wave in this region. Therefore, it should be favourable for the occurrence of QBO nearby the equator in South Asia.

By analysing the transition between the easterly and the westerly at 30 hPa nearby the equator (the data abstracted from Monthly Climatic Data for the World), it is clear that we cannot simply say that the transition between the easterly and the westerly originates in the tropical America. Fig. 1 is an example for the variation of the zonal wind at 30 hPa nearby the equator. We can find that the transition from the easterly to the westerly obviously originates nearby the equator in South Asia. And the transition from the westerly to the easterly seems to originate in tropical America and the equatorial western Pacific. It should be especially pointed out that the difference is generally only 1 — 3 months within the whole globe for the transition between the easterly phase and the westerly phase (a few of them are about 6 months). So the transition between the easterly and the westerly is very rapid within the whole globe.

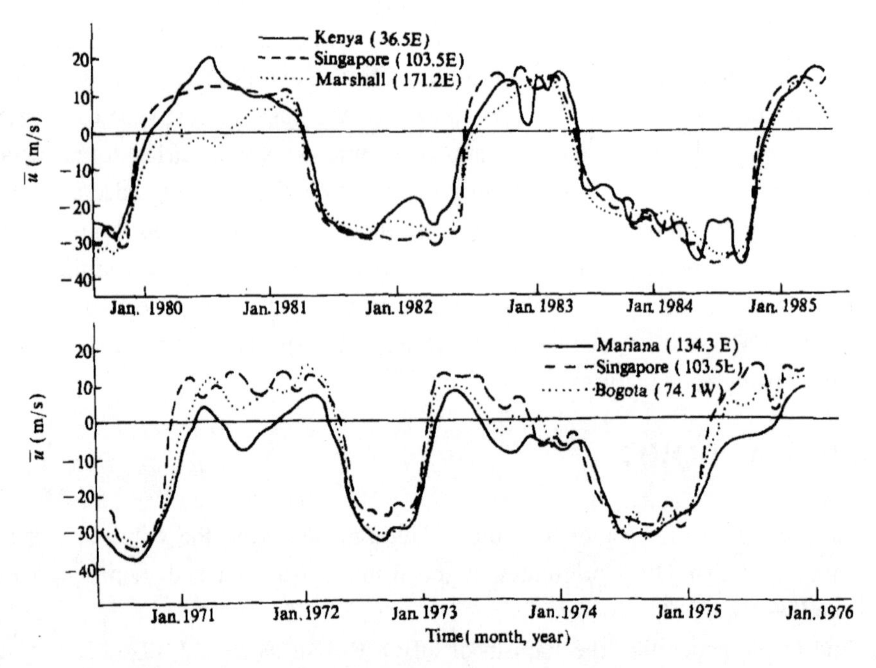

Fig. 1. The variation of the mean zonal wind at 30 hPa nearby the equator.

In order to study further the properties of zonal wind variation, Fig. 2 gives the time-longitude section of pentad mean zonal wind at 30 hPa nearby the equator from 1976 to 1977.Obviously the transition from the easterly to the westerly originated in 90—110°E, next occurring in 70 — 90°E. The former was 15—20 days earlier than the latter, and the latter was 5 — 10 days earlier than the others. The difference in the whole globe for the transition from the westerly to the easterly is smaller, 10 — 15 days or so, and the transition originates in tropical America and the equatorial western Pacific.

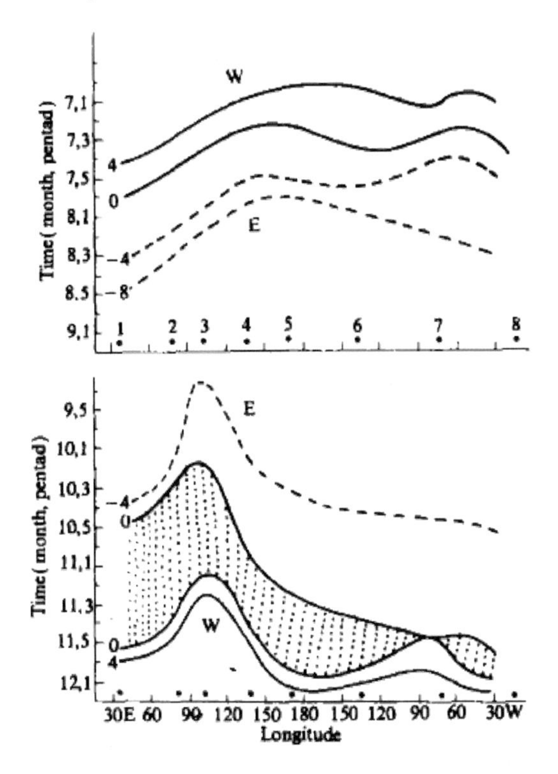

Fig. 2. The time-longitude section of the pentad mean zonal wind at 30 hPa nearby the equator.
E and W represent the easterly and the westerly respectively. The shaded parts are the unstable period of the easterly and the westerly, 1—8 represent the stations in (1°S, 37°E), (8°N, 78°E), (8°N. 101°E), (9°N, 139°E), (8°N, 169°E), (11°S, 139°W), (9°N, 79°W), and (8°S, 14°W) respectively.

According to the planetary wave theory of QBO put forward by Lindzen and Holton, if there is a planetary wave propagating westwards in the upper troposphere (mixed Rossby-gravity wave), then the upward propagation of the planetary wave will lead to the acceleration of mean easterly flow in the lower stratosphere, and the eastward planetary wave (Kelvin wave) will lead to the acceleration of the mean westerly flow. Therefore, the above- mentioned results indicate that the region nearby the equator in South Asia is favourable to the upward propagation of Kelvin wave, leading the westerly wind to be established originally over this zone.

III. ENSO Influence on QBO Variation

It has been pointed out that the mean period of QBO is about 27 months. But, at each observation station, the easterly and the westerly are not changed according to the sine or cosine law. The easterly is stronger than the westerly at most stations. But the duration of the easterly phase is close to the one of the westerly phase.

Recently, a series of studies in relation to ENSO indicate that ENSO event has obvious influence on the general circulation and climatic variation. Then, has ENSO obviously influence on the wind field in the tropical stratosphere? Fig. 3 gives the temporal variation of zonal wind at 30 hPa in Singapore (solid curve) and Mariana (dashed curve) between 1967 and 1988. The El Niño events are marked below the curves with thick transverse in the figure. It can be seen that these El Niño events obviously shorten the westerly period of QBO, which is originated after the El Niño event. Table 1 gives the durations of the zonal westerly at 30 hPa in Singapore and Mariana. For El Niño events, the lasting time of the westerly is 10·3 months and 5.8 months in Singapore and Mariana respectively; but 17.2 months and 12.0 months respectively for non-El Niño events. So the occurrence of El Niño has a great influence upon the westerly phase of QBO. The duration is shortened by about 41%—52%.

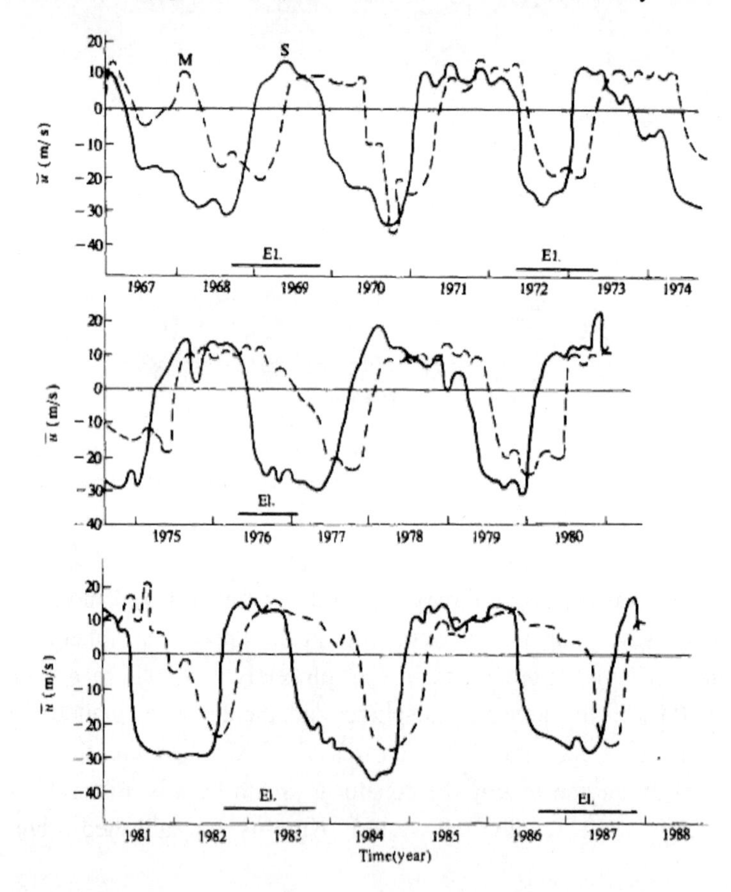

Fig. 3. The temporal variation of the zonal wind at 30 hPa in Singapore (solid curve) and Mariana (dashed curve) between 1967 and 1988.

Table 1. The Mean Lasting Time of the Westerly at 30hPa in Singapore and Mariana (Months)

	El Niño	Non-El Niño
Singapore	10.3	17.2
Mariana	5.8	12.0

A very interesting phenomenon is that El Niño event has no influence on the easterly phase of QBO, including its intensity and lasting time. The problem that El Niño event can lead to the shortening of the westerly lasting time of QBO and its dynamical mechanism will be studied further.

Ⅳ. Conclusion

We have obtained some new results about the evolution of QBO based on the analyses of zonal wind data at 30 hPa.

(1) The transition from the easterly phase of QBO to the westerly phase is very rapid in the whole globe. We have further found that the transition from the easterly to the westerly obviously originates in South Asia. But the transition from the westerly to the easterly originates in tropical America and the equatorial western Pacific.

(2) The occurrence of El Niño event has an obvious influence on the variation of QBO. El Niño event leads to the obvious shortening of duration of the westerly phase of the QBO (about 50%), but it has no obvious influence on the easterly phase of QBO (intensity and duration).

References

[1] Read, R. J. et al., *J. Geophys. Res.*, **66**(1961). 813-818.
[2] Read, R. J. & Rogers, D. G., *J. Atmos. Sci.*, **19**(1962), 127-135.
[3] Tucker, G. B. & Hopwood, J. M., *ibid.* **25**(1968), 293-298.
[4] Angell, J. K. & Korshover, J., *Mon. Wea. Rev.*, **96**(1968), 778-784
[5] Trenberth, K. E., *ibid.*, **108**(1980), 1370-1377.
[6] Lindzen, R. S. & Holton, J. R., *J. Atmos. Sci.*, **25**(1968), 1095-1107.
[7] Holton, J. R. & Lindzen, R. S., *ibid.*, **29**(1972), 1076-1080
[8] Belmont, A. D, & Dartt, D. G., *Mon. Wea. Rcv.*, **96**(1968), 767-777.
[9] Holton, J.R., *J. Atmos. Sci.*, **29**(1972), 368-375.
[10] 李崇银, 中国科学, **B辑**, 1983, 857-864.
[11] ——, 热带气象, **3**(1987), 191-196.

Relationship Between Subtropical High Activities over the Western Pacific and Quasi —Biennial Oscillation in the Stratosphere*

Li Chongyin(李崇银), Long Zhenxia(龙振夏)

LASG, Institute of Atmospheric Physics, Chinese Academy of Sciences, Beijing

Abstract Based on data analyses, the quasi-biennial feature of subtropical high over the western Pacific is studied. The results show that the oscillation is very clear in both the relative intensity and the latitude location of subtropical high ridge. The analyses also show that the vertical shear of zonal wind in the lower stratosphere is related to the subtropical high activities over the western Pacific: the easterly (westerly) wind shear corresponds to the stronger (weaker) subtropical high and the further northward (southward) ridge location. The rising (sinking) motion in the upper troposphere over the equator caused by the easterly (westerly) wind shear in the lower stratosphere will enhance (reduce) the Hadley cell, and it is likely an important mechanism of how the quasi-biennial oscillation in the stratosphere to influence the subtropical high activities over the western Pacific. The simulations with IAP-GCM are also completed in this study and the results are similar to those by observational data analyses.

Keywords subtropical high over the western Pacific, quasi-biennial oscillation, zonal wind shear, Hadley cell.

1. Introduction

The subtropical high over the western Pacific is an important atmospheric circulation system, its activities and anomaly have great influences on the weather and climate in East Asia/ western Pacific regions. Therefore, Chinese meteorologists have paid much attention to it and some studies such as its east-west stretch and retreat, south north movement[1]. its structure characteristics[2], its influences on the climate in China[3], the effect of ENSO on it[4] are completed. Based on these studies, the fundamental understanding of the subtropical high over the western Pacific and its activities is obtained, but some questions still need to be investigated, especially the longer time-scale variation of subtropical high, which is important to the climate anomaly and climate prediction.

The quasi-biennial oscillation (QBO) in the stratosphere has long been an important research project since it was discovered in the 1960s. During recent years, the quasi-biennial

---

Chinese manuscript received February 8, 1996; revised October 11, 1996.

*Projece 49635170 supported by the National Natural Science Foundation of China.

oscillation phenomenon of the tropospheric circulation and climate variation, which is called TBO, is also revealed, and its relationship with the QBO is also studied. For example, the monsoon rainfall in India shows a quasi-biennial oscillation and it is closely related to the QBO, there is less monsoon rainfall in India during the easterly phase of the QBO[5]; the hurricane frequency in the Atlantic is also related to the QBO, there are more hurricanes in the Atlantic during the westerly phase of the QBO[6]. In our previous study, it was shown that the precipitation and temperature variations in the eastern China and the typhoon activities over the western Pacific are related to the QBO[7]. Therefore, connecting the QBO with the TBO and investigating their influence on the climate are an important project in atmospheric sciences.

In the present paper, the interannual variation of subtropical high over the western Pacific will be analyzed and the influence of the QBO on the activity of subtropical high will be studied through observational data and numerical simulation. The results will provide some scientific bases for interannual climate prediction in East Asia.

2. Quasi-Biennial Oscillation of Subtropical High over the Western Pacific

The subtropical high over the western Pacific is a huge warm high system, which exists in the whole troposphere. For simplicity, the intensity and variability of subtropical high over the western Pacific are represented by using the geopotential height anomalies at 500 hPa in (20°-30°N, 120°-150°E) region. Fig.1a shows the power spectrum of the intensity variation of subtropical high, a 25-month spectrum peak is evident, therefore, the intensity of subtropical high over the western Pacific has quasi-biennial fluctuation. In Fig. 1b, the 27-month band-pass filtered (QBO period is about 27-month) results of geopotential height anomalies at 500 hPa in (20°-30°N, 120°-50°E) region during 1967-1980 and the temporal variation of 30 hPa zonal wind at Singapore are shown respectively. It is clear that the quasi-biennial fluctuation of subtropical high intensity over the western Pacific is closely related to the QBO

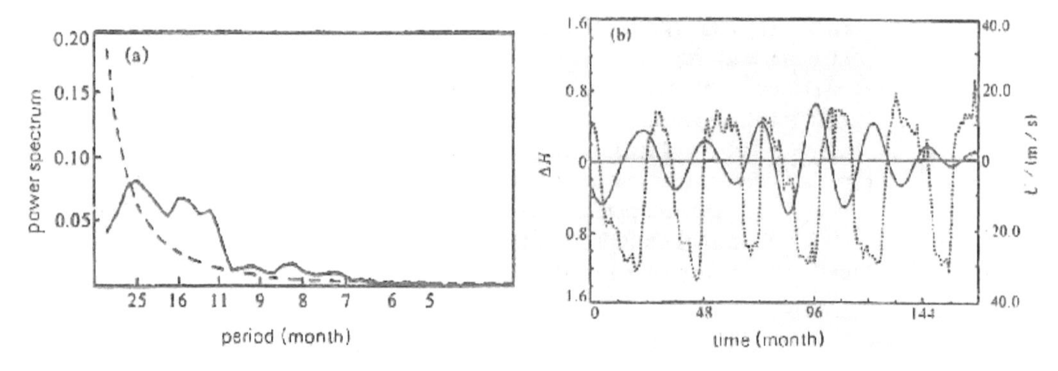

Fig. 1. The power spectrum of geopotcntial height anomalies at 500 hPa in (20°-30°N, 120°-150°E) region (a) and the 27-month band-pass filtered results of geopotential height anomalies (b, solid line) and temporal variation of 30 hPa zonal wind at Singapore (b, dashed line).

in the stratosphere. The subtropical high is weaker during the transforming period from the westerly phase of the QBO into the easterly phase; on the contrary, the subtropical high is stronger during the transforming period from the easterly phase of the QBO into the westerly phase.

Besides the intensity, the zonal location is also an important feature for the subtropical high activities over the western Pacific, which can be represented by the located lalitude of ridge line. The data analysis shows that the zonal location of the ridge line of subtropical high over the western Pacific is also quasi-biennial periodic variation. The interannual variation of the ridge line location at 500 hPa of the subtropical high averaged in 120°-150°E during June-August is given in Fig.2, where the dotted line represents 3-year moving average results and the dashed line shows the ridge line location ascertained by OLR data. The dashed line and the solid line in Fig.2 show similar variation trend, which means that the analysis is reliable. Although the 3-month regional mean is used in Fig.2, so that the amplitude is not great, the interannual variation of the ridge line location of subtropical high over the western Pacific is clear and shows evident quasi-biennial oscillation feature. In Fig.2, the character "w" represents the zonal westerly wind shear in the lower stratosphere during the spring-summer of that year and the easterly wind shear occurs during periods without "w". In general, the zonal wind shear in the lower stratosphere corresponds to the zonal wind direction at 30 hPa. It is shown in Fig.2 that the ridge line location of subtropical high over the western Pacific is related to the zonal wind shear in the lower stratosphere. The easterly wind shear in the lower stratosphere corresponds to the further northward ridge line location of subtropical high over the western Pacific, but the westerly wind shear corresponds to the further southward ridge line location. On the average during 1951-1989 summer, the ridge line of subtropical high is located at 25.7°N corresponding to the easterly wind shear, but at 24.8°N corresponding to the westerly wind shear in the lower stratosphere.

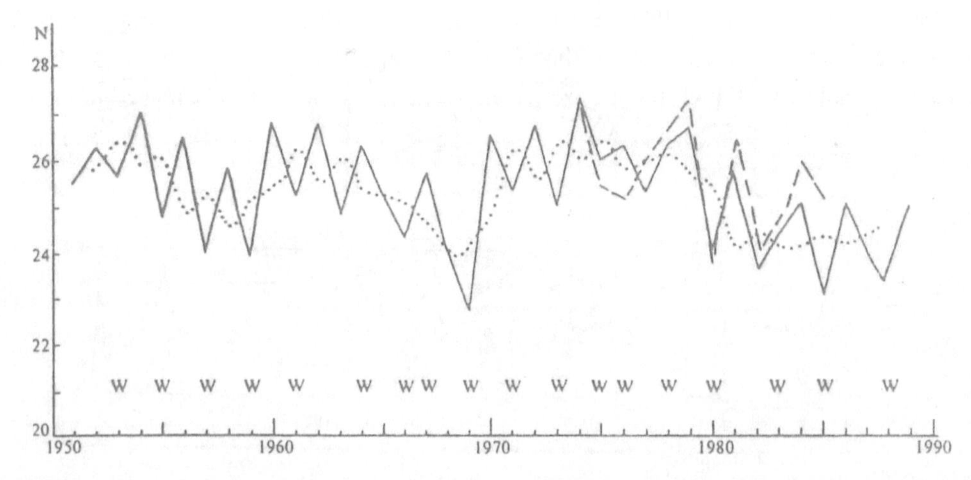

Fig. 2. Interannual variation of the ridge line location of subtropical high at 500 hPa over the western Pacific averaged in (120°-150°E) region during June-August (solid line); the dotted line shows 3-ycar moving average results, and the dashed line shows the ridge line location ascertained by OLR data.

3. Influence of the QBO on the Activities of Subtropical High over the Western Pacific

Above analyses show that both the intensity and the zonal location of subtropical high over the western Pacific have quasi-biennial fluctuations, and this interannual variation is related to the QBO in the stratosphere. In order to investigate the relationship between the QBO and the activities of subtropical high over the western Pacific, especially the influence of the QBO in the stratosphere on the subtropical high, we need to know the feature of zonal wind in the stratosphere over the equator region. The temporal height section of the monthly zonal wind (after Ref. [8], see Fig.3) shows that there are clear easterly (westerly) wind phase and easterly (westerly) wind vertical shear in the stratosphere. Next, we will discuss how the QBO affect the subtropical high over the western Pacific.

The study on the interannual variation of the tropopause over the tropics has indicated that the tropical tropopause has a QBO-like expending or contracting phenomenon in vertical direction[9], and the observations at Singapore have also shown that there is a positive (negative) height anomaly in the upper troposphere corresponding to the easterly (westerly) wind phase of the QBO in the stratosphere. Gray et al. connected the isobaric surface variation and the temperature variation near the tropopause with the vertical sheal of zonal wind in the lower stratosphere. They gave a schematic diagram shown in Fig.4a. The schematic diagram, which represents the influence of the QBO on the subtropical high over the western Pacific, is shown in Fig.4b. For the easterly wind shear in the lower stratosphere, there is anomalous ascending motion in the upper troposphere over the equator, and then the Hadley cell is enhanced; the positive anomaly of Hadley cell will be of great advantage to the enhancement of subtropical high over the western Pacific and leads to further northward location of the ridge line. On the contrary, there is sinking motion in the upper troposphere over the equatorial area corresponding to the westerly wind shear in the lower stratosphere, and then the Hadley cell is weakened; the weaker Hadley cell will lead to a further southward and weak subtropical high.

Above analyses show that the subtropical high over the western Pacific can be influenced by the zonal wind shear in the lower stratosphere, the ridge line of subtropical high shifts further northward and its intensity is stronger corresponding to the easterly wind shear, but the location is further southward and the intensity is weaker when there is the westerly wind shear in the lower stratosphere. In fact, the observations verified this result. It is clearly shown in Fig.2 that the ridge line of subtropical high over the western Pacific shifts further northward (southward) corresponding to the easterly (westerly) wind shear in the lower stratosphere for taking an average. At the same time, it is also shown that for the weather analyses and prediction practices the subtropical high over the western Pacific is stronger when it locates further northward and it is weaker when deviating further southward.

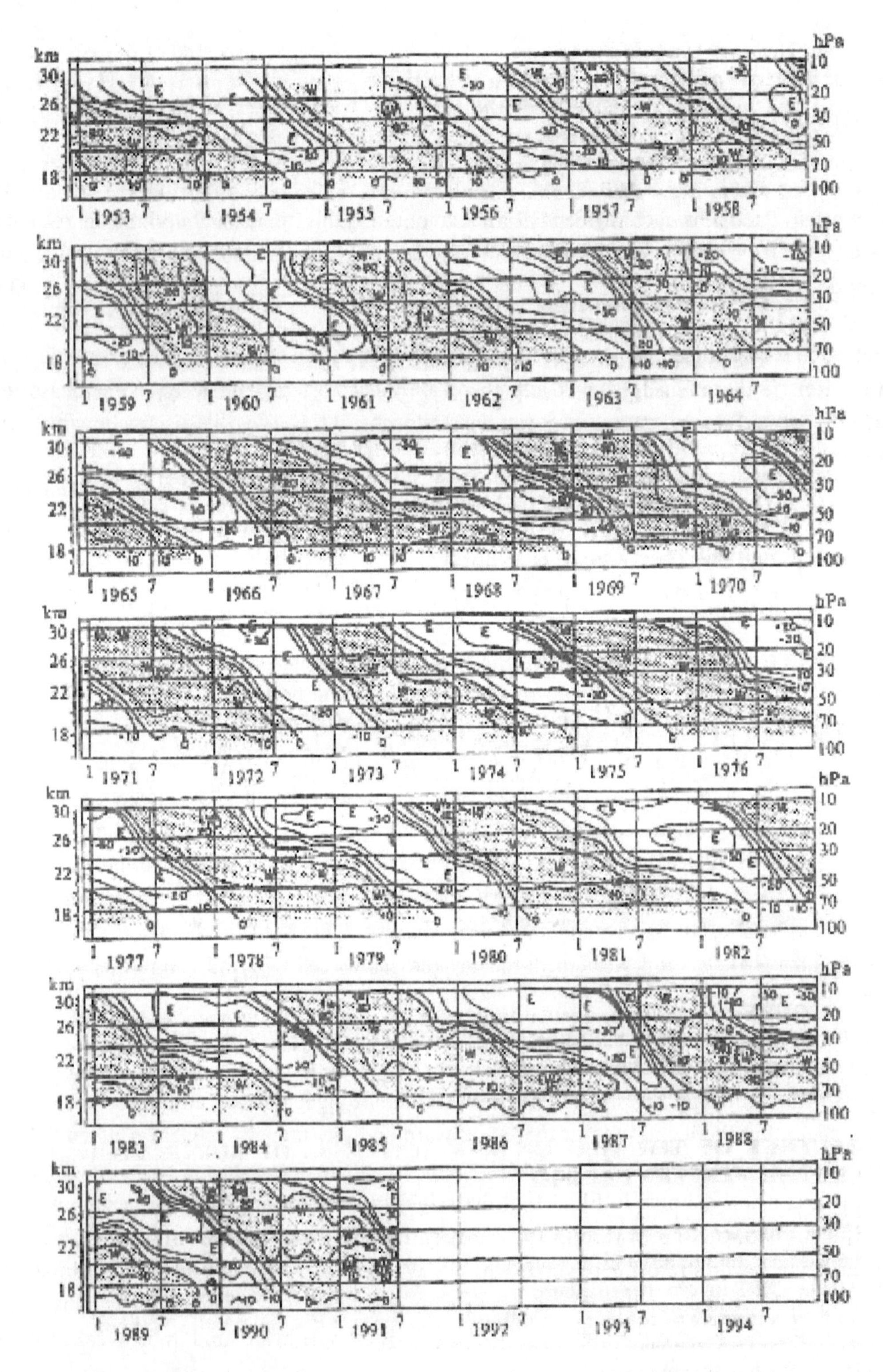

Fig. 3. Time-altitude section of monthly zonal wind (m / s) in the stratosphere in the equatorial area during 1953- 1991 (after Ref. [8]).

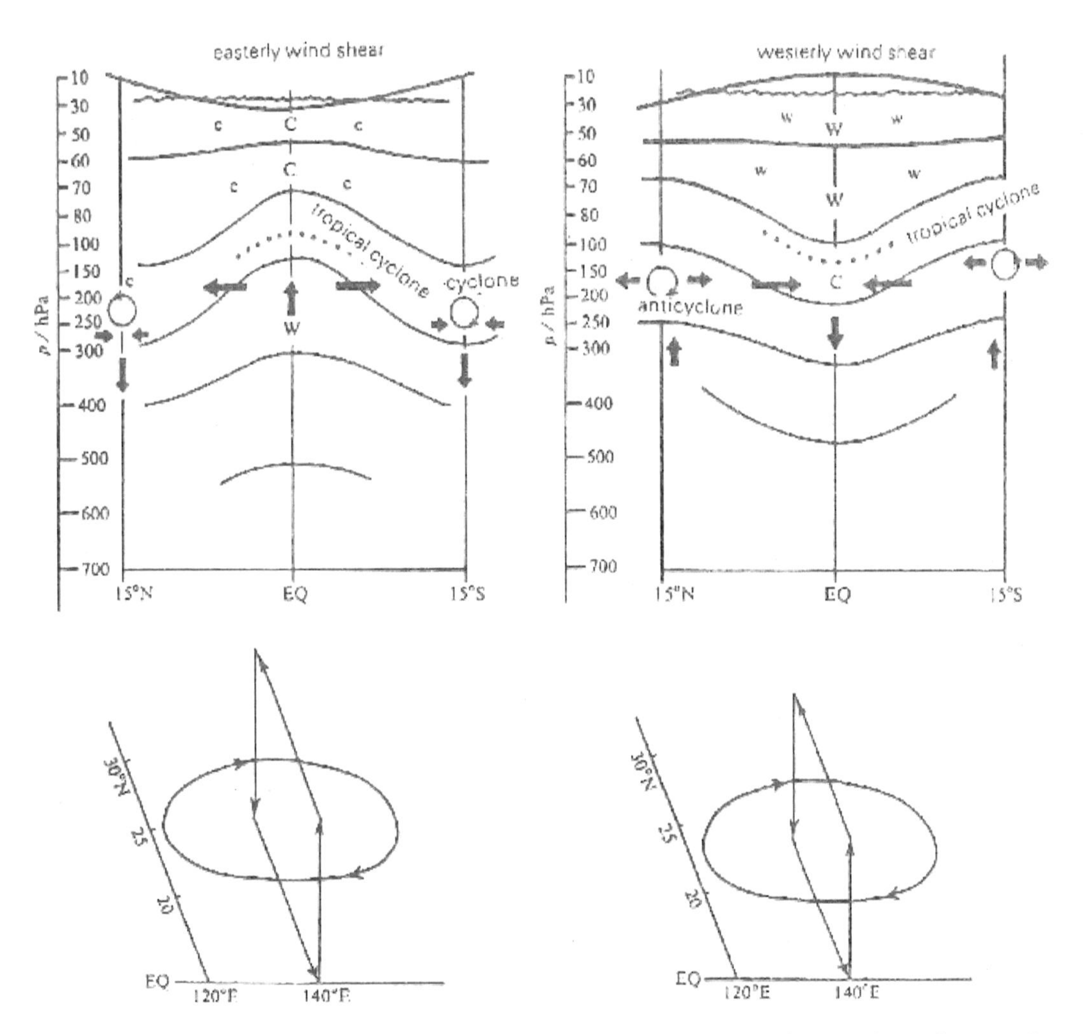

Fig. 4. Schematic diagram of the zonal wind shear in the lower stratosphere and its effect on the subtropical high over the western Pacific, which is represented by the elliptic anticyclonic cell in the lower part.

In order to show the impact of QBO in the stratosphere (especially the zonal wind shear) on the subtropical high, the meridional circulation is studied as follows. The altitude-longitude sections of meridional wind averaged in 120°-150°E region in July are respectively given in Fig.5 for the cases in 1985, 1986 and 1987. The case in July 1985 corresponds to the westerly wind shear in the lower stratosphere, and the cases in July 1986 and 1987 correspond to the easterly wind shear. It is very clear that the Hadley cell locates further southward and weaker when there is the westerly wind shear in July 1985, but the Hadley cell locates further northward and stronger when there is the easterly wind shear in July 1986 and 1987. These results verify the impact of zonal wind shear in the lower stratosphere on the subtropical high over the western Pacific.

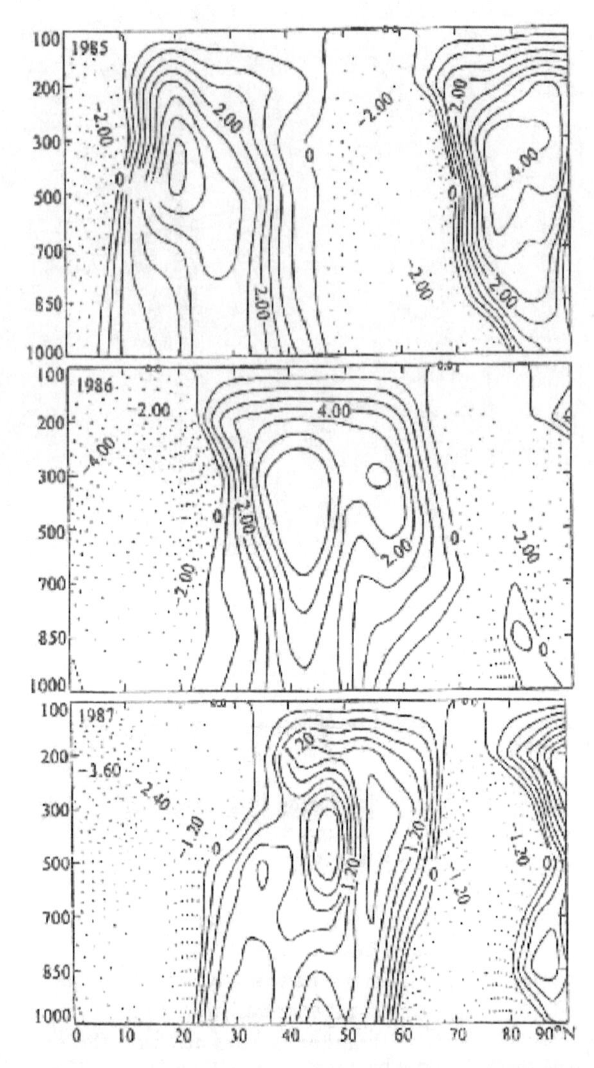

Fig. 5. Altitude-longitude sections of meridional wind averaged in 120°-150°E region in July 1985, 1986 and 1987, respectively. Solid (dashed) line represents the northerly (southerly) wind and the interval is 0.3 m/s.

4. Numerical Simulation Results

By using IAP-GCM, numerical simulation experiments are completed to study the influence of zonal wind shear in the lower stratosphere on the subtropical high over the wstern Pacific. For saving computer time, we used a two-level IAP-GCM, which is a grid - point model (5° longitude × 4° latitude) and the physical processes have been considered in the model completely. Detail information can be referred to the relevant hand book and paper[10, 11]. It can be said that this model has better simulation ability in climatology as well as 9 level GCM, although it has only two levels. As shown in Fig.4, there is anomalous ascending (sinking) motion and anomalous ascent (descent) of isobaric surface in the upper troposphere over the equatorial area when there is the easterly (westerly) wind shear in the lower stratosphere. But the two-level IAP-GCM doesn't include the stratosphere, therefore in order to study the impact of easterly wind shear in the lowel stratosphere, it is assumed that there are positive anomalies of geopotential height at 200 hPa in (2°S-6°N, 100°-180°E) region, and the height anomalies are 100 m in the central area and 50 m along the boundary. The numerical integrations of the GCM are from May to August, and the differences of the model outputs between the perturbation experiment with the height anomalies at 200 hPa and the control run without the height anomalies can represent the influence of the easterly wind shear in the lower stratosphere. Because we focus on discussing the impact of the QBO on the subtropical high over the western Pacific in the perturbation numerical experiment, the height anomalies at 200 hPa are not in the global equatorial region.

The height anomalies at 500 hPa during June-August caused by the easterly wind shear in the lower stratosphere are shown in Fig.6. It is very clear that there are evidently positive anomalies over the northwestern Pacific region and the major anomalies are in (10°-35°N, 120°-165°E) area. This distribution of positive height anomalies means that the easterly wind shear can lead the subtropical high over the western Pacific to enhance and to extend northwards, and that the ridge line of subtropical high deviates also further northward, because the mean location of ridge line over the western Pacific in June-August is about 25°N. The time-latitude section of the simulated height anomalies at 500 hPa averaged in

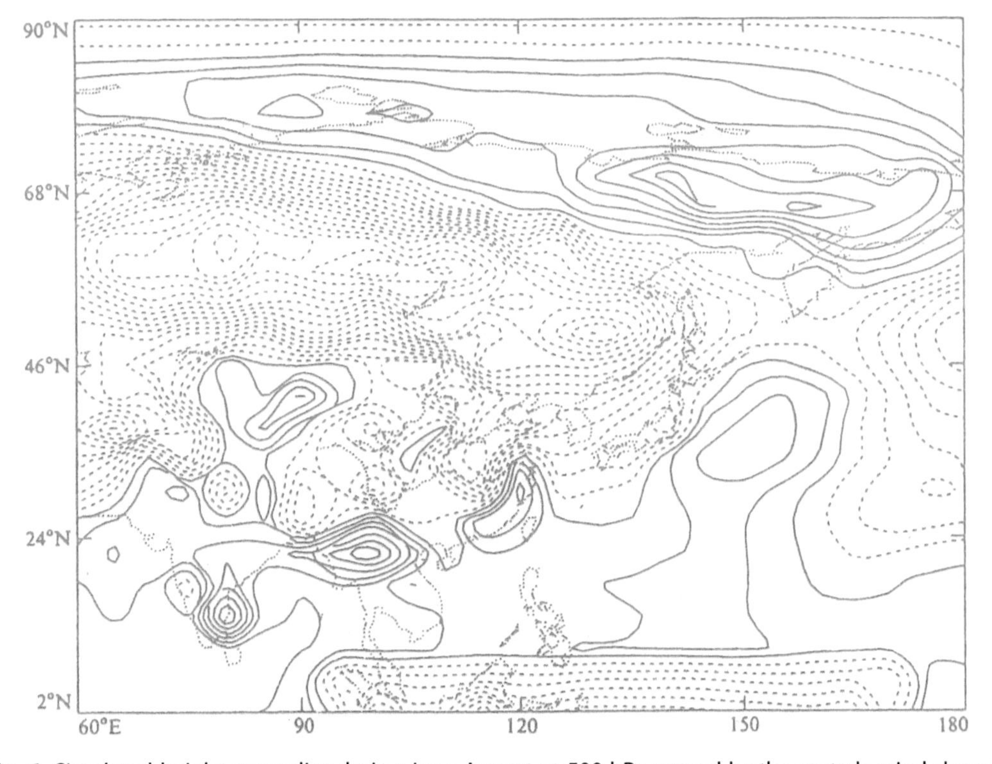

Fig. 6. Simulated height anomalies during June-August at 500 hPa caused by the easterly wind shear in the lower stratosphere. The solid (dashed) lines arc positive (negartive) anomalies, and the interval is ±3 m.

(120°-150°E) area caused by the easterly wind shear in the lower stratosphere is shown in Fig.7. It is also evident that there are continuous positive anomalies of the height at 500 hPa in 16°-38°N latitudes and the evolution of the height anomaly shows low frequency feature similar to the result indicated in our previous research; i.e. the responses in the atmosphere to external forcing are mainly low-frequency (30-60 day) remote responses[12] For the temporal variation of positive anomaly center, we can find that it extends northwards in June and locates at about 38°N in very late June, then it retreats southwards and locates at about 15°N in middle of July; after that, the positive anomaly center propels northward again and it is at about 40°N in August. In other words, the variation of subtropical high over the western Pacific caused by the easterly wind shear in the lower stratosphere shows not only the enhancement of intensity and further northward ridge line, but also the temporal evolution with a low-frequency (30-60 day) oscillation feature.

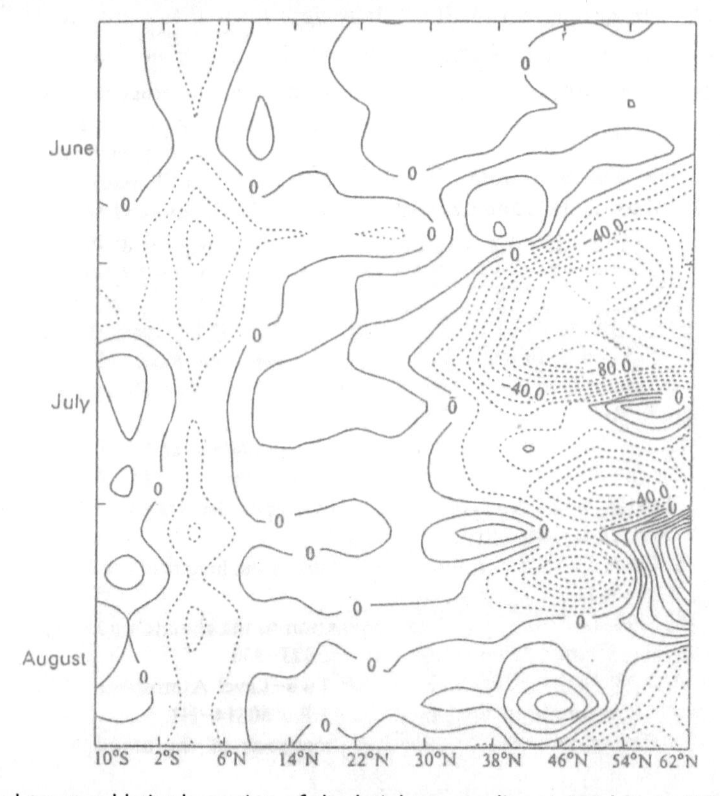

Fig. 7. Simulated temporal-latitude section of the height anomalies at 500 hPa in (120°-150°E) region caused by the easterly wind shear in the lower stratosphere. The solid (dashed) lines are positive (negative) anomalies and the interval is ±10 m

For the westerly wind shear in the lower stratosphere, numerical simulation shows that the subtropical high over the western Pacific will shift weakened and its ridge line will be further southward (figure not given), but the impact of the westerly wind shear is not as evident as that in the easterly wind shear.

5. Conclusions

In the present paper, the activities of subtropical high over the western Pacific and the influence of QBO in the stratosphere on subtropical high are studied through data analyses, and the numerical simulation is also completed with IAP-GCM. The major results can be summarized as follows:

1) The activities of subtropical high over the western Pacific (including the intensity variation and latitude location of ridge line) show evident quasi biennial oscillations.

2) The activities of subtropical high over the western Pacific are related to the zonal wind shear in the lower stratosphere (or the zonal wind at 30 hPa). The subtropical high is stronger and its ridge line deviates further northward, when there is the easterly wind shear in the lower stratosphere, and the subtropical high is weaker and its ridge line deviates further southward in the case of the westerly wind shear.

3) The easterly (westerly) wind shear in the lower stratosphere can produce anomalous ascending (sinking) motion in the upper troposphere over the equatorial area, then the Hadley cell will be enhanced (reduced) and the activities of subtropical high over the western Pacific will also be influenced to have quasi-biennial oscillation phenomenon. This is likely an important mechanism of the stratospheric QBO in influencing the subtropical high over the western Pacific.

4) Numerical simulation experiment in the IAP-GCM obtained similar results to the data analyses. Corresponding to the easterly wind shear in the lower stratosphere (there is anomalous ascending motion or positive height anomalies in the upper troposphere), the simulated subtropical high over the western Pacific is stronger and deviates further northward; on the contrary, the simulated subtropical high is weaker and deviates further southward in the case of the westerly wind shear in the lower stratosphere.

References

[1] Huang Shisong and Tang Mingmin, 1962: Some features of annual variation in the north-south location of subtropical high and their meanings, *Journal of Nanjing University*, 2, 86-95. (in Chinese)

[2] Huang Shisong et al., 1962: Some question studies on the structure of subtropical high and the corresponding circulation, *Acta Meteorologica Sinica*, 32, 339-359. (in Chinese)

[3] Tao Shiyan and Xu Shuyin, 1962: The circulation features of the continued drought and flood in the Yangtzc-Huaihe River valley in summer, *Acta Meteorologica Sinica*, 32, 1-10. (in Chinese)

[4] Chen Licting, 1982: Interaction between subtropical high over the northwestern Pacific and SST in the equatorial western Pacific, *Scientia Atmospherica Sinica*, 6, 148-156. (in Chinese)

[5] Mukherzee, B.K., K. Indira and R.S. Reddy, et al., 1985: Quasi-biennial oscillation in stratospheric zonal wind and Indian summer monsoon, *Mon. Wea. Rev.*, 113, 1421-1424.

[6] Gray, W.M., 1984: Atlantic seasonal hurricane frequency, Part I: El Kiño and 30 mb quasi-biennial oscillation influences, *Mon. Weq. Rev.*, 112, 1649-1668.

[7] Li Chongyin and Long Zhenxia, 1992: Quasi-biennial oscillation and its influence on general at-

mospheric circulation and climate in East Asia, *Chinese J. Atmos. Sci.*, 16, 70-79.

[8] Gray, W.M., J.D. Sheaffer and J.A. Knaff, 1992: Influence of the stratospheric QBO on ENSO variability, *J. Meteor. Soc. Japan*, 70, No.5, 1-21.

[9] Reid, G.C. and K.S. Gage, 1985: Interannual variations in the height of the tropical tropopausc, *J. Geophys. Res.*, 90, 5629-5635.

[10] Zeng Qingcun et al., 1990: IAP-GCM and its application to the climate studies, in: *Climate Change Dynamics and Modelling*, China Meteorological Press, 303-330.

[11] Zeng Qingcun et al., 1989: Documentation of IAP Two-Level Atmospheric General Circulation Model, United States Department of Energy, DOE / ER / 60314-HI.

[12] Xiao Ziniu and Li Chongyin, 1992: Numerical simulation of the atmospheric low-frequency tele-response to external forcing, Part I: Anomalous SST in the equatorial eastern Pacific, *Chinese J. Atmos. Sci.*, 16, 372-382.

On the Differences and Climate Impacts of Early and Late Stratospheric Polar Vortex Breakup

Li Lin[1] (李琳), Li Chongyin*[1,2] (李崇银), Pan Jing[2] (潘静),
TAN Yanke[1] (谭言科)

[1] Institute of Meteorology, PLA University of Science and Technology, Nanjing
[2] LASG, Institute of Atmospheric Physics, Chinese Academy of Sciences, Beijing

Abstract The stratospheric polar vortex breakup (SPVB) is an important phenomenon closely related to the seasonal transition of stratospheric circulation. In this paper, 62-year NCEP/NCAR reanalysis data were employed to investigate the distinction between early and late SPVB. The results showed that the anomalous circulation signals extending from the stratosphere to the troposphere were reversed before and after early SPVB, while the stratospheric signals were consistent before and after the onset of late SPVB. Arctic Oscillation (AO) evolution during the life cycle of SPVB also demonstrated that the negative AO signal can propagate downward after early SPVB. Such downward AO signals could be identified in both geopotential height and temperature anomalies. After the AO signal reached the lower troposphere, it influenced the Aleutian Low and Siberian High in the troposphere, leading to a weak winter monsoon and large-scale warming at mid latitudes in Asia. Compared to early SPVB, downward propagation was not evident in late SPVB. The high-latitude tropospheric circulation in the Northern Hemisphere was affected by early SPVB, causing it to enter a summer circulation pattern earlier than in late SPVB years.

Keywords stratospheric polar vortex breakup (SPVB), stratosphere-troposphere interaction, Arctic Oscillation (AO), season transition

1. Introduction

In recent years, it has been suggested that the stratosphere exerts an important effect on tropospheric climate change at various timescales, as discussed in several data analysis and numerical simulation studies (Baldwin and Dunkerton, 1999; Christiansen, 2001; Thompson et al., 2002; Black, 2002; Chen et al., 2006; Wang and Chen, 2010).

As the most distinctive feature of the stratospheric circulation system, the stratospheric polar vortex is thought to be a leading candidate for the stratosphere- troposphere interaction process (Jung and Barkmeijer, 2006; Limpasuvan et al., 2005; Perlwitz and Graf, 2001). The

Received July 25, 2011; revised January 10, 2012.
*Corresponding author: LI Chongyin, lcy@lasg.iap.ac.cn

stratospheric polar vortex breakup (SPVB) usually occurs at the time of stratospheric final warming (SFW), and it causes stratospheric circulation to change from strong circumpolar westerlies during the winter to weaker circumpolar easterlies during the summer. The persistence of the stratospheric polar vortex has increased significantly since the mid- 1980s, and the interannual variability of SPVB is related to the preexisting stratospheric flow structure and variations in the upward propagation of tropospheric planetary waves (Waugh et al., 1999). Waugh and Rong (2002) found large interannual variability of SPVB, with different characteristics of early and late vortex breakup.

Downward signals and their impact on the troposphere during strong and weak stratospheric polar vortices were investigated by Baldwin and Dunkerton (2001) to demonstrate that stratospheric harbingers may be used as a predictor of tropospheric weather regimes. With the downward propagation of anomalous stratospheric signals, the winter monsoon is enhanced by stratospheric sudden warming (SSW) (Deng et al., 2008). Although comprehensive studies have shown the basic dynamic structure of the circulation evolutions during breakup events (Black et al., 2006; Black and Mcdaniel, 2007; Wei et al., 2007), the features of downward stratospheric signals are not well understood with respect to early and late SPVB. Therefore, in order to understand the downward features and processes of stratospheric signals in early and late SPVB, we looked at the differences in downward propagation between early and late SPVB. Meanwhile, we also investigated the impact of SPVB on tropospheric climate.

The paper is organized as follows. The dataset and approach used are outlined in section 2. The differences in downward propagation and the influence of early and late breakup are presented in sections 3 and 4. Section 5 discusses the impact on climate of early and late SPVB, and conclusions and discussions are provided in section 6.

2. Data and Methods

Daily NCEP/NCAR reanalysis data from 1948 to 2009 were employed in this study, using a 2.5°×2.5° horizontal resolution and extending from 1000 hPa to 10 hPa with 17 pressure layers vertically.

Previously, different methods have been used to determine the time of SPVB (Waugh and Rong, 2002; Black et al., 2006; Black and Mcdaniel, 2007; Wei et al., 2007). Here, we calculated and defined the date of SPVB according to the approach set out previously by Wei et al. (2007), which is the final date when the zonal wind at the polar jet falls below zero and does not recover until the following fall. According to this definition, the average SPVB date over the 62-year study period was 15 April, which is close to the date obtained by Wei et al. (2007) using ERA-40 data.

Figure 1 shows the date of SPVB year-by-year. Large interannual variation in the breakup date can clearly be seen, as reported in previous studies (Waugh et al., 1999; Wei et al., 2007). Early (late) events were categorized as such if they were below (above) one standard deviation from the average, and under this definition there were 11 early cases and 15 late cases identified, with an average breakup date of 19 March and 9 May, respectively (Table 1). The life cycle of SPVB was analyzed following the methods used to explore strong stratosphere-troposphere coupling during SSW by Limpasuvan et al. (2004). Day 0 was defined as the time when the 65°N wind at 10 hPa became negative or easterly.

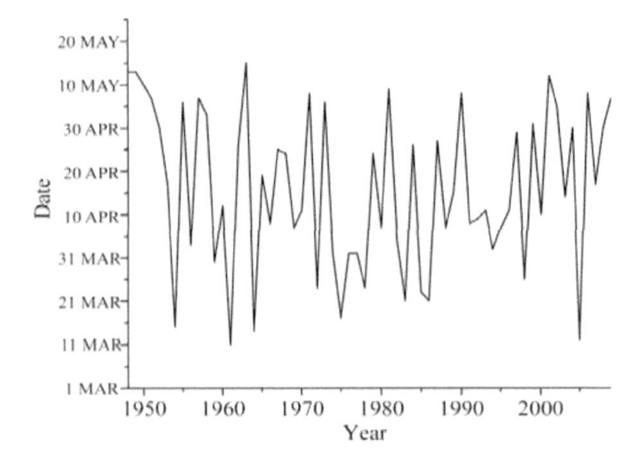

Fig. 1. Stratospheric polar vortex breakup (SPVB) from 1948 to 2009.

Table 1. Early and late breakup from 1948 to 2009.

Early breakups		Late breakups	
1954	15 March	1948	13 May
1961	11 March	1949	13 May
1964	14 March	1950	10 May
1972	24 March	1951	7 May
1975	17 March	1955	6 May
1978	24 March	1957	7 May
1983	21 March	1963	15 May
1985	23 March	1971	8 May
1986	21 March	1973	6 May
1998	26 March	1981	9 May
2005	10 March	1990	8 May
		2001	12 May
		2002	5 May
		2006	8 May
		2009	7 May

3. Features of Early and Late SPVB

The breakup of the stratospheric polar vortex represents the onset of spring, with the boreal stratospheric circulation changing from a winter to a summer state. Figure 2 shows the variation of the zonal mean zonal wind at10 hPa and the geopotential height and temperature gradient inside and outside the arctic vortex at 10 hPa. The gradient was calculated as the geopotential height and temperature difference between the zonal average at 60°N and 90°N. Day 0 refers to the vortex breakup onset, and the days preceding and following Day 0 are labeled as negative and positive, respectively. The linear tendency of the time series of early and late cases shows that late decay happens more gradually than early decay, as suggested by Black et al. (2006). Different to that in SSW, the temperature gradient shifts smoothly (Fig. 2c). As the temperature gradient reverses earlier than zonal mean zonal wind, the breakup date could be 10 days earlier if it is defined by temperature gradient instead of zonal wind. The larger amplitude prior to breakup indicates that stratospheric circulation is unstable in the winter.

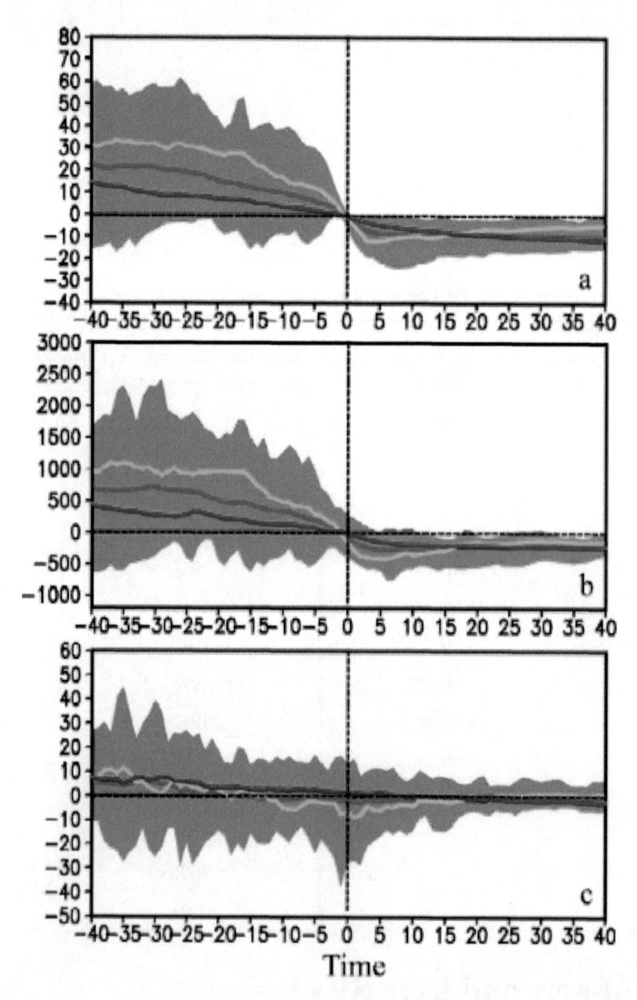

Fig. 2. Time series of (a) zonal average zonal wind; (b) zonal average geopotential height; and (c) temperature gradients between 60° N and 90° N at 10 hPa. Shading indicates the range for 1949 to 2009; the dark blue line indicates the 62-year average; the red line shows late breakups; and the light blue line shows early breakups.

To provide more reliable details about the difference between early and late breakups, a composite analysis was conducted with the 11 early cases and 15 late cases. The time evolutions of the zonal mean temperature anomaly, zonal wind anomaly and geopotential height anomaly at north of 60°N during early and late cases are shown in Fig. 3; anomalies were determined in terms of deviation from climate mean state. Opposite anomalies for zonal mean temperature, zonal wind and geopotential height are noticeable before and after the onset of early breakup. Taking the zonal wind anomaly for instance, there is a positive anomaly before the onset of early breakup, while after the breakup it turns into a negative anomaly. The positive and negative anomalies before and after the onset of early breakup exist from troposphere to stratosphere as a quasi-barotropic structure. The anomalous

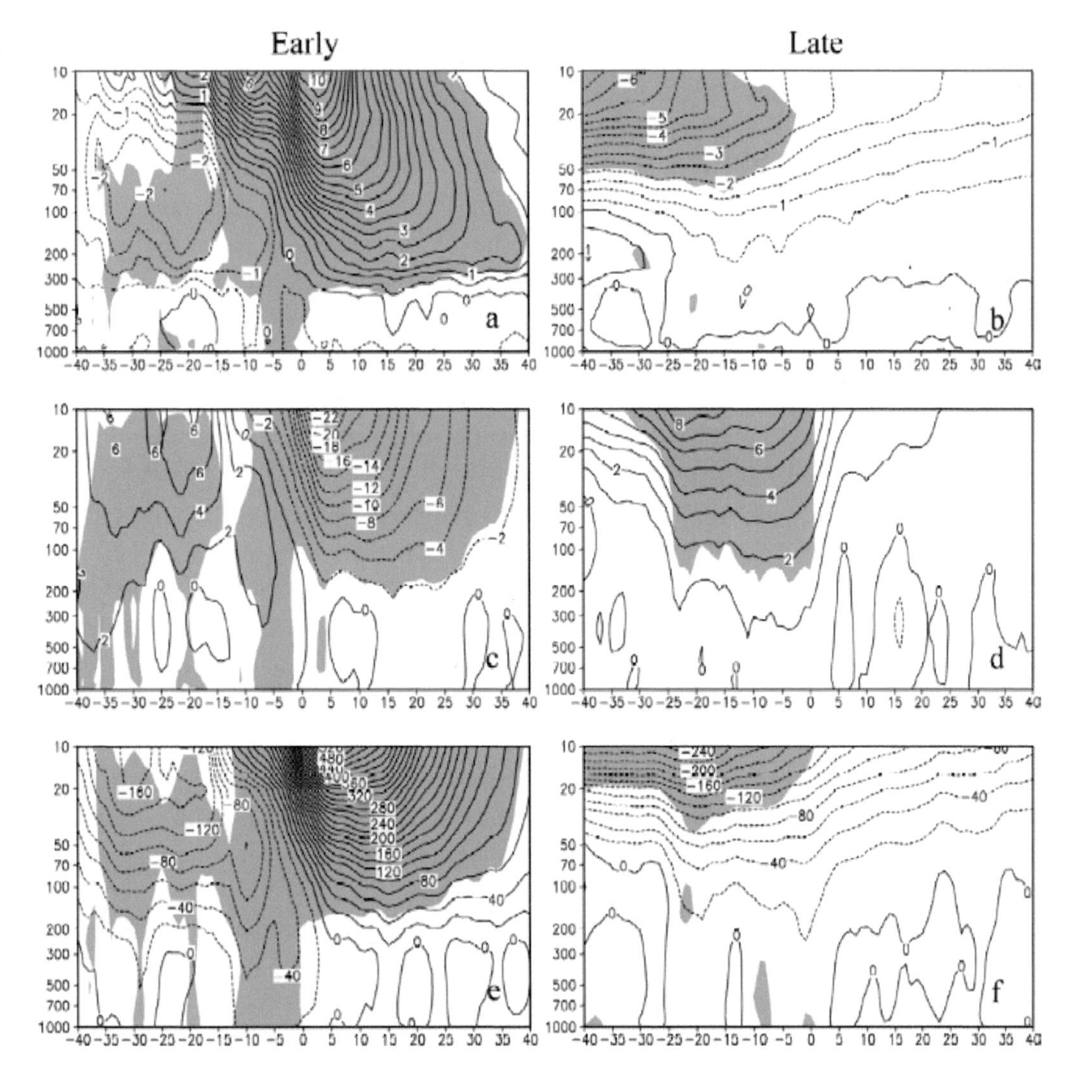

Fig. 3. Time evolutions of (a, b) the zonal average temperature (℃) anomaly; (c, d) zonal wind (m s^{-1}) anomaly; and (e, f) geopotential height (gpm) anomaly north of 60°N during early (left) and late (right) breakups. Shading denotes the region above the 90% confidence level.

distributions of temperature and geopotential height are similar to the zonal wind anomaly in early breakup. The zonal wind anomaly reversal happens earlier in the upper stratosphere than the troposphere in early breakup, and such an earlier signal is even more evident with respect to the temperature anomaly. Unlike in early breakups, the anomalies are consistent in the stratosphere before and after the onset of late breakups, shrinking to the upper stratosphere after late breakups, which again is different to the situation in early breakups.

The stratospheric zonal circulation must be "preconditioned" before a SSW to allow upward waves to exert considerable influence (Labitzke, 1981). Such "preconditioning" can be noticed as an anomalously strong vortex before the onset of a SSW (Limpasuvan et al., 2004). We found that in both early and late SPVB, the strongest precursor emerged approximately before Day–30 with enhanced westerly wind, lower geopotential height and lower temperatures at high latitudes. It seems that stratospheric circulation was also "preconditioned" before the breakups.

4. The Downward Signals in SPVB and Their Influence

The AO typically exhibits its anomaly in the stratosphere and can then propagate downward, reaching the ground in approximately three weeks and affecting the weather and climate in the troposphere (Baldwin and Dunkerton, 2001). To identify down ward signals from the stratosphere to the troposphere during SPVB, the AO index was calculated (Fig.4). The results clearly show a negative AO signature propagating downward from the stratosphere and reaching the ground after the onset of early breakups. The downward picture closely resembles that of weak vortex events as presented by Baldwin and Dunkerton (2001). The AO has a zonal systematical structure: the negative AO with a positive geopotential height anomaly occurs over the polar cap and the opposite anomaly occurs in the middle latitudes. Negative AO signatures appear after the onset of late breakups, but different to that in early breakups, they are limited to the upper stratosphere with less downward transition. Although a negative AO exists in the troposphere on Day 15, it seems not well connected with the negative AO signatures in the stratosphere.

The 81-day breakup life cycle was divided into 16 phases. Figure 5 gives the vertical cross sections of zonal mean geopotential height anomalies during Day -5 to Day -1, Day 6 to Day 10, Day 16 to Day 20 and Day 26 to Day 30 in early and late breakup. The positive and negative geopotential height anomaly appears in high and mid latitudes, forming an AO pattern in the upper stratosphere when early breakup occurs (Figs. 5a). This AO pattern in the geopotential height anomaly then propagates downward to the lower troposphere (Figs. 5c). In contrast to early breakups, the negative geopotential height anomaly controls the high latitudes in the stratosphere during the life cycle of late breakups. No evident downward AO pattern is formed in the geopotential height anomaly in late breakups (Figs.5e-h). The evolution of temperature anomalies shows a similar downward process to that of the geopotential height anomalies. A downward AO pattern can also be seen in the temperature anomaly in early breakups, but not in late breakups (data not shown).

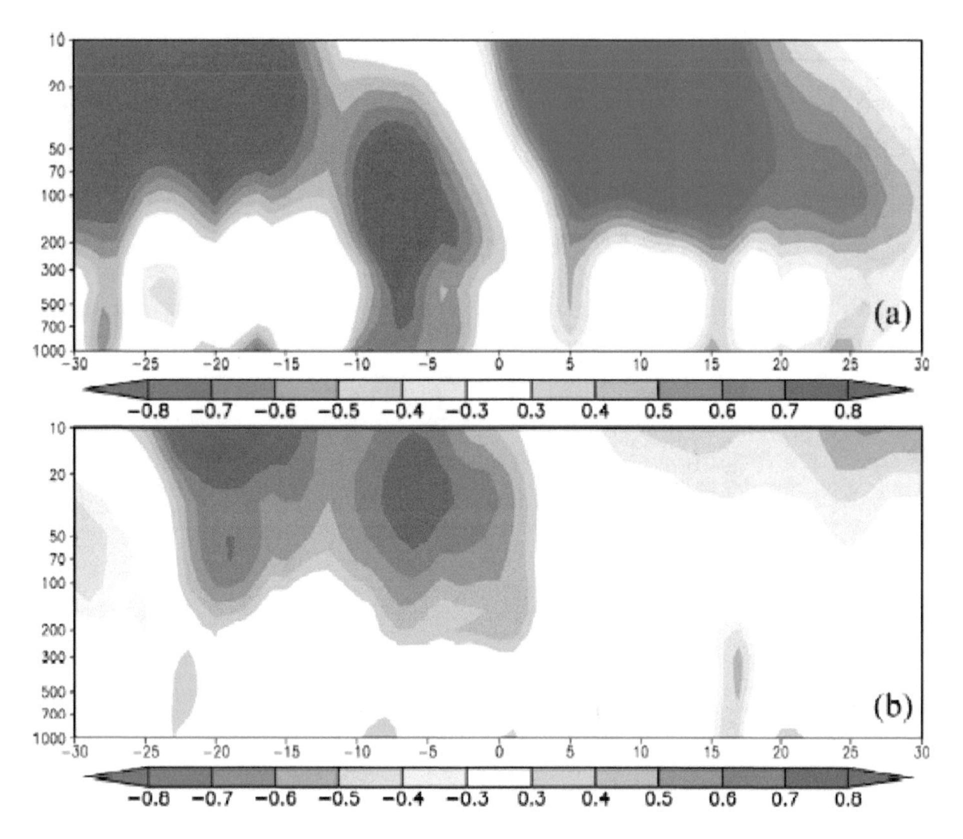

Fig. 4. Time-height evolution of the AO index during (a) early and (b) late breakups.

From the above analysis, the negative AO was identified as operating downward from the stratosphere to the troposphere after early breakups. Therefore, it is necessary to investigate its influence on tropospheric weather. At 1000 hPa, from Day 16 to Day 20 and Day 21 to Day 25, positive geopotential height anomalies occupied the polar region (Figs. 6c and d). At the same time, Eurasia and the North Pacific were covered by negative and positive geopotential height anomalies, respectively. These results indicated that the Siberian High and Aleutian Low, two important components of the East Asian winter monsoon system, weakened after early breakups. Furthermore, winter monsoons were found to be decreasing with a southerly wind anomaly emerging near the east coast of Asia. As the winter monsoon weakened, temperature increased in the mid latitudes and along the east coast of Asia (Figs. 6a and b).

5. Impact of SPVB on the Tropospheric Season Transition Pattern

The polar vortex is a large-scale circulation system extending from the troposphere to the stratosphere in winter. Boreal climate is closely linked with the strength and location of the polar vortex (Frauenfeld and Davis, 2003). The changing persistence of the stratospheric vortex, observed in early and late breakup of the stratospheric polar vortex, should have some impact on tropospheric climate.

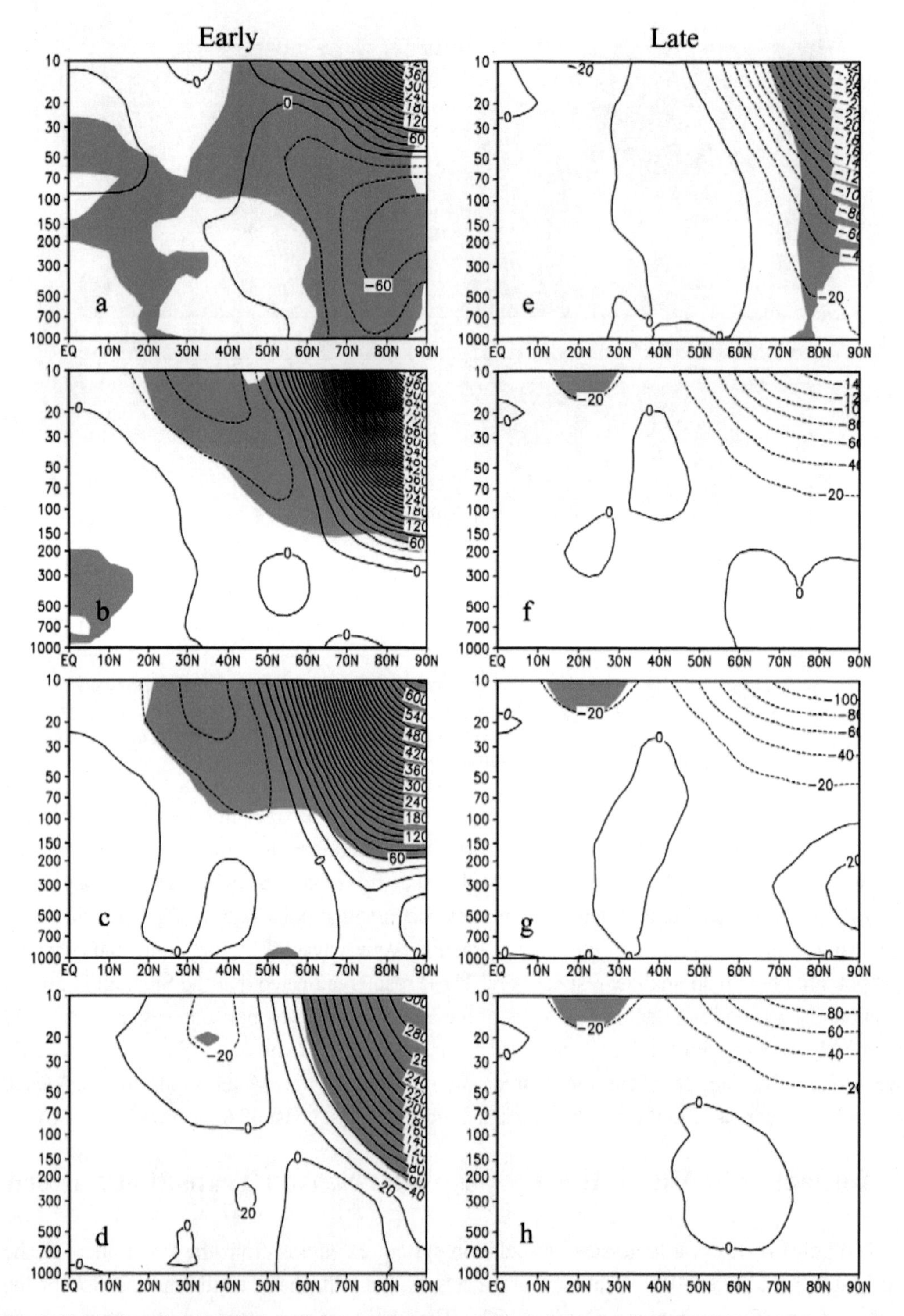

Fig. 5. Vertical sections of the zonal average geopotential height anomaly (gpm) during (a, e) Day -5 to Day -1; (b, f) Day 6 to Day 10; (c, g) Day 16 to Day 20; and (d, h) Day 26 to Day 30 during early (left) and late (right) breakups. Shading denotes the region above the 10% significance level.

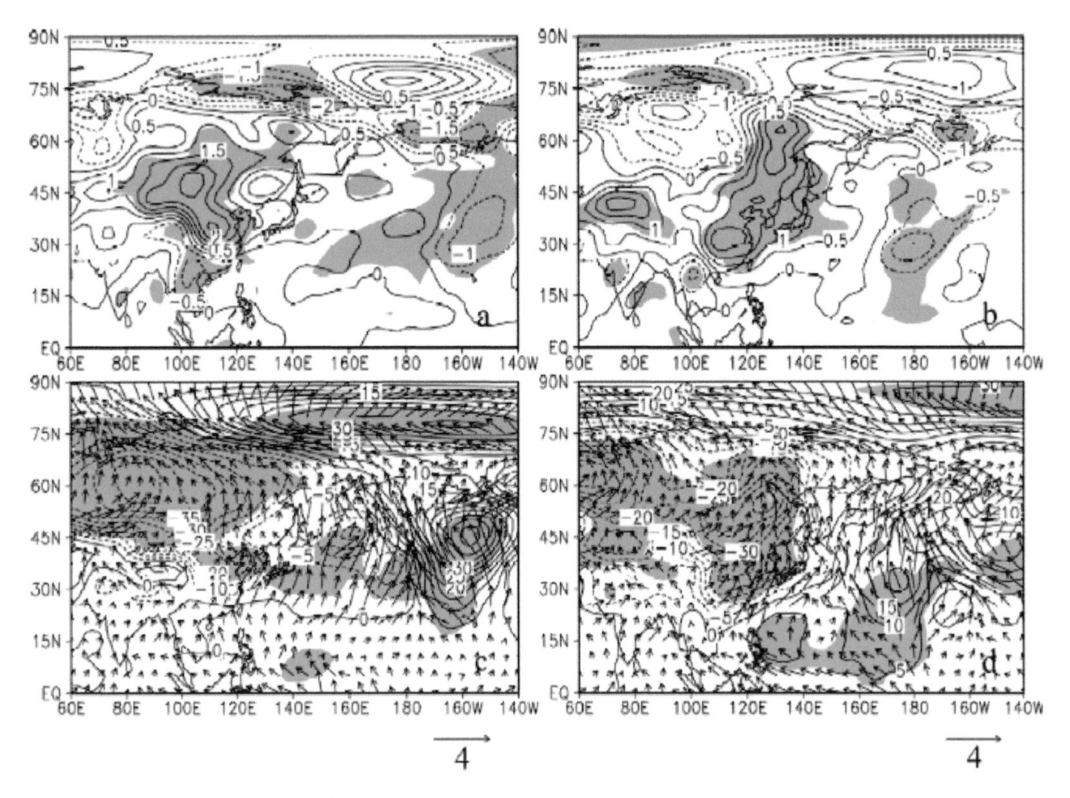

Fig. 6. Temperature anomaly (℃) at 1000 hPa (a, c) and geopotential height anomaly (gpm) and wind anomaly (ms^{-1}) at 850 hPa (b, d) during Day 16 to Day 20 (left) and Day 21 to Day 25 (right) in early breakups. Shading denotes the region above the 90% confidence level.

From the above analysis, stratospheric summer circulation is established in April of early decay years, while the winter circulation was enhanced in April of late breakup years as the circumpolar flue is "preconditioned" before late breakup occurs. Figure 7 shows the zonal averages of geopotential height, temperature and zonal wind in April of early and late breakup years and their differences. An easterly appeared at high latitudes of the stratosphere in April of early breakup years, whereas a westerly still controlled the stratosphere in April of late breakup years (Figs. 7g, h). The differences of zonal averages of geopotential height, temperature and zonal wind between early and late breakups show noticeable distinction in the upper troposphere and stratosphere at high latitudes, while the distinction in the lower troposphere is not evident (Figs. 7c, f, i).

Seasonal transitions can be tracked from the stationary wave changes in westerly trough and ridge variations from winter to summer at high latitudes in the boreal troposphere. The Aleutian Low and Siberian High are two important elements of the East Asian winter monsoon system that weaken in the summer. In Fig. 8, the distribution of high and low pressure centers at high latitudes at 1000 hPa has the same pattern in the April of early and late breakup years, but the Aleutian Low and Siberian High are much weaker in the early breakup years than in the late breakup years. According to the differences between the circulation system in the April of early and late breakup years, the advancement of the stratospheric season transition can cause the tropospheric circulation pattern at high latitudes to appear more summer-like in April in early breakup years, and the opposite in late breakup years. Figure 7 shows no obvious zonal average differences in the lower troposphere, while Fig. 8 gives the opposite result. Such a difference implies that the stratospheric influences on tropospheric season transition may not be zonally symmetric.

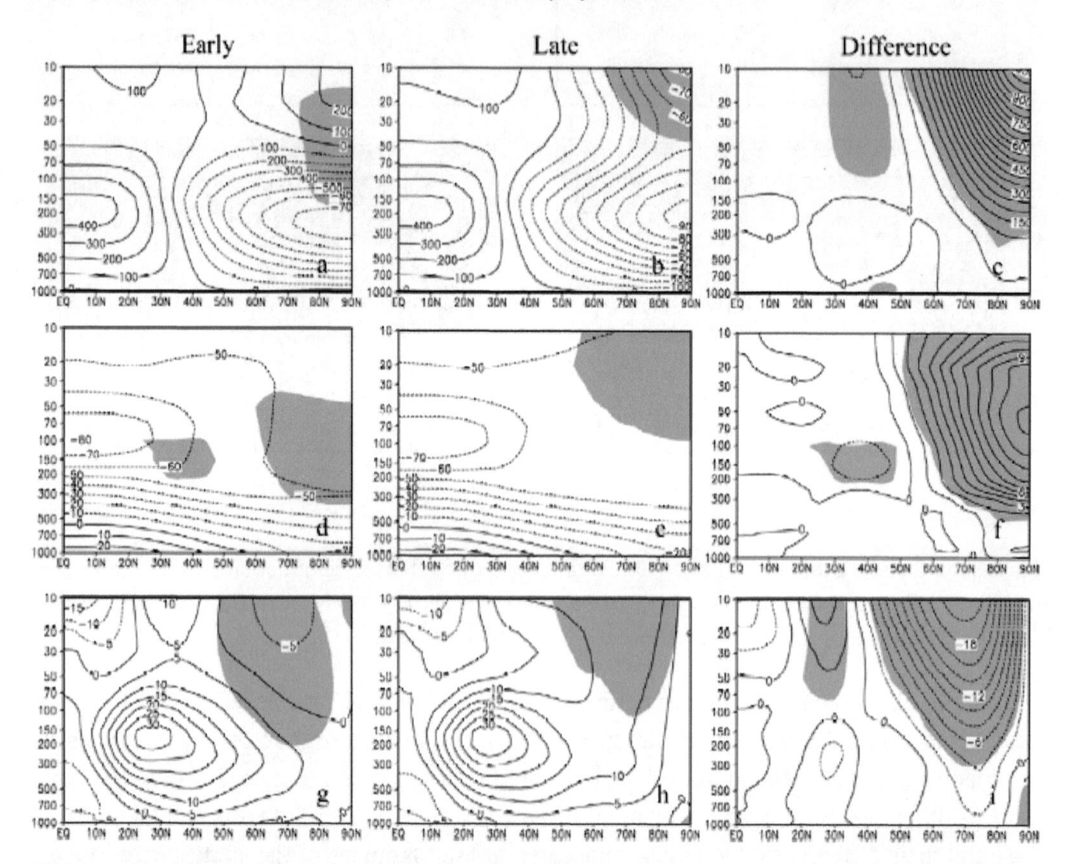

Fig. 7. Zonal average geopotential height (gpm) (a, b, c), temperature (℃) (d, e, f) and zonal wind (m s^{-1}) (g, h, i) in April for early breakup years (top), late breakup years (middle), and their differences (bottom). Shading denotes the region above the 90% confidence level.

6. Discussion and Conclusions

In this study, a composite analysis was applied to investigate the distinction between early and late breakup of the stratospheric polar vortex, with 11 early and 15 late cases. Notable differences in the circulation anomalies were found to occur between the early and late cases. Although circulation prior to breakup onset was "preconditioned" in both early and late cases, stratospheric circulation anomalies reversed after the onset of early breakups, while remaining consistent in late breakups. The circulation anomalies had a quasi-barotropic structure prior to and after early breakups from the lower troposphere to the upper stratosphere, while the consistent anomalies shrank to the stratosphere after late breakups.

The AO connects the stratosphere and troposphere conceptually, and analyzing its evolution helped depict stratosphere-troposphere interactions in early and late breakups. The results showed that a negative AO signal existed after both early and late breakups in the middle and high stratosphere, whereas its downward propagation was only evident after early breakups, and became limited in the stratosphere after late breakups.

A downward AO pattern was also identified in the geopotential height and temperature anomalies in early breakups, but not in late breakups. The signal reached the lower troposphere and affected the weather in early breakups. At 1000 hPa, the Aleutian Low and Siberian High were weakened, resulting in a weak winter monsoon and increased temperature in the mid latitudes and over the east coast of Asia. Previous studies have explored the near-surface anomaly patterns after a SSW in mid winter and suggest that a negative AO downward transmission leads to enhanced Aleutian Low and Siberian High events (Deng et al., 2008). These results revealed different tropospheric patterns from those we explored in early vortex breakups. Black et al. (2006) pointed out that near-surface anomaly patterns followed stratospheric polar vortex breakup and did not correspond well to the AO because the major extratropical anomaly features are retracted or shifted northward toward the pole. We think that a downward stratospheric AO anomaly may have various influences on tropospheric circulation during a SSW and vortex breakup. Owing to the fact it can influence the Aleutian Low and Siberian High and lead to a weak or strong East Asian winter monsoon (Chen et al., 2005; Wang et al., 2009), the stationary planetary wave may play some different roles in the impact after a SSW and SPVB, which needs some further investigation.

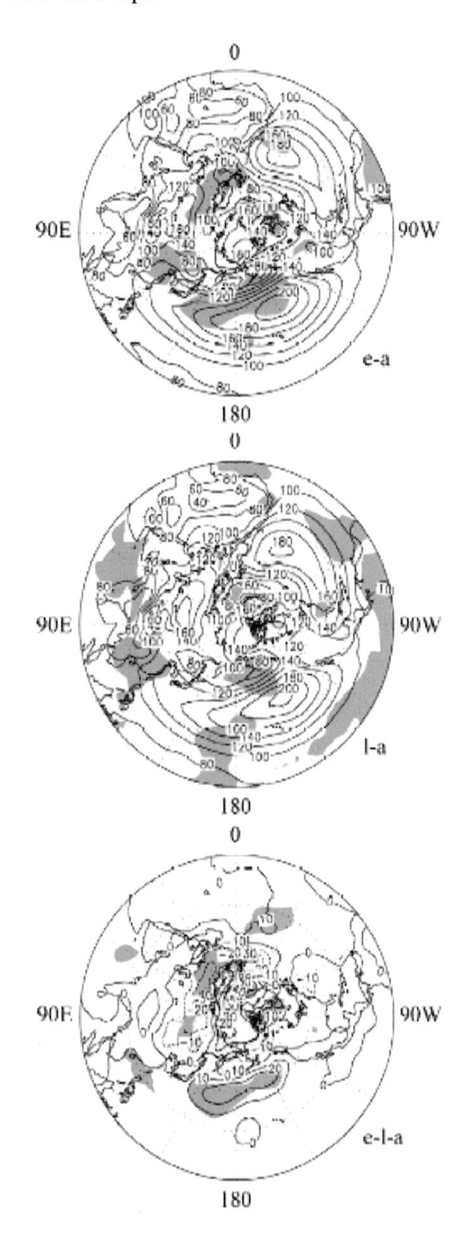

Fig. 8. Geopotential height (gpm) at 1000 hPa in April for early breakup years, late breakup years, and their differences. Shading denotes the region above the 90% confidence level; "e" represents early breakups; "l" represents late breakups; and "e-l" represents difference between early and late breakups.

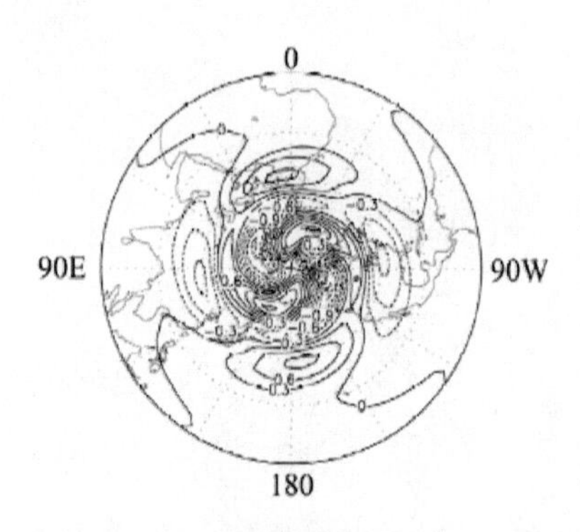

Fig. 9. Differences in the zonal wavenumber-2 distribution of April mean sea level pressure for early and late breakup years.

Ayarzaguena and Serrano (2009) studied the monthly characterization of the tropospheric circulation in early and late SFW. They suggested that late or early SFW may have some effect on the troposphere over the Euro-Atlantic area in April. By comparing the circulation differences in April in early and late breakup years, we found that early and late breakups can affect the process of tropospheric season transition, especially in the Asian-Pacific area. In April of early breakup years, tropospheric circulation at high latitudes was more summer-like, but it retained its winter features in the April of late breakup years. Meanwhile, we noticed that stratospheric influences on tropospheric season transition were not zonally symmetric. Ayarzaguena and Serrano (2009) suggested that planetary waves of wavenumber-2 play an important role in April in both the troposphere and stratosphere. Actually, wavenumber-2 is known to cause a zonally asymmetrical impact on the Siberian high and Aleutian low (Chen et al., 2005; Wang et al., 2009). Figure 9 shows the differences of the zonal wavenumber-2 distribution of April mean sea level pressure for early and late breakup years. The evident differences in Fig. 9 corresponding with those in Figs. 8e-l-a and e-l-b indicate that planetary waves (wavenumber-2) are important in terms of tropospheric season transition in early and late SPVB.

As the variations of stratospheric vortex persistence have a significant impact on interannual variations in polar ozone distribution (Zurek et al., 1996; Shindell et al., 1998), we suggest the climatic impact of SPVB may be more complex and deserves further dynamic analysis and numerical simulations.

Acknowledgements This work was supported by the Chinese Key Developing Program for Basic Sciences (Grant No. 2010CB950400) and the National Natural Science Foundation of China (Grant No. 40705023).

References

Ayarzaguena, B., and E. Serrano, 2009: Monthly characterization of the tropospheric circulation over the Euro-Atlantic area in relation with the timing of stratospheric final warmings. *J. Climate*., 22, 6313-6324.

Baldwin, M. P., and T. J. Dunkerton, 1999: Propagation of the Arctic oscillation from the stratosphere to the troposphere. *J. Geophys. Res*., 104, 30937-30946.

Baldwin, M. P., and T. J. Dunkerton, 2001: Stratospheric Harbingers of Anomalous Weather Regimes. *Science*, 294(5542), 581-584.

Black, R. X., 2002: Stratospheric forcing of surface climate in the Arctic Oscillation. *J. Climate*, 15(3), 268-277.

Black, R. X, and B. A. Mcdaniel, 2007: The dynamics of Northern Hemisphere stratospheric final warming events. *J. Atmos. Sci*., 64, 2934-2946.

Black, R. X., B. A. Mcdaniel, and W. A. Robinson, 2006: Stratosphere-troposphere coupling during spring onset. *J. Climate*, 19, 4891-4901.

Chen, W., S. Yang, and R. H. Huang, 2005: Relationship between stationary planetary wave activity and the East Asian winter monsoon. *J. Geophys. Res*., 110, doi: 10.1029/2004JD005669.

Chen, Y. J., R. J. Zhou, C. H. Shi, and Y. Bi, 2006: Study on the trace species in the stratosphere and their impact on climate. *Adv. Atmos. Sci*., 23(6), 1020-1039.

Christiansen, B., 2001: Downward propagation of zonal mean zonal wind anomalies from the stratosphere to the troposphere: Model and reanalysis. *J. Geophys*. Res., 106(21), 27307-27322.

Deng, S. M., Y. J. Chen, T. Luo, Y. Bi, and H. F. Zhou, 2008: The possible influence of stratospheric sudden warming on East Asian weather. *Adv. Atmos. Sci*.., 25(5), 841-846, doi: 10.1007/s00376-008-0841-7.

Frauenfeld, O. W., and R. E. Davis, 2003: Northern Hemisphere circumpolar vortex trends and climate change implications. *J. Geophys. Res*., 108(D4), 4423-4436.

Jung, T., and J. Barkmeijer, 2006: Sensitivity of the tropospheric circulation to changes in the strength of the stratospheric polar vortex. *Mon. Wea. Rev*., 134, 2191-2207.

Labitzke, K., 1981: The amplification of height wave 1 in January 1979: A characteristic precondition for the major warming in February. *Mon. Wea. Rev*., 109, 983-989.

Limpasuvan, V., D. W. J. Thompson, and D. L. Hartmann, 2004: The life cycle of the northern hemisphere sudden stratospheric warmings. *J. Climate*., 17(13), 2584-2597.

Limpasuvan, V., D. L. Hartmann, D. W. J. Thompson, K. Jeev, and Y. L. Yung, 2005: Stratosphere-troposphere evolution during polar vortex intensification. *J. Geophys. Res*., 110, D24101, doi: 10.1029/2005JD006302.

Perlwitz, J., and H. F. Graf, 2001: Troposphere-stratosphere dynamic coupling under strong and weak polar vortex conditions. *Geophys. Res. Lett*., 28(2), 271-274.

Shindell, D., D. Rind, and P. Lonergan, 1998: Increased polar stratospheric ozone losses and delayed eventual recovery owing to increasing greenhouse-gas concentrations. *Nature*, 392, 589-592.

Thompson, D. W. J., M. P. Baldwin, and J. M. Wallace, 2002: Stratospheric connection to northern hemisphere wintertime weather: Implications for prediction. *J. Climate*., 15, 1421-1428.

Wang, L., and W. Chen, 2010: Downward Arctic Oscillation signal associated with moderate weak stratospheric polar vortex and the cold December 2009. *Geophys. Res. Lett*., 37, L09707, doi: 10.1029/2010GL042659.

Wang, L., R. H. Huang, L. G, W. Chen, and L. H. Kang, 2009: Interdecadal variations of the East Asian winter monsoon and their association with quasi-stationary planetary wave activity. *J. Climate*, 22, 4860-4872.

Waugh, D. W., and P. P. Rong, 2002: Interannual variability in the decay of lower stratospheric Arctic vortices. *J. Meteor. Soc. Japan*, 80(4B), 997-1012.

Waugh, D. W., W. J. Randel, P. Steven, P. A. Newman, and E. R. Nash., 1999: Persistence of the lower stratospheric polar vortices. *J. Geophys. Res*., 104(22), 27191-27201.

Wei, K., W. Chen, and R. H. Huang, 2007: Dynamical diagnosis of the breakup of the stratospheric polar vortex in the Northern Hemisphere. *Science in China* (D), 50(9), 1369-1379.

Zurek, R. W., G. L. Manney., A. J. Miller, M. E. Gelman, and R. M. Nagatani, 1996: Interannual variability of the north polar vortex in the lower stratosphere during the UARS mission. *Geophys. Res. Lett*., 23, 289-292.

Annual and Interannual Variations in Global 6.5DWs from 20 to 110 km During 2002-2016 Observed by TIMED/SABER

Ying Ying Huang[1, 2, 3, 4], Shao Dong Zhang[2, 3, 4*], Chong Yin Li[1, 5], Hui Jun Li[1], Kai Ming Huang[2, 3, 4], and Chun Ming Huang[2, 3, 4]

[1] College of Meteorology and Oceanography, National University of Defense Technology, Nanjing, China,
[2] School of Electronic Information, Wuhan University, Wuhan, China,
[3] Key Laboratory of Geospace Environment and Geodesy, Ministry of Education, Wuhan, China,
[4] State Observatory for Atmospheric Remote Sensing, Wuhan, China,
[5] LASG, Institute of Atmospheric Physics, Chinese Academy of Sciences, Beijing, China

Abstract Using version 2.0 of the TIMED/SABER kinetic temperature data, we have conducted a study on the annual and interannual variations of 6.5DWs at 20-110 km, from 52°S to 52°N for 2002-2016. First, we obtained global annual variations in the spectral power and amplitudes of 6.5DWs. We found that strong wave amplitudes emerged from 25°S/N to 52°S/N and peaked in the altitudes of the stratosphere, mesosphere, and the lower thermosphere. The annual variations in the 6.5DWs are similar in both hemispheres but different at various altitudes. At 40-50 km, the annual maxima emerge mostly in winters. In the MLT, annual peaks occurred twice every half year. At 80-90 km, 6.5DWs appeared mainlyin equinoctial seasons and winters. At 100-110 km, 6.5DWs emerged mainly in equinoctial seasons. Second, we continued the study of the interannual variations in 6.5DW amplitudes from 2002 to 2016. Frequency spectra of the monthly mean amplitudes showed that main dynamics in the long-term variations of 6.5DWs were AO and SAO in both hemispheres. In addition, 4 month period signals were noticed in the MLT of the NH. The amplitudes of SAO and AO were obtained using a band-pass filter and were found to increase with altitude, as do the 6.5DW amplitudes. In both hemispheres, the relative importance of SAO and AO changes with altitude. At 40-50 and 100-110 km, AO play a dominant role, while at 80-90 km, they are weaker than SAO. Our results show that both the annual and interannual variations in 6.5DWs are mainly caused by the combined action of SAO and AO.

Keywords: Three separate altitude peaks of 6.5DW emerge from stratosphere to the thermosphere; Annual variations of 6.5DW are different in stratosphere, mesosphere, and the lower thermosphere; The SAO and AO in 6.5DWs have a long-term variation.

1. Introduction

Planetary waves (PWs) with a period of 5-7 days, traveling westward with zonal wave number1, are known as 5 day wave (5DW). These have been some of the most attractive PWs during recent decades in both numerical and observational studies [*Geisfer and Dickinson*, 1976; *Wu et al.*, 1994; *Riggin et al.*, 2006]. The 5DW has been considered as the gravest symmetric Rossby normal mode, the (1, 1) mode, because of their similar spatial distributions [*Madden and Julian*, 1972; *Williams and Avery*, 1992]. PWs that have similar meridional structures to 5DWs have also been found in the mesosphere and the lower thermosphere (the MLT, approximately 60-110 km). However, the periods of these PWs are usually longer, close to 6.5 days, and they are conventionally accepted as 6.5 day waves (6.5DWs) [*Meyer and Forbes*, 1997; *Talaat et al.*, 2001, 2002; *Lieberman et al.*, 2003; *Liu et al.*, 2004; *Jiang et al.*, 2008].

As a kind of global wave component from the stratosphere to the MLT, 6.5DW traces have been widely detected in several atmospheric parameters, such as zonal wind (10-30 m/s) [*Wu et al.*, 1994; *Meyer and Forbes*, 1997; *Talaat et al.*, 2001; *Kishore et al*, 2004; *Lima et al.*, 2005; *Jiang et al.*, 2008; *Day et al.*, 2012], meridional winds (10-20 m/s) [*Wu et al.*, 1994; *Talaat et al.*, 2001], temperature (10-15 K) [*Talaat et al.*, 2001; *Riggin et al.*, 2006; *Day et al.*, 2012], atomic oxygen ($\sim 4 \times 10^5$ m^{-6}) [*Talaat et al.*, 2001], and occurrences of the sporadic E (Es) layer [*Zuo and Wan*, 2008]. In polar summer mesopause regions, 6.5DWs are even found to modulate short-term variations of polar mesospheric clouds [*Merkel et al.*, 2003, 2008; *von Savignyet al.*, 2007; *Nielsen et al.*, 2010] and polar mesospheric summer echoes [*Kirkwood and Réchou*, 1998; *Kirkwood et al.*, 2002].

Seasonal variations in 6.5DWs have been demonstrated in several studies. However, variation features mentioned in these reports are not very similar. It has been shown that in MLT regions, predominant 6.5DWs are usually enhanced during equinoctial seasons [*Wu et al.*, 1994; *Lieberman et al.*, 2003; *Lima et al.*, 2005; *Jiang et al.*, 2008; *Gan et al.*, 2015] or before and after the equinox [*Talaat et al.*, 2001; *Liu et al.*, 2004] and are minimized at solstices [*Liu et al.*, 2004; *Jiang et al.*, 2008], although active wave events also appear in some winter months [*Jiang et ai*, 2008]. Wave activities in the stratosphere are enhanced at slightly earlier times [*Talaat et al*, 2002] or between fall and the following spring [*Liu et al.*, 2004]. These different characteristics are likely because the data sets in these studies are obtained for different time ranges or latitude coverages or even using different methods. These studies infer interannual variations in 6.5DWs [*Lima et al.*, 2005; *Jiang et al.*, 2008]. However, because the periods of most of these data sets are no longer than 3 years, the exact interannual variation characteristics of 6.5DWs are hard to obtain therein.

The 6.5DWs are thought to be generated at lower levels and propagate upward to the MLT regions [*Taiaat et al.*, 2001, 2002; *Lieberman et al.*, 2003; *Jiang et al.*, 2008]. *Wu et al.* [1994] considered mesospheric 6.5DWs as 5DWs that were Doppler shifted during their upward propagations. However, numerical simulations both in *Meyer and Forbes* [1997] and in *Liu et al.*

[2004] showed no evidence of 6.5DWs resulting from a Doppler shift of the 5DWs. It seems that 6.5DWs are rather unstable modes [*Meyer and Forbes*, 1997] that are amplified by atmospheric instabilities [*Meyer and Forbes*, 1997; *Liu et al.*, 2004]. In addition, the potential generation mechanisms of 6.5DWs have also been put forward, including interhemispheric propagations from winter stratospheres [*Riggin et al.*, 2006; *Day and Mitchell*, 2010; *Day et al.*, 2012]; the effects of heating due to moist convection [*Miyoshi and Hirooka*, 1999]; nonlinear interactions between the second asymmetric normal mode with $m = 2$ (the 7 day wave) and stationary planetary waves (SPWs) with $m = 1$ [*Pogoreitsev et al.*, 2002] or between 4 day $s = 2$ and 10 day $s = 1$ waves [*Taiaat et al.*, 2001]; and a possible Antarctic excitation source of 6.5DWs in troposphere [*Cheong and Kimura*, 1997; *Lieberman et al.*, 2003].

As 6.5DWs traverse the stratosphere and mesosphere, they undergo considerable variations [*Lieberman et al.*, 2003]. The wave source, mean wind structure, instability, and the critical layers of the wave can affect the wave response in the MLT region [*Liu et al.*, 2004]. Therefore, the temporal variations and spatial distributions of 6.5DWs should be determined by assessing vertical and meridional propagations and the accompanying wave-flow interactions. However, the observational evidence of 6.5DWs is still insufficient.

The Thermosphere Ionosphere Mesosphere Energetics Dynamics/Sounding of the Atmosphere Using Broadband Emission Radiometry (TIMED/SABER) mission is able to measure the global kinetic temperature of the neutral atmosphere from the stratosphere to MLT (20-110 km) [*Killeen et al.*, 2006]. Because of its high vertical resolution, acceptable precision, and scarce data gaps [*Huang et al.*, 2006; *Xu et al.*, 2006; *Remsberg et al.*, 2008], in recent decades, data from this mission have been widely used in studies of atmospheric structures [*Xu et al.*, 2007; *Dou et al.*, 2009; *Gan et al.*, 2012; *Jiang et al.*, 2014] and disturbances from the stratosphere to MLT, such as gravity waves [*Ern et al.*, 2011; *John and Kumar*, 2012; *Shuai et al.*, 2014]; PWs [*Garcia et al.*, 2005; *Huang et al.*, 2013]; Kelvin waves [*Forbes et al.*, 2009]; tides [*Oberheide et al.*, 2005; *Zhang et al.*, 2006; *Xu et al.*, 2009]; and even nonlinear interactions between wave couples [*Xu et al.*, 2014; *Gu et al.*, 2015].

In this paper, the annual variations, interannual variations, and latitude distribution characteristics of 6.5DWs are studied by using TIMED/SABERv2.0 Level2A kinetic temperature data for2002-2016, from 52°S to 52°N, at 20-110 km. In section 2, the data employed in this paper and the spectral analysis method are introduced; in section 3, the annual variations and latitudinal distributions of6.5DWs in spectra and amplitudes are studied; in section 4, the interannual variations and their potential controlling processes are studied; finally, discussions and conclusions are provided in sections 5 and 6, respectively.

2. Data and Processing

2.1 Temperature Data

SABER retrieves kinetic temperature for the neutral atmosphere from emission at 15 μm by using LTE (local thermodynamical equilibrium) approximation in the stratosphere (~20-

50 km) [*Giiie and House*, 1971] and non-LTE (nonlocal thermodynamical equilibrium) approximation in the MLT regions (~50 to 120 km) [*Edwards and Lopez-Puertas*, 1993; *Mertens et al.*, 2001]. Approximately 15 evenly spaced points with similar local times are sampled in a given latitude band on each observing day for ascending or descending orbits, respectively. The precession rate is approximately 3° per day in longitude in a backward trajectory, corresponding to 12 min in local time. As a result, SABER takes 60 days to cover 24 h local time completely by considering the data obtained on both ascending and descending orbits. Every60 days, the TIMED satellite yaws to keep the instruments on the antisun side of the spacecraft. Thus, the latitude coverage of SABER flips from 54°S-83°N to 54°N-83°S.

In this paper, v2.0 data for February 2002 to May 2016 between 52°S and 52°N and from 20 to 110 km are used to study the temporal and spatial variations of6.5DWs. The latitude range is divided into 27 bands with a 4° interval, and the vertical interval is interpolated to 1 km. Background temperatures are defined as the linearly fitted temperature in a 60 day window and a 1 day step. The temperature disturbances are obtained by subtracting the background temperature from the data, and they would not be contaminated by any trends longer than 60 days.

2.2 Spectral Analysis

Before conducting spectral analysis, the daily mean values of the disturbances along the ascending and descending tracks of SABER are subtracted, respectively. Because every day SABER samples at a similar local time along its ascending and descending tracks in each latitude band, this step is capable of diminishing the influences of migrating tides in the spectral results. Salby's fast Fourier method [*Salby*, 1982a, 1982b] has been widely used in spectral analysis and is based on asynoptic sampling patterns of satellite observations [*Garcia et al*, 2005; *Merzlyakov and Pancheva*, 2007; *Huang et al.*, 2013]. This method is also used here to calculate the frequency-wave number spectral features of PWs. Similar to *Huang et al.* [2013], each spectral result is determined in a 30 day window, and the spectral results from 52°S to 52°N and 20 to 110 km for a 15 year time span from March 2002 to 2016 are calculated as well to show the annual and spatial variations in 6.5DW spectral power. Sampling points at certain latitude bands are extraordinarily scattered on TIMED's yaw days, so data measured on these days are excluded in spectral analysis here. The time ranges of spectral windows are listed in Table 1. The middle and bottom rows show the days in the year (DoY) and dates of the windows, respectively, based on which abbreviations of the time windows are named and shown in the top row. Although the days between neighboring windows sometimes overlap, natural temporal variations in spectral power are still obtained.

Table 1. Time Ranges of Time Windows in Spectral Analysis

Abbreviation	JF	FM	MA	AM	MJ	JJ	JA	AS	SO	ON	ND	DJ
DoY	019-048	042-071	079-108	104-133	143-172	162-191	199-228	226-255	266-295	288-317	326-355	344-008
Date	1.19-2.17	2.11 - 3.12	3.20-4.18	4.14-5.13	5.23-6.21	6.11 - 7.10	7.18-8.16	8.14-9.12	9.23-10.22	10.15-11.13	11.22-12.21	12.10-1.8

Figure 1 shows the frequency-wave number spectral results at 80 km at 40°N during (a) AM and (b) AS in 2003. Frequency has been transformed into period in this figure. The minimum level and interval of contours are 0.8 and 0.2 K, respectively. Both panels show predominant spectral peaks at 6.5 day period approximately and wave numberW1 (W and E means westward and eastward propagation, respectively). This wave component is the focus of this paper. The spectral power of 6.5DWs attenuates quickly on both sides of the peak period. According to lengths and intervals of the SABER sampling patterns, the period range of 6.5DWs can be determined within 5.5-6.8 days. In addition to 6.5DWs, another spectral peak can be noticed at period an 8.2 day period with wave number W1 in both seasons.

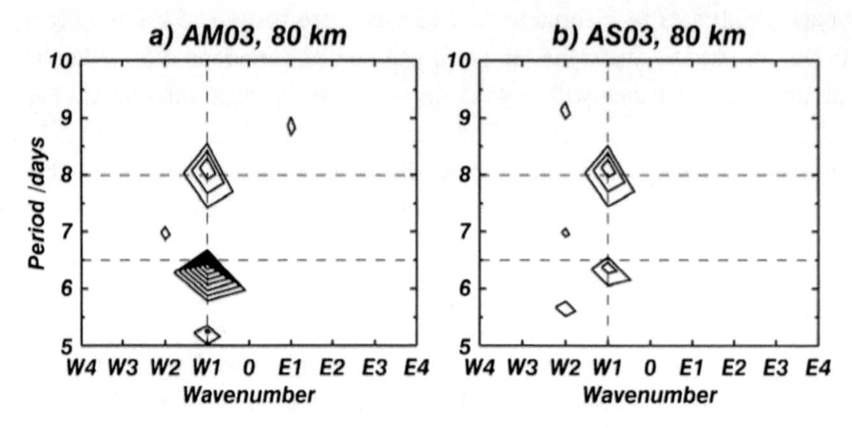

Figure 1. Frequency-wave number power spectra (unit: K) at 80 km at 40°N during (a) AM and (b) AS in 2003. The minimum level of contours is 0.6 K and that of the interval is 0.2 K.

3. Global Annual Variations

3.1 Spectral Results

Planetary waves play important roles in modulating the variations and dynamics of the atmosphere, from the stratosphere to MLT. To show the spatial and temporal variations in 6.5DW spectral peaks, a parameter $\overline{P}_n$ (s, f), normalized spectral power (NSP), is defined. For the ith altitude, the jth latitude, and the kth spectral window, $\overline{P}_n$ (s, f) at each wave number s and frequency f is defined as

$$\overline{P}_n(s,f)=\frac{P_{h_i,\Phi_j,w_k}(s,f)}{\max\left\langle P_{h_i,\Phi_j,w_k}\right\rangle},\tag{1}$$

in which P_{h_i,Φ_j,w_k} (s, f) indicates the spectral power at wave number s and frequency f directly calculated from Salby's method and the denominator max $\langle\rangle$ denotes the maximum spectral power among all s and f. Under this definition, $\overline{P}_n$ (s, f) becomes to be a dimensionless quantity with value between 0 and 1. $\overline{P}_n$ (s, f) shows the relative strengths of PW components. The NSP of 6.5DWs is extracted from each spectral result in all time windows at altitude from 20 to 110 km altitude and between latitudes of 52°S and 52°N latitude ranges from MA 2002 to FM 2016. The 2002-2016 averaged NSP of 6.5DWs is calculated, and their annual variations in the latitude-altitude plane are shown in Figure 2. Figures 2a-2l show the results from JF to DJ Only NSP values greater than 0.5 are colored. This figure provides the statistical results of annual variations in NSPs.

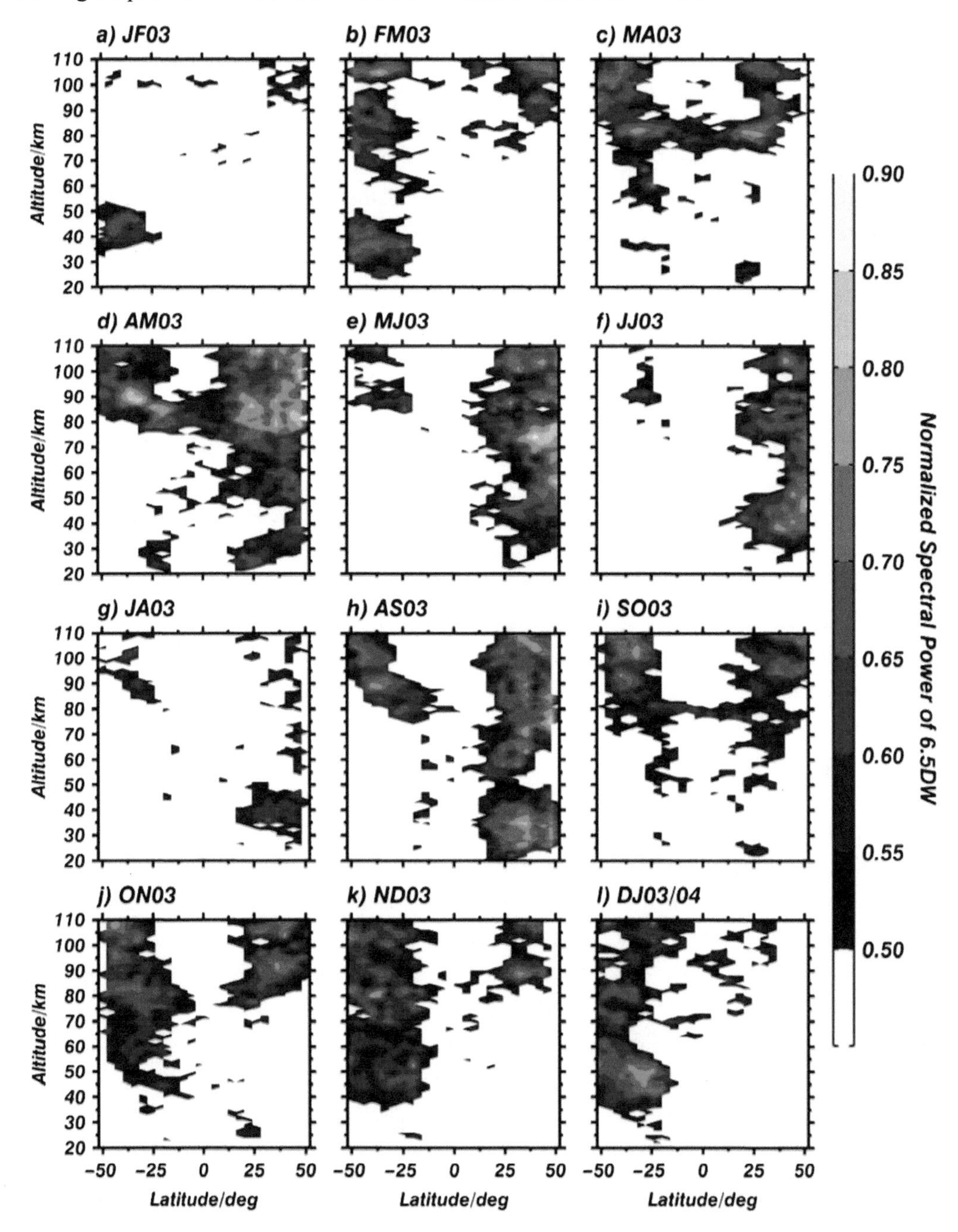

Figure 2. Latitude-altitude distributions of 2002–2016 averaged NSP of 6.5DWs from JF to DJ. Abbreviations shown above each panel correspond with Table 1.

Large NSP values can be mostly noticed in extra tropical regions from 25°S/N to 50°S/N. The annual variations of the averaged NSP during 2002-2016 are related with altitude, and they are similar in both hemispheres. In the stratosphere (30-50 km), dominant NSP peaks arise from late spring to early autumn: during ND-FM in the SH and MJ-AS in the NH. During other seasons, stratospheric NSP is generally invisible. From the upper stratosphere to the mesosphere (50-80 km), large NSP values usually emerge in spring-early summer and early autumn: from AM-JJ and AS in the NH and ON-DJ and FM-MA in the SH. From the mesopause region to the lower thermosphere (80-110 km), the NSP values are symmetrical with the equator during interim seasons between solstitial and equinoctial seasons, i.e., FM-AM and AS-ON. During equinoctial seasons: MJ-JJ and ND-DJ, larger NSP values can be found in the spring.

3.2 Amplitude Results

To determine the absolute strengths of 6.5DWs and their temporal variations, the amplitudes and phases of waves are determined by the least-square fitting method of harmonic functions: $f(A, \varphi)=A \cdot \cos(\omega t - s\lambda - \varphi)$, in which A, φ, ω, and s are the amplitude, phase, angular frequency, and zonal wave number of 6.5DW, respectively. t and λ are the universal time and longitude of each sampling point, respectively. The lifetime of 6.5DWs is found to be ~10-20 days [*Wu et al.*, 1994] or 3-4 weeks [*Talaat et al.*, 2001]. Accordingly, wave parameters are fitted in 1 day steps and 13 day windows, within which wave structures are retained and wave parameters are not changed much. Before fitting, linear trends were removed in each 13 day window to minimize long-term influences on the obtained amplitudes. As a result, the fitted amplitudes represent the average values over a few wave cycles.

Latitude-altitude distributions of the 2002-2016 averaged monthly mean amplitudes of 6.5DW are shown in Figure 3. Figures 3a-3l show the results from January to December, respectively. The latitude-altitude distributions of the monthly mean amplitudes are similar in both hemispheres. The meridional minimum amplitudes are shown around the equator from 20 to 110 km throughout the year. At extratropical latitudes of both hemispheres, large amplitudes appear above 70 km poleward from 25°. In general, these amplitudes increase with altitude. In the MLT, the annual maximum amplitudes (>3.0 K at 80-90 km and >6.0 K at 100-110 km) occur in the late spring and early autumn seasons, from April to May and August to September in the NH and in November and February in the SH. Two altitude peaks emerge at approximately 80-90 km and 100-110 km, respectively. In the stratosphere, the amplitudes are usually less than 1.0 K. However, the amplitudes reach peak values (>1.8 K in the NH and >1.4 K) at 40-50 km poleward from 25° during winters in both hemispheres, from December to February in the NH and July to September in the SH.

Both the NSP and amplitudes of 6.5DWs show large values in middle to high latitudes, from 25°S/N to 52°S/N. In the MLT of both hemispheres, annual variations in NSP are usually similar with those of the amplitudes. The annual variations in NSP reach the annual maxima in late spring and early autumn. Therefore, in the MLT, 6.5DWs must be one of the most predominant PWs, especially during late spring and early autumn. In the stratosphere, the NSP appears to be larger in summers, but amplitudes are larger during winters. Between these latitude bands, it is usually westerlies/easterlies in the winter/summer from troposphere to stratosphere [*Dickinson*, 1968]. PWs can penetrate into the stratosphere during winters, and stationary planetary waves (SPWs) with zonal wave numbers 1 or 2 are usually large [*Smith*, 1983]; thus, the relative strength of 6.5DWs compared to such dominant PW spectra becomes weak. However, in summer, vertical propagations of PWs should be restrained by easterlies, so even small amplitudes of 6.5DWs can dominate PWs' spectra.

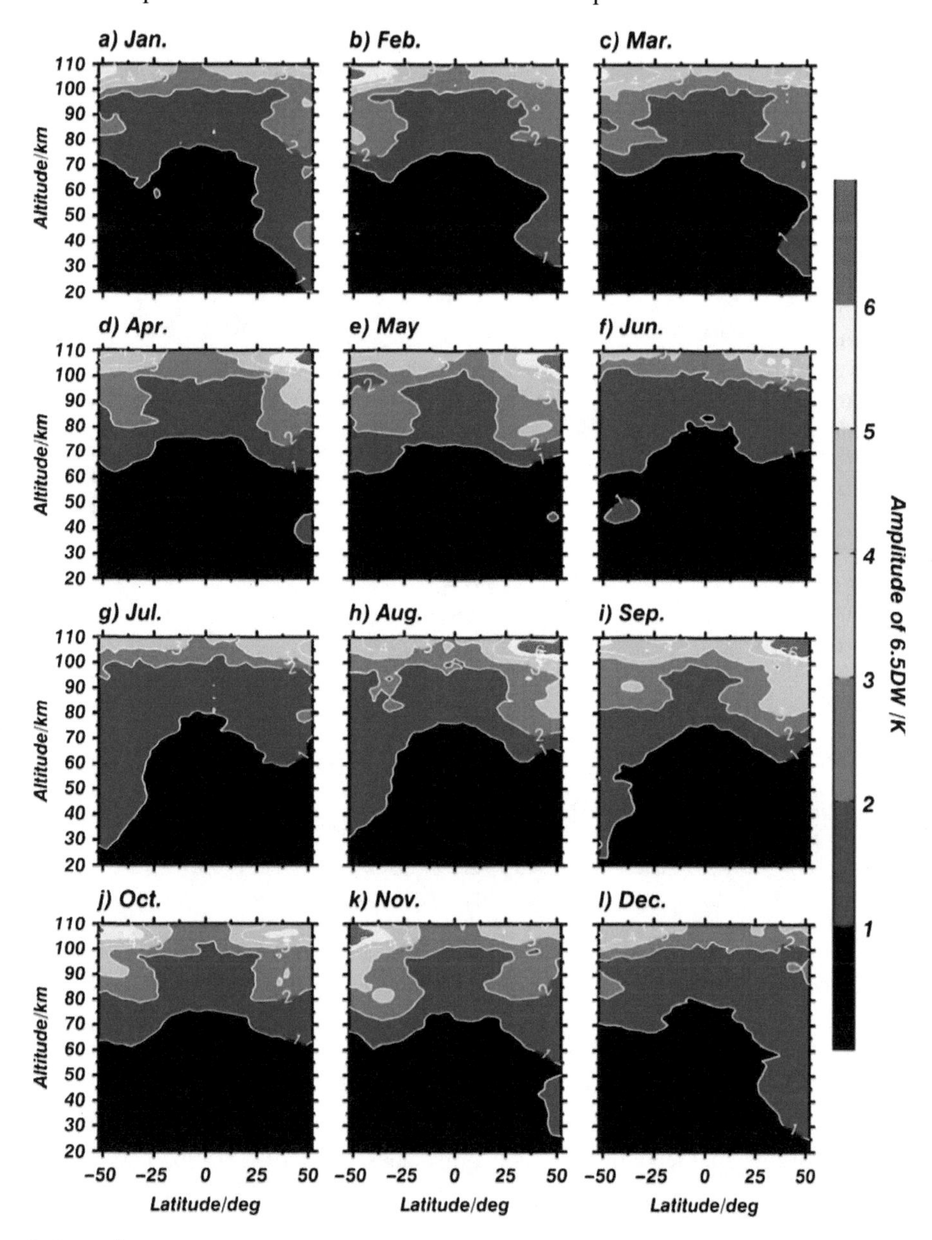

Figure 3. Annual variations in 2002–2016 averaged monthly mean amplitudes. (a–l) Results from January to December.

To show the general annual variations of 6.5DWs, the amplitudes are averaged for 2002-2016. Day-to-day variations of the composite amplitudes from 20 to 110 km are shown in Figure 4. Figures 4a and 4b show the results at 40°N and 40°S, respectively. At these latitudes, the composite amplitudes are less than 2.0 K below 70 km. Above 70 km, the amplitudes increase with altitude and become prominent during certain seasons. The vertical maxima appear at 100-110 km. It is clear that the annual variations in 6.5DW amplitudes are similar at both latitudes but are different below and above 70 km. Below 70 km, the amplitudes vary annually, with the annual maxima/minimums emerging during winters/summers. However, above 70 km, the amplitudes reach peak values twice a year in the late spring and early autumn seasons. The annual variations in the composite amplitudes shown in Figure 4 are similar with those of the monthly mean amplitudes shown in Figure 3.

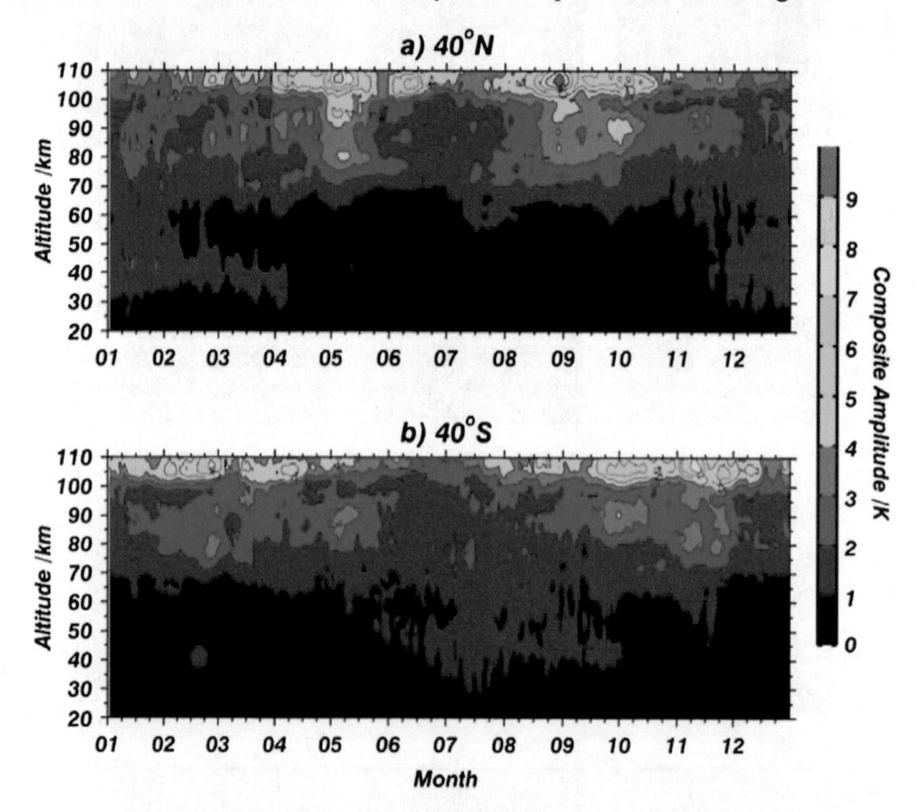

Figure 4. Day-to-day variations in the composite 6.5DW amplitudes averaged for 2002-2016 at (a) 40°N and (b) 40°S.

The altitude transitions of 6.5DW amplitudes from the stratosphere to the MLT during March 2002 to April 2016 are shown in Figure 5. The black, blue, and red curves depict the day-to-day variations in the daily average amplitudes at 40-50, 80-90, and 100-110 km, respectively. Altitude ranges are marked at the left top corner of each panel. Figures 5a-5c and 5d-5f show the results at 40°N and 40°S, respectively. In each panel, the vertical dashed lines separate each year. The letters "M, ""J, " "S, " and "D" below the x axes indicate the first day of March, June, September, and December, respectively. The ranges of the y axes change with altitude. These ranges are 0-6, 0-8, and 0-16 K at 40-50, 80-90, and 100-110 km, respectively.

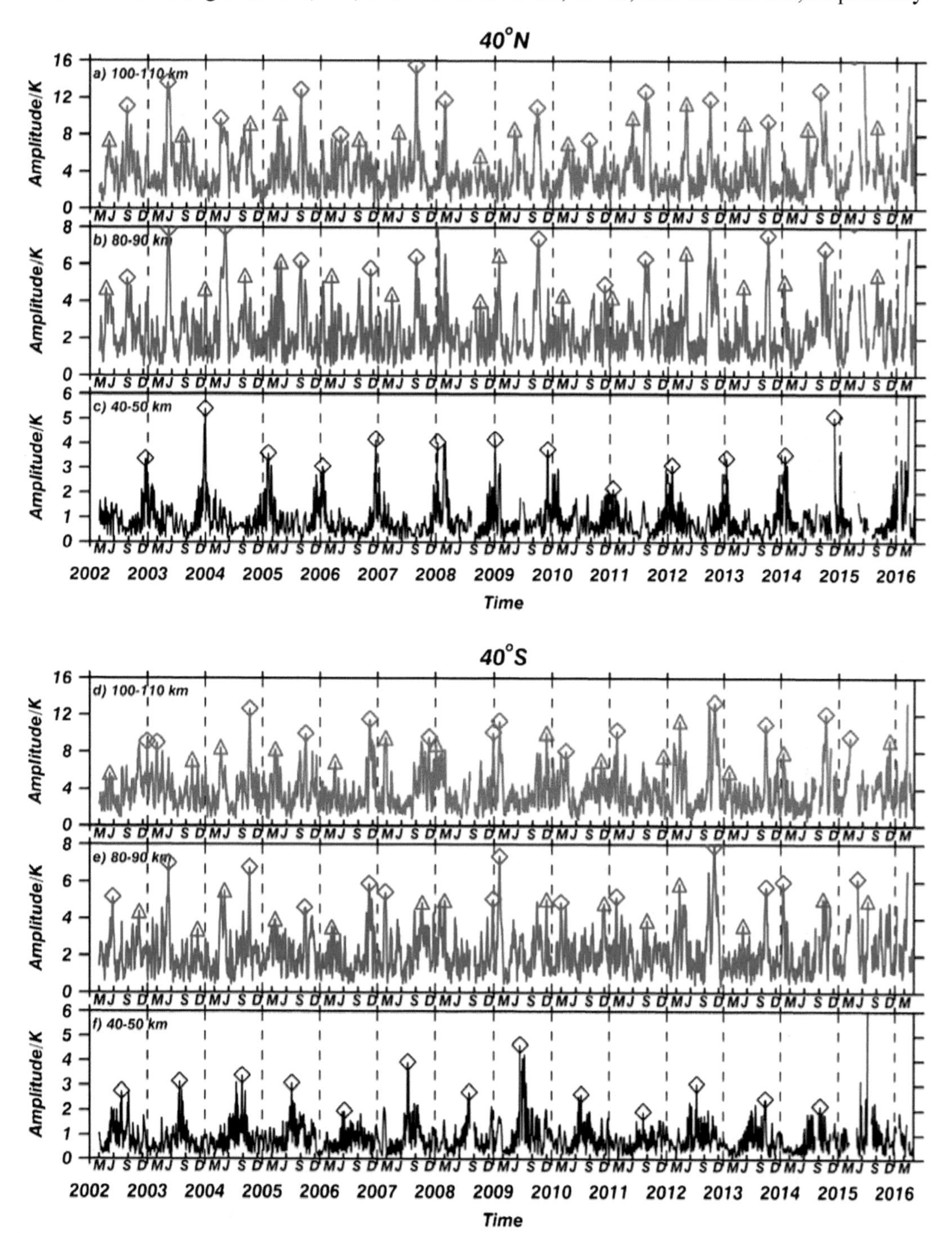

Figure 5. Day-to-day variations of mean amplitudes from March 2002 to April 2016 at 40-50 km (black), 80-90 km (blue), and 100-110 km (red).

From March 2002 to April 2016 in each of the three altitude ranges, the annual variations are similar at both latitudes. The amplitudes at 40-50 km vary periodically within a year. The annual maximum amplitudes mainly emerge in winter seasons. In Figures 5c and 5f, the peaks of every winters are marked as diamonds. The dates and amplitudes of winter peaks are listed in Table 2. The annual variations in the amplitudes at 80-90 and at 100-110 km are similar but are different from the variation characteristics at 40-50 km. During each year, the amplitudes at the upper levels reach peak values in the first and the second half of year, respectively. The larger and smaller one of peaks in each half of the year are considered to be the first and the second peaks of the year and are marked as diamonds and triangles, respectively, in Figures 5a and 5band 5d and 5e. The dates and amplitudes of these peaks at 80-90 and 100-110 km are listed in Tables 3 and 4, respectively. Within each data frame, the corresponding results of the first and the second peaks are shown to the left and right of the slash, respectively. The bottom row in Tables 2-4 shows the average amplitudes and standard deviations of the peak amplitudes.

Table 2. Dates and Amplitudes of the Maxima During Winters at 40-50 km From 2002 to 2016 at 40°S and 40°N. Average Amplitudes and Standard Deviations Are Shown in the Bottom Row of This Table

40°S		40°N	
Dates	Amplitudes/K	Dates	Amplitudes/K
21 Jul 2002	2.8	16 Dec 2002	3.4
21 Jul 2003	3.2	30 Dec 2003	5.4
26 Aug 2004	3.4	4 Feb 2005	3.6
6 Jul 2005	3.1	15 Jan 2006	3.1
7 Jun 2006	2.0	18 Dec 2006	4.2
13 Jul 2007	3.9	12 Jan 2008	4.1
1 Aug 2008	2.7	4 Jan 2009	4.2
16 Jun 2009	4.6	1 Dec 2009	3.8
6 Jul 2010	2.6	21 Jan 2011	2.2
6 Aug 2011	1.9	28 Jan 2012	3.1
12 Jul 2012	3.1	14 Jan 2013	3.4
25 Sep 2013	2.5	24 Jan 2014	3.5
9 Sep 2014	2.2	28 Nov 2014	5.1
6 Jul 2015	9.2	21 Mar 2016	6.8
Average	3.4 ± 1.8	-	3.9 ± 1.2

Both the amplitudes and dates of peaks vary interannually. From March 2002 to April 2016, the times of peaks in each month are recorded. Because amplitudes are usually remarkable from 30°-50° in both hemispheres, and monthly variations of records are similar between 30°S-50°S and 30°N-50°N, the total counts in these two latitude ranges are summed, and their monthly variations of them are shown in Figure 6. Figures 6a-6c and 6d-6f show results for

Table 3. Dates and Amplitudes of the First and the Second Maxima of Mean Amplitudes at 80–90 km From 2002 to 2015[a]

Year	40°S		40°N	
	Date	Amplitude/K	Date	Amplitude/K
2002	26 May/9 Nov	5.2/4.4	22 Aug/12 Apr	5.3/4.7
2003	14May/14 Nov	7.0/3.4	12May/27Dec	7.9/4.7
2004	11 Oct/5 May	6.8/5.6	5 May/8 Sep	8.0/5.4
2005	25 Sep/22 Mar	4.7/4.0	28 Aug/23 Apr	6.2/6.2
2006	9 Nov/17Mar	5.9/3.6	13 Nov/13 Mar	5.8/5.4
2007	20 Feb/12Oct	5.5/4.9	30 Aug/28 Mar	6.4/4.4
2008	27 Dec/28 Feb	5.1/5.0	12 Jan/1 Oct	8.6/4.1
2009	5 Feb/28 Nov	7.4/5.1	2 Oct/28 Jan	7.4/6.5
2010	26 Feb/29 Nov	4.9/4.9	27 Nov/5 Mar	4.9/4.4
2011	16Feb/29 Aug	5.2/3.9	10Aug/14Jan	6.3/4.2
2012	2 Nov/20 Mar	8.0/5.9	27 Sep/26 Apr	8.3/6.7
2013	28 Sep/6 May	5.7/3.7	2 Oct/1 May	7.6/4.9
2014	17 Jan/25 Sep	6.0/5.1	29 Sep/19 Jan	6.9/5.1
2015	1 May/6 July	6.2/5.0	30 Mar/29 Aug	8.3/5.5
Average	-	6.0 ± 1.0/4.6 ± 0.8	-	7.0 ± 1.2/5.2 ± 0.8

[a]Left and right sides of the slash within each data frame show values of the first and the second maxima of each year, respectively. Average amplitudes and standard deviations are shown in the bottom row of this table.

Table 4. Similar as Table 3, but for Results at 100–110 km

Year	40°S		40°N	
	Date	Amplitude/K	Date	Amplitude/K
2002	29 Dec/7 May	9.2/5.8	23 Aug/2 May	11.2/7.5
2003	2 Mar/11 Oct	9.1/7.2	8 May/6 Aug	13.7/8.1
2004	13 Oct/10Apr	12.7/8.6	8 Apr/13 Oct	9.8/9.3
2005	2 Oct/23 Mar	10.0/8.4	29 Aug/19 Apr	12.9/10.4
2006	14 Nov/7 Apr	11.6/7.0	8 May/4 Sep	8.0/7.6
2007	24 Nov/22 Feb	9.6/9.6	30 Aug/11 May	15.5/8.4
2008	28 Dec/5 Jan	10.2/8.9	25 Feb/30 Sep	11.8/5.8
2009	4 Feb/27 Nov	11.4/10.1	24 Sep/5 May	11.0/8.7
2010	30 Mar/9 Nov	8.1/7.2	19 Aug/5 Apr	7.5/7.2
2011	19Feb/9Dec	10.4/7.7	12 Aug/22 May	12.8/10.0
2012	4 Nov/23 Mar	13.4/11.4	29 Sep/27 Apr	11.9/11.5
2013	25 Sep/6 Feb	11.1/6.0	1 Oct/5 May	9.5/9.4
2014	12 Oct/19Jan	12.2/8.1	29 Aug/11 Jun	12.8/8.8
2015	14 Mar/28 Nov	9.7/9.4	30 Mar/29 Aug	16.7/9.1
Average	-	10.6 ± 1.5/8.2 ± 1.6	-	11.8 ± 2.6/8.7 ± 1.4

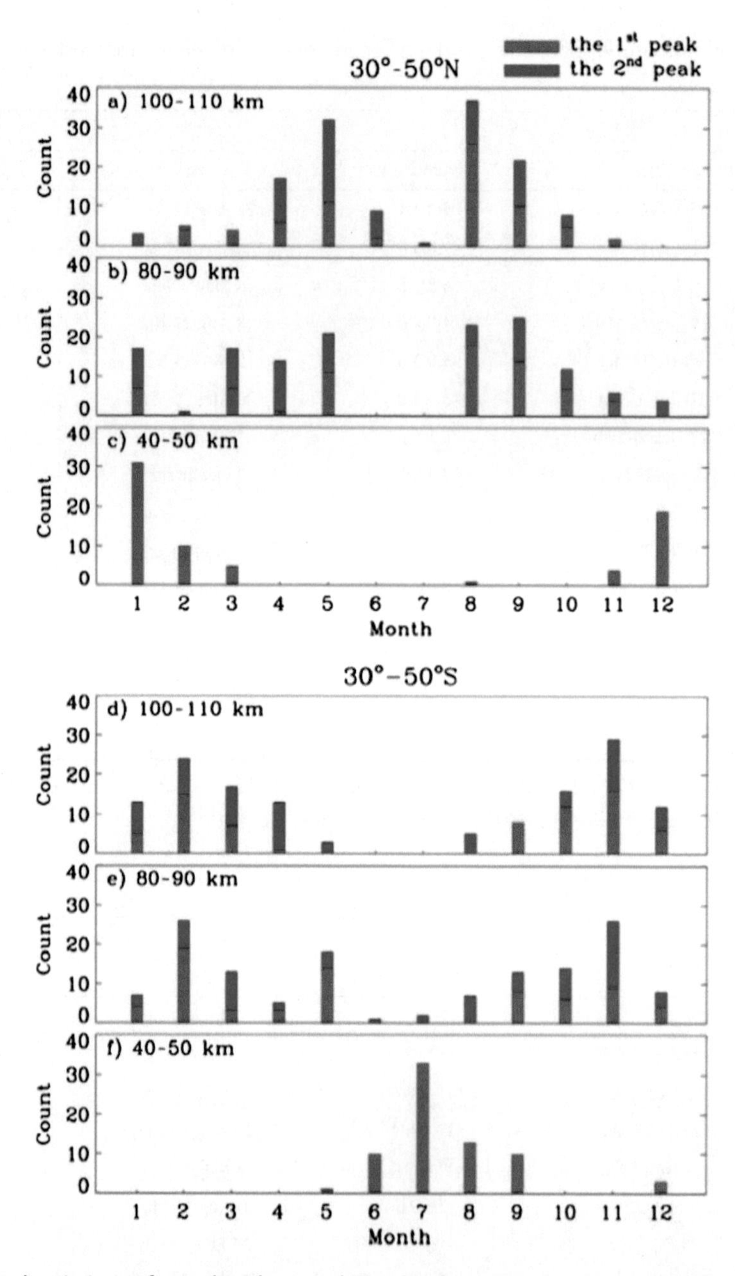

Figure 6. Annual variations of months when peak amplitudes emerge as counted during March 2002 to April 2016. Values are summed over (a–c) 30°N–50°N and are summed over (d–f) 30°S–50°S. Values are presented for 100–110 km (Figures 6a and 6d), 80–90 km (Figures 6b and 6e), and 40–50 km (Figures 6c and 6f).

100-110, 80-90, and 40-50 km at 30°-50°N and 30°-50°S, respectively. At 40-50 km, the red bars represent counts of winter peaks; at 80-90and 100-110km, the red and blue bars represent counts of the first and the second peaks of years, respectively. From 30°N-50°N and 30°S-50°S at 40-50 km, winter peaks occur in January and July, respectively. At each latitude in the 80-90 and 100-110 km altitude ranges, the month distributions of the first and the second peaks are similar. Provided that at 80-90 and 100-110 km the amplitudes of the first and second peaks in each year are comparable, the counts of the first and second peaks are added together and their monthly variations are shown. At 80-90 km from 30°N to 50°N, these peaks usually emerge in January, March to May, and August and September, while from 30°S to 50°S, the peaks often emerge in February, May, and November. In both latitude ranges, the monthly distributions of counts at 100-110 km are more concentrated than at 80-90 km. From 30°N to 50°N, the majority of peaks occur in April and May and August and September, while at 40°S, the peaks clearly occur in February and March and October and November.

It can be concluded from Figure 6 that annual peaks of 6.5DWs at 40-50 km usually emerge in winter seasons in both hemispheres, mostly during January from 30° to 50°N and in July from 30° to 50°S. At 80-90 and 100-110 km, wave peaks emerge twice during each half of the year. At 80-90 km, peaks usually emerge in equinoctial seasons and winters in both hemispheres, in March to May, August and September, and January from 30° to 50°N and in February, November, and May from 30° to 50°S. At 100-110 km, peaks usually emerge during equinoctial seasons, in April and May and August and September from 30° to 50°N and in February and November from 30° to 50°S. Annual variations in wave peaks can provide necessary information about wave sources and the propagation features of wave energy. For 6.5DWs, the propagation features are similar in both hemispheres but are completely different at 40-50, 80-90, and 100-110 km. Both *Talaat et al.* [2001] and *Jiang et al.* [2008] have suggested that 6.5DWs in the MLT are propagated from lower levels. Thus, different annual variations of 6.5DWs in the MLT and in the stratosphere potentially result from their modulation by background winds at lower heights.

4. Interannual Variations

Figure 5 shows clear interannual variations in the amplitudes of 6.5DWs at 40°S/N. To show the general interannual variations and latitude distributions, the wave amplitudes are averaged over every 3 months from 2002 to 2016. The latitude and temporal variations of these waves are shown in Figure 7. The results for 100-110 km, 80-90 km, and 40-50 km are shown in Figures 7a-7c, respectively. From 40-50 km to 100-110 km altitude levels, remarkable amplitudes propagate poleward from 25° in both hemispheres. Therefore, these latitude ranges in the NH and SH are denoted as NH and SH for short in the following descriptions of this section.

It is inferred that latitude centers of strong 6.5DW activities are around or even poleward from 40° to 50° in both hemispheres. The annual maximum amplitudes and their latitude locations change interannually. At 40-50 km, the amplitudes are greater than 1.6 K in every winter during 2003-2016 in the NH except for winters of 2006/2007, 2010/2011, 2012/2013, and 2014/2015 and in the winters of 2002, 2004-2005, and 2007 in the SH. At 80-90 km, larger amplitudes (>3.0 K) emerge during the spring-summer of 2003-2004, the summer-autumn of 2005, 2011-2012 and 2014-2015, and winter of 2015/2016 in the NH while during spring-summer of 2006-2007 and 2012 and summer-autumn of 2003-2004 in the SH. In 100-110 km, larger 6.5DW amplitudes (>6.0 K) emerge during springs of 2003-2004 and 2011 and autumns of 2007 and 2011 in the NH, while they emerge during the springs of 2002, 2006-2008, and 2012 in the SH. It seems that 6.5DW amplitudes show higher values in winters in both hemispheres at 40-50 km. Both amplitudes at 80-90 km and 100-110 km demonstrate higher values during the springs of 2003-2004 and during the summer- autumn in 2011 in the NH and during the springs of 2006-2007 and 2012 in the SH.

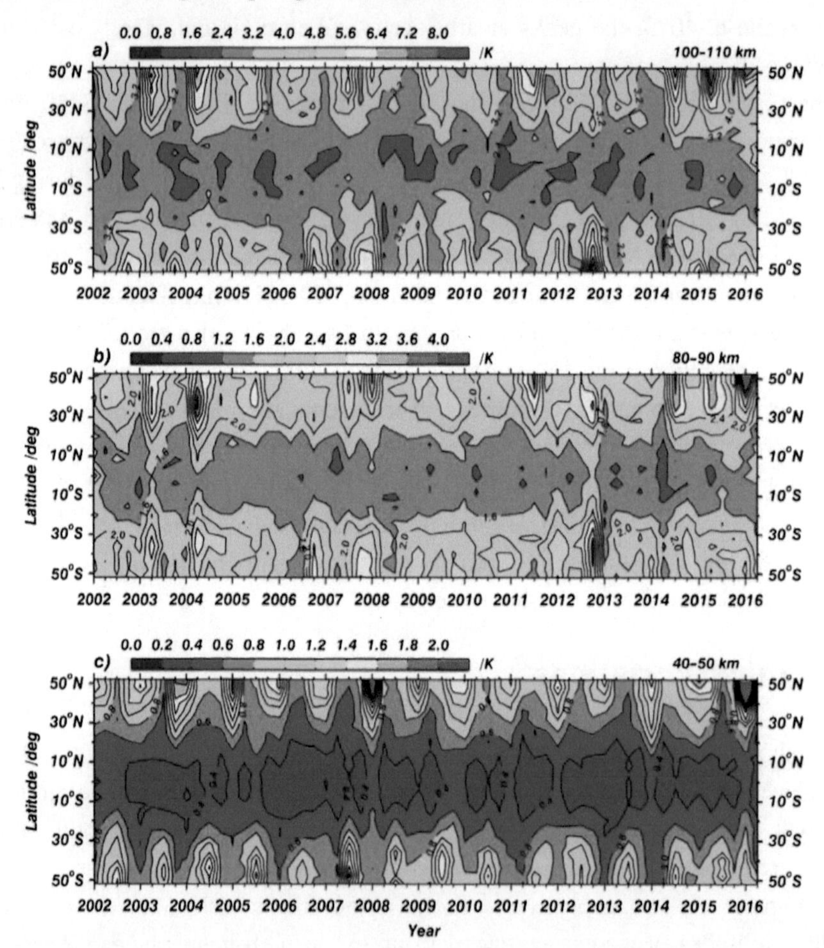

Figure 7. Latitude and temporal variations of 3 month averaged 6.5DW amplitudes from 2002 to 2016 in (a) 100-110 km, (b) 80-90 km, and (c) 40-50 km.

To further diagnose the long-term variations in 6.5DW amplitudes and the possible modulation processes of atmospheric dynamics with longer periods, frequency spectra of monthly mean amplitudes are calculated. Figure 8 shows the latitude variations of spectra at 100-110 (Figure 8a), 80-90 (Figure 8b), and (Figure 8c) 40-50 km. The spectra of periods longer than 3.5 months are shown here. At 40-50 km, the main periods in the long-term variations of

6.5DW amplitudes are 12 months and 6 months. These periods are related to AO (annual oscillations) and SAO (semiannual oscillations). In both hemispheres, the spectral amplitudes of AO are larger than those of SAO. At 80-90 km, the most prominent spectral peak in the NH is at approximately 4 months, with an amplitude of approximately 0.6 K at 52°N. In addition, the AO and SAO peaks are also remarkable. Spectra at approximately 28 months are also visible, and they could be considered as QBO (quasi-biennial oscillation) signals. In both hemispheres, the spectral amplitudes of the QBO are less than 0.2 K. At 100-110 km, the spectral peaks of the AO, SAO and 4 month periods are prominent in both hemispheres. Among them the most remarkable signal is that of the AO. In addition, the spectral peaks of the QBO can be noticed as well, but these signals are much weaker.

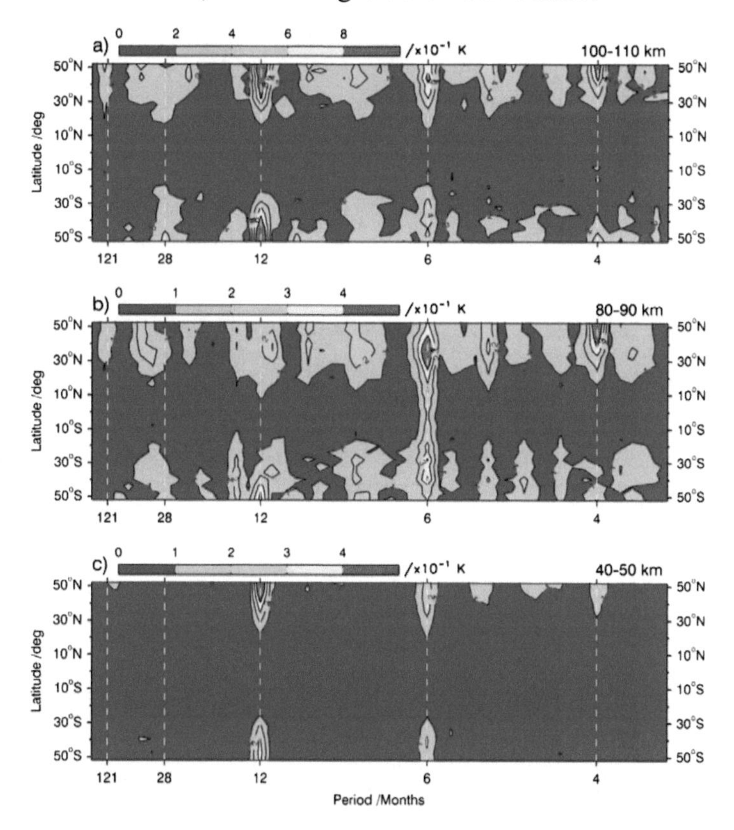

Figure 8. Spectral power of monthly averaged 6.5DW amplitudes at (a) 100-110 km, (b) 80-90 km, and (c) 40-50 km.

Temporal variations in the long-term dynamics of the monthly averaged 6.5DW amplitudes can be extracted by a finite impulse response digital band-pass filter between 5.6-6.5 and 10.9-13.3 months for SAO and AO, respectively. Latitude and temporal variations of SAO and AO in 6.5DWs are shown in Figures 9 and 10, respectively. In Figure 9, crests/troughs of SAO dynamics usually emerge in solstitial/equinoctial seasons in both hemispheres at 40-50 km. The SAO amplitudes are stronger in the NH and are greater than 2.0 K during the winters of 2003-2006, 2007-2009, and 2013/2014 in the NH and during the winters of 2007-2009 in the

SH. The interannual variations of SAO activities are similar at 80-90 and 100-110 km and are comparable in both hemispheres. Crests/troughs usually emerge during equinoctial/ solstitial seasons. At 80-90 km, large amplitudes (>4.0 K) emerge during 2004-2005, 2007-2008, and 2012-2015 in the NH and during 2012-2015 in the SH. At 100-110 km, large amplitudes (>6.0 K) emerge during 2003-2005, 2008, 2010, 2012, and 2014-2015 in the NH and during 2004-2005, 2012, and 2014-2015in the SH.

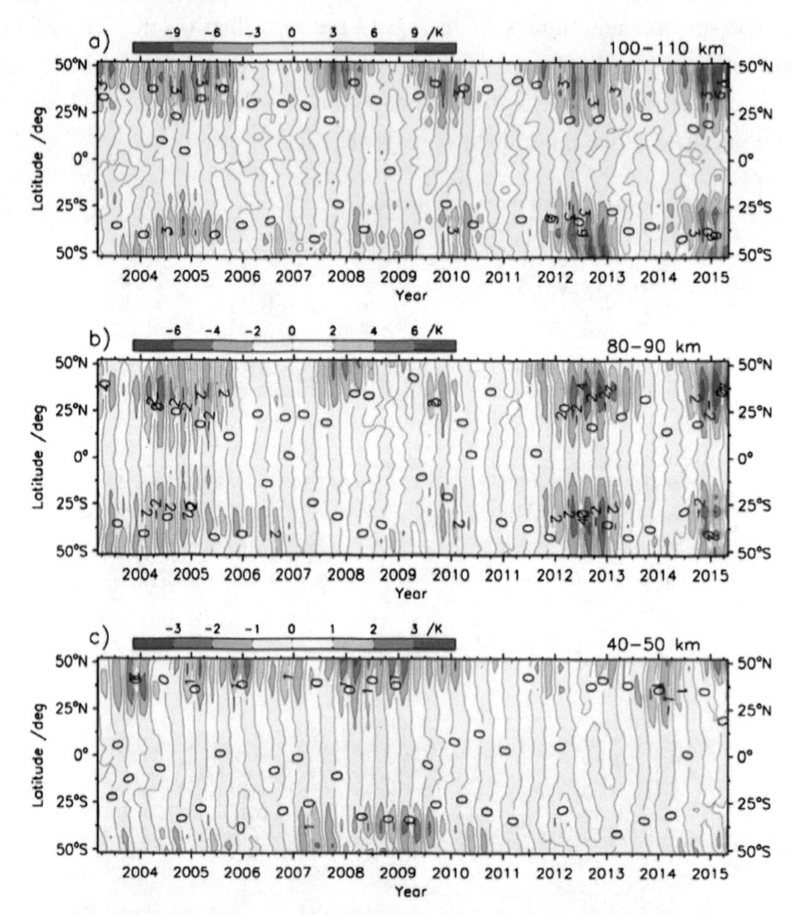

Figure 9. Latitude and temporal variations of SAO amplitudes obtained from monthly mean 6.5DW amplitudes using a digital band-pass filter.

Crests/troughs of the AO amplitudes usually emerge during the autumn-winter/spring-summer seasons at 40-50 km and during the spring-summer/autumn-winter seasons in the upper altitude levels in both hemispheres. At 40-50 km, the AO dynamics are stronger in the NH and are larger than those of SAO in both hemispheres. At 80-90 and 100-110 km, the AO amplitudes have interannual variations. At 80-90 km, these variations are comparable in both hemispheres during most years, while from 2006 to 2009, they are more prominent in the SH. It is also indicated in the middle panel of Figure 8 that AO peaks in the SH are stronger than those in the NH. At 100-110 km, the AO amplitudes are relative larger than those of SAO.

Generally, the AO in both hemispheres are alternating, presenting seesaw-like variation patterns.

It has been shown in Figure 8 that at 40-50 km, AO and SAO are the dominant and secondary signals in the long-term variations of 6.5DW amplitudes, respectively. Figure 10 shows that the annual variations of AO are very stable. It can be inferred from these results that the annual variations of6.5DWs at this altitude level are mainly controlled by AO. It can also be noticed that interannual variations of AO are less significant than those of SAO. Both Figures 7 and 9 show that the tendency of SAO activities is similar to that of the 6.5DW amplitudes. Therefore, the interannual variations of 6.5DWs at 40-50 km are probably caused by the interannual variations of SAO in 6.5DWs. At 80-90 km, the SAO amplitudes are usually larger than those of AO in both hemispheres. This result can explain the double dominant annual peaks of 6.5DWs shown in Figure 6. However, the intervals between these peaks are not exactly 6 months. Considering the results shown in the middle panel of Figure 8, this finding may be caused by AO (in both hemispheres) and 4 month oscillations (in the NH). At 100-110 km, the AO/SAO amplitudes became larger/weaker than at 80-90 km. This result may lead a greater concentration of months during which annual peaks of 6.5DWs arise.

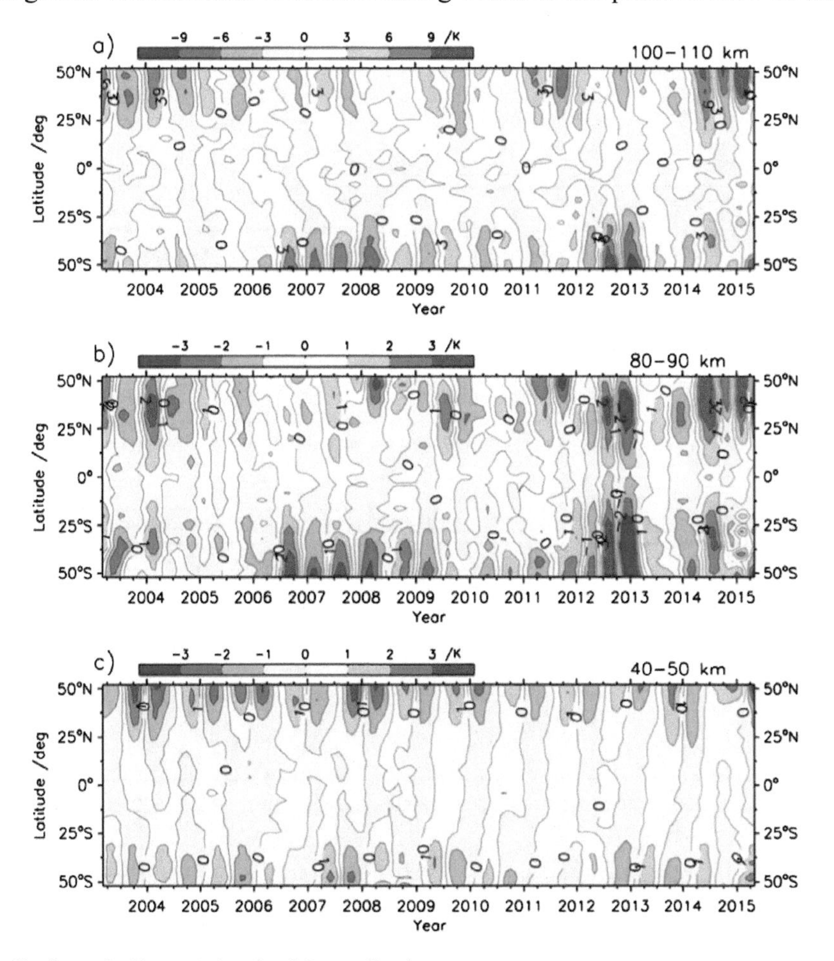

Figure 10. Similar as in Figure 9, but for AO amplitudes.

Vertical variations in SAO and AO amplitudes during the 2003-2015 periods are also studied here. Figures 11 and 12 demonstrate the results at 40°N and 40°S, respectively. The SAO and AO results are shown in panels (a) and (b), respectively. In both hemispheres, SAO are nearly out of phase below and above 70 km during most years, except for in 2008. In the lower levels, the crests/troughs of SAO usually emerge during equinoctial/solstice seasons. AO amplitudes show opposite phases below and above 70 km in most years, as their crests/troughs emerge during the autumn-winter/spring-summer and spring-summer/autumn-winter seasons in the lower and upper levels, respectively. At 40°N in 2008, AO phases are the same below and above 70 km. The amplitudes of both SAO and AO increase with altitude. It is reasonable that the vertical peaks of SAO and AO emerge at 40-50, 80-90, and 100-110 km altitudes, where peaks of 6.5DW amplitudes occur. Increase of 6.5DW in altitude may be related to increase in the SAO amplitude. Vertical structures of AO have little interannual changes in general, and AO signals are obvious from 40-50 to 100-110 km in both hemispheres. At 40-50 km, AO amplitudes are larger than those of SAO in general, and interannual variations of SAO amplitudes are clearer than AO. At 80-90 km, SAO amplitudes exceed those of AO. At 100-110 km, AO became larger than SAO, especially in the SH. This result implies that the relative strengths of SAO and AO in 6.5DWs change with altitude.

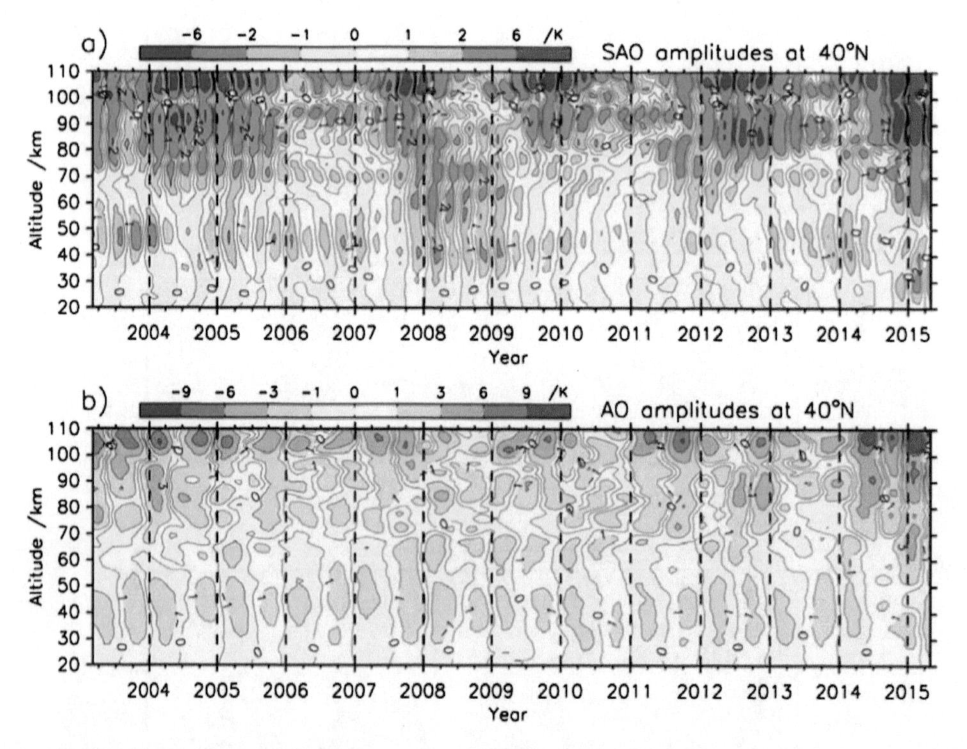

Figure 11. Altitude and temporal variations of (a) SAO and (b) AO amplitudes, respectively, filtered from 6.5DW amplitudes at 40°N.

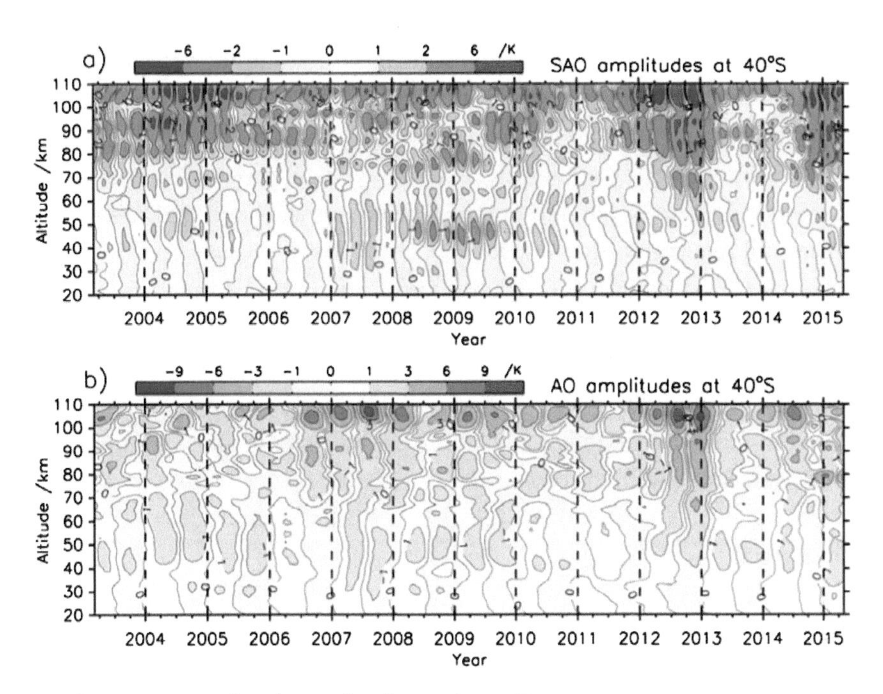

Figure 12. Similar to Figure 11, but for results obtained at 40°S.

5. Discussions

In this paper the global annual variations of temperature in 6.5DWs within a latitude range of 52°S-52°N are obtained using statistical analyses. Unlike 6.5DWs in both zonal and meridional wind fields, whose peak events usually occur above equatorial regions [*Wu et al.*, 1994; *Talaat et al.*, 2001; *Lieberman et al.*, 2003; *Riggin et al*, 2006], our results show that from the stratosphere to MLT, 6.5DW temperature activities in temperature are more notable poleward from 25° in both hemispheres. Both the altitude and annual variations are similar in both hemispheres. In these latitude ranges, three isolated amplitude peaks appear in the stratosphere at 40-50 km, in the mesosphere (80-90 km) and in the LT (100-110 km), respectively.

The annual variations obtained in this paper show different characteristics in the stratosphere and MLT, a finding that has also been mentioned for waves in zonal winds [*Liu et al.*, 2004]. In the stratosphere, wave activities reach a maximum during winter and are weak during other seasons. These results are similar as those in *Liu et al.* [2004], although their results are obtained from zonal winds. In the MLT, 6.5DW peaks usually emerge during equinoctial seasons. This conclusion has been drawn in a number of previous studies [*Wu et al.*, 1994; *Lieberman et al.*, 2003; *Liu et al.*, 2004; *Lima et al.*, 2005; *Jiang et al.*, 2008; *Gan et al.*, 2015]. This finding shows that although latitude distributions of 6.5DWs in horizontal wind and temperature fields are different, their annual variations are similar. In addition, considerable mesospheric peaks occasionally appeared during the winter months as well during the 2002- 2016 periods, but this phenomenon seems to be uncommon in the LT. This slight difference between the annual variations of 6.5DWs in the mesosphere and the LT has been seldom been mentioned.

The vertical propagations of 6.5DWs can be deduced from altitude shifts in their temporal variations [*Talaat et al.*, 2002]. As a result, it can be inferred from our results that 6.5DWs in the LT may originate from the mesosphere. However, background atmospheric conditions, especially zonal wind, have important effects on the vertical propagations of PWs [*Charney and Drazin*, 1961]. Therefore, the question of whether mesospheric 6.5DWs propagate from the stratosphere still requires further study.

The interannual variations of 6.5DWs have been mentioned in former studies [*Lima et al.*, 2005; *Jiang et al.*, 2008]. In *Lima et al.* [2005], these variations were related to phases of the equatorial stratospheric QBO. In our spectral results shown in Figure 8, the peak spectral power in the QBO period can be noticed in the MLT regions in both hemispheres. This finding shows the possible effect of the QBO on the interannual variations of 6.5DWs. However, more evidence should be provided.

6. Conclusions

The annual and interannual variations of 6.5DWs from 20 to 110 km from March 2002 to April 2016 are studied in this paper based on TIMED/SABER observations. Both the composite spectral power and composite amplitudes averaged for 2002-2016 demonstrate predominant wave strengths in the 30°S/N-50°S/N latitude bands in both hemispheres. Our study shows that annual variations of 6.5DW amplitudes from 30°S/N to 50°S/N are similar in both hemispheres. At 40-50 km, the annual maxima emerge in winters, in December and January in the NH and July and August in the SH. In the MLT, annual peaks arise twice in each half of the year. At 80-90 km, annual peaks mainly emerge in equinoctial seasons and winters, in March-May, August and September, and January in the NH and in February, November, and May in the SH. At 100-110 km, the annual peaks mainly emerge in equinoctial seasons, in April and May and August and September in the NH and February and November in the SH. The different annual variations of 6.5DWs in the stratosphere and in the MLT may indicate different wave sources.

Interannual variations in the strengths of 6.5DW strengths show possible modulations of AO and SAO at all three altitude levels. Dynamics with 4 month periods are also clear in the MLT regions and are much stronger in the NH. Using a digital band-pass filter, the amplitudes of AO and SAO are obtained from the monthly mean averaged 6.5DW amplitudes for 2002-2016. The vertical structures of both AO and SAO are very stable and increase with altitude, and their altitude variations are similar to the 6.5DW amplitudes. The relative importance of the SAO and AO changes with altitude plays an important role. At 40-50 and 100-110 km, AO plays a dominant role, while at 80-90 km, AO are relative weaker than SAO. Our results show that combined actions of SAO and AO play major roles in both annual and interannual variations in 6.5DWs.

Acknowledgments

This work was jointly supported by the National Basic Research Program of China (grant

2012CB825605), the National Natural Science Foundation of China (through grant 41221003, 41375045, 41504118, and 40904048), and the Science and Technology Projects of Jiangsu Province (through grant BK20150709). Website of data used is at ftp: //saber.gats-inc.com/custom/Temp_O3/v2.0/. The authors acknowledge the efforts of the TIMED/SABER team in making the data available and freely downloadable. Also thanks to the reviewers for the recommendations of this article and the AGU data policy.

References

Charney, J. G., and P. G. Drazin (1961), Propagation of planetary-scale disturbances from the lower into the upper atmosphere, *J. Geophys. Res.*, *66*(1), 83-109.

Cheong, H. B., and R. Kimura (1997), Excitation of the 5-day wave by Antarctica, *J. Atmos. Sci.*, *54*, 87-102.

Day, K. A., and N. J. Mitchell (2010), The 5-day wave in the Arctic and Antarctic mesosphere and lower thermosphere, *J. Geophys. Res.*, *115*, D01109, doi: 10.1029/2009JD012545.

Day, K. A., M. J. Taylor, and N. J. Mitchell (2012), Mean winds, temperatures and the 16- and 5-day planetary waves in the mesosphere and lower thermosphere over Bear Lake Observatory (42°N, 111°W), *Atmos. Chem. Phys.*, *12*(3), 1571-1585, doi: 10.5194/acp-12-1571-2012.

Dickinson, R.E.(1968), Planetary Rossby waves propagating vertically through weak westerly wind wave guides, *J. Atmos. Sci.*, *25*, 984-1002.

Dou, X., et al. (2009), Seasonal oscillations of middle atmosphere temperature observed by Rayleigh lidars and their comparisons with TIMED/SABER observations, *J. Geophys. Res.*, *114*, D20103, doi: 10.1029/2008JD011654.

Edwards, D. P., and M. Lopez-Puertas (1993), Non-local thermodynamic equilibrium studies of the 15-μm bands of CO2 for atmospheric remote sensing, *J. Geophys. Res.*, 98(D8), 14, 955-14, 977.

Ern, M., P. Preusse, J. C. Gille, C. L. Hepplewhite, M. G. Mlynczak, J. M. Russell, and M. Riese (2011), Implications for atmospheric dynamics derived from global observations of gravity wave momentum flux in stratosphere and mesosphere, *J. Geophys. Res.*, *116*, D19107, doi: 10.1029/2011JD015821.

Forbes, J. M., X. Zhang, S. E. Palo, J. Russell, C. J. Mertens, and M. Mlynczak (2009), Kelvin waves in stratosphere, mesosphere and lower thermosphere temperatures as observed by TIMED/SABER during 2002-2006, *Earth Planets Space*, *61*, 447-453.

Gan, Q., J. Yue, L. C. Chang, W. B. Wang, S. D. Zhang, and J. Du (2015), Observations of thermosphere and ionosphere changes due to the dissipative 6.5-day wave in the lower thermosphere, *Ann. Geophys.*, *33*(7), 913-922, doi: 10.5194/angeo-33-913-2015.

Gan, Q., S. D. Zhang, and F. Yi (2012), TIMED/SABER observations of lower mesospheric inversion layers at low and middle latitudes, *J. Geophys. Res.*, *117*, D07109, doi: 10.1029/2012JD017455.

Gille, J., and F. B. House (1971), On the inversion of limb radiance measurements I: Temperature and thickness, *J. Atmos. Sci.*, *28*, 1427-1442.

Garcia, R. R., R. Lieberman, J. M. Russell III, and M. G. Mlynczak (2005), Large-scale waves in the mesosphere and lower thermosphere observed by SABER, *J. Atmos. Sci.*, *62*, 4384-4395.

Geisler, J. E., and R. E. Dickinson (1976), The five-day wave on a sphere with realistic zonal winds, *J. Atmos. Sci.*, *33*, 632-641.

Gu, S.-Y., H.-L. Liu, T. Li, X. Dou, Q. Wu, and J. M. Russell (2015), Evidence of nonlinear interaction between quasi 2 day wave and quasi-stationary wave, *J. Geophys. Res. Space Physics*, *120*, 1256-1263, doi: 10.1002/2014JA020919.

Huang, F.T., H. G. Mayr, C. A. Reber, T. Killeen, J. Russell, M. Mlynczak, W. Skinner, and J. Mengel (2006), Diurnal variations of temperature and winds inferred from TIMED and UARS measurements, *J. Geophys. Res.*, *111*, A10S04, doi: 10.1029/2005JA011426.

Huang, Y.Y., S. D. Zhang, F.Yi, C. M. Huang, K. M. Huang, Q. Gan, and Y. Gong (2013), Global climatological variability of quasi-two-day waves revealed by TIMED/SABER observations, *Ann. Geophys.*, *31*(6), 1061-1075, doi: 10.5194/angeo-31-1061-2013.

Jiang, G., J. Xiong, W. Wan, B. Ning, and L. Liu (2008), Observation of 6.5-day waves in the MLT region over Wuhan, *J. Atmos. Sol. Terr. Phys.*, *70*(1), 41 -48, doi: 10.1016/j.jastp.2007.09.008.

Jiang, Y., Z. Sheng, and H. Q. Shi (2014), Modes ofzonal mean temperature variability 20-100 km from the TIMED/SABER observations, *Ann. Geophys.*, *32*(3), 285-292, doi: 10.5194/angeo-32-285-2014.

John, S. R., and K. K. Kumar (2012), TIMED/SABER observations of global gravity wave climatology and their interannual variability from stratosphere to mesosphere lower thermosphere, *Clim. Dyn.*, *39*(6), 1489-1505, doi: 10.1007/s00382-012-1329-9.

Killeen, T. L., Q.Wu, S.C. Solomon, D. A.Ortland, W. R.Skinner, R.J. Niciejewski, and D.A. Gell (2006), TIMED Doppler interferometer: Overview and recent results, *J. Geophys. Res.*, *111*, A10S01, doi: 10.1029/2005JA011484.

Kirkwood, S., andA. Réchou (1998), Planetary-wave modulation of PMSE, *Geophys. Res. Lett.*, *25*(24), 4509-4512, doi: 10.1029/1998GL900198.

Kirkwood, S., V. Barabash, B. U. E. Brändström, A. Moström, K. Stebel, N. Mitchell, and W. Hocking (2002), Noctilucent clouds, PMSE and 5-day planetary waves: A case study, *Geophys. Res. Lett.*, *29*(10), 50-51-50-54, doi: 10.1029/2001GL014022.

Kishore, P., S. P. Namboothiri, K. Igarashi, S. Gurubaran, S. Sridharan, R. Rajaram, and M. Venkat Ratnam (2004), MF radar observations of 6.5-day wave in the equatorial mesosphere and lower thermosphere, *J. Atmos. Sol. Terr. Phys.*, *66*(6-9), 507-515, doi: 10.1016/ j.jastp.2004.01.026.

Lieberman, R. S., D. M. Riggin, S. J. Franke, A. H. Manson, C. Meek, T. Nakamura, T. Tsuda, R. A. Vincent, and I. Reid (2003), The 6.5-day wave in the mesosphere and lower thermosphere: Evidence for baroclinic/barotropic instability, *J. Geophys. Res.*, *108*(D20), 4640, doi: 10.1029/ 2002JD003349.

Lima, L. M., P. P. Batista, B. R. Clemesha, and H. Takahashi (2005), The 6.5-day oscillations observed in meteor winds over Cachoeira Paulista (22.7°S), *Adv. Space Res.*, *36*(11), 2212-2217, doi: 10.1016/ j.asr.2005.06.005.

Liu, H. L., E. R.Talaat, R. G. Roble, R. S. Lieberman, D. M. Riggin, and J. H.Yee (2004), The 6.5-day wave and its seasonal variability in the middle and upper atmosphere, *J. Geophys. Res.*, *109*, D21112, doi: 10.1029/2004JD004795.

Madden, R., and P. Julian (1972), Further evidence of global scale, 5-day pressure waves, *J. Atmos. Sci.*, *29*, 1464-1469.

Merkel, A. W., R. R. Garcia, S. M. Bailey, and J. M. Russell (2008), Observational studies of planetary waves in PMCs and mesospheric temperature measured by SNOE and SABER, *J. Geophys. Res.*, *113*, D14202, doi: 10.1029/2007JD009396.

Merkel, A. W., G. E. Thomas, S. E. Palo, and S. M. Bailey (2003), Observations of the 5-day planetary wave in PMC measurements from the Student Nitric Oxide Explorer Satellite, *Geophys. Res. Lett.*, *30*(4), 1196, doi: 10.1029/2002GL016524

Miyoshi, Y., and T. Hirooka (1999), A numerical experiment of excitation of the 5-day wave by a GCM, *J. Atmos. Sci.*, *56*, 1698-1707.

Mertens, C. J., M. G. Mlynczak, M. L. Puertas, P. P. Wintersteiner, R. H. Picard, J. R. W. L. L. Gordley, and J. M. Russell III (2001), Retrieval of mesospheric and lower thermospheric kinetic temperature from measurements of 15μm Earth limb emission under non-LTE conditions, *Geophys. Res. Lett., 28*(7), 1391-1394.

Merzlyakov, E. G., and D. V. Pancheva (2007), The 1.5-5-day eastward waves in the upper stratosphere-mesosphere as observed by the Esrange meteor radar and the SABER instrument, *J. Atmos. Sol. Terr. Phys., 69*(17-18), 2102-2117, doi: 10.1016/j.jastp.2007.07.002.

Meyer, C. K., and J. M. Forbes (1997), A 6.5-day westward propagating planetary wave: Origin and characteristics, *J. Geophys. Res., 102*(D22), 26, 173-26, 178, doi: 10.1029/97JD01464.

Nielsen, K., D. E. Siskind, S. D. Eckermann, K. W. Hoppel, L. Coy, J. P. McCormack, S. Benze, C. E. Randall, and M. E. Hervig (2010), Seasonal variation of the quasi 5 day planetary wave: Causes and consequences for polar mesospheric cloud variability in 2007, *J. Geophys. Res., 115*, D18111, doi: 10.1029/2009JD012676.

Oberheide, J., Q. Wu, D. A. Ortland, T. L. Killeen, M. E. Hagan, R. G. Roble, R. J. Niciejewski, and W. R. Skinner (2005), Nonmigrating diurnal tides as measured by the TIMED Doppler interferometer: Preliminary results, *Adv. Space Res., 35*, 1911 -1917, doi: 10.1016/j.asr.2005.01.063.

Pogoreltsev, A. I., I. N. Fedulina, N. J. Mitchell, H. G. Muller, Y. Luo, C. E. Meek and A. H. Manson (2002), Global free oscillations of the atmosphere and secondary planetary waves in the mesosphere and lower thermosphere region during August/September time conditions, *J. Geophys. Res., 107*(D24), 4799, doi: 10.1029/2001JD001535.

Remsberg, E. E., et al. (2008), Assessment of the quality of the version 1.07 temperature-versus-pressure profiles of the middle atmosphere from TIMED/SABER, *J. Geophys. Res., 113*, D17101, doi: 10.1029/2008JD010013.

Riggin, D. M., et al. (2006), Observations of the 5-day wave in the mesosphere and lower thermosphere, *J. Atmos. Sol. Terr. Phys., 68*(3-5), 323-339, doi: 10.1016/j.jastp.2005.05.010.

Smith, A. K. (1983), Stationary waves in the winter stratosphere: Seasonal and interannual variability, *J. Atmos. Sci., 40*, 245-261.

Salby, M. L. (1982a), Sampling theory for asynoptic satellite observations. Part 1: Space-time spectra, resolution, and aliasing, *J.Atmos. Sci., 39*, 2577-2600.

Salby, M. L. (1982b), Samling theory for asynoptic satellite observations. Part 2: Fast Fourier synoptic mapping, *J. Atmos. Sci., 39*, 2601 -2614.

Shuai, J., S. Zhang, C. Huang, F.Yi, K. Huang, Q. Gan, and Y. Gong (2014), Climatology of global gravity wave activity and dissipation revealed by SABER/TIMED temperature observations, *Sci. China Technol. Sci., 57*(5), 998-1009, doi: 10.1007/s11431-014-5527-z.

Talaat, E. R., J. H. Yee, and X. Zhu (2001), Observations of the 6.5-day wave in the mesosphere and lower thermosphere, *J. Geophys. Res., 106*(D18), 20, 715-20, 723, doi: 10.1029/2001JD900227.

Talaat, E. R., J.-H. Yee, and X. Zhu (2002), The 6.5-day wave in the tropical stratosphere and mesosphere, *J. Geophys. Res., 107*(D12), 4133, doi: 10.1029/2001JD000822.

von Savigny, C., C. Robert, H. Bovensmann, J. P. Burrows, and M. Schwartz (2007), Satellite observations of the quasi 5-daywave in noctilucent clouds and mesopause temperatures, *Geophys. Res. Lett., 34*, L24808, doi: 10.1029/2007GL030987.

Williams, C. R., and S.K. Avery (1992), Analysis of long-period waves using the mesosphere-stratosphere-troposphere radar at Poker Flat, Alaska, *J. Geophys. Res., 97*(18), 20, 588-520, 594.

Wu, D. L., P. B. Hays, and W. R. Skinner (1994), Observations of the5-daywave in the mesosphere and lower thermosphere, *Geophys. Res. Lett., 21*(24), 2733-2736.

Xu, J., H. L. Liu, W. Yuan, A. K. Smith, R. G. Roble, C. J. Mertens, J. M. Russell, and M. G. Mlynczak (2007), Mesopause structure from. Thermosphere, Ionosphere, Mesosphere, Energetics, and Dynamics (TIMED)/

Sounding of the Atmosphere Using Broadband Emission Radiometry (SABER) observations, *J. Geophys. Res.*, *112*, D09102, doi: 10.1029/2006JD007711.

Xu, J., C. Y. She, W. Yuan, C. Mertens, M. Mlynczak, and J. Russell (2006), Comparison between the temperature measurements by TIMED/SABER and lidar in the midlatitude, *J. Geophys. Res.*, *111*, A10S09, doi: 10.1029/2005JA011439.

Xu, J., A. K. Smith, H. L. Liu, W. Yuan, Q. Wu, G. Jiang, M. G. Mlynczak, J. M. Russell, and S. J. Franke (2009), Seasonal and quasi-biennial variations in the migrating diurnal tide observed by Thermosphere, Ionosphere, Mesosphere, Energetics and Dynamics (TIMED), *J. Geophys. Res.*, *114*, D13107, doi: 10.1029/2008JD011298.

Xu, J., A. K. Smith, M. Liu, X. Liu, H. Gao, G.Jiang, and W.Yuan (2014), Evidence for nonmigrating tides produced by the interaction between tides and stationary planetary waves in the stratosphere and lower mesosphere, *J. Geophys. Res. Atmos.*, *119*, 471-489, doi: 10.1002/ 2013JD020150.

Zhang, X., J. M. Forbes, M. E. Hagan, J. M. Russell, S. E. Palo, C. J. Mertens, and M. G. Mlynczak (2006), Monthly tidal temperatures 20-120 km from TIMED/SABER, *J. Geophys. Res.*, *111*, A10S08, doi: 10.1029/2005JA011504.

Zuo, X., and W. Wan (2008), Planetary wave oscillations in sporadic E layer occurrence at Wuhan, *Earth Planets Space*, *60*, 647-652.

Appendix

1. Mak Mankin, **Li Chongyin**: On the overstability convection, *Scientia Sinica* (series B), 1982, 25, 1326-1340.
2. **Li Chongyin**: The CISK-overstability convection, *Scientia Sinica* (series B), 1984, 27, 501-510.
3. **Li Chongyin**: The scale selectivity of oscillation convection, *Kexue Tongbao*, 1984, 27(5), 505-507.
4. **Li Chongyin**: On the CISK with shearing basic current, *Advances in Atmospheric Science*, 1984, 1, 256-262.
5. **Li Chongyin**: Actions of summer monsoon troughs (ridges) and tropical cyclone over South Asia and the moving CISK mode, *Scientia Sinica* (B), 1985, 28, 1197-1206.
6. **Li Chongyin**: El Niño and typhoon action over the western Pacific, *Chinese Science Bulletin*, 1986, 31, 538-542.
7. **Li Chongyin**: A Numerical Simulation of the Typhoon Genesis, *Adv. Atmos. Sci.*, 1985, 2(1), 72-80.
8. **Li Chongyin**: An investigation on forecasting typhoon actions over South China Sea based upon El Niño events, *<Proceedings of the International Conference on Oil Developing in South China Sea>*, 1985, 3238.
9. **Li Chongyin**: Nonlinear influences of horizontal distribution of zonal wind on the large-scale meridional motion in the atmosphere, *<Proceeding of International Summer Colloquium on Nonlinear Dynamics of the Atmosphere>*,1986, 10-20, New York, Science Press.
10. **Li Chongyin**, Hu Ji: A analysis of interaction between the atmospheric circulation over East Asia/Northwest Pacific and El Niño, *Chinese J. Atmos. Sci.*, 1987, 11, 411-420.
11. **Li Chongyin**: Actions of typhoon over the western Pacific (including the South China Sea) and El Niño, *Advan. Atmos. Sci.*, 1988, 5(1), 107-116.
12. **Li Chongyin**: Frequent activities of stronger aerotroughs in East Asia in wintertime and the occurrence of the El Niño event, *Science in China* (B), 1989, 32(8), 976-985.
13. **Li Chongyin**: On the feedback role of tropical convection, *<Tropical Rainfall Measurements>*, , 1988, 141-146, A. Deepak Pubishing, USA.
14. **Li Chongyin**: Warmer winter in eastern China and El Niño, *Chinese Science Bulletin*, 1989, 34, 1801-1805.
15. **Li Chongyin**: An important mechanism producing 30-50 day oscillation in the tropical atmosphere-Feedback of the cumulus convection, *Annual Report, Institute of Atmospheric Physics*, 1989, 8, 33-47, Beijing, Science Press.
16. **Li Chongyin**, Chen Yuxiang, Yuan Chongguang: Important factor cause of El Niño-the frequent activities of stronger cold waves in East Asia, *<Fronters in Atmospheric Sciences>*, 1989, 156-165, New York, Allenton Press INC..
17. **Li Chongyin**: Intraseasonal oscillation in the atmosphere, *Chinese J. Atmos. Sci.*, 1990, 14, 35-52.
18. **Li Chongyin**: A dynamical study on the 30-50 day oscillation in the tropical atmosphere outside the equator, *Chinese J. Atmos. Sci.*, 1990, 14, 101-112.
19. **Li Chongyin**: The influences of sensible heating and vertical wind shear on extratropical CISK disturbance, *Chinese J. Atmos. Sci.*, 1990, 14(2), 191-2002.
20. **Li Chongyin**: Interaction between anomalous winter monsoon in East Asia and El Niño events, *Advances in Atmos. Sci.*, 1990, 7, 36-46.
21. **Li Chongyin**, Wu Peili: An observational study of the 30-50 day atmospheric oscillations. Part Ⅰ:

structure and Propagation, *Advances in Atmos. Sciences*, 1990, 7, 294-304.

22. **Li Chongyin**, Wu Peili: A further inquiry on 30-60 day oscillation in the tropical atmosphere, *Acta Meteor. Sinica*, 1990, 4(5), 525-535.
23. **Li Chongyin**: On interaction between anomalous circulation/climate in East Asia and El Niño event, < *Climate Change, Dynamics and Modelling*>, 1990, 101-126, Beijing, China Meteor. Press.
24. **Li Chongyin**, Wu Peili, Zhong Qin: Characteristics of 30-60 day oscillation of general circulation in northern hemisphere, *Science in China* (B), 1991, 34, 457-468.
25. **Li Chongyin**, Long Zengxia: Evolution of QBO and the influence of ENSO, *Chinese Science Bulletin*, 1991, 36(12), 1016-1020.
26. **Li Chongyin**, Zhou Yaping: An observational study of the 30-50 day atmospheric oscillations, Part II: Temporal evolution and hemispheric interaction across the equator, *Advan. Atmos. Sci.*, 1991, 8(4), 399-406.
27. **Li Chongyin**: The Global Characteristics of 30-60 day atmospheric oscillation, *Chinese J. Atmos. Sci.*, 1991, 15(2), 130-140.
28. **Li Chongyin**, Xiao Ziniu: The 30-60 day oscillations in the global atmosphere excited by warming in the equatorial eastern Pacific, *Chinese Science Bulletin*, 1992, 37(6), 484-489.
29. **Li Chongyin**, Zhang Qin: Global atmospheric low-frequency teleconnection, *Progress in Natural Science*, 1991, 1, 447-452.
30. **Li Chongyin**, Yan Jinghua: A study of numerical simulation on the development of depressions in the South China Sea, *Acta Meteor. Sin.*, 1992, 6(3), 265-274.
31. **Li Chongyin**, Long Zhenxia: Quasi-biennial oscillation and its influence on general atmospheric circulation and climate in East Asia, *Chinese J. Atmos. Sci.*, 1992, 16, 70-79.
32. Xiao Ziniu, **Li Chongyin**: Numerical simulation of the atmospheric low-frequency teleresponse to external forcing, Part I: Anomalous sea surface temperature in the equatorial eastern Pacific Ocean, *Chinese J. Atmos. Sci.*, 1992, 16(4), 372-382.
33. **Li Chongyin**, Xiao Ziniu: Numerical simulation of atmospheric low frequency teleresponse to the external forcing, Part II: Response to anomalous "cold wave" over middle-high latitudes in Eurasian Area, *Chinese J. Atmos. Sci.*, 1993, 17(3), 287-296.
34. **Li Chongyin**: A further inquiry on the mechanism of 30-60 day oscillation in the tropical atmosphere, *Adv. Atmos. Sci.*, 1993, 10, 41-53.
35. **Li Chongyin**, Long Zhenxia, Xiao Ziniu: On low-frequency remote responses in the atmosphere to external forcings and their influences on climate, <*Climate Variability*>, 1993, 177-190, Beijing, China Meteor. Press.
36. **Li Chongyin**: Some differences of the 30-60 day atmospheric oscillation between mid-high latitudes and tropics, <*Climate, Environment and Geophysical Fluid Dynamics*>, 1993, 99-110,, Beijing, China Meteor. Press.
37. Luo Dehai, **Li Chongyin**: The resonant interaction of periodic external forced Rossby waves and low-frequency oscillations in the mid-hig latitudes, <*Climate, Environment and Geophysical Fluid Dynamics*>, 1993, 111-122, Beijing, China Meteor. Press.
38. **Li Chongyin**, Zhou Yaping: Relationship between Intraseasonal oscillation in the tropical atmosphere and ENSO, *Chinese J. Geophysics*, 1994, 37(2), 213-223.
39. **Li Chongyin**, Yan Jinghua: Numerical simulation study of the occurrence and development of a mid-tropospheric cyclone over the South China Sea, *Acta. Meteor. Sin.*, 1994, 8, 150-160.
40. **Li Chongyin**, Ian Smith: Numerical simulation of the tropical intraseasonal oscillation and the effect of warm SSTs, *Acta Meteor. Sin.*, 1995, 9(1), 1-12.

41. **Li Chongyin**, Cao Wenzhong, Li Guilong: Influences of basic flow on unstable excitation of intraseasonal oscillation in mid-high latitudes, *Science in China* (series B), 1995, 38(9), 1135-1145.
42. Liao Qinghai, **Li Chongyin**: CISK-Rossby wave and the 30-60 day oscillation in the tropics , *Adv. Atmos. Sci.*, 1995, 12, 1-12.
43. **Li Chongyin**: The further studies of intraseasonal oscillation in the tropical atmosphere, <*Proceedings of the International Scientific Conference on the TOGA Programme*>, 1995, (717), 260-264, WCRP-91-WMO/TD.
44. **Li Chongyin**: Westerly anomalies over the equatorial western Pacific and Asian winter monsoon, <*Proceedings of the International Scientific Conference on the TOGA Programme*>, 1995, (717), 557-561, WCRP-91-WMO/TP.
45. **Li Chongyin**: Kinetic energy transfer of tropical atmospheric system associated with the occurrence of El Niño event, <*Proceedings of the Second International Study Conference on GEWEX in Asia* >, 1995, 87-90, Pattaya, Thailand.
46. **Li Chongyin**, Huang Ronghui, Yang Dasheng, Ni Yunqi: Atmospheric dynamics in recent four years in China, <*China National Report on Meteorology and Atmospheric Sciences*>, 1995, 69-81, Beijing, China Meteor. Press.
47. **Li Chongyin**, Li Guilong: A dynamical study of influence of El Niño on intraseasonal oscillation in the tropical atmosphere, *Chinese J. Atmos. Sci.*, 1996, 20, 148-159.
48. **Li Chongyin**, Liao Qinghai: Behaviour of coupled modes in a simple nonlinear air-sea interaction model, *Adv. Atmos. Sci.*, 1996, 13(2), 183-195.
49. **Li Chongyin**: Further studies on evaporation-wind feedback, *J. Tropical Meteorology*, 1996, 3(1), 11-17.
50. **Li Chongyin**: Dynamic mechanism of intraseasonal oscillation in the tropical atmosphere, <*From Atmospheric Circulation to Global Change*>, 1996, 351-364, Beijing, China Meteor. Press.
51. **Li Chongyin**: Quasi-two weeks Oscillation in the tropical atmosphere, *Theoretical and Applied Climatology*, 1996, 55, 121-128.
52. **Li Chongyin**: A further study on the interaction between anomalous winter monsoon in East Asia and El Niño, *Acta Meteor. Sin.*, 1996, 10(3), 309-320.
53. **Li Chongyin**: ENSO cycle and anomalous East-Asian winter monsoon, *Workshop on El Niño/Southern Oscillation and Monsoon*, 1996.SMR/930 – 18, ICTP, Trieste.
54. **Li Chongyin**, Long Zhenxia: Relationship between subtropical high activities over the western Pacific and Quasi Biennial Oscillation in the stratosphere, *Chinese J. Atmos. Sci.*, 1997, 21(4), 343-352.
55. **Li Chongyin**, Li Guilong: Evolution of intraseasonal oscillation over the tropical western Pacific/South China Sea and its effect to the summer precipitation in Southern China, *Advances Atmos., Sci.*, 1997, 14, 246-254.
56. **Li Chongyin**, Li Guilong: Activities of quasi-stationary waves in the tropical atmosphere an El Niño/Southern Oscillation, *Progress in Natural Science*, 1998, 8, 321-325.
57. **Li Chongyin**, Liao Qinghai: The exciting mechanism of tropical intraseasonal oscillation to El Niño event, *J. Tropical Meteor.*, 1998, 4, 113-121.
58. **Li Chongyin**, Mu Mingquan: ENSO cycle and anomalies of winter monsoon in east Asia, <*East Asia and Western Pacific Meteorology and Climate*>, C. P. Chang, J. C. L. Chan, J. T. Wang, 1998, 60-73, Word Scientific, Singapore.
59. **Li Chongyin**: The quasi-decadal oscillation of air-sea system in the northwestern Pacific region, *Adv. Atmos. Sci.*, 1998, 15(1), 31-40.
60. **Li Chongyin**, Li Guilong: Activities of low-frequency waves in the tropical atmosphere and ENSO, *Adv. Atmos. Sci.*, 1998, 15, 1993-203.

61. **Li Chongyin**, Mu Mingquan: Numerical simulations of anomalous winter monsoon in east Asia Exciting ENSO, *Chinese J. Atmos. Sci.*, 1998, 22, 393-403.
62. **Li Chongyin**, Li Guilong: The relationship between low-frequency atmospheric oscillation and SST in the equatorial Pacific, *Chinese Science Bulletin*, 1999, 44, 1126-1129.
63. Long Zhenxia, **Li Chongyin**: Numerical simulations on the sensitivity of summer climate over east Asia to the duration of sea surface temperature anomalies over eastern equatorial Pacific, *Chinese J. Atmos. Sci.*, 1999, 23, 88-103.
64. Mu Mingquan, **Li Chongyin**: ENSO signals in interannual variability of East Asian winter monsoon, Part I: Obeserved data analyses, *Chinese J. Atmos. Sci.*, 1999, 23, 139-149.
65. **Li Chongyin**, Zhang Liping: Summer monsoon activities in South China Sea and Its impacts, *Chinese J. Atmos.Sci.*, 1999, 23, 111-120.
66. **Li Chongyin**, Mu Mingquan, Zhou Guangqin: The variation of warm pool in the equatorial western Pacific and Its impacts to Climate, *Advance Atmos. Sci.*, 1999, 16, 378-394.
67. **Li Chongyin**, Huang Ronghui: On interaction between ENSO and East-Asian monsoon, *<1995-1998 China National Report on Meteorology and Atmospheric Sciences>*, 1999, 68-78, Beijing, China Meteor. Press.
68. **Li Chongyin**, Mu Mingquan, Zhou Guangqing: Subsurface ocean temperature anomalies in the Pacific warm pool and ENSO occurrence, *<Programme on Weather Prediction Research Report>*, 1999, 979 (series13), 232-240, WMO/TD, Geneve.
69. **Li Chongyin**, Li Guilong: Variation of the NAO and NPO associated with climate jump in the 1960s, *Chinese Science Bulletin*, 1999, 44, 1983-1986.
70. **Li Chongyin**, Mu Mingquan: ENSO occurrence and sub-surface ocean temperature anomalies in the equatorial warm pool, *Chinese J. Atmos.Sci.*, 1999, 23, 217-225.
71. Mu Mingquan, **Li Chongyin**: Numerical simulations on the relationship between Indian summer monsoon and El Niño, *Chinese J. Atmos. Sci.*, 1999, 23, 377-386.
72. Cho Han-Ru, **Li Chongyin**: Equatorial Kelvin waves and intraseasonal oscillation in the equatorial troposphere, Chinese. *J. Atmos. Sci.*, 1999, 23(3), 237-244.
73. **Li Chongyin**, Wu Jingbo: On the onset of the South China Sea summer monsoon in 1998 , *Advance Atmos. Sci.*, 2000, 17, 193-204.
74. **Li Chongyin,** Li Guilong: The NAO/NPO and interdecadal climate variation in China, *Advance Atmos. Sci.*, 2000, 17, 555-561.
75. **Li Chongyin**, Qu Xin: Atmospheric circulation evolutions associated with summer monsoon onset in the South China Sea, *Chinese J. Atmos. Sci.*, 1999, 23, 311-325.
76. Mu Mingquan, **Li Chongyin**: Interaction between subsurface ocean temperature anomalies in the western Pacific warm pool and ENSO cycle, *Chinese J. Atmos. Sci.*, 2000, 24, 107-121.
77. **Li Chongyin**, Mu Mingquan, Bi Xunqiang: Interdecadal variation of atmospheric circulation Part Ⅱ； A numerical simulations with GCM. *Chinese J. Atmos. Sci.*, 2000, 24, 333-343.
78. **Li Chongyin**, Mu Mingquan: Relationship between East-Asian winter monsoon, warm pool situation and EVSO cycle, *Chinese Science Bulletin*, 2000, 45, 1448-1455.
79. **Li Chongyin**, Long Zhenxia: El Niño occurrence in 1997 and intraseasonal oscillation anomalies in the Tropical atmosphere, *<Proceedings of the Second international Symposium on Asian Monsoon System>*, 2000, 106-111, Cheju, Korea.
80. Mu Mingquan, **Li Chongyin**: Interdecadal variations of atmoapheric sciculation, Part Ⅰ: Obsrevational data analyses. *Chinese J. Atmos. Sci.*, 2000, 24(3), 270-278.
81. **Li Chongyin**, Mu Mingquan: Influence of the Indian Ocean dipole on Asian monsoon circulation, *Exchanges*, 2001, 6, 11-14.

82. **Li Chongyin**, Sun Shuqing, Mu Mingquan: Origin of the TBO-Interaction between anomalous East-Asian winter monsoon and ENSO cycle, *Advance Atmos. Sci.*, 2001, 18, 554-566.
83. Long Zhenxia, **Li Chongyin:** Simulating influences of positive sea surface temperature anomalies over the eastern equatorial Pacific on subtropical high over the western Pacific, *Chinese J. Atmos. Sci.*, 2001, 25, 1-16.
84. **Li Chongyin**, Long Zhenxia: Intraseasonal oscillation anomalies in the Tropical atmosphere and the 1997 El Niño occurrence, *Chinese J. Atmos. Sci.*, 25(4), 337-345.
85. **Li Chongyin**, Mu Mingquan: The influence of the Indian Ocean dipole on atmospheric circulation and climate, *Advance Atmos. Sci.*, 2001, 18, 831-843.
86. Long Zhenxia, **Li Chongyin**: Numerical simulation of lag influence of ENSO on East-Asian Monsoon, *Acta Meteor. Sin.*, 2001, 15(1), 59-70.
87. Long Zhenxia, **Li Chongyin**: Interannual variation of tropical atmospheric 30-60 day low-frequency oscillation and ENSO cycle, *Chinese J. Atmos. Sci.*, 2002, 26(1), 51-62.
88. **Li Chongyin**, Long Zhenxia, Zhang Qingyun: Strong/weak summer monsoon activity over the South China Sea and atmospheric intraseasonal oscillation, *Advance Atmos. Sci.*, 2001, 18, 1146-1160.
89. **Li Chongyin**, Mu Mingquan, Zhou Guangqing: Sub-surface ocean temperature anomalies in the Pacific warm pool and ENSO occurrence, *<Dynamics of Atmospheric and Oceanic Circulations and Climate >*, 2001, 601-620, Beijing, China Meteor. Press.
90. **Li Chongyin**, Mu Mingquan, Pan Jing: Indian Ocean temperature dipole and SSTA in the equatorial Pacific Ocean, *Chinese Science Bulletin*, 2002, 47(3), 236-239.
91. **Li Chongyin**, Cho Han-Ru, Wang Jough-Tai: CISK Kelvin wave with evaporation wind feedback and air-sea interaction-A further study of tropical intraseasonal oscillation mechanism, *Adv. Atmos. Sci.*, 2002, 19, 379-390.
92. **Li Chongyin**, Long Zhenxia: intraseasonal oscillation anomalies in the tropical atmosphere and El Niño events, *Exchanges*, 2002, 7(2), 12-15.
93. Mu Mingquan, **Li Chongyin**: Indian Ocean dipole and its relationship with ENSO mode, *Acta Meteor. Sinica*, 2002, 16(4), 489-497.
94. **Li Chongyin**, Mu Mingquan: A further study of the essence of ENSO, *Chinese J. Atmos. Sci.*, 2002, 26, 309-328.
95. **Li Chongyin**, Mu Mingquan, Long Zhenxia: Influence of intraseasonal oscillation on East-Asian summer monsoon, *Acta Meteor. Sin.*, 2003, 17, 130-142.
96. Long Zhenxia, Pan Jing, **Li Chongyin**: Influences of anomalous summer monsoon over the South China Sea on climate variation, *Acta Meteor. Sin.*, 2003, 17, 118-129.
97. Xian Peng, **Li Chongyin**: Interdecadal modes of sea surface temperature in the North Pacific Ocean and its evolution, *Chinese J. Atmos. Sci.*, 2003, 27(2), 118-126.
98. Yang Hui, **Li Chongyin**: The relation between atmospheric intraseasonal oscillation and summer severe flood and drought in the Changjiang-Huaihe basin, *Adv. Atmos. Sci.*, 2003, 20, 540-553.
99. **Li Chongyin**, Xian Peng: Atmospheric Anomalies related to Interdecadal Variability of SST in the North Pacific, *Adv. Atmos. Sci.*, 2003, 20, 859-874.
100. Yang Hui**, Li Chongyin**: The Relation between Atmospheric Intraseasonal Oscillation and Summer Severe Flood and Drought in the Changjiang-Huaihe River Basin[J]. Advances in Atmospheric Sciences, 2003, 20(4):540-553.
101. Weng Hengyi, Sumi Akimasa, Takayabu Yukari N., Kimoto Masahide, **Li Chongyin**: Interannual-Interdecadal variation in large-scale atmospheric circulation and extremely wet and dry summers in China/Japan during 1951-2000, Part I: Spatial patterns, *J. Meteor. Soc. Japan*, 2004, 82,

775-788.

102. Weng Hengyi, Sumi Akimasa, Takayabu Yukari N., Kimoto Masahide, **Li Chongyin:** Interannual-Interdecadal variation in large-scale atmospheric circulation and extremely wet and dry summers in China/Japan during 1951-2000, Part II: Dominant timescales, *J. Meteor. Soc. Japan*, 2004, 82, 789-804.
103. **Li Chongyin**, He Jinhai, Zhujinhong: A review of decadal/interdecadal climate variation studies in China, *Adv. Aqtmos. Sci.*, 2004, 21, 425-436.
104. Ding Yihui, **Li Chongyin**, Liu Yanju: Overview of the South China Sea monsoon experiment, *Adv. Atmos. Sci.*, 2004, 21, 343-360.
105. Chan J C L, **Li Chongyin**: The East Asian winter monsoon, <*East Asian Monsoon*>, C P Chang C P, 2004, 54-106, World Scientific Publisher, Singapore.
106. **Li Chongyin**, Wang Jough-Tai, Lin Shi-Zhei, Cho Han-Ru, The relationship between East Asian summer monsoon activity and northward jumps of the upper air westerly jet location, *Chinese J. Atmos. Sci.*, 2005, 29(1), 1-20.
107. **Li Chongyin**, Yang Hui: Atmospheric intraseasonal oscillation and summer rainfall over the Yangtze-Huai river valley, <*Climate Chang and Yangtze Floods*>, 2004, 46-62, Jiang et al., Beijing, Science Press.
108. **Li Chongyin**, Pei Shunqiang, Pu Ye, Dynamical impact of anomalous East-Asian winter monsoon on zonal wind over the equatorial western Pacific, *Chinese Science Bulletin*, 2005, 50(14), 1520-1526.
109. Yang Hui, **Li Chongyin**: Lasting Time of El Niño and Circulation Anomaly, *Chinese J. Geophysics*, 2005, 48, 821-830.
110. **Li Chongyin**, Hu Ruijin, Yang Hui: Intraseasonal oscillation in the tropical Indian Ocean, *Adv. Atmos. Sci.,* 2005, 22, 617-624.
111. Zhou wen, Chan J C L, **Li Chongyin**: South China Sea summer monsoon onset in relation to the off-equatorial ITCZ, *Adv. Atoms. Sci.*, 2005, 22, 605-676.
112. Gu Wei, **Li Chongyin**, Yang Hui: An analysis on interdecadal variation and trend of summer rainfall over East China, *Acta Meteor. Sin.*, 2005, 63(5), 728-739.
113. Jia Xiaolong, **Li Chongyin**: Dipole Oscillation in the Southern Indian Ocean and Its Impacts on Climate, *Chinese J. Geophysics*, 2005, 48, 1348-1356.
114. Hu Ruijin, Liu Qinyu, **Li Chongyin**: A heat budget study on the mechanism of SST variation in the Indian Ocean Dipole region, *Journal of Ocean University of China*, 2005, 4(4), 334-342.
115. Wang Xin, **Li Chongyin**, Zhou Wen: Interdecadal variation of the relationship between Indian rainfall and SSTA modes in the Indian Ocean, *Inter. J. Climate*, 2006, 26, 595-606.
116. Zhou Wen, Li **Chongyin**, Chan J C L: The interdecadal variations of the summer monsoon rainfall over South China, *Meterol. Atmos. Phys.*, 2006, DOI: 10.1007/s00703-006-0184-9.
117. Ding Yihui, **Li Chongyin**, He Jinhai, et al.: South China Sea monsoon experiment (SCSMEX) and East Asian Monsoon, *Acta Meteor. Sin.*, 2006, 20(2), 159-190.
118. Song Jie, Zhou Wen, Pan Jing, **Li Chongyin**: The global influence of the Northern hemisphere second mode of the zonal average of the zonal wind, *Geophys. Res. Lett.*, 2006, 33(18), L18703, DOI: 10.1029/ 2006 GL026380.
119. **Li Chongyin**, Pan Jing, The atmospheric circulation characteristics accompanied with the Asian summer monsoon onset, *Adv. Atmos. Sci.*, 2006, 23(6), 925-939.
120. **Li Chongyin**, Zhou Wen, Jia Xiaolong, Wang Xin: Decadal/interdecadal variations of the Ocean temperature and its impacts on the climate. *Adv. Atmos. Sci.*, 2006, 23(6), 964-981.
121. Yang Hui, Jia Xiaolong, **Li Chongyin**: The tropical Pacific-Indian Ocean temperature anomaly mode

and its effect. *Chinese Science Bulletin*, 2006, 51(23), 2878-2884.

122. Wang Xin, **Li Chongyin,** Zhou Wen: Interdecadal Mode and Its Propagating characteristics of SSTA in the South Pacific, *Meteorol. Atmos. Phys.*, 2006, DOI: 10.1007/S00703-006-0235-2.
123. Zhou Wen, **Li Chongyin,**Wang Xin: Possible connection between Pacific Oceanic Interdecadal Pathway and East Asian Winter Monsoon, *Geophys. Res. Lett.*, 2006, DOI: 10.1029/2006GL027809.
124. Jia Xiaolong, **Li Chongyin,**Zhou Zingfang: A GCM atudy on the tropical intraseasonal oscillation, *Acta Meteor. Sin.*, 2006, 20(3), 352-366.
125. Chen Zhe, **Li Chongyin**, Fu Zuntao: Periodic structures of Rossby wave under influence of dissipation, *Commun. Theor. Phys.*, 2007, 47(1), 35-40.
126. **Li Chongyin**, Ling Jian, Jia Xiaolong, Dong Min: Numerical simulation and comparison study of the atmospheric intraseasonal oscillation, *Acta Meteorological Sinica*, 2007, 21, 1-8.
127. Zhou Wen, Wang Xin, Zhou Tianjun, **Li Chongyin**, Chan J C L: Interdecadal variability of the relationship between the East Asian winter monsoon and ENSO. *Meteorol Atmos. Phys.*, 2007, DOI: 10.1007/s00703-007-0263-6.
128. Jia Xiaolong, **Li Chongyin**: Sensitivity of simulated tropical intraseasonal oscillation to cumulus parameterizations, *Acta Meteor. Sin.*, 2008, 22(3), 257-276.
129. Jia Xiaolong, **Li Chongyin**: Seasonal variations of the tropical intraseasonal oscillation and its reproduction in SAMIL-R42L9. *Journal of Tropical Meteorology*, 2007, 13(2), 173-176.
130. Jia Xiaolong, **Li Chongyin,** Ling Jian, Chidong Zhang: Impacts of the GCM's resolution on the MJO Simulation. *Advances in Atmos. Sci.*, 2008, 25(1), 139-156.
131. Yuan Yuan, Zhou Wen, Yang Hui, **Li Chongyin**: Warming in the Northwestern Indian Ocean Associated with the El Niño event. *Advance in Atmos. Sci.*, 2008, 25(2), 246-252.
132. Yuan Yuan, Zhou Wen, Chan J C L, **Li Chongyin**: Impacts of the basin-wide Indian Ocean SSTA on the South China Sea summer monsoon onset, *International Journal of Climatology*, 2008, DOI: 10.1002/JOC.1671.
133. Yuan Yuan, Yang Hui, Zhou Wen, **Li Chongyin**: Influnces of the Indian Ocean dipole on the Asian summer monsoon in the following year, *International Journal of Climatology*, 2008, 28, 1849-1859, DOI: 10.1002/JOC.1678.
134. Yuan Yuan, Chan J C L, Zhou Wen, **Li Chongyin**: Decadal and interannual variability of the Indian Ocean dipole, *Advance in Atmos. Sci.*, 2008, 25, 856-866.
135. Yuan Yuan, **Li Chongyin**: Decadal variability of the IOD-ENSO relationship. *Chinese Science Bulletin*, 2008, 53(11), 1745-1752.
136. **Li Chongyin**, Gu Wei, Pan Jing: Mei-Yu, Arctic oscilation and statospheric circulation anomalies, *Chinese J. Geophysics*, 2008, 51 (6), 1127-1135.
137. **Li Chongyin**, Huang Ronghui: El Niño and the Southern Oscillation-monsoon interaction and interannual climate, *<Changes in the Human-Monsoon System of East Asia in the Context of Global Change>*, 2008, 75-88, Congbin Fu, et al., World Scientific.
138. Zhang Liang, Zhang Lifeng, **Li Chongyin**, et al.: Some new exact solutions of Jacobian elliptic function about the generalized Boussinesq equation and Boussinesq-Burgers equation, *Chinese Physics* B, 2008, 17(2), 403-410.
139. **Li Chongyin**, Jia Xiaolong, Ling Jian, Zhou Wen, Zhang Chidong: Sensitivity of MJO simulations to diabatic heating profiles, *Climate Dynamics*, 2009, 32, 167-187, DOI: 10.1007/s00382-008-0455.x.
140. Jia Xiaolong, **Li Chongyin**, Ling Jian: Impacts of cumulus parameterization and resolution on the MJO simulation, *Journal of Tropical Meteorology*. 2009, 15(1), 106-110.
141. Gu Wei, **Li Chongyin**, Wang Xin, Zhou Wen,Li Weijing: Linkage between Mei-yu precipitation and

North Atlantic SST on the decadal timescale, *Adv. Atmos. Sci.*, 2009, 26(1), 101-108.

142. Song Jie, **Li Chongyin**, Zhou Wen, Pan Jing: The Linkage between the Pacific-North American teleconnection pattern and the North Atlantic Oscillation, *Adv. Atmos. Sci.*, 2009, 26(2), 229-239.
143. Gu Wei, **Li Chonhyin,** Li Weijing, Zhou Wen, Chan J C L: Interdecadal unstationary relationship between NAO and East China's summer precipitation patterns, *Geophys. Res. Lett.*, 2009, 36, L13702, DOI: 10.1029/2009GL038843.
144. Ling Jian, **Li Chongyin**, Jia Xiaolong: Impacts of cumulus momentum transport on MJO simulation, *Adv. Atmos. Sci.*, 2009, 26(5), 864-876.
145. Song Jie, Zhou Wen, **Li Chongyin**, Qi Lixin: Signature of the Antarctic Oscillation in the Northern Hemisphere, *Meteorol. Atmos. Phys.*, 2009, 105, 55-67, DOI: 10. 1007/s00703-009-0036-5.
146. **Li Chongyin**, Ling Jian: Physical essence of the "predictability barrier", *Atmos. Oce. Sci. Let.*, 2009, 2(5), 290-294.
147. Tian Hua, **Li Chong-yin,** Yang Hui: Modulation of TC Genesis over the Northwestern Pacific by Atmospheric Intraseasonal Oscillation, *Journal of Tropical Meteorology*, 2012, 18(1), 9-16.
148. Xiao Ziniu, Liang Hongli, **Li Chongyin**: Relationship between the number of summer typhoons engendered over the northwest Pacific and South China Sea and main climate conditions in the preceding winter and spring, *Acta Meteor. Sin.*, 2010, 24(4), 441-451.
149. **Li Chongyin**, Pan Jing, Que Zhiping: Variation of the East Asian monsoon and the Tropospheric biennial oscillation, *Chinese Science Bulletin*, 2011, 56(1), 70-75.
150. Jia Xialong, **Li Chongyin**, Zhou Ningfang, Ling Jian: The MJO in an AGCM with three different cumulus parameterization schemes, *Dyn. Atmos. Oceans*, 2010, 49, 141-163.
151. Jia Xiaolong, Chen Lijuan, Ren Fumin, **Li Chongyin**: Impacts of the MJO on winter rainfall and Circulation in China, *Adv. Atmos. Sci.*, 2011, 28(3), 521-533.
152. **Li Chongyin**, Li Lin, Tan Yanke: Structure of South Asia High in the stratosphere and influence of ENSO, *J. Trop. Meteor.* 2011, 27(3), 193-201.
153. Song Jie, Zhou Wen, Wang Xin, **Li Chongyin**: Zonal asymmetry of the annular mode and its downstream subtropical jet: an idealized model study, *J. Atoms. Sci.*, 2011, 68, 1946-1973; DOI: 10.1175/2011 JAS3656.1
154. Huang Yong, **Li Chongyin**, Wang Yin: Numerical simulations of the Pacific meridional mode impacts on tropical cyclones activity over the western North Pacific, *J. Tropical Meteorology*, 2012, 18(4), 428-442.
155. Huang Yong, **Li Chongyin**, Wang Ying: Numerical simulations of the pacific meridional mode impacts on tropical cyclones activity over the western north pacific. *Journal of Tropical Meteorology*, 2012, 18(4), 512-520.
156. Bai Xuxu, **Li Chongyin** , Tan Yanke: The impacts of Madden-Julian oscillation on spring rainfall in East China. *Journal of Tropical Meteorology*, 2013, 19(3), 214-222.
157. Li Lin, **Li Chongyin**, Song Jie: Arctic oscillation anomaly in winter 2009/2010 and its impacts on weather and climate, *Science China*-Earth Sciences, 2012, 55(4), 567-579.
158. Yan Hongming, Yang Hui, Yuan Yuan, **Li Chongyin**: Relationship Between East Asian Winter Monsoon and Summer Monsoon, *Adv. Atmos. Sci.*, 2011, 28(6), 1345-1356.
159. Pan Jing, **Li ChongYin**: Low-Frequency Vortex Pair over the Tropical Eastern Indian Ocean and the South China Sea Summer Monsoon Onset, *Atmospheric and Oceanic Science Letters*, 2011, 4(6), 304-308.
160. Wang Xin, Wang Dongxiao, Zhou Wen, **Li Chongyin**: Interdecadal modulation of the influence of La Niña events on mei-yu rainfall over the Yangtze River Valley. *Adv. Atmos. Sci.*, 2012, 29(1), 157-168,

DOI: 10. 1007/s00376-011-1021-8.
161. Wang Xin, Zhou Wen, **Li Chongyin,** Wang Dongxiao: Effects of the East Asian summer monsoon on tropical cyclones genesis over the South China Sea on an interdecadal timescales. *Adv. Atmos. Sci.*, 2012, 29, 249-262, DOI: 10.1007/s00376-011-1080-x
162. Li Lin, **Li Chongyin**, Pan Jing, Tan Yanke: On the different and climate impacts of early and late stratospheric polar vortex breakup, *Adv. Atmos. Sci.*, 2012, 29(5),119-128.
163. Li Gang, **Li Chongyin,** Tan Yanke, Bai Tao: Seasonal Evolution of Dominant Modes in South Pacific SST and Relationship with ENSO. *Adv. Atmos. Sci*, 2012, 29, 1238-1248, DOI: 10.1007/s00376-012-1191-z.
164. Liu Peng, **Li Chongyin**, Wang Yu, Fu Yunfei: Climatic characteristics of convective and stratiform precipitation over the Tropical and Subtropical areas as derived from TRMM PR, *Science China*-Earth Science, 2013, 56(3), 375-385.
165. **Li Chongyin**, Li Lin, Pan Jing: Spatial and temporal variations of stratospheric atmospheric circulation in the Northern Hemisphere during the boreal summer, *China Science Bulletin*, 2012, 57(24), DOI: 10.1007/s11434-012-5606-0.
166. **Li Chongyin**, Li Lin, Que Zhiping: Further research on mechanism of TBO in south Asian monsoon region, *J. Trop. Meteor.*, 2014, 20(3), 202-207.
167. Li Gang, **Li Chongyin**, Tan Yanke, Bai Tao: Impacts of central Pacific and eastern Pacific types of ENSO on sea surface temperature variability over the South Pacific. *Theor Appl Climatol.*, 2013, 111, DOI: 10.1007/s00704-013-0840-1.
168. Ling Jian, **Li Chongyin,** Zhou Wen, Jia Xiaolong: To begin or not to begin? A case study on the MJO initiation problem, *Theor Appl Climatol.*, 2013, 112, DOI: 10.1007/s00704-013-0889-x.
169. Ling Jian, **Li Chonhyin**, Zhou Wen, Jia Xiaolong, Zhang Chidong: Effect of boundary layer latent heating on MJO simulations. *Adv. Atoms. Sci.*, 2013, 30(30), 101-115.
170. Yan Hongming, Li Qingquan, Yuan Yuan, **Li Chongyin**: Circulation variation over western North Pacific and its association with tropical SSTA over Indian Ocean and the Pacific, *Chinese J. Geophys.* 2013, 56(8), 2542-2557, DOI: 10.6038/cjg20130805
171. Song Jie, **Li Chongyin**, Zhou Wen: High and low latitude types of the downstream influences of the North Atlantic Oscillation, *Clim. Dyn.*, 2013, (12), DOI: 10.1007/s 00382-013-1844-3.
172. Li Xiuzhen, Zhou Wen, **Li Chongyin**, Song Jie: Comparison of the annual cycles of moisture supply over southwest and southeast China, *Journal of Climate*, 2013, 26(13), DOI: 10.1175/JCLI-D-13-00057.1.
173. Yang Guang, **Li Chongyin,** Tan Yanke: The interdecadal variation of the intensity of South Asian High and its possible causes. *Journal of Tropical Meteorology*, 2016, 22(1), 66-76, DOI: 10.16555/j.1006-8775.2016.01.003.
174. Yuan Yuan, **Li Chongyin**, Song Yang: Decadàl anomalies of winter precipitation over Southern China in association with E1 Nifio and La Nina, *J. Meteor. Res.*, 2014, 29(1), 91-110.
175. Li Xin, **Li Chongyin**: Occurrence of two types of El Niño events and the subsureface ocean temperature anomalies in the equatorial Pacific, *Chin. Sci. Bull.*, 2014, 59(27), 3471-3483, DOI: 10. 1007/s 11434-014-0365-8.
176. Li Gang, **Li Chongyin**, Tan Yanke, Wang Xin: Observed relationship of boreal winter South Pacific tripole SSTA with eastern China rainfall during the following spring, *Journal of Climate*, 2014, 27: 8094-8106, DOI: 10. 1175/JCLI-D-14-00074.1.
177. Li Gang, **Li Chongyin**, Tan Yanke: The interdecadal changes of the South Pacific sea surface temperature in the mid-1990s and their connections with ENSO, *Adv. Atoms. Sci.*, 2014, 31(1), 66-84, DOI:

10.1007/s00376-013-2280-3.

178. **Li Chongyin**, Ling Jian, Song Jie, Pan Jing, Tian Hua and Chen Xiong: Studying progresses in China on tropical atmospheric intraseasonal oscillation, *J. Meteor. Res.*, 2014, 28(5), 671-692.
179. Wang Dongxiao, Wang Xin, Zhou Wen, **Li Chongyin**: Comparisons of two types of El Niño impacts on TC genesis over the South China Sea, *Chapter 17 in Typhoon Impact and Crisis Management*, 2014, DOI: 10.1007/978-3-642-40695-9_17, Springer Verlag Berlin Heidelberg.
180. Holbrook Neil J, Li Jianping, Collins M, Lorenzo E D, Jin Feifei, Knutson Thomas, Latif M, **Li Chongyin**, Power Scott, Huang Rhonghui, Wu Guoxiong: Decadal climate variability and cross-scale interactions, *Bull. Amer. Meteor. Soc.*, 2014, October, DOI: 10.1175/BAMS-D-13-00201.1.
181. Ling Jian, **Li Chongyin**: Impact of convective momentum transport by deep convection on simulation of tropical intraseasonal oscillation, *Journal of Ocean University of China*, 2014, 13(5), DOI: 10.1007/s11802-014-2295-0
182. Wang Xin, Zhou Wen, **Li Chongyin**, Wang Dongxiao: Comparison of the impact of two types of El Niño on tropical cyclone genesis over the South China Sea. *Int. J. Climatol.*, 2014, 34, 2651-2660, DOI: 10.1002/joc.3865.
183. Zheng Chongwei, Pan Jing, **Li Chong-yin**: Prospect and suggestions on the development of wave energy resource in the South China Sea. *Applied Mechanics and Materials*, 2014, 672-674, 459-466.
184. Yang Shenggao, Zhang Beichen, Fang Hanxian, Liu Junming, Zhang Qinghe, Hu Hongqiao, Liu Ruiyuan, **Li Chongyin**: F-lacuna at cusp latitude and its associated TEC variation, *J. Geophys. Res.* Space Physics, 2014, 119, DOI: 10.1002/2014JA020607.
185. Song Jie, **Li Chongyin**: Contrasting Relationship between Tropical Western North Pacific Convection and Rainfall over East Asia during Indian Ocean Warm and Cold Summers, *J. Climate*, 2014, 27, 2562-2576. DOI: 10.1175/JCLI-D-13-00207.1
186. Chen Xiong, **Li Chongyin**, Tan Yanke: The influence of El Niño on atmospheric MJO over the equatorial Pacific, *Journal of Ocean University of China*, 2015, 14(1), 1-8, DOI: 10.107/s11082-015-2381_.
187. Zheng Chongwei, **Li Chongyin**: Variation of the wave energy and significant wave height in the China Sea and adjacent waters, *Renewable and Sustainable Energy Reviews* (RSER), 2015, 43(3),381-387.
188. Xue Feng, Zeng Qingcun, Huang Ronghui, **Li Chongyin**, Lu Riyu,Zhou Tianjun: Recent advances in monsoon studies in China, *Adv. Atoms. Sci.*, 2015, 32(2),206-229.
189. Chen Xiong, **Li Chongyin**, Tan Yanke, Guo Wenhua: The contrasts between strong and weak MJO activity over the equatorial western Pacific in winter. *Journal of Tropical Meteorology*; 2017, 23(2), 133-145.
190. Chen Xiong, **Li Chongyin**, Ling Jian, Tan Yanke: Impact of East Asian winter monsoon on MJO over the equatorial western Pacific, *Theor. Appl. Climatol.*, 2015, 127(3), 551-561, DOI: 10.1007/s00704-015-1649-x.
191. Li Xin, **Li Chonhyin**, Ling Jian, Tan Yanke: The Relationship between Contiguous El Niño and La Niña Revealed by Self-Organizing Maps., *J. Climate*, 2015, 28, 8118-8134, DOI: 10.1175/JCLI-D-15-0123.1.
192. Li Gang, Tan Yanke, **Li Chongyin**, et al.: The distribution characteristics of total ozone and its relationship with stratospheric temperature during boreal winter in the recent 30 years, *Chinese J. Geo.*, 2015, 58(5), 1475-1491, DOI: 10.6038/cjg 20150502.
193. Yang Shenggao, Zhang Beichen, Fang Hanxian, Kamide Y., **Li Chongyin**, Liu Junming, Zhang Shunrong, Liu Ruiyuan, Zhang Qinghe, Hu Hongqiao: New evidence of dayside plasmatransportation over the polar cap to theprevailing dawn sector in the polarupper atmosphere for solar-maximumwinter, *J. Geophys.*

Res. Space Physics, 2016, 121, DOI: 10.1002/2015JA022171.

194. Zheng Chongwei, Pan Jing, **Li Chongyin**: Global oceanic wind speed trends, *Ocean & Coastal Management*, 2016, 129, 15-24. SCI referral: 000379371800003.
195. Zheng Chongwei, **Li Chongyin**, Pan Jing, Liu Mingyang, Xia Linlin: An overview of global ocean wind energy resources evaluatio, *Renewable and Sustainable Energy Reviews*, 2016, 53, 1240-1251. SCI referral: 000367758100087.
196. Zheng ChongWei, **Li ChongYin**, Chen Xuan, Jing Pan: Numerical Forecasting Experiment of the Wave Energy Resource in the China Sea, *Advances of Meteorology*, 2016, Article ID 5692431, 1-12.
197. Xia Linlin, Tan Yanke, **Li Chongyin**, Cheng Cheng: The Classification of Synoptic-Scale Eddies at 850hPa over the North Pacific in Wintertime, *Advances of Meteorology*, 2016, Article ID 4797103, 1-8, DOI: 10.1155/2016/4797103.
198. Jue Zhiping, Wu Fan, Bi Cheng, **Li Chongyin**: Impacts of monthly anomalies of intraseasonal oscillation over south China sea and south Asia on the activity of summer monsoon and rainfall in eastern China, *Journal of Tropical Meteorology*, 2016, 22(2), 145-158.
199. Gui Fayin, **Li Chongyin**, Tan Yanke, et al.: The warming mechanism in the southern Arabian sea during the development of Indian Ocean dipole events, *Journal of Tropical Meteorology*, 2016, 22(2), 159-171.
200. Xiao Ziniu, Liao Yunchen, **Li Chongyin**: Possible impact of solar activity on the convection dipole over the tropical pacific ocean. *Journal of Atmospheric and Solar-Terrestrial Physics*, 2016, 140, 94-107, DOI: 10.1016/j. jastp. 2016.02.0.
201. Chen Xiong, Ling Jian, **Li Chongyin**: Evolution of Madden-Julian Oscillation in two types of El Niño. *Journal of Climate*. 2016, 29, 1919-1934.
202. Ling Jian, **Li Chongyin**, Li Tiam, Jia Xiaolong, Khouides Boualem, Maloney Eric, Vitart Frederic, Xiao Ziniu, Zhang Chidong: Challenges and Opportunities in MJO Studies, *Bulletin of Amer. Meteor. Soc.*, 2017, 98(2), ES53-ES56, DOI: 10.1175/BAMS-D-16-0238.1.
203. Zheng Chongwei, **Li Chongyin**, Li Xin: Recent decadal trend in the North Atlantic wind energy resources. *Advances in Meteorology*, 2017, Article ID 7257492, 1-8, DOI: 10.1155/2017/7257492.
204. Zheng Chongwei, **Li Chongyin**, Gao Chengzhi, Liu Mingyang: A seasonal grade division of the global offshore wind energy resource, *Acta Ocean. Sinica*, 2017, 36(3), 109-114.
205. ling Jian, **Li Chongyin**: A New Interpretation of the Ability of Global Models to Simulate the MJO, *Geophysical Research Letters*, 2017, DOI: 10.1002/2017GL073891.
206. Chen Xiong, **Li Chongyin,** Ling Jian: Further inqurry into characteristics of MJO in boreal winter, *International Journal of Climatology*, 2017, 37, 4451-4462, DOI: 10.1002/joc.5098..
207. Wang Gongjie, Cheng Lijing, Boyer Timothy, **Li:Chongyin** Halosteric Sea Level Changes during the Argo Era, *Water*, 2017, 9(7), 1-13, DOI: 10.3390/w9070484.
208. Chen Xiong, **Li Chongyin**, Tan Yanke, Guo Wenhua: The contrasts between strong and weak MJO zctivity over the quatorial western Pacific in winter. *Journal of Tropical Meteorology*, 2017, 23(2), 133-145.
209. Zheng Chongwei, **Li Chonhyin**: Propagation characteristic and intraseasonal oscillation of the swell energy of the Indian Ocean. *Applied Energy*, 2017, 197, 342-353, DOI: 10.1016/j.apenergy.2017.04.052 .
210. Zheng Chongwei, **Li Chongyin**: Analysis of temporal and spatial characteristics of waves in the Indian Ocean based on ERA-40 wave reanalysis. *Applied Ocean Research*, 2017, 63, 217-228.
211. Zheng Chongwei, Wang Qing, **Li Chongyin**: An overview of medium- to long-term predictions of global wave energy resources. *Renewable and Sustainable Energy Reviews*, 2017, 79, 1492-1502.

212. Huang Yingying, Zhang Shaodong, **Li Chongyin,** Li Huijun, Huang Kaiming, and Huang Chunming: Annual and interannual variations in global 6.5DWs from 20 to 110 km during 2002–2016 observed by TIMED/SABER, *J. Geophys. Res.* Space Physics, 2017, 122, 8985-9002, DOI: 10.1002/2017JA023886.

213. Li Huijun, **Li Chongyin**, Feng Xueshang, Xiang Jie, Huang Yingying, Zhou Shudao: Data completion with Hilbert transform over plane rectangle: technique renovation for the Grad-Shafranov reconstruction, *J. Geophys. Res.* Space Physics, 2017, 122(4), 3949-3960.

214. Li Xin, **Li Chongyin**: The tropical Pacific–Indian Ocean associated mode simulated by LICOM2.0. *Adv.Atmos. Sci.*, 2017, 34(12), 1426-1436.

215. Wang Gongjie, Cheng Lijing, Abraham John, **Li Chongyin**: Consensuses and discrepancies of basin-scale ocean heat content changes in different ocean analyses, *Climate Dynamics*, 2018, 50(7-8), 2471-2487, DOI: 10.1007/s00382- 017-3751-5.

216. Li Huijun, **Li Chongyin,** Feng Xueshang, et al.: Corner singularity and its application in regular parameters optimization: technique renovation enovation for grad-shafranov reconstraction, *AASTX Technique Renovation for the GS-Reconstruction*, 2018, 8, arXv: 1712.02479v1.

217. Li Gang, Chen Jiepeng, Wang Xin, **Li Chongyin**, et al.: Remote impacts of North Atlantic sea surface temperature on rainfall in southwestern China. *Climate Dynamics*, 2017, DOI: 0.1007/s00382-017-3625-x.

218. Zheng Chongwei, **Li Chongyin**, Pan Jing: Propagation route and speed of swell in the Indian Ocean. *Journal of Geophysical Research:* Oceans, 2018, 123, https: //doi.org/10.1002/2016JC012585

219. Chen Xiong, **Li Chongyin**, Li Xin, Liu Mingyang: The Northern and Southern Modes of East Asian Winter Monsoon and their Relationships with ENSO. *International Journal of Climatology*, 2018, DOI: 10.1002/joc.5683.

220. Zheng Chongwei, Xiao Ziniu, Peng Yuehua, **Li Chongyin**, Du Zhi-bo: Rezoning global offshore wind energy resources. *Renewable Energy*, 2018, 129, 1-11. WOS: 000439745700001.

221. Yin Ming, Li Xin, **Li Chongyin**, Tan Yanke: Relationships between intensity of the Kuroshio current in the East China Sea and the East Asian winter monsoon, *Acta Oceanologica Sinica*, 2018, 37(7), 8-19, DOI: 10.1007/s13131-018-1240-2.